HANDBOOK OF RENEWABLE ENERGY TECHNOLOGY

HANDBOOK OF RENEWABLE ENERGY TECHNOLOGY

editors

Ahmed F. Zobaa
Brunel University, U.K.

Ramesh C. Bansal
The University of Queensland, Australia

World Scientific

NEW JERSEY · LONDON · SINGAPORE · BEIJING · SHANGHAI · HONG KONG · TAIPEI · CHENNAI

Published by

World Scientific Publishing Co. Pte. Ltd.

5 Toh Tuck Link, Singapore 596224

USA office: 27 Warren Street, Suite 401-402, Hackensack, NJ 07601

UK office: 57 Shelton Street, Covent Garden, London WC2H 9HE

British Library Cataloguing-in-Publication Data
A catalogue record for this book is available from the British Library.

HANDBOOK OF RENEWABLE ENERGY TECHNOLOGY

ISBN-13 978-981-4289-06-1
ISBN-10 981-4289-06-X

Typeset by Stallion Press
Email: enquiries@stallionpress.com

Printed in Singapore by B & Jo Enterprise Pte Ltd

Dedicated to
Lord Sun, source of all kinds of energies

Preface

Effects of environmental, economic, social, political and technical factors have led to the rapid deployment of various sources of renewable energy-based power generation. The incorporation of these generation technologies have led to the development of a broad array of new methods and tools to integrate this new form of generation into the power system network. This book, arranged into six sections, tries to highlight various renewable energy based generation technologies.

Section 1 provides a general overview of the wind power technology, where the classification of wind turbines based on generators, power electronic converters, and grid connection is described in detail. In Chapter 1, the fundamentals of wind power systems and their design aspects are presented; the modeling methods of the wind phenomenon and turbine mechanical system are described in Chapter 2. Chapter 3 presents modeling and integration of wind power systems to the grid, while a literature review on the technologies and methods used for wind resource assessment (WRA) and optimum wind turbine location is presented in Chapter 4. In the next chapter, the descriptions of the different types of economic analysis methods are presented with case studies. The operation and control of a line side converter used in variable-speed wind energy conversion systems under balanced and unbalanced grid voltages conditions is discussed in Chapter 6, and lastly, the wake effect from wind turbines on overhead lines and, in particular, a tower line close to wind farms is analyzed in Chapter 7.

Section 2 is on solar energy. Although sun charts are widely used, there are situations where charts are inadequate and precise computations are preferred. This is discussed in Chapter 8 as a computational approach that is applicable to both thermal collection/conversion processes and photovoltaics (PV) systems. In Chapter 9, the different types of PV systems, grid-connected and stand-alone, designing of stand-alone PV system, both for electricity supply to remote homes and solar water pumping systems, are presented. Concentrated solar power appears to be a method of choice for large capacity, utility-scale electric generation in the near future, in particular, distributed trough systems, which represent a reasonably mature approach. The power tower configuration is also a viable candidate. Both these technologies have the possibility of energy storage and auxiliary heat production during the unavailability of sunlight, and a discussion on this is carried out in Chapter 10. Chapter 11 presents various overviews on battery-operated solar energy storage, its charging technologies and performance, and maximum power point tracking (MPPT). Non-grid solar thermal technologies like water heating systems, solar cookers, solar drying applications and solar thermal building designs are simple and can be readily adopted, as can be seen in Chapter 12. The solar tunnel dryer is one of the promising technologies for large scale agricultural and industrial processes. In this technology, the loading and unloading of material in process is relatively easy and thus, more quantity can be dried at lower cost, as discussed in Chapter 13.

Section 3 focuses on bio-mass energy. Chapter 14 presents biomass as a source of energy which stores solar energy in chemical form in plant and animal materials. It is one of the most commonly used, but precious and versatile resource on earth, and has been used for energy purposes since the Stone Age. Biomass energy can be sustainable, environmentally benign and an economically sound source. Chapter 15 presents a resource known as forest biomass. An analysis of its potential energy, associated to its two sources, forest residue and energy crops, is carried out. It discusses the collection and transportation systems and their performance. Chapter 16 discusses different aspects of the production and utilization of bioethanol. It also presents the technical fundamentals of various manufacturing systems, depending on the raw material used. Biodiesel and its use as fuel could help to reduce world dependence on petrol. In Chapter 17, the main characteristics that make biodiesel an attractive biofuel are discussed, with Chapter 18 discussing the raw materials used to obtain biodiesel and their principal advantages and disadvantages.

Section 4 is based on small hydro and ocean-based energies. Chapter 18 focuses on some of the key challenges faced in the development of marine energy. It presents a prototype form of marine energy being widely deployed as a contributor to the world's future energy supply. Chapter 19 describes electrical circuits and operations of low power hydro plants. Grid connection issues and power quality problems are explained with some examples. In the case of small hydro power plants, operational problems and solutions through the strengthening of grid connection codes are presented in Chapter 19. In the case of the isolated small hydro power plant, frequency is generally maintained constant either by dump load/load management or by input flow control. Frequency control by using a combination of dump load and input flow control is discussed in Chapter 20.

Section 5 is devoted to the simulation tools for renewable energy systems, distributed generation (DG) and renewable energy integration in electricity markets. In Chapter 21, a review is undertaken of the main capabilities of the most common software packages for feasibility studies of renewable energy installations. Here, the chapter details the models implemented in these tools for representing loads, resources, generators and dispatch strategies, and summarizes the approaches used to obtain the lifecycle cost of a project. A short description of a methodology for estimating greenhouse gas (GHG) emission reductions is also included. Chapter 22 reviews the distributed generation from a power system's point of view. A detailed analysis on DG allocation in a distribution system for loss reduction is presented in Chapter 23, while the next chapter (Chapter 24) describes the aggregation of DG plants which gives place to a new concept: the Virtual Power Producer (VPP). VPPs can reinforce the importance of these generation technologies by making them valuable in electricity markets. Thus, DG technologies are using various power electronics based converters.

Section 6 covers a range of assorted topics on renewable energy, such as power electronics, induction generators, doubly-fed induction generators (DFIG), power quality instrumentation for renewable energy systems and energy planning issues. Chapter 25 describes the power-electronic technology for the integration of renewable energy sources like wind, photovoltaic and energy-storage systems, with grid interconnection requirements for the grid integration of intermittent renewable energy sources discussed in detail. Chapter 26 provides an analysis of an induction generator and the role of DFIG-based wind generators; their control is presented in Chapter 27. Chapter 28 presents power quality instrumentation and measurements in a distributed and renewable energy-based environment. The gap in the demand and supply of energy can only be met by an optimal allocation of energy resources and the need of the day for developing countries like India. For the socio-economic development of India, energy allocation at the rural level is gaining in importance. Thus, a detailed analysis of such cases and scenarios is presented in Chapter 29.

We are grateful to a number of individuals who have directly (or indirectly) made contributions to this book. In particular, we would like to thank all the authors for their contributions, and the reviewers for reviewing their book chapters, thus improving the quality of this handbook.

We would also like to thank the Authorities and staff members of Brunel University and The University of Queensland for being very generous and helpful in maintaining a cordial atmosphere, and for leasing us the facilities required during the preparations of this handbook. Thanks are due to World Scientific Publishing, especially to Gregory Lee, for making sincere efforts for the book's timely completion.

Lastly, we would like to express our thanks and sincere regards to our family members who have provided us with great support.

Ahmed F. Zobaa and Ramesh C. Bansal
Editors

About the Editors

Ahmed Faheem Zobaa received his B.Sc. (Hon.), M.Sc. and Ph.D. degrees in Electrical Power and Machines from the Faculty of Engineering at Cairo University, Giza, Egypt, in 1992, 1997 and 2002. He is currently a Senior Lecturer in Power Systems at Brunel University, UK. In previous postings, he was an Associate Professor at Cairo University, Egypt, and a Senior Lecturer in Renewable Energy at University of Exeter, UK.

Dr. Zobaa is the Editor-In-Chief for the *International Journal of Renewable Energy Technology,* and an Editorial Board member, Editor, Associate Editor, and Editorial Advisory Board member for many other international journals. He is a registered Chartered Engineer, and a registered member of the Engineering Council, UK, and the Egyptian Society of Engineers. Dr. Zobaa is also a Fellow of the Institution of Engineering and Technology, and a Senior Member of the Institute of Electrical and Electronics Engineers. He is a Member of the Energy Institute (UK), International Solar Energy Society, European Society for Engineering Education, European Power Electronics & Drives Association, and IEEE Standards Association.

His main areas of expertise are in power quality, photovoltaic energy, wind energy, marine renewable energy, grid integration, and energy management.

Ramesh C. Bansal received his M.E. degree from the Delhi College of Engineering, India, in 1996, his M.B.A. degree from Indira Gandhi National Open University, New Delhi, India, in 1997, and his Ph.D. degree from the Indian Institute of Technology (IIT)-Delhi, India, in 2003. He is currently a faculty member in the School of Information Technology and Electrical Engineering, The University of Queensland, St. Lucia Campus, Qld., Australia. In previous postings, he was with the Birla Institute of Technology and Science, Pilani, the University of the South Pacific, Suva, Fiji, and the Civil Construction Wing, All India Radio.

Dr. Bansal is an Editor of the *IEEE Transactions on Energy Conversion and Power Engineering Letters*, an Associate Editor of the *IEEE Transactions on Industrial Electronics* and an Editorial Board member of the *IET, Renewable Power Generation, Electric Power Components and Systems Energy Sources*. He is also a Member of the Board of Directors of the International Energy Foundation (IEF), Alberta, Canada, a Senior Member of IEEE, a Member of the Institution of Engineers (India) and a Life Member of the Indian Society of Technical Education.

Dr. Bansal has authored or co-authored more than 125 papers in national/ international journals and conference proceedings. His current research interests include reactive power control in renewable energy systems and conventional power systems, power system optimization, analysis of induction generators, and artificial intelligence techniques applications in power systems.

Contents

Section 2.   Solar Energy Systems

13.   Solar Tunnel Dryer — A Promising Option for Solar Drying   289

Mahendra S. Seveda, Narendra S. Rathore and Vinod Kumar

Section 3.   Bio Fuels

14.   Biomass as a Source of Energy   323

Mahendra S. Seveda, Narendra S. Rathore and Vinod Kumar

Section 4. Ocean and Small Hydro Energy Systems

Section 5. Simulation Tools, Distributed Generation and Grid Integration

Section 6. Induction Generators, Power Quality, Power Electronics and Energy Planning for Renewable Energy Systems

25. Modern Power Electronic Technology for the Integration of Renewable Energy Sources   673

Vinod Kumar, Ramesh C. Bansal, Raghuveer R. Joshi, Rajendrasinh B. Jadeja and Uday P. Mhaskar

26. Analysis of Induction Generators for Renewable Energy Applications   717

Kanwarjit S. Sandhu

Section 1

Wind Energy and Their Applications

Chapter 1

Wind Energy Resources: Theory, Design and Applications

Fang Yao

School of Electrical, Electronic and Computer Engineering,
Faculty of Engineering, Computer and Mathematics,
University of Western Australia
20852621@student.uwa.edu.au

Ramesh C. Bansal

School of Information Technology and Electrical Engineering,
The University of Queensland, Australia
rcbansal@ieee.org

Zhao Yang Dong

Department of Electrical Engineering,
The Hong Kong Polytechnic University, Hong Kong

Ram K. Saket

Department of Electrical Engineering, Institute of Technology,
Banaras Hindu University, Varanasi (U.P.), India

Jitendra S. Shakya

Samrat Ashok Technological Institute, Vidisha, M.P., India

The technology of obtaining wind energy has become more and more important over the last few decades. The purpose of this chapter is to provide a general discussion about wind power technology. The fundamental knowledge of wind power systems and their design aspects are presented. The description of the fundamental topics which are essential to understand the wind energy conversion and its eventual use is also provided in the chapter. This chapter discusses the wind farms and hybrid power systems as well.

1.1   Introduction

Wind power is one of the renewable energy sources which has been widely developed in recent years. Wind energy has many advantages such as no pollution, relatively low capital cost involved and the short gestation period. The first wind turbine for electricity generation was developed at the end of the 19th century. From 1940 to 1950, two important technologies, i.e., three blades structure of wind turbine and the AC generator which replaced DC generator were developed.[1] During the period of 1973 to 1979, the oil crises led to lots of research about the wind generation. At the end of 1990s, wind power had an important role in the sustainable energy. At the same time, wind turbine technologies were developed in the whole world, especially in Denmark, Germany, and Spain. Today, wind energy is the fastest growing energy source. According to the Global Wind Energy Council (GWEC), global wind power capacity has increased from 7600 MW at the end of 1997 to 195.2 GW by 2009. However wind power accounts for less than 1.0% of world's electrical demand. It is inferred that the wind power energy will develop to about 12% of the world's electrical supply by 2020.[2]

A lot of developments have been taken place in the design of wind energy conversion systems (WECS). Modern wind turbines are highly sophisticated machines built on the aerodynamic principles developed from the aerospace industry, incorporating advanced materials and electronics and are designed to deliver energy across a wide-range of wind speeds. The following sections will discuss the different issues related to wind power generation and wind turbines design.

The rest of the chapter is organized as follows. A number of important topics including aerodynamic principle of wind turbine, power available in the wind, rotor efficiency, factors affecting power in the wind, wind turbine power curve, optimizing rotor diameter and generator rated power have been presented in Sec. 2. Section 3 discusses a number of design considerations such as choice between two and three blades turbine, weight and size considerations. Grid connected wind farms, problems related with grid connections and latest trends of wind power generation are described in Sec. 4. Section 5 discusses hybrid power system and economics of wind power system. The conclusion is presented in Sec. 7, followed by references at the end of chapter.

Classification of wind turbine rotors, different types of generators used in the wind turbines, types of wind turbines, dynamic models of wind turbine will be discussed in detail in Chap. 2 of the book.

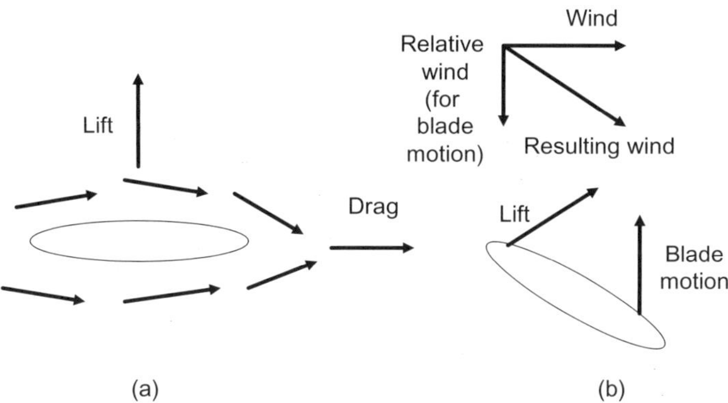

Fig. 1.1.   The lift in (a) is the result of faster air sliding over the top of the wind foil. In (b), the combination of actual wind and the relative wind due to blade motion creates a resultant that creates the blade lift.[3]

1.2   Power in the Wind

1.2.1   *Aerodynamics principle of wind turbine*

Figure 1.1(a) shows an airfoil, where the air moving the top has a greater distance to pass before it can rejoin the air that takes the short cut under the foil. So the air pressure on the top is lower than the air pressure under the airfoil. The air pressure difference creates the lifting force which can hold the airplane up.

In terms of the wind turbine blade, it is more complicated than the aircraft wing. From Fig. 1.1(b) we can find that a rotating turbine blade sees air moving toward it not only from the wind itself, but also from the relative motion of the blade. So the combination of the wind and blade motion is the resultant wind which moves toward the blade at a certain angle.

The angle between the airfoil and the wind is called the angle of attack as shown in Fig. 1.2. Increasing the angle of attack can improve the lift at the expense of increased drag. However, if we increase the angle of attack too much the wing will stall and the airflow will have turbulence and damage the turbine blades.

1.2.2   *Power available in the wind*

The total power available in wind is equal to the product of mass flow rate of wind m_w, and $V^2/2$. Assuming constant area or ducted flow, the continuity equation states

                    F. Yao et al.

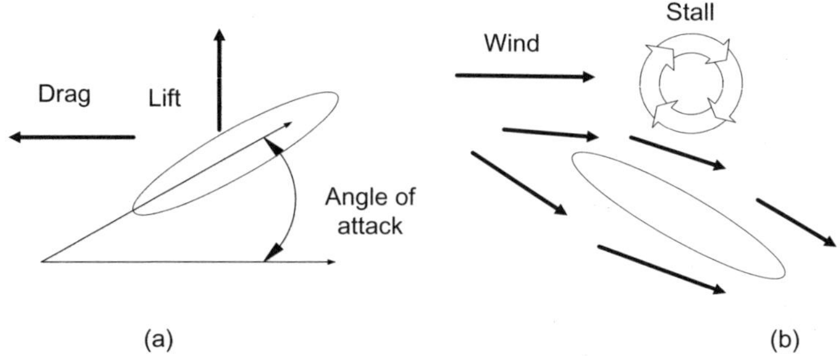

Fig. 1.2.   An increase in the angle of attack can cause a wing to stall.[3]

that $m_w = \rho A V$, where ρ is the density of air in kg/m^3, A is the blades area in m^2, and V is velocity in m/s.

Thus, the total wind power,

$$P_w = (m_w V^2)/2 = (\rho A V^3)/2. \tag{1.1}$$

Here, the ρ is a function of pressure, temperature and relative humidity. Let us assume the inlet wind velocity is V_i and the output velocity is V_o, then the average velocity is $(V_i + V_o)/2$.

The wind power recovered from the wind is given as

$$P_{\text{out}} = m_w(V_I^2 - V_O^2)/2 = (\rho A/4)(V_i + V_O)(V_i^2 - V_o^2)$$
$$= (P_w/2)(1 + x - x^2 - x^3), \tag{1.2}$$

where $x = V_o/V_i$. Differentiating Eq. (1.2) with respect to x and setting it to zero gives the optimum value of x for maximum power output

$$d(P_{\text{out}})/dx = 0 = (1 - 2x - 3x^2) \tag{1.3}$$

and then we can get $x_{\text{max}\,p} = 1/3$.

Substituting the value of $x_{\text{max}\,p}$ in Eq. (1.2), the maximum power recovered is

$$P_{\text{out max}} = 16/27 P_w = 0.593 P_w. \tag{1.4}$$

It can be found that the maximum power from a wind system is 59.3% of the total wind power.

The electrical power output is,

$$P_e = C_p \eta_m \eta_g P_w, \tag{1.5}$$

where C_p is the efficiency coefficient of performance when the wind is converted to mechanical power. η_m is mechanical transmission efficiency and η_g is the electricity transmission efficiency.[4] The optimistic values for these coefficients are $C_p = 0.45$, $\eta_m = 0.95$ and $\eta_g = 0.9$, which give an overall efficiency of 38%. For a given system, P_w and P_e will vary with wind speed.

1.2.3 *Rotor efficiency*

For a given wind speed, the rotor efficiency is a function of the rotor turning rate. If the rotor turns too slowly, the efficiency drops off because the blades are letting too much wind pass by unaffected. However, if the rotor turns too fast, efficiency will reduce as the turbulence caused by one blade increasingly affects the blade that follows. The tip-speed ratio (TSR) is a function which can illustrate the rotor efficiency. The definition of the tip-speed-ratio is:

$$\text{TSR} = \text{rotor tip speed/wind speed} = (\pi dN)/60v, \qquad (1.6)$$

where N is rotor speed in rpm, d is the rotor diameter (m); and v is the wind speed (m/s) upwind of the turbine.

1.2.4 *Factors affecting wind power*

1.2.4.1 *Wind statistics*

Wind resource is a highly variable power source, and there are several methods of characterizing this variability. The most common method is the power duration curve.[5] Another method is to use a statistical representation, particularly a Weibull distribution function.[6] Long term wind records are used to select the rated wind speed for wind electric generators. The wind is characterized by a Weibull density function.

1.2.4.2 *Load factor*

There are two main objectives in wind turbine design. The first is to maximize the average power output. The second one is to meet the necessary load factor requirement of the load. The load factor is very important when the generator is pumping irrigation water in asynchronous mode.[7] Commonly assumed long-term average load factors may be anywhere from 25% to 30%.

1.2.4.3 *Seasonal and diurnal variation of wind power*

It is clear that the seasonal and diurnal variations have significant effects on wind. The diurnal variation can be reduced by increasing the height of the wind power generator tower. In the early morning, the average power is about 80% of the long term annual average power. On the other hand, in early afternoon hours, the average power can be 120% of the long term average power.

1.2.5 *Impact of tower height*

Wind speed will increase with the height because the friction at earth surface is large.[8] The rate of the increase of wind speed that is often used to characterize the impact of the roughness of the earth's surface on wind speed is given as:

$$\left(\frac{v}{v_o}\right) = \left(\frac{H}{H_o}\right)^{\alpha},\tag{1.7}$$

where v is the wind speed at height H, v_o is the nominal wind speed at height H_o, and α is the friction coefficient. This can be translated into a substantial increase in power at greater heights. Table 1.1 gives the typical values of friction coefficient for various terrain characteristics.

It is known that power in the wind is proportional to the cube of wind speed, so even the modest increase in wind speed will cause significant increase in the wind power. In order to get higher speed winds, the wind turbines will be mounted on a taller tower. The air friction is also an important aspect to be considered, in the first few hundred meters above the ground, wind speed is greatly affected by the friction that air experiences. So smoother is the surface, lesser is the air movement friction.

1.2.6 *Wind turbine sitting*

The factors that should be considered while installing wind generator are as follow:

(1) Availability of land.
(2) Availability of power grid (for a grid connected system).
(3) Accessibility of site.
(4) Terrain and soil.
(5) Frequency of lighting strokes.

Once the wind resource at a particular site has been established, the next factor that should be considered is the availability of land.[10–12] The area of the land required depends upon the size of wind farm. In order to optimize the power output from a

Table 1.1.   Friction coefficient for various terrain characteristics.[9]

Terrain characteristics	Friction coefficient α
Smooth hard ground, calm water	0.10
Tall grass on ground	0.15
High crops and hedges	0.20
Wooded countryside, many trees	0.25
Small town with trees	0.30
Large city with tall buildings	0.40

given site, some additional information is needed, such as wind rose, wind speeds, vegetation, topography, ground roughness, etc. In addition other information such as convenient access to the wind farm site, load bearing capacity of the soil, frequency of cyclones, earthquakes, etc., should also be considered. A detailed discussion on technologies and methods used in wind resource assessment is presented in Chap. 4 of the book.

1.2.7 *Idealized wind turbine power curve*

The power curve is an important item for a specific wind turbine. The wind power curve also shows the relationship between wind speed and generator electrical output.

1.2.7.1 *Cut-in wind speed*

When the wind speed is below the cut-in wind speed (V_C) shown in Fig. 1.3, the wind turbines cannot start.[13,14] Power in the low speed wind is not sufficient to overcome friction in the drive train of the turbine. The generator is not able to generate any useful power below cut in speed.

1.2.7.2 *Rated wind speed*

We can see from Fig. 1.3 that as the wind speed increases, the power delivered by the generator will increase as the cube of wind speed. When the wind speed reached V_R the rated wind speed, the generator can deliver the rated power. If the wind speed exceeds V_R, there must be some methods to control the wind power or else the generator may be damaged. Basically, there are three control approaches for large wind power machines: active pitch-control, passive stall-control, and the combination of the two ways.

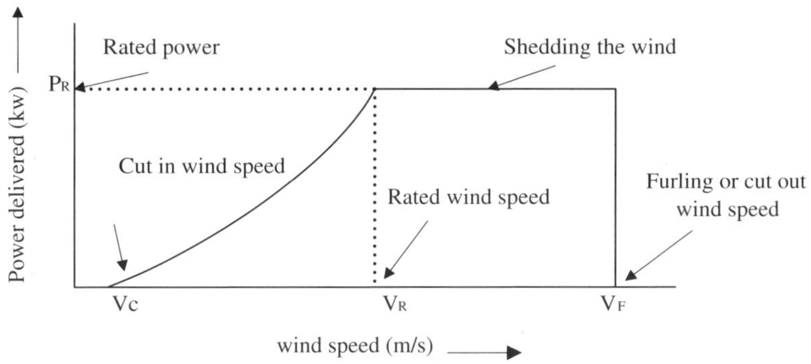

Fig. 1.3.   Idealized power curve.

In pitch-control system, an electronic system monitors the generator output power. If the power exceeds the rated power, the pitch of the turbine blades will adjust to shed some wind. The electronic system will control a hydraulic system to slowly rotate the blades about the axes, and turn them a few degrees to reduce the wind power. In conclusion, this strategy is to reduce the blade's angle of attack when the wind speeds over the rated wind speed.

For the stall-controlled machines, the turbine blades can reduce the efficiency automatically when the winds exceed the rated speed. In this control method, there are no moving parts, so this way is a kind of passive control. Most of the modern, large wind turbines use this passive, stall-controlled approach.

For large (above 1.0 MW), when the wind speed exceed the rated wind speed, the turbine machine will not reduce the angle of attack but increases it to induce stall.

For small size wind turbines, there are a variety of techniques to spill wind. The common way is the passive yaw control that can cause the axis of the turbine to move more and more off the wind. Another way relies on a wind vane mounted parallel to the plane of the blades. As winds get strong, the wind pressure on the vane rotate the machine away from the wind.

From Fig. 1.3 we can see that there is no power generated at wind speeds below V_C; at wind speeds between V_R and V_F, the output is equal to the rated power of the generator; above V_F the turbine is shut down.[13,14]

1.2.7.3   *Cut-out or furling wind speed*

Sometimes, the wind is too strong to damage the wind turbine. In Fig. 1.3 this wind speed is called as cut-out or the furling wind speed. Above V_F, the output power is zero. In terms of active pitch-controlled and passive stall-controlled machines, the rotor can be stopped by rotating the blades about their longitudinal axis to create a stall. However, for the stall-controlled machines, there will be the spring-loaded on the large turbine and rotating tips on the ends of the blades. When it is necessary, the hydraulic system will trip the spring and blade tips rotate 90° out of the wind and stop the turbine.

1.2.7.4   *Optimizing rotor diameter and generator rated power*

Figure 1.4 shows the trade-offs between rotor diameter and generator size as methods to increase the energy delivered by a wind turbine. In terms of Fig. 1.4(a), increasing the rotor diameter and keeping the same generator will shift the power curve upward. In this situation, the turbine generator can get the rated power at a lower wind speed. For Fig. 1.4(b), keeping the same rotor but increasing the generator size will allow

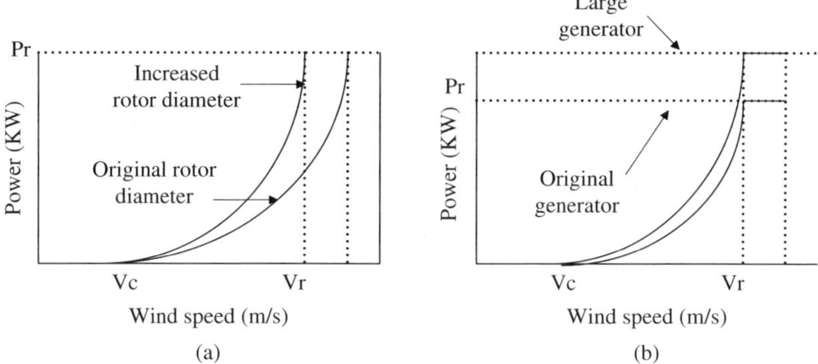

Fig. 1.4. (a) Increasing rotor diameter gives the rate power at lower wind speed, (b) increasing the generator size increases rate power.[9]

the power curve to continue upward to the new rated power. Basically, for the lower speed winds, the generator rated power need not change, but for the high wind speed area, increasing the rated power is a good strategy.[9,15,16]

1.2.8 *Speed control for maximum power*

It is known that the rotor efficiency C_p depends on the tip-speed ratio (TSR). Modern wind turbines operate optimally when their TSR is in the range of around 4–6.[15] In order to get the maximum efficiency, turbine blades should change their speed as the wind speed changes. There are different ways to control the rotor blades speed:

1.2.8.1 *Pole-changing induction generators*

In terms of the induction generator, the rotor spins at a frequency which is largely controlled by the number of poles. If it is possible for us to change the number of poles, we can make the wind turbine spin at different operating speeds. The stator can have external connections that switch the number of poles from one value to another without change in the rotor.

1.2.8.2 *Variable slip induction generators*

It is known that the speed of a normal induction generator is around 1% of the synchronous speed. The slip in the generator is a function of the dc resistance in the rotor conductors. If we add a variable resistance to the rotor, then the slip can range up to about 10%.[15]

1.3 Wind Turbine Design Considerations

A wind turbine consists of rotor, power train, control and safety system, nacelle structure, tower and foundations, etc.; the wind turbine manufacturer must consider many factors before selecting a final configuration for development.

First of all, the intended wind location environment is the most important aspect. The turbines for high turbulent wind sites should have robust, smaller diameter rotors. The International Electro-technical Commission (IEC) specified design criteria, which are based on the design loads on the mean wind speed and the turbulence level.

Secondly, minimizing cost is the next most important design criteria. In fact electricity generated by wind is more expensive than the electrical power from fuel-based generators. So cost is a very important factor that restrains the wind power generation from diversifying. If the cost of wind energy could be reduced by an additional 30% to 50%, then it could be globally competitive. In order to reduce the cost of wind energy, the wind energy designers can increase the size of the wind turbine, tailor the turbines for specific sites, explore new structural dynamic concepts, develop custom generators and power electronics.[16]

1.3.1 *Basic design philosophies*

There are three wind turbine design principles for handing wind loads: (i) withstanding the loads, (ii) shedding or avoiding loads and (iii) managing loads mechanically and/or electrically.[17] For the first design philosophy, the classic Danish configuration was originally developed by Paul La Com in 1890. These kinds of designs are reliability, high solidity but non-optimum blade pitch, low tip speed ratio (TSR) and three or more blades. For the wind turbines based on the second design philosophy, these turbines have design criteria such as optimization for performance, low solidity, optimum blade pitch, high TSR, etc. In terms of the designs based on the third philosophy, these wind turbines have design considerations like optimization for control, two or three blades, moderate TSR, mechanical and electrical innovations.

1.3.2 *Choice between two and three blade rotors*

Wind turbine blades are one of the most important components of a wind turbine rotor. Nowadays, fiber glass rotor blades are very popular. Rotor moment of inertia is the main difference between two and three blades. For the three bladed rotors mass movement has polar symmetry, whereas the two bladed rotor mass movements do not have the same, so the structural dynamic equations for the two bladed turbine system are more complex and have periodic coefficients.[17] In terms of the three bladed systems, the equations have constant coefficients which make them easier to solve. In conclusion, the three blade turbines are more expensive than the two blades. However, three blades can provide lower noise and polar symmetry.

1.3.3　*Weight and size considerations*

Wind tower is the integral component of the wind system. In order to withstand the thrust on the wind turbine, the wind tower must be strong enough. In addition, the wind tower must also support the wind turbine weight. It is common to use the tall wind towers because they can minimize the turbulence induced and allow more flexibility in siting. The ability of a wind tower to withstand the forces from the high wind is an important factor of a wind tower. The durability of the wind tower depends on the rotor diameter of wind turbine and its mode of operation under such conditions. In terms of the wind tower cost, the cost of operation and maintenance (O&M) and the cost of major overhauls and repairs also needed to be considered.

1.4　Grid Connected Wind Farms

1.4.1　*Wind farms*

Nowadays, a single wind turbine is just used for any particular site, such as an off-grid home in rural places or in off-shore areas. In a good windy site, normally there will many wind turbines which are often called as a wind farm or a wind park. The advantages of a wind farm are reduced site development costs, simplified connections to transmission lines, and more centralized access for operation and maintenance.

How many wind turbines can be installed at a wind site? If the wind turbines are located too close, it will result in upwind turbine interfering with the wind received by those located downwind. However, if the wind turbines are located too far, it means that the site space is not properly utilized.

When the wind passes the turbine rotor, the energy will be extracted by the rotor and the power which is available to the downwind machines will be reduced. Recent studies show that the wind turbine performance will degrade when the wind turbines are too close to each other. Figure 1.5 shows the optimum spacing of towers is estimated to be 3–5 rotor diameters between wind turbines within a row and 5–9 diameters between rows.[9,15]

1.4.2　*Problems related with grid connections*

For wind power generation, there must be a reliable power grid/transmission network near the site so that the wind generated power can be fed into the grid. Generally, the

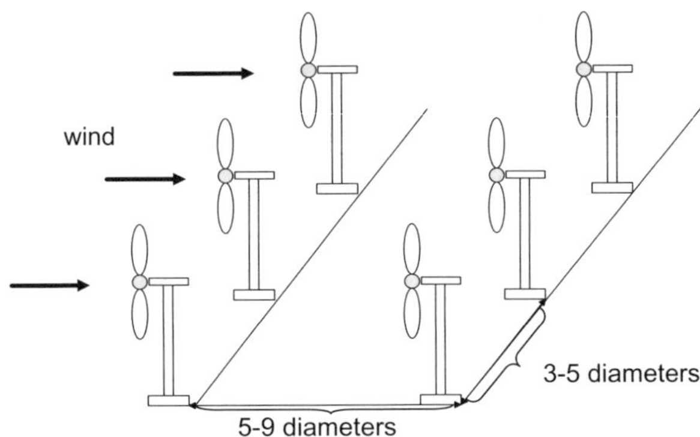

Fig. 1.5. Optimum spacing of towers in wind farm.

wind turbine generates power at 400 V, which is stepped up to 11–110 kV, depending upon the power capacity of the wind system. If the wind power capacity is up to 6 MW, the voltage level is stepped up to 11/22 kV; for a capacity of 6–10 MW, the voltage level is increased up to 33 kV; and for capacity higher than 10 MW, it is preferred to locate a 66 or 110 kV substation at the wind farm site.[18] An unstable wind power generation system may have the following problems:

1.4.2.1 *Poor grid security and reliability*

From economic point of view, the poor grid stability may cause 10–20% power loss,[18] and this deficiency may be the main reason for low actual energy output of wind power generation.

In China, many wind farms are actually not connected to the power grid because of the stability issues and difficulties in dispatching by the system operators. Major wind power researches are being conducted in aspects of dispatch issues, and long distance transmission issues.

In the Australian National Electricity Market (NEM), before the connection of a wind farm to a power grid, the (wind) generation service provider must conduct connectivity studies by itself and/or with the transmission network service provider for which the wind farm is to be connected. The connectivity study needs to check if the proposed wind generator can be hosted by the existing power grid in view of stability as well as reliability aspects. Depending on the study results conducted by the transmission network service provider, the cost associated and the suitability of the connection point of the proposed wind farm will be given for the generation company to make further decisions regarding its investment.

Table 1.2.   Offshore wind farms in Europe.[21]

Country	Project name	Capacity (MW)	Number of turbines	Wind turbine manufacturer
Denmark	Horns Rev 1	160	80	Vestas
Denmark	Nysted	165.6	72	Siemens
Denmark	Horns Rev 2	209	91	Siemens
Netherlands	Egmond Aan zee	108	38	Vestas
Netherlands	Prinses Amaila	120	60	Vestas
Sweden	Lillgund	110.4	48	Siemens
Gunfleet sands 1 and 2	Clacton-on Sear	104.4	29	Siemens

1.4.2.2 *Low frequency operation*

There is no doubt that the low frequency operation of the wind generation will affect the output power. Normally, when the frequency is less than 48 Hz, many wind power generations do not cut in. The power output loss could be around 5–10% on account of low frequency operation.[18]

1.4.2.3 *Impact of low power factor*

A synchronous generator can supply both active and reactive power. However, reactive power is needed by the wind power generation with induction generator for the magnetization. However, in terms of a wind power generator with induction generators, instead of supplying reactive power to the grid, they will absorb reactive power from grid. As a result, a suitable reactive power compensation device is required to supply the reactive power to wind generator/grid.[19,20]

1.4.3 *Latest trend of wind power generation*

In Europe, offshore projects are now springing up off the coasts of Denmark, Sweden, UK, France, Germany, Belgium, Irelands, Netherlands, and Scotland. The total offshore wind farm installed capacity in 2009 has reached 2055 MW. Table 1.2 shows operational offshore wind farms having installed of more than 100 MW in Europe till 2009.[21]

1.5 **Hybrid Power Systems**

There are still many locations in different parts of the world that do not have electrical connection to grid supply. A power system which can generate and supply power to such areas is called a remote, decentralized, standalone, autonomous, isolated power system, etc. It is a common way to supply electricity to these loads by diesel power plants. The diesel system is highly reliable which has been proved for many years. The main problems of diesel systems are that the cost of fuel, transportation, operation and maintenance are very high.

The cost of electricity can be reduced by integrating diesel systems with wind power generation. This system has another advantage of reductions in size of diesel engine and battery storage system, which can save the fuel and reduce pollution. Such systems having a parallel operation of diesel with one or more renewable energy based sources (wind, photovoltaic, micro hydro, biomass, etc.) to meet the electric demand of an isolated area are called autonomous hybrid power systems. Figure 1.6 shows a typical wind-diesel hybrid system with main components.[22] A hybrid system can have various options like wind-diesel, wind-diesel-photovoltaic, wind-diesel-micro hydro, etc.

The operation system of a diesel engine is very important. Normally there are two main modes of system operation which are running diesel engine either continuously or intermittently. The continuous diesel system operation has the advantage of technical simplicity and reliability. The main disadvantage of this approach is low utilization of renewable energy sources (wind) and not very considerable fuel savings. Basically, the minimum diesel loading should be 40% of the rated output, and then minimum fuel consumption will be around 60% of that at full load.[23] In order to get large fuel savings, it is expected that the diesel engine runs only when wind energy is lower than the demand. Nevertheless unless the load is significantly

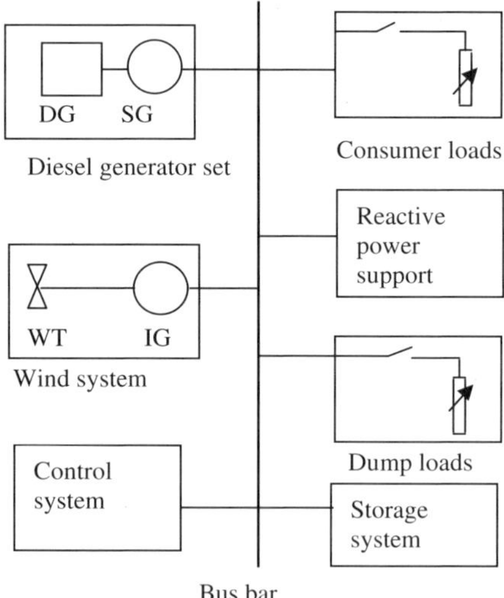

Fig. 1.6.   Schematic diagram of general isolated wind-diesel hybrid power system.

less than the energy supplied by the wind turbine, the diesel generator will not be able to stay off for a long time. The start-stop can be reduced by using the energy storage methods. To make the supply under these circumstances continuous, it is required to add complexity in the architecture or control strategy.

As wind is highly fluctuating in nature, and it will affect the supply quality considerably and may even damage the system in the absence of proper control mechanism. Main parameters to be controlled are the system frequency and voltage, which determine the stability and quality of the supply. In a power system, frequency deviations are mainly due to real power mismatch between the generation and demand, whereas voltage mismatch is the sole indicator of reactive power unbalance in the system. In the power system active power balance can be achieved by controlling the generation, i.e., by controlling the fuel input to the diesel electric unit and this method is called automatic generation control (AGC) or load frequency control (LFC) or by scheduling or management of the output power. The function of the load frequency controller is to generate, raise or lower command, depending upon the disturbance, to the speed-gear changer of the diesel engine which in turn changes the generation to match the load. Different methods of controlling the output power of autonomous hybrid power systems are dump load control, priority load control, battery storage, flywheel storage, pump storage, hydraulic/pneumatic accumulators, super magnetic energy storage, etc.[24]

It is equally important to maintain the voltage within specified limits, which is mainly related with the reactive power control of the system.[9,10] In general, in any hybrid system there will be an induction generator for the wind/hydro system. An induction generator offers many advantages over a synchronous generator in an autonomous hybrid power system. Reduced unit cost, ruggedness, brushless (in squirrel cage construction), absence of separate DC source for excitation, ease of maintenance, self-protection against severe overloads and short circuits, etc., are the main advantages.[25]

However the major disadvantage of the induction generator is that it requires reactive power for its operation. In the case of the grid-connected system, the induction generator can get the reactive power from grid/capacitor banks, whereas in the case of the isolated/autonomous system, reactive power can only be supplied by capacitor banks. In addition, most of the loads are also inductive in nature, therefore, the mismatch in generation and consumption of reactive power can cause serious problem of large voltage fluctuations at generator terminals especially in an isolated system. The terminal voltage of the system will sag if sufficient reactive power is not provided, whereas surplus reactive power can cause high voltage spikes in the system, which can damage the consumer's equipment and affect the supply quality. To take care of the reactive power/voltage control an appropriate reactive power compensating device is required.[19,22,24] Another approach available from ENERCON[27] consists of a wind turbine based on an annular generator connected to a diesel generator with energy storage to form a stand-alone power system.

1.6   Economics of Wind Power Systems

There is no doubt that the purpose of all types of energy generation ultimately depends on the scale of economics. Wind power generation costs have been falling over recent years. It is estimated that wind power in many countries is already competitive with fossil fuel and nuclear power if social/environmental costs are considered.[26]

The installation cost of a wind system is the capital cost of a wind turbine (see Fig. 1.7 for the normalized contribution of an individual sub-system towards total capital cost of a wind turbine), land, tower, and its accessories, and it accounts for less than any state or federal tax credits.

The installation cost of a wind system is the cost of wind turbine, land, tower, and its accessories and it accounts for less than any state or federal tax credits. The maintenance cost of the wind system is normally very small and annual maintenance cost is about 2% of the total system cost. The cost of financing to purchase the wind system is significant in the overall cost of wind system. Furthermore the extra cost such as property tax, insurance of wind system and accidents caused from the wind system. One of the main advantages of generating electricity from the wind system is that the wind is free. The cost of the wind system just occurs once. On the

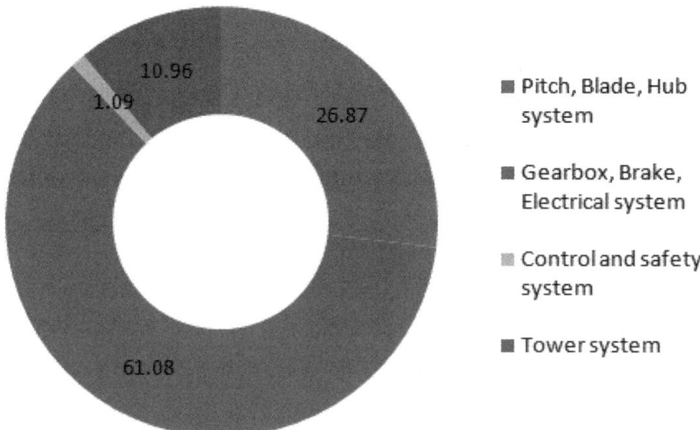

Fig. 1.7.   Contribution of various sub-systems towards capital cost of wind turbine.

other hand, the cost of non-renewable energies is more and more expensive, which is required for renewable energies such as wind power.

Nowadays, research and development make the wind power generation competitive with other non-renewable fuels such as fossil fuel and nuclear power. Lots of efforts have been done to reduce the cost of wind power by design improvement, better manufacturing technology, finding new sites for wind systems, development of better control strategies (output and power quality control), development of policy and instruments, human resource development, etc.[20]

1.7 Conclusion

Wind power generation is very essential in today's society development. Lots of wind power technologies have been researched and numbers of wind farms have been installed. The performance of wind energy conversion systems depends on the subsystems such as wind turbine (aerodynamic), gears (mechanical), and generator (electrical). In this chapter a number of wind power issues, such as power in the wind, impact of tower height, maximum rotor efficiency, speed control for maximum power, some of the design considerations in wind turbine design, wind farms, latest trend of wind power generation from off shore sites, problems related with grid connections and hybrid power systems have been discussed.

References

1. R.C. Bansal, T.S. Bhatti and D.P. Kothari, "On some of the design ascpects of wind energy conversion systems," *Energy Conversion and Management* **43** (2002) 2175–2187.
2. "Global wind scenario," *Power Line* **7** (2003) 49–53.
3. S. Muller, M. Deicke and R.W. De Doncker, "Double fed induction generator systems," *IEEE Industry Application Magazine* **8** (2002) 26–33.
4. B. Singh, "Induction generator — A prospective," *Electric Machines and Power Systems* **23** (1995) 163–177.
5. P.K. Sandhu Khan and J.K. Chatterjee, "Three-phase induction generators: A discussion on performance," *Electric Machines and Power Systems* **27** (1999) 813–832.
6. P. Gipe, *Wind Power* (Chelsea Green Publishing Company, Post Mills, Vermount, 1995).
7. G.D. Rai, *Non Conventional Energy Sources*, 4th edition (Khanna Publishers, New Delhi, India, 2000).
8. A.W. Culp, *Principles of Energy Conversion*, 2nd Edition (McGraw Hill International Edition, New York, 1991).
9. G.M. Masters, *Renewable and Efficient Electrical Power Systems* (John Wiley & Sons, Inc., Ho, New Jersey, 2004).
10. T.S. Jayadev, "Windmills stage a comeback," *IEEE Spectrum* **13** (1976) 45–49.
11. G.L. Johnson, "Economic design of wind electric generators," *IEEE Trans. Power Apparatus Systems* **97** (1978) 554–562.

12. K.T. Fung, R.L. Scheffler and J. Stolpe, "Wind energy — a utility perspective," *IEEE Trans. Power Apparatus Systems* **100** (1981) 1176–1182.

13. G.L. Johnson, *Wind Energy Systems* (Prentice-Hall, Englewood Cliffs, New Jersey, 1985).

14. J.F. Manwell, J.G. McGowan and A.L. Rogers, *Wind Energy Explained Theory, Design and Application* (John Wiley & Sons, Inc., Ho, New Jersey, 2002).

15. M.R. Patel, *Wind and Solar Power Systems* (CRC Press LLC, Boca Raton, Florida, 1999).

16. D.C. Quarton, "The evolution of wind turbine design analysis — A twenty years progress review," *Int. J. Wind Energy* **1** (1998) 5–24.

17. R.W. Thresher and D.M. Dodge, "Trends in the evolution of wind turbine generator configurations and systems," *Int. J. Wind Energy* **1** (1998) 70–85.

18. R.C. Bansal, T.S. Bhatti and D.P. Kothari, "Some aspects of grid connected wind electric energy conversion systems," *Interdisciplinary J. Institution on Engineers (India)* **82** (2001) 25–28.

19. Z. Saad-Saund, M.L. Lisboa, J.B. Ekanayka, N. Jenkins and G. Strbac, "Application of STATCOMs to wind farms," *IEE-Proc. Generation, Transmission and Distribution* **145** (1998) 511–516.

20. J. Bonefeld and J.N. Jensen, "Horns rev-160 MW offshore wind," *Renewable Energy World* **5** (2002) 77–87.

21. European Wind Energy Association, http://www.ewea.org.

22. R.C. Bansal and T.S. Bhatti, *Small Signal Analysis of Isolated Hybrid Power Systems: Reactive Power and Frequency Control Analysis* (Alpha Science International U.K. & Narosa Publishers, New Delhi, 2008).

23. R. Hunter and G. Elliot, *Wind-Diesel Systems, A Guide to the Technology and Its Implementation* (Cambridge University Press, 1994).

24. R.C. Bansal, "Automatic reactive power control of isolated wind-diesel hybrid power systems," *IEEE Trans. Industrial Electronics* **53** (2006) 1116–1126.

25. A.A.F. Al-Ademi, "Load-frequency control of stand-alone hybrid power systems based on renewable energy sources," Ph.D Thesis, Centre for Energy Studies, Indian Institute of Technology, Delhi (1996).

26. J. Beurskens and P.H. Jensen, "Economics of Wind Energy Prospects and Directions," *Renewable Energy World* **4** (2001) 103–121.

27. Enercon Wind Diesel Electric System, http://www.enercon.de.

Chapter 2

Wind Turbine Systems: History, Structure, and Dynamic Model

S. Masoud Barakati

Faculty of Electrical and Computer Engineering,
University of Sistan and Baluchestan
Zahedan, Iran
smbaraka@ece.usb.ac.ir

This chapter focuses on wind turbine structure and modeling. First, a brief historical background on the wind will be presented. Then classification of the wind turbine based on generators, power electronic converters, and connecting to the grid will be discussed. The overall dynamic model of the wind turbine system will be explained in the end of the chapter.

2.1   Wind Energy Conversion System (WECS)

A wind energy conversion system (WECS) is composed of blades, an electric generator, a power electronic converter, and a control system, as shown in Fig. 2.1. The WECS can be classified in different types, but the functional objective of these systems is the same: converting the wind kinetic energy into electric power and injecting this electric power into the electrical load or the utility grid.

2.1.1   *History of using wind energy in generating electricity*

History of wind energy usage for the generation of electricity dates back to the 19th century, but at that time the low price of fossil fuels made wind energy economically unattractive.[1] The research on modern Wind Energy Conversion Systems (WECS) was put into action again in 1973 because of the oil crisis. Earlier research was on making high power modern wind turbines, which need enormous electrical generators. At that time, because of technical problems and high cost of manufacturing, making huge turbines was hindered.[1,2] So research on the wind turbine turned to

                                    S. M. Barakati

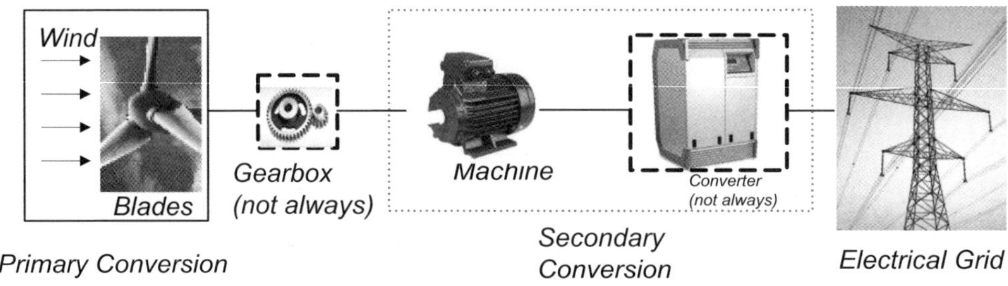

Fig. 2.1.   Block diagram of a WECS.

making low-price turbines, which composed of a small turbine, an induction generator, a gearbox and a mechanical simple control method. The turbines had ratings of at least several tens of kilowatts, with three fixed blades. In this kind of system, the shaft of the turbine rotates at a constant speed. The asynchronous generator is a proper choice for this system. These low-cost and small-sized components made the price reasonable even for individuals to purchase.[3]

As a result of successful research on wind energy conversion systems, a new generation of wind energy systems was developed on a larger scale. During the last two decades, as the industry gained experience, the production of wind turbines has grown in size and power rating. It means that the rotor diameter, generator rating, and tower height have all increased. During the early 1980s, wind turbines with rotor spans of about 10 to 15 meters, and generators rated at 10 to 65 kW, were installed. By the mid-to late 1980s, turbines began appearing with rotor diameters of about 15 to 25 meters and generators rated up to 200 kW. Today, wind energy developers are installing turbines rated at 200 kW to 2 MW with rotor spans of about 47 to 80 meters. According to the American Wind Energy Association (AWEA), today's large wind turbines produce as much as 120 times more electricity than early turbine designs, with Operation and Maintenance (O&M) costs only modestly higher, thus dramatically cutting O&M costs per kWh. Large turbines do not turn as fast, and produce less noise in comparison to small wind turbines.[4]

Another modification has been the introduction of new types of generators in wind systems. Since 1993, a few manufacturers have replaced the traditional asynchronous generator in their wind turbine designs with a synchronous generator, while other manufacturers have used doubly-fed asynchronous generators.

In addition to the above advances in wind turbine systems, new electrical converters and control methods were developed and tested. Electrical developments include using advanced power electronics in the wind generator system design, and introducing the new concept, namely variable speed. Due to the rapid advancement of power electronics, offering both higher power handling capability and lower price/kW,[5] the application of power electronics in wind turbines is expected to increase further. Also, some control methods were developed for big turbines with the variable-pitch blades in order to control the speed of the turbine shaft. The pitch control concept has been applied during the last fourteen years.

A lot of effort has been dedicated to comparison of different structures for wind energy systems, as well as their mechanical, electrical and economical aspects. A good example is the comparison of variable-speed against constant-speed wind turbine systems. In terms of energy capture, all studies come to the same result that variable speed turbines will produce more energy than constant speed turbines.[6] Specifically, using variable-speed approach increases the energy output up to 20% in a typical wind turbine system.[7] The use of pitch angle control has been shown to result in increasing captured power and stability against wind gusts.

For operating the wind turbine in variable speed mode, different schemes have been proposed. For example, some schemes are based on estimating the wind speed in order to optimize wind turbine operation.[8] Other controllers find the maximum power for a given wind operation by employing an elaborate searching method.[9–11] In order to perform speed control of the turbine shaft, in an attempt to achieve maximum power, different control methods such as field-oriented control and constant Voltage/frequency (V/f) have been used.[12–15]

As mentioned in the previous section, in the last 25 years, four or five generations of wind turbine systems have been developed.[16] These different generations are distinguished based on the use of different types of wind turbine rotors, generators, control methods and power electronic converters. In the following sections, a brief explanation of these components is presented.

2.1.2 *Classification of wind turbine rotors*

Wind turbines are usually classified into two categories, according to the orientation of the axis of rotation with respect to the direction of wind, as shown in Fig. 2.2[17,18]:

- Vertical-axis turbines
- Horizontal-axis turbines.

2.1.2.1 *Vertical-axis wind turbine (VAWT)*

The first windmills were built based on the vertical-axis structure. This type has only been incorporated in small-scale installations. Typical VAWTs include the Darrius rotor, as shown in Fig. 2.2(a). Advantages of the VAWT[20,21] are:

- Easy maintenance for ground mounted generator and gearbox,
- Receive wind from any direction (no yaw control required), and
- Simple blade design and low cost of fabrication.

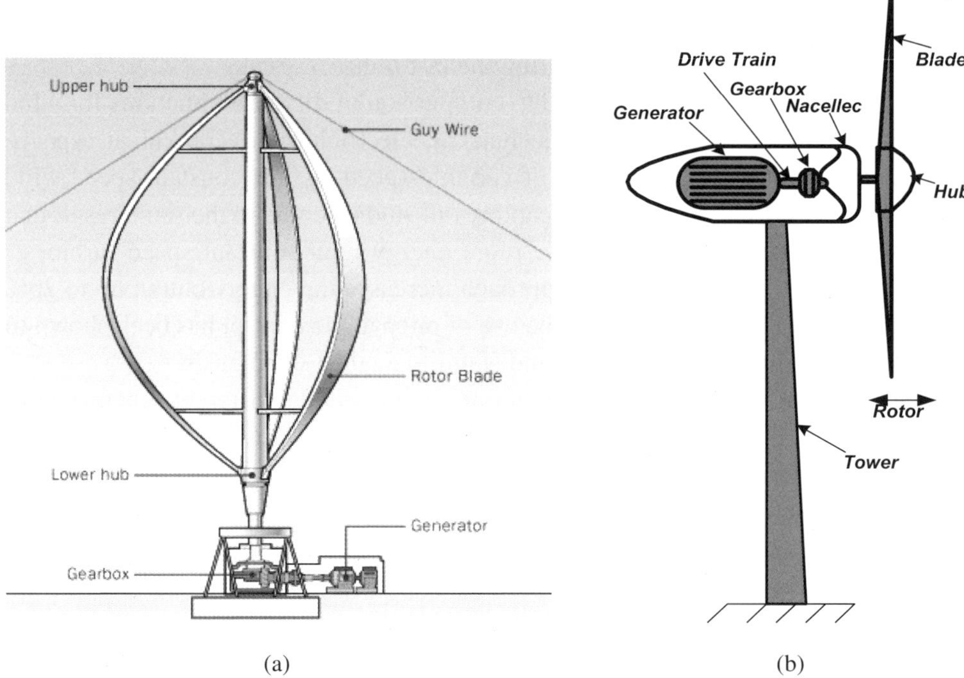

(a)   (b)

Fig. 2.2.   (a) A typical vertical-axis turbine (the Darrius rotor),[19] (b) a horizontal-axis wind turbine.[1]

Disadvantages of a vertical-axis wind turbine are:

- Not self starting, thus, require generator to run in motor mode at start,
- Lower efficiency (the blades lose energy as they turn out of the wind),
- Difficulty in controlling blade over-speed, and
- Oscillatory component in the aerodynamic torque is high.

2.1.2.2   *Horizontal-axis wind turbines (HAWT)*

The most common design of modern turbines is based on the horizontal-axis structure. Horizontal-axis wind turbines are mounted on towers as shown in Fig. 2.2(b). The tower's role is to raise the wind turbine above the ground to intercept stronger winds in order to harness more energy.

Advantages of the HAWT:

- Higher efficiency,
- Ability to turn the blades, and
- Lower cost-to-power ratio.

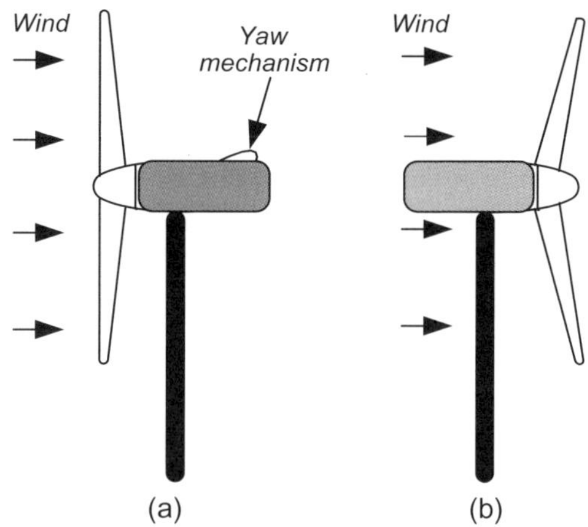

Fig. 2.3.   (a) Upwind structure, (b) downwind structure.[1]

Disadvantages of the horizontal-axis:

- Generator and gearbox should be mounted on a tower, thus restricting servicing, and
- More complex design required due to the need for yaw or tail drive.

The HAWT can be classified as upwind and downwind turbines based on the direction of receiving the wind, as shown in Fig. 2.3.[22,23] In the upwind structure the rotor faces the wind directly, while in downwind structure, the rotor is placed on the lee side of the tower. The upwind structure does not have the tower shadow problem because the wind stream hits the rotor first. However, the upwind needs a yaw control mechanism to keep the rotor always facing the wind. On the contrary, the downwind may be built without a yaw mechanism. However, the drawback is the fluctuations due to the tower shadow.

2.1.3   *Common generator types in wind turbines*

The function of an electrical generator is providing a means for energy conversion between the mechanical torque from the wind rotor turbine, as the prime mover, and the local load or the electric grid. Different types of generators are being used with wind turbines. Small wind turbines are equipped with DC generators of up to a few kilowatts in capacity. Modern wind turbine systems use three-phase AC generators.[21] The common types of AC generator that are possible candidates in modern wind turbine systems are as follows:

- Squirrel-Cage rotor Induction Generator (SCIG),
- Wound-Rotor Induction Generator (WRIG),

- Doubly-Fed Induction Generator (DFIG),
- Synchronous Generator (with external field excitation), and
- Permanent Magnet Synchronous Generator (PMSG).

For assessing the type of generator in WECS, criteria such as operational characteristics, weight of active materials, price, maintenance aspects and the appropriate type of power electronic converter, are used.

Historically, the induction generator (IG) has been extensively used in commercial wind turbine units. Asynchronous operation of induction generators is considered an advantage for application in wind turbine systems, because it provides some degree of flexibility when the wind speed is fluctuating.

There are two main types of induction machines: squirrel-cage (SC), and wound-rotor (WR). Another category of induction generator is the DFIG; the DFIG may be based on the squirrel-cage or wound-rotor induction generator.

The induction generator based on SCIG is a very popular machine because of its low price, mechanical simplicity, robust structure, and resistance against disturbance and vibration.

The wound-rotor is suitable for speed control purposes. By changing the rotor resistance, the output of the generator can be controlled and also speed control of the generator is possible. Although the WRIG has the advantage described above, it is more expensive than a squirrel-cage rotor.

The DFIG is a kind of induction machine in which both the stator windings and the rotor windings are connected to the source. The rotating winding is connected to the stationary supply circuits via power electronic converter. The advantage of connecting the converter to the rotor is that variable-speed operation of the turbine is possible with a much smaller, and therefore much cheaper converter. The power rating of the converter is often about 1/3 the generator rating.[24]

Another type of generator that has been proposed for wind turbines in several research articles is a synchronous generator.[25–27] This type of generator has the capability of direct connection (direct-drive) to wind turbines, with no gearbox. This advantage is favorable with respect to lifetime and maintenance. Synchronous machines can use either electrically excited or permanent magnet (PM) rotor.

The PM and electrically-excited synchronous generators differ from the induction generator in that the magnetization is provided by a Permanent Magnet pole system or a dc supply on the rotor, featuring providing self-excitation property. Self-excitation allows operation at high power factors and high efficiencies for the PM synchronous.

It is worth mentioning that induction generators are the most common type of generator use in modern wind turbine systems.[5]

2.1.3.1 *Mechanical gearbox*

The mechanical connection between an electrical generator and the turbine rotor may be direct or through a gearbox. In fact, the gearbox allows the matching of the generator speed to that of the turbine. The use of gearbox is dependent on the kind of electrical generator used in WECS. However, disadvantages of using a gearbox are reductions in the efficiency and, in some cases, reliability of the system.

2.1.3.2 *Control method*

With the evolution of WECS during the last decade, many different control methods have been developed. The control methods developed for WECS are usually divided into the following two major categories:

- Constant-speed methods, and
- Variable-speed methods.

2.1.3.2.1 Variable-speed turbine versus constant-speed turbine

In constant-speed turbines, there is no control on the turbine shaft speed. Constant speed control is an easy and low-cost method, but variable speed brings the following advantages:

- Maximum power tracking for harnessing the highest possible energy from the wind,
- Lower mechanical stress,
- Less variations in electrical power, and
- Reduced acoustical noise at lower wind speeds.

In the following, these advantages will be briefly explained.

Using shaft speed control, higher energy will be obtained. Reference 28 compares the power extracted for a real variable-speed wind turbine system, with a 34-m-diameter rotor, against a constant-speed wind turbine at different wind speeds. The results are illustrated in Fig. 2.4. The figure shows that a variable-speed system outputs more energy than the constant-speed system. For example, with a fixed-speed system, for an average annual wind speed of 7 m/s, the energy produced is 54.6 MWh, while the variable-speed system can produce up to 75.8 MWh, under the same conditions. During turbine operation, there are some fluctuations related to mechanical or electrical components. The fluctuations related to the mechanical parts include current fluctuations caused by the blades passing the tower and various current amplitudes caused by variable wind speeds. The fluctuations related to the electrical parts, such as voltage harmonics, is caused by the electrical converter. The electrical harmonics can be conquered by choosing the proper electrical filter.

                                        S. M. Barakati

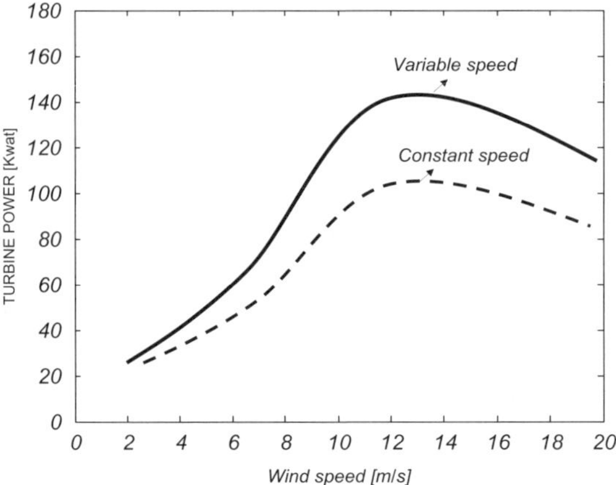

Fig. 2.4.   Comparison of power produced by a variable-speed wind turbine and a constant-speed wind turbine at different wind speeds.

However, because of the large time constant of the fluctuations in mechanical components, they cannot be canceled by electrical components. One solution that can largely reduce the disturbance related to mechanical parts is using a variable-speed wind turbine. References 6 and 28 compare the power output disturbance of a typical wind turbine with the constant-speed and variable-speed methods, as shown in Fig. 2.5. The figure illustrates the ability of the variable-speed system to reduce or increase the shaft speed in case of torque variation. It is important to note that the disturbance of the rotor is related also to the mechanical inertia of the rotor.

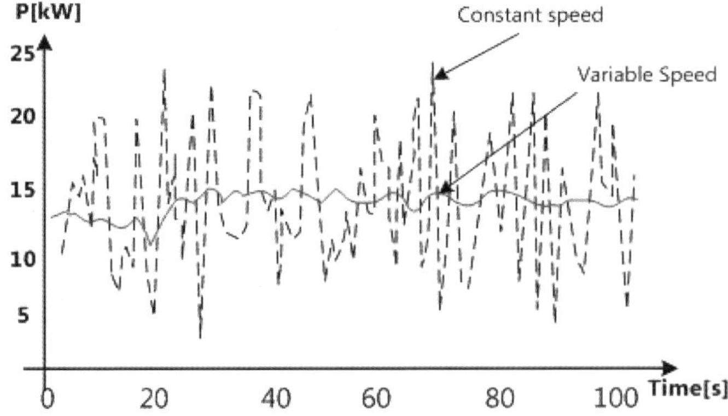

Fig. 2.5.   Power output disturbance of a typical wind turbine with constant-speed method and variable-speed methods.[1,5,27]

Although a variable-speed operation is adopted in modern wind turbines, this method has some disadvantages, such as additional cost for extra components and complex control methods.[9,30]

2.1.4 *Power electronic converter*

The power electronic (PE) converter has an important role in modern WECS with the variable-speed control method. The constant-speed systems hardly include a PE converter, except for compensation of reactive power. The important challenges for the PE converter and its control strategy in a variable-speed WECS are[31]:

- Attain maximum power transfer from the wind, as the wind speed varies, by controlling the turbine rotor speed, and
- Change the resulting variable-frequency and variable-magnitude AC output from the electrical generator into a constant-frequency and constant-magnitude supply which can be fed into an electrical grid.

As a result of rapid developments in power electronics, semiconductor devices are gaining higher current and voltage ratings, less power losses, higher reliability, as well as lower prices per kVA. Therefore, PE converters are becoming more attractive in improving the performance of wind turbine generation systems. It is worth mentioning that the power passing through the PE converter (that determines the capacity the PE converter) is dependent on the configuration of WECS. In some applications, the whole power captured by a generator passes through the PE converter, while in other categories only a fraction of this power passes through the PE converter.

The most common converter configuration in variable-speed wind turbine system is the rectifier-inverter pair. A matrix converter, as a direct AC/AC converter, has potential for replacing the rectifier-inverter pair structure.

2.1.4.1 *Back-to-back rectifier-inverter pair*

The back-to-back rectifier-inverter pair is a bidirectional power converter consisting of two conventional pulse-width modulated (PWM) voltage-source converters (VSC), as shown in Fig. 2.6. One of the converters operates in the rectifying mode, while the other converter operates in the inverting mode. These two converters are connected together via a dc-link consisting of a capacitor. The dc-link voltage will be maintained at a level higher than the amplitude of the grid line-to-line voltage, to achieve full control of the current injected into the grid. Consider a wind turbine system including the back-to-back PWM VSC, where the rectifier and inverter are connected to the generator and the electrical grid, respectively. The power flow is controlled by the grid-side converter (GSC) in order to keep the dc-link

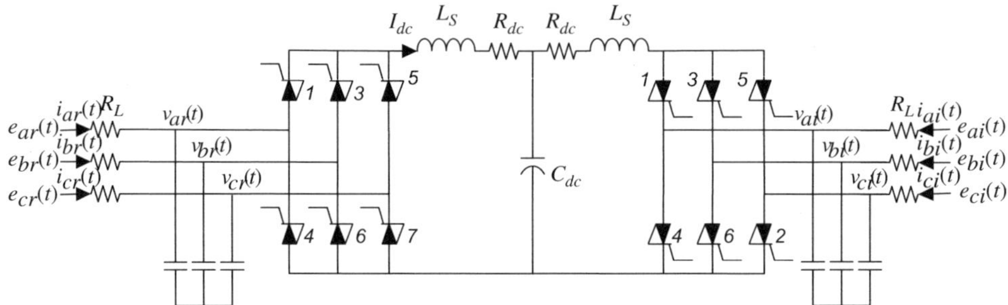

Fig. 2.6.   The back-to-back rectifier-inverter converter.

voltage constant, while the generator-side converter is responsible for excitation of the generator (in the case of squirrel-cage induction generator) and control of the generator in order to allow for maximum wind power to be directed towards the dc bus.[31] The control details of the back-to-back PWM VSC structure in the wind turbine applications has been described in several papers.[32-35]

Among the three-phase AC/AC converters, the rectifier-inverter pair structure is the most commonly used, and thus, the most well-known and well-established. Due to the fact that many semiconductor device manufacturers produce compact modules for this type of converter, the component cost has gone down. The dc-link energy-storage element provides decoupling between the rectifier and inverter. However, in several papers, the presence of the dc-link capacitor has been considered as a disadvantage. The dc-link capacitor is heavy and bulky, increases the cost, and reduces the overall lifetime of the system.[36-39]

2.1.4.2   *Matrix converter*

Matrix converter (MC) is a one-stage AC/AC converter that is composed of an array of nine bidirectional semiconductor switches, connecting each phase of the input to each phase of the output. This structure is shown in Fig. 2.7.

The basic idea behind the matrix converter is that a desired output frequency, output voltage and input displacement angle can be obtained by properly operating the switches that connect the output terminals of the converter to its input terminals. The development of MC configuration with high-frequency control was first introduced in the work of Venturini and Alesina in 1980.[40,41] They presented a static frequency changer with nine bidirectional switches arranged as a 3 × 3 array and named it a matrix converter. They also explained the low-frequency modulation method and direct transfer function approach through a precise mathematical analysis. In this method, known as direct method, the output voltages are obtained from multiplication of the modulation transfer matrix by input voltages.[42] Since then, the research

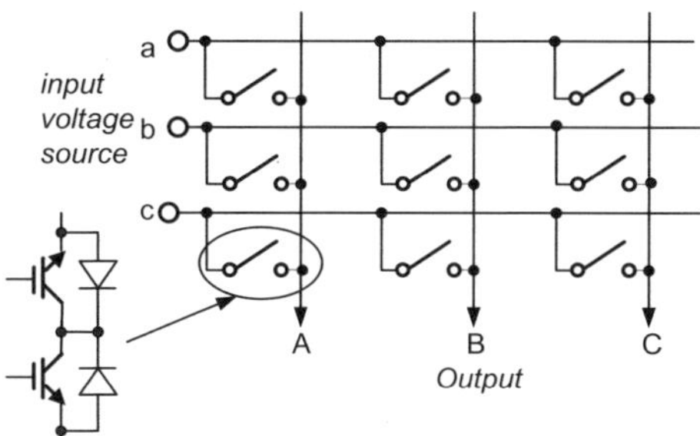

Fig. 2.7.   Matrix converter structure, the back-to-back rectifier-inverter converter.

on the MC has concentrated on the implementation of bidirectional switches, as well as modulation techniques.

As in case of comparison MC with the rectifier-inverter pair under PWM switching strategy, MC provides low-distortion sinusoidal input and output waveforms, bi-directional power flow, and controllable input power factor.[43] The main advantage of the MC is in its compact design which makes it suitable for applications where size and weight matter, such as in aerospace applications.[44]

The following drawbacks have been attributed to matrix converters: The magnitude of the MC output voltage can only reach 0.866 times than that of the input voltage, input filter design for MC is complex, and because of an absence of a dc-link capacitor in the MC structure the decoupling between input and output and ride-through capability do not exist, limiting the use of MC.[45]

2.1.5   *Different configurations for connecting wind turbines to the grid*

The connection of the wind turbine to the grid depends on the type of electrical generator and power electronic converter used. Based on the application of PE converters in the WECS, the wind turbine configurations can be divided into three topologies: directly connected to the grid without any PE converter, connected via full-scale the PE converter, and connected via partially-rated PE converter. In the following, the generator and power electronic converter configurations most commonly used in wind turbine systems are discussed.

As a simple, robust and relatively low-cost system, a squirrel-cage induction generator (SCIG), as an asynchronous machine, is connected directly to the grid, as depicted in Fig. 2.8. For an induction generator, using a gearbox is necessary in order to interface the generator speed and turbine speed. The capacitor bank (for reactive power compensation) and soft-starter (for smooth grid connection) are also required. The speed and power are limited aerodynamically by stall or pitch control. The variation of slip is in the range of 1–2%, but there are some wind

     S. M. Barakati

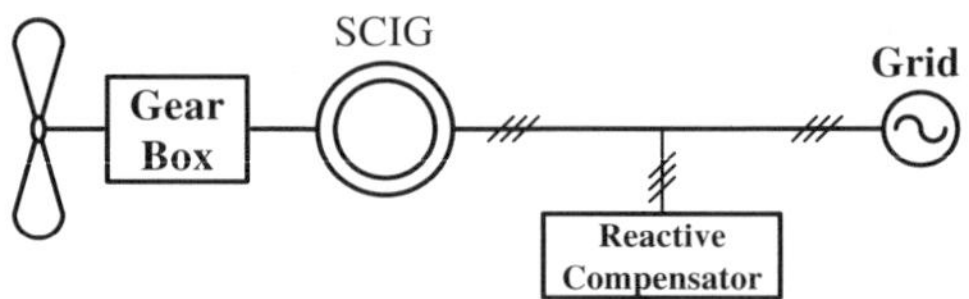

Fig. 2.8.   Wind turbine system with SCIG.

turbines based on SCIG in industry with increased rotor resistance and, therefore, increased slip (2–3%). This scheme is used to allow a little bit of speeding up during wind gusts in order to reduce the mechanical stresses. However, this configuration based on an almost fixed speed is not proper for a wind turbine in a higher power range and also for locations with widely varying wind velocity.[5,46]

Three wind turbine systems based on induction generators, with the capability of variable-speed operation are shown in Fig. 2.9.[5,16] The wind turbine system in Fig. 2.9(a) uses a wound-rotor induction generator (WRIG). The idea of this model

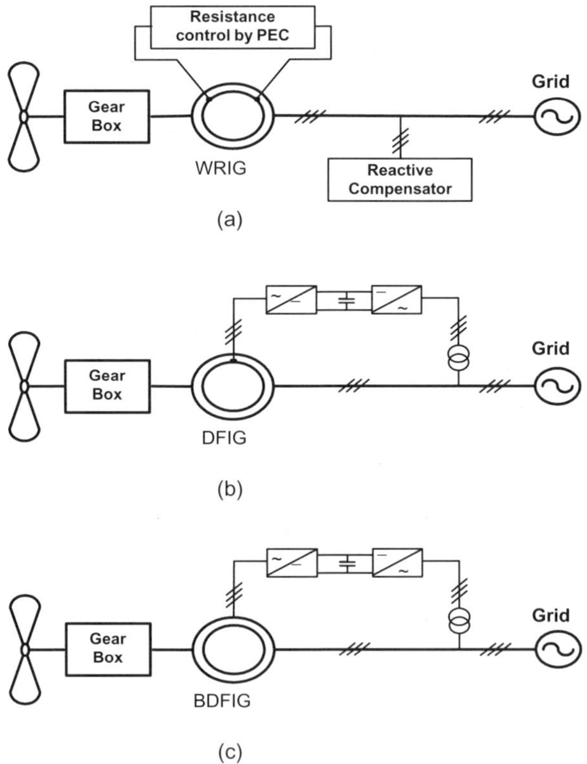

Fig. 2.9.   Wind turbine systems based on the induction generator with capability of variable-speed operation: (a) Wound-Rotor, (b) Doubly-Fed, and (c) Brushless Doubly-Fed induction generators.

is that the rotor resistance can be varied electronically using a variable external rotor resistance and a PE converter. By controlling the rotor resistance, the slip of the machine will be changed over a 10% range (speed range 2–4%).[5] In normal operation, the rotor resistance is low, associated with low slip, but during wind gusts the rotor resistance is increased to allow speeding up.

Figure 2.9(b) shows a configuration employing a Doubly-Fed Induction Generator (DFIG) and a power electronic converter that connects the rotor winding to the grid directly. With this configuration, it is possible to extend the speed range further without affecting the efficiency. The reason for speed control without loss of efficiency is that slip power can be fed back to the grid by the converter instead of being wasted in the rotor resistance. Note that the power rating of the power converter is sP_{nom}, where "s" is the maximum possible slip and P_{nom} is the nominal power of the machine. The rotor slip (s) can be positive or negative because the rotor power can be positive or negative, due to the bidirectional nature of the power electronic converter. For example, if the power rating of the converter is 10% of the power rating of the generator, the speed control range is from 90% to 110% of the synchronous speed. It means that at 110% speed, $s = -0.1$ and power is fed from the rotor to the grid, whereas at 90% speed, the slip is $s = +0.1$, and 10% of the power is fed from the grid to the rotor through the converter. With these attributes, i.e., a larger control range and smaller losses, the configuration in Fig. 2.9(b) is more attractive than the configuration in Fig. 2.9(a).

In the configurations shown in Figs. 2.9(a) and 2.9(b), with the wound-rotor induction generator, the access to the rotor is possible through the slip rings and brushes. Slip rings and brushes cause mechanical problems and electrical losses. In order to solve the problems of using slip rings and brushes, one alternative is by using the Brushless Doubly-Fed induction generator (BDFIG), shown in Fig. 2.9(c). In this scheme, the stator windings (main winding) are directly connected to the grid, while the three-phase auxiliary winding is connected to the electrical grid through a PE converter. By using the appropriate control in the auxiliary winding, it is possible to control the induction machine at almost any speed. Also, in this configuration, a fraction of the generator power is processed in the converter.

In the third category, the electrical machine is connected to the electrical grid via a fully-rated converter. It means that the whole power interchanged between the wind turbine and the electrical grid must be passed through a PE converter. This implies extra losses in the power conversion. However this configuration will improve the technical performance. In this configuration, as an electrical machine, it is possible to use an induction machine or synchronous machine, as shown in Fig. 2.10.[5,16,31] Note that the system of Fig. 2.10(a) uses a gearbox together with a SCIG. The systems of Figs. 2.10(b) and 2.10(c) use synchronous generators without a gear box.

                          S. M. Barakati

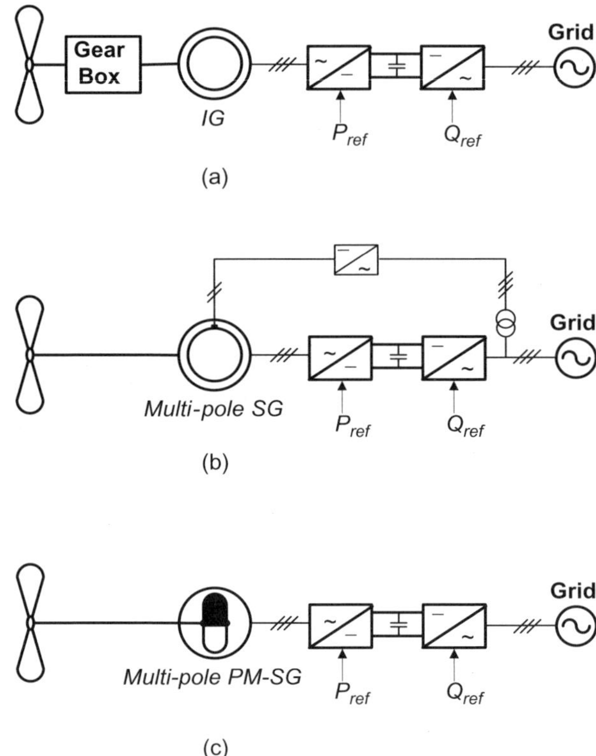

Fig. 2.10.   Wind turbine systems with a fully-rated power converter between generator terminals and the grid: (a) induction generator with gearbox, (b) synchronous and (c) PM synchronous.

In the configuration in Fig. 2.10(b), the synchronous generator needs a small power electronic converter for field excitation, and slip rings. An advantage of using the synchronous generator is the possibility of eliminating the gearbox in the wind turbine (direct-drive wind turbine). Direct drive generators essentially have a large diameter because of the high torque. In gearless drives, induction machines cannot be used because of the extreme excitation losses in these large machines due to the large air gap. However, synchronous machines can be used in direct-drive wind turbines, with either electrically excited or permanent-magnet rotor structures (Fig. 2.10(c)). Direct-drive systems with permanent magnet excitation are more expensive, because of the high price of magnets, but have lower losses. Nowadays, the price of permanent magnets is decreasing dramatically. Another disadvantage of using the permanent magnet synchronous machine is the uncontrollability of its excitation.

All configurations shown in Fig. 2.10 have the same control characteristics since the power converter between the generator and the grid enables fast control of active and reactive power. Also, the generator is isolated from the grid by a dc-link capacitor. But, using a fully-rated power electronic converter is the disadvantage of these configurations.

Different wind turbine manufacturers produce different configurations. Comparing different systems from different points of view shows a trade-off between cost and performance.

2.1.6　*Starting and disconnecting from electrical grid*

When wind velocities reach approximately 7 miles per hour, the wind turbine's blades typically start rotating, but at 9 to 10 mph, they will start generating electricity. To avoid damage, most turbines automatically shut themselves down when wind speeds exceed 55 to 65 mph. When the wind turbines are connected to or disconnected from the grid, voltage fluctuation and transient currents can occur. The high current can be limited using a soft-start circuit.[20]

2.2　Overall Dynamic Model of the Wind Turbine System and Small Signal Analysis

2.2.1　*Dynamic model of the wind turbine system*

In this section, a nonlinear dynamic model of a grid-connected wind-energy conversion system is developed in *qdo* reference frame. Dynamic models of the mechanical aerodynamic conversion, drive train, electrical generator, and power electronic converter are presented.

Different components of a wind turbine system model and the interactions among them are illustrated in Fig. 2.11.[47] The figure shows model blocks for wind speed, the aerodynamic wind turbine, mechanical components, electrical generator, power electronic converter, and utility grid. The system may also contain some mechanical parts for blades angle control. In the following sections, detailed discussions of

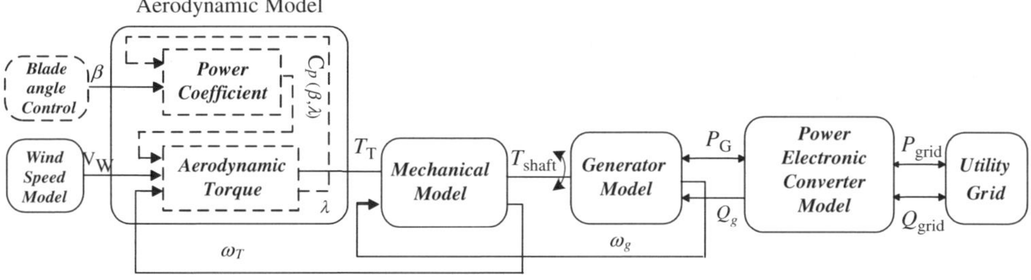

Fig. 2.11.　Block diagram of the overall wind turbine system model.

the building blocks of the overall model are presented. Note that, in the modeling, a wind turbine system with constant blade angle, without blade angle control, is considered.

2.2.1.1 *Aerodynamic model*

As illustrated in Fig. 2.11, the output of the aerodynamic model block is the mechanical torque on the wind-turbine shaft, that is a function of the wind-turbine characteristics, wind speed, shaft speed, and the blade angle. In the following, a formula for the turbine output power and torques will be introduced.

2.2.1.2 *Wind turbine output torque*

As the wind blows, it turns the wind turbine's blades, which turns the generator rotor to produce electricity. The output power of the wind turbine is related to two parameters: wind speed and rotor size. This power is proportional to the cubic wind speed, when all other parameters are assumed constant. Thus, the output power of wind turbines will increase significantly as the wind speed increases. In addition, larger rotors allow turbines to intercept more wind, increasing their output power. The reason is that the rotors sweep a circular surface whose area is a function of the square of the blade length. Thus, a small increase in blade length leads to a large increase in the swept area and energy capture. But, for economical and technical reasons, the size of the blades in wind turbines has limitations.

The mechanical power and mechanical torque on the wind turbine rotor shaft are given by Eqs. (2.1) and (2.2), respectively.[60–63]

$$P_T = \frac{1}{2}\rho A_r C_p(\beta, \lambda) V_w^3, \tag{2.1}$$

$$T_T = \frac{1}{2\omega_T}\rho A_r C_p(\beta, \lambda) V_w^3, \tag{2.2}$$

where

P_T = mechanical power extracted from turbine rotor,

T_T = mechanical torque extracted from turbine rotor,

A_r = area covered by the rotor = ΠR^2 where R is turbine rotor radius [m],

V_W = velocity of the wind [m/s],

C_p = performance coefficient (or power coefficient),

ρ = air density [kg/m^3],

λ = tip-speed-ratio (TSR),

β = rotor blade pitch angle [rad.],

ω_T = angular speed of the turbine shaft [rad/s].

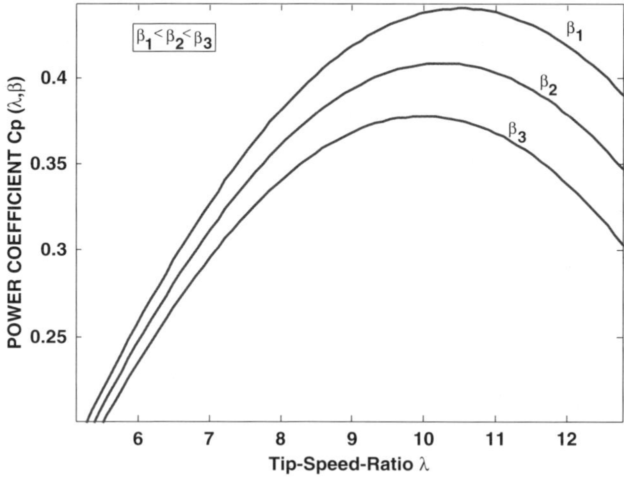

Fig. 2.12.   A typical C_p versus λ curve.

The blade tip-speed-ratio is defined as follows:

$$\lambda = \frac{\text{blade tip speed}}{\text{wind speed}} = \frac{\omega_T \times R}{V_w}. \tag{2.3}$$

The power coefficient C_p is related to the tip-speed-ratio λ, and rotor blade pitch angle, β. Figure 2.12 shows a typical C_p versus tip-speed-ratio curve. C_p changes with different values of the pitch angle, but the best efficiency is obtained for $\beta = 0$.[18] In the study, it is assumed that the rotor pitch angle is fixed and equal to zero.

The power coefficient curve has been described by different fitted equations in the literature.[9,18,63] In this study, the C_p curve is approximated analytically according to:[61,62]

$$C_p(\lambda, \beta) = (0.44 - 0.0167\beta) \sin \left[\frac{\pi(-3 + \lambda)}{15 - 0.3\beta} \right] - 0.00184(-3 + \lambda)\beta. \tag{2.4}$$

The theoretical upper limit for C_p is 0.59 according to Betz's Law, but its practical range of variation is 0.2–0.4.[18,64]

Equations (2.1)–(2.4) give a model for the transfer of wind kinetic energy to mechanical energy on the shaft of wind turbine. The block diagram of this model is shown in Fig. 2.13.

2.2.1.3   *Tower-shadow effect*

The tower-shadow effect is caused by the periodical passing by of the wind turbine blades past the wind tower.[65,66] This gives a drop in the mechanical torque which is transferred to the generator shaft and subsequently felt as a drop in the output

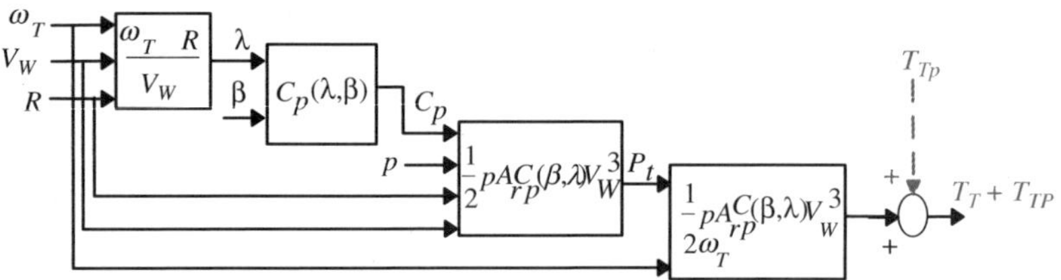

Fig. 2.13.   Block diagram of the aerodynamic wind turbine model.

voltage. Usually the tower-shadow effect has a frequency proportional to the number of blades, for example, three per revolution for a three blade turbines.

To account for the tower-shadow effect, a periodic torque pulse with frequency f_{TP} is added to the output torque of the aerodynamic model. The frequency of the periodic torque is:[46]

$$f_{TP} = N \times f_r, \tag{2.5}$$

where N is the number of blades and f_r the rotor angular speed (in Hz).

The magnitude of the torque depends on the type of wind turbine. As mentioned, based on the direction of wind received by the wind turbine, there are two structures: upwind and downwind. The tower-shadow effect is more significant in the downwind turbine. For this case, as a rule of thumb, the magnitude of this torque pulse equals 0.1 p.u., based on the rated torque of the wind turbine. The magnitude of the torque pulse for the upwind rotor is smaller in comparison with that for the downwind rotor.[20,21] The tower-shadow torque should be considered as a disturbance at the output of block diagram Fig. 2.13.

2.2.1.4 *Mechanical model*

In this subsection, a complete mechanical model for the wind turbine shaft dynamics is presented. Since the time constants of some mechanical parts are large in comparison with those of the electrical components, and detailed information on all mechanical parameters is not available,[67] the mechanical model has been developed based on reasonable time constant values and the data available. The model of a wind turbine drive train is fundamentally a three-mass model corresponding to a large mass for the wind turbine rotor, a mass for the gearbox and a mass for the generator. The moments of inertia of the shafts and gearbox can be neglected because they are small compared with the moments of inertia of the wind turbine and the generator.[68,69] Therefore, the mechanical model is essentially a two-mass model of rotor dynamics, consisting of a large mass and a small mass, corresponding to the wind turbine rotor inertia J_T and generator rotor inertia J_G, respectively,[8,63,66−70] as shown in Fig. 2.14. The low-speed shaft is modeled as an inertia, a spring with

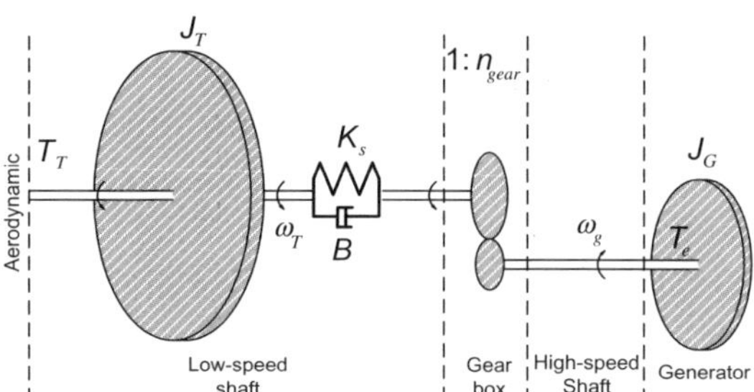

Fig. 2.14.   A complete mechanical model of the wind turbine shaft.

Table 2.1.   Mechanical model parameters.

Parameter	Description	Parameter	Description
J_T	Wind turbine inertia [kg.m^2]	ω_T	Wind turbine shaft speed [rad/s]
J_G	Generator inertia [kg.m^2]	ω_g	Generator shaft speed [rad/s]
K_s	Stiffness coefficient [N.m/rad]	θ_T	Wind turbine shaft angle [rad]
B	Damper coefficient [N.m/rad./s]	θ_g	Generator shaft angle [rad]
T_T	Wind turbine torque [N.m]	$1:n_{gear}$	Gear ratio
T_e	Generator electromechanical torque [N.m]		

stiffness coefficient K_s, and a damper with damping coefficient B. An ideal gear box with the gear ratio $1 : n_{gear}$ is included between the low-speed and high-speed shafts. Also, the parameters of the mechanical model are defined in Table 2.1.

The drive train converts the aerodynamic torque T_T on the low-speed shaft to the torque on the high-speed shaft T_e. The dynamics of the drive train are described by the following three differential equations:

$$\frac{d}{dt}\omega_T = \frac{1}{J_T}[T_T - (K_s\delta\theta + B\delta\omega)], \tag{2.6}$$

$$\frac{d}{dt}(\delta\theta) = \delta\omega, \tag{2.7}$$

$$\frac{d}{dt}\omega_g = \frac{1}{J_T}\left[\frac{1}{\eta_{gear}}(K_s\delta\theta + B\delta\omega) - T_e\right], \tag{2.8}$$

where $\delta\theta = \theta_T - \theta_g/n_{gear}$, $\delta\omega = \omega_T - \omega_g/n_{gear}$, T_T is the turbine mechanical torque from Eq. (2.2) and T_e is the generator electromechanical torque which will be introduced in the next section.

It is worth mentioning that as a simple dynamic model, one can consider a single mass model, i.e., one lumped mass accounting for all the rotating parts of the wind turbine. In fact, the stiffness and damping of shaft are used for the sake of completeness and can be removed in case they are not important in a specific application. This removal simplifies the dynamic model and reduces system order, but the completeness of the dynamic model will be compromised.

2.2.1.5 *Induction machine model*

Figure 2.15 shows an idealized three-phase induction machine consisting of a stator and a rotor.[71,72] Each phase in stator and rotor windings has a concentrated coil structure. The balanced three-phase ac voltages in the stator induce current in the short-circuited rotor windings by induction or transformer action. It can be shown that the stator current establishes a spatially sinusoidal flux density wave in the air gap which rotates at synchronous speed given by:

$$\omega_s = \frac{2}{P}\omega_e, \tag{2.9}$$

where ω_s is the synchronous speed in rad/sec, ω_e stator angular electrical frequency in rad/sec, and P the number of poles. If the mechanical shaft speed of the machine is defined as ω_r (in rad/sec), at any speed ω_s, the speed difference $\omega_s - \omega_r$ creates slip (s). The slip is defined as follows:

$$s = \frac{\omega_s - \omega_r}{\omega}. \tag{2.10}$$

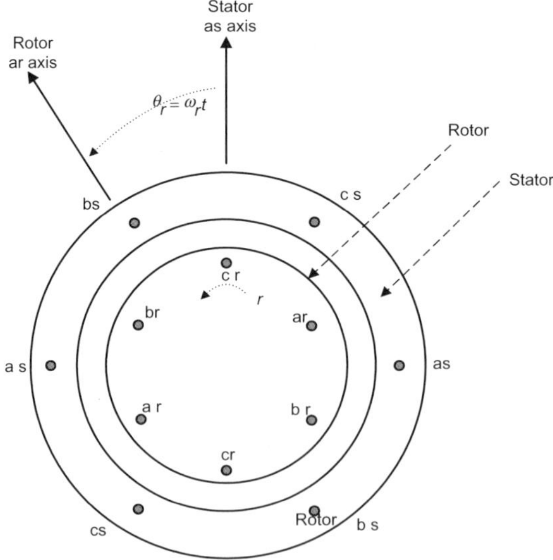

Fig. 2.15. Equivalent circuit for the induction machine.[69]

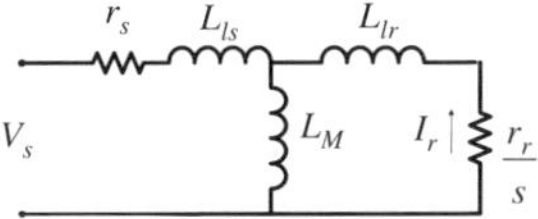

Fig. 2.16. A per-phase equivalent circuit for induction machine.

In the induction generator, at steady-state operating point, ω_r ($= \omega_g$) is slightly higher than ω_s (i.e., $s < 0$), while in induction motor, ω_r is slightly lower than ω_s (i.e., $s > 0$).

A transformer-like per-phase equivalent circuit for induction machine, in steady-state, is shown in Fig. 2.16.

In this equivalent circuit, r_s is the stator resistance, L_{ls} the stator inductance, L_M the magnetizing inductance, L_{lr} the rotor inductance (referred to stator circuit) and r_r the rotor resistance (referred to stator circuit). In the generator mode, the resistance r_r/s has a negative value. This negative resistance implies the existence of a source and therefore, the direction of power in the generator mode is from the rotor circuit to the stator circuit.

In steady-state, the electromechanical torque on the shaft is a function of the rotor current, rotor resistance and slip, as expressed by Eq. (2.11).[71,72]

$$T_e = \frac{3}{\omega_s} I_r^2 \frac{T_r}{s}. \tag{2.11}$$

If the terminal voltage and frequency are constant, T_e can be calculated as a function of slip (s) from Eq. (2.11). Figure 2.17 shows the torque-speed curve, where the value of slip is extended beyond the region $0 \leq s \leq 2$. In Fig. 2.17 two distinct zones can be identified: generating mode ($s < 0$) and motoring mode ($0 \leq s \leq 1$). The sign of the torque in the motoring and generating regions has been specified based on the convention that: $T_{\text{motor}} > 0$ and $T_{\text{generator}} < 0$. The magnitude of the counter torque that is developed in the induction generator as a result of the load connected at the machine's stator terminals is then $T_c = -T_e$. The theoretical range of operation in the generator mode is limited between the synchronous angular speed ω_s and the ω_r corresponding to the pushover torque.

It is worth noting that, as shown in the equivalent circuit of Fig. 2.17, the induction machines have inductive nature, and therefore, the induction generator (similar to induction motor) absorbs reactive power from its terminals. The reactive power essentially sustains the rotating magnetic field in the air gap between the cage rotor and the stator winding. This reactive power should be supplied by the grid, in grid-connected mode, or by the capacitor-bank that is connected at the stator terminals, in stand-alone mode. Moreover, it is possible to add a power electronic converter, acting as a dynamic Var compensation device, at the stator terminals, for additional and smoother Var control.[73]

                                   S. M. Barakati

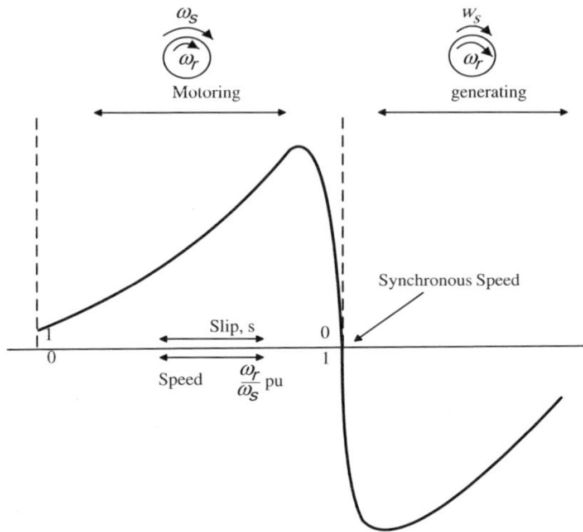

Fig. 2.17.   Torque-speed curve of induction machine.[69,70]

The output voltage of the generator, in stand-alone operation, can be estimated from the intersection point of the magnetization curve of the machine and the impedance line of the capacitor. This intersection point defines the operating point. Also the output frequency, in a grid connection, is dictated by the grid, while in stand-alone operation, it is a function of the load, rotor speed and excitation capacitance.[74]

2.2.1.5.1   Dynamic model of the induction machine

A commonly-used induction machine model is based on the flux linkages.[72] The dynamic equivalent circuit of the induction machine in *qdo* frame is illustrated in Refs. 72 and 75.

Note that in the *qdo*-equivalent circuit all the rotor parameters are transferred to the stator side. The machine is described by four differential equations based on flux linkage in the *qdo* frame and one differential equation based on rotor electrical angular speed, as follows:

$$\frac{d\psi_{qs}}{dt} = C_1\psi_{qs} - \omega_e\psi_{ds} + C_2\psi_{qr} + \omega_b v_{qs}, \tag{2.12}$$

$$\frac{d\psi_{ds}}{dt} = \omega_e\psi_{qs} + C_1\psi_{ds} + C_2\psi_{dr} + \omega_b v_{ds}, \tag{2.13}$$

$$\frac{d\psi_{qr}}{dt} = C_3\psi_{qs} + C_4\psi_{qr} - (\omega_e - \omega_{re})\psi_{dr}, \tag{2.14}$$

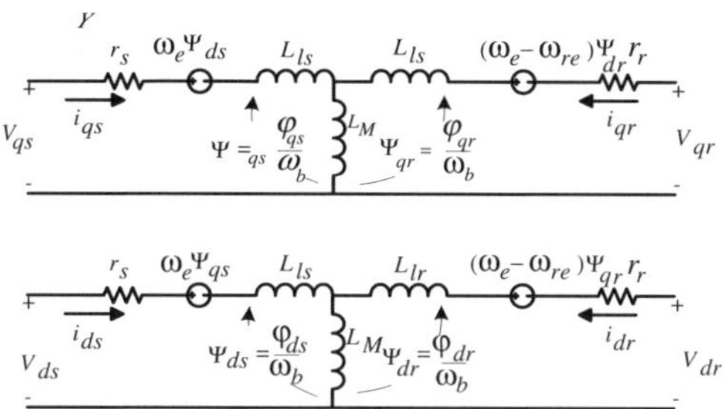

Fig. 2.18.   Qdo-equivalent circuit of an induction machine.[70,73]

$$\frac{d\psi_{dr}}{dt} = C_3\psi_{ds} + (\omega_e - \omega_{re})\psi_{qr} + C_4\psi_{dr}, \tag{2.15}$$

$$\frac{d\omega_{re}}{dt} = \left(\frac{p}{2J}\right)(T_e - T_L), \tag{2.16}$$

where

$$C_1 = \omega_b \frac{r_s}{x_{ls}}\left(\frac{x_{ml}^*}{x_{ls}} - 1\right), \quad C_2 = \omega_b \frac{r_s}{x_{ls}}\frac{x_{ml}^*}{x_{lr}}, \quad C_3 = \omega_b \frac{r_r}{x_{lr}}\frac{x_{ml}^*}{x_{ls}},$$

$$C_4 = \omega_b \frac{r_r}{x_{lr}}\left(\frac{x_{ml}^*}{x_{lr}} - 1\right) x_{ml}^* = \left(x_m^{-1} + x_{ls}^{-1} + x_{lr}^{-1}\right)^{-1},$$

ψ_{ds}, ψ_{qs}, ψ_{dr}, and ψ_{qr}: d-axis and q-axis stator and rotor flux linkages, r_r and r_s: rotor and stator resistances, $X_{ls} = \omega_e L_{ls}$ and $X_{lr} = \omega_e L_{lr}$: stator and rotor leakage reactances, $X_m = \omega_e L_{ml}$: magnetization reactance, ω_e, ω_b: stator and base electrical angular speeds, ω_{re}: rotor electrical angular speed, v_{qs}, v_{ds}: q and d-axis stator voltages, v_{qr}, v_{dr}: q and d-axis rotor voltages, T_e and T_L: electromechanical and load torque.

The stator and rotor currents, in the qdo-equivalent circuit of Fig. 2.18, can be found as follows:

$$i_{qs} = \frac{1}{\chi_{ls}}(\psi_{qs} - \psi_{mq}), \tag{2.17}$$

$$i_{ds} = \frac{1}{\chi_{ls}}(\psi_{ds} - \psi_{md}), \tag{2.18}$$

$$i_{qr} = \frac{1}{\chi_{lr}}(\psi_{qr} - \psi_{mq}), \tag{2.19}$$

$$i_{dr} = \frac{1}{\chi_{lr}}(\psi_{dr} - \psi_{md}), \tag{2.20}$$

44   *S. M. Barakati*

where

$$\psi_{mq} = x_{ml}^* \left[\frac{\psi_{qs}}{x_{ls}} + \frac{\psi_{qr}}{x_{lr}} \right] \quad \text{and} \quad \psi_{md} = x_{ml}^* \left[\frac{\psi_{ds}}{x_{ls}} + \frac{\psi_{dr}}{x_{lr}} \right].$$

In addition, the electromechanical torque of the machine can be written as follows:

$$T_e = \frac{3}{2} \left(\frac{p}{2} \right) \frac{1}{\omega_b} (\psi_{ds} i_{qs} - \psi_{qs} i_{ds}) = C_5 (\psi_{dr} \psi_{qs} - \psi_{qr} \psi_{ds}), \tag{2.21}$$

where $C_5 = \frac{3}{2} \frac{P}{2} \frac{1}{\omega_b} \frac{x_{ml}^*}{x_{ls} x_{lr}}$.

2.2.1.5.2   Constant V/f speed control method

To avoid saturation of the induction machine when the stator frequency changes, the stator terminal voltage is also adjusted using a constant V/f strategy. This method is well known for the induction machine speed control.[76] A power electronic converter should be employed at the terminals of the induction generator to implement the constant V/f strategy. In the study, this strategy is implemented for adjusting the speed of the turbine shaft to achieve maximum power point tracking.

2.2.1.6   *Gearbox model*

The duty of a mechanical gearbox is transforming the mechanical power from the slow turning rotor shaft to a fast-turning shaft, which drives the generator. The gearbox is mostly used in the wind turbines with induction generators. The need for this transmission arises from the problem that an induction generator cannot be built for very low speeds with good efficiency.

In order to model the gearbox, it is only needed to consider that the generator torque can simply be transferred to the low speed shaft by a multiplication. For example, for the gearbox of Fig. 2.14, one can write[77]:

$$\frac{T_T}{T_e} = \frac{\omega_g}{\omega_T} = \eta_{\text{gear}}. \tag{2.22}$$

Note that for a non ideal gearbox, the efficiency of the gearbox should be considered in the model.

2.2.1.7   *Grid model*

The grid model consists of an infinite bus. The infinite-bus model can be used when the grid power capacity is sufficiently large such that the action of any one user or generator will not affect the operation of the power grid. In an infinite bus, the system frequency and voltage are constant, independent of active and reactive power flows.

2.2.1.8 *Wind speed model*

Although the wind model is not part of the wind turbine model, the output power calculation in the wind turbine rotor requires the knowledge of instantaneous wind speed.

Wind is very difficult to model because of its highly-variable behavior both in location and time. Wind speed has persistent variations over a long-term scale. However, surface conditions such as buildings, trees, and areas of water affect the short-term behaviour of the wind and introduce fluctuations in the flow, i.e., wind speed turbulence.

A brief review of the literature reveals different wind speed models. For example, a wind model based on superposition of components is proposed in Ref. 78. In this method, the wind speed is modeled by four components: mean wind speed, ramp wind component, gust wind component and noise wind component. However, determining all four components is a difficult task.

In this study, wind speed is modeled with a random process. The model is based on Van Der Hoven and Von Karman's models.[60,79] The instantaneous value of wind speed, $v_W(t)$, can be described as the wind speed average value plus fluctuations in the wind speed, as follows:

$$v_w(t) = V_{WM} + \sum_{i=1}^{N} A_i \cos(\omega_i t + \psi_i), \qquad (2.23)$$

where V_{WM} is the mean value of wind speed, typically determined as a 10-minute average value, A_i is the amplitude of the wind speed fluctuation at discrete frequency of w_i ($i = [1, N]$), N is the number of samples, and ψ_i is a random phase angle with a uniform distribution in the interval $[-\pi, \pi]$.

The amplitudes A_i are based on a spectral density function $S(\omega)$ that is empirically fit to wind turbulence. The function $S(\omega)$ can be determined using Van Der Hoven's spectral model.[79] The independence of the model from the mean wind speed is a drawback of the model. Therefore, it cannot model the low frequency components, and it is not proper for a complete description of the wind speed over a short time scale, i.e., seconds, minuets, or hours.[79] Von-Karman's distribution given by Eq. (2.24),[79] a commonly-used turbulence spectral density function, is a solution to this problem.

$$S(w) = \frac{0.475\sigma^2 \frac{L}{V_{WM}}}{\left[1 + \left(\frac{\omega L}{V_{WM}}\right)^2\right]^{5/6}}. \qquad (2.24)$$

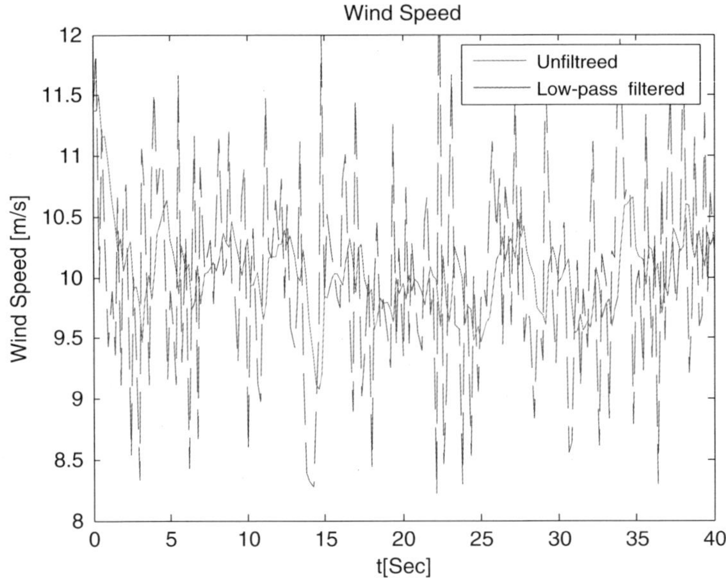

Fig. 2.19.   Wind speed fluctuation: Unfiltered and low pass filtered.

In Eq. (2.24), σ is the standard deviation of the wind speed, and L is the turbulence length scale [m]. The parameter L equals:

$$\begin{cases} 20h, & \text{if } h \leq 30\,\text{m} \\ 600, & \text{if } h > 30\,\text{m} \end{cases} \tag{2.25}$$

where h is the height at which the wind speed signal is of interest [m], which normally equals the height of the wind turbine shaft.

The amplitude of the ith harmonic, A_i, based on the spectral density function of Eq. (2.24), can be defined as:

$$A_i(\omega_i) = \frac{2}{\pi}\sqrt{\frac{1}{2}[S(\omega_i) + S(\omega_{i+1})](\omega_{i+1} - \omega_i)}. \tag{2.26}$$

Figure 2.19 shows a spectral density function based on Eq. (2.24). The parameters chosen for the simulation were: $V_{WM} = 10$ [m/s], $L = 180$ [m], $\sigma = 2$, $N = 55$. The instantaneous wind speed fluctuation, based on Von-Karman's spectral density over time is shown in Fig. 2.20.

2.2.1.8.1   High-frequency damping effect

For wind power calculations, the instantaneous wind speed model should be augmented with complex wind effects on the wind turbine blades, including

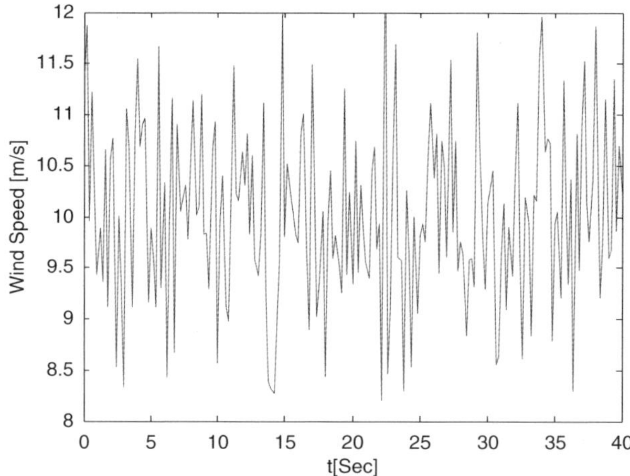

Fig. 2.20.   Instantaneous wind speed as a function of time.

high-frequency damping effects and tower-shadow effects. In this section, the high-frequency damping effect is discussed.

The phenomenon of damping the high-frequency wind speed variations over the blades surface, namely high-frequency damping effect, should be included in the aerodynamic model of the wind turbine.[63] To approximate this effect, a low-pass filter with the following transfer function is employed.

$$H(s) = \frac{1}{1 + \tau s}. \tag{2.27}$$

The filter time constant τ depends on the turbine radius, average wind speed at hub height, and the intensity of wind turbulence. Figure 2.19 shows the instantaneous wind speed and corresponding low-pass filtered signal.

It is worth mentioning that the low-pass filtered wind speed data can be saved in a memory and used later for simulation, instead of using the instantaneous wind speed data and the low-pass filter dynamic equation.

References

1. S. Masoud Barakati, *Applications of Matrix Converters for Wind Turbine Systems* (VDM Verlag Dr. Muller, Germany, 2008).
2. S. Krohn, "The wind energy pioneer — Poul la Cour," The Danish Wind Turbine Manufacturers Association, http://webasp.ac-aix-marseille.fr/rsi/bilan/action_0405/13LduRempart/Doc/Doc_page/ windpower.pdf (2008).
3. R. Hoffmann and P. Mutschler, "The influence of control strategies on the energy capture of wind turbines," *Application of Electrical Energy* (2000), pp. 886–893.

4. P. Migliore, J. van Dam and A. Huskey, "Acoustic Tests of Small Wind Turbinesat," http://www.wind.appstate.edu/reports/NRELAcousticTestsofSmallWindTurbines.pdf (2007).

5. L.H. Hansen, L. Helle, F. Blaabjerg, E. Ritchie, S. Munk-Nielsen, H. Bindner, P. Sørensen and B. Bak-Jensen, "Conceptual survey of generators and power electronics for wind turbines," Risø National Laboratory, Roskilde, Denmark, December 2001.

6. O. Carlson, J. Hylander and K. Thorborg, "Survey of variable speed operation of wind turbines," *European Union Wind Energy Conf.*, Goeteborg, Sweden, May 1996, pp. 406–409.

7. M. Idan, D. Lior and G. Shaviv, "A robust controller for a novel variable speed wind turbine transmission," *J. Solar Energy Engineering* **120** (1998) 247–252.

8. S. Bhowmik and R. Spee, "Wind speed estimation based variable speed wind power generation," *Proc. 24th Annual Conf. of the IEEE*, vol. 2, 31 August–4 September 1998, pp. 596–601.

9. M.G. Simoes, B.K. Bose and R.J. Spiegel, "Fuzzy logic based intelligent control of a variable speed cage machine wind generation system," *IEEE Tran. Power Electronics* **12** (1997) 87–95.

10. Q. Wang and L. Chang, "An intelligent maximum power extraction algorithm for inverter-based variable speed wind turbine systems," *IEEE Trans. Power Electronics* **19** (2004) 1242–1249.

11. A. Miller, E. Muljadi and D.S. Zinger, "A variable speed wind turbine power control," *IEEE Trans. Energy Conversion* **12** (1997) 181–186.

12. L. Xu and C. Wei, "Torque and reactive power control of a doubly fed induction machine by position sensorless scheme," *IEEE Trans. Industry Applications* **31** (1995) 636–642.

13. W. Lu and B.T. Ooi, "Multiterminal LVDC system for optimal acquisition of power in wind-farm using induction generators," *IEEE Trans. Power Electronics* **17** (2002) 558–563.

14. J.L. Rodriguez-Amenedo, S. Arnalte and J.C. Burgos, "Automatic generation control of a wind farm with variable speed wind turbines," *IEEE Trans. Energy Conversion* **17** (2002) 279–284.

15. D. Seyoum and M.F. Rahman, "The dynamic characteristics of an isolated self-excited induction generator driven by a wind turbine," *Rec. in Proc. IAS Conf., Industry Applications Conf., Annual Meeting*, vol. 2, October 2002, pp. 731–738.

16. F. Blaabjerg, Z. Chen and S.B. Kjaer, "Power electronics as efficient interface in dispersed power generation systems," *IEEE Trans. Power Electronics* **19** (2004) 1184–1194.

17. L.L. Freries, *Wind Energy Conversion Systems* (Prentice Hall, 1990).

18. S. Heier, *Grid Integration of Wind Energy Conversion Systems*, Chap. 1 (John Wiley & Sons Ltd, 1998).

19. Live Journal, www.windturbine.livejournal.com.

20. J.F. Manwell, J.G. McGowan and A.L. Rogers, *Wind Energy Explained, Theory, Design and Application* (Wiley, 2002).

21. S. Mathew, *Wind Energy Fundamentals, Resource Analysis, and Economics* (Springer, 2006).

22. P.G. Casielles, J. Sanz and J. Pascual, "Control system for a wind turbine driven self excited asynchronous generator," *Proc. Mediterranean Electrotechnical Integrating Research, Industry and Education in Energy and Communication Engineering, MELECON'89 Conf.*, April 1989, pp. 95–98.

23. D.A. Torrey, "Variable-reluctance generators in wind-energy systems," *Rec. in IEEE PESC '93, Power Electronics Specialists Conf.*, June 1993, pp. 561–567.

24. J. Matevosyan, "Wind power in areas with limited export capability," Licentiate Thesis Royal Institute of Technology Department of Electrical Engineering Stockholm (2004).

25. A.J.G. Westlake, J.R. Bumby and E. Spooner, "Damping the power-angle oscillations of a permanent-magnet synchronous generator with particular reference to wind turbine applications," *IEE Proc. Electric Power Applications*, vol. 143, May 1996, pp. 269–280.

26. S. Grabic, N. Celanovic and V. Katie, "Series converter stabilized wind turbine with permanent magnet synchronous generator," *IEEE Conf. PESC'04, Power Electronics Specialists Conf.*, vol. 1, June 2004, pp. 464–468.

27. L. Dambrosio and B. Fortunato, "One step ahead adaptive control technique for a wind turbine-synchronous generator system," *Proc. IECEC-97, Intersociety Energy Conversion Engineering Conf.*, vol. 3, July 1997, pp. 1970–1975.

28. D.S. Zinger and E. Muljadi, "Annualized wind energy improvement using variable speeds," *IEEE Trans. Industry Applications* **33** (1997) 1444–1447.

29. P. Bauer, S.W.H. De Haan, and M.R.Dubois, "Wind energy and offshore windparks: State of the art and trends", *EPE-PEMC Dubrovnik & Cavat* (2002), pp 1–15.

30. P. Bauer, S.W.H. De Haan, C.R. Meyl and J.T.G. Pierik, "Evaluation of electrical systems for offshore wind farms," *Rec. in IEEE Industry Applications Conf.*, vol. 3, October 2000, pp. 1416–1423.

31. R. Teodorescu and F. Blaabjerg, "Flexible control of small wind turbines with grid failure detection operating in stand-alone and grid-connected mode," *IEEE Trans. Power Electronics* **19** (2004) 1323–1332.

32. E. Bogalecka and Z. Krzeminski, "Control systems of doubly-fed induction machine supplied by current controlled voltage source inverter," *6th Int. Conf. on Electrical Machines and Drives, Conf. Publ. No. 376*, September 1993, pp. 168–172.

33. C. Knowles-Spittle, B.A.T. Al Zahawi and N.D. MacIsaac, "Simulation and analysis of 1.4 MW static Scherbius drive with sinusoidal current converters in the rotor circuit," *Proc. IEE 7th Int. Conf. on Power Electronics and Variable Speed Drives, Conf. Publ. No. 456*, September 1998, pp. 617–621.

34. R. Pena, J.C. Clare and G.M. Asher, "A doubly fed induction generator using back-to-back PWM converters supplying an isolated load from a variable speed wind turbine," *Electric Power Applications, IEE Proc.*, vol. 143, September 1996, pp. 380–387.

35. T. Yifan and L. Xu, "A flexible active and reactive power control strategy for a variable speed constant frequency generating system," *IEEE Trans. Power Electronics* **10** (1995) 472–478.

36. W.-S. Chienand and Y.-Y. Tzou, "Analysis and design on the reduction of DC-link electrolytic capacitor for AC/DC/AC converter applied to AC motor drives," *Proc. IEEE PESC 98, Power Electronics Specialists Conf.*, vol. 1, May 1998, pp. 275–279.

37. J.S. Kim and S.K. Sul, "New control scheme for AC-DC-AC converter without DC link electrolytic capacitor," *IEEE Proc. PESC'93, Power Electronics Specialists Conf., 1993,* June 1993, pp. 300–306.

38. K. Siyoung, S. Seung-Ki and T.A. Lipo, "AC to AC power conversion based on matrix converter topology with unidirectional switches," *Proc. APEC'98, Applied Power Electronics Conf. and Exposition, 1998*, vol. 1, February 1998, pp. 301–307.

39. M. Salo and H. Tuusa, "A high performance PWM current source inverter fed induction motor drive with a novel motor current control method," *Proc. IEEE PESC'99, Power Electronics Specialists Conf., 1999*, vol. 1, 27 June–1 July 1999, pp. 506–511.

40. M. Venturini, "A new sine wave in sine wave out, conversion technique which eliminates reactive elements," *Proc. POWERCON 7* (1980), pp. E3_1–E3_15.

41. M. Venturini and A. Alesina, "The generalized transformer: A new bidirectional sinusoidal waveform frequency converter with continuously adjustable input power factor," *Proc. IEEE PESC'80* (1980), pp. 242–252.

42. J. Rodriguez, "A new control technique for AC–AC converters," *Proc. IFAC Control in Power Electronics and Electrical Drives Conf.*, Lausanne, Switzerland (1983), pp. 203–208.

43. L. Helle, K.B. Larsen, A.H. Jorgensen, S. Munk-Nielsen and F. Blaabjerg, "Evaluation of modulation schemes for three-phase to three-phase matrix converters," *IEEE Trans. Industrial Electronics* **51** (2004) 158–171.

44. M. Aten, C. Whitley, G. Towers, P. Wheeler, J. Clare and K. Bradley, "Dynamic performance of a matrix converter driven electro-mechanical actuator for an aircraft rudder," *Proc. 2nd Int. Conf. Power Electronics, Machines and Drives, PEMD 2004*, vol. 1, April 2004, pp. 326–331.

45. B. Ooi and M. Kazerani, "Elimination of the waveform distortions in the voltage-source-converter type matrix converter," *Conf. Rec. IEEE-IAS Annu. Meeting*, vol. 3 (1995), pp. 2500–2504.

46. J. Wu, P. Dong, J.M. Yang and Y.R. Chen, "A novel model of wind energy conversion system," *Proc. DRPT2004, IEEE Int. Conf. on Electric Utility Deregulation and Power Technologies*, April 2004, Hong Kong, pp. 804–809.

47. S.M. Barakati, M. Kazerani and J.D. Aplevich, "A dynamic model for a wind turbine system including a matrix converter," *Proc. IASTED, PES2007*, January 2006, pp. 1–8.

48. T. Satish, K.K. Mohapatra and N. Mohan, "Speed sensorless direct power control of a matrix converter fed induction generator for variable speed wind turbines," *Proc. Power Electronics, Drives and Energy Systems*, PEDES'06, December 2006, pp. 1–6.

49. S.F. Pinto, L. Aparicio and P. Esteves, "Direct controlled matrix converters in variable speed wind energy generation systems power engineering," *Proc. Energy and Electrical Drives, POWERENG 2007*, April 2007, pp. 654–659.

50. K. Huang and Y. He, "Investigation of a matrix converter-excited brushless doubly-fed machine wind-power generation system," *Proc. Power Electronics and Drive Systems, PEDS 2003*, November 2003, vol. 1, pp. 743–748.

51. Q. Wang, X. Chen and Y. Ji, "Control for maximal wind energy tracing in matrix converter AC excited brushless doubly-fed wind power generation system," *Proc. IEEE Industrial Electronics, IECON 2006*, November 2006, pp. 718–723.

52. L. Zhang and C. Watthanasarn, "A matrix converter excited doubly-fed induction machine as a wind power generator," *Proc. Power Electronics and Variable Speed Drives*, September 1998, pp. 532–537.

53. D. Aouzellag, K. Ghedamsi and E.M. Berkouk, "Network power flow control of variable speed wind turbine," *Proc. Power Engineering, Energy and Electrical Drives, POWERENG 2007*, April 2007, pp. 435–439.

54. L. Zhang, C. Watthanasarn and W. Shepherd, "Application of a matrix converter for the power control of a variable-speed wind-turbine driving a doubly-fed induction generator," *Proc. Industrial Electronics, Control and Instrumentation, IECON'97*, vol. 2, November 1997, pp. 906–911.

55. M. Kazerani, "Dynamic matrix converter theory development and application," Ph.D. thesis, Department of Electrical Eng., McGill University, Canada (1995).

56. B.T. Ooi and M. Kazerani, "Application of dyadic matrix converter theory in conceptual design of dual field vector and displacement factor controls," *Proc. IEEE Industry Applications Society Annual Meeting*, vol. 2, October 1994, pp. 903–910.

57. H. Nikkhajoei, A. Tabesh and R. Iravani, "Dynamic model of a matrix converter for controller design and system studies," *IEEE Trans. Power Delivery* **21** (2006) 744–754.

58. A. Alesina and M. Venturini, "Solid-state power conversion: A Fourier analysis approach to generalized transformer synthesis," *IEEE Trans. Circuits and Systems* **28** (1981) 319–330.

59. S.M. Barakati, M. Kazerani and J.D. Aplevich, "An overall model for a matrix converter," *IEEE Int. Symp. Industrial Electronics, ISIE08*, June 2008.

60. S.M. Barakati, M. Kazerani and J.D. Aplevich, "A dynamic model for a wind turbine system including a matrix converter," *Proc. IASTED, PES2007*, January 2006, pp. 1–8.

61. S.M. Barakati, M. Kazerani and X. Chen, "A new wind turbine generation system based on matrix converter," *Proc. IEEE PES General Meeting*, June 2005, pp. 2083–2089.

62. E.S. Adin and W. Xu, "Control design and dynamic performance analysis of a wind turbine-induction genrator unit," *IEEE Trans. Energy Conversion* **15** (2000) 91–96.

63. J.G. Slootweg, H. Polinder and W.L. Kling, "Representing wind turbine electrical generating systems in fundamental frequency simulations," *IEEE Trans. Energy Conversion* **18** (2003) 516–524.

64. L.S.T. Ackermann, "Wind energy technology and current status. A review," *Renewable and Sustainable Energy Review* **4** (2000) 315–375.

65. E. Muljadi and C.P. Butterfield, "Power quality aspects in a wind power plant," *IEEE Power Engineering Society General Meeting Montreal*, Quebec, Canada, June 2006.

66. R. Grünbaum, "Voltage and power quality control in wind power," ABB Power Systems AB, http://library.abb.com/GLOBAL/SCOT/scot221.nsf/VerityDisplay/6B5A44E46BAD5D3DC12 56FDA003B4D11/$File/PowerGenEurope2001.pdf (2008).

67. S. Kim, S.-K. Sul and T.A. Lipo, "AC/AC power conversion based on matrix converter topology with unidirectional switches," *IEEE Tran. Industry Applications* **36** (2000) 139–145.

68. A.D. Hansen, C. Jauch, P. Sørensen, F. Iov and F. Blaabjerg, "Dynamic wind turbine models in power system simulation tool DIgSILENT," Risø National Laboratory, Roskilde, http://www.digsilent.de/Software/Application_Examples/ris-r-1400.pdf (2003).

69. S. Seman, F. Iov, J. Niiranen and A. Arkkio, "Comparison of simulators for variable speed wind turbine transient analysis," *Int. J. Energy Research* **30** (2006) 713–728.

70. L. Mihet-Popa, F. Blaabjerg and I. Boldea, "Wind turbine Generator modeling and simulation where rotational speed is the controlled variable," *IEEE Trans. Industry Application* **40** (2004) 3–10.

71. B.K. Bose, *Power Electronics And Ac Drives* (Prentice-Hall, 1986).

72. P.C. Krause, O. Wasynczuk and S.D. Sudhoff, *Analysis of Electric Machinery* (IEEE Press, 1994).

73. T.F. Chan and L.L. Lai, "Capacitance requirements of a three-phase induction generator self-excited with a single capacitance and supplying a single-phase load," *IEEE Trans. Energy Conversion* **17** (2002) 90–94.

74. Electric Systems Consulting ABB Inc., "Integration of wind energy into the alberta electric system — stages 2 & 3: Planning and Interconnection Criteria," *Alberta Electric System Operator Report Number*: 2004-10803-2.R01.4.

75. B. Ozpineci and L.M. Tolbert, "Simulink implementation of induction machine model — a modular approach," *Proc. Electric Machines and Drives Conference, IEMDC'03*, vol. 2, June 2003, pp. 728–734.

76. A. Munoz-Garcia, T.A. Lipo and D.W. Novotny, "A new induction motor V/f control method capable of high-performance regulation at low speeds," *IEEE Tran. Industry Applications* **34** (1998) 813–821.

77. Powersim Technologies INC, Psim user manual, version 4.1, Set. 1999, http://www.powersim tech.com (2008).

78. P.M. Anderson and A. Bose, "Stability simulation of wind turbine systems," *IEEE Trans. Power Apparatus and system* **PAS-102** (1983) 3791–3795.

79. C. Nichita, D. Luca, B. Dakyo and E. Ceanga, "Large band simulation of the wind speed for real time wind turbine simulators," *IEEE Trans. Energy Conversion* **17** (2002) 523–529.

80. Starting (and stopping) a turbine, http://www.windpower.org/en/tour/wtrb /electric.htm (2008).

81. L. Neft and C.D. Shauder, "Theory and design of a 30-hp matrix converter," *IEEE Trans. Industrial Applications* **28** (1992) 546–551.

82. P. Wheeler and D. Grant, "Optimized input filter design and low loss switching techniques for a practical matrix converter," *Proc. of Industrial Electronics*, vol. 144, January 1997, pp. 53–60.

83. D.G. Holmes and T.A. Lipo, "Implementation of a controlled rectifier using AC-AC matrix converter theory," *Conf. Rec. PESC'89, Power Electronics Specialists Conf.*, June 1989, vol. 1, pp. 353–359.

84. H.W. van der Broeck, H.-C. Skudelny and G.V.Stanke, "Analysis and realization of a pulsewidth modulator based on voltage space vectors," *IEEE Tran. Industry Applications* **24** (1988) 142–150.

Chapter 3

Wind Turbine Generation Systems Modeling for Integration in Power Systems

Adrià Junyent-Ferré[*,‡] and Oriol Gomis-Bellmunt[†]

*Centre d'Innovació Tecnològica en Convertidors
Estàtics i Accionaments (CITCEA-UPC),
Departament d'Enginyeria Elèctrica,
Universitat Politècnica de Catalunya,
ETS d'Enginyeria Industrial de Barcelona,
Av. Diagonal, 647, Pl. 2. 08028 Barcelona, Spain*

†*IREC Catalonia Institute for Energy Research,
Josep Pla, B2, Pl. Baixa. E-08019 Barcelona, Spain*
‡*adria.junyent@citcea.upc.edu*

This chapter deals with the modeling of wind turbine generation systems for integration in power systems studies. The modeling of the wind phenomenon, the turbine mechanical system and the electrical machine, along with the corresponding converter and electrical grid is described.

3.1 Introduction

Wind power generation has grown in the last three decades and is considered one of the most promising renewable energy sources. However, its integration into power systems has a number of technical challenges concerning security of supply, in terms of reliability, availability and power quality.

Wind power impact mainly depends on its penetration level, but depends also on the power system size, the mix of generation capacity, the degree of interconnections to other systems and load variations. Since the penetration of wind power generation is growing, system operators have an increasing interest in analyzing the impact of wind power on the connected power system. For this reason, grid connection requirements are established. In the last few years, the connection requirements have incorporated, in addition to steady state problems, dynamic requirements, like voltage dip ride-through capability. This leads to the need for detailed modeling

53

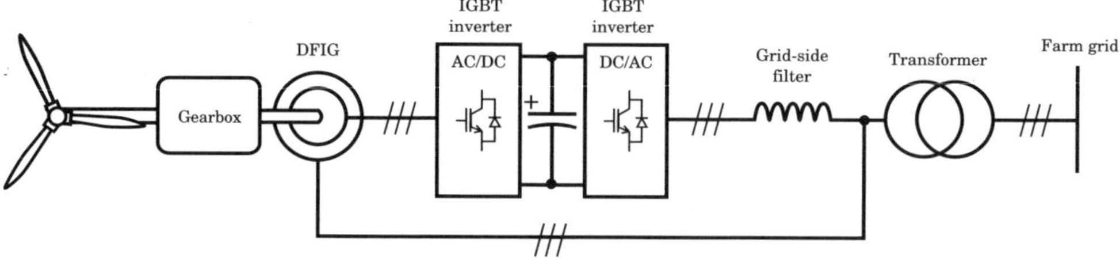

Fig. 3.1.   The DFIG wind generator concept.

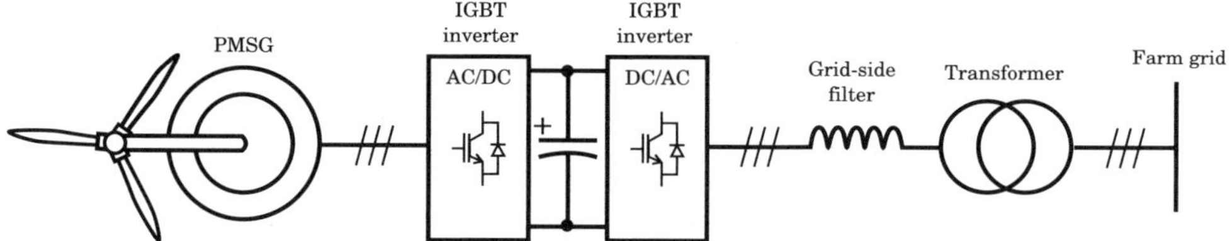

Fig. 3.2.   The gearless PMSG wind generator concept.

of wind turbine systems in order to analyze the dynamic phenomena in the power grid.

Moreover, new wind turbine technology integrates power electronics and control making it possible for wind power generation to participate in active and reactive power control. Nowadays most of the installed variable speed wind generators are based on the doubly-fed induction generator (DFIG) but new types of wind generators based on permanent magnet synchronous generators (PMSG) are expected to gain market popularity in the following years. The DFIG configuration consists of a wound rotor induction generator with the stator windings directly connected to the grid and the rotor windings connected to a voltage-source back-to-back power converter which transfers a fraction of the extracted power (see Fig. 3.1). The PMSG configuration needs a full power converter and allows the use of multipolar generators making it possible to suppress the gearbox (see Fig. 3.2).

In this chapter we discuss a model for the induction and synchronous type generators, the back-to-back converter and the electrical network. Including the detailed description of the reactive power, DC bus voltage and torque controllers, along with its corresponding current loops, simulation results are analyzed and discussed.

3.2 Wind Turbine Modeling

Wind turbine electrical generation systems' power comes from the kinetic energy of the wind, thus it can be expressed as the kinetic power available in the stream of air multiplied by a C_P factor called power coefficient. C_P mainly depends on the relation between the average speed of the air across the area covered by the wind wheel and its angular speed and geometric characteristics of the turbine (including the instantaneous blade pitch angle configuration).[1] The power extracted by the wind turbine has the following expression:

$$P_{ww} = c_P P_{\text{wind}} = c_P \frac{1}{2} \rho A v_w^3, \tag{3.1}$$

where P_{wind} is kinetic power of the air stream that crosses the turbine rotor area, ρ is the air density assumed to be constant, A is the surface covered by the turbine and v_w is the average wind speed.

There have been different approaches to model the power coefficient ranging from considering it to be constant for steady state and small signal response simulations to using lookup tables with measured data. Another common approach is to use an analytic expression originated from Ref. 1 of the form:

$$c_P(\lambda, \theta_{\text{pitch}}) = c_1 \left(c_2 \frac{1}{\Lambda} - c_3 \theta_{\text{pitch}} - c_4 \theta_{\text{pitch}}^{c_5} - c_6 \right) e^{-c_7 \frac{1}{\Lambda}}, \tag{3.2}$$

where λ is the so-called tip speed ratio and it is defined as:

$$\lambda \triangleq \frac{\omega_t R}{v_1} \tag{3.3}$$

and

$$\frac{1}{\Lambda} \triangleq \frac{1}{\lambda + c_8 \theta_{\text{pitch}}} - \frac{c_9}{1 + \theta_{\text{pitch}}^3}, \tag{3.4}$$

where $[c_1 \cdots c_9]$ are characteristic constants for each wind turbine and θ_{pitch} is the blade pitch angle.

Thus by knowing the wind speed, the angular speed of the wind turbine and the blade pitch angle, the mechanical torque on the turbine shaft can be easily computed:

$$\Gamma_t = c_P(v_w, \omega_t) \frac{1}{2} \rho A v_w^3, \tag{3.5}$$

where Γ_t is the turbine torque.

3.3 Wind Modeling

Wind speed usually varies from one location to another and also fluctuates over the time in a stochastic way. As it has been previuously seen, it maintains a direct relation to the torque over the turbine axis and therefore it may also have some direct effect on the power output of the wind turbine generation system (WTGS) hence its evolution must be taken into account to properly simulate the WTGS dynamics.

One possible approach to generate the wind speed signal on simulations may be to use logs of real measurements of the speed on the real location of the WTGS. This approach has some evident limitations because it requires a measurement to be done on each place to be simulated. Another choice, proposed by Ref. 2 is to use a mathematical model which takes some landscape parameters to generate a wind speed sequence for any location. This wind speed expression has the form:

$$v_w(t) = v_{wa}(t) + v_{wr}(t) + v_{wg}(t) + v_{wt}(t), \tag{3.6}$$

where $v_{wa}(t)$ is a constant component, $v_{wr}(t)$ is a common ramp component, v_{wg} is a gust component and v_{wt} is a turbulence component.

The gust component may be useful to simulate an abnormal temporary increase of the speed of the wind and its expression is:

$$v_{wg}(t) = \begin{cases} 0, & \text{for } t < T_{sg} \\ \hat{A}_g \left(1 - \cos \left[2\pi \left(\dfrac{t - T_{sg}}{T_{eg} - T_{sg}} \right) \right] \right), & \text{for } T_{sg} \leq t \leq T_{eg} \\ 0, & \text{for } T_{eg} < t \end{cases} \tag{3.7}$$

where A_g is the amplitude of the gust and T_{sg} and T_{eg} are the start and the end time of the gust.

Finally, as discussed in Ref. 3, the turbulence component is a signal which has a power spectral density of the form:

$$P_{Dt}(f) = \frac{l\hat{v}_w \left[\ln \left(\frac{h}{z_0} \right) \right]^{-2}}{\left[1 + 1.5 \frac{fl}{\hat{v}} \right]^{5/3}}, \tag{3.8}$$

where $\hat{v}_w$ is the average wind speed, h is the height of interest (the wind wheel height), l is the turbulence scale which is twenty times h and has a maximum of 300 m and z_0 is a roughness length parameter which depends on the landscape type as shown in Table 3.1

Table 3.1. Values of the z_0 for different types of landscapes.

Landscape type	Range of z_0 (m)
Open sea or sand	0.0001–0.001
Snow surface	0.001–0.005
Mown grass or steppe	0.001–0.01
Long grass or rocky ground	0.04–0.1
Forests, cities and hilly areas	1–5

Source: Panofsky and Dutton, 1984; Simiu and Scanlan, 1986.

By knowing the height of the wind turbine, the average wind speed and the kind of landscape where the WTGS is, the power spectral density of the wind speed turbulence is known. The next step is to generate a signal, function of time, which has the desired power spectral density. There are many ways of doing this. One approach suggested in Ref. 4 is to sum a large number of sines with random initial phase and amplitude according to the P_{Dt}. The suggested method here is to use a linear filter designed to shape a noise signal to give it the desired spectral density. Provided that the P_{Dt} is very close to the response of a first order filter, a possible filter that accomplishes this goal is:

$$H(s) = \frac{K}{s + p}, \tag{3.9}$$

where

$$p = \frac{2\pi \left((K_1^2)^{3/5} - 1 \right)}{K_2 \sqrt{K_1^2 - 1}}, \quad K = K_1 p. \tag{3.10}$$

3.4 Mechanical Transmission Modeling

The drive-train of a WTGS comprises the wind wheel, the turbine shaft, the gearbox and the generator's rotor shaft. The gearbox usually has a multiplication ratio between 50 and 150 and the wind wheel inertia usually is about the 90% of the inertia of the whole system.

Because of the high torque applied to the turbine shaft, its deformation must not be neglected and its elastic behavior should be taken into account because of its filtering properties. A common way to model the drive-train is to treat it as a series of masses connected through an elastic coupling with a linear stiffness, a damping ratio and a multiplication ratio between them. On this paper a model with two masses,

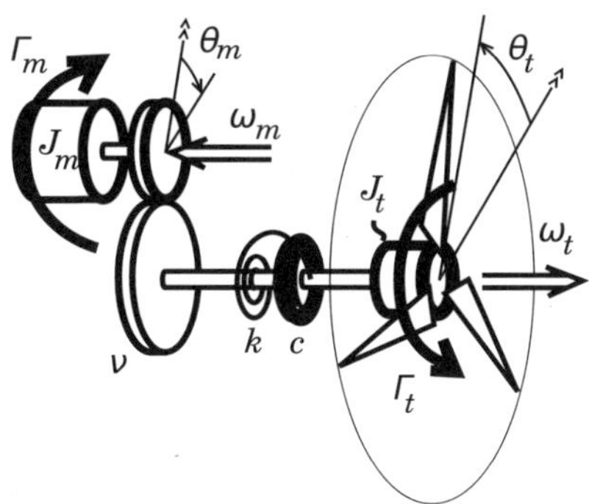

Fig. 3.3.   Two mass drive-train model.

graphically presented in Fig. 3.3, is used treating the wind wheel as one inertia J_t and the gearbox and the generator's rotor as another inertia J_m connected through the elastic turbine shaft with a k angular stiffness coefficient and a c angular damping coefficient. Applying the Newton's laws, the dynamics of the resulting system can be described as:

$$
\begin{bmatrix} \dot{\omega}_m \\ \dot{\omega}_t \\ \omega_m \\ \omega_t \end{bmatrix} =
\begin{bmatrix}
\frac{-v^2 c}{J_m} & \frac{vc}{J_m} & -\frac{v^2 k}{J_m} & \frac{vk}{J_m} \\
\frac{vc}{J_t} & -\frac{c}{J_t} & \frac{vk}{J_t} & -\frac{k}{J_t} \\
1 & 0 & 0 & 0 \\
0 & 1 & 0 & 0
\end{bmatrix}
\begin{bmatrix} \omega_m \\ \omega_t \\ \theta_m \\ \theta_t \end{bmatrix} +
\begin{bmatrix}
\frac{1}{J_m} & 0 \\
0 & \frac{1}{J_t} \\
0 & 0 \\
0 & 0
\end{bmatrix}
\begin{bmatrix} \tau_m \\ \tau_t \end{bmatrix},
\tag{3.11}
$$

where θ_t and θ_m are the angles of the wind wheel and the generator shaft, ω_t and ω_m are the angular speed of the wind wheel and the generator, τ_t is the torque applied to the turbine axis by the wind wheel and τ_m is the generator torque.

3.5 Electrical Generator Modeling

3.5.1 *Induction machine*

The generator of a doubly-fed WTGS is a wounded rotor asynchronous machine. We will assume the stator and rotor windings to be placed sinosuidally and symetrical and the magnetical saturation effects and the capacitance of all the windings neglectable. Taking as positive the currents flowing towards the machine, the relations between the voltages on the machine windings and the currents and its first derivative can be written as:

$$
v_s^{abc} = r_s i_s^{abc} + \frac{d}{dt}\lambda_s^{abc},
\tag{3.12}
$$

$$
v_r^{abc} = r_r i_r^{abc} + \frac{d}{dt}\lambda_r^{abc},
\tag{3.13}
$$

where v_s^{abc} and i_s^{abc} is the stator abc voltage and current vectors, v_r^{abc} and i_r^{abc} is the rotor abc voltage and current vector, λ_s^{abc} and λ_r^{abc} are the stator and rotor flux linkage abc vectors defined as:

$$
\begin{bmatrix} \lambda_s^{abc} \\ \lambda_r^{abc} \end{bmatrix} =
\begin{bmatrix} L_{ss}^{abc} & L_{sr}^{abc} \\ L_{rs}^{abc} & L_{rr}^{abc} \end{bmatrix}
\begin{bmatrix} i_s^{abc} \\ i_r^{abc} \end{bmatrix},
\tag{3.14}
$$

where

$$
L_{ss}^{abc} =
\begin{bmatrix}
L_{ls} + L_{ms} & -\frac{1}{2}L_{ms} & -\frac{1}{2}L_{ms} \\
-\frac{1}{2}L_{ms} & L_{ls} + L_{ms} & -\frac{1}{2}L_{ms} \\
-\frac{1}{2}L_{ms} & -\frac{1}{2}L_{ms} & L_{ls} + L_{ms}
\end{bmatrix},
\tag{3.15}
$$

$$
L_{rr}^{abc} = \begin{bmatrix} L_{lr} + L_{mr} & -\frac{1}{2}L_{mr} & -\frac{1}{2}L_{mr} \\ -\frac{1}{2}L_{mr} & L_{lr} + L_{mr} & -\frac{1}{2}L_{mr} \\ -\frac{1}{2}L_{mr} & -\frac{1}{2}L_{mr} & L_{lr} + L_{mr} \end{bmatrix},
\tag{3.16}
$$

$$
L_{sr}^{abc} = \left\{ L_{rs}^{abc} \right\}^{t} = L_{sr} \begin{bmatrix} \cos(\theta_r) & \cos(\theta_r + \frac{2\pi}{3}) & \cos(\theta_r - \frac{2\pi}{3}) \\ \cos(\theta_r - \frac{2\pi}{3}) & \cos(\theta_r) & \cos(\theta_r + \frac{2\pi}{3}) \\ \cos(\theta_r + \frac{2\pi}{3}) & \cos(\theta_r - \frac{2\pi}{3}) & \cos(\theta_r) \end{bmatrix},
\tag{3.17}
$$

also, the mechanical torque can be written as a function of the machine current as:

$$
\Gamma_m = \frac{P}{2} \begin{bmatrix} i_s^{abc} \\ i_r^{abc} \end{bmatrix}^{t} \begin{bmatrix} 0 & N_{sr}^{abc} \\ N_{rs}^{abc} & 0 \end{bmatrix} \begin{bmatrix} i_s^{abc} \\ i_r^{abc} \end{bmatrix},
\tag{3.18}
$$

where

$$
N_{sr}^{abc} = \left\{ N_{rs}^{abc} \right\}^{t} = -L_{sr} \begin{bmatrix} \sin(\theta_r) & \sin(\theta_r + \frac{2\pi}{3}) & \sin(\theta_r - \frac{2\pi}{3}) \\ \sin(\theta_r - \frac{2\pi}{3}) & \sin(\theta_r) & \sin(\theta_r + \frac{2\pi}{3}) \\ \sin(\theta_r + \frac{2\pi}{3}) & \sin(\theta_r - \frac{2\pi}{3}) & \sin(\theta_r) \end{bmatrix}
\tag{3.19}
$$

for further details on the formulation of the voltage equations of the asynchronous machine, the reader is referred to Ref. 5.

As these equations have a hard dependency on the rotor angle position, it is not recommended its use in simulation. Instead, it is preferred to introduce the Park variable transformation to the equations and use the transformated variables to integrate the dynamical equations of the machine. The Park transformation matrix is a non-singular matrix defined as:

$$
T(\theta) = \frac{2}{3} \begin{bmatrix} \cos(\theta) & \cos(\theta - \frac{2\pi}{3}) & \cos(\theta + \frac{2\pi}{3}) \\ \sin(\theta) & \sin(\theta - \frac{2\pi}{3}) & \sin(\theta + \frac{2\pi}{3}) \\ \frac{1}{2} & \frac{1}{2} & \frac{1}{2} \end{bmatrix},
\tag{3.20}
$$

where θ is the so called Park reference angle which may be choosen as constant or linear time-varying for different purposes.

The $qd0$ transformed variables are defined as:

$$
x^{qd0} \equiv T(\theta)x^{abc}.
\tag{3.21}
$$

By choosing $\theta = \omega_s t$ where ω_s is the nominal grid frequency for the stator variables transformation and $\theta = \omega_s t - \theta_r$ where θ_r is the rotor angular position multiplied by the number of pole pairs of the machine for the rotor variables, the machine current and voltage variables written in qd become constant in steady state, which benefits the numerical integration methods used by the simulation software.

By doing this, the machine equations can be written as:

$$
\begin{bmatrix} v_{sq} \\ v_{sd} \\ v_{rq} \\ v_{rd} \end{bmatrix} = \begin{bmatrix} L_s & 0 & M & 0 \\ 0 & L_s & 0 & M \\ M & 0 & L_r & 0 \\ 0 & M & 0 & L_r \end{bmatrix} \frac{d}{dt} \begin{bmatrix} i_{sq} \\ i_{sd} \\ i_{rq} \\ i_{rd} \end{bmatrix}
$$

$$
+ \begin{bmatrix} r_s & L_s\omega_s & 0 & M\omega_s \\ -L_s\omega_s & r_s & -M\omega_s & 0 \\ 0 & sM\omega_s & r_r & sL_r\omega_s \\ -sM\omega_s & 0 & -sL_r\omega_s & r_r \end{bmatrix} \begin{bmatrix} i_{sq} \\ i_{sd} \\ i_{rq} \\ i_{rd} \end{bmatrix} \tag{3.22}
$$

and

$$
V_{s0} = L_{ls}\frac{di_{s0}}{dt} + r_s i_{s0}, \tag{3.23}
$$

$$
V_{r0} = L_{lr}\frac{di_{r0}}{dt} + r_r i_{r0}, \tag{3.24}
$$

where $L_s \equiv \frac{3}{2}L_{ms} + L_{ls}$ and $L_r \equiv \frac{3}{2}L_{mr} + L_{lr}$ are the stator and rotor windings self-inductance coefficient, $M = \frac{3}{2}L_{sr}$ is the coupling coefficient between stator and rotor windings and s is the slip defined as the relation between the mechanical speed and the reference frame angular speed ($s \triangleq \frac{\omega_s - \omega_r}{\omega_s}$).

Also the torque expression and the stator reactive power, which are the control objectives of the rotor-side converter control, have the following form:

$$
\Gamma_m = \frac{3}{2}PM\left(i_{sq}i_{rd} - i_{sd}i_{rq}\right), \tag{3.25}
$$

where P is the number of pairs of poles of the generator.

Also, according to the so-known *pq-theory*[6] the instantaneous reactive power can be written as:

$$
Q_s = \frac{3}{2}(v_{sq}i_{sd} - v_{sd}i_{sq}). \tag{3.26}
$$

3.5.2 *Permanent magnet synchronous machine*

To model the dynamical behavior of the permanent magnet synchronous machine, we will assume again the stator windings to be placed sinosuidally and symetrical and the magnetical saturation effects and the capacitance of all the windings neglectable. Taking as positive the currents flowing towards the machine, the relations between the voltages on the machine windings and the currents and its first derivatives can be written as[7] :

$$
v_s^{abc} = r_s i_s^{abc} + \frac{d}{dt}\lambda_s^{abc}, \tag{3.27}
$$

where the stator flux linkage in *abc* can be written as:

$$\lambda_s^{abc} = ([L_1] + [L_2(\theta_r)]) \, i_s^{abc} + \lambda_m \begin{bmatrix} \sin(\theta_r) \\ \sin(\theta_r - \frac{2\pi}{3}) \\ \sin(\theta_r + \frac{2\pi}{3}) \end{bmatrix} \tag{3.28}$$

and

$$[L_1] = \begin{bmatrix} L_{ls} + L_A & -\frac{1}{2}L_A & -\frac{1}{2}L_A \\ -\frac{1}{2}L_A & L_{ls} + L_A & -\frac{1}{2}L_A \\ -\frac{1}{2}L_A & -\frac{1}{2}L_A & L_{ls} + L_A \end{bmatrix}, \tag{3.29}$$

$$[L_2(\theta_r)] = -L_B \begin{bmatrix} \cos 2(\theta_r) & \cos 2(\theta_r - \frac{\pi}{3}) & \cos 2(\theta_r + \frac{\pi}{3}) \\ \cos 2(\theta_r - \frac{\pi}{3}) & \cos 2(\theta_r + \frac{\pi}{3}) & \cos 2(\theta_r) \\ \cos 2(\theta_r + \frac{\pi}{3}) & \cos 2(\theta_r) & \cos 2(\theta_r - \frac{\pi}{3}) \end{bmatrix}, \tag{3.30}$$

where λ_m is the flux due to the rotor magnet, L_A is the constant fraction of the stator linkage inductance and L_B is a rotor position dependent inductance term due to rotor asymmetry.

The mechanical torque can be written as:

$$\Gamma_m = P \left(\{i_s^{abc}\}^T \frac{d}{d\theta_r} [L_2(\theta_r)] i_s^{abc} + \lambda_m \{i_s^{abc}\}^T \begin{bmatrix} \cos(\theta_r) \\ \cos(\theta_r - \frac{2\pi}{3}) \\ \cos(\theta_r + \frac{2\pi}{3}) \end{bmatrix} \right). \tag{3.31}$$

By differentiating the stator flux linkage variables over time, we get:

$$\frac{d}{dt} \lambda_s^{abc} = ([L_1] + [L_2(\theta_r)]) \frac{d}{dt} i_s^{abc} + \omega_r \frac{d}{d\theta_r} [L_2(\theta_r)] i_s^{abc}$$

$$+ \lambda_m \omega_r \begin{bmatrix} \cos(\theta_r) \\ \cos(\theta_r - \frac{2\pi}{3}) \\ \cos(\theta_r + \frac{2\pi}{3}) \end{bmatrix}. \tag{3.32}$$

By substituting the flux linkage expression in the Eq. (3.27), the explicit relation between current and voltage is obtained:

$$v_s^{abc} = \left(r_s [I_3] + \omega_r \frac{d}{d\theta_r} [L_2(\theta_r)] \right) i_s^{abc} + ([L_1] + [L_2(\theta_r)]) \frac{d}{dt} i_s^{abc}$$

$$+ \lambda_m \omega_r \begin{bmatrix} \cos(\theta_r) \\ \cos(\theta_r - \frac{2\pi}{3}) \\ \cos(\theta_r + \frac{2\pi}{3}) \end{bmatrix}. \tag{3.33}$$

As in the case of the induction machine, we see a strong dependency on the rotor position which is not desirable for the numerical stability of the integration algorithm used in simulation. We introduce a Park variable transformation for the stator variables taking $\theta = \theta_r$ which gives constant values to the voltage and current variables in steady state. The system equations in this reference frame can be written as:

$$v_s^{qd} = \begin{bmatrix} r_s & \omega_r \left(L_{ls} + \frac{3}{2}\left(L_A + L_B\right)\right) \\ -\omega_r \left(L_{ls} + \frac{3}{2}\left(L_A - L_B\right)\right) & r_s \end{bmatrix} i_s^{qd}$$
$$+ \begin{bmatrix} L_{ls} + \frac{3}{2}\left(L_A - L_B\right) & 0 \\ 0 & L_{ls} + \frac{3}{2}\left(L_A + L_B\right) \end{bmatrix} \frac{d}{dt}i_s^{qd} + \lambda_m \omega_r \begin{bmatrix} 1 \\ 0 \end{bmatrix}$$

$$(3.34)$$

and

$$v_s^0 = r_s i_s^0 + L_{ls}\frac{d}{dt}i_s^0,\tag{3.35}$$

also the torque can be rewritten as:

$$\Gamma_m = \frac{3}{2}P\left(\{i_s^{qd0}\}^T \begin{bmatrix} 0 & 3L_B & 0 \\ 3L_B & 0 & 0 \\ 0 & 0 & 0 \end{bmatrix} i_s^{qd0} + \lambda_m \{i_s^{qd0}\}^T \begin{bmatrix} 1 \\ 0 \\ 0 \end{bmatrix}\right).\tag{3.36}$$

3.6 Converter Modeling

The most common converter topology used in variable speed wind turbines is the forced commutation voltage source back-to-back converter with insulated-gate bipolar transistors (IGBT). The structure of this type of converter is shown in Fig. 3.4. The AC side on the left, which we will call the machine side, is connected to the rotor of the machine in the DFIG configuration and to the stator of the machine in the PMSG while the right AC side, grid side from now on, is connected to the wind turbine transformer (see also Figs. 3.1 and 3.2). Notice that the current direction is depicted according to the machine equations discussed in the previous section.

The dynamics of the converter involve continuous state variables corresponding to voltages and currents and discrete states corresponding to the commutation state of the IGBTs. These dynamics are complex to model and simulate and different levels of detail may be achieved by doing some assumptions. Usually an averaged converter model is used, neglecting the commutation effects assuming they are filtered by the low-pass dynamics of the rest of the system.

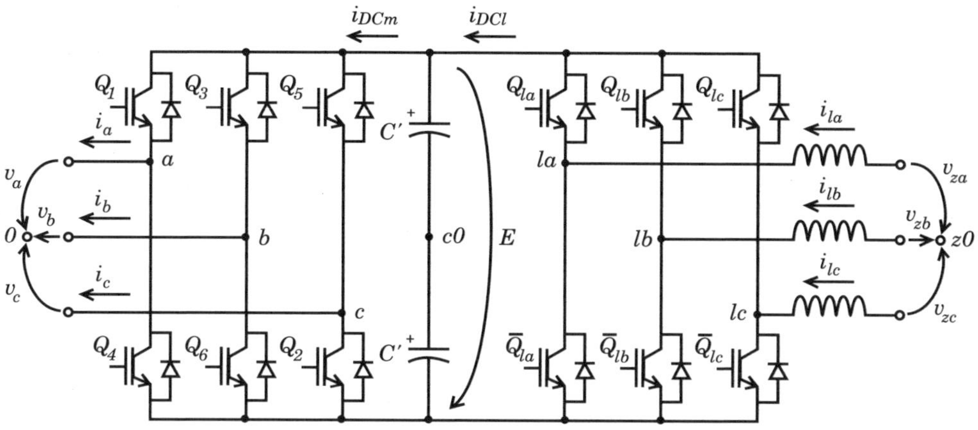

Fig. 3.4.   The IGBT voltage source back-to-back converter.

Making this assumptions, the dynamics of the grid-side electrical circuit between the grid voltage and the voltage applied on the AC side of the converter assuming the currents are positive when flowing towards the machine can be described as:

$$v_z^{abc} - v_l^{abc} - (v_{c0} - v_{z0}) \begin{bmatrix} 1 \\ 1 \\ 1 \end{bmatrix} = r_l i_l^{abc} + L_l \frac{d}{dt} i_l^{abc}, \tag{3.37}$$

also when no neutral conductor is present, it can be stated that:

$$v_{c0} - v_{z0} = \frac{1}{3} \begin{bmatrix} 1 & 1 & 1 \end{bmatrix} \cdot (v_z^{abc} - v_l^{abc}), \tag{3.38}$$

where v_z^{abc} and v_l^{abc} are the *abc* voltages on the grid side of the grid converter filter and the AC side of the converter, r_l is the resistance of the filter inductors and L_l is the inductance of the filter.

The dynamics of the voltage of the DC bus can be described as:

$$\frac{dE}{dt} = \frac{1}{C}(i_{DCl} - i_{DCm}), \tag{3.39}$$

where E is the voltage of the DC bus, i_{DCl} is the current through the DC side of the grid-side inverter, i_{DCm} is the current through the machine side inverter and both currents can be computed by doing a power balance on each inverter:

$$E i_{DCm} = v_a i_a + v_b i_b + v_c i_c, \tag{3.40}$$

$$E i_{DCl} = v_{la} i_{la} + v_{lb} i_{lb} + v_{lc} i_{lc}. \tag{3.41}$$

3.7 Control Modeling

In this section the control of the wind turbine is briefly explained. First we introduce a simple speed control that gives the torque reference that the generator must follow. Then the basics of the electrical control of the DFIG, the PMSG and the grid side converter are presented.

3.7.1 *Speed control*

The purpose of the speed controller is to maximize the power extracted by the turbine. The power available in the wind is a function of the wind speed thus extracting the maximum power available would require, at least, knowing exactly the value of the wind speed. To avoid the dependency on wind speed measurements, a series of open-loop control algorithms have been developed, the most common of them being the constant tip speed ratio control.[8] This control scheme is based on the fact that given a fixed wind speed, it can be proved that the optimal operation point can be achieved by controlling the generator so that the torque follows a function of the square of the wind speed. This function can be expressed as:

$$\Gamma_m^* = \frac{1}{v} K_{C_P} \omega_t^2, \tag{3.42}$$

where Γ_m^* is the desired generator torque and K_{C_P} is a parameter which depends on the characteristics of the turbine:

$$K_{C_P} = \frac{1}{2} \rho A R^3 \frac{c_1 \left(c_2 c_7 c_9 + c_6 c_7 + c_2\right)^3 e^{-\frac{c_6 c_7 + c_2}{c_2}}}{c_2^2 c_7^4}. \tag{3.43}$$

3.7.2 *DFIG current dynamics decoupling and linearization*

DFIG is controlled by applying voltages to the rotor of the machine in order to obtain the desired rotor currents. The desired rotor current values can be calculated given a grid voltage and the desired torque and stator reactive power values in steady state. From the equations of the DFIG machine we see that:

$$i_{rq}^* = -\frac{2L_s}{3PMv_{sq}} \Gamma_m^*, \tag{3.44}$$

$$i_{rd}^* = -\frac{2L_s}{3Mv_{sq}} Q_s^* + \frac{v_{sq}}{\omega_e M}, \tag{3.45}$$

where i_{rq}^* and i_{rd}^* are the rotor current reference values and Γ_m^* and Q_s^* are the desired torque and reactive power.

To be able to design a current controller a linearization and decoupling feedback is introduced:

$$\begin{bmatrix} v_{rq} \\ v_{rd} \end{bmatrix} = \begin{bmatrix} \hat{v}_{rq} + s\omega_s \left(Mi_{sd} + L_r i_{rd} \right) \\ \hat{v}_{rq} - s\omega_s \left(Mi_{sq} + L_r i_{rq} \right) \end{bmatrix}, \tag{3.46}$$

where $\hat{v}_{rq}$ and $\hat{v}_{rd}$ are the new auxiliar voltage inputs.

This way the voltage-current relations of the resulting system become equivalent to a circuit with a resistor in series with a inductor:

$$\begin{bmatrix} i_{rq}(s) \\ i_{rd}(s) \end{bmatrix} = \begin{bmatrix} \frac{1}{L_r s + r_r} & 0 \\ 0 & \frac{1}{L_r s + r_r} \end{bmatrix} \begin{bmatrix} \hat{v}_{rq}(s) \\ \hat{v}_{rd}(s) \end{bmatrix}. \tag{3.47}$$

3.7.3 *PMSG current dynamics decoupling and linearization*

PMSG is controlled by applying voltages to the stator of the machine to obtain the desired stator current evolution. As in the DFIG, the stator current reference values can be calculated to obtain the desired steady state torque and to achive different objectives like minimizing the stator current or obtaining the maximum torque with a given stator voltage modulus. From the steady state equations of the PMSG we see that to control the desired torque we need:

$$i_{sq}^* = -\frac{2}{3P\lambda_m}\Gamma_m^*. \tag{3.48}$$

Also to simplify the design of the current controllers, we introduce a linearization and decoupling feedback[9] :

$$\begin{bmatrix} v_{sq} \\ v_{sd} \end{bmatrix} = \begin{bmatrix} \hat{v}_{sq} + \omega_r \left(L_{ls} + \frac{3}{2} \left(L_A + L_B \right) \right) i_{sd} + \omega_r \lambda_m \\ \hat{v}_{sq} - \omega_r \left(L_{ls} + \frac{3}{2} \left(L_A - L_B \right) \right) i_{sq} \end{bmatrix}, \tag{3.49}$$

where $\hat{v}_{sq}$ and $\hat{v}_{sd}$ are the new auxiliar voltage inputs.

This way the voltage-current relations of the resulting system become:

$$\begin{bmatrix} i_{sq}(s) \\ i_{sd}(s) \end{bmatrix} = \begin{bmatrix} \frac{1}{\left(L_{ls} + \frac{3}{2}(L_A - L_B) \right)s + r_s} & 0 \\ 0 & \frac{1}{\left(L_{ls} + \frac{3}{2}(L_A + L_B) \right)s + r_s} \end{bmatrix} \begin{bmatrix} \hat{v}_{sq}(s) \\ \hat{v}_{sd}(s) \end{bmatrix}. \tag{3.50}$$

3.7.4 *Grid-side converter control*

The goal of the grid side converter control is to maintain the DC bus voltage in the desired nominal value and also to produce the desired reactive power through the grid side converter. Taking the v_z grid voltage angle as the reference angle it can be seen that the output reactive power is proportional to i_{ld} while the active power is proportional to i_{lq}. Thus, from the steady state equations, a reference value can be deducted for i_{ld} to obtain the desired reactive power:

$$i_{ld}^* = \frac{2}{3v_{zq}}Q_z^*.$$

(3.51)

The DC bus regulation is done through i_{lq} by feeding the voltage error to a proportional-integrator (PI) controller to obtain a 0 voltage steady state error:

$$i_{lq}^* = \frac{K_p s + K_i}{s}(E^* - E).$$

(3.52)

Given the current reference values, current controllers can be designed with ease introducing the following linearization and decoupling feedback:

$$\begin{bmatrix} v_{lq} \\ v_{ld} \end{bmatrix} = \begin{bmatrix} \hat{v}_{lq} + \omega_s L_l i_{ld} + v_{zq} \\ \hat{v}_{lq} - \omega_s L_l i_{lq} \end{bmatrix},$$

(3.53)

where $\hat{v}_{lq}$ and $\hat{v}_{ld}$ are the new auxiliary voltage inputs.

This way the voltage-current relations of the resulting system become:

$$\begin{bmatrix} i_{lq}(s) \\ i_{ld}(s) \end{bmatrix} = \begin{bmatrix} \frac{1}{L_l s + r_l} & 0 \\ 0 & \frac{1}{L_l s + r_l} \end{bmatrix} \begin{bmatrix} \hat{v}_{lq}(s) \\ \hat{v}_{ld}(s) \end{bmatrix}.$$

(3.54)

3.7.5   *Current controller design*

We have shown that for each system, we can introduce a linearization and decoupling feedback from measured magnitudes and transform a multivariable system with time-varying parameters into a circuit with a resistor in series with a inductor. Thus the same current controller design procedure can be applied to each system. Here we briefly present a PI design for a RL circuit.

For a given RL circuit, the current dynamics can be expressed in the Laplace domain as a function of the applied voltages as:

$$i(s) = \frac{1}{Ls + r}v(s).$$

(3.55)

This corresponds to a first order system, hence it can be controlled with a PI controller achieving a zero steady state error and a desired closed loop settling time. This controller can be expressed as

$$v(s) = \frac{K_i + K_p s}{s}\left(i^*(s) - i(s)\right)$$

(3.56)

and

$$K_p = \frac{L}{\tau}, \quad K_i = \frac{r}{\tau},$$

(3.57)

where K_p and K_i are the proportional and integral parameters of the PI controller and τ is the desired closed loop time constant.[10]

The closed loop transfer function of the current as a function of the current reference becomes:

$$i(s) = \frac{1}{\tau s + 1} i^*(s).$$

(3.58)

3.8 Electrical Disturbances

We introduce here a general three phase grid voltage expression:

$$v_z^{abc}(t) = \frac{\sqrt{2}}{\sqrt{3}} V_D(t) \begin{bmatrix} \cos(\varphi_D(t)) \\ \cos(\varphi_D(t) - \frac{2\phi}{3}) \\ \cos(\varphi_D(t) + \frac{2\phi}{3}) \end{bmatrix} + \frac{\sqrt{2}}{\sqrt{3}} V_I(t) \begin{bmatrix} \cos(\varphi_I(t)) \\ \cos(\varphi_I(t) + \frac{2\phi}{3}) \\ \cos(\varphi_I(t) - \frac{2\phi}{3}) \end{bmatrix}$$

(3.59)

and

$$\varphi_D(t) = \varphi_D(0) + \int_0^t \omega_s(t)dt,$$

(3.60)

$$\varphi_I(t) = \varphi_I(0) + \int_0^t \omega_s(t)dt,$$

(3.61)

where $V_D(t)$ and $V_I(t)$ are the phase-to-phase positive and negative sequence voltage rms values as functions of time, $\varphi_D(t)$ and $\varphi_I(t)$ are the phase angles of the positive and negative sequences, and $\omega_s(t)$ is the derivative of the phase angles.

The wind turbine system's response to a series of grid fault types can be simulated by inputting the right grid voltages to the presented model, the function definitions to accomplish this can be seen on Table 3.2

3.9 Conclusions

In this chapter we have presented a dynamical simulation model for the DFIG and the PMSG wind generators which are the main types of variable speed wind generators that we will see in wind farms in following years, DFIG being the most common nowadays. First we presented a wind able to simulate the evolution of its speed for a given wind turbine location. Then we have introduced a simple model that describes the dynamic behavior of the wind turbine mechanical components and the generator electrical dynamics. Finally we have briefly discussed the control of both

Table 3.2. Grid voltage function definition for different grid faults simulation.

Fault type	Function	Value	Comments
Symmetrical voltage sag	$V_D(t)$	$V^N\,(1 - \alpha u(t - t_0))$	V_N is the nominal voltage, α is the per-unit dip amplitude, $u(t - t_0)$ is a t_0 delayed step function
	$V_I(t)$	0	
	$\omega_s(t)$	ω_s^N	ω_s^N is the nominal grid frequency
Asymmetrical voltage sag	$V_D(t)$	$V^N\,(1 - \alpha_D u(t - t_0))$	α is the per-unit positive sequence dip amplitude
	$V_I(t)$	$V_I u(t - t_0)$	V_I is the negative sequence amplitude during the asymmetrical dip
	$\omega_s(t)$	ω_s^N	ω_s^N is the nominal grid frequency
Frequency step change	$V_D(t)$	V^N	
	$V_I(t)$	0	
	$\omega_s(s)$	$\omega_s^N\,(1 + \alpha_\omega)$	α_ω is the per-unit grid frequency increase

wind turbine topologies and how we can model a series of electrical disturbances of interest with the model.

References

1. S. Heier, *Grid Integration of Wind Energy Conversion Systems* (John Wiley and Sons, 1998).
2. J.G. Slootweg, "Reduced-order modelling of wind turbines," in *Wind Power in Power Systems*, (Wiley, 2005), pp. 555–585.
3. H.A. Panofsky and J.A. Dutton, *Atmospheric Turbulence*. (Wiley-Interscienc, 1984).
4. M. Shinozuka and C.-M. Jan, "Digital simulation of random processes and its applications," *J. Sound and Vibration* **25** (1972) 111–128.
5. P.C. Krause, *Analysis of Electric Machinery* (McGraw-Hill, 1986).
6. H. Akagi, E. Watanabe and M. Aredes, *Instantaneous Power Theory and Applications to Power Conditioning* (Wiley, 2007).
7. P.D. Chandana-Perera, "Sensorless control of permanent-magnet synchronous motor drives," PhD thesis, Faculty of Engineering & Science at Aalborg University (2002).
8. D. Goodfellow and G. Smith, "Control strategy for variable speed of a fixed-pitch wind turbine operating in a wide speed range," *Proc. 8th BWEA Conf.*, Cambridge (1986), pp. 219–228.
9. M. Chinchilla, S. Arnaltes and J.C. Burgos, "Control of permanent-magnet generators applied to variable-speed wind-energy systems connected to the grid," *IEEE Trans. Energy Conversion* **21** (2006) 130–135, doi: 10.1109/TEC.2005.853735.
10. L. Harnefors and H.-P. Nee, "Model-based current control of ac machines using the internal model control method," *IEEE Trans. Industry Applications* **34** (1999) 133–141, doi: 10.1109/28.658735.

Chapter 4

Technologies and Methods used in Wind Resource Assessment

Ravita D. Prasad

College of Engineering, Science and Technology,
Fiji National University,
P.O. Box 3722, Samabula, Fiji Islands
ravita.prasad@fnu.ac.fj

Ramesh C. Bansal

School of Information Technology and Electrical Engineering,
The University of Queensland, St. Lucia, Brisbane, QLD 4072, Australia
rcbansal@ieee.org

Wind energy is one of the fastest growing energy sources in the world today. The reason for this can be due to the fact that there have been vast improvements in wind energy technology which has led to lower cost. For any wind energy project to be successful there should be a thorough wind resource assessment (WRA) carried out. This chapter presents literature review on the technologies and methods used in WRA. The chapter is organized as follows. Section 4.1 gives an overview on wind energy. Section 4.2 presents the literature review on technologies, software and methods used in WRA. Section 4.3 describes a method for finding the optimum wind turbine for a site. Section 4.4 presents the uncertainty involved in predicting wind speed using different methods. Finally, some conclusions are drawn.

4.1   Introduction

At present the world is consuming much higher energy than it used in the past. This is mainly due to the fact that industrialization is on the rise and people are introducing new technologies. There are concerns from government as well as non-government organizations about the increase in pollution and escalating cost of fuel that has led industries to look for alternative fuel sources.

Some of the widely used renewable energies are solar, hydropower, geothermal, wind, tidal, wave, biomass and many more. The technologies involved in each are always under constant research and development so that the cost of energy generated from each can be further reduced and that the technology is efficient, reliable, and safe.

Wind is a form of solar energy and it is caused by the uneven heating of the atmosphere by the sun, the irregularities of the earth's surface, and the rotation of the earth.[1] Wind flow patterns are modified by the earth's terrain, bodies of water, and vegetative cover. This wind flow or motion of energy when harvested by modern wind turbines can be used to generate electrical energy. The terms "wind energy" or "wind power" describe the process by which wind is used to generate mechanical power or electricity. Wind turbines convert the kinetic energy in the wind into mechanical power. This mechanical power can be used for specific tasks (such as grinding grain or pumping water) or a generator is used is convert this mechanical energy into electrical energy to power homes, businesses, schools, etc.[2,3] The total wind power, P_w, available to wind turbine is given by[4–15]

$$P_w = \frac{1}{2}\rho A v^3, \tag{4.1}$$

where ρ is the density of air in kg/m^3, A is the swept area in m^2, v is the wind speed in m/s. The maximum wind power that can be harnessed by a wind turbine is 59.3% (which is known as the Betz coefficient) of the total wind power. The electrical output power (P_e) from a wind turbine is given by[16]

$$P_e = C_{op}\frac{1}{2}\rho A v^3, \tag{4.2}$$

where C_{op} is the overall power coefficient of the wind turbine which is the product of the mechanical efficiency (η_m), electrical efficiency (η_e) and the aerodynamic efficiency (Betz coefficient).

However, before wind energy can be harnessed from a particular site, the wind resource assessment (WRA) and analysis is critical to estimate the economic feasibility of a wind turbine at a site and the annual energy yield. Since wind is an intermittent source of energy in WRA the wind distribution for the site is determined.

4.2 Literature Review, Methods and Software used in WRA

With nearly 90% of all the life cycle costs (LCC) of a wind power plant being upfront, the financial and economic viability of electricity generation from wind energy is dependent on the level and extent of energy content in winds prevalent at a particular site[17] and also on the payment expected for power generated. Prevalent wind at any location is both site specific and very much dependent on the terrain and topographic features around the location. For proper and beneficial development of wind power at any site wind data analysis and accurate wind energy potential assessment are the key requirements. An accurate WRA is an important and critical factor to be well understood for harnessing the power of the wind. The reason is that if one looks at Eq. (4.1) an error of 1% in wind speed measurement can lead to a 3% error in energy output since energy is proportional to cube of wind speed.[6,18,19]

It is well known that wind resource is seldom consistent and it varies with the time of the day, season of the year, height above the ground, type of terrain, and from year to year.[5,6,14,15,20] All of these factors lead to the reason why WRA and analysis should be done carefully and completely. The surface roughness and the obstacle in the vicinity of wind measuring tower are also important factors to be considered for WRA.[21,22] The following subsections give an overview of steps carried out to choose a site for wind turbine installation from a set of potential sites. Preliminary wind survey is the initial stage of WRA.

4.2.1 *Preliminary wind survey*

Preliminary wind survey includes surveying a number of sites and choosing the best site for installing a wind speed monitoring instruments. Wind resource is the best factor to be looked at before installing the wind monitoring system at the site. However, there are some other factors that have to be considered before significant time and energy have been invested at a particular site. One purpose of such visits is to look for physical evidence to support the wind resource estimate development in the large-area screening. For example, consistently bent trees and vegetation (flagging) is a sure sign of strong winds. Another purpose is to check for potential siting constraints. A third purpose of the site visit is to select a possible location for a wind monitoring station. Furthermore, access to the site is another factor, since, construction of roads will add to the overall cost of wind energy. The factors that are studied in preliminary wind survey are as follows.

(1) Instantaneous wind speed measurement: This is done using wind watch. The measurement is taken at the site during site visits. The wind speed measurement is taken every 5 minutes for 2 hours and at 3 different times of a day and 3 different times in a year. This is done basically to see how much potential a site has for wind speed measuring instrument to be installed.
(2) Interview people: This is not always a formal interview. Sometimes it can be just an informal chat. The main focus of interview/chat is to find out if the area has some endangered flora or fauna whose habitat would probably be destroyed in installing a wind monitoring equipment.
(3) Study of meteorological information on wind speed and wind direction: This would be used to compare the wind speed value obtained from the wind watch.

Also this information can be used to find out how the wind speed is varying in different years in the long term.

(4) Land availability: One has to find out which kind of tenure the land has, i.e., freehold or leased. At some parts of the world land issue is very sensitive, hence memorandum of understanding (MOU) has to be made before a wind project can commence.

(5) Terrain features: Some of the features that have to be studied are:

- Buildings and trees height around the site. The size of these will influence the speed of wind available at a site.
- Soil conditions. It is important for the foundation of a tower for wind turbine if it gets selected.
- Accessibility of site. One question that has to be answered is that "is there a proper road leading to the site where the wind monitoring system is to be installed?" This is considered because when equipment is brought for installation then it can be easily transported to the site. No proper road would increase the cost of wind project.
- Vegetation flagging. This indicates the presence of prevalent strong wind. Figure 4.1[23] can be used to determine the wind speed values at a site by observing vegetation flagging. Nevertheless, it should be noted that the

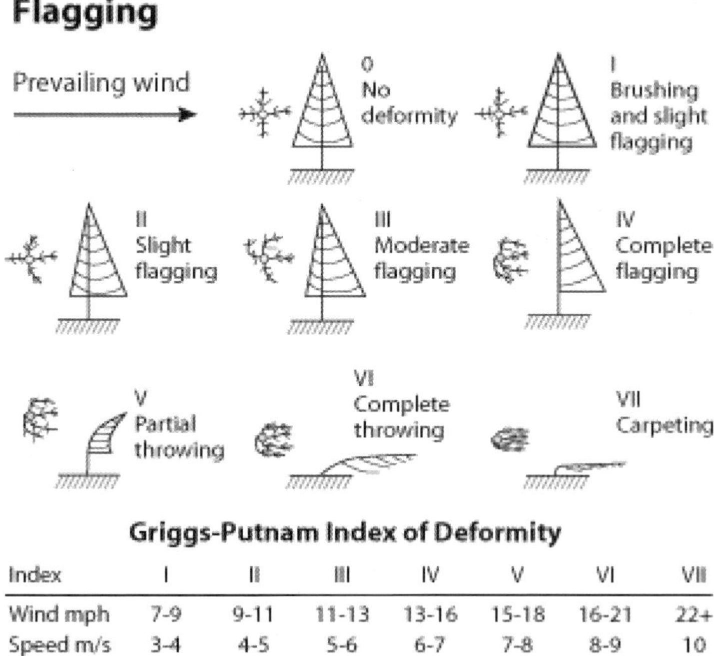

Griggs-Putnam Index of Deformity

Index	I	II	III	IV	V	VI	VII
Wind mph	7-9	9-11	11-13	13-16	15-18	16-21	22+
Speed m/s	3-4	4-5	5-6	6-7	7-8	8-9	10

Fig. 4.1. Flagging illustrations and a Griggs–Putnam index of deformity.

absence of flagging does not mean that this site does not experience strong wind motion. It may simply be that there is no prevalent wind.

Once a site is selected by a preliminary wind survey, a detailed WRA is carried out at the site which is explained in the following subsections.

4.2.2 *Direct wind measurement*

After a preliminary wind survey, a site is selected and wind monitoring equipment is installed. The most reliable approach to site assessment is to directly measure the wind speed, ideally at the hub height of the proposed wind turbines so as to remove any uncertainties arising from predicting wind shear (the way in which wind speed increases with height). It should always be kept in mind that wind resource determines;

- Project location and size.
- Tower height.
- Turbine selection and layout.
- Energy output (annual, seasonal and capacity factor).
- Cost of energy/cash flow.
- Size of emissions credits.

There are two types of wind measurement program:

 (i) Short-term study: a short-term (1–4 weeks) measurement program may be con-ducted for the purpose of verifying model estimates made. The objective would be to obtain data from periods of 1–3 hours of relatively constant wind direction (maximum range 10–20 degrees in a series of, for example, 10-minutes means) for as many directions as are of interests.
(ii) Long-term study: in some cases long-term measurement program may be required. The objective would be to estimate the wind climatology more directly by obtaining, for example, hourly measurements over a period of 6 months to 2 years.[22] The shorter period may be sufficient when there are no significant seasonal variations, which will generally not be the case at middle or high latitudes.

For direct wind measurement program the following are some of the parameters that are monitored:

- Wind speed — cup anemometer is used. These are oriented in such a way to minimize any wake effects.[24] Please refer to Chap. 7 for a detailed study of wake effects.

- Wind direction — wind vane is used. Wind direction will help to better layout a wind turbine at the most optimum location. It will be used to construct wind rose which indicates the prevalent wind.
- Temperature. This is measured because temperature affects the air density which is directly proportional to the power available in the wind speed. Temperature measurement may be made at 2 to 3 meters above ground. Measuring at this height minimizes the effects of surface heating during daylight hours. Also at temperatures near freezing, precipitation can collect and freeze on the sensors, affecting their performance. There may be periods of time when the anemometer is measuring high wind speeds while the wind vane is immobile, or the wind vane may be indicating direction changes while the anemometer is not measuring a wind speed. Correlating temperatures to these data can verify icing conditions.[25] Incorrect data should be removed or corrected before analysis is conducted.

All the parameters measured will be stored in a data logger from where data can be downloaded after certain time. Wide operating temperature ranges and weather-proof enclosures are needed[26] for data loggers to ensure reliable data collection even in adverse conditions. The stored parameters in the data logger will be used to determine wind shear, turbulence intensity and air density.

It is always an intricate decision to determine at what height the wind speed should be measured. It is preferred that the best height for wind speed measurement is the hub-height of the wind turbine to reduce uncertainty in energy output estimation from a location. However, it is not always known what wind turbine will be installed at the site, so wind speed and other parameters can be measured at different heights. This would aid in determining the surface roughness coefficient for the site and correcting measured wind speed at turbines hub-height for determining annual energy output. For a 50-meter tower, measurements at 10, 25 and 50 meters are normal and for a 60-meter tower; measurements are at 10, 30 and 60 meters.[27] Ten-meter data is the standard height for wind measurements. In areas that contain obstructions or vegetation, particularly within forest canopies, the lowest wind sensor is placed at a height that minimizes effects of surface roughness or obstructions.

Usually it is not cost effective to measure wind data in long term. The following subsections discuss methods used to rectify this issue.

4.2.3 *Derivation of long-term wind speed*

As it is not feasible or financially viable to measure the wind resource at a potential wind farm site for number of years in order to gather enough data for long term resource prediction. The data measured over 6 months (minimum) must be further processed in order to estimate the long term resource.

One method of achieving this is to use a measure-correlate-predict method (MCP). MCP is a statistical technique used for predicting the long-term wind resource at a candidate site by relating measurements from a short-term measurement campaign at the candidate site to long-term measurement at a reference site.[28] The measured data (candidate site) is matched with a meteorological station (reference site) for which high quality, long term records are available. Ideally, the meteorological station should be as close to the wind farm site as possible and have a similar exposure. Concurrent data sets for the wind farm and the meteorological station are compared and correlations derived. These correlations are then applied to the long term meteorological station data, to construct an estimate of the wind resource at the wind farm site would have been over the period of the long term data.

Measure-correlate-predict methods take into account the fact that the wind resource will vary from year to year — the period of measurement is unlikely to be representative of the long term wind resource without this manipulation.

4.2.4 *Forecasting wind speed*

Sometimes due to lack of time or high cost the wind speed at a particular site is recorded only for a short duration (4–6 months[29]) for WRA. Many researchers focus on providing a forecasting tool in order to predict wind power production with good accuracy. Depending on the input, these tools are classified as physical (which uses meteorological, topographical information and technical characteristics of wind turbines) or statistical (which uses explanatory variables and online measurements like recursive least squares) or artificial neural networks (ANN) approaches or a combination of all three. Accurate short-term or long-term forecasting can aid in

- Improved marketing trading.
- Optimized scheduled maintenance.
- Enhanced plant scheduling by system operators.[30]

4.2.4.1 *ANN approach to WRA*

Bechrakis *et al.*[31] developed a model to simulate the wind speed to estimate the wind power of an area, in which wind speed at another site is given. This method takes into account the evolution of the sample cross correlation function (SCCF) of wind speed in time domain and uses ANN to perform the wind speed simulation. The tests showed that the higher is the SCCF value between the two sites, the better is the simulation achieved. Some researchers use spatial correlation models as functions of time to improve wind speed forecasting at a specific site using data from two or more stations.[32] Short term wind predictability[33] is the ability to foresee hourly wind energy one or several days in advance.

In Ref. 34 one year's measured wind speeds of one site have been used to extrapolate the annual wind speed at a new site using ANN. After derivation of the simulated wind speed time series for the target site, its mean value and corresponding Weibull distribution parameters were calculated to make an assessment of the annual wind energy resource in the new area with respect to a particular wind turbine model. Results indicated that only a short time period of wind data acquisition in a new area might provide the information required for a satisfactory assessment of the annual wind energy resource. Barbounis *et al.*[35] used local recurrent neural network models to perform long-term wind speed and power forecasting. Owing to the large time-scale, they were based three days ahead meteorological predictions (wind speed and direction) provided at four nearby sites of the park.

Nichita *et al.*[36] proposed two modeling procedures for wind speed simulation. These simulations could be implemented on the structure of a wind turbine simulator during studies concerning stand-alone or hybrid wind systems. The turbulence component is assumed to be dependent on the medium and long term wind speed evolution. Authors[37–41] simulated wind speeds using site correlation for wind resource assessment.

4.2.4.2 *Models for wind speed forecasting*

Riso National Laboratory and the Technical University of Denmark have used detailed area specific, three-dimensional weather models and have worked with numerical weather prediction (NWP) models, such as HIRLAM (High Resolution Limited Area Model), the UK MESO (UK Meteorological Office Meso-scale model) or the LM (Lokal–Model of the German Weather Service). These systems work like a weather forecast, predicting wind speeds and directions for all the wind farms in a given area.[42] Some of the systems use statistics, ANN or fuzzy logic (FL) instead of physical equations since in this method they are able to learn from experience. The disadvantage is that they need a large data set to be trained before the system works properly.[42] The current wind power prediction tool (WPPT) "Prediktor" uses physical models. Other WPPT are Zephyr, Previento, eWind and Sipreolico. WPPT applies statistical methods for predicting the wind power production in larger areas. It uses online data that cover only a subset of the total population of wind turbines in the area. In general, WPPT uses statistical methods to determine the optimal weight between online measurements and meteorological forecasted variables.

4.2.5 *Softwares used in WRA*

Prediction of the wind resource at a given site is a crucial stage in the development of a commercial (large-scale) wind energy installation. This is because the energy which can be harvested from a given site and the project economics are both highly dependent on the wind resource at the site.[43] The energy output of a wind farm is a function of the cube of the wind speed — so if the wind speed doubles, the available power will increase by a factor of eight. The more energy produced, the better the return on investment made.

Wind farms vary in size and scale depending on the limitations of the land available and the type of terrain. The exact location of each turbine in the wind farm must be at such a place so that there is maximum energy output from the wind turbines. In order to determine this one needs to be very accurate and thorough in deciding where each turbine has to be located. Hence softwares such as Garrad Hassan WindFarmer, WindSim, RESoft Windfarm, etc., are used by commercial companies to make right decisions for a successful wind energy project. These softwares helps the decision-maker by using all the minute details (such as the type of terrain, turbulence intensity, wake effect, etc.) for the location and finding the optimum location for each turbine. Brief description of some of the commercially available softwares is as follows.

4.2.5.1 *Garrad Hassan WindFarmer (GH WindFarmer)*

GH WindFarmer software is used for wind farm design and it combines all aspects of data processing, wind farm assessment and wind farm layout into one integrated easy-to-use program making fast and accurate calculations. GH WindFarmer is technically advanced and powerful. It enables the user to automatically and efficiently optimize the wind farm layout for maximum energy yield, whilst meeting environmental, technical and constructional constraints.[30] The latest version of GH WindFarmer has complete uncertainty calculations with standard deviations, historical and future uncertainties and exceedance levels for the net energy yield, compares turbine design parameters with estimates of Design Equivalent Turbulence, detailed shadow flicker calculations with greater ease of use and new options such as user-defined rotor orientation and creates a turbine ranking table as part of the site conditions report to rapidly identify the least productive turbines.

4.2.5.2 *WindSim*

WindSim software is used for simulation of wind resources in complex terrain. WindSim is based on Computational Fluid Dynamics (CFD). It combines advanced numeric processing with 3D visualization in a user-friendly user interface.

Customers use WindSim to optimize park layouts by finding turbine locations with the highest wind speeds, but with low turbulence. WindSim is also used for calculation of loads on turbines.[44] The first PC based version of the Wind–Sim software was launched in 2003. Since then there has been a steady growth in the number of users, including the leading companies within the wind industry such as Enercon, Gamesa, Siemens and Vestas.

4.2.5.3 *RESoft WindFarm*

RESoft WindFarm calculates the energy yield of a wind farm simultaneously including topographic and wake effects. The turbine layout can be optimized for maximum energy yield or minimum cost of energy whilst subject to natural, planning (including noise) and engineering constraints. The energy yield analysis is somewhat more sophisticated than other software, and includes numerous advanced wake options. WindFarm has advanced graphics, and can perform wind flow calculations, noise calculations (showing the noise contours) and measure-correlate-predict analysis of wind speed data. The powerful visualisation tools create 3D visualizations of the landscape (a virtual World), planning quality photomontages including animation, display wire frame views of the wind farm, analyze shadow flicker and create zone-of-visual-influence maps including cumulative impact.[45]

4.2.5.4 *Limitations of GH WindFarmer, RESoft WindFarm, WAsP and WindSim*

All these models have limitations due to linearization of the model equations. This restricts their applicability to low terrain slopes (e.g., <0.3). These models are also limited by the fact that they do not take into account thermal effects such as sea breezes or mountain-valley winds. Though these models have some limitations, they can give good results if "handled" carefully.[27] Each of the softwares have uncertainty involved in estimating the annual energy output from wind farm and predicting wind speed.

4.2.6 *WRA by topographical and numerical modeling*

On a more refined scale, wind speeds at a particular site can be modeled using information on the elevation, topography and ground surface cover. There are a number of models that are used by wind planners for estimating wind speed variations in simple and complex terrains, namely Wind Atlas Analysis and Application Program (WAsP), Mixed Spectral Finite-Difference (MSDF), MS-Micro/3, Guidelines for Windows (GLW), etc. In simple terrain models roughness coefficient and turbulence intensity are calculated. If more detail is required on wind flow in complex terrain near proposed wind turbine sites, then more sophisticated models may be appropriate.[28]

4.2.6.1 *WAsP (Wind Atlas Analysis and Application Program)*

WAsP is a PC program for predicting wind climates, wind resources and power productions from wind turbines and wind farms. The predictions are based on wind data measured at stations in the same region. The program includes a complex terrain flow model, a roughness change model and a model for sheltering obstacles.[46] It is used for vertical and horizontal extrapolation of wind climate statistics. It contains several physical models to describe the wind flow over different terrains and close to sheltering obstacles.

4.2.6.2 *MS-Micro/3*

MS-Micro/3 is a numerical model for estimating wind speed variations in complex terrain. It runs on a microcomputer using disk operating system (DOS). In general, it can run on Windows in a DOS compatibility box or command line mode. Its principal uses are wind energy site selection, wind resource estimates, wind loading estimates, local climatological studies and evaluation of representativeness of measured wind data.

4.2.6.3 *MSDF (Mixed Spectral Finite-Difference)*

MSDF is a numerical model used for estimating wind speed variation in complex terrains. Its main features include, 3D steady-state surface boundary-layer flow, spatial variations in terrain height and surface roughness, high spatial resolution, results at any height or height above the ground and turbulence closure. This model is used for wind energy site selection, wind resource and turbulence estimates, local climate studies and evaluation of representativeness of measured wind data.[47] This model retains many of the advantages of its parent MS-Micro/3.

4.2.6.4 *GLW (Guidelines for windows)*

This is a graphical implementation for Windows of the simple Guidelines for estimating wind flow in simple complex terrain situations. Its main features include, estimation of wind speed profiles for complex terrains characterized by heterogeneous surface roughness and topography. Wind speed input measurements can be defined at remote "reference" site, instantaneous computations allow immediate graphical feed-back — useful for teaching and sensitivity analysis and topographic features treated: 2D ridges, valleys, escarpments; 3D hills, basins; 2D rolling topography; 3D rolling topography; user-defined features.[47]

Once wind data has been analysed for a site, it is vital to determine the exact location to install wind turbine. This is known as micrositing.

4.2.7 *Micrositing*

Micrositing is the process of choosing the type of wind turbine generator (WTG) and its exact position in planning work of a wind park. Proximity to transmission lines and roads, terrain, and land use are just a few of the other factors that require consideration when siting a wind project. The other factors that have to be regarded during micrositing are:

- Wind conditions (statistic data concerning wind speed and wind direction).
- Building requirements (e.g., distance to residences).
- Ownership structure of the area.
- Accessibility (existing roads).
- Influence of the WTG on the environment (e.g., shadow flickering, noise emission).
- Distances between the individual turbines in a park.

Micrositing report for a wind farm consists of the following[27]:

- Description and maps of the site (in terms of orography and roughness).
- Characteristics of the turbine and wind farm (turbine dimensions, farm layout, power curve, and control systems).
- Description of the long-term reference site and internal wind speed variations.
- Temporal variations of the wind (seasonal, diurnal, and turbulence intensity).
- Spatial variations of wind over the project area.
- Air density adjustment for the site and site-corrected power curve.
- Wind turbine array efficiency.
- Wind turbine expected availability.
- Electrical losses.
- High wind hysteretic losses (losses near the turbine cutout speed).
- Frequency of forced shutdowns by the utility.
- Discussion of the uncertainties of all measurements and estimated.
- Discussion of the uncertainty of the energy output estimate.

Siting a turbine within a wind project involves careful consideration of an array of factors relating to wind flow, terrain, equipment access, environmental and land-use issues, and visual impact. Maximizing production is the most important factor, but without attention to aesthetics, the project may not make it past the permitting phase. To maximize production, careful attention must be paid to the prevailing wind direction(s), wind obstructions from man-made structures or vegetation, and terrain effects. The impact of the wind disturbance caused by one turbine on another is another important factor. Below is a list of general rules for siting turbines.

- On a site with multiple wind turbines, the turbines should be placed at least two rotor diameters apart in the plane perpendicular to the prevailing wind direction, and at least ten rotor diameters apart in the plane parallel to the prevailing wind direction. This will prevent the turbines from experiencing reduced wind speeds and increased turbulence due to the other turbines.
- To avoid turbulence, turbines should be placed at a distance twenty or more times the height of any man-made structure or vegetation upwind of the project. The turbulent wind flow created by a structure generally extends vertically to twice the height of the structure, so small structures may not have any impact on the tall turbines used today.
- Avoid areas of steep slope. The wind on steep slopes tends to be turbulent and has a vertical component that can affect the turbine. Also, the construction costs for a steep slope are greatly increased.
- On ridgelines and hilltops, set the turbines back from the edge to avoid the impacts of the vertical component of the wind.

Along with the rules stated above, it is important to take into consideration the visual impact of the site. Grids of turbines tend to be much less visually appealing than turbines placed along the curves created by natural features. It is also sometimes possible, on a large site, to place turbines where they will not be as visible to residents in the proximity. Softwares such as Windfarmer and RESoft windfarm can be used to aid micrositing.

4.3 Wind Characteristics for Site

Wind speed varies in both time and space. Space variations are generally dependent on height above the ground and global and local geographical conditions. After the wind resource data such as wind speed, direction and temperature are known at a particular site then the following parameters are determined using the above information.

4.3.1 *Annual average wind speed*

This is the mean of wind speed recorded during a year.

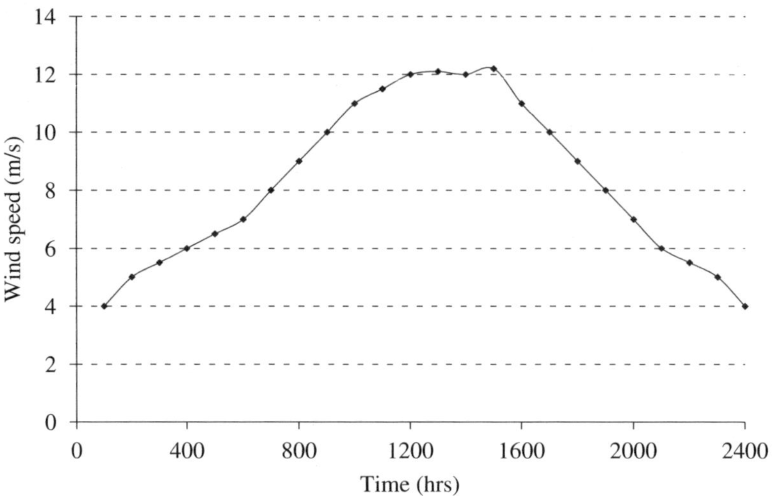

Fig. 4.2.   Annual daily variation in wind speed.

4.3.2   *Annual diurnal wind speed*

This shows how the wind speed varies during a day by considering the whole year's wind speed data. The different hourly wind speeds data is averaged for the year and a graph is plotted, Fig. 4.2. Daily variation in wind speed is due to differential heating of the earth's surface during the daily radiation cycle.

4.3.3   *Monthly diurnal wind speed*

This shows the daily variation in wind speed in different months throughout a year. This will be similar to Fig. 4.2 but instead of the graph for a whole year, it would be diurnal variation in wind speed for different months. This information will help the wind planner to match the daily load in different months to the daily wind resource.

4.3.4   *Monthly variation in wind speed*

This shows how wind speeds are changing in different months in a year, Fig. 4.3. It would aid in knowing which months have high wind speeds and which has low, consequently helping to decide the type of storage system that would be needed.

4.3.5   *Frequency distribution of wind speed*

This is the most important of all wind characteristics. Using this information the annual wind energy for a site is estimated. To obtain this graph, wind speed is binned,

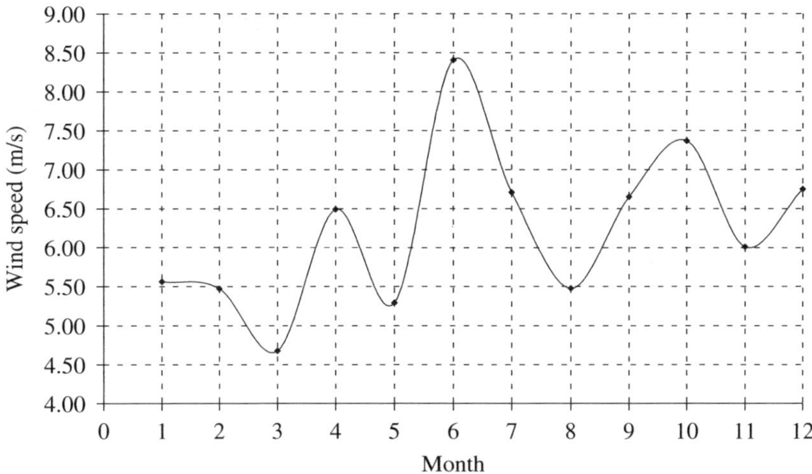

Fig. 4.3.   Monthly variation in wind speed for a site.

i.e., bins 0–1 m/s, 1–2 m/s, 2–3 m/s, etc., are created and different wind speed in year are binned in the respective bins. The frequency (or number of hours in a year) is then determined. This can be done in Microsoft Excel or in other software such as Windographer, where this is done automatically and the wind speed frequency curve is obtained.

Table 4.1, shows the wind speed bins and frequency and Fig. 4.4 is obtained using Table 4.1.

4.3.6   *Weibull distribution of wind speed*

When the wind speed frequency (distribution) for a site is unknown then the Weibull distribution can be used to estimate the wind speed distribution for a site by putting in the shape parameter (k) and the scale parameter (c). The Weibull probability density function (PDF) of wind speed, v, $f(v)$ is given by Eq. (4.3).

$$f(v) = \left(\frac{k}{c}\right)\left(\frac{v}{c}\right)^{k-1} \exp\left(-\left(\frac{v}{c}\right)^k\right). \tag{4.3}$$

The area under the PDF gives the probability of wind speeds occurring at a particular site. Probability density function gives the probability of occurrence of wind speed between certain intervals. Once the probability is found, this probability is then multiplied with the 8760 hours (that is number of hours in a year). This product gives the frequency of wind speed in a year. The frequency of wind speed is then multiplied with the power output from a wind turbine at that particular wind speed.

Table 4.1. Wind speed frequency bins.

Wind speed bins (m/s)	Class mark (m/s)	Frequency (hours)
0–1.0	0.5	69
1.0–2.0	1.5	175
2.0–3.0	2.5	387
3.0–4.0	3.5	1061
4.0–5.0	4.5	1518
5.0–6.0	5.5	1530
6.0–7.0	6.5	1300
7.0–8.0	7.5	930
8.0–9.0	8.5	640
9.0–10.0	9.5	400
10.0–11.0	10.5	325
11.0–12.0	11.5	280
12.0–13.0	12.5	97
13.0–14.0	13.5	20
14.0–15.0	14.5	15
15.0–16.0	15.5	7
16.0–17.0	16.5	5
17.0–18.0	17.5	1
18.0–19.0	18.5	0
19.0–20.0	19.5	0

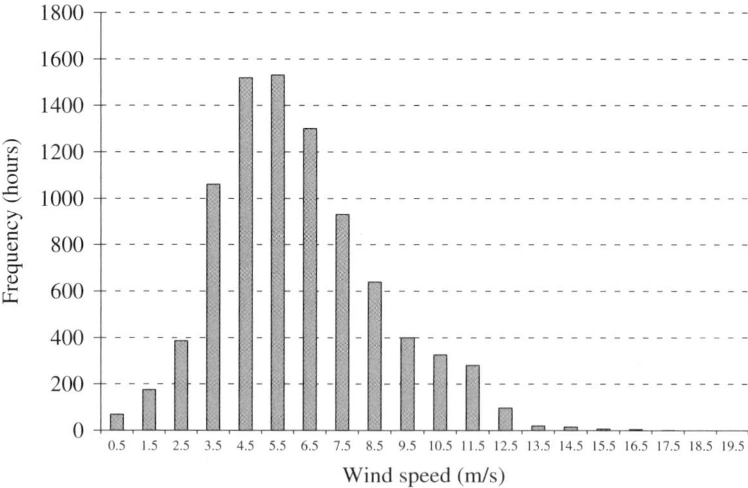

Fig. 4.4. Wind speed frequency graph.

The probability that the wind speed is between v_1 and v_2 is given by

$$P(v_1 < v < v_2) = \int_{v_1}^{v_2} f(v) \cdot dv$$

$$= \int_{v_1}^{v_2} \left(\frac{k}{c}\right) \left(\frac{v}{c}\right)^{k-1} e^{-\left(\frac{v}{c}\right)^k} dv$$

$$\text{let } x = -\left(\frac{v}{c}\right)^k$$

$$\frac{dx}{dv} = -k \left(\frac{v}{c}\right)^{k-1} \cdot \frac{1}{c}$$

$$-dx = \left(\frac{k}{c}\right) \left(\frac{v}{c}\right)^{k-1} \cdot dv.$$

Limits of integration are: lower limit $x_1 = -\left(\frac{v_1}{c}\right)^k$ and upper limit $x_2 = -\left(\frac{v_2}{c}\right)^k$
Therefore,

$$P(x_1 < x < x_2) = \int_{x_1}^{x_2} -e^x dx$$

$$= -\left[e^x\right]_{x_1}^{x_2}$$

$$= -(e^{x_2} - e^{x_1})$$

$$= e^{x_1} - e^{x_2}$$

$$\text{Hence, } P(v_1 < v < v_2) = e^{-\left(\frac{v_1}{c}\right)^k} - e^{-\left(\frac{v_2}{c}\right)^k}. \tag{4.4}$$

If the k value for a site is 2.5 and the c value is 6 m/s, then the probability density for different wind speeds and the frequency for the wind speeds are given in Table 4.2.

It is observed that there is not much difference between Figs. 4.4 and 4.5. If one wants to estimate the wind speed distribution for a site then the Weibull parameters k and c value are required to match the site.

4.3.7 *Wind rose*

This shows the frequency with which the wind direction falls within each direction sector, Fig. 4.6. The wind direction and changes in it are determined by geography, global and local climatic conditions and by the rotation of earth. Locally, the wind direction will vary with the lateral turbulence intensity and for coast near locations; in particular, the wind direction can vary between day and night.[16] Though the yaw system of the wind turbine will hold the rotor in the direction of the mean wind direction, short-term fluctuations in the wind direction give rise to fatigue loading. At high wind speeds sudden changes in the wind direction during production can give rise to extreme loads.

Table 4.2.   Probability for different wind speeds.

v_1	v_2	$P(v_1 < v < v_2)$	Freq
0	1	0.0113	99
1	2	0.0509	446
2	3	0.0999	875
3	4	0.1423	1247
4	5	0.1652	1447
5	6	0.1626	1425
6	7	0.1380	1209
7	8	0.1015	889
8	9	0.0648	568
9	10	0.0359	314
10	11	0.0171	150
11	12	0.0071	62
12	13	0.0025	22
13	14	0.0008	7
14	15	0.0002	2
15	16	0.0000	0
16	17	0.0000	0
17	18	0.0000	0
18	19	0.0000	0
19	20	0.0000	0

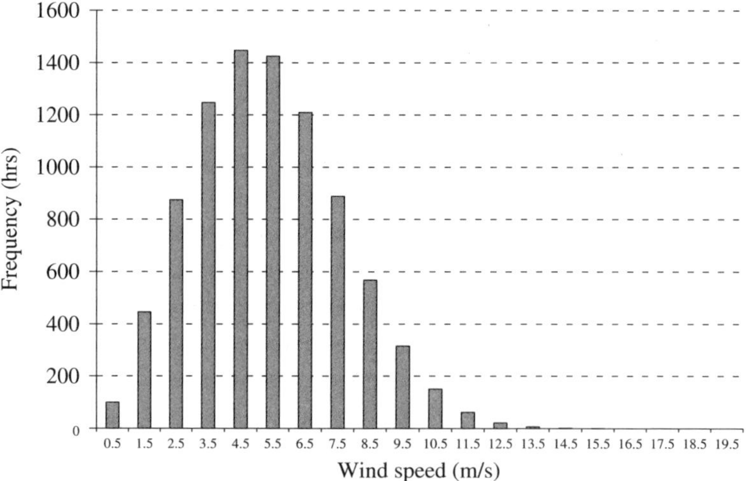

Fig. 4.5.   Wind speed frequency using Weibull distribution.

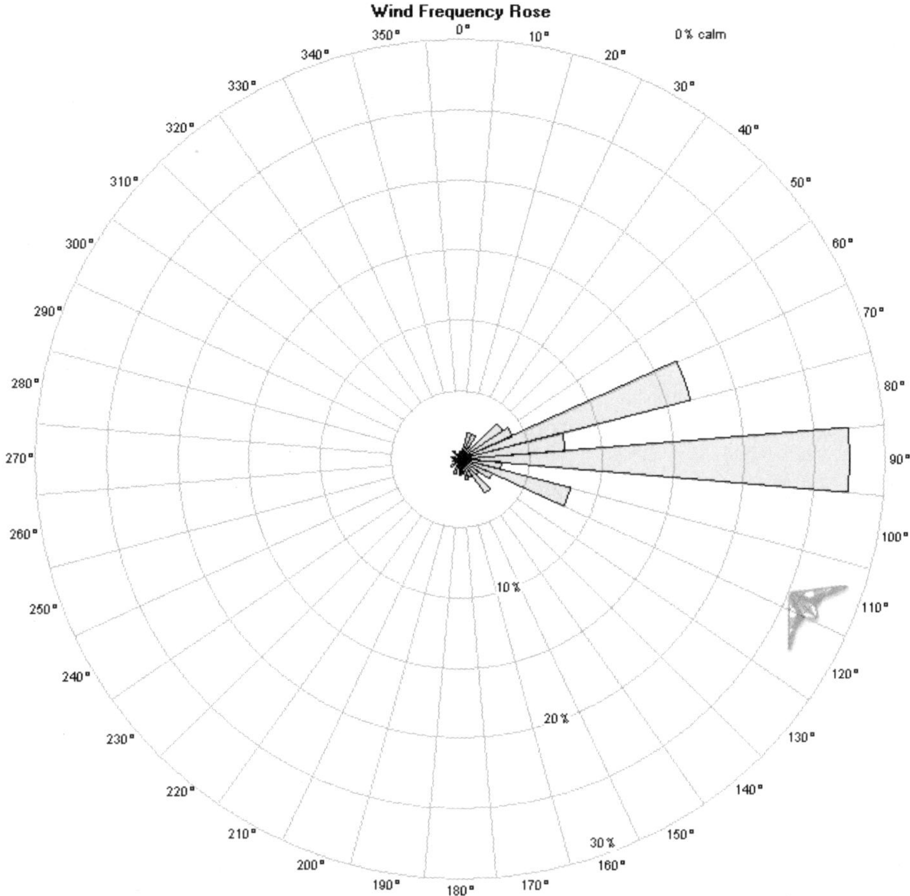

Fig. 4.6.   Wind rose for a site using Windographer.

The other factors such as the slope of the terrain, the type of soil and other topographical features that are obtained during preliminary wind survey can be used to optimize the wind turbine location. Softwares that are mentioned in Secs. 4.2.5 and 4.2.6 can be used to locate the wind turbines at a particular location.

4.4   To Find the Optimum Wind Turbine which Yields High Energy at High Capacity Factor

The optimum turbine that can be installed at the site can be found using the normalized power curve method. Jangamshetti *et al.*[48] presented a method of matching wind turbine generators to a site using normalized power curves. The site matching was based on identifying optimum speed parameters v_C, v_R, and v_F from the turbine performance index (TPI) curve, which was obtained from the normalized curves, so as to yield higher energy at higher CF. The wind speeds are obtained using cubic mean cube root and statistically modeled Weibull PDF. Usefulness of these normalized curves for identifying optimum wind turbine generator parameters for a site is presented by means of two illustrative case studies. It was shown that there exists a unique TPI curve for every site from which speed parameters of a turbine that will optimally match a site. The following are some of the formulae to find TPI for a site.

The average electrical power output (W) of wind turbine system is

$$P_{e,\text{ave}} = P_{eR}(\text{CF}), \tag{4.5}$$

where P_{eR} is the rated power output at rated wind speed v_R and it is given by

$$P_{eR} = C_{pR} \cdot \eta_{mR} \cdot \eta_{gR} \cdot \frac{1}{2}\rho A v_R^3. \tag{4.6}$$

In Eq. (4.6) C_{pR} = coefficient of performance at v_R; η_{mR} = mechanical interface efficiency at rated power; η_{gR} = generator efficiency at rated power; ρ = air density (kg/m^3) and A = turbine swept area (m^2).

CF in Eq. (4.5) is the capacity factor given by

$$\text{CF} = \frac{1}{v_R^3} \int_{v_c}^{v_R} v^3 f(v)dv + \int_{v_R}^{v_F} f(v)dv, \tag{4.7}$$

where $f(v)$ is the Weibull PDF Eq. (4.3).

By using integration by substitution, the definitions of Gamma function (Γ) and incomplete gamma function (γ), the capacity factor in Eq. (4.7) can be expressed as

$$\text{C.F} = p^3 \varepsilon^{-(v_c/c)^k} + \frac{3\Gamma\left(\frac{3}{k}\right)}{k}\left(\frac{v_R}{c}\right)^3$$
$$\cdot \left[\gamma\left(\left(\frac{v_R}{c}\right)^k, \frac{3}{k}\right) - \gamma\left(\left(\frac{v_c}{c}\right)^k, \frac{3}{k}\right)\right] - \varepsilon^{-(v_F/c)^k}, \tag{4.8}$$

where $\left(\frac{v_R}{c}\right)$ normalized rated speed $v_c = p \cdot v_R$ and $v_F = q \cdot v_R$, where $p < 1.0$ and $q > 1.0$[48]; and the values of k and c values are approximated by the formula[13]

$$k = \left(\frac{\sigma}{\bar{x}}\right)^{-1.086} \quad \text{and} \quad \frac{c}{\bar{x}} = \left(0.568 + \frac{0.433}{k}\right)^{\frac{-1}{k}}. \tag{4.9}$$

The average electrical power is then given by

$$P_{e,\text{ave}} = \eta_{oR}\frac{1}{2}\rho A v_r^3(\text{CF}), \tag{4.10}$$

where overall efficiency, $\eta_{oR} = C_{pR}\eta_{mR}\eta_{gR}$.

Hence, the normalized power is given by Eq. (4.11)

$$P_N = \frac{P_{e,\text{ave}}}{\eta_{oR}\frac{1}{2}\rho A c^3} = (\text{CF})\left(\frac{v_R}{c}\right)^3.$$ (4.11)

To yield a total energy production closer to the maximum, at a much better capacity factor for a given wind regime, $P_N = r \cdot P_{N,\max}$ where $0.5 \leq r \leq 1.0$. Thus, turbine performance index (TPI) is defined as

$$\text{TPI} = \frac{P_N \times \text{C.F}}{P_{N,\max} \times \text{C.F}_{\max}}.$$ (4.12)

Example 4.1. A site has annual average wind speed of 7 m/s with a standard deviation of 2.5 m/s. Find the estimated k and c values for the site.

Solution:

$$k = \left(\frac{\sigma}{\bar{x}}\right)^{-1.086}$$

$$= \left(\frac{2.5}{7}\right)^{-1.086}$$

$$= \underline{2.70}$$

$$\frac{c}{\bar{x}} = \left(0.568 + \frac{0.433}{k}\right)^{\frac{-1}{k}}$$

$$\frac{c}{7} = \left(0.568 + \frac{0.433}{2.70}\right)^{\frac{-1}{2.70}}$$

$$c = \underline{\underline{7.87\,m/s.}}$$

Example 4.2. Consider that the normalized rated speed, $\left(\frac{v_R}{c}\right)$, is varied in intervals of 0.1 til 3 and Table 4.3 is obtained for the capacity factor by using $k = 2.5$ and $c = 6$ m/s.

(i) Complete the values for the normalized power and the turbine performance index, TPI for the respective normalized rated speed and comment on

    (a) the values of capacity factor and TPI when normalized power is maximum,
    (b) the values of capacity factor and normalized power when TPI is maximum.

(ii) Using the completed Table 4.3, draw the TPI, normalized power and capacity factor on one graph.

(iii) Via the completed Table 4.3 and graph, one can find the turbine performance index of the site and also find the wind turbine speed specification which would best suit the site or use the specifications of different turbines and find the TPI.

Table 4.4 gives some wind turbine specifications.

   For each wind turbine;

(a) find the normalized rated speed,

(b) find the normalized power at these rated speed and,

(c) find the TPI and capacity factor.

Table 4.3.   Capacity factor.

Normalized rated speed $\left(\frac{v_R}{c}\right)$	C.F	P_N	TPI
0	0.0000		
0.1	0.0139		
0.2	0.0755		
0.3	0.1911		
0.4	0.3422		
0.5	0.4941		
0.6	0.6117		
0.7	0.6746		
0.8	0.6829		
0.9	0.6506		
1	0.5951		
1.1	0.5303		
1.2	0.4647		
1.3	0.4025		
1.4	0.3459		
1.5	0.2956		
1.6	0.2520		
1.7	0.2146		
1.8	0.1829		
1.9	0.1564		
2	0.1341		
2.1	0.1155		
2.2	0.0999		
2.3	0.0868		
2.4	0.0757		
2.5	0.0662		
2.6	0.0581		
2.7	0.0511		
2.8	0.0450		
2.9	0.0398		
3	0.0352		

Table 4.3. (Continued)

Normalized rated speed $\left(\frac{v_R}{c}\right)$	C.F	P_N	TPI
0	0.0000	0.0000	0.0000
0.1	0.0139	0.0000	0.0000
0.2	0.0755	0.0006	0.0001
0.3	0.1911	0.0052	0.0013
0.4	0.3422	0.0219	0.0102
0.5	0.4941	0.0618	0.0415
0.6	0.6117	0.1321	0.1099
0.7	0.6746	0.2314	0.2123
0.8	0.6829	0.3497	0.3248
0.9	0.6506	0.4743	0.4198
1	0.5951	0.5951	0.4818
1.1	0.5303	0.7059	**0.5093**
1.2	0.4647	0.8030	0.5076
1.3	0.4025	0.8843	0.4842
1.4	0.3459	0.9491	0.4465
1.5	0.2956	0.9977	0.4012
1.6	0.2520	1.0320	0.3537
1.7	0.2146	1.0542	0.3077
1.8	0.1829	1.0669	0.2655
1.9	0.1564	**1.0725**	0.2281
2	0.1341	1.0730	0.1958
2.1	0.1155	1.0699	0.1681
2.2	0.0999	1.0640	0.1446
2.3	0.0868	1.0560	0.1247
2.4	0.0757	1.0461	0.1077
2.5	0.0662	1.0345	0.0932
2.6	0.0581	1.0211	0.0807
2.7	0.0511	1.0059	0.0699
2.8	0.0450	0.9889	0.0606
2.9	0.0398	0.9700	0.0525
3	0.0352	0.9493	0.0454

Table 4.4. Wind turbine specification.

Model	Rating (kW)	v_c (m/s)	v_r (m/s)	v_f (m/s)
1	25	2.5	10	20
2	30	3	12	25

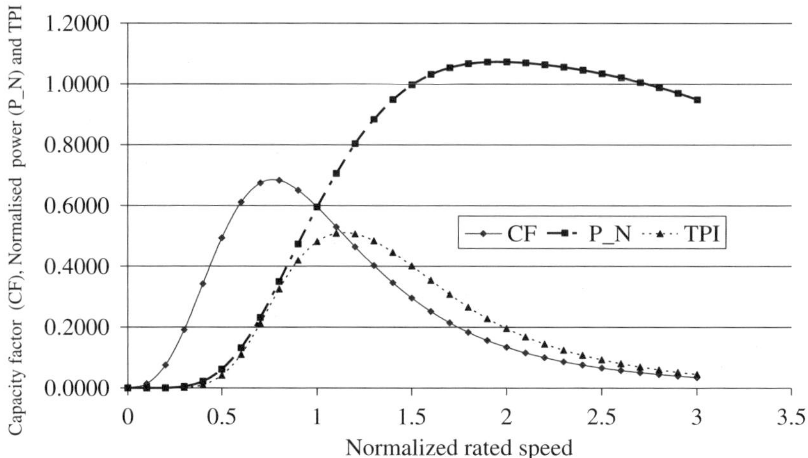

Fig. 4.7.   CF, P_N and TPI.

Solution

(i) Using Eqs. (4.11) and (4.12), Table 4.3 is completed as

   (a) when normalized power is maximum, the capacity factor is very low 0.16 and the TPI is also low of 0.23,

   (b) when TPI is maximum, the capacity factor is high but the normalized power is low (Fig. 4.7).

This implies that when some percentage of maximum normalized power is taken so that the TPI and capacity factor both are high.

- For Model 1, the normalized rated speed, $\left(\frac{v_R}{c}\right)$, is $\left(\frac{10}{6}\right) = 1.67$. At this $\left(\frac{v_R}{c}\right)$, the capacity factor and TPI are found from the completed Table 4.3, and the values respectively are; 0.2146 and 0.3537.
- For Model 2, the normalized rated speed, $\left(\frac{v_R}{c}\right)$, is $\left(\frac{12}{6}\right) = 2$. At this $\left(\frac{v_R}{c}\right)$, the capacity factor and TPI are found from the completed Table 4.3, and the values respectively are; 0.1341 and 0.1958.

Hence, just by comparing these two values of capacity factor and the TPI, a wind planner would choose Model 1 wind turbine, since it has high capacity factor as well as high TPI.

Note: *Other factors such as load demand at the site, the cost of wind turbine would also need to be considered.*

Once the wind turbine is selected for the site, then the annual energy output can be determined. This will help to gain some insight as to how much annual energy is produced from a wind turbine or a wind farm. This information can aid in finding the cost of energy and also carry out some economic analysis for the wind energy project. This is because the economic viability of a wind project is also another factor that has to be considered when installing a wind energy plant. The uncertainty involved in estimating the annual energy output should also be included in the wind resource assessment report.

4.5　Uncertainties Involved in Predicting Wind Speeds using the Different Approaches of WRA

In WRA, wind speed at a particular site and height play a major role in establishing the estimate annual energy yield from that site. Due to the relationship between wind speed (v), power output from a turbine (P_e) and energy output from a turbine (E), a 1% error in wind speed leads to a 3% error in energy output.[6,18] Table 4.5 gives the uncertainty in wind speed when the annual mean wind speed is estimated using

Table 4.5.　Uncertainty in wind speed using different approaches.

Different methods	Uncertainty in wind speed (%)
1. Methods in predicting annual mean wind speed	
o Observational wind atlas (Mesoscale modeling)[49]	10–30
o Numerical wind atlas (Microscale modeling)[49]	1–15
WAsP[35]	2.0–5.9
ANN[35]	1.7–6.8
MCP[50]	5–10
2. Monitoring periods (months) for on-site wind data collection[51]	
o 1	6.4–11.8
o 3	4.9–10.3
o 6	3.5–7.8
o 12	1.2–2.8
o 24	0.6–1.5

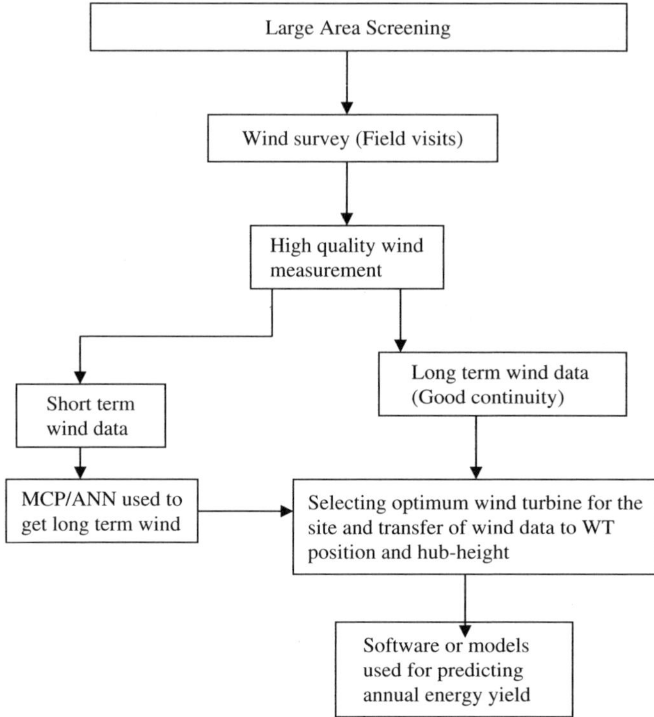

Fig. 4.8.   Wind resource assessment techniques.

the different approaches mentioned and also the uncertainty in mean wind speed for on-site data collection for different durations.

Considering Table 4.5, minimum uncertainty in wind speed is using WAsP or ANN to predict wind speeds. However, each of the different methods of predicting wind speeds at a site has its pros and cons. It depends upon the wind planner to decide which method would be best to predict the wind speeds at a particular site. For instance, WAsP may have a large uncertainty in wind speed prediction when the topography of a site is complex.[52] MCP or ANN method would be better for complex topographies since these methods do not require the topographical details of the site.[53] It is also worthwhile to note that the uncertainty for MCP method would decrease if the duration of wind speed measurement at reference site increases.[54] Mesoscale modeling offers a number of advantages for WRA, such as the ability to simulate, with reasonable accuracy, complex wind flows in areas where surface measurements are scant or non-existent[55] whereas microscale modeling is best suited to estimating the wind resource in areas of simple to moderate terrain slopes with distances up to tens of kilometers from the reference mast.[56,57]

It is also seen from Table 4.5 that the uncertainty in wind speed decreases as the duration of wind data collection increases. The factors that are considered for the estimation of wind speed uncertainty are: anemometer measurement uncertainty; the vertical spacing on the tower; the monitoring period; the temporal period used in the measure-correlate-predict analysis; and the r-squared of the monitoring/reference station relationship.[54] If wind speed is measured at the site then the temperature and atmospheric pressure must also be measured since these two factors have great influence over the air density.[58] Air density is used to estimate the wind power density at a particular site. Hence the uncertainty in wind power density at a site will depend on the uncertainty in wind speed measurement and also the uncertainty in temperature and atmospheric pressure measurement.

The whole procedure of carrying out a wind resource assessment using the techniques mentioned above may be presented in the flow diagram as shown in Fig. 4.8.

4.6 Concluding Remarks

A proper wind resource assessment would lead to a successful wind energy project. After a WRA, consultants can be hired to give professional advice as to where exactly wind turbine(s) is/are to be installed by looking at the terrain features and the wind resource as the site. This can be done by using software or models that are described in Sec. 4.2 of this chapter. Once a WRA is done, the optimum wind turbine for the site is found which is used to estimate the annual energy yield.

References

1. BWEA, "The economics of wind energy," http://www.bwea.com.html (2008).
2. T. Burton, D. Sharpe, N. Jenkins and E. Bossanyi, *Wind Energy Handbook* (John Wiley & Sons Ltd., UK, 2001).
3. E. Hau and H. Renouard, *Wind Turbines: Fundamentals, Technologies, Applications and Economics* (Springer-Verlag Berlin Heidelberg, Germany, 2006).
4. J.W. Twidel and A.D. Weir, *Renewable Energy Sources* (Taylor and Francis, New York, 2006).
5. R.C. Bansal, T.S. Bhatti and D.P. Kothari, "On some of the design aspects of wind energy conversion systems," *Energy Conversion and Management* **43** (2002) 2175–2187.
6. P. Gipe, *Wind Power: Renewable Energy for Home, Farm and Business* (Chelsea Green Publishing Company, Vermont, 2004).
7. R.W. Thresher and D.M. Dodge, "Trends in the evolution of wind turbine generator configurations and systems," *Int. J. Wind Energy* **1** (1998) 70–85.
8. D.C. Quarton, "The evolution of wind turbine design analysis — A twenty year progress review," *Int. J. Wind Energy* **1** (1998) 1–24.
9. A.W. Culp, *Principles of Energy Conversion* (McGraw Hill Int., New York, 1991).
10. M.R. Patel, *Wind and Solar Power Systems: Design, Analysis and Operation* (CRC Press, Boca Raton, FL, 2005).
11. G.M. Masters, *Renewable and Efficient Electric Power Systems* (John Wiley & Sons, New Jersey, 2004).
12. Ackermann, T. (ed.), *Wind Power Systems* (John Wiley & Sons, New Jersey, 2005).

13. J.F. Manwell, J. McGowan and G. Rogers, *Wind Energy Explained: Theory, Design and Application* (John Wiley & Sons Ltd., UK, 2002).

14. R.C. Bansal, A.F. Zobaa and R.K. Saket, "Some issues related to power generation using wind energy conversion systems: An overview," *Int. J. Emerging Electric Power Systems*, vol. 3, article 1070 (2005).

15. R.C. Bansal and T.S. Bhatti, *Small Signal Analysis of Isolated Hybrid Power Systems* (Narosa Publishing House Pvt. Ltd., New Delhi, 2008).

16. Det Norske Veritas and Risø National Laboratory, "Guidelines for design of wind turbines," Jydsk Centraltrykkeri, Denmark (2002).

17. B. Rajsekhar, "Wind resource assessment using Mapinfo GIS software tool, GIS development," http://www.gisdevelopment.net/application/utility/power/utilityp0004.htm (2007).

18. J. Park, "Common sense: Wind energy," California Office of Appropriate Technology, USA, May 1983.

19. AWEA, "Wind Energy," http://www.awea.org/faq/basicwr.html (2007).

20. J. Allen and R.A. Bird, "The prospects for the generation of electricity from wind energy in the United Kingdom," U.S. Department of Energy, July 1997.

21. S. Rehman, I.M. El-Amin, F. Ahmad, S.M. Shaahid, A.M. Al-Shehri and J.M. Bakhashwain, "Wind power resource assessment for Rafha, Saudi Arabia," *Renewable and Sustainable Energy Reviews* **11** (2007) 937–950.

22. R. Hunter and G. Elliot, *Wind — Diesel Systems: A Guide to the Technology and its Implementation* (Cambridge University Press, New York, 2004).

23. "Small wind electric system resource evaluation," in *The Encyclopedia of Alternative Energy and Sustainable Living*, http://www.daviddarling.info/encyclopedia/S/AE_small_wind_electric_system_resource_evaluation.html (2009).

24. R.S. Hunter, "Wind speed measurement and use of cup anemometry," www.ieawind.org/Task_11/RecommendedPract/11%20Anemometry_secondPrint.pdf (2008).

25. Northwest Community Energy, "Resource validation," http://www.nwcommunityenergy.org/wind/resource-assessment/resource-validation (2008).

26. "Specifications for wind resource assessment system components," www.energy.gov.lk/pdf/specifications_for_wind_loggers10.pdf (2008).

27. Wind Resource Assessment Unit, "Wind resource assessment techniques," *Pavan: A News Bulletin from the Centre for Wind Energy Technology*, Issue 6, July–September 2005, www.cwet.tn.nic.in/ images/Newsletter-PDF/Issue6.pdf (2008).

28. R.D. Prasad, R.C. Bansal and M. Sauturaga, "Some of the design and methodology considerations in wind resource assessment," *IET-Renewable Power Generation* **3** (2009) 53–64.

29. D. Maunsell, T.J. Lyons and J. Whale, "Wind resource assessment for a site in Western Australia," www.rise.org.au/pubs/Maunsell_pape_Wind.pdf (2007).

30. GH WindFarmer, "The wind farm design software," http://www.garradhassan.com/products/ghwindfarmer/ (2008).

31. D.A. Bechrakis and P.D. Sparis, "Correlation of wind speed between neighboring measuring stations," *IEEE Trans. Energy Conversion* **19** (2004) 400–406.

32. M.C. Alexiadis, P.S. Dokopoulos and H.S. Sahamanohlou, "Wind speed and power forecasting based on spatial correlation models," *IEEE Trans. Energy Conversion* **14** (1999) 836–842.

33. L.A.F.M. Ferreira, "Evaluation of short-term wind predictability," *IEEE Trans. Energy Conversion* **7** (1992) 409–417.

34. D.A. Bechrakis, J.P. Deane and E.J. McKeogh, "Wind resource assessment of an area using short term data correlated to a long term data set," *Solar Energy* **76** (2004) 725–732.

35. T.G. Barbounis, J.B. Theocharis, M.C. Alexiadis and P.S. Dokopoulos, "Long term wind speed and power forecasting using local recurrent neutral network models," *IEEE Trans. Energy Conversion* **21** (2006) 273–284.

36. C. Nichita, D. Luca, B. Dakyo and E. Ceanga, "Large bands simulation of the wind speed for real time wind turbine simulators," *IEEE Trans. Energy Conversion* **17** (2002) 523–529.

37. D.A. Bechrakis and P.D. Sparis, "Simulation of wind speed at different heights using artificial neutral networks," *Wind Engineering* **24** (2000) 127–136.
38. J.R. Salmon and J.L. Walmsley, "A two-site correlation model for wind speed, direction and energy estimates," *J. Wind Engineering Industrial Aerodynamics* **79** (1999) 233–268.
39. A. Joensen, L. Landberg and H. Madsen, "A new measure-correlate predict approach for resource assessment," *Proc. Eur. Wind Energy Conf.* (1999), pp. 1157–1160.
40. A.D. Sahin and Z. Sen, "Wind energy directional spatial correlation functions and application for prediction," *Wind Engineering* **24** (2000) 223–231.
41. M. Mohandes, S. Rehman and T.O. Halawani, "A neutral network approach for wind speed prediction," *Renewable Energy* **13** (1998) 345–354.
42. B. Ernst, "Wind power forecast for the German and Danish networks," *Wind Power in Power Systems* (John Wiley and Sons Ltd, New York, 2005).
43. K. Syngellakis and H. Traylor, "Urban wind resource assessment in the UK," www.urban-wind.org/pdf/Reports_UrbanWindResourceAssessment_UK.pdf (2008).
44. WindSim: When the terrain gets rough, www.windsim.com/documentation/0803_prospect.pdf (2008).
45. Wind Farm analysis, design and optimization, http://www.resoft.co.uk/html/details.htm (2008).
46. WAsP, http://www.wasp.dk/ (2008).
47. Zephyr North, "Products and software," http://www.zephyrnorth.com/products.html (2008).
48. S.H. Jangamshetti and V.G. Rau, "Normalized power curves as a tool for identification of optimum wind turbine generator parameters," *IEEE Trans. Energy Conversion* **16** (2001) 283–288.
49. J.C. Hansen, N.G. Mortensen, J. Badger, N.E. Clausen and P. Hummelshoj, "Opportunities for WRS using numerical and observational wind atlases: Modeling, verification and application," Riso National Laboratory, Techinical University of Denmark (2007), www.risoe.dk/vea/projects/nimo/download/WP%20Shanghai%202007%-%20Risoe%20-%20jchal5Oct 2007.ppt (2008).
50. E. Walls, A.L. Rogers and J.F. Manwell, "Eastham MA-SODAR based wind resource assessment," Renewable Energy Research Laboratory, University of Massachusetts, July 2007, www.ceere.org/rerl/publicatons/resource_data/Eastham/Reports/Eastham_SODAR_Site_Assessment_Report_2007.pdf (2008).
51. M. Taylor, M. Filippelli and F. Kreikebaum, "Assessments reduce prediction uncertainty," North American Wind Power (2007), www.awstruewind.com/files/NAW_AWSTruewind0602.pdf.
52. D. Maunsell, T.J. Lyons and J. Whale, "Wind resource assessment for a site in Western Australia," www.rise.org.au/pubs/Maunsell_pape_Wind.pdf.
53. K. Klink, "Atmospheric circulation effects on wind speed variability at turbine height," *J. Applied Meteorology and Climatology* **46** (2007) 445–456.
54. M. Taylor, P. Mackiewicz, M.C. Brower and M. Markus, "An analysis of wind resource uncertainty in energy production estimates," www.awstruewind.com/files/EWEC_2004_WRA_Uncertainty_EPR.pdf.
55. M. Brower, J.W. Zack, B. Bailey, M.N. Schwartz and D.L. Elliot, "Mesoscale modeling as a tool for wind resource assessment and mapping," http://ams.confex.com/ams/pdfpapers/72138.pdf.
56. A.J. Bowen and N.G. Mortensen, "Exploring the limits of WAsP: The Wind Atlas Analysis and Application Program," *Proc. 1996 European Union Wind Energy Conf. and Exhibition*, Göteborg, Sweden, 20–24 May 1996, pp. 584–587.
57. J.L. Walmsley, I. Troen, D.P. Lalas and P.J. Mason, "Surface-layer flow in complex terrain: Comparison of models and full-scale observations," *Boundary Layer Meteorology* **52** (1990) 259–281.
58. S.Y.C. Catunda, J.E.O. Pessanha, J.V. FonsecoNeto, N.J. Camelo and P.R.M. Silva, "Uncertainty analysis for defining a wind power density measurement system structure," *Proc. 21st IEEE on Instrumentation and Measurement Technology Conference*, Italy, 18–20 May 2004, pp. 1043–1047.

Chapter 5

Economic Analysis of Wind Systems

Ravita D. Prasad

College of Engineering, Science and Technology,
Fiji National University,
P.O. Box 3722, Samabula, Fiji Islands
ravita.prasad@fnu.ac.fj

Ramesh C. Bansal

School of Information Technology and Electrical Engineering
The University of Queensland, St. Lucia,
Brisbane, QLD 4072, Australia
rcbansal@ieee.org

Wind energy is one of the fastest growing energy sources in the world today. The reason for this can be due to the fact that there have been vast improvements in wind energy technology which has led to lower cost. This chapter presents an overview on the economic analysis of a wind energy project. The chapter is organized as follows. Section 5.1 gives an overview on wind energy and reasons for carrying out the economic analysis. Section 5.2 describes the wind system economic components, Sec. 5.3 presents the different types of economic analysis methods and Sec. 5.4 presents two case studies on economic analysis. Finally, some conclusions are drawn.

5.1 Introduction

Generating electricity from the wind makes economic as well as environmental sense; the wind is a free, clean and renewable fuel which will never run out. Even though wind is free its cost of electricity however, is not free. There are initial capital cost of purchasing wind turbines, towers, transportation of materials, labor charge, expertise charge, operation and maintenance cost, etc. Wind turbines are becoming cheaper and more powerful, with larger blade lengths which can utilize more wind and therefore produce more electricity, bringing down the cost of renewable power generation.[1–4] There are two main factors which affect the cost of electricity generated from the wind and therefore its final price which depends upon: (i) technical factors, such as wind speed and the nature of the turbines and (ii) the financial perspective of those that commission the projects, e.g., what rate of return is required on the capital, and the length of time over which the capital is repaid.[1] To be economically viable the cost of making the electricity has to be less than its selling price. It is extensively known that the cost of energy will be low if the site has a high wind speed (Fig. 5.1), the wind turbine optimally matches the wind characteristics for the site and cost of wind turbine and installation is low.

The unit cost of electrical energy can be determined easily by knowledge of capital investment and operating cost. Before investment decisions are made it is vital to determine the electrical energy output from the site.

This chapter presents economic analysis for wind energy systems. The chapter is organized as follows. Section 5.2 presents an overview of the components of wind energy economics and Sec. 5.3 describes the methods of economic analysis such as levelized cost of energy (LCOE), cost of energy (COE), benefit to cost ratio, simple payback period (SPB), internal rate of return (IRR), discount rate and net present value (NPV). Section 5.4 offers two case studies of economic analysis; (1) wind turbine alone and (2) wind diesel hybrid system at Vadravadra, Gau Island in Fiji. Finally some conclusions are drawn.

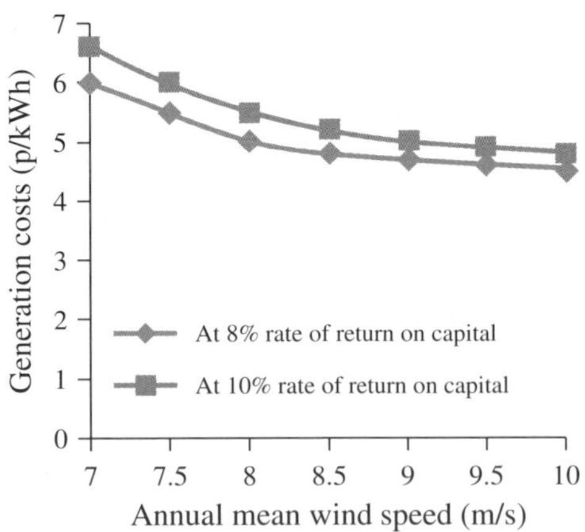

Fig. 5.1.　Generation cost against annual wind speed.[1]

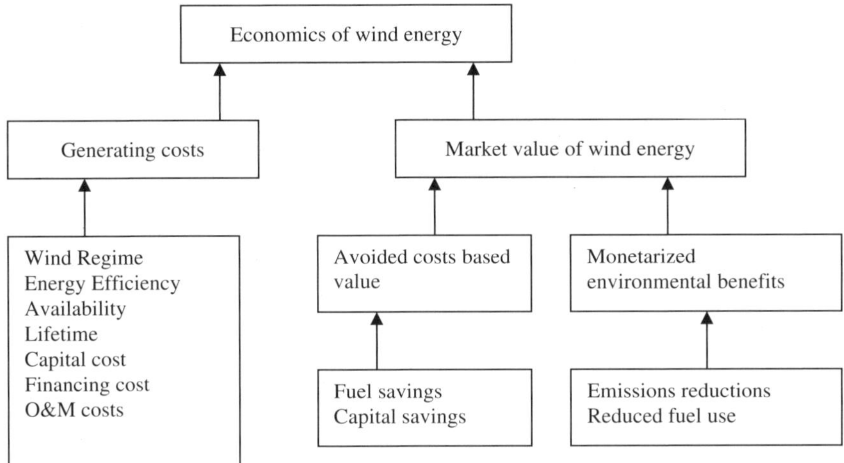

Fig. 5.2.   Components of wind system economics.[5]

5.2   Wind System Economic Components

If someone has designed a wind energy conversion system (WECS) that can reliably produce energy, one should be able to predict its annual energy production. With this result and the determination of the manufacturing, installation, operation and maintenance, and financing costs, the cost-effectiveness can be addressed. Figure 5.2 shows the economic aspects of wind energy systems, which are discussed briefly.

5.2.1   *Availability*

The availability is the percentage of the time in a year that the wind turbine is able to generate electricity. The times when a wind turbine is not available includes downtime for periodic maintenance or unscheduled repairs.

5.2.2   *Lifetime of the system*

It is common practice to equate the design lifetime with the economic lifetime of a wind energy system. In Europe, an economic lifetime of 20 years is often used for the economic assessment of wind energy systems.[5]

5.2.3   *Energy efficiency*

The efficiency of the wind energy conversion system also affects the economics of the wind system. The theoretical maximum efficiency of wind turbine is 59.3% (known as Betz coefficient). Low overall efficiency means low return on investments.

5.2.4 *Wind regime*

Wind regime is the distribution of wind speed throughout a year where the wind turbine is supposed to be commissioned. It can be presented in a wind speed duration curve or wind speed frequency curve. A wind speed duration curve shows the number of hours that wind speed exceeds a particular value and wind speed frequency curve shows the number of hours in a year that a particular wind speed will occur. For instance, the wind regime for Vadravadra site in Gau Island in Fiji is shown in Fig. 5.3 (wind speed duration curve) and Fig. 5.4 (wind speed–frequency curve).

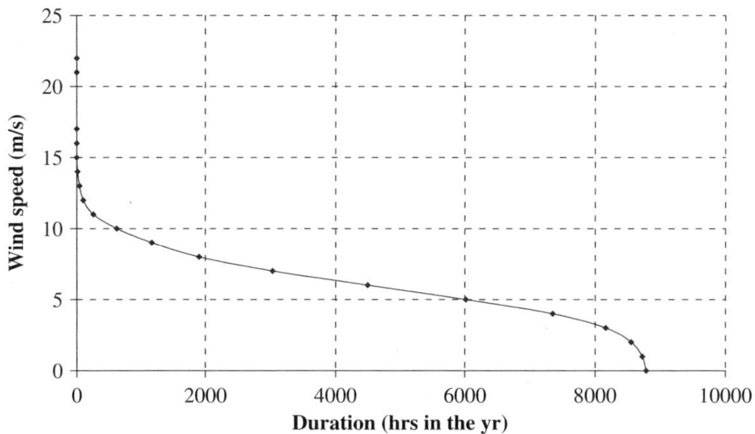

Fig. 5.3.   Wind speed duration curve for Vadravadra.

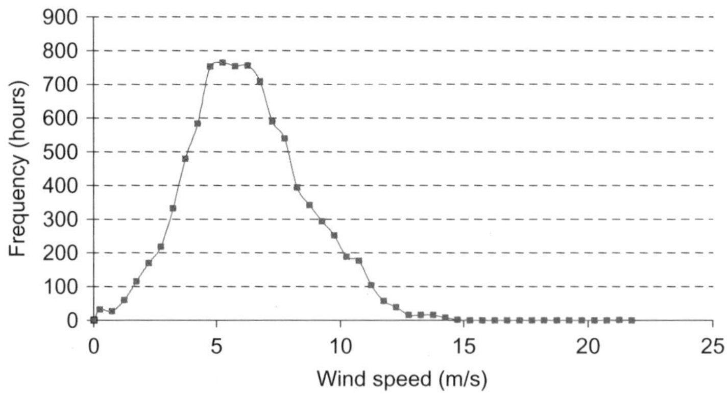

Fig. 5.4.   Wind speed frequency curve for Vadravadra.

5.2.5 *Investment costs*

Any cost from the start of the idea until the date of operation which includes land preparation, site, equipment, transport, design, consultancy, project management, etc., are "written off" over the life time of WECS.

5.2.5.1 *Capital costs*

The determination of the capital or total investment cost generally involves the cost of wind turbines and its auxiliaries, i.e., tower, wiring, utility interconnection or battery storage equipment, power conditioning unit, etc., and delivery and installation charges, professional fees and sales tax.[6–8] Figure 5.5 shows the typical installation costs of wind turbines in remote areas which could be typically varying between two curves depending upon siting conditions. One way to estimate the capital costs of a wind turbine is to use cost data for smaller existing machines normalized to a machine size parameter. Here, the usual parameters that are being used are unit cost per kW of rated power or unit cost per area of the rotor diameter. Remote systems with operating battery storage typically cost more, averaging between US$4000–5000/kW. Individual batteries cost from US$150–300 for a heavy-duty, 12 V, 220 Ah deep-cycle type. Larger capacity batteries, those with higher amp-hour ratings, cost more. A 110 V, 220 Ah battery storage system, which includes a charge controller, costs at least US$2000.[6] The cost of wind turbines has increased from US$1200/kW to US$1600/kW.[9,10] This is why the cost of wind turbine for this project was taken as F$4000/kW. Cost of transmission line is US$20000–40000/mile, but costs can be higher in some cases.[6] Similar to a grid-connected system, a remote system has

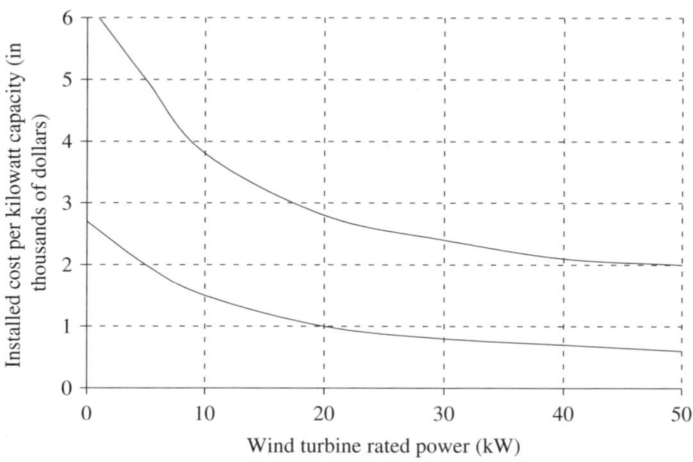

Fig. 5.5.   Installation cost of wind turbines.[6]

initial and lifetime operating costs. Initial costs include equipment components such as batteries, control systems and an inverter to supply AC loads. A more fundamental way to determine the capital cost of a wind turbine is to divide the machine into its various components and to determine the cost of each component.

5.2.5.2 *Financing costs*

Wind energy projects have intensive amount of money to be invested in the beginning so that the purchase and installation costs are met. For this reason, the developer or purchaser will pay a limited down payment of 10–20% and borrow the rest. The source of capital may be a bank or investors where the lenders will expect a return. The return in the case of a bank is referred to as the interest. Over the lifetime of the project, the cumulative interests can add up to a significant amount of the total costs.

5.2.6   *Recurring costs*

Recurring cost includes operation and maintenance (O&M) cost (administration, labor, spare parts, consumables, lubrication), fuel cost, and capital cost (interest on outstanding capital and transaction costs).

5.2.6.1 *Operation and maintenance costs*

According to Danish Wind Industry Association,[8] O&M costs are very low when the turbines are brand new but increase as the turbine gets old. The O&M costs generally range from 1.5% to 3% of the original turbine cost. Annual operating costs also include battery replacement every 3 to 10 years, depending on the battery type and the number of discharges.[6]

5.2.6.2 *Avoided cost based value of wind energy*

The traditional way to assess the value of wind energy is to equate it to the direct savings that would result due to the use of the wind rather than the most likely alternative. These savings are often referred to as "avoided costs". The avoided costs include fuel and capacity costs.

5.2.7   *Environmental value of wind energy*

The primary environmental value of electricity generated from wind energy systems is that the wind offsets emissions that would have been caused by conventional fossil fueled power plants. These emissions include sulphur dioxide (SO_2), nitrogen oxides (NO_x), carbon dioxide (CO_2), particulates, slag and ash. The amount of emissions saved via the use of energy depends on the types of power plant that are replaced by the wind system, and the particular emissions control systems currently installed on the various fossil-fired plants.

5.2.8 *Market value of wind energy*

The market value of wind energy is the total amount of revenue one will receive by selling wind energy or will avoid paying through its generation and use. The value that can be "captured" depends strongly on three considerations; the market application, the project owner or developer and the types of revenues available. In remote areas the environmental, social and legal factors are least affected.[7,10–12]

5.3 Economic Analysis Methods

There are two types of economic analysis:

- Absolute analysis: Are the costs higher or lower than the benefits? Is the project viable?
- Relative analysis: For projects (such as the wind turbine alone or those with hybrid system) are the benefits higher and at what cost? How do the projects rank in terms of costs and benefits?

  (I) *Cost-benefit analysis*: A time period is chosen and the sum of all costs and benefits in that period is determined. The net benefit is determined by subtracting total benefits and total cost in that time period.

  $$\text{Net benefit} = \sum (\text{benefits}) - \sum (\text{costs}).$$

  (II) *Benefit to cost ratio (BCR)*: A time period is chosen then the sum of all costs and benefits in that period is determined. The ratio of benefit to cost gives the benefit to cost ratio.

  $$\text{BCR} = \frac{\sum (\text{benefits})}{\sum (\text{costs})}.$$

  (III) *Simple payback period (SPB)*: This is one of the most common ways of finding the economic value of a wind energy project. Payback considers the initial investment costs and the resulting annual cash flow. The payback time (period) is the length of time needed before an investment makes enough to recoup the initial investment.

  $$\text{SPB (in years)} = \frac{\sum (\text{investment costs})}{(\text{yearly benefits} - \text{yearly costs})}.$$

However, the payback does not account for savings after the initial investment is paid back from the profits (cash flow) generated by the investment (project). This method is a "first cut" analysis to evaluate the viability of investment. It does not include anything about the longevity of the system. For example, two wind turbines may both have the same 5-year payback periods, but even though one lasts for 20 years and the other one falls apart after 5 years, the payback period makes absolutely no distinction between the two.

For example the investment cost of a wind turbine project comes to $100000 but a net annual operational saving is $10000. When the net annual savings is divided into the initial investment, the sample payback period is calculated as follows:

$$\text{SPB} = \frac{\sum (\text{investment costs})}{(\text{yearly benefits} - \text{yearly costs})} = \frac{100000}{10000} = 10\,\text{years}.$$

(IV) *Initial rate of return*: This is the opposite of simple payback period. The value makes the investment look too good.

$$\text{Initial rate of return} = \frac{(\text{yearly benefits} - \text{yearly costs})}{\sum(\text{investment costs})} \times 100\%.$$

For example in the previous example, the initial rate of return can be calculated as

$$\frac{10000}{100000} \times 100\% = 10\%.$$

This initial rate of return acts as a minimum threshold indicator for the investment. If the internal rate of return is below this minimum threshold there is no need to proceed with the investment.

(V) *Levelized cost of energy (LCOE)*: All the costs are added during a selected time period which is divided by units of energy. A net present value (NPV) calculation is performed and solved in such a way that for the value of the LCOE chosen, the project's NPV becomes zero. This means that the LCOE is the minimum price at which energy must be sold for an energy project to break even.

$$\text{LCOE} = \frac{\sum \text{costs/no. of years}}{\text{annual yeild (kWh)}}.$$

For example if the total cost a wind energy project for a duration of 20 years is $500000 and the annual wind energy yield is 60000 kWh, then the LCOE is calculated as follows:

$$\text{LCOE} = \frac{\sum \text{costs/no. of years}}{\text{annual yeild (kWh)}} = \frac{500000/20}{60000}$$

$$= \$0.417/\text{kWh} \quad \text{or} \quad \approx 42\,\text{cents/kWh}$$

(VI) *Cash flow analysis*: One of the most flexible and powerful way to analyze an energy investment is the cash-flow analysis. This technique easily accounts for complicating factors such as fuel escalation, tax-deductible interest, depreciation, periodic maintenance costs, and disposal or salvage value of the equipment at the end of its lifetime. In a cash flow analysis, rather than using increasingly complex formulas to characterize these factors, the results are computed numerically using a spreadsheet. Each row of the resulting table corresponds to one year of operation, and each column accounts for a contributing factor. Simple formulas in each cell of the table enable detailed information to be computed for each year along with very useful summations. Cash flow is always positive.

$$\sum \text{cash flow}_n = \sum \text{benefits}_n - \sum \text{costs}_n$$

where n is the number of years of operation from the start system operation.

(VII) *Discounted cash flow (DCF)*: DCF analysis uses future free cash flow projections and discounts them to arrive at a present value, which is used to evaluate the potential for investment. If the value arrived through DCF analysis is higher than the current cost of the investment, the opportunity may be a good one. The purpose of DCF analysis is to estimate the money one would receive from an investment and to adjust for the time value of money.

$$\text{Discount cash flow}_n = \frac{\sum \text{benefits}_n - \sum \text{costs}_n}{(1+i)^n}$$

where $i = $ the discount rate which is the interest rate used in calculating the present value of future cash flows and $n = $ the years from the system starts operation. The present worth factor in the above formula is $\frac{1}{(1+i)^n}$.

The value that is chosen for i can often "weigh" the decision towards one option or another, so the basis for choosing the discount must clearly be carefully evaluated. The discount rate depends on the cost of capital, including the balance between debt-financing and equity-financing, and an assessment of the financial risk.

(VIII) *Net present value (NPV)*: NPV compares the value of a dollar today to the value of that same dollar in the future, taking inflation and returns into account. If the NPV of a prospective project is positive, it should be accepted.

However, if NPV is negative, the project should probably be rejected because cash flows will also be negative.[13] To calculate NPV; choose the time period for the project and sum all the discounted cash flows in that time period.

$$\sum \text{Discount cash flow}_n = \frac{\sum \text{benefits}_n - \sum \text{costs}_n}{(1+i)^n} = \text{NPV}$$

(IX) *Internal rate of return* (*IRR*): This is perhaps the most persuasive measure of the value of a wind energy project. The IRR allows the energy investment to be directly compared with the return that might be obtained for any other competing investment. IRR is the discount rate that makes the NPV of the energy investment equal to zero. When the IRR is less than discount rate, it is a good indicator for the project.

$$\text{IRR} \Rightarrow \text{NPV} = 0; \qquad \text{i.e.,} \quad \frac{\sum \text{benefits}_n - \sum \text{costs}_n}{(1+i)^n} = 0.$$

5.4   Case Study for the Economic Analysis of a Wind Turbine

For a site in Vadravadra village in Gau Island in Fiji the average annual wind speed recorded for the site was 6.24 m/s. For 68% of the time in a year the wind speed is more than 5 m/s. A 30 kW Fuhrlaender (FL-30) wind turbine was found to yield 60000 kWh of electrical energy annually when coefficient of performance (C_{op}) was taken as 25%. Vadravadra village has 174 individuals but for economic analysis 200 individuals are assumed. When 200 individuals are taken the annual electrical energy consumption from wind is about 300 kWh/capita.

Case 1: Wind turbine is the only source of electricity generation

For purchasing and installing the wind monitoring system in Vadravadra the total cost was F$21642.

Assumptions:

- It is assumed that no subsidies are given from the government or any other organizations and there are no tax and CO_2 saving incentives.
- The revenue that is generated from a FL-30 wind turbine installed at the site is the product of annual energy yield and the cost of energy (COE) in cents/kWh. The only source of revenue in the analysis is through the production of annual energy from the wind turbine.
- The lifetime of the wind turbine is taken as 20 years.

Table 5.1.   Costs breakdown at Vadravadra for wind turbine installation.

Parameters	Costs (F$)
Investments (one-time costs)	
Wind turbine	120000.00
Tower	60000.00
Foundation and site preparation	20000.00
Transport and freight (from overseas and to Gau Island)	100000.00
Labor	12800.00
Expert hiring	150000.00
Wind survey	21642.00
Total Investment	***484442.00***
Recurring costs (occurs every year)	
Land lease	1000.00
O&M	12111.05
Insurance	2422.21
Total costs	***15533.26/year***

Note: Wind turbine is purchased at F$4000/kW, labor cost consists of 20 people working for 320 hours at F$2.00/hour rate and O&M is 2.5% of total investment cost.

The cash flow analysis was carried out and the discount rate was taken as 10% (this choice of discount rate was taken so that it matched the interest rate for borrowing). The cost of FL-30 wind turbine, installation cost, labor, foundation preparation and other costs are shown in Table 5.1.

Table 5.1 gives the cost breakdown of the wind turbine that is going to be installed at the site. High prices are considered for the wind turbine, tower and other items because Vadravadra village is in a remote island and the cost for tower purchase, transportation and the labor cost would be high. An additional cost of hiring specialists to install the wind turbine at the site is taken as F$150000. This cost is high since the specialists would be from overseas to help in the installation of the wind turbine at the remote island site. The annual O&M cost is taken as 2.5% of the fixed cost. The annual revenue generated will be from the energy sold to the customers in F$/kWh.

In Fiji the cost of one unit of electricity is 21 Fcents/kWh. In the discounted cash flow analysis when COE is taken as 21 Fcents/kWh and the discount rate is taken as 10% then the NPV is negative value. This negative value of NPV means that the project is unprofitable at the current discount rate and the COE. Hence, the COE should be increased.

When NPV is zero, it means that no profit is made over the 20 years. From cash flow analysis the COE when the NPV is zero is F$1.20/kWh. Table 5.2 shows the discounted cash flow analysis.

Table 5.2.   Discounted cash flow analysis at 10%.

Year	Benefits ($)	Costs ($)	Benefit − Cost ($)	Present worth factor	Discounted cash flow (Present value = benefit − cost) ($)
0		484442	−484442	1.00	−484442
1	72435.64	15533.26	56902.38	0.91	51729.43
2	72435.64	15533.26	56902.38	0.83	47026.76
3	72435.64	15533.26	56902.38	0.75	42751.6
4	72435.64	15533.26	56902.38	0.68	38865.09
5	72435.64	15533.26	56902.38	0.62	35331.9
6	72435.64	15533.26	56902.38	0.56	32119.91
7	72435.64	15533.26	56902.38	0.51	29199.92
8	72435.64	15533.26	56902.38	0.47	26545.38
9	72435.64	15533.26	56902.38	0.42	24132.16
10	72435.64	15533.26	56902.38	0.39	21938.33
11	72435.64	15533.26	56902.38	0.35	19943.94
12	72435.64	15533.26	56902.38	0.32	18130.85
13	72435.64	15533.26	56902.38	0.29	16482.59
14	72435.64	15533.26	56902.38	0.26	14984.17
15	72435.64	15533.26	56902.38	0.24	13621.98
16	72435.64	15533.26	56902.38	0.22	12383.61
17	72435.64	15533.26	56902.38	0.20	11257.83
18	72435.64	15533.26	56902.38	0.18	10234.39
19	72435.64	15533.26	56902.38	0.16	9303.99
20	72435.64	15533.26	56902.38	0.15	8458.18
Total	**1448712.71**	**795107.2**			**0**

In Table 5.2, the discounted cash flow column is filled by the product of present worth factor and the (Benefit − Cost) for that particular year. When the NPV is zero, the SPB is 20 years, i.e., end of the project life. The LCOE is found to be

$$\text{LCOE} = \frac{\sum \text{costs/no. of years}}{\text{annual yeild (kWh)}} = \frac{\$795107.20\,/20}{60000} = \text{F}\$0.66/\text{kWh}.$$

This LCOE is less than the COE because in the calculation for LCOE the total cost occurring over the wind turbine lifetime does not take into account the discount factor. The benefit to cost ratio is calculated to be

$$\text{BCR} = \frac{\sum (\text{benefits})}{\sum (\text{costs})} = \frac{\$1448712.71}{\$795107.20} = 1.82.$$

Table 5.3.   Effect of varying COE on NPV, SPB and IRR.

COE (cents/kWh)	NPV ($)	SPB (yrs)	IRR (%)
115	−24652	–	9.25
120	889	20	10.03
125	26429	17	10.79
130	**51970**	**15**	**11.54**
135	**77511**	**14**	**12.28**
140	103051	13	13.01
145	128592	11.5	13.73
150	154133	11	14.44

This value is more than 1 which is a good indicator that the wind project will be profitable. However, BCR does not take into account the discount factor; it just uses the present value of cost and benefit.

Table 5.3 is obtained when the discount rate of 10% is considered. From Table 5.3, it is observed that as the COE is increased the NPV increases and the SPB period decreases.

When the COE to the consumers is taken as F$1.30/kWh then IRR is 11.54% which is a good indicator since IRR is the discount rate which gives NPV equal to zero. In this case for COE equal to F$1.30/kWh the discount rate (10%) is less than IRR (11.54%). At this COE the SPB period is 15 years and NPV is approximately F$50000. The SPB period is found from the discounted cash flow graph, Fig. 5.6.

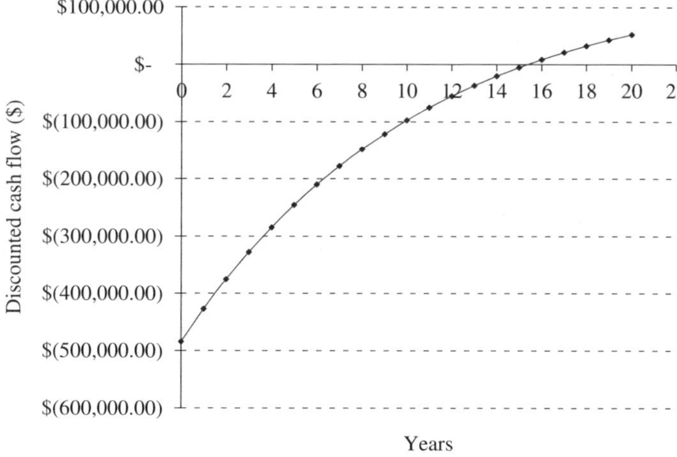

Fig. 5.6.   Discounted cash flow curve for discount rate of 10% and COE of F$1.30/kWh.

                    R. D. Prasad and R. C. Bansal

Table 5.4.   Effect of discount rate on NPV and SPB.

Discount rate (%)	NPV (F$)	SPB (years)
0	775692.80	7.6
2	545808.51	8.4
4	371840.16	9.6
6	238240.34	10.5
8	134167.46	12.4
10	51969.90	15.4
12	−13816.71	>20
14	−67140.14	>20

The place where the curve cuts the x-axis is the payback period. From Fig. 5.6 the SPB period is approximately 15 years.

Table 5.4 shows the values of SPB and NPV when the discount rate is changed. The cost of energy is taken as F$1.30/kWh. This COE is taken because if one sees Table 5.3 then at this COE the IRR is 11.54%. If the discount rate is increased above this IRR then loss would be made. To determine the IRR of the wind energy project the discount rate can be varied and its corresponding NPV calculated using the discounted cash flow analysis. It is seen from Table 5.4 that as the discount rate increases the SPB period also.

The point where the curve in Fig. 5.7 (obtained from Table 5.4) cuts the x-axis (discount rate axis), i.e., NPV $= 0$, it is known as IRR. The IRR is found to be 11.54%. This value basically means is that if the wind project is going to be profitable then the interest rate must be less than this. If the interest rate is more than IRR then the project will be going in at a loss.

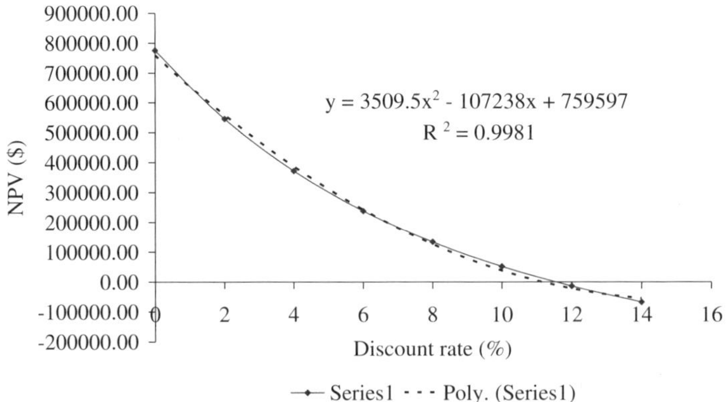

Fig. 5.7.   NPV as a function of discount rate.

Table 5.5. Effect of NPV on the COE.

NPV (F$)	COE (F$/kWh)	SPB (yrs)	IRR (%)
5000	1.21	19	10.15
10000	1.22	19	10.30
20000	1.24	18	10.60
30000	1.26	17	10.90
40000	1.28	16	11.19
50000	1.30	15.5	11.49
60000	**1.32**	**15**	**11.78**
70000	1.34	14.5	12.07
80000	**1.35**	**14**	**12.35**
90000	1.37	13	12.64

Table 5.5 shows how the COE changes when a desired NPV is required. The discount rate is taken as 10%.

From Table 5.5 it is seen that as the NPV increases the SPB period decreases and IRR increases but at a slower rate. It is noticed that for every $10000 increase in NPV the IRR increases by 0.3.

Therefore, from all these analyses for Case 1 the COE can be taken as F$1.32/kWh because then the IRR will approximately be 12% and at this IRR, COE is lowest. Always the IRR value is sought to be more than the discount factor because then the NPV would be positive. This COE is high for the village people and they will surely not be able to afford this high cost, which is why if some kind of overseas aid should be required to carry out wind energy project in a small country like Fiji, the Government or the state should consider giving more grants for renewable energy projects.

Case 2: Wind-diesel hybrid configuration

A 15 kW diesel generator (DG) set is considered. HOMER software was used to simulate the optimum configuration of wind-diesel hybrid based on the COE. When 15 kW DG set and 30 kW wind turbine are used, it is found using HOMER software that 107653 kWh of electrical energy would be produced annually. This gives energy consumption of an amount of 538 kWh/capita annually. Since the diesel generator is already in operation at Vadravadra village, the installation cost of the generator is taken as F$0.00 with the following assumptions:

- The diesel generator is operated 2500 hours annually
- The O&M costs is F$1500 annually
- Current price of diesel fuel is taken as F$2.5/L.

- The diesel price is assumed to increase at 5% per year.
- The fuel usage is 0.4 L/kWh.
- No grant or subsidy is available for the project.
- Wind turbine life time is 20 years.

All the other costs are the same as mentioned in Table 5.1.

Since the diesel price is assumed to be increasing at 5% per year and this factor is taken into account in the cash flow analysis under the column of cost (Table 5.6). For hybrid configuration, when the discount rate is 10% and the NPV is taken as zero, it gives the COE as F$1.36/kWh, LCOE as F$1.08/kWh, the SPB is 20 years and IRR is 10%.

It is noted that when the DG set is also in operation the recurring cost is high due to the diesel cost and the O&M cost of DG set. The benefit to cost ratio is 1.37 which is more than 1. However, this value is 30.5% less than the one obtained in

Table 5.6. Discounted cash flow analysis for wind-diesel hybrid system.

Year	Benefits ($)	Costs ($)	Benefit − Cost ($)	Present worth factor	Discounted cash flow (present value = benefit − cost) ($)
0		518192	(518,192.00)	1.00	−518192
1	128316.15	52470.76	75845.39	0.91	68950.35
2	128316.15	54242.64	74073.52	0.83	61217.78
3	128316.15	56103.1	72213.05	0.75	54254.73
4	128316.15	58056.6	70259.55	0.68	47988.22
5	128316.15	60107.76	68208.39	0.62	42352.04
6	128316.15	62261.49	66054.66	0.56	37286.13
7	128316.15	64522.9	63793.25	0.51	32736.02
8	128316.15	66897.38	61418.77	0.47	28652.31
9	128316.15	69390.59	58925.56	0.42	24990.19
10	128316.15	72008.45	56307.7	0.39	21709.05
11	128316.15	74757.21	53558.94	0.35	18772.08
12	128316.15	77643.41	50672.74	0.32	16145.9
13	128316.15	80673.92	47642.23	0.29	13800.26
14	128316.15	83855.95	44460.2	0.26	11707.76
15	128316.15	87197.09	41119.06	0.24	9843.58
16	128316.15	90705.28	37610.87	0.22	8185.22
17	128316.15	94388.88	33927.27	0.20	6712.33
18	128316.15	98256.66	30059.49	0.18	5406.46
19	128316.15	102317.83	25998.32	0.16	4250.93
20	128316.15	106582.06	21734.09	0.15	3230.63
	2566323.01	2030631.95			0

Case 1, the reason is that the cost is much high when DG set is used even though there is increase in the revenue. Revenue is generated by selling the energy produced in F\$/kWh.

When the discount rate is 10%, Table 5.7 is obtained. From Table 5.7 it is seen that when the COE is F\$1.44/kWh, the IRR is 12% which is more than 10% (discount rate). This is a good indicator because the NPV would be positive which means one would be making profit. Since, the wind energy project is not a profit making scheme then the COE can be taken as F\$1.38/ kWh.

To have a 12% IRR which is the same as Case 1, the COE is taken as F\$1.44/kWh for Case 2 which is an 11% increase in COE when comparing it to the value F\$1.30/kWh for Case 1 in Table 5.3. This difference is due to the fact that the total cost is increased in Case 2 because of the installation of the DG set and the purchasing of diesel fuel regularly and also, the cost of diesel fuel increases every year.

Table 5.8 shows the values of SPB, NPV, and the IRR when the discount rate is changed. The cost of energy is taken as F\$1.44/kWh. This COE is taken because if one sees Table 5.7 then at this COE the IRR is 12% and it is the lowest COE at this

Table 5.7. NPV, SPB and IRR variation with variation with variation in the cost of energy.

COE (\$/kWh)	NPV (F\$)	SPB (yrs)	IRR (%)
1.34	−18314	−0	9.37
1.36	−2351	20	9.92
1.38	13612	17	10.46
1.40	29575	15	10.98
1.42	45538	14	11.49
1.44	**61500**	**13**	**12.00**
1.46	**77464**	**12**	**12.50**
1.48	93427	11	12.99

Table 5.8. Effect of discount rate on NPV, SPB and IRR.

Discount rate (i) %	NPV (F\$)	SPB (yrs)
8	136513	10.5
9	97094	11.5
10	61500	13
11	29265	15
12	−14	20

                           R. D. Prasad and R. C. Bansal

Table 5.9.   Effect NPV on the COE.

NPV (F$)	COE (F$/kWh)	SPB (yrs)	IRR (%)
5000	1.37	19	10.17
10000	1.38	17.5	10.34
20000	1.39	16	10.67
30000	1.40	15	10.99
40000	1.41	14	11.32
50000	1.43	13.5	11.64
60000	**1.44**	**13**	**11.95**
70000	1.45	12.5	12.27
80000	1.46	12	12.58
90000	1.48	11.5	12.88

IRR. If the discount rate is increased above this IRR then loss would be made. From Table 5.8 it is seen that as the discount rate increases the NPV decreases and the SPB period increases. It is expected because if NPV decreases then it would take more time to recover the cost.

Table 5.9 shows how the cost of energy changes when a desired NPV is required. The discount rate is taken as 10%. From Table 5.9 it is seen that as NPV increases SPB period decreases. For this case of hybrid of wind turbine and diesel generator the COE comes out to be F$1.44/kWh which is a high value.

Overall, it is seen that the COE for either the wind turbine alone installed at the site or a hybrid of wind-diesel at the site has COE approximately F$1.40/kWh. This COE is very high for the consumers of Vadravadra village. The village's source of income is by selling copra, dalo and mats. This high COE is due to the large capital cost involved in buying, transporting and installing a wind turbine at the site. The only way to decrease the COE to the consumers is by getting some kind of grant or subsidy from the government, or overseas aid in terms of technical and financial aid. By technical assistance it is implied that trained and expert overseas personnel are involved in installing the wind turbine at the site.

In Fiji, the electrical energy that is sold to the consumers by Fiji Electricity Authority (FEA) is at a rate of 21 Fcents/kWh. However, on islands in Fiji where the electricity is not provided to consumers by FEA, it is provided by some private company or some business persons. When this is the case, the COE at the consumers pay on these islands it very high. For example, in Taveuni Island the COE that the electricity consumers pay is F$1.55/kWh. Hence, it is seen that people are paying huge amounts of money for electricity because they need electricity in their daily lives.

5.5 Conclusions

A proper wind resource assessment would lead to a successful wind energy project. After a WRA, consultants can be hired to give professional advice as to where exactly wind turbine(s) is/are to be installed by looking at the terrain features and the wind resource as the site. This can be done by using software or models that are described in Sec. 5.2 of this chapter. Once a WRA is done and the wind turbine is selected to get the annual energy yield, an economic analysis of the wind project needs to be carried out. One of the main reasons why the economic analysis is done is because the start of wind project requires huge sums of money to be invested and hence its economic viability needs to be determined. From the key findings of the economic analysis the COE to the consumers is determined, the NPV of the wind project is known, and its IRR is known. These are just some of the vital factors that need to be known before the start of any wind energy project.

References

1. BWEA, "The economics of wind energy," http://www.bwea.com/ref/econ.html (2008).
2. T. Burton, D. Sharpe, N. Jenkins and E. Bossanyi, *Wind Energy Handbook* (John Wiley & Sons Ltd., UK, 2001).
3. E. Hau and H. Renouard, *Wind Turbines: Fundamentals, Technologies, Applications and Economics* (Springer-Verlag Berlin Heidelberg, Germany, 2006).
4. R. Harrison, E. Hau and H. Snel, *Large Wind Turbines: Design and Economics* (Wiley & Sons Ltd., UK, 2000).
5. J.F Manwell, J. McGowan and G. Rogers, *Wind Energy Explained: Theory, Design and Application* (John Wiley & Sons Ltd., UK, 2002).
6. Iowa Energy Centre, "Wind energy manual-wind energy economics," http://www.energy.iastate.edu/renewable/wind/wem/wem-13_econ.html (2006).
7. P. Gipe, *Wind Power: Renewable Energy for Home, Farm and Business* (Chelsea Green Pub. Co, White River Junction, 2004).
8. Danish Wind Industry Association, http://www.windpower.org/ (2007).
9. S. Mathew, *Wind Energy: Fundamentals, Resource Analysis and Economics* (Springer-Verlag Berlin Heidelberg, Netherlands, 2006).
10. S. Ghahremanian, "Wind energy: Economical aspects and project development with real best practice projects data," http://www.exergy.se/goran/hig/ses/06/wind%202.pdf (2008).
11. Wikipedia, "Wind power," http://en.wikipedia.org/wiki/Wind_power# Small_scale (2007).
12. A.H. Marafia and H.A. Ashour, "Economics of off-shore/on-shore wind energy systems in Qatar," *Renewable Energy* **28** (2003) 1953–1963.
13. Investopedia: A Forbes media company, http://www.investopedia.com/terms/d/discountrate.asp (2008).

Chapter 6

Line Side Converters in Wind Power Applications

Ana Vladan Stankovic

Electrical and Computer Engineering Department,
Cleveland State University, Cleveland, Ohio 44115, U.S.A.
a.stankovic@csuohio.edu

Dejan Schreiber

Semikron Elektronik GmbH & Co KG
Sigmundstr. 200, 90431 Nürnberg, Germany
dejan.schreiber@semikron.com

This chapter describes the operation and control of a line side converter used in variable-speed wind energy conversion systems. Control techniques used for the line side converter under balanced and unbalanced grid voltages are presented. Several simulation examples illustrate the effectiveness of the control methods presented in the chapter.

6.1 Introduction

Wind is a free, renewable and clean energy resource. Wind energy is one of the fastest-growing forms of electricity generation in the world. Worldwide installed wind power development is shown in Fig. 6.1. The United States has more than 8000 MW of installed wind power. In 2008, the Department of Energy's report concluded that "the U.S. possesses sufficient and affordable wind resources to obtain at least 20% of its electricity from the wind by 2030".[1] The economic stimulus bill passed in February 2009 contains various provisions to benefit the wind industry. The development of wind power technology improves not only the economy but also the environment. Using wind instead of coal and gas reduces carbon dioxide emission by 99% and 98%, respectively.[1]

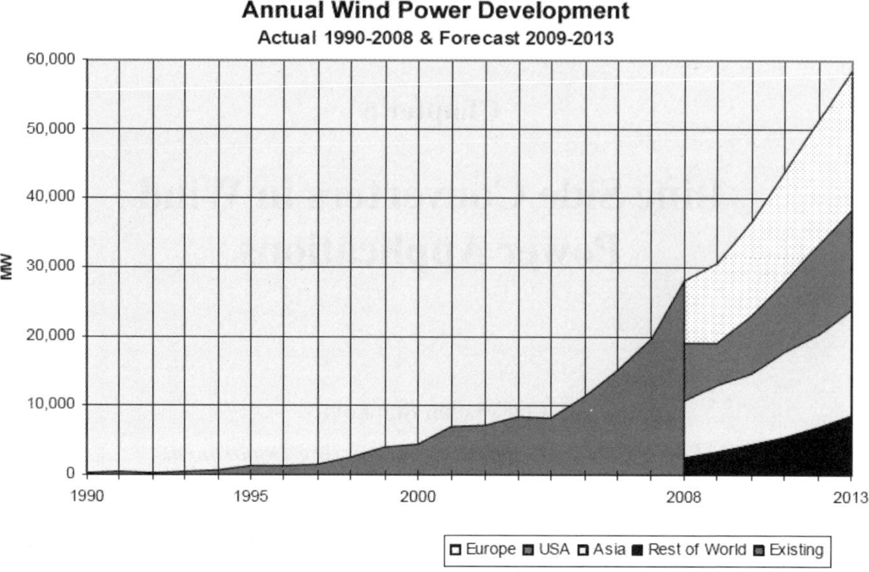

Fig. 6.1.   Annual installed wind power development.

Recent work in wind power technology includes the wind turbine design, the maximum power acquired from variable speed wind turbine, power flow control between the grid and the wind power system, filter design, and the operation under unbalanced grid voltages.[2–10] In the past decade the design and control of power electronics converters for variable — speed wind energy conversion systems was of growing interest. In comparison with the constant speed wind systems, the variable-speed wind energy conversion systems can deliver approximately 20% more power to the grid. In this chapter the analysis and control of the line side converter used in variable-speed wind energy conversion systems under balanced and unbalanced grid voltages is presented.[11–14]

6.2   Line Side Converters

Commonly used circuits for variable speed wind energy conversion systems are shown in Figs. 6.2 to 6.4.[15] In all circuits the PWM line side converter's objective is to export the active power to the grid and to allow the reactive power exchange between the grid and the converter as required by the application specifications.

PWM converter used as a line side converter in wind power applications has been extensively developed and analyzed in recent years.[16–18] It offers advantages because of its capability for nearly instantaneous reversal of power flow, power factor management and reduction of input/output harmonic distortion.

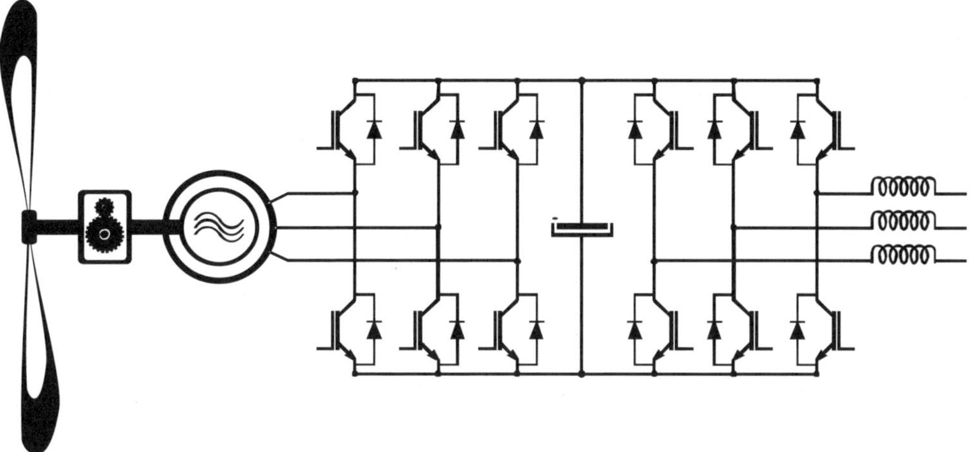

Fig. 6.2.   Induction or synchronous generator with two back-to-back PWM converters.

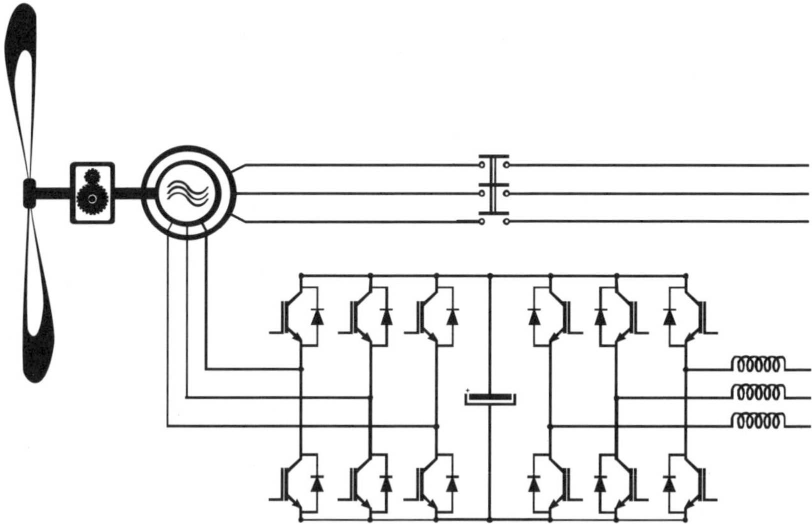

Fig. 6.3.   Doubly-fed induction generator with two PWM converters.

6.3   Principle of Operation

Figure 6.5 shows the structure of the PWM line side converter.

Power flow in the PWM converter is controlled by adjusting the phase angle δ between the source voltage U_1 and the respective converter reflected input voltage V_{s1}.[19] When U_1 leads V_{s1} the real power flows from the ac source into the converter. Conversely, if U_1 lags V_{s1}, power flows from the converter's dc side into the ac

          A. V. Stankovic and D. Schreiber

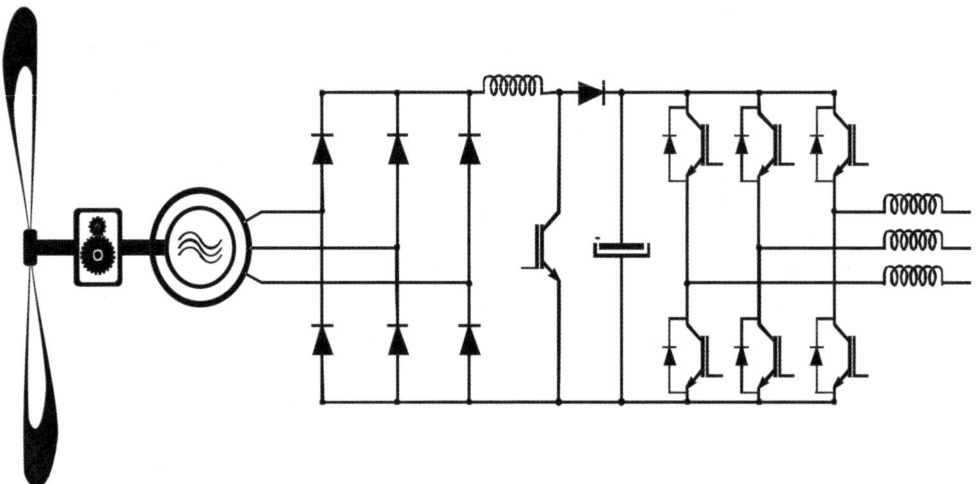

Fig. 6.4.   Synchronous generator with the rectifier, boost chopper, and the PWM line-side converter.

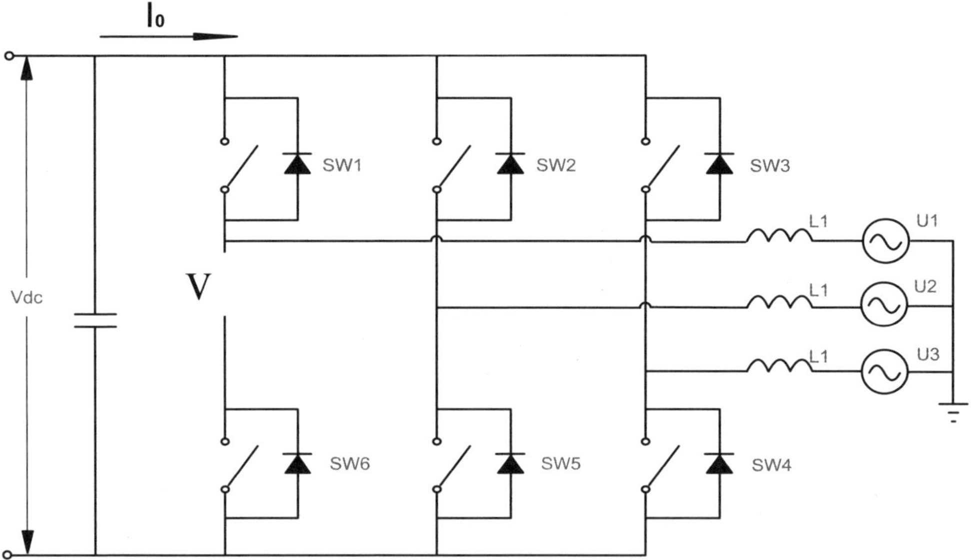

Fig. 6.5.   PWM converter.

source. The real power transferred is given by the Eq. (6.1).

$$P = \frac{U_1 V_{s1}}{X_1} \sin(\delta). \tag{6.1}$$

The ac power factor is adjusted by controlling the amplitude of the converter synthesized voltage V_{s1}. The per phase equivalent circuit and phase diagrams of the leading,

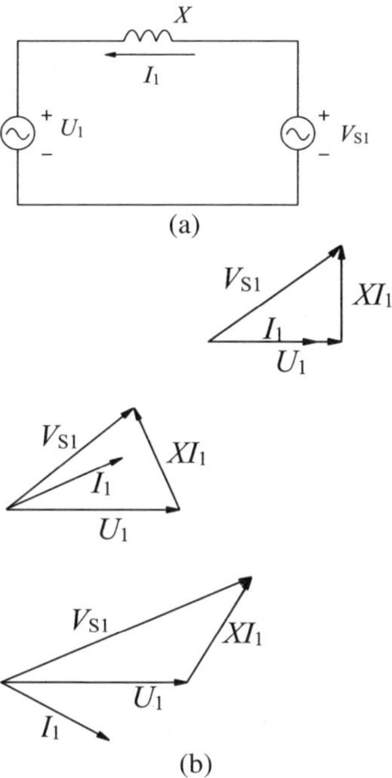

Fig. 6.6.   (a) Per-phase equivalent circuit of the line side converter, (b) phasor diagrams for unity, leading and lagging power factor operation.

lagging and unity power factor operation is shown in Fig. 6.6(a). The phasor diagram in Fig. 6.6(b) shows that to achieve a unity power factor, V_{s1} has to be,

$$V_{s1} = \sqrt{U_1^2 + (X_1 I_1)^2}. \tag{6.2}$$

6.4   Control of a Line-Side Converter under Balanced Operating Conditions

Two control methods of a line side converter under balanced operating conditions are explained in detail.

6.4.1   *Control of a line-side converter in the abc reference frame*

The line side converter controller is responsible for maintaining the dc link voltage at the reference value by exporting the active power to the grid. It is also designed to exchange the reactive power between the converter and the grid when required. In order to control the dc link voltage of the PWM converter, the line currents must be regulated.[20–25] In typical converter controllers presented to date, the dc bus voltage error is used to synthesize a line current reference. Specifically, the line

                    A. V. Stankovic and D. Schreiber

current reference is derived through the multiplication of a term proportional to the bus voltage error by a template sinusoidal waveform. The sinusoidal template is directly proportional to the phase shifted grid voltage, resulting in the desired power factor. The line current is then controlled to track this reference. Current regulation is accomplished through the use of hysteresis controllers.[20] A proposed control method[17] is shown in Fig. 6.7. In order to explain the closed-loop operation of the PWM Boost type converter, the switch matrix theory is used. The current I_0 of the matrix converters is a function of the converter transfer function vector T and the line current vector i is given by,

$$I_0 = Ti. \tag{6.3}$$

The converter transfer function vector T is composed of three independent line to neutral switching functions: SW_1, SW_2, SW_3

$$T = \begin{bmatrix} SW_1 & SW_2 & SW_3 \end{bmatrix}. \tag{6.4}$$

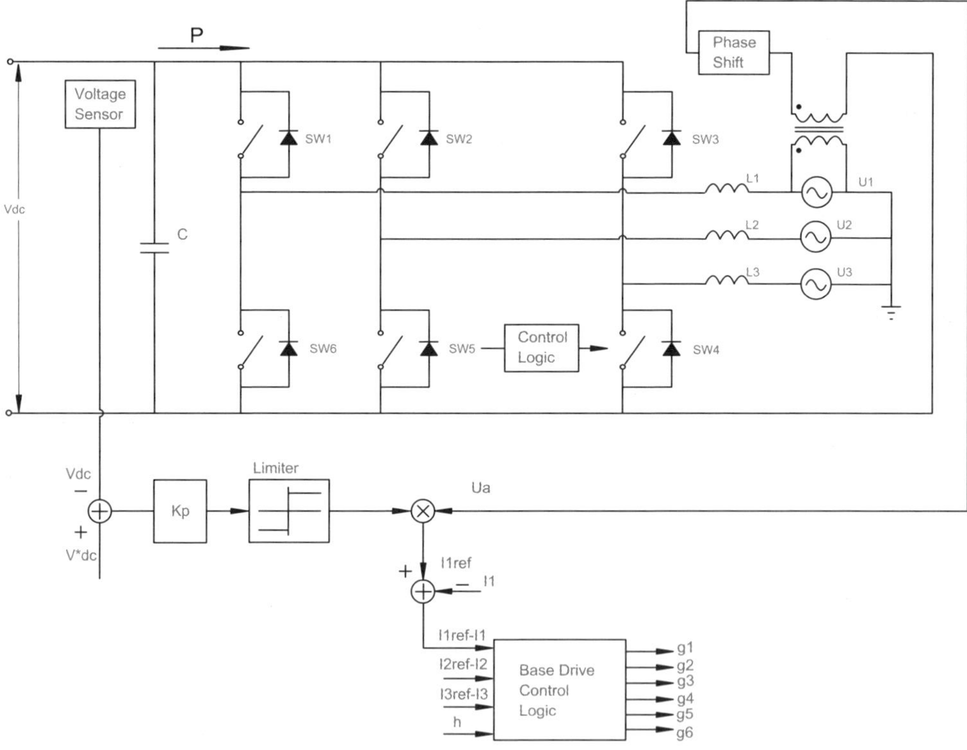

Fig. 6.7.   Control of the PWM line side converter in the abc reference frame.

The line current vector is given by,

$$i = \begin{bmatrix} i_1 \\ i_2 \\ i_3 \end{bmatrix}. \tag{6.5}$$

The line-to-neutral switching functions are balanced and are represented by their fundamental components only.

$$SW_1(t) = S_1 \sin(wt - \Theta),$$
$$SW_2(t) = S_1 \sin(wt - \Theta - 120^0), \tag{6.6}$$
$$SW_3(t) = S_1 \sin(wt + 120^0 - \Theta).$$

Therefore, the converter synthesized line to neutral voltages can be expressed as,

$$V_{s1} = \frac{1}{2} V_{dc} S_1 \sin(wt - \Theta),$$

$$V_{s2} = \frac{1}{2} V_{dc} S_1 \sin(wt - 120^0 - \Theta), \tag{6.7}$$

$$V_{s3} = \frac{1}{2} V_{dc} S_1 \sin(wt + 120^0 - \Theta).$$

Equation (6.7) shows the converter synthesized voltages. V_{dc} represents the dc link voltage. In the time domain, the fundamental components of the three phase line currents are given by,

$$i_1(t) = I_1 \sin(wt - \phi_1),$$
$$i_2(t) = I_1 \sin(wt - 120^0 - \phi_1), \tag{6.8}$$
$$i_3(t) = I_1 \sin(wt + 120^0 - \phi_1).$$

By combining Eqs. (6.3), (6.6) and (6.8) the current $I_0(t)$ is obtained and given by,

$$I_0(t) = I_1 \sin(wt - \phi_1)S_1 \sin(wt - \Theta) + I_1 \sin(wt - 120^0 - \phi_1)S_1$$
$$\times \sin(wt - 120 - \Theta) + I_1 \sin(wt + 120^0 - \phi_1)S_1 \sin(wt + 120^0 - \Theta). \tag{6.9}$$

By using a trigonometric identity, $I_0(t)$ becomes,

$$I_0(t) = \frac{3}{2} I_1 S_1 \cos(\Theta - \phi_1). \tag{6.10}$$

Since the angle, $(\Theta - \phi_1)$ is constant for any set value of the power factor, the dc current, $I_0(t)$, is proportional to the magnitude of the line current, $I_1(t)$ and so is the dc link voltage, V_{dc}. For unity power factor control, angle φ_1 is equal to zero.

The dc link voltage V_{dc} is given by,

$$(V_{dc\,ref} - V_{dc}) = KI_1. \tag{6.11}$$

Figure 6.7 shows that the dc bus error $(V_{dc\,ref} - V_{dc})$ is used to set the reference for the line current magnitude. The line voltage, U_a, is multiplied by the dc bus error and it becomes a reference for the input current in phase 1. The reference value for current in phase 2 is phase-shifted by 120^0 with respect to the current in phase 1 since this section considers control under balanced operating conditions. Since the sum of three input currents is always zero, the reference for current in phase 3 is obtained from the following equation,

$$i_{3\,ref}(t) = -i_{1\,ref}(t) - i_{2\,ref}(t). \tag{6.12}$$

The line currents, $i_1(t)$, $i_2(t)$, $i_3(t)$ are measured and compared with the reference currents, $i_{1\,ref}(t)$, $i_{2\,ref}(t)$, $i_{3\,ref}(t)$. The error is fed to a comparator having a prescribed hysteresis band $2\Delta I$. Switching of the leg of the rectifier (SW1 off and SW4 on) occurs when the current attempts to exceed a set value corresponding to the desired current $i_{ref} + \Delta I$. The reverse switching (SW1 on and SW4 off) occurs when the current attempts to become less than $i_{ref} - \Delta I$. The hysteresis controller produces a very good quality waveform and is simple to implement. Unfortunately, with this type of control (hysteresis controller) the switching frequency does not remain constant but varies along different portions of the desired current.

6.4.2 *Control of a line side converter in d-q reference frame*

Control of the line-side converter in d-q reference frame is shown in Fig. 6.8.[3] In synchronous reference frame d and q components of the line voltages are given by,

$$u_d = u_{id} - Ri_d - L\frac{di_d}{dt} + \omega Li_q, \tag{6.12.1}$$

$$u_q = u_{iq} - Ri_q - L\frac{di_q}{dt} - \omega Li_d, \tag{6.12.2}$$

where L, R are grid inductance and resistance, respectively, and u_{iq}, u_{id} are d and q components of the inverter synthesized voltage.

$$P = \frac{3}{2}u_d i_d, \tag{6.12.3}$$

$$Q = \frac{3}{2}u_d i_q. \tag{6.12.4}$$

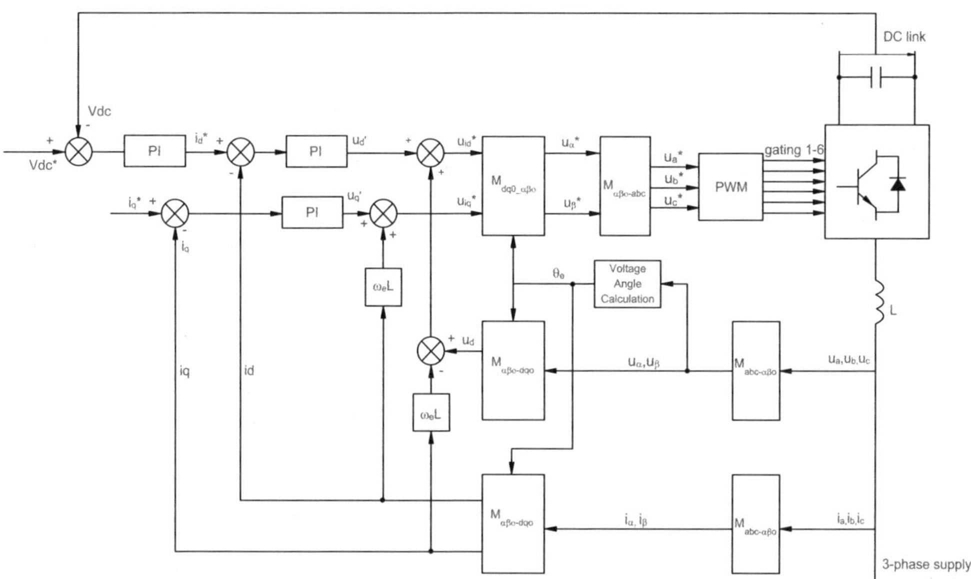

Fig. 6.8.   Line side converter control in *d-q* reference frame.

From Eqs. (6.12.3) and (6.12.4) it follows that the active and reactive power control is achieved by controlling d and q components of the line currents. The active power is instantaneously transferred to the grid via inverter by controlling the d axis current i_d. As shown in Fig. 6.8, the outer dc link voltage controller is used to set the reference for the current i_d. The reactive power is controlled by setting the reference for the q axis current.

6.5   Line Side Converters under Unbalanced Operating Conditions

Unfortunately, the features that PWM converter offers are fully realized only when the grid voltages are balanced. Under an unbalanced grid voltages there is a deterioration of the converter input and output characteristics. The imbalance in grid voltages may occur frequently, especially in weak systems. Non-uniformly distributed single phase loads, faults or unsymmetrical transformer windings could cause imbalance in the three-phase voltages both in magnitude and in phase. Regardless of the cause, unbalanced voltages have a severe impact on the performance of the PWM converter. Actually, the huge harmonics of lower frequencies, not present in the PWM switching functions appear at both the input and output ports of the converter. The problems include a significant distortion in line current waveforms and increase in the dc capacitor ripple current and voltage. These additional low frequency components cause additional losses and should be considered in the filter design of these converters. An analysis of the PWM converter under unbalanced operating conditions will be presented next. Special emphasis is given to the evaluation of control method for input output harmonic elimination of the PWM converter under unbalanced operation conditions. The proposed technique maintains a high quality sinusoidal line currents and dc link voltages even though the grid voltages remain unbalanced. In addition, severe unbalanced operating conditions have been considered. Under severe fault conditions in the distribution system, not only line voltages, but also line impedances must be considered as unbalanced. The generalized control method maintains high quality line currents and dc link voltage with adjustable power factor under severe fault conditions is described.[17]

6.6 Analysis of the PWM Converter under Unbalanced Operating Conditions

The unbalanced input voltages cause an abnormal second harmonic at the dc link voltage which reflects back to the ac side causing the third-order harmonic current to flow. Next, the third harmonic current causes the fourth-order harmonic voltage at the output, and so on. This results in the appearance of even harmonics at the dc link voltage and odd harmonics in the line currents. These additional components should be considered in the filtering design of these converters. The dc link current I_0 of the matrix converters is a function of the converter transfer function vector T and the line current vector i and is given by,

$$I_0 = Ti. \tag{6.13}$$

The converter transfer function vector T is composed of three independent line to neutral switching functions and is given by,

$$T = \begin{bmatrix} SW_1 & SW_2 & SW_3 \end{bmatrix}. \tag{6.14}$$

The line current vector is given by,

$$i = \begin{bmatrix} i_1 \\ i_2 \\ i_3 \end{bmatrix}. \tag{6.15}$$

The line to neutral switching functions are balanced and can be represented only by their fundamental components.

$$SW_1(t) = S_1 \sin(wt - \Theta),$$
$$SW_2(t) = S_1 \sin(wt - \Theta - 120^0), \quad (6.16)$$
$$SW_3(t) = S_1 \sin(wt + 120^0 - \Theta).$$

Therefore, the converter synthesized line to neutral voltages can be expressed as,

$$V_{s1} = \frac{1}{2} V_{dc} S_1 \sin(wt - \Theta),$$

$$V_{s2} = \frac{1}{2} V_{dc} S_1 \sin(wt - 120^0 - \Theta), \quad (6.17)$$

$$V_{s3} = \frac{1}{2} V_{dc} S_1 \sin(wt + 120^0 - \Theta).$$

The equation shows that the converter synthesized voltages will always be balanced when the switching functions are balanced. For this reason, there will be no negative sequence voltage component present at its terminals. It follows that the line currents are unbalanced and given by,

$$\begin{bmatrix} I_1 \\ I_2 \\ I_3 \end{bmatrix} = \begin{bmatrix} 1 & 1 & 1 \\ 1 & a^2 & a \\ 1 & a & a^2 \end{bmatrix} \begin{bmatrix} I^0 \\ I^+ \\ I^- \end{bmatrix}, \quad (6.18)$$

where I^0, I^+, I^- are 0, positive and negative sequence currents.

Since $I_1 + I_2 + I_3 = 0$, the zero sequence current never flows in this circuit. $I^0 = 0$.

The line to neutral voltages are unbalanced and given by,

$$\begin{bmatrix} U_1 \\ U_2 \\ U_3 \end{bmatrix} = \begin{bmatrix} 1 & 1 & 1 \\ 1 & a^2 & a \\ 1 & a & a^2 \end{bmatrix} \begin{bmatrix} U^0 \\ U^+ \\ U^- \end{bmatrix}, \quad (6.19)$$

where U^0, U^+ and U^- are zero, positive and negative sequence line voltages. In the time domain, the fundamental components of the three phase currents are given by,

$$i_1(t) = I_1 \sin(wt - \phi_1), \quad i_2(t) = I_2 \sin(wt - 120^0 - \phi_2),$$
$$i_3(t) = I_3 \sin(wt + 120^0 - \phi_3). \quad (6.20)$$

According to Eq. (6.13), the dc output current $I_0(t)$ is given by,

$$I_0(t) = I_1 \sin(wt - \varphi_1)S_1 \sin(wt - \Theta) + I_2 \sin(wt - 120^0 - \varphi_2)$$
$$\times \, S_1 \sin(wt - 120 - \Theta) + I_3 \sin(wt + 120^0 - \varphi_3)S_1 \sin(wt + 120^0 - \Theta).$$
$$(6.21)$$

By using a trigonometric identity, $I_0(t)$ becomes,

$$I_0(t) = \frac{1}{2}I_1 S_1[\cos(\Theta - \varphi_1) - \cos(2wt - \Theta - \varphi_1)]$$
$$+ \frac{1}{2}I_2 S_1[\cos(\Theta - \varphi_2) - \cos(2wt - 240^0 - \Theta - \varphi_2)]$$
$$+ \frac{1}{2}I_3 S_1[\cos(\Theta - \varphi_3) - \cos(2wt + 240^0 - \Theta - \varphi3)]. \quad (6.22)$$

The dc link current consists of a dc and a harmonic current.

$$I_0(t) = I_{dc} + I_{sh}(2wt), \quad (6.23)$$

where $I_{sh}(2wt)$ is the second-order harmonic current and is given by,

$$I_{sh}(2wt) = -\frac{S_1 I_1}{2}\cos(2wt - \Theta - \varphi_1) - \frac{I_2 S_1}{2}\cos(2wt - 240^0 - \Theta - \varphi_2)$$
$$- \frac{I_3 S_1}{2}[\cos(2wt + 240^0 - \Theta - \varphi_3)]. \quad (6.24)$$

Therefore, the dc link voltage will also contain the second-order harmonic, which will reflect back to the output causing the third-order harmonic current to flow. The third harmonic current will reflect back to the input causing the fourth-order harmonic to flow. As the literature indicates,[21] even harmonics will appear at the input and odd at the output of the converter under unbalanced voltages. The second- and third-order harmonics are of the primary concern.

6.7   Control Method for Input-Output Harmonic Elimination of the PWM Converter under Unbalanced Operating Conditions

Recently, Stankovic and Chen[17] proposed a generalized method for input–output harmonic elimination of the three-phase PWM converters under unbalanced operating conditions. The method is related to harmonic elimination of the PWM Boost type converters under severe fault conditions. Under severe fault conditions both line voltages and line impedances have to be considered unbalanced.

The circuit in Fig. 6.5 is analyzed with unbalanced line voltages and unbalanced line impedances. It is assumed that the converter is lossless. Harmonic elimination can be achieved by generating unbalanced reference commands for three line currents under unbalanced voltages and impedances. The following equations are derived for unbalanced grid voltages U_1, U_2, U_3 and unbalanced line impedances z_1, z_2, and z_3. From the circuit shown in Fig. 6.5 it follows,

$$V_{s1} = U_1 + z_1 I_1, \tag{6.25}$$

$$V_{s2} = U_2 + z_2 I_2, \tag{6.26}$$

$$V_{s3} = U_3 + z_3 I_3, \tag{6.27}$$

$$I_1 = -I_2 - I_3, \tag{6.28}$$

$$S^* = -(U_1^* I_1 + U_2^* I_2^* + U_3^* I_3), \tag{6.29}$$

$$SW_1 I_1 + SW_2 I_2 + SW_3 I_3 = 0. \tag{6.30}$$

Where U_1, U_2, U_3, I_1, I_2, I_3, z_1, z_2, z_3, V_{s1}, V_{s2}, V_{s3}, S, SW_1, SW_2 and SW_3 are grid voltages, line currents, line impedances, synthesized voltages at the input of the converter, apparent power and switching functions, respectively, represented as phasors.

Equation (6.30) represents the condition for the second harmonic elimination. Synthesized voltages V_{s1}, V_{s2} and V_{s3} can be expressed as,

$$V_{s1} = SW_1 \frac{V_{dc}}{2\sqrt{2}}, \tag{6.31}$$

$$V_{s2} = SW_2 \frac{V_{dc}}{2\sqrt{2}}, \tag{6.32}$$

$$V_{s3} = SW_3 \frac{V_{dc}}{2\sqrt{2}}, \tag{6.33}$$

where V_{dc} is the dc voltage.

By substituting Eqs. (6.31) to (6.33) into Eqs. (6.25) to (6.27) the following set of equations is obtained,

$$U_1 = SW_1 \frac{V_{dc}}{2\sqrt{2}} - z_1 I_1, \tag{6.34}$$

$$U_2 = SW_2 \frac{V_{dc}}{2\sqrt{2}} - z_2 I_2, \tag{6.35}$$

$$U_3 = SW_3 \frac{V_{dc}}{2\sqrt{2}} - z_3 I_3, \tag{6.36}$$

$$I_1 = -I_2 - I_3, \tag{6.37}$$

$$S^* = -(U_1^* I_1 + U_2^* I_2 + U_3^* I_3), \tag{6.38}$$

$$SW_1 I_1 + SW_2 I_2 + SW_3 I_3 = 0. \tag{6.39}$$

For given input power, S grid voltages, U_1, U_2, U_3 and line impedances z_1, z_2 and z_3, line currents, I_1, I_2 and I_3, can be obtained from the above set of equations. By multiplying Eqs. (6.34) to (6.36) by I_1, I_2 and I_3, respectively, and adding them up the following equation is obtained,

$$U_1 I_1 + U_2 I_2 + U_3 I_3 = -z_1 I_1^2 - z_2 I_2^2 - z_3 I_3^2,$$
$$+ \frac{V_{\text{dc}}}{2\sqrt{2}}(SW_1 I_1 + SW_2 I_2 + SW_3 I_3). \tag{6.40}$$

The set of six equations with six unknowns, Eqs. (6.34) to (6.39), reduces to three equations with three unknowns.

By substituting Eq. (6.39) into (6.40) the following equation is obtained,

$$U_1 I_1 + U_2 I_2 + U_3 I_3 = -z_1 I_1^2 - z_2 I_2^2 - z_3 I_3^2, \tag{6.41}$$

$$I_1 = -I_2 - I_3, \tag{6.42}$$

$$S^* = -(U_1^* I_1 + U_2^* I_2 + U_3^* I_3). \tag{6.43}$$

Equations (6.41) to (6.43) represent a set of three equations with three unknowns.

By substituting Eq. (6.42) into Eqs. (6.41) and (6.43), the following set of equations is obtained and given by,

$$U_1(-I_2 - I_3) + U_2 I_2 + U_3 I_3 = -z_1(-I_2 - I_3)^2 - z_2 I_2^2 - z_3 I_3^2, \tag{6.44}$$

$$S^* = -(-U_1^* I_2 - U_1^* I_3 + U_2^* I_2 + U_3^* I_3). \tag{6.45}$$

Equation (6.44) can be simplified as,

$$I_2(U_2 - U_1) + I_3(U_3 - U_1) = -(z_1 + z_2)I_2^2 - (z_1 + z_3)I_3^2 - 2z_1 I_2 I_3. \tag{6.46}$$

From Eq. (6.43) current, I_2, can be expressed as,

$$I_2 = \frac{-S^* - I_3(U_3^* - U_1^*)}{U_2^* - U_1^*}. \tag{6.47}$$

Finally by substituting Eq. (6.47) into Eq. (6.46),

$$\frac{-S^* - I_3(U_3^* - U_1^*)}{U_2^* - U_1^*}(U_2 - U_1) + I_3(U_3 - U_1)$$

$$= -(z_1 + z_2)\frac{S^{*2} + 2S^* I_3(U_3^* - U_1^*) + I_3^2(U_3^* - U_1^*)^2}{(U_2^* - U_1^*)^2}$$

$$- (z_1 + z_2)I_3^2 - 2z_1 \frac{-S^* - I_3(U_3^* - U_1^*)}{U_2^* - U_1^*} I_3, \tag{6.48}$$

$$\left[-\frac{2z_1(U_3^* - U_1^*)}{U_2^* - U_1^*} + \frac{(z_1 + z_2)(U_3^* - U_1^*)^2}{(U_2^* - U_1^*)^2} + (z_1 + z_3) \right] I_3^2$$

$$+ \left[(U_3 - U_1) - \frac{(U_3^* - U_1^*)(U_2 - U_1)}{U_2^* - U_1^*} - \frac{2z_1 S^*}{U_2^* - U_1^*} \right.$$

$$\left. + \frac{2S^*(z_1 + z_2)(U_3^* - U_1^*)}{(U_2^* - U_1^*)^2} \right] I_3$$

$$- \frac{S^*(U_2 - U_1)}{U_2^* - U_1^*} + \frac{(z_1 + z_2)S^{*2}}{(U_2^* - U_1^*)^2} = 0. \tag{6.49}$$

Currents I_2 and I_1 can be obtained from Eqs. (6.47) and (6.42).

Equations (6.42), (6.47) and (6.49) represent the steady state solution for line currents under both unbalanced grid voltages and unbalanced line impedances. An analytical solution represented by Eq. (6.49) always exists unless all the coefficients of the quadratic equations are equal to zero.

Critical Evaluation

The analytical solution that has been obtained is general. The only constraint that exists, as far as the level of unbalance is concerned, is governed by constraints of the operation of the PWM Converter itself.

The proposed generalized method for input–output harmonic elimination is valid if and only if $U_i, z_i \neq 0$, where $i = 1, 2, 3$. In other words the solution exists for all levels of unbalance in line voltages and impedances, except for cases where both voltage and impedance in the same phase are equal to zero. Therefore, the maximum level of voltage imbalance with balanced line impedances, for which the proposed solution is still valid is given as,

$$U_1 \neq 0$$
$$U_2 = U_3 = 0$$
$$z_1 = z_2 = z_3 \neq 0.$$

The maximum level of imbalance in both line voltages and impedances for which the proposed solution is still valid is given as,

$$U_1 \neq 0$$
$$U_2 = U_3 = 0$$
$$z_1 = 0$$
$$z_2 \neq z_3 \neq 0.$$

Based on the analysis of the open loop configuration presented above, a feed forward control method is proposed. The line voltages as well as line impedances

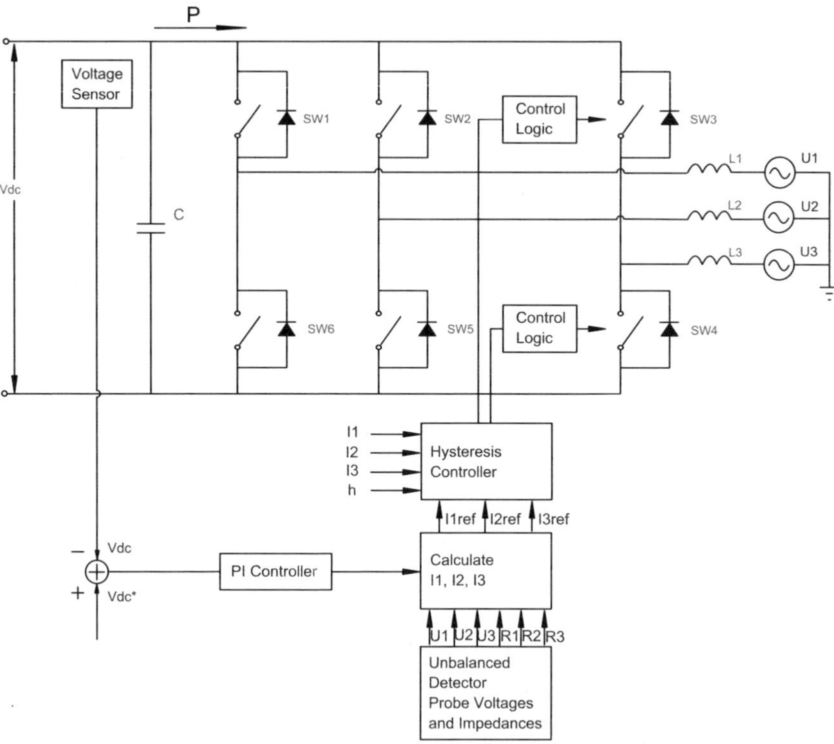

Fig. 6.9.   Control of a line side converter under unbalanced operating conditions.

have to be measured as shown in Fig. 6.9. In Fig. 6.9, block "unbalanced detector" is used to measure unbalanced voltages and unbalanced impedances. Based on this information and a dc bus error, reference currents are calculated (block "calculate I_1, I_2, I_3) according to Eqs. (6.42), (6.47) and (6.49) which become reference signals for the hysteresis controller[16] shown in Fig. 6.9. Only one PI controller is utilized, which has been shown to be sufficient for good regulation. The proposed control method is shown in more detail in Fig. 6.9.

6.8   Examples

Wind turbine with two PWM converters shown in Fig. 6.10 was simulated in Simulink under balanced and unbalanced operating conditions.

Example 1

In this example control method shown in Fig. 6.8, under balanced grid voltages was used in simulation.

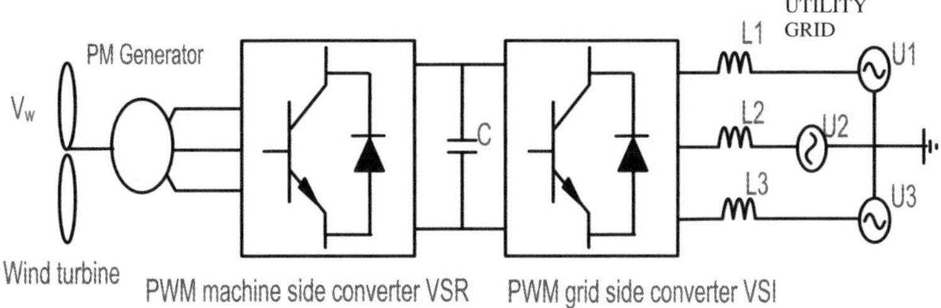

Fig. 6.10.   Wind turbine with two PWM converters.

The following parameters were used in simulation:

3 kW PM synchronous generator
The rotor speed w_m = 77.78 rad/s
The electromagnetic torque $T_B = 30$ Nm
dc link capacitor $C = 1$ mF
Line inductances $L = 10$ mH
The grid voltages are balanced $v_{an} = v_{bn} = v_{cn} = 220$ V
The power $P = 2318$ W, $Q = 0$.

Figures 6.11 to 6.17 show rotor speed, electromagnetic torque, stator voltages, stator currents, dc link voltage, grid side voltages and line currents for one wind speed.

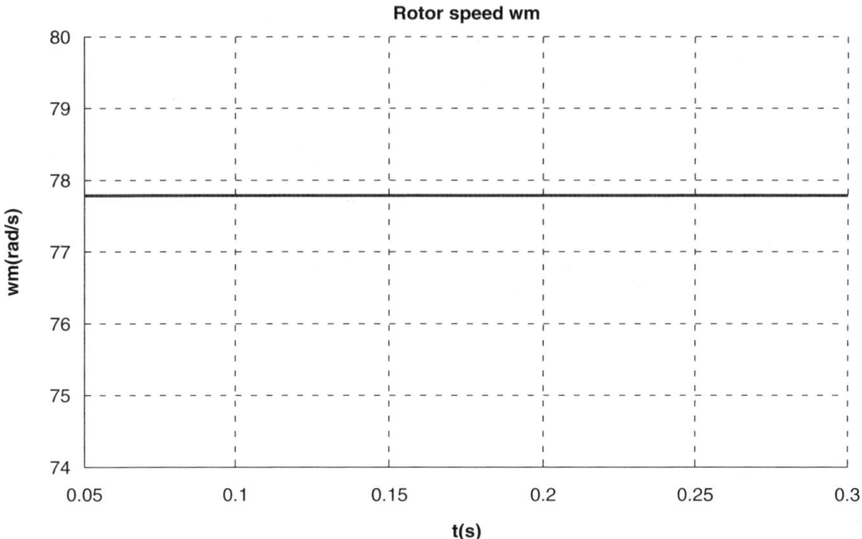

Fig. 6.11.   Rotor speed for one wind speed (Example 1).

A. V. Stankovic and D. Schreiber

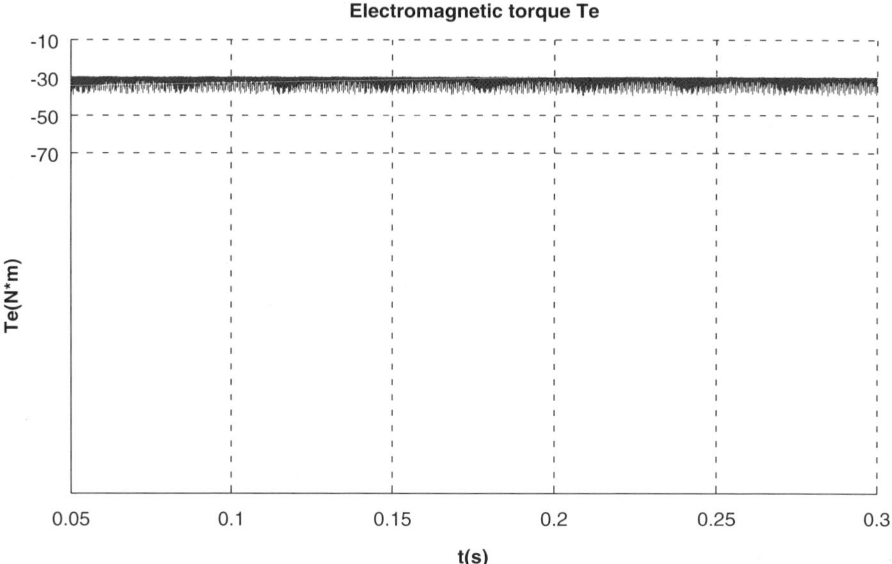

Fig. 6.12.   Electromagnetic torque for one wind speed (Example 1).

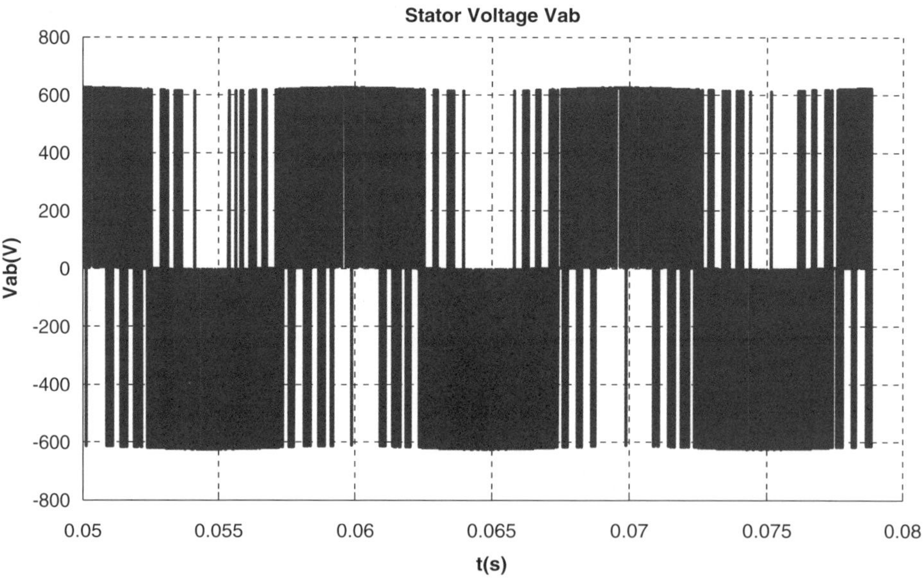

Fig. 6.13.   Stator voltage for one wind speed (Example 1).

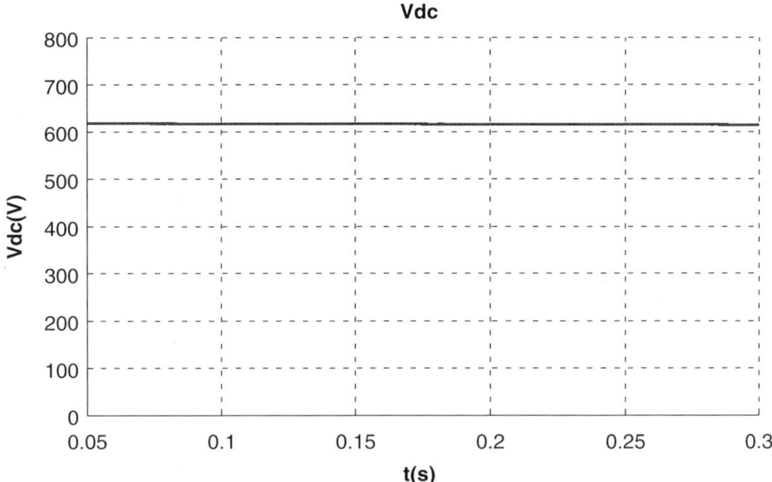

Fig. 6.14.   DC link voltage for one wind speed (Example 1).

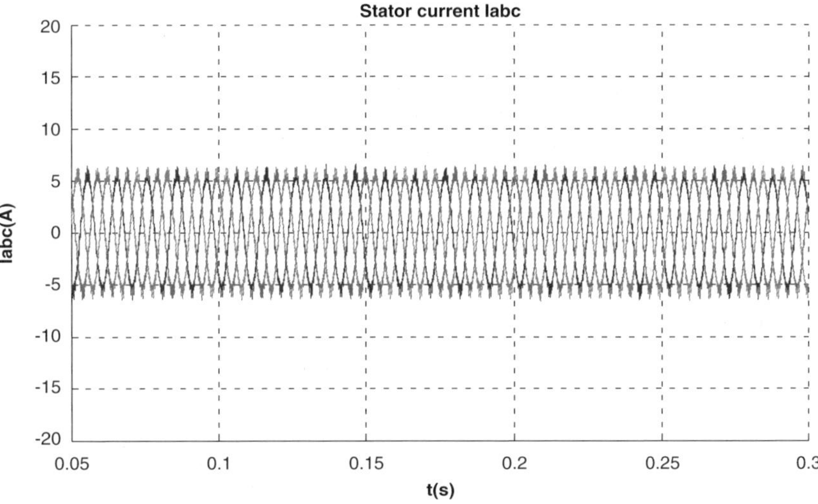

Fig. 6.15.   Stator currents for one wind speed (Example 1).

Example 2

In this example the wind turbine system shown in Fig. 6.10 was simulated under unbalanced grid voltages. The control method shown in Fig. 6.9 was used in the simulation.

The following parameters were also used in the simulation:

3 kW PM synchronous generator

The rotor speed $w_m = 77.78$ rad/s

     A. V. Stankovic and D. Schreiber

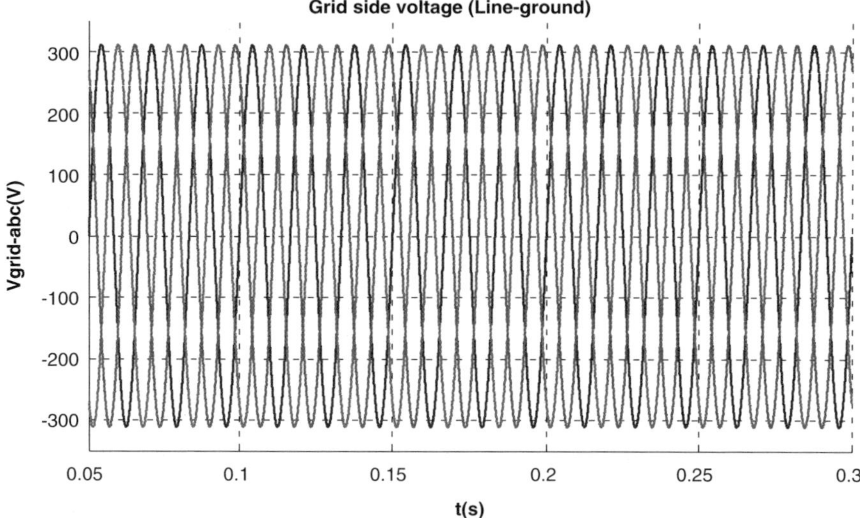

Fig. 6.16.   Grid side voltages for one wind speed (Example 1).

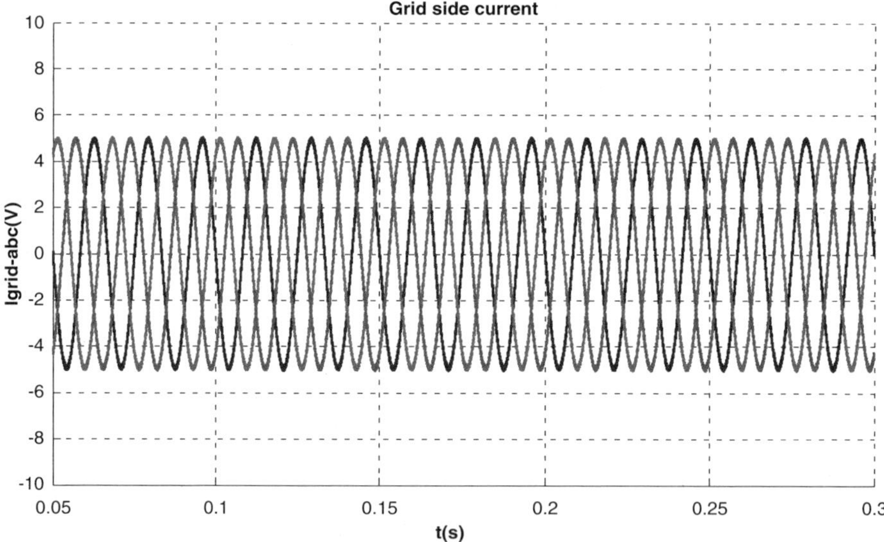

Fig. 6.17.   Grid side currents for one wind speed (Example 1).

The electromagnetic torque $T_B = 30\,\mathrm{Nm}$
dc link capacitor $C = 1\,\mathrm{mF}$
Line inductances $L = 10\,\mathrm{mH}$
The grid voltages are unbalanced $v_{an} = 150\angle 0\, v_{bn}\angle -120 = v_{cn} = 220V\angle 120$
The power $P = 2318\,\mathrm{W}$.

Figures 6.18 to 6.24 show rotor speed, electromagnetic torque, stator voltages, stator currents, dc link voltage, unbalanced grid voltages and line currents for one wind speed. In spite of unbalance in grid voltages, line currents do not contain low order harmonics. This was achieved by using the control method shown in Fig. 6.9.

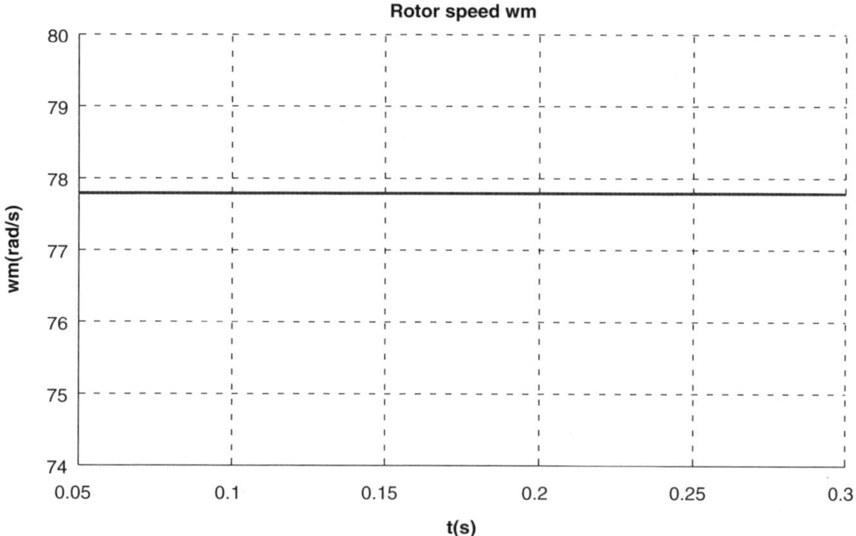

Fig. 6.18.   Rotor speed for one wind change (Example 2).

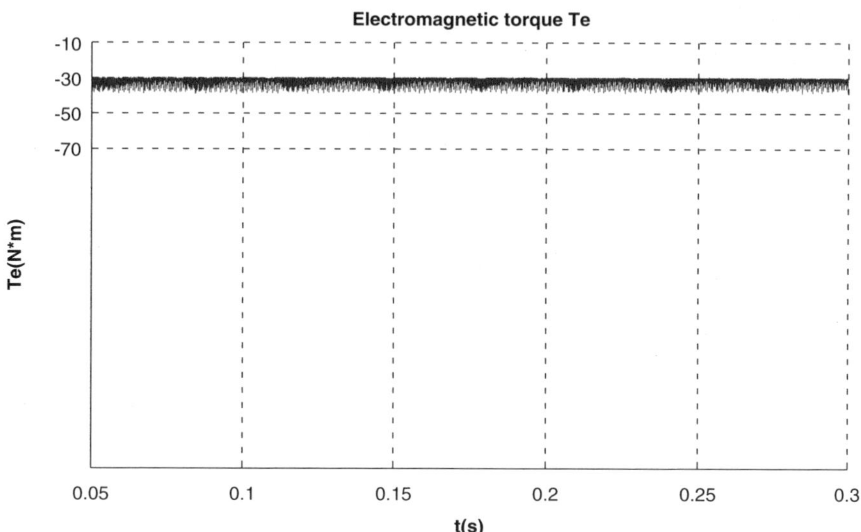

Fig. 6.19.   Electromagnetic torque for one wind speed (Example 2).

          A. V. Stankovic and D. Schreiber

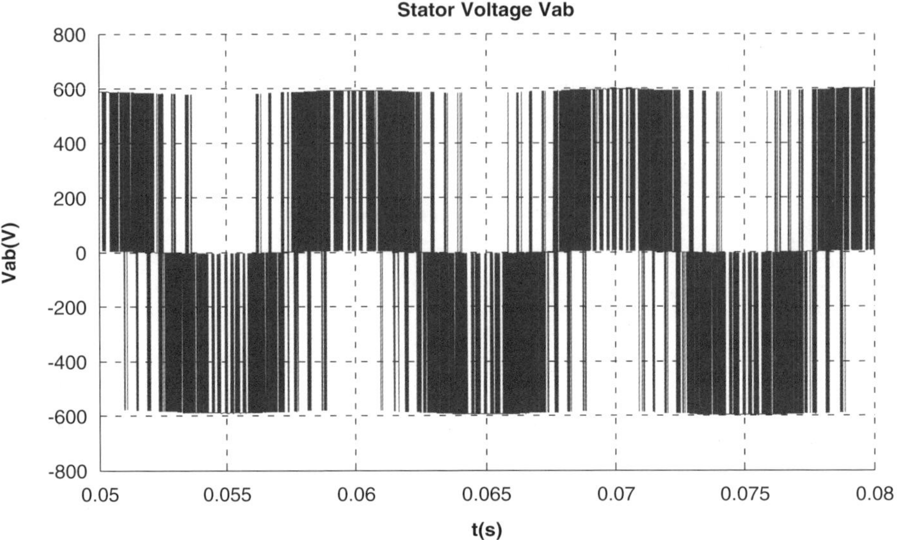

Fig. 6.20.   Stator voltages for one wind speed (Example 2).

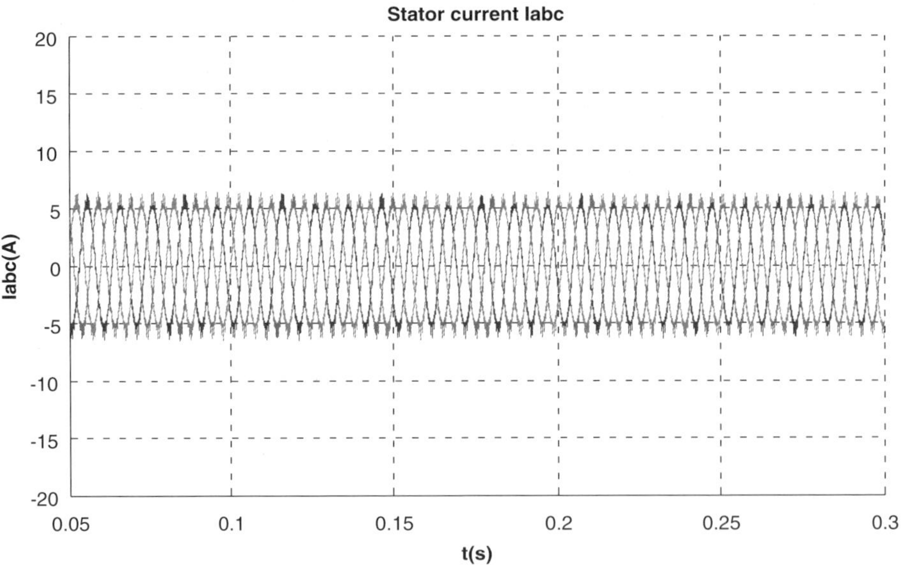

Fig. 6.21.   Stator currents for one wind speed (Example 2).

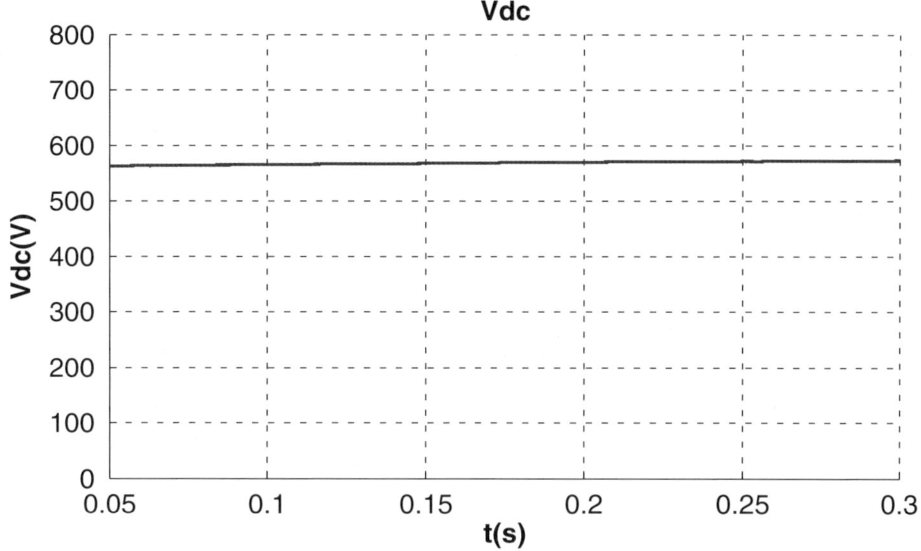

Fig. 6.22. DC link voltage for one wind speed (Example 2).

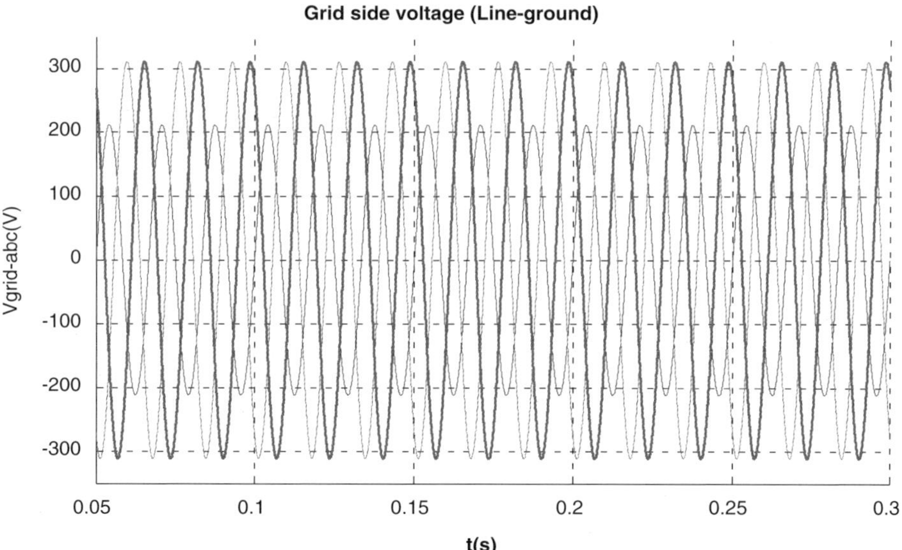

Fig. 6.23. Grid side voltage for one wind speed (Example 2).

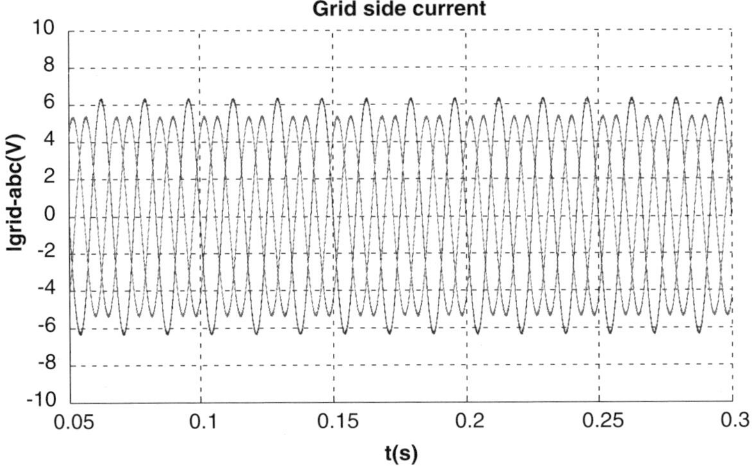

Fig. 6.24.   Grid side currents for one wind speed (Example 2).

Example 3

In this example the wind turbine system shown in Fig. 6.10 is simulated. The grid voltages are unbalanced and the variable speed of a wind turbine is incorporated. The grid voltages are unbalanced.

$$v_{an} = 150\angle 0 \, v_{bn} \angle -120 = v_{cn} = 220\,\text{V}\angle 120.$$

Figures 6.25 to 6.30 show rotor speed, electromagnetic torque, stator voltage, stator currents, unbalanced grid voltages and line currents. In spite of unbalanced

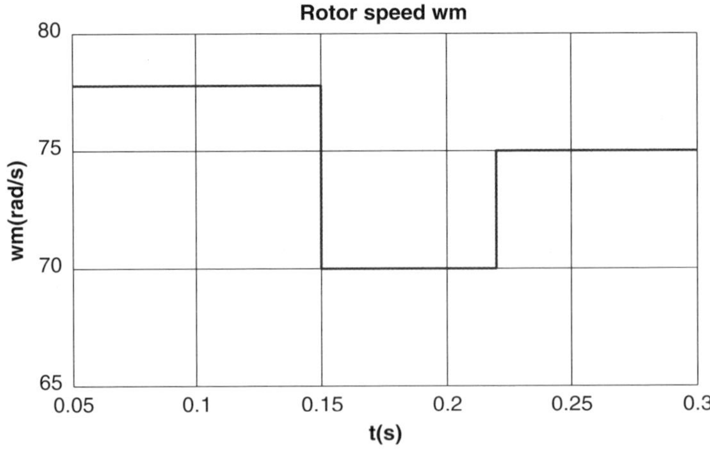

Fig. 6.25.   Rotor speed (Example 3).

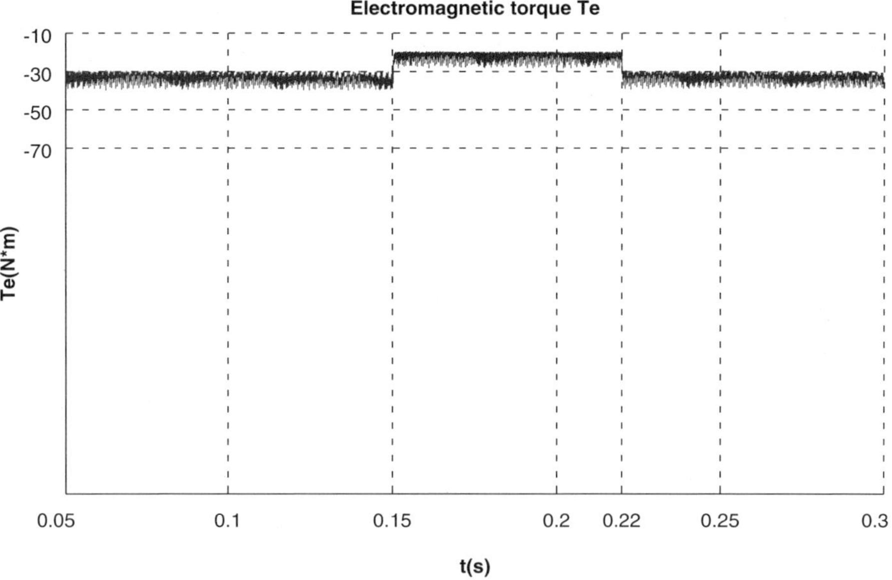

Fig. 6.26. Electromagnetic torque (Example 3).

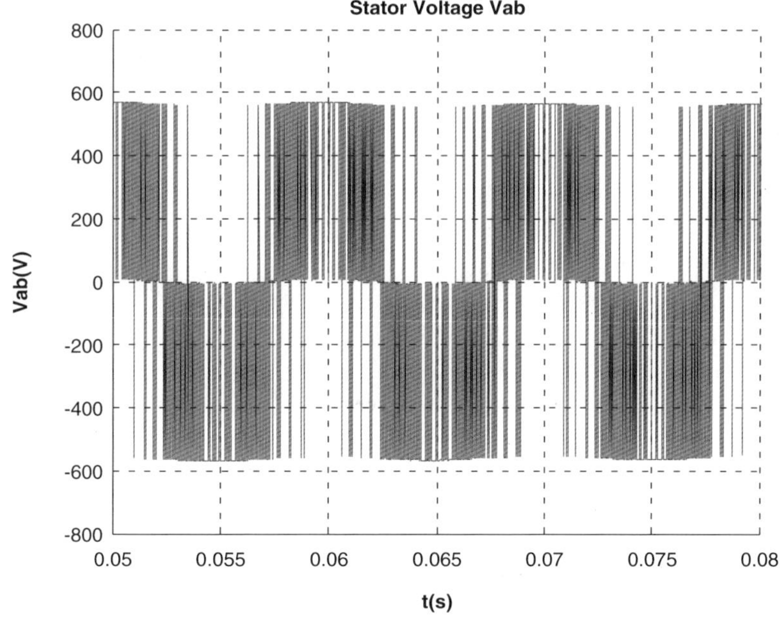

Fig. 6.27. Stator voltage (Example 3).

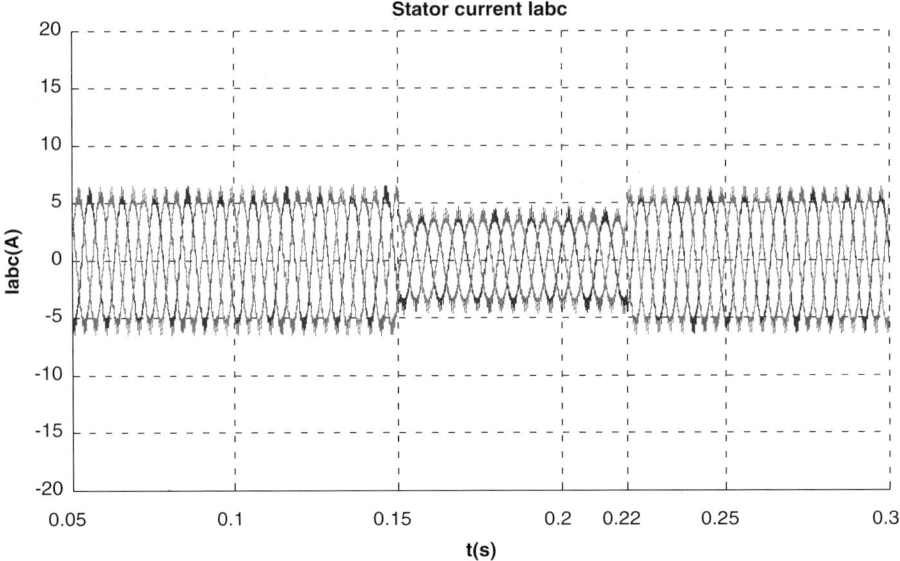

Fig. 6.28.   Stator currents (Example 3).

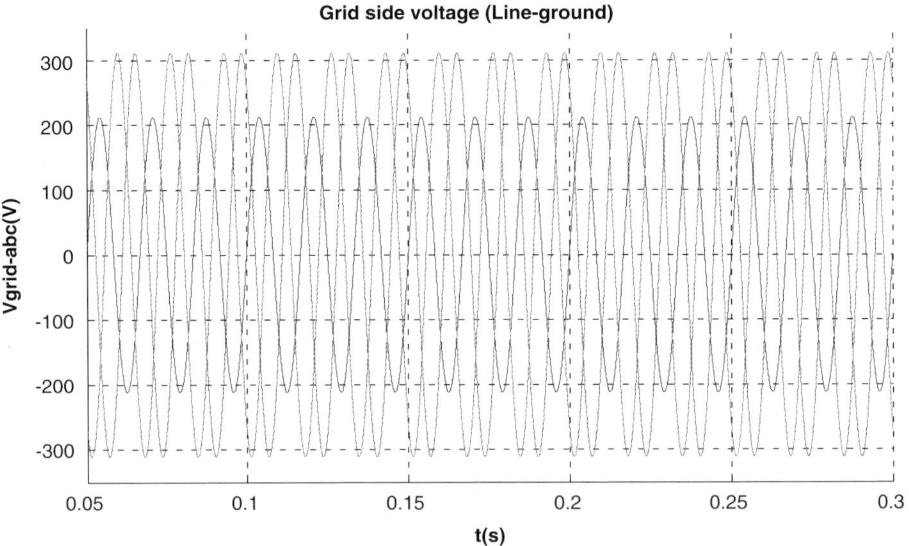

Fig. 6.29.   Grid side voltages (Example 3).

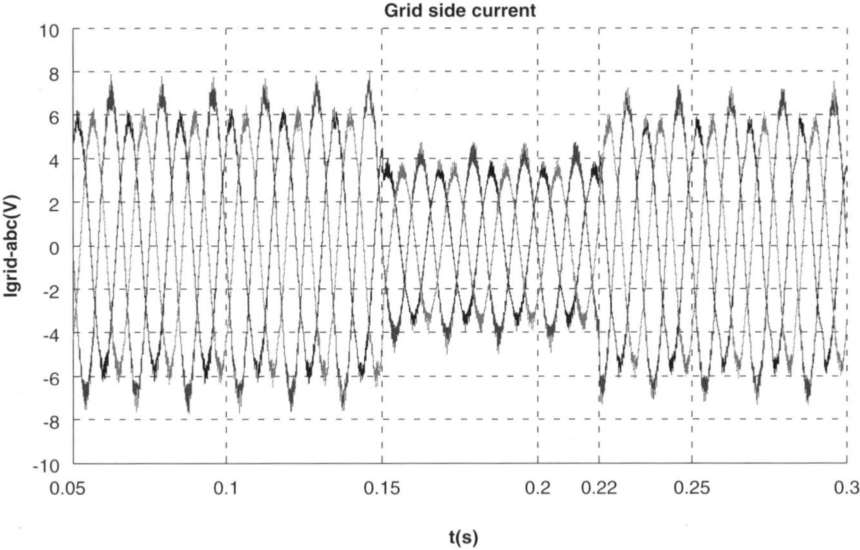

Fig. 6.30.   Grid side currents (Example 3).

voltages, line currents do not contain low order harmonics which was achieved by using the control method shown in Fig. 6.9.

6.9   Concluding Remarks

In this chapter, operation of a line side converter used in variable-speed wind energy conversion systems under balanced and unbalanced grid voltages was analyzed. Control methods under balanced and unbalanced grid voltages were described and simulated. It has been shown that the PWM line side converter can operate under unbalanced grid voltages without injecting harmonic currents into the grid.

References

1. U.S Department of Energy, "20% Wind energy by 2030 report," July 2008, http://www1.eere. energy.gov/windandhydro/pdfs/41869.pdf.
2. J.M. Carrasco, L.G. Franquelo, J.T. Bialasiewicz, E. Galvan, R.C.P. Guisado, A.M. Prats, J.I. Leon and N. Moreno-Alfonso, "Power-electronic systems for the grid integration of renewable energy source: A survey, *IEEE Trans. Industrial Electronics* **53** (2006) 1002–1016.
3. M. Chinchilla, S. Arnaltes and J.C. Burgos, "Control of permanent-magnet generator applied to variable-speed wind-energy systems connected to the grid," *IEEE Trans. Energy Conversion* **21** (2006) 130–135.
4. A. Grauers, "Efficiency of three wind energy generator systems," *IEEE Trans. Energy Conversion* **11** (1996) 650–657.

5. T. Ahmed, M. Nakaoka and K. Nishida, "Advanced control of a boost AC-DC PWM rectifier for variable-speed induction generator," *Applied Power Electronics Conf. and Exposition*, March 2006.

6. J.-I. Jang, Y.-S. Kim and D.-C. Lee, "Active and reactive power control of DFIG for wind energy conversion under unbalanced grid voltage," *Power Electronics and Motion Control Conf.* (2006).

7. T. Sürgevil and E. Akpınar, "Modeling of a 5-kw wind energy conversion system with induction generator and comparison with experimental results," *Int. J. Renewable Energy* **30** (2004) 913–929.

8. I. Schiemenz and M. Stiebler, "Control of a permanent magnet synchronous generator used in a variable speed wind energy system," *IEEE Int. Electric Machines and Drive Conf.* (2001).

9. G. Johnson, *Wind Energy Systems* (Prentice-Hall, Inc., 1985).

10. S. Morimoto, T. Nakamura and Y. Takeda, "Power maximization control of variable-speed wind generation system using permanent magnet synchronous generator," *Electrical Engineering in Japan* **150** (2005) 1573–1579.

11. A.V. Stankovic and K. Chen, "A new control method for input-output harmonic elimination of the PWM boost type rectifier under extreme unbalanced operating conditions," *IEEE Trans. Industrial Electronics* **56** (2009) 2420–2430.

12. A.V. Stankovic and T.A. Lipo, "A novel generalized control method for input output harmonic elimination of the PWM boost type rectifier under simultaneous unbalanced input voltages and input impedances," *32nd Annual Power Electronics Specialists Conf.* (2001).

13. A.V. Stankovic and T.A. Lipo, "A novel control method for input output harmonic elimination of the PWM boost type rectifiers under unbalanced operating conditions," *IEEE APEC 2000* (2000), pp. 413–419.

14. A.V. Stankovic and T.A. Lipo, "A novel control method for input output harmonic elimination of the PWM boost type rectifiers under unbalanced operating conditions," *IEEE Trans. Power Electronics* **16** (2001) 603–611.

15. D. Schreiber, "State of the art of variable speed wind turbines," *11th Int. Symp. Power Electronics — Ee 2001*, Novi Sad, Yugoslavia.

16. A. V. Stankovic and T. A. Lipo, "A novel control method for input-output harmonic elimination of the PWM boost type rectifier under unbalanced operating conditions," *IEEE Trans. Power Electronics* **16** (2001) 603–611.

17. J.W. Dixon and B.T. Ooi, "Indirect current control of a unity power factor sinusoidal current boost type three-phase rectifier," *IEEE Trans. Industry Applications* **35** (1988) 508–515.

18. T.A. Lipo, "Recent progress and development of solid state AC motor drives," *IEEE Trans. on Power Electronics* **3** (1988) 105–117.

19. J.W. Wilson, "The forced-commutated inverter as a regenerative rectifier," *IEEE Trans. Industry Applications* **IA-14** (1978) 335–340.

20. D.M. Brod and D.W. Novotny, "Current Control of VSI-PWM Inverters," *IEEE Trans. Industry Applications* **IA-21** (1984) 769–775.

21. L. Moran, P.D. Ziogas and G. Joos, "Design aspects of synchronous pwm rectifier-inverter system under unbalanced input voltages conditions," *IEEE Trans. Industry Applications* **28** (1992) 1286–1293.

22. A.V. Stankovic, "Unbalanced operation of three-phase boost type rectifiers," in *Handbook of Automotive Power Electronics Motor Drives* (CRC Press, 2005).

23. D. Schreiber, "State of the art of variable speed wind turbines," *11th Int. Symp. Power Electronics — Ee 2001*, Novi Sad, Yugoslavia.

24. R.C. Bansal, "Three-phase self-excited induction generator: An overview," *IEEE Trans. Energy Conversion* **20** (2005) 292–299.

25. http://www.gwec.net/index.php?id=153.

Chapter 7

Wake Effects from Wind Turbines
on Overhead Lines

Brian Wareing

Brian Wareing Tech Ltd,
Overhead Lines and Lightning Protection Consultancy,
Rosewood Cottage, Vounog Hill, Penyffordd,
Chester CH4 0EZ, UK
bwareing@theiet.org

This chapter discusses the effect of wind turbine wake eddies on overhead lines (OHLs) and in particular tower lines close to wind farms. The overall effect of the wake eddies would be to shorten the lifetime of OHL conductors by increased levels of vibration and sub-span oscillations. Since the rapid growth in wind farms has occurred in recent years only, it is likely that conductor fatigue damage has yet to be seen in actual failures but is a potential future problem. The work reported here was carried out on behalf of Scottish Power, a UK utility. It looks at the possibility of conductor damage from wave eddies from a wave mechanics perspective, provides a literature survey of relevant measurements in the field and describes the current state of the investigation. The information in general indicates that the problem is less severe at distances of 300 m from a 3 MW turbine. However, wind turbines are proposed to be installed at just over falling distance (135 m) from the line and the situation is therefore likely to be more severe.

7.1 Introduction

7.1.1 *General*

It has been known for many years that turbulence and wind enhancement and reduction effects mean that wind turbines should not be installed within 2D (where D = rotor diameter) of other turbines across the prevailing wind direction and 8D along the prevailing wind direction, otherwise economic power loss and possible blade problems can occur.[1] Turbulence intensity (which is defined as the relationship of the standard deviation to the mean wind speed) is likely to be greater in the wake eddies of turbines than in classical air flow (flow over the terrain without the wind turbine). It has also been suggested in a study in 1999,[2] that excitation of vibrations in a power line is possible due to wake eddies from a wind turbine. At that time wind turbines were much smaller machines than today. It was found that winds normal to the line in the speed range 1 to 7 m/s are most likely to cause damage.

7.1.2 *Damage potential*

Overhead line (OHL) conductors can be affected by wind induced oscillations in two basic ways:

High frequency (5–60 Hz) vibrations caused by vortex shedding of the wind flow downstream from the conductor (Aeolian vibration). This normally occurs on single conductors at relatively low wind speeds (<5 m/s) normal to the conductor. Amplitudes can be up to the conductor diameter and fatigue often occurs just within the conductor clamp where a "forced" node is present.

Low frequency ($\sim$1 Hz) oscillations caused in bundled conductors. This is known as sub-span oscillation and is caused by the wake induced by the windward conductor(s) of the bundle inducing elliptical movements with amplitudes that can reach the amount of sag in the conductor span or sub-span. It generally requires moderate to strong winds (>8 m/s).

The first problem is likely to occur in the earth wire of the tower whilst the other is common in quad or twin bundled phase conductors. The problem with the wind turbine wake is that consecutive sub-spans may be in totally different wind patterns. The basic situation is that overhead lines use vibration dampers on their conductors in normal "classical" wind flow. Most of the problems due to vibration, however, occur in relatively light winds and so a tower line would normally have vibration dampers installed already and the additional presence of a wind turbine would possibly require additional damping. With a wind turbine nearby there are two basic wind patterns — the relatively undisturbed flow around but away from the turbine blades and the disturbed flow (wave eddies) which passes through the turbine blade envelope. This second area can bring about a major reduction in wind speed when the OHL is downwind of the turbine. This means that for a higher percentage of the time, the OHL conductors will be in an air flow of <5 m/s. This may require the specified further damping for the conductors to survive. A second feature is the boundary between the two air flows where turbulence will occur. This will be at a frequency similar to the rotation speed of the turbine blades (0.5 to 1 Hz) but over a relatively short length of conductor ($\sim$50–70 m). This could mean that, for a line with bundled conductors using spacers, one section (sub-span) could be in a major turbulent flow whilst the next section may not be. This could cause conductor damage at the spacer clamp. The investigation started by Scottish Power aimed to determine where and over what areas these two effects could be significant for the OHL.

7.1.3   *Reasons for concern*

So there are strong concerns that the wake from land based wind farms can induce damaging vibration or low frequency oscillations on overhead lines. This topic has been discussed in many countries world-wide and Germany has put specific recommendations for 132 kV tower and wood pole lines in their CENELEC NNAs EN50341-3-4 and EN50423-3-4. These concern lines within 3D of a wind turbine. Currently, the 3MW turbines being installed in the UK have rotor diameters of around 90 m and the current UK government position is to install 7000 more wind turbines by 2020 with 3000 land based.

7.2   Literature Survey and Review of any Modeling or Field Test Work

7.2.1   *European Wind Energy Conferences (EWEC)*

Smith

A paper by Gillian Smith of Garrad Hassan & Partners (UK)[3] presented at the EWEC 2006 in Athens in March, 2006, covered an advance wake model for very closely spaced turbines. Although directed at determining the appropriate spacing for wind turbines, it did provide a basic modeling and wind tunnel investigation of the wake from a turbine. Figure 7.1 shows the wake of a single turbine in schematic form.

Although this paper concentrates on wake effects on wind turbines, it does note that closely spaced turbines (2-4D apart) lead to significant reductions in the

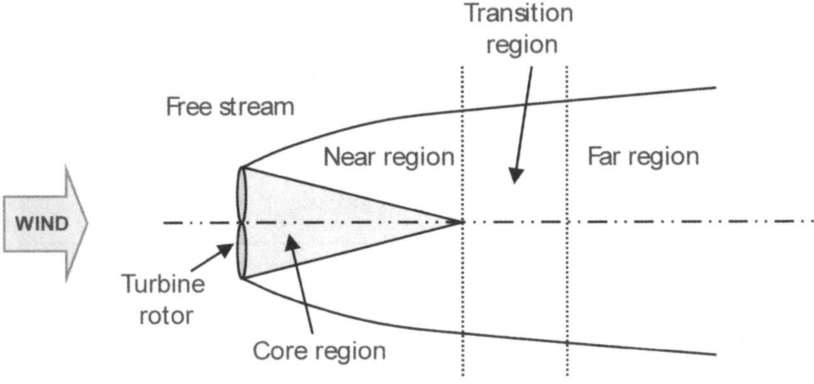

Fig. 7.1.   Wake from a single turbine.[3]

horizontal wind velocity, reductions in turbulence intensity across the wake (see Sec. 3 for an explanation of how turbulence affects OHL conductors) and changes in the wind velocity profile across the wake. All these factors can lead to problems for an OHL conductor at these distances.

Turbulence is generated by:

— Ambient effects
— Momentum gradients in the wake
— The rotor

In Fig. 7.1, the core region generally has stable but reduced wind speeds. However, at the boundary of this with the near wake region there are major disturbances from tip vortices. This combines with the free stream turbulence. The core region is between 1 and 3D long. After this region, there is a narrow "transition" region before the far wake takes effect. In this area the ambient turbulence dominates and the effect of other turbines can reduce wind speeds significantly. The paper suggests that average wind speeds can be 9 to 13% lower at 8D from a row of turbines.

The presentation associated with the paper gives wind velocity profiles at distances of 2.5D, 5D and 10D from a turbine. These are shown in Figs. 7.2 to 7.4. It can be seen that for a 2 MW turbine with D = 80 m, a dip in wind speed occurs over a width of around 100 m at 2.5D and over 200 m at 10D. The maximum dip at 2.5D is a reduction in wind speed of over 60%, compared with 35% at 5D and only 20% at 10D. The significance of this is that there will be an increase in the periods of wind with velocities below 5 m/s in which region vibration effects occur.

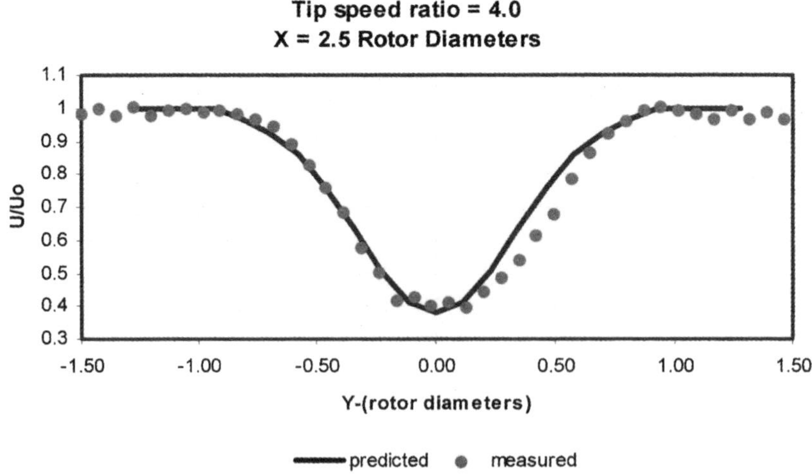

Fig. 7.2. Wind speed reduction at 2.5D behind a wind turbine.[3]

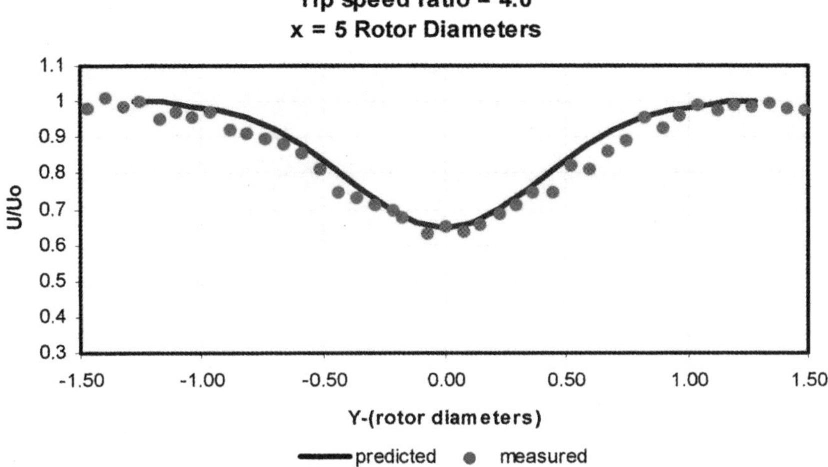

Fig. 7.3.   Wind speed reduction at 5D behind a wind turbine.[3]

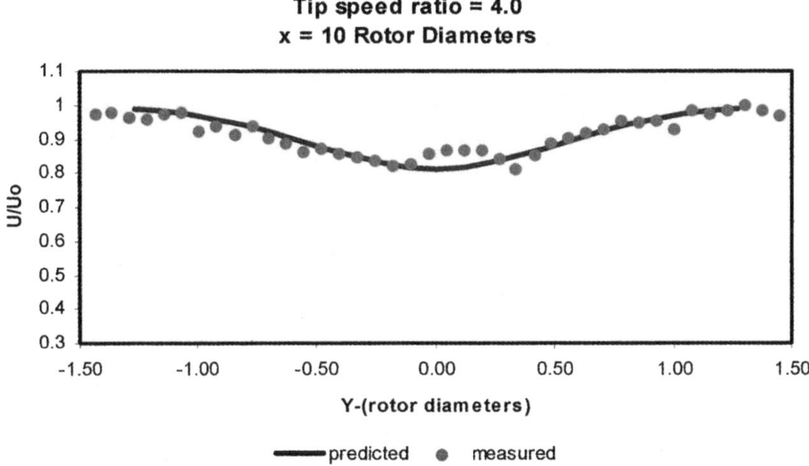

Fig. 7.4.   Wind speed reduction at 10D behind a wind turbine.[3]

A further point is that adjoining sub-spans (distance between spacers on twin or bundled conductors) could have substantially different wind speeds.

Turbulence

A paper by Wessel[4] calculated that turbulence factors behind 2MW turbines spaced at 8.1D and 9.1D at between 5.8 and 7.2%. These are lower than the levels assumed by Lilien.[5] Jiminez[6] looked at eddy simulations and compared his calculations with measurements at a Netherlands wind farm. He argues that the turbulence situation

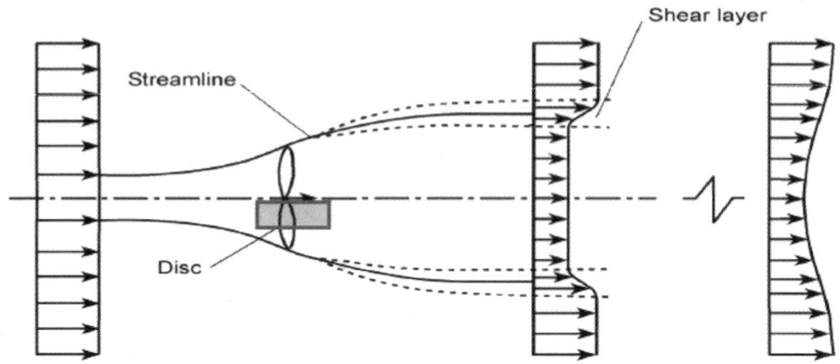

Fig. 7.5.   Schematic representation of a wind turbine wake.[6]

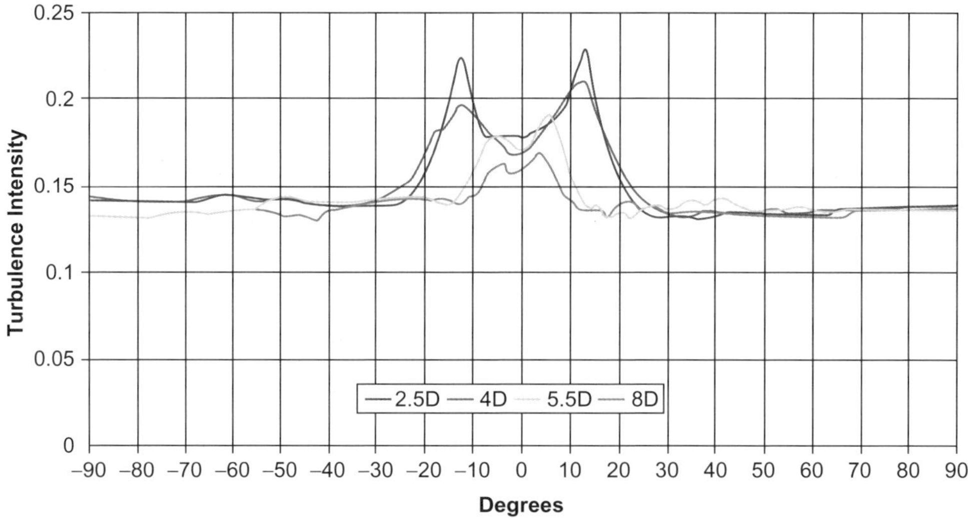

Fig. 7.6.   Turbulence intensity in a horizontal plane at hub height.[6]

arises only in the area between the flow outside the wake and inside the wake where the flow is significantly retarded (Fig. 7.5). He estimates that the wake effect is significant at 8D and does not become completely negligible until 15–20D behind the turbine. Practical measurements were made behind a wind turbine with a 30 m diameter rotor with anemometers at 75, 165 and 200 m behind the turbine. The data is shown in Fig. 7.6 in a horizontal profile at hub height. It can be seen that the turbulence is quite high and wide at 4D but becomes narrower (10° angle from hub) at 5.5D.

Figure 7.7 shows the same data as Fig. 7.6 but in a vertical configuration. The characteristic changes dramatically between 2.5D and 4D. This is the effect of

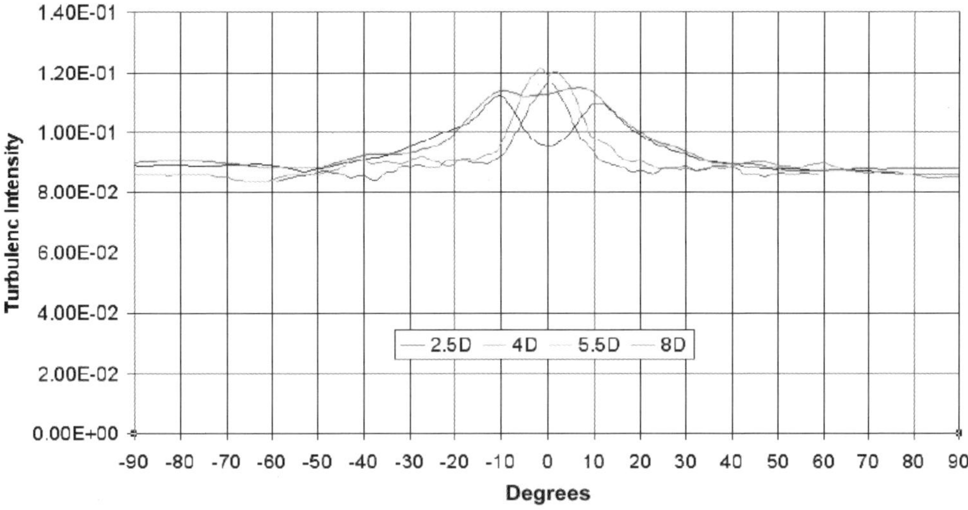

Fig. 7.7.   Turbulence intensity in a vertical direction at hub height.[6]

moving out of the "near wake" zone. The turbulence then changes significantly again between 4D and 5.5D, exhibiting the same narrowing of the wake effect zone.

Turbulence intensity remains high at hub height but only in a decreasing cone angle (10° above and below hub height at 8D). At 4D the conical turbulence envelope has an angle of over 30° around the hub height. The significance of this is in the angle found in this data compared with that used in the Lilien paper[5] and aids the knowledge of how far away from the turbine the overhead line is subject to turbulence.

The horizontal and vertical widths of a significant wake decrease as the distance from the turbine increases, but this decrease does not occur in a linear manner. The data from these graphs has been put into an Excel spreadsheet and re-drawn as a distance profile related to the spread angle.

7.2.2   *North Rhine-Westphalian study*

A study was initiated into the effect of wake eddies from wind turbines on overhead lines by the North Rhine-Westphalian Administration. The results were published in *Elektrizitäts Wirkshaft*.[2] The paper uses a mean wind speed of 7.7 m/s and looks at the wind effects at 50 m height (assumed nacelle height) above ground. For an average 400 kV tower, this point would be around 11 m above the average height of the earth wire and 19 m above the upper phase conductors. It is interesting to note that using an angle of 6° quoted in the paper for the angle, β, shown in Fig. 7.8, the lower edge of the wake for a GE Energy 1.5 MWsle turbine (nacelle at 61.4 m, rotors 80 m diameter)[7] would reach the ground after 3D distance. This is in contradiction

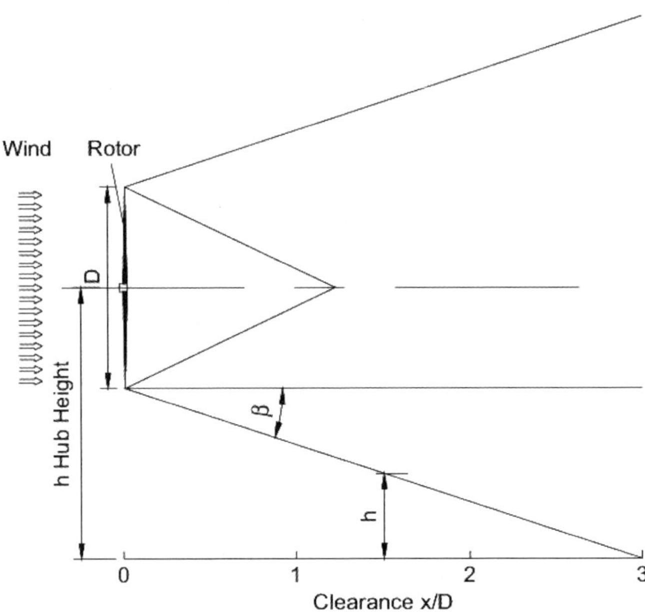

Fig. 7.8.   Cone of wake eddy.[2]

Table 7.1.   Reduction in wind speed, additional turbulence intensity and damage potential in the wake.

Distance (x/D)	Fall in wind speed (V_w/V_o)	Additional turbulence intensity (I_{add})	Damage potential	
			without wind turbine	with wind turbine
1.5	0.4	5%	12	45
2.5	0.4	5%	12	45
3.5	0.5	5%	12	32

to the data measured in Sec. 7.2.1 but it is used to generate the recommendations in the CENELEC standard EN50423-3-4. The paper estimates that in the center of the wake the mean wind speed could be reduced to between 0.4 and 0.6 times the classical air flow velocity. This is in agreement with other data.

The wind speeds relative to classical air flow velocity were estimated and are shown in Table 7.1 as a function of distance. The worst conductor oscillation amplitudes are observed at low turbulence intensity since this allows a steady laminar air flow. Since there will always be some turbulence over a "normal" landscape (i.e., not a flat desert or expanse of water), a minimum value additional turbulence intensity of $I_{add} = 5\%$ has been assumed.

The table also includes a value for damage potential. This is defined as being inversely proportional to time to fatigue failure of conductors. The paper states that this has been evaluated assuming a linear degradation and use of the Cigré S-N curve. The implications in this data and the damage potential values are that at 3.5D distant, the lifetime of the conductors at any height would be significantly reduced. This does not equate with the final recommendations that OHLs would be safe beyond this point.

The paper also does not attempt to determine at what frequency the wind induced oscillations would be. It does imply that wind speeds in the wake of a wind turbine are lower in comparison to classical wind speeds. If this pushes mean speeds of 7.7 m/s into the <5 m/s region then conductor vibration will increase. It is therefore possible that OHLs will experience greater stresses due to oscillation generated by eddies than when wind turbines are not present. This is excluding the problem recognized by Lilien[5] about low frequency (<3 Hz) oscillations. Accordingly, this means that even at a distance of 3D from the wind turbine rotors the damage potential is greater than if the wind turbine was not there. The paper concedes that there will be a shortening of the expected life span of overhead conductors in the wake of a turbine even at 3.5D, but declines to be more specific, stating that line design would affect the situation.

Two things can be noted from Figs. 7.9 and 7.10, which show the comparison of measured and calculated wind speed and turbulence intensity from Stefanatos[8] in

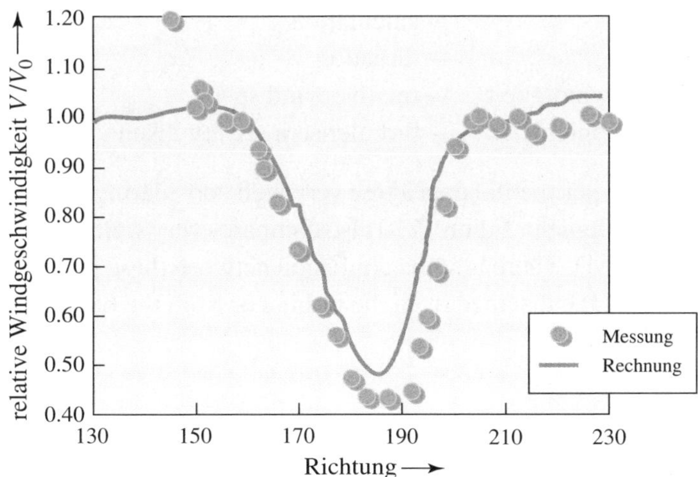

Fig. 7.9.   Horizontal profile of the wind speed at the height of the rotor hub at a distance of 2D.[8]

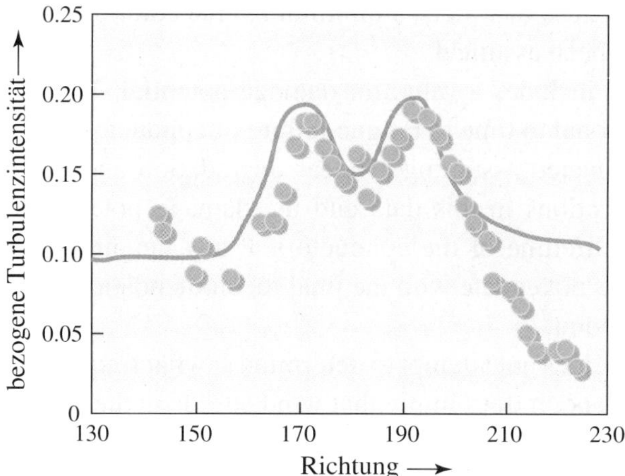

Fig. 7.10.   Horizontal profile of the turbulence intensity at the height of the rotor hub at a distance of 2D.[8]

EWES 1996. The first is that at 2D the turbulence is indeed at a peak at the interface of the wake and the undisturbed air mass. This is at a width of $2 \times \tan 10°$ to $30°$. The wind speed drops to 0.45 of the classical flow at hub height center (in agreement with earlier data) and is reduced over an overall width of $2 \times \tan 20°$ (where $x = 2D$) or 116 m for an 80 m rotor.

In Figs. 7.9 and 7.10 the German parameters are:

Messung   — measurement
Rechnung   — calculation
Richtung   — direction
relative Windgeschwindigkeit   — relative wind speed
bezogene Turbulenzintensität   — turbulence intensity taken

Modeling and experimental data agree very well considering the normal fluctuations in wind measurement. Figure 7.10 also compares very well with Fig. 7.3 which is at a distance of 2.5D. There is some confusion between these papers whether they are referring to D as the distance from the turbine or from the blade tip in its closest position.

7.2.3   *University of Liège paper*[5]

Lilien gives the equation for the mechanical power from a wind turbine in watts:

$$P = \frac{1}{2}\rho_{\mathrm{air}} C_x V^3 A$$

where A is the swept area (m^2),

V the wind speed (m/s),

C_x the power coefficient,

ρ_{air} the air density (kg/m^3).

He states that C_x is around maximum (0.6) when the downstream wind speed is 20 to 40% of the upstream wind speed. Data from Denmark indicates that the minimum operational wind speed is 4 m/s and the average is 8 m/s. This implies that for lines behind a stationary turbine, winds <4 m/s will go through mainly undisturbed whereas wind speeds around the average will be reduced to around 2–4 m/s. This will significantly increase the amount of time that an OHL downstream of the turbine will spend in winds <4 m/s. As vibration in OHL conductors occurs mainly below 5 m/s, a significant increase in the number of periods (not necessarily intensity) of damaging vibration cycles will occur, leading to reduced conductor lifetime before fatigue causes the conductor to become fatally weak and break at the fittings. As the rotor speed is around 30 to 120 rpm, there is also the possibility of low frequency oscillations at <3 Hz. Lilien looks at three areas:

— The frequency profile of the wind speed.
— The transient response of an OHL conductor.
— The consequences of low frequency oscillations on the OHL conductor.

The frequency response of the wind speed was found to be mainly below 0.1 Hz with very little energy above 1 Hz in the turbulent flow. The wake size downstream of the turbine was estimated to expand at an angle of around 3°, about half the value used in the North Rhine-Westphalian study.

Lilien then looked at the OHL response based on a 32 mm diameter, 61 strand ACSR conductor with a weight of 1.7 kg/m. There is no exact equivalent in the UK sizes but this is equivalent to a 500 mm^2 conductor. An undisturbed wind speed (10 minute average) of 20 m/s was used which effectively reduced to 10 m/s downstream of the turbine. Based on a 200 m span and a sag of 2.8 m, conductor oscillations of up to 150 mm amplitude (5 × diameter) at 0.5 Hz were calculated from SAMCEF software. The conductor tension was 28.6 kN. Lilien then used the basic Cigré fatigue expressions which relate the total number of cycles of vibration before 10% of the strands suffer fatigue failure.[9–11] The bending constraint at the failure point (normally a few mm into the clamp) is obtained from the standard Poffenberger–Swartz formula. This shows that the damage potential is directly related to the conductor diameter, the frequency, amplitude and the square root of the mass. Using the data provided the maximum number of cycles to failure was estimated at 8.10^5 when one would generally expect values 100 times this figure. The actual lifetime depends on how much time the line spends in such conditions (the wind direction and speed will have an effect) but it is certainly a low enough figure to cause concern.

In this paper, Lilien[5] did not consider the increased Aeolian vibration levels due to the wind speed reduction downstream. This is discussed later in this chapter. In order to obtain the effect of this, the data from the other papers can be applied to the locally measured wind profile at any particular OHL site (or close to) with no turbines present.

7.2.4　*Data comparison*

Comparing the Stefanatos[8] and Jiminez[6] data, both give turbulence intensity at hub height against the angle of the measured/modeled point from the turbine. Figure 7.11 shows that turbulence intensity can be significantly high (50% above the undisturbed flow) at 4D over an angular width of 30° (equivalent to around 170 m for an 80 m blade). However, at 5.5D the effect is negligible. The Stefanatos measured data at 2D looks fairly close to the Jiminez data and so the Stefanatos wind data (shown in the same figure) should be trustworthy.

The wind data at a distance of 2D shows a major reduction (>50% on the undisturbed flow) over an angle of 15° (around 43 m for an 80 m blade) and a significant reduction (>30% on the undisturbed flow) over an angle of 25° (around 75 m). It is necessary here to decide how much of a reduction is significant in increasing the amount of wind in the vibration region (<5 m/s) compared with the undisturbed flow. If vibration dampers are required in normal undisturbed

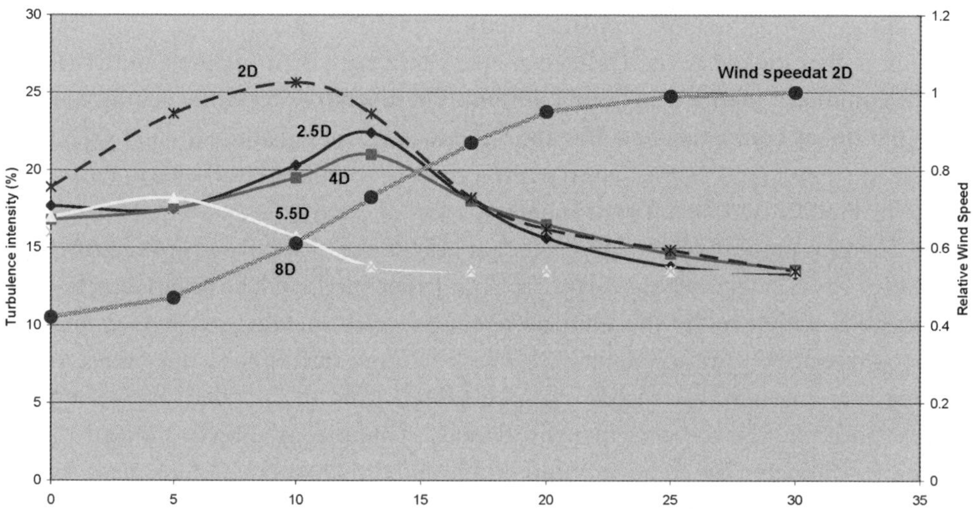

Fig. 7.11.　Comparison of the Jiminez data at distances of 2.5D, 4D, 5.5D and 8D with the Stefanatos data at 2D for turbulence intensity against angle from the turbine.

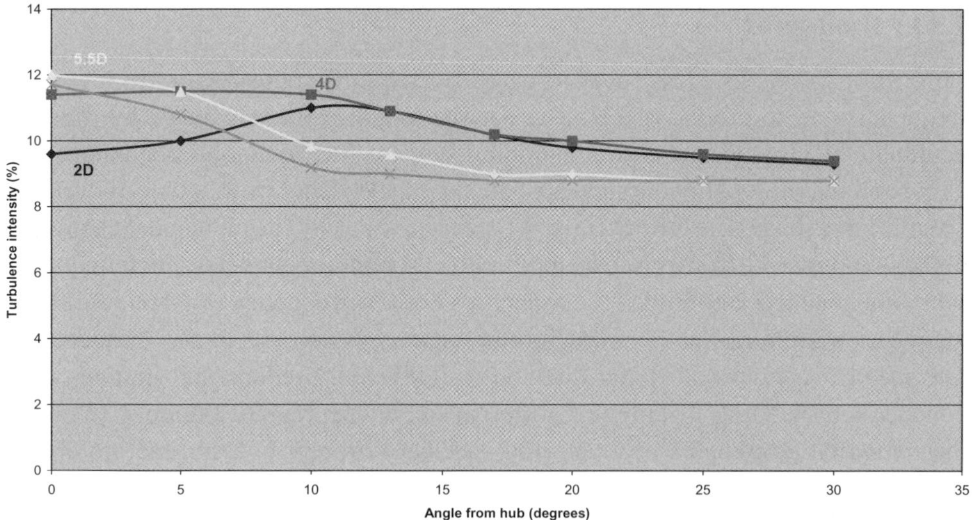

Fig. 7.12.   Vertical turbulence intensity against angle from hub.

flow then how much more damping is required if the average wind speed is reduced by 30% or 50%. This will require some basic validation in order to make sure that the wake-induced frequencies are fully damped out by the damping system.

A second feature is the height at which these wind effects occur. Jiminez[6] has measured vertical wind profiles (Fig. 7.7). Figure 7.12 looks at these in more detail.

Noting from the previous data that there are likely to be problems up to turbine-OHL distances up to around 4D (320 m for an 80 m blade), the turbulence is 30% above background at $61.4 \times (1 - \tan 10°) = 50.6$ m above the ground (approximate height of an L6 tower) and 11% above background at a height of 36 m (approximate height of the upper phase on an L2 tower). The turbulence width at this level would be around 270 m. The middle phase on an L12 tower is around 31 m which would have an increase in turbulence intensity of around 8%. However, looking at the data for a distance of 5.5D, the turbulence width drops down to virtually zero and the wind reduction (from Smith[3]) is around 30% over a width of around 40 m and 20% over a width of around 60 m. It appears, therefore that whilst turbulence (low frequency effects) is significant around 4D, it virtually disappears by 5.5D. The main effect will then be wind reduction over a limited width of 20–30%. This may still require additional vibration dampers.

7.3   Effect of Wind Speed and Turbulence on Overhead Lines

7.3.1   *Wind speed*

Low wind speeds (<5 m/s) tend to cause overhead line conductors to vibrate at frequencies up to around 60 Hz. The damage done to the conductor by repeated bending close to and just within clamping systems has been well documented by Cigré and the limits of endurance are calculated according the number of cycles at specific stress levels, i.e., the S-N (or Wöhler) curves.[11,12] The movement changes at higher wind speeds, from this low amplitude (∼ conductor diameter), high frequency vibration to a high amplitude (∼ conductor sag), low frequency (<3 Hz) oscillation known as galloping. In the wind turbine case any increase in the frequency of low speed winds normal to the line could significantly reduce the lifetime of the conductor by leading to fatigue failures in the clamp region. Figure 7.13 shows that vibration amplitudes peak at wind speeds of around 1–2 m/s and are almost insignificant at wind speeds above 4 m/s.

Figure 7.14 shows the same conditions close to a clamp or fitting. The forced "node" causes the amplitude peak to move to between 2 and 3 m/s wind speeds but the damage effect is still insignificant above 5 m/s wind speed.

Cigré calculations on conductor lifetime are based on vibration monitor data and the S-N curve in the range of normally present wind speeds in the area. If this range of wind speeds is significantly altered then the lifetime could be affected. Changing from a scenario of a "normal" spread of wind speeds with only a small proportion

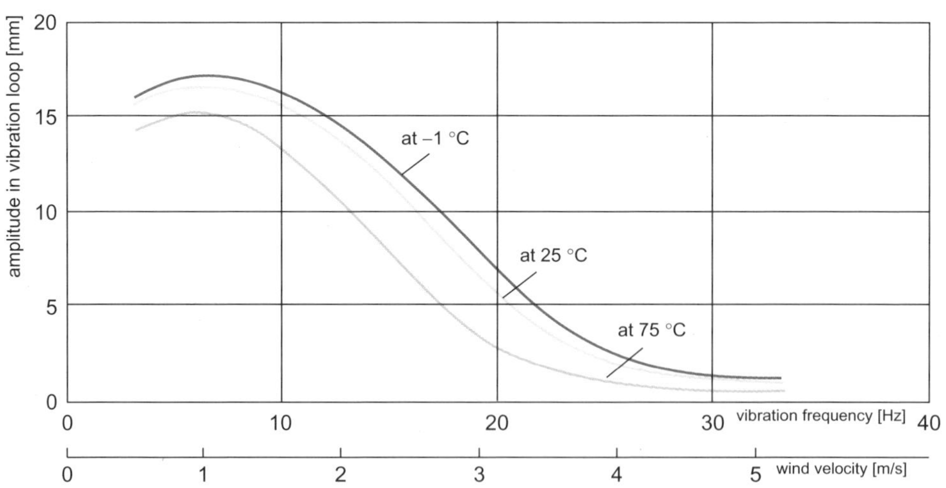

Vibration Amplitude of the undamped conductor

Fig. 7.13.   Vibration amplitude against wind speed.[6]

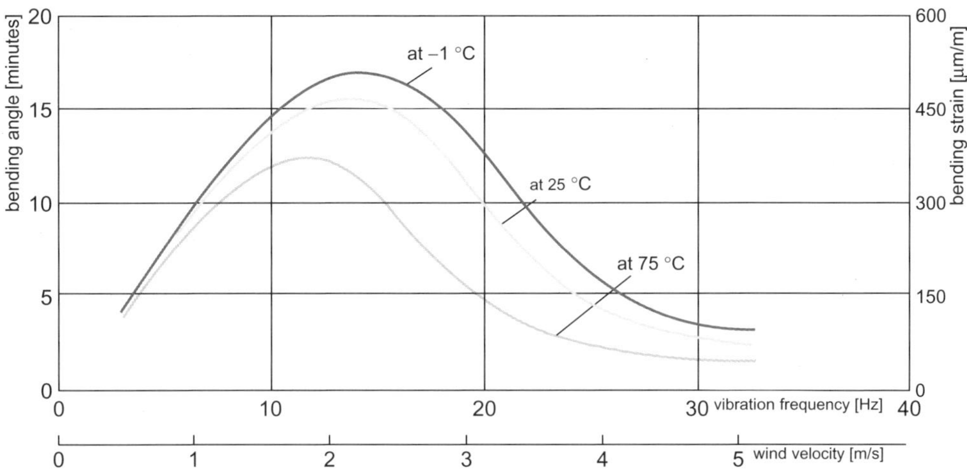

Fig. 7.14.   Bending angle and strain in a conductor against wind speed (Courtesy: Pfisterer).

below 5 m/s to a regime where this proportion is significantly increased is likely to increase damage levels in the OHL conductor. The amount can only be determined by monitoring the vibration levels for a period long enough to be typical (normally around 3 months).

7.3.2 *Turbulence*

Turbulence occurs when objects disrupt the normal laminar flow of wind in the area of the (in this case) OHL conductor. Turbulence occurs naturally close to the ground due to the presence of trees, hedgerows, houses, etc. It is normally classified in five types (0 to 4) according to terrain types in IEC standards (Table 7.2). In the UK, the most common terrain type is 2–3 where trees and isolated houses disrupt the air flow in the lower 50 m or so height above ground level.[13]

Table 7.3 translates these terrain factors into turbulence factors. Obviously more turbulence means less consistent flow over a conductor and so less vibration. The conductor movement becomes erratic and not regular and this is less damaging.

Cigré data[14] indicates that a reduction of turbulence from the 15–22% range to 8% can increase vibration levels by up to 20% and thereby have a significant effect on conductor life. In some countries where snow cover is present on hills for long periods during the winter, the terrain roughness can reduce from a typical type 2 to type 1 terrain.

Table 7.2.   Terrain categories.

Terrain category	Terrain description
0	Smooth, flat terrain
1	Open, flat, no trees, no obstruction or near/across water or desert, snow cover
2	Open, flat rural areas with no obstructions and few and low obstacles, e.g., farmland
3	Open, flat or undulating with very few obstacles, e.g., open grass or farmland with few trees, hedgerows or other barriers, prairie, tundra.
4	Built-up areas with some trees and buildings, e.g., suburbs, small towns, woodlands and shrubs, broken country with large trees, small fields with hedges.

Table 7.3.   Turbulence intensity against terrain roughness.[13]

Terrain factor	Terrain description	Turbulence (%)	Turbulence intensity
1	Open water or desert, snow cover, no trees, no obstructions	8	0.11
2	Open, flat rural areas with no obstructions and few and low obstacles	15	0.18
3	Open, flat or undulating. Low density housing, open woodland with hedgerows and small trees, prairie, tundra.	22	0.25
4	Built-up areas with some trees and buildings, e.g., suburbs, small towns, woodlands and shrubs, broken country with large trees, small fields with hedges.	30	0.35

7.3.3 *Vibration damage*

General

Overhead line conductors may be subject to severe Aeolian vibrations arising from the alternating forces caused by vortex shedding. The conductor vulnerability to damage increases with conductor tension and this damage generally occurs at suspension or damper clamps where the conductor is forced into a node.[15] This can result in many bending cycles of high enough amplitude to initiate fractures in the conductor strands. Once these develop sufficiently (normally about 1/3 diameter) the strands tend to break under elastic tensile stress. To avoid this scenario, the historical technique has been to limit the conductor tension to below a specific value related to the conductor material.

The fatigue behavior of stranded conductors is highly complex. It does result from dynamic stresses developed by the alternate bending of the conductors at clamps but the effects of these stresses are highly aggravated by fretting at interfaces between the strands themselves or between strands and clamps. A thorough review of this whole issue is presented in Chap. 2, "Fatigue of Overhead Conductors" of the EPRI Transmission Line Reference Book, *Wind-Induced Conductor Motion*.[16]

The fatigue characteristics of stranded conductors are highly scattered. Hence, a statistical treatment to conductor fatigue test data has to be applied to get a better quantitative insight into the phenomenon. The results take the form of fatigue endurance curves (the so-called S-N curves).

Test data[16] together with the corresponding log-mean S-N curve and the 95% confidence interval have been presented.[17] Safe limit lines for both multi-layer and single-layer ACSR conductors were also proposed by Cigré Working Group 22.04.[12] Identified as safe border lines, they are also expressed in terms of stress as a function of number of cycles to strand failure. Figure 7.15 depicts the log-mean S-N curve and the safe limit line[17] as well as the Cigré safe border line corresponding to multi-layer ACSR conductors. It may be appreciated that the Cigré safe border line is not

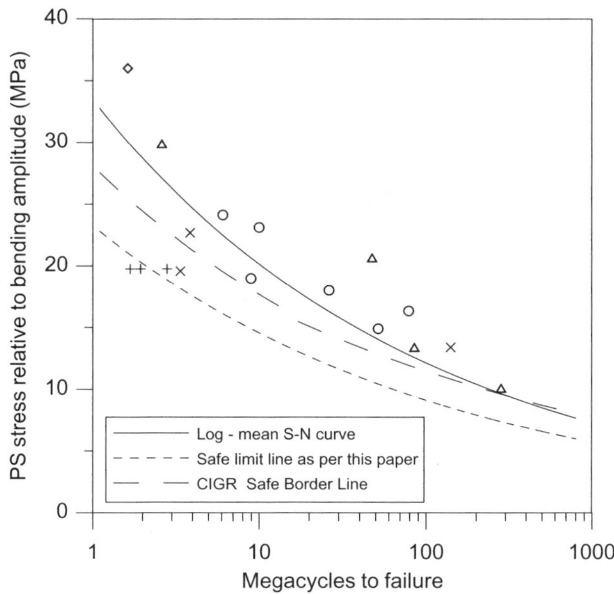

Fig. 7.15.   Fatigue characteristics of multi-layer ACSR relative to bending amplitude.

really safe as it stands well on the right-hand side of the safe limit line derived from the paper.[17] It actually lies close to the log-mean S-N curve at the lower stress levels, which translates into a probability of survival of 50% only.

Assessing vibration severity

There are two ways of evaluating vibration severity:

Comparing the maximum bending stress with the fatigue endurance limit of the conductor.

Miner's rule[15]

The first method assumes that, as long as the fatigue endurance limit is not reached, then the conductor will last forever! The second method is more commonly applied. Here, the accumulation of fatigue damage over a conductor lifetime (calculated to first strand breakage) is evaluated against conductor type. This leads to the expression:

$$D_L = \sum_i \frac{n_i}{N_i},\tag{7.1}$$

where D_L = total damage per year

n_I = number of cycles at a stress level, σ_I, during one year

N_I = number of cycles to failure at stress, σ_i.

$D_L = 1$ is the end-of-life situation. The value of σ_I is dependent on the vibrational frequency (f) and amplitude (Y) For AAC and ACSR conductor D_L is proportional to f^8 and Y^7. The frequency (f) is directly related to the wind speed by the Strondal expression:

$$St = fD/V,\tag{7.2}$$

where V is the wind speed.

Cigré method

The Cigré method[18] of determining a conductor lifetime is based on a combination of the vibration amplitude, the frequency at which the various amplitudes occur and the total number of vibration cycles. This has been evaluated for the most common conductor types (ACSR, AAAC, etc.) and has resulted in the "safe" borderline S-N (stress against number of cycles) curve as shown in Figs. 7.15 and 7.16. The Cigré "safe" borderline is thus applicable to all stranded conductors and all types of clamps. It is given by:

$$\sigma F = C \times N^Z$$

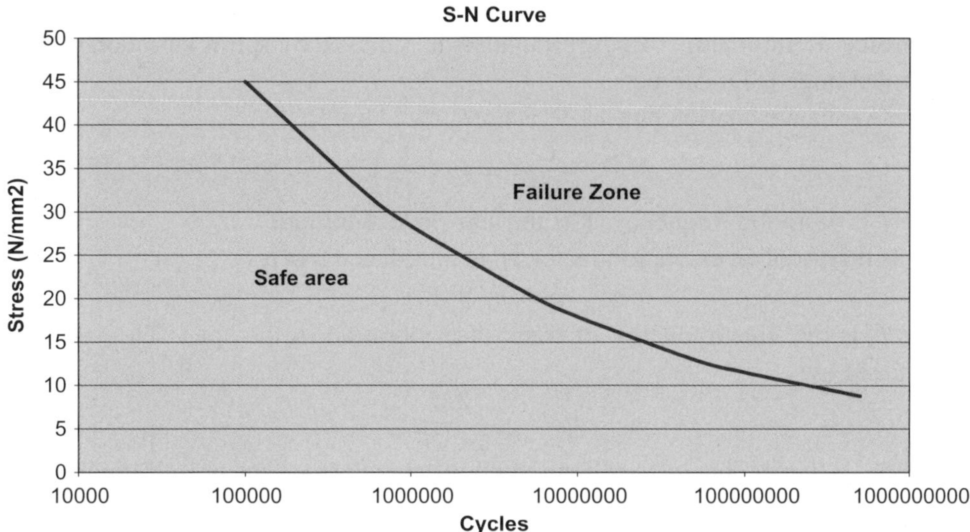

Fig. 7.16. Typical S-N curve showing the 50% failure probability limit for conductor vibration.

where

σ_F is the dynamic stress (N/mm^2) at the frequency, F in Hz,

N is the number of cycles to failure

and

$$C = 450, \quad z = -0.2, \quad \text{for } N < 2 \times 10^7,$$
$$C = 264, \quad z = -0.17, \quad \text{for } N > 2 \times 10^7.$$

As mentioned, this curve represents the 50% probability of failure and the measured S-N curve for a conductor should be well to the left of the curve in the "safe area". After monitoring the above parameters as well as meteorological data over a specified test period, the basic vibration performance is extrapolated to provide a life time value for the conductor at the specified tension level.

7.3.4 *Modeling Aeolian vibration*

The energy put into the conductor by vortex shedding can be balanced in the conductor damping to result in a steady state oscillation with a range of amplitudes and frequencies. Conductors possess self-damping properties where small relative movement between the strands loses energy by frictional effects. This has been applied to similar strand conductors, such as AAC (All Aluminium) and AAAC (All

Aluminium Alloy) and also to dissimilar strands, such as in ACSR (Aluminium Conductor Steel Reinforced).[19] Energy can also be removed by spiral vibration (SVD) and Stockbridge (SD) dampers.

The Aeolian excitation power, P, is normalized by:

$$\frac{P}{L} \cdot \frac{1}{f^3 D^4} = fn\left(\frac{Y}{D}\right), \tag{7.3}$$

where f is vibration frequency, Y is the anti-node amplitude.

A further Aeolian excitation factor, η, is introduced where:

$$\eta = P/W.E, \tag{7.4}$$

where E is the kinetic energy of transverse vibration in the span. The factor is normalized by:

$$\eta_n = \eta \cdot \frac{2M}{\rho_{air} D^2}, \tag{7.5}$$

where ρ_{air} is the air density. Aeolian vibrations are related to the normal component of the wind intensity of turbulence, I_o, and so η_n can be expressed as a family of curves of Y/D for specific values of I_o. Combining these with the terrain factors, a family of curves that allow the safe working tension to be determined can be obtained. These allow the correct type and/or number of SD dampers to be recommended.

7.4 CENELEC Standards

7.4.1 *EN50341*

EN50341 (2004) is the European CENELEC design standard for OHLs operated above 45kV. Germany has put the following paragraph as §5.4.5 DE.2 in their annexe EN50341-3-4 2004:

5.4.5 DE.2 Clearance to wind energy converters

Between the rotor blade tip of the wind energy converter in the most unfavorable position and the closest conductor of the overhead line in still air the following clearances shall be obeyed:

- More than or equal to three times the rotor diameter if the conductors are not damped against wind induced vibrations.
- More than one times rotor diameter if the conductors are damped against wind induced vibrations.
- If it is ensured that the overhead transmission line is outside the wake of the wind energy converter and the distance between the rotor blade tip in the most unfavorable position and the closest conductor is more than one times rotor diameter, it is possible to avoid the damping of the conductors.

Furthermore, the rotor blade tip is not allowed to project over the right of way along the overhead line.

The above clause refers to "wind induced vibrations" and is rather open in stating that damping should be provided. It is standard practice to damp OHL transmission conductors against Aeolian vibration. The devices commonly used are Stockbridge dampers and spacer dampers. These dampers are efficient only within specific vibration bands — typically around 30 and 60 Hz depending on the damper and conductor. What is not clear in this standard is whether **additional** damping because of the presence of the wind turbine is required. It also does not refer to any low frequency oscillations.

7.4.2 *EN50423*

EN50423 (2004) is the European CENELEC design standard for OHLs operated below 45 kV. Germany has put the following paragraph[20] as §5.4.5 DE.2 in their annexe EN50423-3-4 2004:

5.4.5 DE.2 Clearances to wind energy converters

In addition to the clearances specified in EN 50341-3-4, Clause 5.4.5/DE.2, the following stipulations apply to overhead electrical lines less than AC 45 kV:

- The rotor blade tip shall not penetrate into the right-of-way of the overhead electrical line.
- The clearances specified for overhead electrical lines exceeding AC 45 kV apply also to overhead electrical lines less than AC 45 kV if they are installed on structures which are typical for overhead electrical lines for AC 110 kV (see EN 50341-3-4, Clause 5.4.5/DE.2).

If it is guaranteed that the overhead electrical line is situated outside the wake of the wind energy converter, a horizontal clearance according to EN 50423-1, Table 5.4.5.2 (overhead electrical line adjacent to buildings) shall be met between the swung outermost conductor of the overhead electrical line not exceeding AC 45 kV and the most unfavorable position of the rotor blade. The wake of the wind energy converter is determined by its lower edge (see Fig. 7.8 earlier in this chapter):

$$-h = h_{\mathrm{hub}} - D/2 - 0, \quad 1x$$

where h_{hub} height of rotor hub above ground, D rotor blade diameter, x distance from energy converter axis.

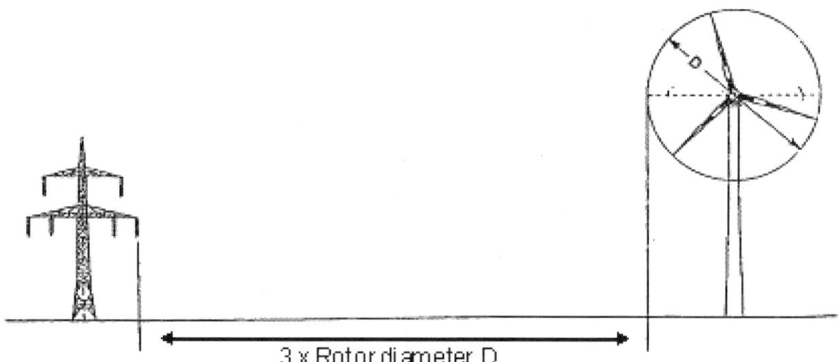

Fig. 7.17.   Example of 3D for undamped lines.[21]

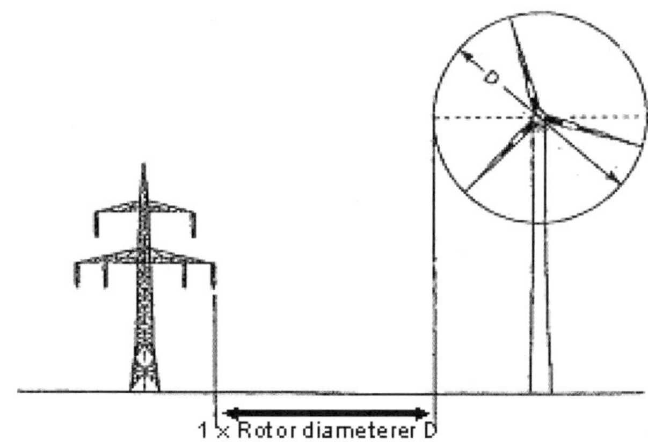

Fig. 7.18.   Example of 1D for damped lines.[21]

7.4.3 *German standard S(1) 14*

Section S(1) 14[21] of the German standard covers wind effects from wind turbines. Figures 7.17 and 7.18 come from this standard. No mention is given concerning transmission tower heights and the figures merely confirm the paragraph in EN50341-3-4.

In effect, this paper adds nothing to the current knowledge based on the North Rhine-Westphalian Administration study.

7.5   Wind Tunnel Results

7.5.1 *Wind tunnel tests using a model turbine*

A wind tunnel was used to investigate the wave eddy and wind speed reduction phenomena of a wind turbine over a range of wind speeds and modes of operation

Fig. 7.19.   The Milan Polytechnic wind tunnel and the model turbine.

and orientation.[22] The tests were carried out at Milan Polytechnic which has a world standard wind tunnels test facility (Fig. 7.19). The scale of the wind turbine model used in the tests was 1:50, based on a Vesta V90-3.0MW turbine.[23] Flow patterns were measured at various distances from the turbine at different wind velocities. Visualization techniques using fog and helium bubbles were also employed. Tests were performed on the aeroelastic response a 1:50 model of a simulated OHL conductor (Araucaria), reproducing the dynamic characteristics for a 400 m long span. The motion of the cable was measured in 5 points along the line using infrared cameras.

7.5.2 *Wind speed and turbulence intensity tests*

Turbulent flow

A series of tests were carried out for smooth and turbulent flow. Turbulent flow is more significant for the practical aspects of the wind turbine/OHL interaction as the turbulence intensity (TI) values will be comparable to those experienced at land based wind farm locations.

Effect of wind speed

Figures 7.20 and 7.21 show the effect of wind speed (by changing the tip speed ratio (TSR), higher TSR = lower wind speed) at tip height on wind speed and TI.

The wind velocity variation is lower in the turbulent flow (TF) condition than smooth flow (SF). The percentage reduction in wind speed at tip height appears to be independent of the wind speed. However, there is a significant drop in TI at tip height at the higher wind speed. Effectively, therefore, an increase in wind speed of

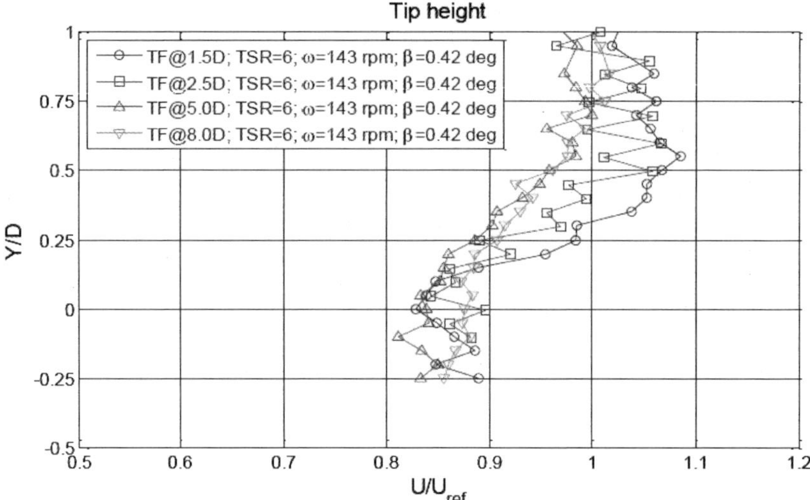

Fig. 7.20.   Wind speed effect on wind profile at tip height in turbulent flow.

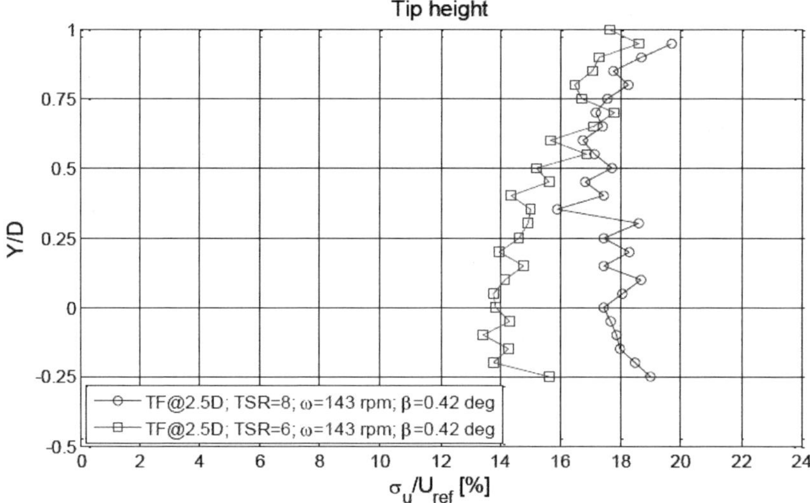

Fig. 7.21.   Wind speed effect on wind profile at tip height in turbulent flow.

33% before the turbine is reflected in a 30% decrease in TI after the turbine. So, as wind speed increases, the wind turbine effectively keeps a section of the OHL (equal to the rotor diameter of 90 m) in a much lower, and possibly damaging, turbulence intensity region. This is in fact contrary to what was seen in the smooth flow case. The turbulence intensity base level at tip height is much higher than the value measured at hub height because of the presence of the boundary layer vertical profile.

Effect of height from ground

A comparison between the wind velocity profiles at 4 different heights from the ground (hub height (HH), tip height (TH), 1/4 blade chord (1/4 C) and overhead line (OHL)), is shown in Figs. 7.22 and 7.23 for one wind turbine operating condition

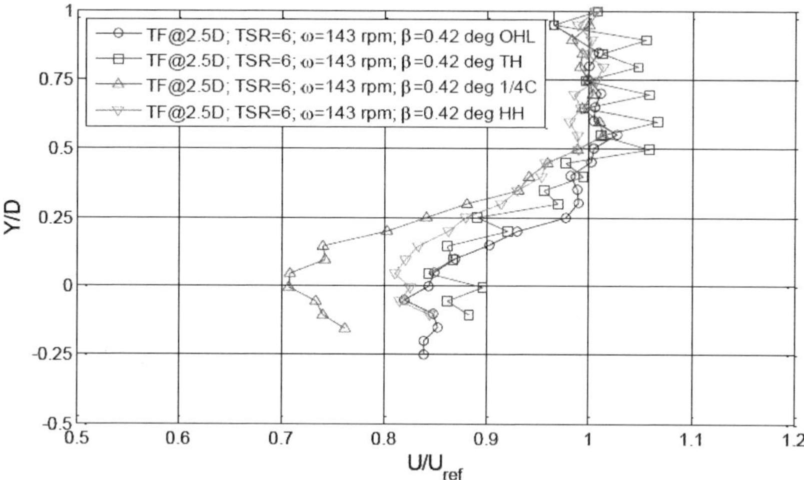

Fig. 7.22.   Effect of height above ground on wind speed profile at 2.5D in turbulent flow.

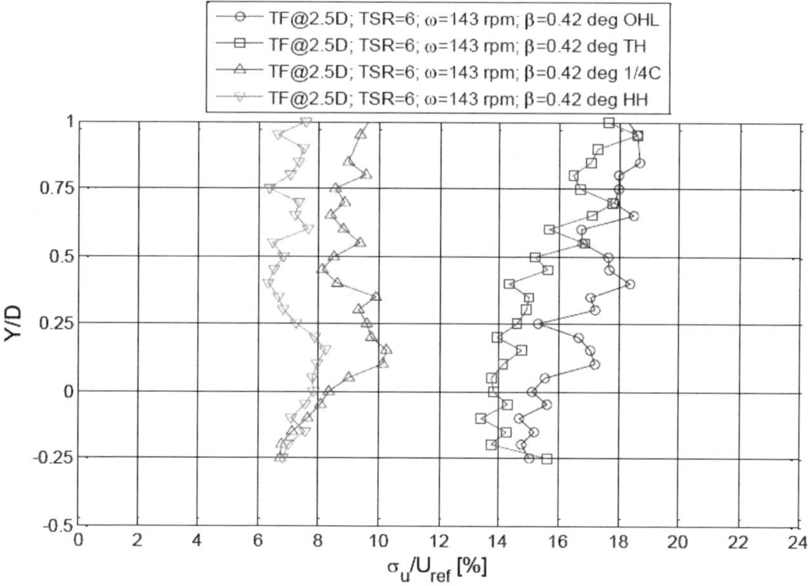

Fig. 7.23.   Effect of height above ground on turbulence intensity profile at 2.5D in turbulent flow.

(TSR $= 6$; $= 143$ rpm, $= 0.42$ deg) at a distance of 2.5D downstream the wind turbine.

Due to the presence of the ground, a different reference wind speed has been assumed to make the value non dimensional at different heights. In this manner a value lower than 1 means a decrease in the mean wind speed if compared to the incoming wind speed at the same height.

The wind speed profile is similar for all heights (apart from the 80 m height) with the wake width around one blade diameter (90 m). The tip effect has virtually disappeared in the turbulent flow. At 2.5D the turbulence intensity at 80–85 m above ground is significantly reduced but at 35–40 m the drop is only around 30%. This could be due again to the modification of the boundary layer profile because of the presence of the wind turbine.

Effect of distance downwind of turbine

Figures 7.24 to 7.27 show how the wake structure along the wind direction by comparing the mean wind speed and turbulence intensity profiles. These have been measured at hub height (85 m) and at tip height (40 m) at 1.5D, 2.5D, 5D and 8D downstream the wind turbine position.

The effects induced by the increase in the distance from the wind turbine are clearly visible. In these turbulent flow (TF) conditions the wind speed drop at 1.5D is significant (35%) but the values at 2.5D are lower and very similar of the 5D ones (15–20%). At 8D the effect is very small. At tip height (close to earth wire

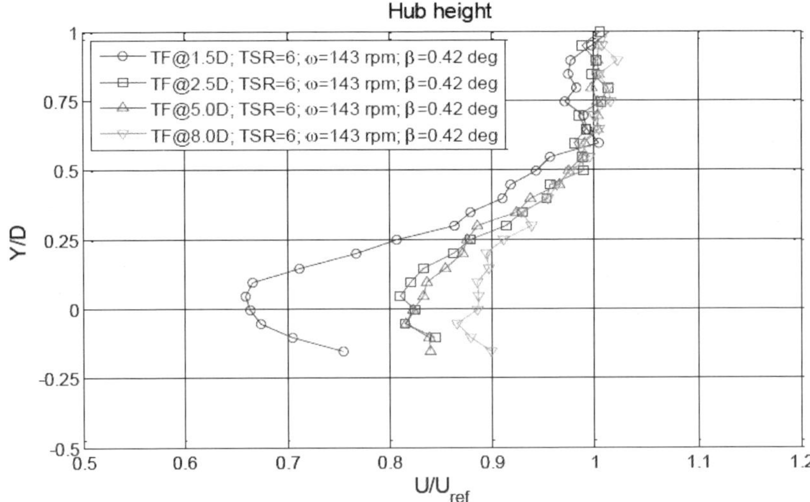

Fig. 7.24.   Effect of height above ground on wind speed profile at hub height in turbulent flow.

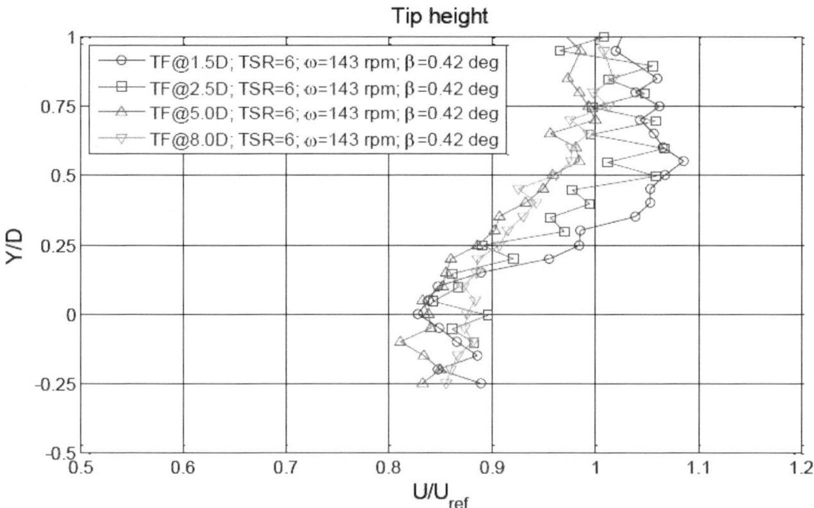

Fig. 7.25.   Effect of height above ground on wind speed profile at tip height in turbulent flow.

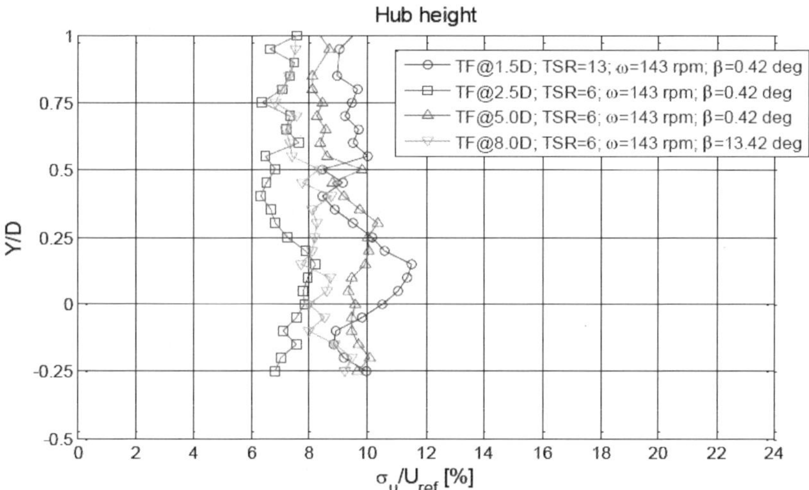

Fig. 7.26.   Effect of height above ground on turbulent intensity profile at hub height in turbulent flow.

height) there is little to choose between the values at 1.5 to 8D (~15%) but the wake tends to spread out as the distance increases. Whilst there is little to choose between the turbulence intensity values at hub height, there is a distinct increase in turbulence intensity for the 5 and 8D distances compared to 1.5 and 2.5D. The turbulence intensity at 40 m above ground (tip height) without the turbine present is

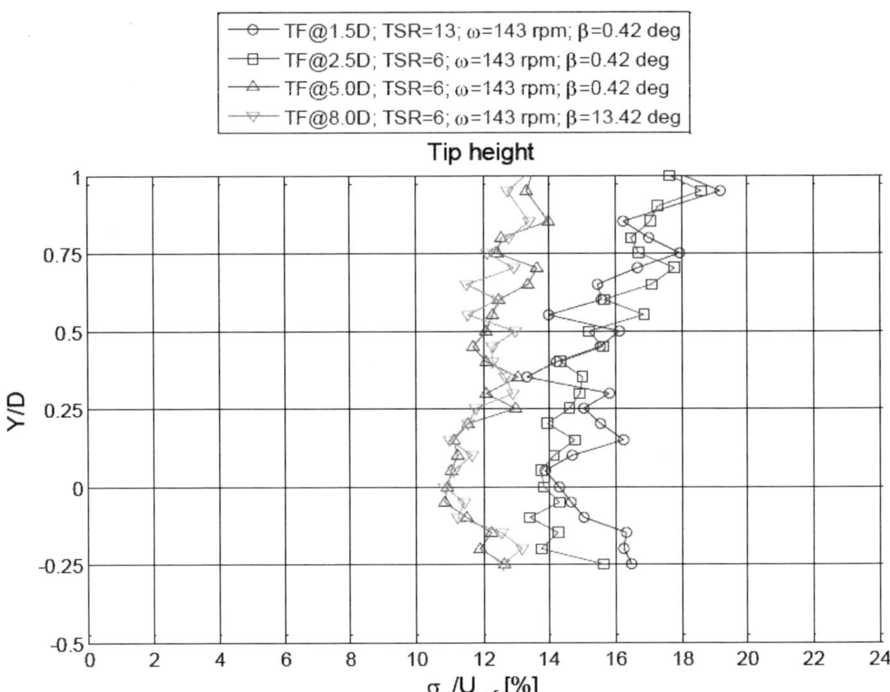

Fig. 7.27.   Effect of height above ground on turbulent intensity profile at tip height in turbulent flow.

around 21%, so at 1.5/2.5D the turbulence intensity is dropping to half this value and in the category 1 classification for H in the H/w scale for conductor vibration. Under Cigré recommendations[12] this would require a tension reduction of around 12% to maintain conductor lifetimes.

7.5.3 *Scaled overhead line*

Set-up

After the completion of all the above tests, a scaled aeroelastic representation of an Araucaria conductor on a 400 m span was set up. The model of an OHL was designed and built in the same geometry scale of the wind turbine (1:50) and represents an Araucaria conductor in terms of dynamic and aerodynamic characteristics. The major parameters of the line are shown in Table 7.4. The line was fitted with reflective spots along the central section and observed with no turbine present. Figure 7.28 shows the span position in relation to the turbine.

The tests were performed in turbulent wind conditions so as to more accurately represent the real world. The dynamic response of the cable was studied by means of

Table 7.4. Full scale Araucaria OHL characteristics.

Span length	400 m
Maximum sag	15 m
Diameter	37.3 mm
Number of strands	61
Wire diameter	4.14 mm
Aluminium equivalent area	700 mm^2
Area full	821 mm^2
Axial force	34 kN
Mass per unit length	2.7 kg/m

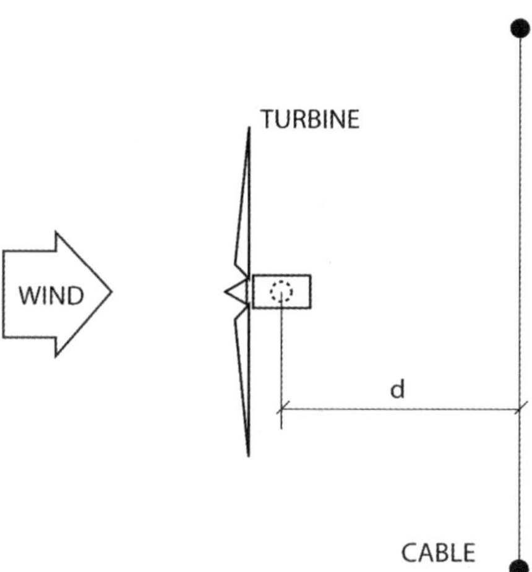

Fig. 7.28. Relative position of the turbine and the OHL model.

an optical system able to define the spatial motion of markers along the conductor. The positions of the markers (labeled "a", "b", "c", "d" and "e") and the reference frame adopted for the data post processing are shown in Fig. 7.29.

The runs were at wind speeds up to 20 m/s, at 1.5 and 2.5D and at various line heights, both with and without the turbine present. Recordings were made with four video cameras. The video data is then processed to provide 3-D locations for all the reflective dots and hence the line movement. The mechanical parameters of the model line. using the scaling rules of aeroelastic models' are given in Table 7.5.

The tower dimensions are shown in Table 7.6 together with the sag of the conductors at the heights for a typical L12 tower in the UK. Since only a section of the OHL is in the wake of the wind turbine, the height to be considered depends on

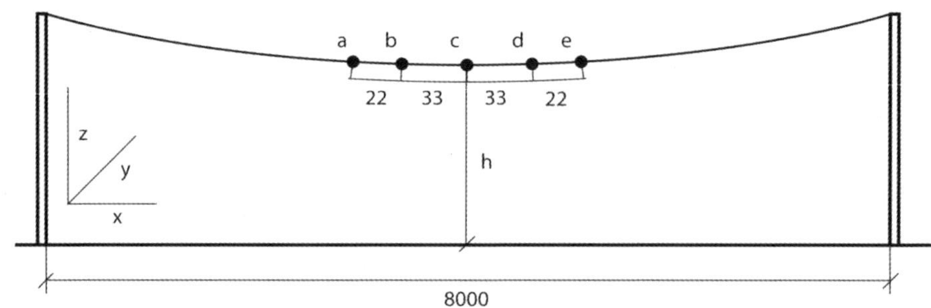

Fig. 7.29.   Markers and reference frame (dimensions in mm).

Table 7.5.   Modeled mechanical parameters of the Araucaria line.

Span length	8 m
Maximum sag	315 mm
Diameter	0.75 mm
Axial force	0.38 N
Mass per unit length	1.1 g/m

Table 7.6.   Tower conductor heights.

Conductor	Height at tower	Minimum height	Medium height
Earth wire	46.5 m	34.5 m	38.5 m
Upper phase	42.1 mm	21 m	31 m
Middle phase	32.8 mm	17.5 m	21.5 m

the relative position of the OHL and the wind turbine. Average heights have been therefore been used for each conductor during the wind tunnel tests and these are shown in Table 7.6.

A special structural scheme was adopted using a steel wire with diameter of 0.3 mm to reproduce the cable stiffness and some plastic cylinders with a diameter of 2 mm and a length of 25 mm to reproduce the mass and the aerodynamic properties of the cable. The plastic cylinders contribute only to transmit the aerodynamic forces and to add the correct amount of mass uniformly distributed along the structure in order to have the correct reproduction of the modal shapes and of the natural frequencies. The OHL aeroelastic model was positioned during the wind tunnel tests downwind the wind turbine at a distance d = 1.5D or 2.5D and tests were performed for wind speeds of U $=$ 1.4 m/s, 2.12 m/s and 2.8 m/s for the following conditions:

— without wind turbine (H $=$ 0.62 m $\sim$31 m high)
— with the wind turbine at 1.5D (H $=$ 0.62 m $\sim$31 m high)
— with the wind turbine at 2.5D (H $=$ 0.62 m (31 m), 0.77 m (38.5 m) and 0.43 m (21.5 m))

In Fig. 7.30, the mean y displacements (along the wind direction) of the 5 markers are shown for different operating conditions at U $=$ 2.12 m/s. It can be seen that:

1. The displacements along the wind direction are considerably higher (over 5 m as opposed to 3 m) when the wind turbine is put upwind the conductor model.
2. Considering the 0.62 m (31 m) height of the OHL from the floor and two different distances from the wind turbine, slightly larger displacements are recorded when the conductor is closer to the wind turbine.
3. Comparing results at d $=$ 2.5D, the mean lateral displacement are significantly larger when the OHL is at the lower height from the floor.

Looking at Fig. 7.31, it can be seen that:

1. The oscillation along the wind direction approximately doubles in amplitude when the wind turbine is put upwind the cable model.

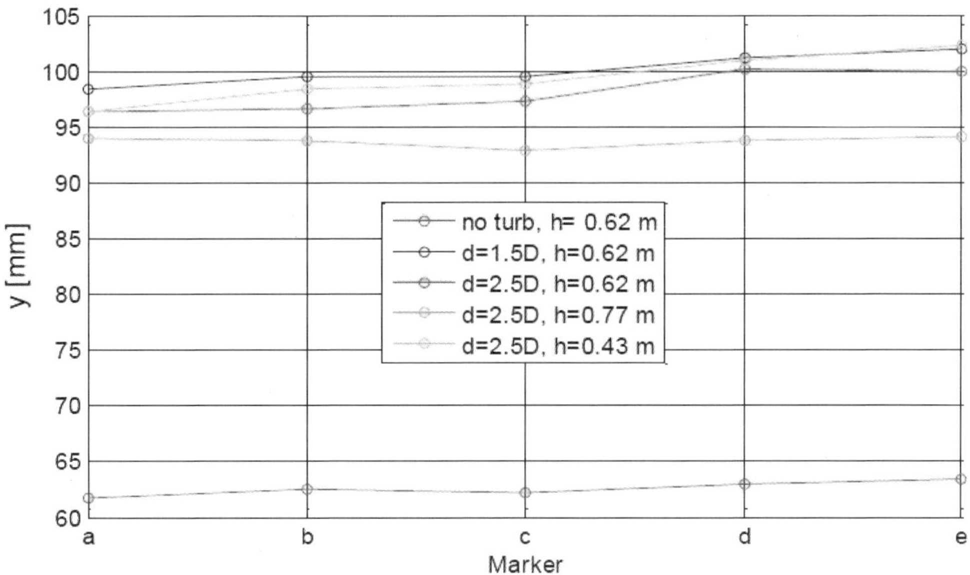

Fig. 7.30.   Mean y displacement of markers at 2.12 m/s wind speed.

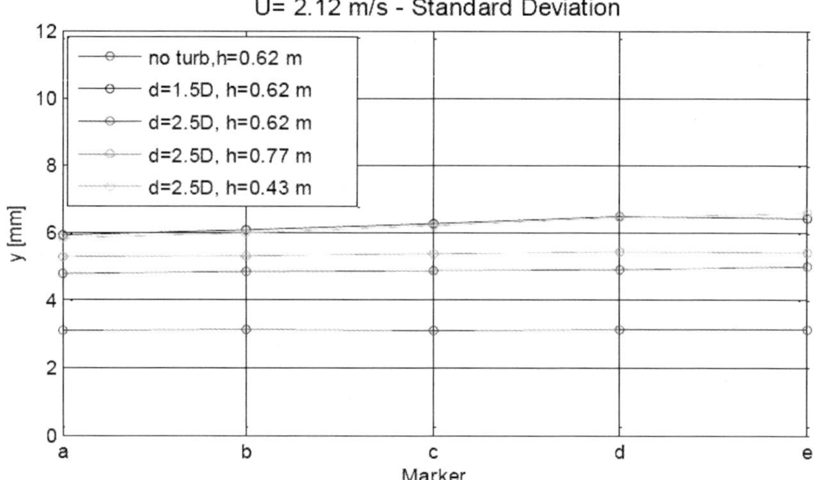

Fig. 7.31.   Standard deviation of y displacement of markers at 2.12 m/s wind speed.

2. The worst case is again at the lowest height (equivalent to 21.5 m above ground) at 2.5D where the amplitude is 300 mm equivalent amplitude (around 10× the Araucaria diameter).
3. Considering the same height of the OHL from the floor and two different relative distances between the wind turbine and the OHL, larger oscillations are recorded when the cable is closer to the wind turbine.

In Fig. 7.32 the standard deviation along the vertical z direction of the 5 markers are shown for U = 2.12 m/s. It can be seen that:

1. The oscillation along the vertical direction is more than double that without the wind turbine.
2. The amplitudes are about 40% of the horizontal (windward) values.
3. Larger oscillations are recorded when the conductor is closer to the wind turbine.
4. The fluctuation of the vertical displacement are larger when the OHL is at the lower height from the floor.

7.5.4 *Concerns*

The simulated OHL data appears to suggest that the German NNA in EN50341-3 is actually insufficient to protect their lines. The basic idea is that the wake will flow over the lower profile 132 kV (and wood pole) lines and so minimal vibration protection is required. However, the Milan data indicates greater movement at distances close to the ground — the lowest height considered being 21.5 m — which

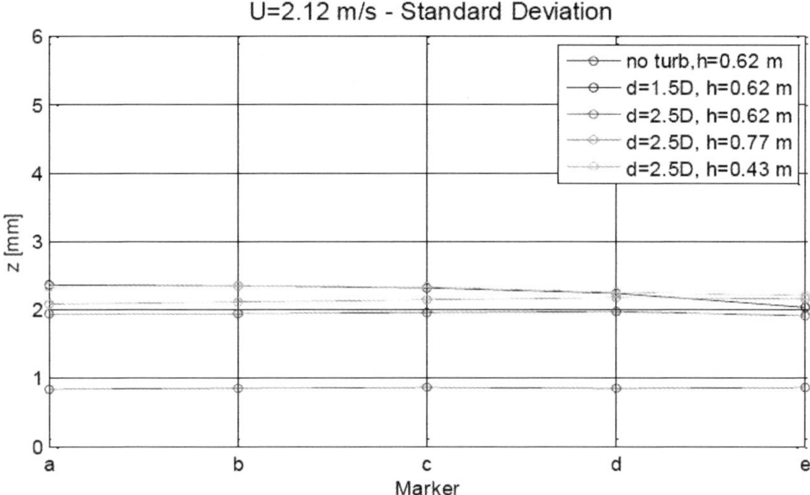

Fig. 7.32. Standard deviation of z displacement of markers at 2.12 m/s wind speed.

would raise concerns for 132 kV lines as well as 275/400 kV tower lines. There is a distinct and significant increase in amplitude of the conductor movement when the turbine is introduced and this is likely to be at the frequency of the rotor blade movement — around 1–2 Hz. The limitations of the wind tunnel data include the fact that sub-span lengths were not covered. However, the data raises sufficient questions and concerns that field data is required to determine the extent of the problem more accurately.

7.6 Comparison with Other Data

7.6.1 *General*

An initial investigation by Scottish Power[24] covered data from other sources in the field and wind tunnels although none had looked at this specific problem except J-L Lillien[5] and the German reviewers.[20,21] This section therefore picks out some of this reported data and other work that has been located more recently.

7.6.2 *Other wind tunnel data*

Dahlberg[25] dealt with the sideways extent of the wake with distance from the turbine. The wake edge is identified by the vortices from the blade tips and its sideways motion by radial velocity sensors. A clear result was that the position wake edge already, at 0.5D distance, moved out with respect to the blade tip. The wake is first expanding, identified as a radial velocity directed outwards due to the effect of the reduction of speed through the turbine, but soon the entrainment of flow from the free-stream to the wake takes over. The position corresponding to 2D is probably when the radial velocity changes sign, i.e., the wake ceases to expand but levels off before starting to contract. The vortex positions correspond to a very high peak in the rms velocity values and a high velocity gradient in the wake. They merge at a distance of 2 to 3D. Dahlberg observed, both from the smoke visualization and from a spectral analysis,that this distance corresponds with no free-stream turbulence. This basically contradicts earlier thoughts that the wake kept expanding until fading out and combining with the free stream flow.

Medici[26] performed smooth and turbulent flow wind tunnel tests using a small, 2-bladed wind turbine model (blades 80 mm long compared with 900 mm in Milan). All the data in the tests was freely provided by the Royal Institute of Technology in Stockholm and one profile is shown in Fig. 7.33. This confirms the maximum spread of the wake after around 2D but does indicate the wind flow changes by almost an order of magnitude on a scale of less than a blade diameter at 1.5D. Another feature of the data was the appearance of a low frequency fluctuation both in the wake and

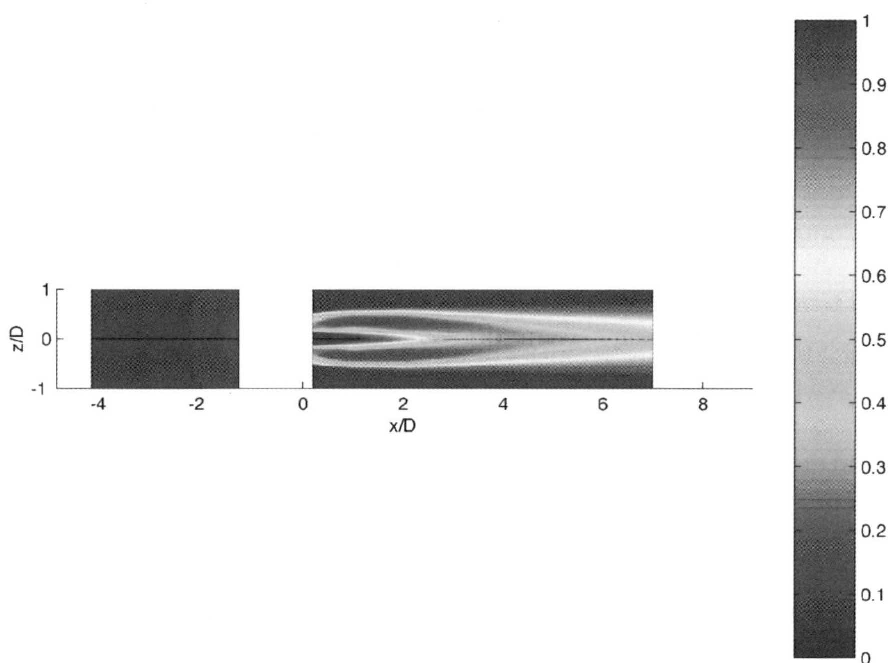

Fig. 7.33.   Medici profile. The scale is relative flow to free flow (U/U$_o$).[26]

in the flow outside. This fluctuation was found both with and without free stream turbulence.

7.7   Effect of Multiple Turbines on the OHL

This chapter has looked at the effect of a single turbine on a stretch of OHL. However, wind farms require many turbines mounted as close as possible with regard to economics, power efficiency and other factors. In particular, a wind farm can have several turbine positions paralleling the route of the OHL. Such a situation would apply an almost sinusoidal wind effect along the line. With wind farms often set up with 2D between them normal to the prevailing wind and 4D along the wind direction, it can happen that the turbines could be spaced around 2D apart along the line. So velocity profiles could occur along each span such as that shown in Fig. 7.34. Figure 7.35[27] shows the two turbines separated by 2D and a third turbine at which the power output is measured as it is traversed along and away from the first two.

The configuration at several wind farms is that a string of turbines follow the OHL route at falling distance (1.5D). The power output (which is related to the wind speed at the third turbine) is seen to vary from 100% to 5% in a cyclical fashion at a distance of 2D. Each unit on the x-axis is 1D (90 m for a 3 MW turbine) and so a 375 m span would be just over 4 units (-2 to $+2$) and suffer two complete cycles overlaid with the 3 Hz rotor blade circulation frequency.

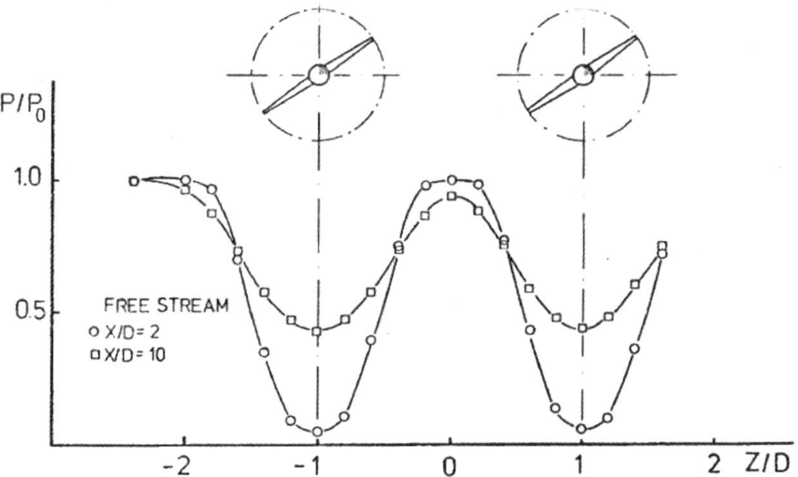

Fig. 7.34.   Power output from a downstream turbine traversed in the spanwise direction at 2D and 10D downstream.[27]

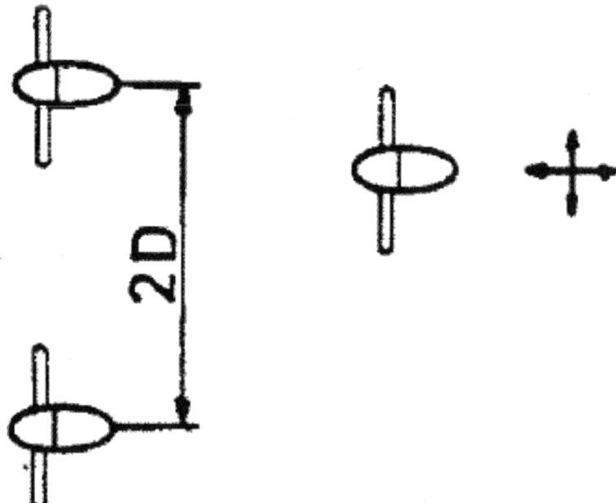

Fig. 7.35.   Configuration for Fig. 7.34. Wind flow is from LHS.

7.8 Solutions

Possible solutions are:

1. Halt all wind turbine installation closer than an agreed distance (minimum 3.5D) from any tower line.
2. Install extra and/or different damping mechanisms on susceptible lines and structures.
3. Possibly changing the positions or use of spacer dampers as the length of susceptible OHL conductor may be critical.
4. More expensive solutions may include the use of anti-galloping conductors or low profile towers.

At this stage it is not possible to speculate on other possible solutions until field data has confirmed the possibility of damage to OHL conductors.

7.9 Summary

1. A Belgian paper had indicated that wind turbines could cause problems on nearby OHL conductors due to oscillations from wake eddy effects.
2. German investigations have indicated that this is indeed a concern and recommendations were included in a national standard and CENELEC standards EN50341 and 50423. These standards cover lines within $3.5 \times D$ (305 m for a 3 MW turbine) of the turbine position. Damage potential was identified but the effect on conductor vibrations and other oscillations were never fully evaluated.

3. The effect of the wake eddies are two-fold

   a. A major reduction in wind speed occurs when the OHL is downwind of the turbine, causing the OHL to be in damaging low wind speed conditions for a higher percentage of their life time. This may require further damping for the conductors to survive.
   b. Turbulence will occur at a frequency similar to the rotation speed of the turbine blades (0.5 to 1 Hz) but over a relatively short length of conductor ($\sim$50–70 m) possibly causing cause conductor damage at the spacer damper.

4. A literature survey of modeling and wind measurements showed that significant wind reduction and turbulence can occur at 4D from the turbine but that this was substantially reduced by 5.5D. This implied that the damage potential could be worse than assumed in the German standard.

5. Wind tunnel results show a maximum reduction of the mean values of the wind up to 54% at hub height at 2.5D downwind of the wind turbine, in smooth flow conditions.

6. The velocity reduction becomes lower as the distance from the wind turbine increases and at 8D reaches 20%. The global dimension of the wake is larger than a rotor diameter due to the deflection of the incoming flow and can be estimated as >1.5D up to 2.5D downwind of the wind turbine whilst it decreases to <1D at 8D downwind.

7. Comparing results at different heights, the cylindrical structure of the wake is a zone of increased velocity before the profile reaches the undisturbed flow velocity. This feature is present also at the OHL height, i.e., outside the rotor area, probably due to interaction with the ground. This caused an increased static deflection measured on the aeroelastic model of the OHL conductor positioned behind the wind turbine compared with its behavior with no turbine present.

8. The cylindrical shape of the wake and therefore the cylindrical distribution of the zone of increased wind speed appear to lead to high aerodynamic loads on the OHL.

9. Results in turbulent flow have shown the largest reduction in the wind flow velocity at 1.5D equal to 65%. Due to the large levels of turbulence intensity already present in the incoming flow the variation of this parameter due to the wind turbine interaction is less evident.

10. The simulated OHL data appears to suggest that the German NNA in EN50341-3-4 is actually insufficient to protect their lines. The basic idea is that the wake will flow over the lower profile 132 kV (and wood pole) lines and so minimal vibration protection is required. However, the wind tunnel data indicates greater movement at distances close to the ground — the lowest height considered being 21.5 m — which would raise concerns for 132 kV lines as well as 275/400 kV

tower lines. There is a distinct and significant increase in amplitude of the conductor movement when the turbine is introduced and this is likely to be at the frequency of the rotor blade movement — around 1–2 Hz. The data raises sufficient questions and concerns that field data is required to determine the extent of the problem more accurately.

Acknowledgment

The author is indebted to Scottish Power, UK, for supporting this work and kindly providing permission for publication.

References

1. T. Hahm and J. Kröning, "In the wake of a wind turbine," *Fluent News Spring* (2002).
2. T. Degner, F. Kiessling and J. Tzchoppe, "Minimum distance between wind energy plants and overhead lines," *Elektrizitätswirtschaft Jg.*, vol. 98 (1999), in German.
3. G. Smith, "Advance wake model for very closely spaced turbines," EWEC 2006 Paper, Athens, 27 February–2 March 2006.
4. A. Wessel *et al.*, "Verification of a new model to calculate turbulence intensity inside a wind farm," EWEC 2006 Paper, Athens, 27 February–2 March 2006.
5. J.L. Lilien, R. Keutgen and N. Raimarckers, "Wind Turbine turbulence effects on cable structures in their vicinity," Rev AIM Liège 1-2/2001.
6. A. Jiminez *et al.*, "Large eddy simulation of a wind turbine wake," EWEC 2006 Paper, Athens 27 February–2 March, 2006.
7. GE Energy literature 1.5 and 2.5 MW wind turbine data (2005).
8. N. Stefanatos, E. Morfiadakis and G. Glinou, "Wake measurements in complex terrain," EWEC 1996, Paper, pp. 773–7.
9. Cigré SC22 WG01 "Aeolian Vibrations," *Electra*, vol. 124 (1989).
10. Cigré SCB2 WG11 "Fatigue endurance capability of conductor/clamp systems," Technical Brochure (2007).
11. Cigré SCB2 WG11 "Guide to vibration measurements on overhead lines," *Electra*, vol. 163 (1995).
12. CIGRE Study Committee No. 22, Working Group 04, "Endurance capability of conductors," Final Report, July 1988.
13. J.B. Wareing, "The work of Cigre SCB2 WG11 (Conductor Dynamics) 2004-5," EATL Report 5804, July 2005.
14. J.B. Wareing, "New Vibration limits for overhead lines," EATL Report 5675, November 2003.
15. M.A. Miner, "Cumulative damage due to fatigue," *Proc. ASME J. Applied Mechanics* (1945).
16. C.B. Rawlins, "Fatigue of Overhead Conductors," in *Transmission Line Reference Book, Wind-Induced Conductor Motion* (EPRI, Palo Alto, 1979), pp. 51–81.
17. C. Hardy and A. Leblond, "Statistical analysis of stranded conductor fatigue endurance data," *Proc. 4th Int. Symp. Cable Dynamics*, Montreal, Canada, 28–30 May 2001, pp. 195–202.
18. Cigré SC22 WG04, "Recommendations for the lifetime evaluation of transmission line conductors," *Electra*, vol. 63 (1979), pp. 103–145.
19. C. Hardy *et al.* "Review of models of self-damping of stranded cables," *Proc. 1st Symp. Cable Dynamics*, Liège, 19–21 October 1995, pp. 61–68.

20. German Annexe, EN50341-3-4 (2001).
21. German Standard S(1)14 Deutschen Elektrotechnischen Kommission (in German).
22. J.B. Wareing, "Investigation of the Wake effects from wind turbines on OHL conductors and structures," BWTR_132, June 2008.
23. Vesta wind power solutions, http://www.vestas.com.
24. J.B. Wareing, "Investigation of the Wake effects from wind turbines on OHL conductors and structures — Phase 2," BWTR_132, July 2009.
25. J.-A. Dahlberg *et al.*, "Potential improvement of wind turbine array efficiency by active wake control," *Proc. EWEC*, Madrid (2003).
26. D. Medici, "Wind Turbine Wakes — Control and Vortex Shedding," Thesis, Royal Institute of Technology, Stockholm, Sweden (2004).
27. P.-H. Alfredsson and J.-A. Dahlberg, "Measurements of wake interaction effects on the power output from small wind turbine models," Technical note HU-2189 Pt5, National Swedish Board for Energy Source Development (1981).

Section 2

Solar Energy Systems

Chapter 8

Solar Energy Calculations

Keith E. Holbert

School of Electrical, Computer and Energy Engineering,
Arizona State University, P.O. Box 875706
Tempe, AZ 85287-5706
keith.holbert@asu.edu

Devarajan Srinivasan

ViaSol Energy Solutions, 1705 W. University Drive
Tempe, AZ 85281
devarajan.srinivasan@viasolenergy.com

This chapter presents calculations of the position of the sun as a function of location, date and time. In addition, direct, diffuse and reflected components of solar insolation on a tilted surface can be determined. This computational approach is applicable to both thermal collection/conversion processes and photovoltaics (PV). Although sun charts are widely used, there are situations where charts are inadequate and precise computations are preferred. For example, consider fixed PV modules atop residential or industrial buildings in which the value of the electricity varies with date and time.

8.1 Introduction

Three energy conversion processes are normally associated with the sun[1]:

(i) heliochemical — which is principally the photosynthesis process,
(ii) helioelectrical — which is commonly exploited in solar cells (i.e., photo-voltaics), and
(iii) heliothermal — which is a conversion of sunlight into thermal heat as employed within concentrating solar power (CSP) plants.

189

In the case of the latter two methods, it is oftentimes necessary to calculate the solar energy incident to the energy collection apparatus. This chapter describes general-purpose computational approaches to determine optimal solar panel positions and expected energy collection. Solar energy is dilute and intermittent. Hence solar energy harvesting facilities have to be optimized to improve their economic return. Although the solar irradiance is dilute, estimates are that the annual extraterrestrial radiation to the contiguous U.S. is 4.43×10^{16} kWh.[2]

The fundamental goal of such calculations is to determine the solar insolation on a given area, so solar collectors can be oriented to maximize collected energy. The total incident solar energy on the collectors is a function of several variables — including the day of the year, time of day, latitude, longitude, orientation of the collectors, and atmospheric conditions. In general, one begins by computing angles based on the particular day of the year, and then time of the day. This is followed by considering the specific collector-panel surface position and orientation with respect to the incident sunlight. The sections of this chapter are arranged with this procedure in mind.

8.2   Earth's Orbit

The earth, as a member of the solar system, circles the sun once a year. The orbital path of the earth around the sun, as shown in Fig. 8.1, is slightly elliptical and not a perfect circle.[3-5] The imaginary plane of earth's elliptical orbit is called the orbital plane (also known as the ecliptic plane). The direction of earth's orbit is counterclockwise when the orbital plane is viewed from above the North Pole (looking down at the North Pole).

The sun is slightly offset from the center of the ellipse. At perihelion, on January 3, the earth reaches its closest approach to the sun of around 147.09 million km. The earth is farthest from the sun at aphelion, on July 4, at around 152.10 million km. On average, the earth is about 149.60 million km from the sun. The earth travels about 940 million km during a year in its orbit around the sun.

In addition to orbiting the sun, the earth also spins, every 24 hours, on an axis that passes through the North and South Poles. Looking down at the earth from above the North Pole, earth's rotation is counterclockwise, from the west to the east. Earth's axis is not perpendicular to the orbital plane, but is tilted at 23.45° to the perpendicular. The Northern hemisphere tilts towards the sun during summer and away from the sun during winter. The tilt of the earth's axis and its motion around the sun causes variation:

 (i) in the intensity and duration of sunlight received at different places on the earth, which results in the seasons, and

(ii) in the apparent motion of the sun across the sky.

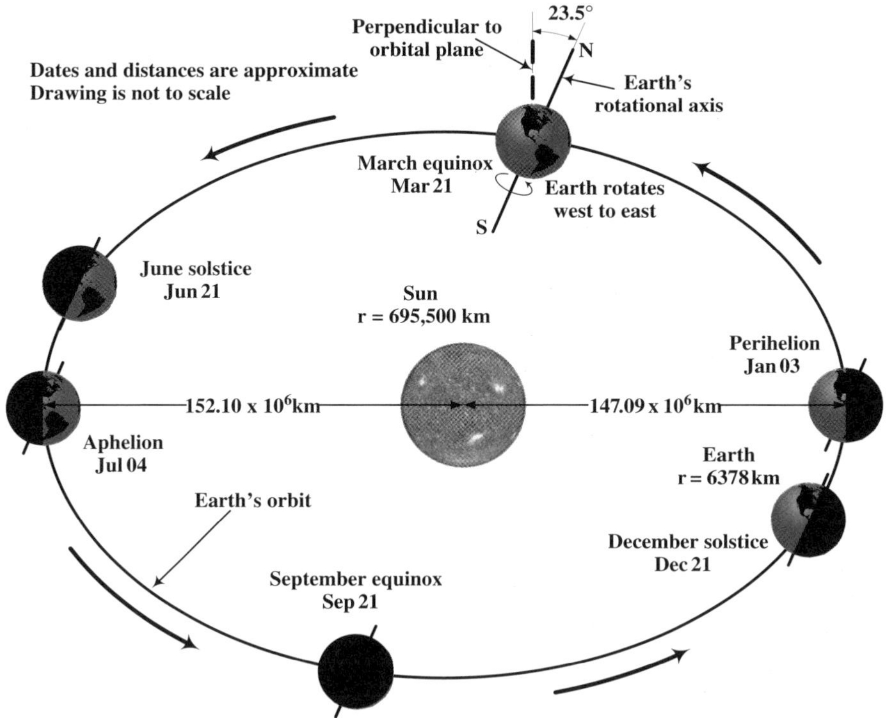

Fig. 8.1.   Earth's orbit around the sun.

Four associated astronomical events are the vernal and autumnal equinoxes, and the summer and winter solstices. During the March equinox ($\sim$March 21) and September equinox ($\sim$September 21), the sun is directly over the equator and the lengths of day and night are equal ($\sim$12 hours). During the June solstice, the sun is directly over the Tropic of Cancer (23.45°N latitude), and thus at the equator, the sun appears at its northernmost position (23.45°). In June, it is summer in the Northern Hemisphere and the day is longer than the night; and it is winter in the Southern Hemisphere and the day is shorter than the night. During the December solstice, the sun is directly over the Tropic of Capricorn (23.45°S latitude), and hence, at the equator, the sun appears at its southernmost position ($-23.45°$).

8.3   Solar Constant and Solar Spectra

The orbit of the earth is elliptical, hence the distance between the earth and sun varies over the year, leading to apparent solar irradiation values throughout the year approximated by[6]:

$$I_0 = I_{SC}\left[1 + 0.033\cos\left(\frac{N}{365} \times 360°\right)\right], \tag{8.1}$$

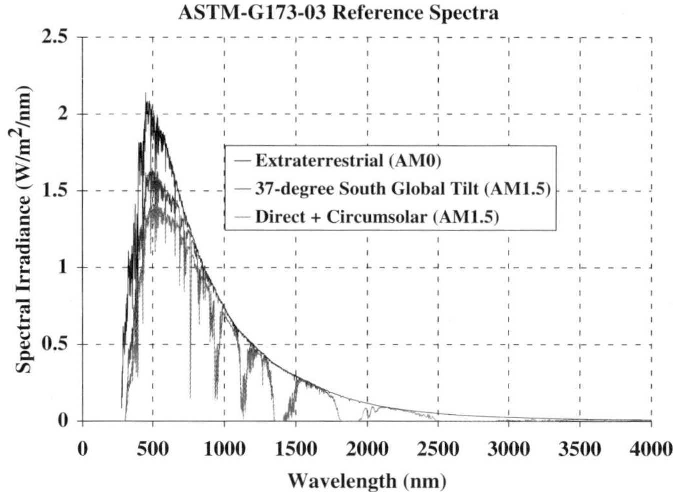

Fig. 8.2.   Extraterrestrial and photovoltaic reference solar spectra. The direct plus circum-solar spectrum is the direct normal component of the 37° south global tilt (AM1.5) spectrum. The diffuse component (not shown) is the difference between the two AM1.5 spectra. Data are from Ref. 7.

where I_{SC} is the solar constant, and N is the day number of the year, with January 1 equal to 1. The amount of solar radiation, just above the earth's atmosphere, measured in a plane perpendicular to the rays of the sun, is called the *solar constant*. The solar constant generally accepted by the space community is approximately 1.3661 kW/m². Solar radiation consists of several frequencies, in addition to visible light.[7,8] The extraterrestrial solar spectrum is called the Air Mass Zero (AM0) spectrum and is defined by the ASTM E490 standard.[9] The AM0 spectrum is shown in Fig. 8.2.

The spectrum of sunlight reaching the earth's surface is affected by atmospheric conditions and varies by location and time of day. The photovoltaic (PV) industry accepts a standard spectrum for testing and rating PV devices. This standard (AM1.5) spectrum is defined in the ASTM G173-03 standard[10] for specific atmospheric conditions on a south-facing surface tilted at 37° (i.e., the average latitude for the contiguous USA). The selected conditions are considered an average for the contiguous USA over a one year period.

8.4   Solar Angles[11,12]

The declination is the angular distance of the sun north or south of the earth's equator. The *declination angle*, δ, for the Northern Hemisphere (reverse the declination angle sign for the Southern Hemisphere) is[13]

$$\delta = 23.45° \sin \left[\frac{N + 284}{365} \times 360° \right]. \tag{8.2}$$

The earth is divided into latitudes and longitudes. For longitudes, the global community has defined $0°$ as the prime meridian which is located at Greenwich, England. The longitudes are described in terms of how many degrees they lie to the east or west of the prime meridian. A 24-hr day has 1440 mins, which when divided by $360°$, means that it takes 4 mins to move each degree of longitude.

The apparent solar time, *AST* (or local solar time) in the western longitudes is calculated from[14]

$$AST = LST + (4 \min /\deg)(LSTM - Long) + ET, \tag{8.3}$$

where

$LST =$ Local standard time or clock time for that time zone (may need to adjust
        for daylight saving time, *DST*, that is $LST = DST - 1$ hr),
$Long =$ local longitude at the position of interest, and
$LSTM =$ local longitude of standard time meridian, which is

$$LSTM = 15° \times \left(\frac{\text{Long}}{15°} \right)_{\text{round to integer}}. \tag{8.4}$$

The difference between the true solar time and the mean solar time changes continuously day-to-day with an annual cycle. This quantity is known as the *equation of time*. The equation of time, *ET* in minutes, is approximated by[15]

$$ET = 9.87 \sin(2D) - 7.53 \cos(D) - 1.5 \sin(D)$$
$$\text{where} \quad D = 360° \frac{(N - 81)}{365}. \tag{8.5}$$

As illustrated in Fig. 8.3, there are several angles of interest in describing the position of the sun at a particular time instant. The *hour angle, H,* is the azimuth angle of the sun's rays caused by the earth's rotation, and H can be computed from Ref. 16:

$$H = \frac{AST - 720 \text{ mins}}{4 \text{ min} /\deg}. \tag{8.6}$$

The hour angle as defined here is negative in the morning and positive in the afternoon ($H = 0°$ at noon).

The solar *altitude angle* (β_1) is the apparent angular height of the sun in the sky if you are facing it. The *zenith angle* (θ_z) and its complement the altitude angle (β_1) are given by

$$\cos(\theta_z) = \sin(\beta_1) = \cos(L) \cos(\delta) \cos(H) + \sin(L) \sin(\delta), \tag{8.7}$$

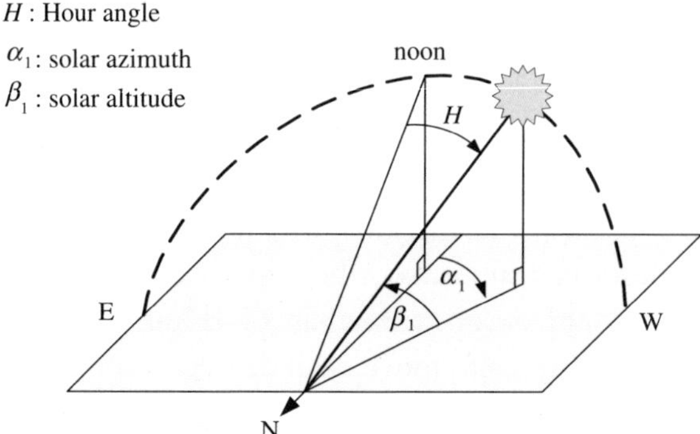

Fig. 8.3.   Important solar angles.

where L is the latitude (positive in either hemisphere) [$0°$ to $+90°$]; δ is the declination angle (negative for Southern Hemisphere) [$-23.45°$ to $+23.45°$]; and H is
the hour angle. The noon altitude is $\beta_N = 90° - L + \delta$.

The sun rises and sets when its altitude is $0°$, not necessarily when its hour angle
is $\pm90°$. The hour angle at sunset or sunrise, H_S, can be found from using Eq. (8.7)
when $\beta_1 = 0$

$$\cos(H_S) = -\tan(L)\tan(\delta), \tag{8.8}$$

where H_S is negative for sunrise and positive for sunset. The apparent sunrise
(apparent sunset) occurs earlier (later) than the value computed above since atmospheric refraction bends the sunlight. In addition, since the sun is about $0.5°$ wide
as viewed from the earth, whether one considers the first appearance sunlight as
sunrise also plays a role. Absolute values of $\cos(H_S)$ greater than unity can occur in
the arctic zones when the sun neither rises nor sets. The length of a day is twice the
absolute value of H_S.

The *solar azimuth*, α_1, is the angle away from south in the Northern Hemisphere
(referenced to north in the Southern Hemisphere)

$$\cos(\alpha_1) = \frac{\sin(\beta_1)\sin(L) - \sin(\delta)}{\cos(\beta_1)\cos(L)}, \tag{8.9}$$

where α_1 is positive toward the west (afternoon), and negative toward the east
(morning), and therefore, the sign of α_1 should match that of the hour angle.

If $\delta > 0$, the sun can be north of the east-west line in the Northern Hemisphere.
The time at which the sun is due east and west can be determined from

$$t_{E/W} = 4\frac{\min}{\deg}\left\{180° \mp \arccos\left[\frac{\tan(\delta)}{\tan(L)}\right]\right\}, \tag{8.10}$$

where these times are given in minutes from midnight AST.

Using the above equations, the apparent route of the sun through the sky can
be graphically represented by a sun path chart as shown in Fig. 8.4. The elevation

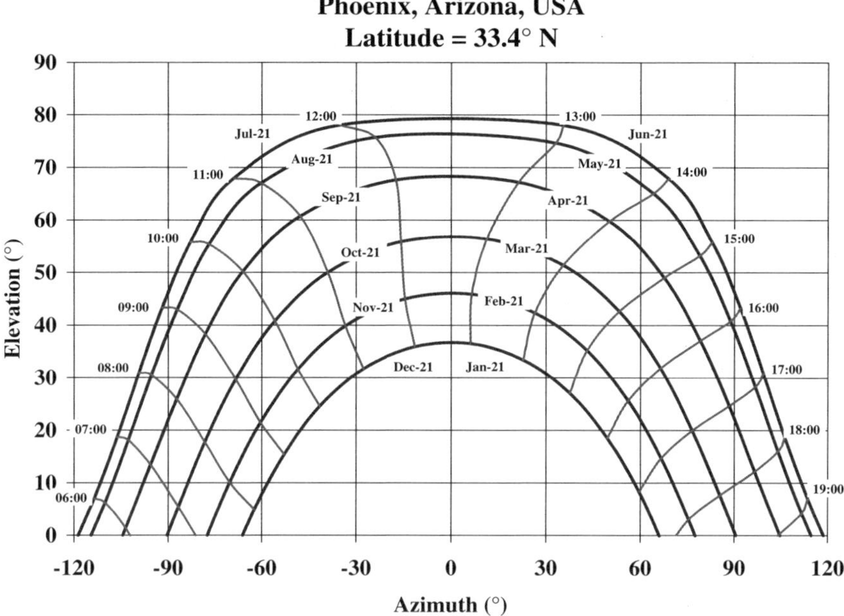

Fig. 8.4.   Elevation and azimuth angles of the sun at various dates and times during the year for 33.4°N latitude.

(or altitude angle) of the sun is plotted against the azimuth for various times during the year. This type of chart is typically used for shading studies. By plotting the elevation versus azimuth of various obstructions like buildings, trees, and mountains on the same chart, the days and times of potential shading can be identified.

8.5   Collector Angles

With knowledge of the position of the sun with respect to a particular location on earth, the solar collector orientation may be determined. The collector angle (θ) between the sun and normal to the surface is

$$\cos(\theta) = \sin(\beta_1)\cos(\beta_2) + \cos(\beta_1)\sin(\beta_2)\cos(\alpha_1 - \alpha_2), \qquad (8.11)$$

where α_2 is the azimuth angle normal to the collector surface, and β_2 is the tilt angle from the ground, as shown in Fig. 8.5. If θ is greater than 90°, then the sun is behind the collector. Some collector panel angles of interest are given in Table 8.1.

Given the collector angle, the total daily radiation to a surface is determined from integrating the solar irradiance, $I_0 \cos(\theta)$, from sunrise to sunset. For a horizontal surface, the collector and zenith angles are identical ($\theta = \theta_z$). If the extraterrestrial

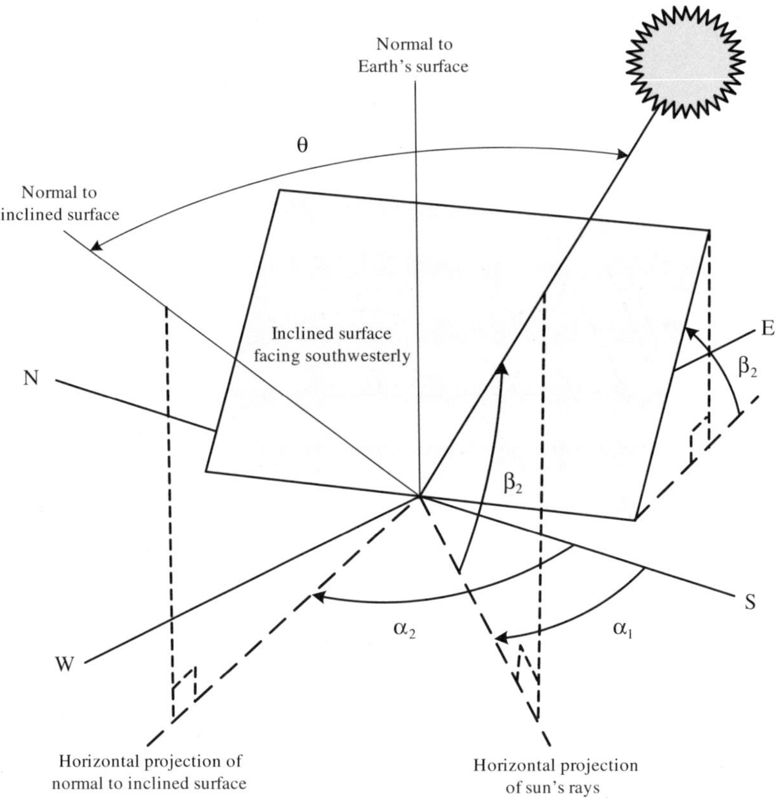

Fig. 8.5.   Solar angles on a tilted flat surface; diagram derived from Ref. 16.

Table 8.1.   Solar collector angles of special interest.

Azimuth, α_2	Tilt, β_2	θ	Orientation
n/a	$0°$	$90° - \beta_1$	Horizontal (flat)
—	$90°$	varies	Vertical wall
$0°$	$90°$	varies	South facing vertical
$-90°$	$90°$	varies	East facing wall
$+90°$	$90°$	varies	West facing wall
α_1	$90° - \beta_1$	$0°$	Tracking system

irradiance were to reach such a horizontal panel, then the total radiation received is[17]

$$H_O = \int_{-H_S}^{H_S} I_0[\cos(L)\cos(\delta)\cos(H) + \sin(L)\sin(\delta)]\frac{24}{2\pi}dH$$

$$= I_0\frac{24}{\pi}[\cos(L)\cos(\delta)\sin(H_S) + H_S\sin(L)\sin(\delta)]. \tag{8.12}$$

In similar fashion, the total extraterrestrial radiation to an equator facing tilted surface can be determined as

$$H_O = I_0 \frac{24}{\pi} [\cos(L - \beta_2) \cos(\delta) \sin(\tilde{H}_S) + \tilde{H}_S \sin(L - \beta_2) \sin(\delta)], \qquad (8.13)$$

where $\tilde{H}_S$ is the time of sunset on the tilted surface

$$\tilde{H}_S = \min[H_S, \arccos(-\tan(L - \beta_2)\tan(\delta))]. \qquad (8.14)$$

The sunrise and sunset hours on a panel that is not facing directly toward the equator are different when the collector is shadowed by itself. The collector panel sunrise and sunset hours may be computed from[18]

$$H_{SRS} = \mp \min \left\{ H_S, \arccos \left[\frac{ab \pm \sqrt{a^2 - b^2 + 1}}{a^2 + 1} \right] \right\}, \qquad (8.15)$$

where

$$a = \frac{\cos(L)}{\sin(\alpha_2)\tan(\beta_2)} + \frac{\sin(L)}{\tan(\alpha_2)}$$

$$b = \tan(\delta) \left[\frac{\cos(L)}{\tan(\alpha_2)} - \frac{\sin(L)}{\sin(\alpha_2)\tan(\beta_2)} \right]. \qquad (8.16)$$

For sunrise the negative sign of the leading "$\mp$" sign is selected, and for sunset this sign is positive. The choice of sign with respect to the "$\pm$" sign depends on α_2: if sunrise is being calculated, the sign used is the same as that for α_2; for sunset, the sign is opposite to that of α_2.

8.6 Solar Irradiance

As depicted in Fig. 8.6, there are three basic components to the solar irradiance on a surface:

(i) direct (beam),
(ii) diffuse-scattered, and
(iii) reflected short wavelength (solar) radiation from other surfaces.

Solar irradiance is the amount of solar energy incident on a surface per unit time, per unit area. The irradiance is typically expressed in kW/m^2. Solar insolation is the integration of the irradiance with respect to time, and may be quantified in kWh or kWh/m^2.

The direct normal irradiance at the earth's surface is[6]

$$I_{DN} = A \exp \left(-\frac{p}{p_0} \frac{B}{\sin(\beta_1)} \right), \qquad (8.17)$$

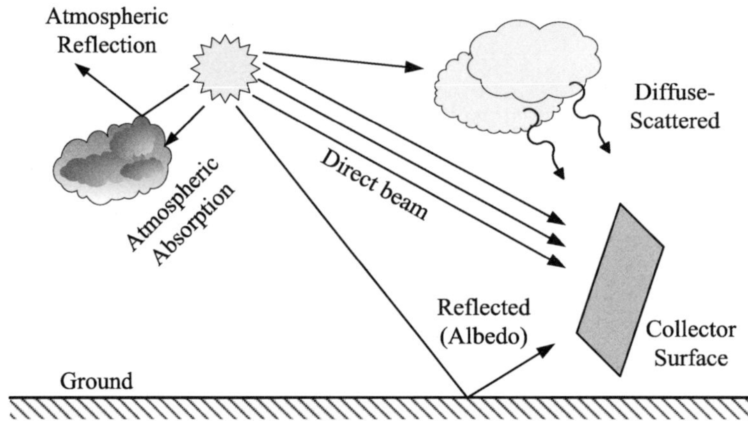

Fig. 8.6.   Components of solar radiation.

where A is the apparent extraterrestrial solar intensity, B is the atmospheric extinction coefficient (mainly due to changes in atmospheric moisture), and p/p_0 is the pressure at the location of interest relative to a standard atmosphere, given by

$$p/p_0 = \exp(-0.0001184z), \tag{8.18}$$

where z is the elevation in meters above sea level. The values for A and B are given in Table 8.2. The direct normal radiation at sea level then is

$$I_{DN}(0\,\text{m}) = A \exp\left[\frac{-B}{\sin(\beta_1)}\right]. \tag{8.19}$$

As noted earlier, calculation of the energy received must account for the position of the panel relative to the sun. The direct radiation flux onto the collector is

$$I_D = I_{DN}\cos(\theta). \tag{8.20}$$

The diffuse-scattered radiation, also referred to as sky radiation, flux onto the collector is

$$I_{DS} = CI_{DN}\left[\frac{1+\cos(\beta_2)}{2}\right], \tag{8.21}$$

where C is the ratio of diffuse radiation on a horizontal surface to the direct normal irradiation. The reflected, or albedo, irradiance for a non-horizontal surface may be approximated by

$$I_{DR} = I_{DN}\rho(C + \sin(\beta_1))\left[\frac{1-\cos(\beta_2)}{2}\right], \tag{8.22}$$

Table 8.2. Apparent solar irradiation (A), atmospheric extinction coefficient (B), and ratio of diffuse radiation on a horizontal surface to the direct normal irradiation (C) for 1964 in the Northern Hemisphere.[13]

Date	Day of Year	$A(\text{W/m}^2)$	B	C
Jan 21	21	1230	0.142	0.058
Feb 21	52	1215	0.144	0.060
Mar 21	80	1186	0.156	0.071
Apr 21	111	1136	0.180	0.097
May 21	141	1104	0.196	0.121
June 21	172	1088	0.205	0.134
July 21	202	1085	0.207	0.136
Aug 21	233	1107	0.201	0.122
Sept 21	264	1152	0.177	0.092
Oct 21	294	1193	0.160	0.073
Nov 21	325	1221	0.149	0.063
Dec 21	355	1234	0.142	0.057

For the Southern Hemisphere: to determine A, divide the A value by I_0 on the date of interest, and then multiply it by I_0 for half a year later (I_0 is computed using Eq. (8.1)); similarly, for B and C, use the values six months after the date of interest.[6]

where ρ is the foreground reflectivity with some values given below

ρ	Surroundings condition
0.2	Ordinary ground or vegetation
0.8	Snow cover
0.15	Gravel roof

The total radiation flux, also referred to as the global irradiance, is then

$$I_{\text{Tot}} = I_D + I_{DS} + I_{DR}. \tag{8.23}$$

However, for concentrating solar power, only the direct component is of importance since the mirrored surfaces are positioned to re-direct and concentrate the direct beam sunlight; and at the mean sun-earth distance, the sun subtends an angle of 32 min. The intensity of the direct beam, as compared to the diffuse component, can be seen in Fig. 8.7 for a collector tilted $33.4°$ (at latitude) in Phoenix, Arizona where the direct beam constitutes 83% of the $7.1\,\text{kWh/m}^2/\text{day}$ of energy received that day. Further illustrated in the graph is the increased energy collection made possible by tracking the sunlight, which in this case increases the total irradiance to $10.6\,\text{kWh/m}^2/\text{day}$.

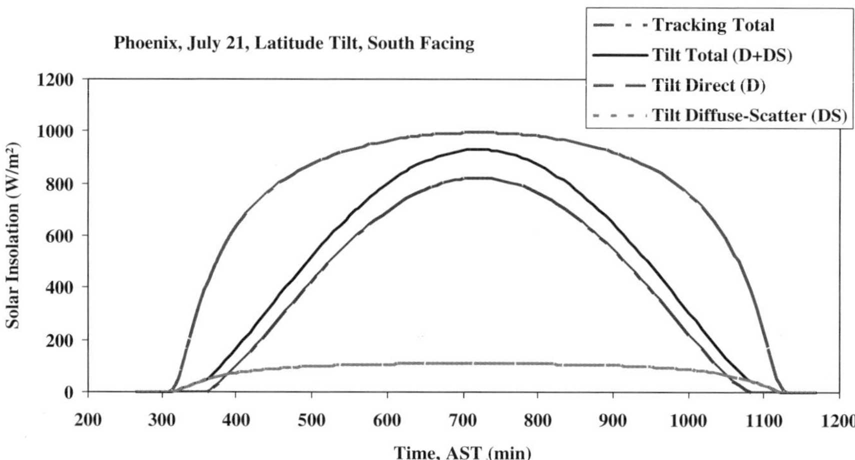

Fig. 8.7. Solar irradiance for a summer day for a collector tilted at 33.4° compared to a panel utilizing dual-axis tracking. For the tilted panel, early in the morning and late in the evening, the total irradiance is from the diffuse-scattered component, but during the remainder of the daytime, the direct component provides the majority of the insolation.

Small variations in tilt from horizontal ($\pm 10°$) and orientation from south ($\pm 20°$) do not reduce performance significantly.[13] The overall annual incident energy for tilts ranging from horizontal (0°) to vertical (90°) and collector azimuth angles from south facing to west facing are presented in Fig. 8.8 for a latitude of 33.4°.

In some localities electricity pricing differs during the day, that is, higher rates are charged during peak usage periods. Whereas maximum energy production occurs for panels oriented directly south at a fixed angle near latitude, maximum economic return depends on the electricity rate structure which may include different charges due to seasonal and time-of-use energy demand. In such situations, fixed PV panels may be more optimally oriented to maximize power production during on-peak times.[19] The total annual return for electricity generation is computed from

$$R = A \sum_{i=1}^{365} \int \eta_i(t) I_i(t) r_i(t) dt, \qquad (8.24)$$

where A is the active panel area; η is the overall conversion efficiency of the PV system; $I_i(t)$ and $r_i(t)$ are the solar insolation and value of the electricity, respectively, for the i-th day of the year at time t; and the integration is carried out over the hours in which the panel is illuminated.

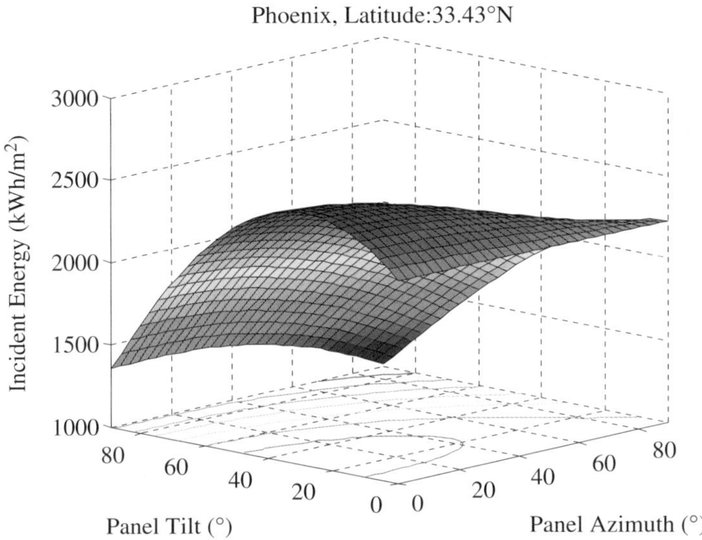

Fig. 8.8. Annual incident energy as a function of collector position. The maximum of 2528 kWh/m² occurs at a tilt of 29° and azimuth of 0°.

Not all the incident solar insolation is absorbed by a particular body. The emissivity (ε) quantifies the energy absorption

$$E_{abs} = \varepsilon_{\text{surf}} I_{\text{Tot}}. \tag{8.25}$$

Representative surface ε at solar energies[16] include

- asphalt pavement: $\varepsilon = 0.82 - 0.88$.
- white (flat) paint: $\varepsilon = 0.12 - 0.25$.
- red brick, tile, concrete: $\varepsilon = 0.65 - 0.77$.

8.7 Comparison to Measured Data

The solar energy received is also affected by varying atmospheric conditions. For example, weather conditions — such as rain, wind, pressure, temperature, humidity and pollution — inhibit the solar to energy conversion processes. Therefore, the availability of long-term measured data can also be useful; however, such data are typically acquired for a limited number of collector orientations.

For instance, solar radiation data for flat-plate collectors facing south are available for Phoenix, Arizona, from 30 years of measured data, and are plotted in Fig. 8.9 for a fixed tilt of 0° (horizontal) and 90° (vertical). The annual average

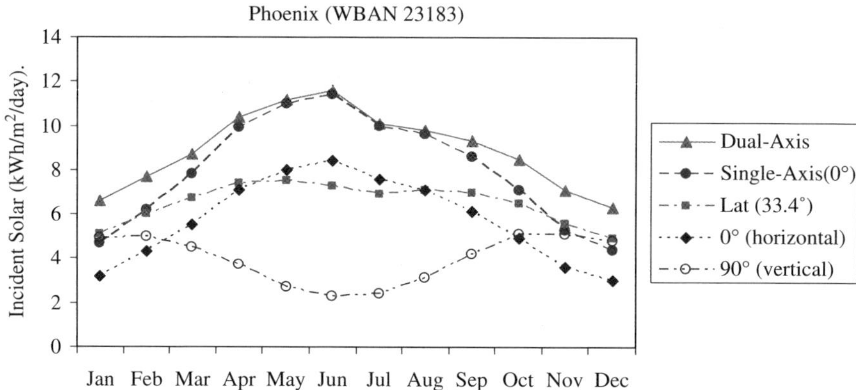

Fig. 8.9.   Solar radiation for flat-plate collectors facing south at various tilts. Data are from Ref. 20.

solar radiation for horizontal and south-facing vertical flat-plate collectors is 5.7 and 4.0 kWh/m²/day, respectively. The graph also shows the solar radiation for the more optimal configuration of a south-facing tilt at latitude (33.4°), which gathers an annual average of 6.5 kWh/m²/day. Figure 8.9 also demonstrates that by using either single or dual axis tracking, further increases in the total annual energy collection are achieved. In particular, non-tilted single-axis and dual-axis trackers received 8.0 and 8.9 kWh/m²/day, respectively. Such measured data may be used to refine analytical predictions in situations where measured data are unavailable for unique collector geometries.

8.8   Photovoltaic Energy Conversion

The electrical output of PV modules depends on the electrical, thermal, solar spectral, and optical properties of the module (or array) as well as the angle and amount of incident radiation.[21] The amount of input radiation has a linear (direct) effect on the current output of PV modules. The effect of variation in the solar spectrum on module output is relatively small for air mass values between 1 and 2 for crystalline PV modules. Empirical methods are available to estimate the effect of changing solar spectrum on PV module currents.

For surfaces not normal to the incident sunlight, the density of incident radiation is reduced, due to geometry, by a factor equal to the cosine of the angle of incidence as shown in Fig. 8.10 (see Eq. (8.20)).

In addition to this typical cosine loss of input radiation, large collector angles lead to optical losses due to reflectance from the surface of PV modules. These optical losses result in a lower incident radiation on the PV cell under the glass, than

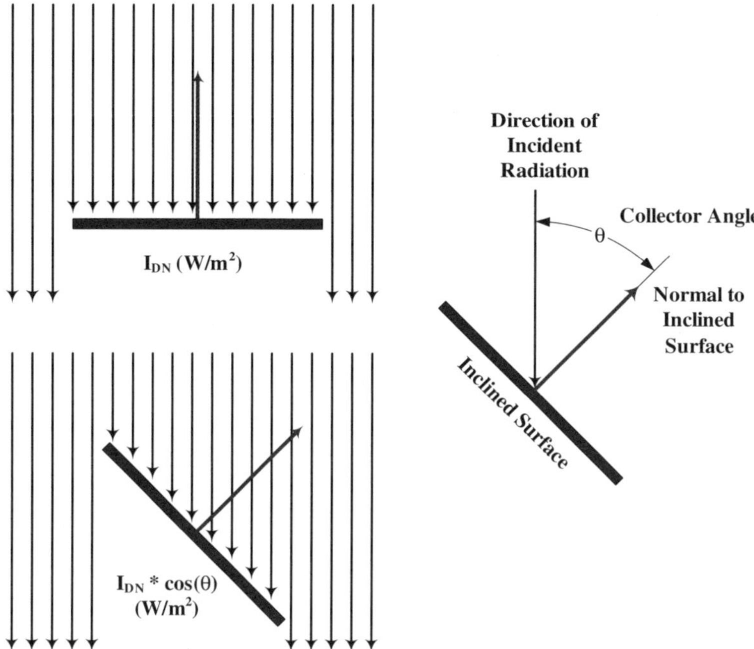

Fig. 8.10. An inclined surface, not normal to the incident radiation, receives a reduced radiation density, due to geometric effects of the collector angle.

what is suggested by Eq. (8.20). For crystalline silicon PV modules, the effect of optical losses is significant for angles of incidence greater than $55°$.[21]

8.9 Concluding Remarks

This chapter has provided straightforward methods to determine the position of the sun with respect to an earth-based solar collector. Such algorithms can be utilized for studies of solar collector performance and for active solar tracking equipment. There are more precise, and correspondingly more complex, computational approaches. For example, an algorithm to compute the solar zenith and azimuth angles within an uncertainty of just $\pm 0.0003°$ for Julian calendar years from -2000 to 6000 is described in Ref. 22. It is also noteworthy that passive solar tracking devices have also been developed.

References

1. G. Zhao, H. Kozuka, H. Lin and T. Yoko, "Solar energy conversion," *Wiley Encyclopedia of Electrical and Electronics Engineering* (John Wiley & Sons, New York, 1999), pp. 638–649.
2. M.M. El-Wakil, *Powerplant Technology* (McGraw-Hill, New York, 1984).

3. M. Pidwirny, *Fundamentals of Physical Geography*, 2nd edition, Chap. 6 (University of British Columbia, Okanagan, 2006), http://www.physicalgeography.net/fundamentals/contents.html.

4. Oklahoma Climatological Survey, Norman, OK (1997), http://okfirst.mesonet.org/train/meteorology/Seasons.html.

5. D.R. Williams, "Earth Fact Sheet," National Space Science Data Center, NASA Goddard Space Flight Center, Greenbelt, MD (2001), http://nssdc.gsfc.nasa.gov/planetary/factsheet/earthfact.html.

6. P.J. Lunde, *Solar Thermal Engineering: Space Heating and Hot Water Systems* (John Wiley & Sons, New York, 1980), pp. 62–100.

7. "Solar Spectra," Renewable Resource Data Center, National Renewable Energy Laboratory, Golden, CO, http://rredc.nrel.gov/solar/spectra/.

8. "Simple Model of the Radiative Transfer of Sunshine," National Renewable Energy Laboratory, Golden, CO, http://www.nrel.gov/rredc/smarts/.

9. ASTM E490-00a, Standard solar constant and zero air mass solar spectral irradiance tables, ASTM International, West Conshohocken, PA (2006).

10. ASTM G173-03, Standard tables for reference solar spectral irradiances: Direct normal and hemispherical on 37° tilted surface, ASTM International, West Conshohocken, PA (2003).

11. I. Reda and A. Andreas, Solar position algorithm for solar radiation applications, National Renewable Energy Laboratory, Technical Report, NREL/TP-560-34302 (2005).

12. "Solar Position and Intensity," Renewable Resource Data Center, National Renewable Energy Laboratory, Golden, CO, http://rredc.nrel.gov/solar/codesandalgorithms/solpos/.

13. "Solar energy use," in *ASHRAE Handbook — HVAC Applications* (*SI*), Chap. 33 (American Society of Heating, Refrigerating, and Air-Conditioning Engineers, Atlanta, GA, 2007).

14. A. Rabi, *Active Solar Collectors and Their Applications* (Oxford University Press, New York, 1985), p. 30.

15. D.S. Chauhan and S.K. Srivastava, *Non-conventional Energy Resources*, 2nd edition (New Age International, New Delhi, 2006), p. 47.

16. A.W. Culp, *Principles of Energy Conversion*, 2nd ed. (McGraw-Hill, New York, 1991), pp. 97–107.

17. B.Y.H. Liu and R.C. Jordan, "Daily insolation on surfaces tilted toward the equator," *ASHRAE J.* **3** (1961) 53–59.

18. S.A. Klein, "Calculation of monthly average insolation on tilted surfaces," *Solar Energy* **19** (1977) 325–329.

19. K.E. Holbert, "An analysis of utility incentives for residential photovoltaic installations in Phoenix, Arizona," *Proc. 39th North American Power Symp.*, Las Cruces, NM (2007), pp. 189–196.

20. W. Marion and S. Wilcox, "Solar radiation data manual for flat-plate and concentrating collectors," National Renewable Energy Laboratory, NREL/TP-463-5607 (1994).

21. D.L. King, W.E. Boyson and J.A. Kratochvil, "Photovoltaic array performance model," Sandia National Laboratories, Technical Report, SAND2004–3535 (2004).

22. I. Reda and A. Andreas, "Solar position algorithm for solar radiation applications," *Solar Energy* **76** (2004) 577–589; Corrigendum in **81** (2007) 838.

Chapter 9

Photovoltaic Systems

Ravita D. Prasad

College of Engineering, Science and Technology,
Fiji National University,
P.O. Box 3722, Samabula, Fiji
ravita.prasad@fnu.ac.fj

Ramesh C. Bansal

School of Information Technology and Electrical Engineering,
The University of Queensland, St. Lucia, Brisbane, QLD 4072, Australia
rcbansal@ieee.org

Generation of electricity from light began in the 19th century which led to a lot of developments in solar energy since then. This chapter focuses on photovoltaic (PV) systems which generate electricity. The first section gives an overview of how electricity is generated in a solar cell while the second section presents PV modules — the different types available, interconnection and the performance of modules. The third section describes the different types of PV systems; grid-connected and stand alone. Designing a stand-alone PV system is also presented; both for electricity supply to remote homes and solar water pumping systems. Finally, in the fourth section some conclusions are drawn.

9.1 Introduction

Solar radiation received on earth in just one hour is more than what the whole world's population consumes in one year. In 1839, a young French physicist named Edmund Bacquerel discovered the photovoltaic effect. While working with two metal electrodes in an electricity-conducting solution, he noted that the apparatus generates voltage when exposed to light.[1] It was not until 1904, when Albert Einstein published a paper on the photoelectric effect, that the general scientific community stopped looking at photovoltaic as some type of scientific hoax.

205

     R. D. Prasad and R. C. Bansal

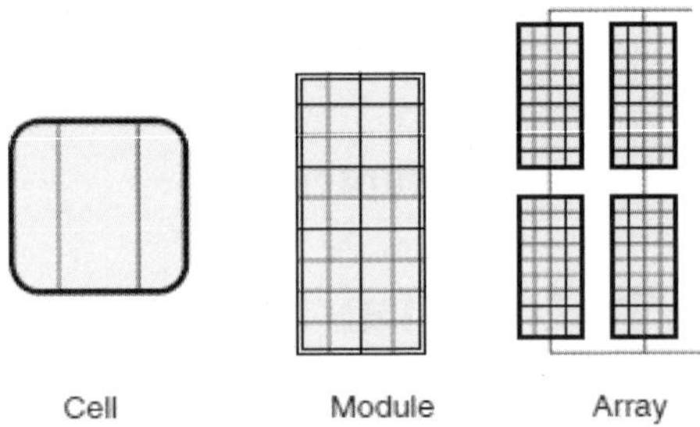

Fig. 9.1.   The photovoltaic hierarchy.[3]

Photovoltaic is the conversion of light directly into electricity through semiconductor materials. The basic component of a PV system is solar cell. If light with adequate energy falls onto silicon arranged to form a *p-n* junction and penetrates to a point near the junction, then, because of the photo-electric effect, it will create free electrons near the junction. These electrons immediately move under the influence of the *p-n* junction's electric field. The electrons continue to move through the cell to the surface of the cell. On the way towards the surface of the cell some of the electrons may be re-absorbed by the silicon atoms, but many electrons still reach the surface of the cell. These electrons can be collected by a metallic grid and an electric current will flow if the grid is connected to the metal contact on the other side of the cell by an external circuit.

The PV hierarchy is shown in Fig. 9.1. One solar cell has output voltage of around 0.5–0.6 V and very few appliances work at this voltage so solar cells are connected in series in a module to increase the output voltage of the module.[2] The number of cells in a module is governed by the voltage of the module. Photovoltaic module manufacturers make modules which can work with 12 V batteries. In allowing for some over-voltage to charge the battery and to compensate for lower output under non standard test conditions (STC), modules usually have 33–36 solar cells in series to ensure reliable operation. To increase the modules output current the series strings of solar cells are connected in parallel. Based on the desired current-voltage output of the module, solar cells are connected in both parallel and series combination. The modules can then in turn be connected in series and parallel to have the desired PV system voltage and current. Such combinations of modules are referred to as arrays.

9.2   PV Modules

PV modules or solar modules have d.c. electrical output power even though there are no moving parts and no pollutants emitted. PV systems are modular which gives

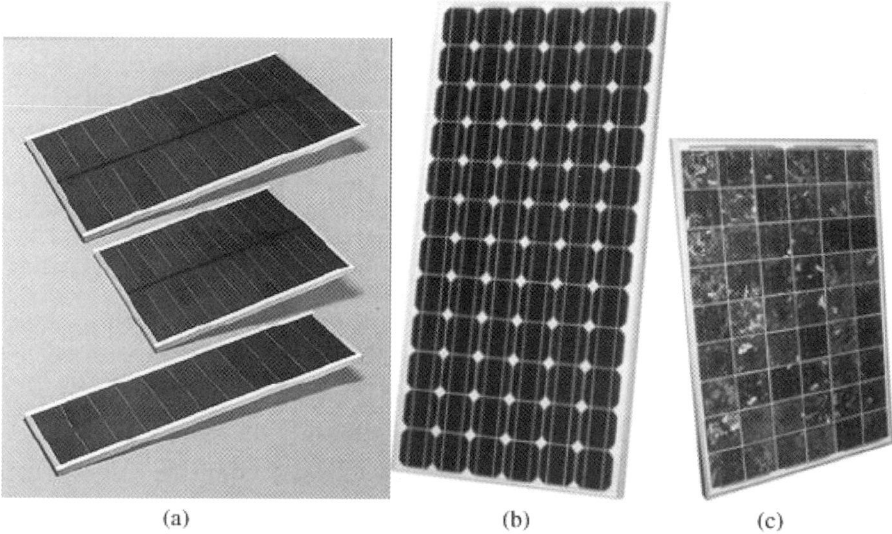

Fig. 9.2. Examples of the types of commercially available PV modules: (a) amorphous (b) monocrystalline and (c) polycrystalline.[4]

it an advantage of being able to increase its size even after it has been installed. It is flexible, easily and quickly coordinated and constructed into an array or a PV plant.

There are three types of PV modules (Fig. 9.2) on the market and they differ in the form of silicon used to manufacture them.

(i) Amorphous silicon modules: these are made from uncrystallized forms of silicon. They are often called thin film silicon (TFS) modules as the silicon is deposited in a thin layer or film on a variety of surfaces, such as glass. This type of module is in a dark matt colour and performs well in low light conditions. However, this has low efficiencies (typically 5–8%)[5] and thus requires a much larger roof area than all other technologies.

(ii) Monocrystalline solar cells: these are thin wafer cut from a large single crystal of silicon to form the individual cells and are bluish black in color. According to Ref. 6 this type of cell has the best efficiency for a given module area and well made modules have a proven long life.

(iii) Polycrystalline solar cells: these are thin wafers cut from a block of multiple crystal silicon. They are easily recognized by its color (usually blue), but there are other colors also and this is the most common panels available from a range of manufacturers.[7]

                    R. D. Prasad and R. C. Bansal

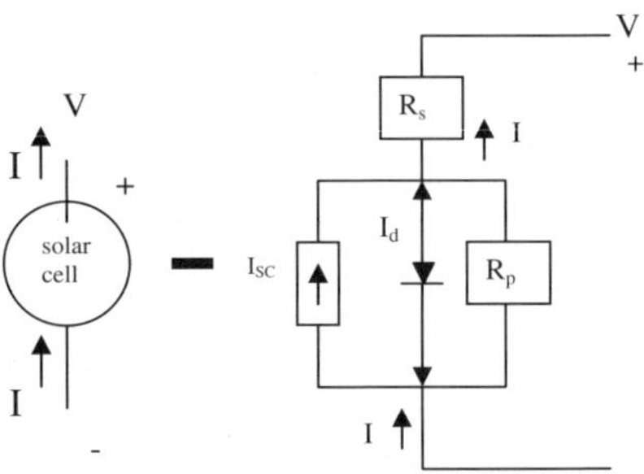

Fig. 9.3.   Equivalent circuit for solar cell in full sunlight.

9.2.1 *Shading effects on PV modules*

Since PV modules have around 33–36 cells in series, shading on just one cell in this series of cells can have reduction in power output from the PV module. The equivalent circuit for the PV cell is shown in Fig. 9.3 where the cell is in full sunlight.

Figure 9.4 shows one of the cells shaded in a PV module of n cells. R_p and R_s are the parallel leakage resistance and series resistance of a solar cell respectively. When one of the cells in PV module is shaded the short circuit current (I_{SC}) through the shaded cell is zero and the diode becomes reverse biased hence the current through

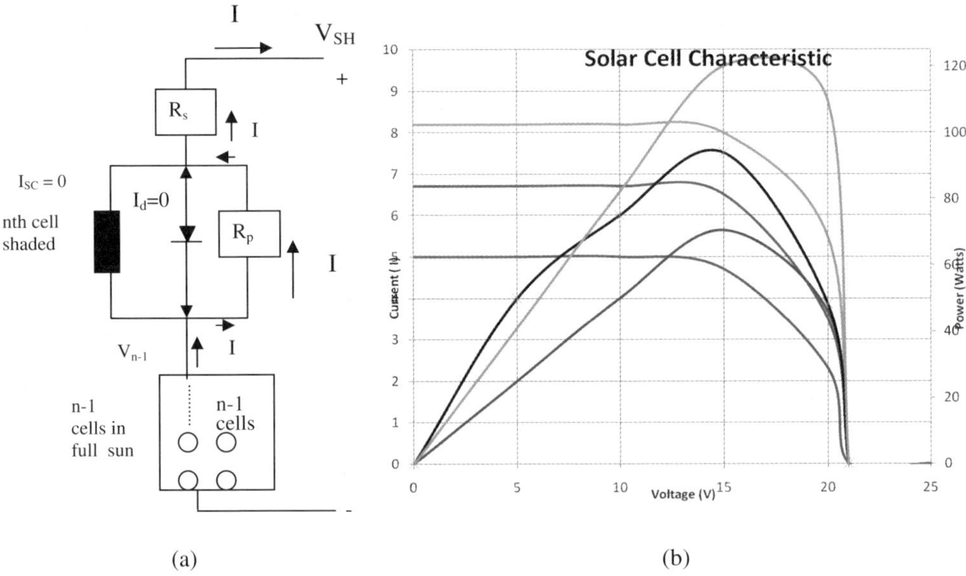

(a)   (b)

Fig. 9.4.   (a) Equivalent circuit for nth cell in solar module to be shaded. The figure shows typical solar output as a function of irradiation, (b) a typical power curve from a solar array. The efficiency of the solar module reaches its peak at a unique point and this point is called the maximum power point of the solar cell.

the diode (I_d) is zero. This results in the current produced by $n - 1$ cells in the PV module to pass through R_p and R_s which results in loss of voltage. According to Ref. 3 the drop in voltage (ΔV) by the shaded cell is given by

$$\Delta V = \frac{V}{n} + I(R_p + R_s), \tag{9.1}$$

where V is the voltage output of PV module when all cells are in full sunlight.

Since R_p is much greater than R_s, Eq. (9.1) simplifies to

$$\Delta V \cong \frac{V}{n} + IR_p. \tag{9.2}$$

9.2.2 *Interconnection of PV modules*

When modules are connected in series and parallel combination it contains by-pass and blocking diodes. These diodes protect the modules and prevent it from acting as a load in the dark or during shading.

9.2.3 *Performance of solar cells and modules*

The efficiency of a solar cell is

$$\eta_{\text{solar cell}} = \frac{P_{\text{out of solar cell}}}{\textit{Solar power incident on solar cell}}. \tag{9.3}$$

The losses are due to

- Grid coverage: The surface of the cell has to be covered with a metallic grid to collect the electrons produced by the photoelectric effect.
- Reflection loss: Some of the incoming solar radiation is reflected from the front surface of the cell.
- Spurious absorption: Some of the electrons ejected from their electron shell will be absorbed by impure atoms in the crystal.
- Some of the incoming solar radiation do not have sufficient energy to eject an electron from its electron shell.
- Some of the incoming solar radiation has more than enough energy to eject an electron from its electron shell. The extra energy is dissipated as heat in the crystal.
- Quantum efficiency: Of the photons with the correct energy to eject an electron from its electron shell only 90% will actually strike an electron and eject it.
- Absorption not near junction. Some photons are absorbed by the crystal far from the junction. These photons create electron hole pairs which do nothing but immediately recombine, leaving only a little heat as their legacy.
- Shading of solar cell: When part of solar cell or module is shaded the output voltage decreases.

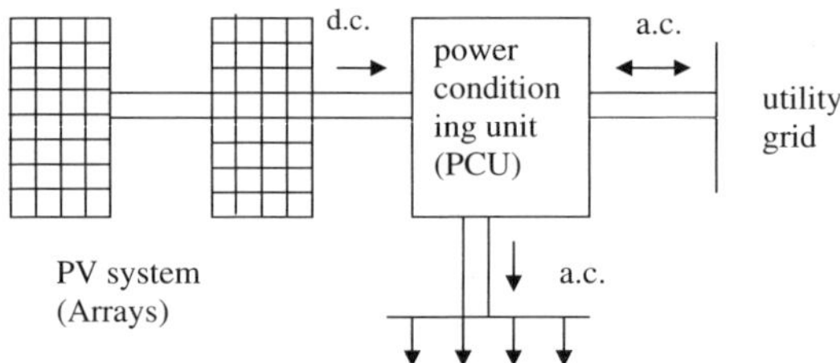

Fig. 9.5.   Grid connected PV system.

9.3   Types of PV Systems

There are two types of PV systems.

9.3.1   *Grid-connected system or utility interface (UI)*

In a grid connected system, the grid acts as a back-up and there is no need for battery storage unless there is a power outage problem. This makes grid connected PV systems relatively simple. However, PV systems have to compete with cheap grid supply which makes it hard to justify the PV system unless some kinds of subsidies are provided.

In PV systems, PV arrays may be pole mounted or attached on the roof or may be an integral part of the building known as Building Integrated Photovoltaics (BIPV). Sometimes batteries may be attached for backup if the utility grid is problematic. Before power from the PV array is distributed to the a.c. loads and the utility grid it goes through a power conditioning unit which changes d.c. to a.c., which maintains quality of supply and helps to operate PV system most efficiently (Fig. 9.5).

PV systems can be economical because a PV array can supply power to homes and businesses during mid day, when cost of electricity drawn from grid may be high and when PV systems are an integral part of building.

PVs mounted on buildings are becoming popular as the price of PV systems is decreasing and technology is getting more mature. Principal components in grid connected home size PV systems are PV arrays with two leads from each string sent to a combiner box that includes blocking diodes and individual fuses for each string. The strings are attached to the lightning surge arrestor.

9.3.1.1   *Net-metering*

When PV systems delivers more power, the electric meter runs backward, building credit with utility. This arrangement with a single electric meter running in both directions is called net-metering. In net-metering, power produced from the solar system is connected to the electric grid and credited in real time. All the surplus electrical power produced by the solar system that a home is not using is fed back into the power grid. As this surplus electrical power is fed back into the power grid, the meter at the home spins in reverse. This means the power produced by the solar system which is not used in the home is "sold" to the local utility company so the consumer's electricity bill is reduced.[8] This reduction of electricity bill has a positive impact on the economics of investment; since the bill reduces, the payback period for the investment would also reduce.

9.3.1.2 *Estimation of actual a.c. output power from PV systems*

Grid connected PV systems consists of an array of modules and a power conditioning unit which includes an inverter to convert d.c. from PV to a.c. required by grid or load. A good starting point to estimate system performance is rated d.c power under standard test conditions (STC), i.e., 1 sun (i.e., $1\,kW/m^2$), air mass ratio of 1.5 (AM1.5) and 25°C cell temperature. Then to estimate the actual a.c. power output, $P_{a.c.}$, Eq. (9.4) is used.[3]

$$P_{a.c.} = P_{d.c.,STC} \times \eta_{\text{conversion}},\qquad(9.4)$$

where $P_{d.c.,\,STC}$ the d.c. power of array is obtained by simply adding individual module ratings under STC and $\eta_{\text{conversion}}$ is the conversion efficiency which accounts for inverter efficiency, dusty collectors, mismatched modules and difference in ambient conditions. Even in full sunlight the impact of these losses can de-rate the power output by 10–20%.

Mismatch of modules

Three cases of mismatch of modules are explained below:

- If there are two modules in parallel combination in a PV system both rated at 180 W, but one module produces 180 W at 30 V and 6 A and the other produces 180 W at 36 V and 5 A. Then the maximum power output from the parallel combination is 330 W (11 A and 30 V) instead of 360 W.
- All modules may not have same power output even if they are made by the same manufacturer. For example, 100 W modules may actually be 96 W or 104 W.
- Different operating conditions may exist in different parts of the PV array, such as cleanliness of different parts of the PV array or some modules may be obscured by a cloud which is covering only part of the array.

These mismatches can easily drop the array output by several per cent.

Cell temperature

When the module operates under standard conditions of irradiance ($0.8\,kW/m^2$), spectral distribution of AM 1.5, ambient temperature of 20°C and wind speed (>1 m/s), the cell is operating at normal operating cell temperature (NOCT). As cell temperature increases above 25°C, the power produced decreases.

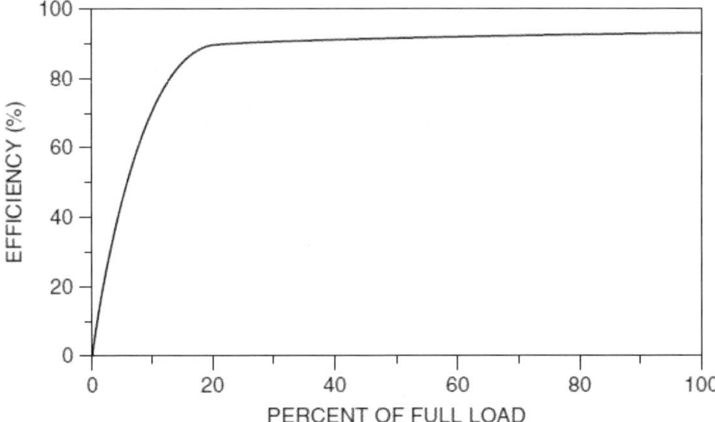

Fig. 9.6. Efficiency of inverter as a function of percentage of full load.[3]

The cell temperature is given by the following

$$T_c = T_a + \left[\frac{NOCT - 20}{0.8} \times S \right], \tag{9.5}$$

where T_c and T_a are the cell temperature and ambient temperature respectively in °C and S is the solar irradiance in kW/m^2.

Inverter efficiency

Since the power output from solar cell is d.c., the inverter is usually used to convert d.c. to a.c power. The power output from the inverter will depend on the efficiency of inverter. Efficiency of the inverter depends upon the load (Fig. 9.6). To improve the combined efficiency of the PV-based system, MPPT types of algorithms are deployed.

9.3.1.3 *Grid connected system sizing*

Grid connected system sizing is not critical as that of the standalone system because of back-up power arrangement, hence, trying to match supply to demand is not that important. Grid connected system sizing is more a matter of how much area is available on the building and budget of the buyer.

9.3.2 *Stand-alone system or off grid system*

In case of standalone systems such as the PV-battery system, it has to compete with a diesel generator (DG) or it competes with the cost of bringing the grid to the site. Off grid systems must be designed with great care to assure satisfactory performance. Users must be willing to check and maintain batteries; they must be willing to adjust

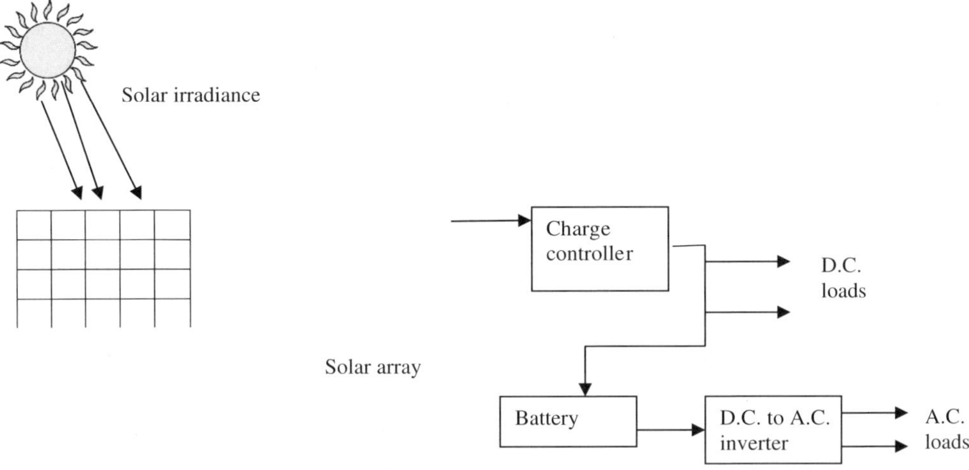

Fig. 9.7.   Simple schematic of stand-alone solar with battery storage.

their energy demands and must take responsibility of safe operation of the system for the reward of getting electricity which is truly valued.

For stand alone solar systems with battery storage charge controllers, inverters are important components apart from solar panel and battery (Fig. 9.7).

Charge controller as the name implies, controls the amount of charge that the battery would receive. When the battery has reached its charging limit, the charge controller will withhold further charging of the battery, sometimes known as *voltage regulating*. The absence of this would lead to overcharging of batteries, consequently damaging batteries and even causing fires. The majority of charge controllers have *low-voltage disconnect* (*LVD*) which prevent the batteries from over-discharging. Some charge controllers also include *maximum power point tracking* (*MPPT*) which optimize the solar PV array's energy production. Furthermore, charge controllers offer *battery temperature compensation* (*BTC*) where the charge controller regulates the charge rate based on the basis of the temperature of the battery since batteries are sensitive to temperature variations above or below 75°F.[9] A detailed discussion about MPPT controllers is prescribed in Chap. 11.

Solar batteries are usually *lead-acid deep cycle batteries*. This kind of batteries are tolerant to the way of charging and discharging going on in a solar system and they have a long life time compared to regular car batteries. Deep cycle batteries can discharge 80% of their rated capacity. In PV deep cycle batteries are favored so that during no sunlight periods the battery is able to supply the load demand. Another kind of batteries using Nickel cadmium (NiCd) plates can be used in PV systems. NiCd batteries are expensive but can withstand harsh weather conditions, and can be completely discharged without damage and the electrolyte will not freeze.

Inverters convert d.c. voltage from the battery to a.c. voltage so that household appliances can be powered since most of electrical appliances used in homes are a.c. There are two types of inverters used in solar systems; modified sine wave (MSW) and true sine wave (TSW). MSW inverters are simple and turn d.c. into approximately 120 V, 60 cycle a.c. but do not produce a true sine wave. Some electrical appliances, such as computers, immediately turn a.c. power into d.c. and so work perfectly fine with MSW inverters. However, other appliances do not work well with MSW. Audio equipment, for example, can produce an annoying hum when used with MSW inverters.[10] TSW inverters are more expensive, but for home use, they are almost always a better choice than MSW inverters. According to Ref. 9, a high quality inverter will include:

- An auto start system which allows an inverter to switch to a low power consumption standby mode when nothing is connected and turned on. This will save a lot of manual switching and/or wasted power.
- Tweaking ability which adjusts parameters such as auto-start and battery depth discharge.
- High quality heavy-duty power transformer.

Application of stand-alone PV systems

- Space: solar cells are used to power spacecraft, satellites and remotely-controlled vehicles on Mars.
- Marine navigational aids: PV systems are used to power marine navigational aids.
- Telecommunications: PV systems are used to power telecommunication stations which are in remote area and without grid support.
- Health stations: PV systems provide power to health stations which are in remote areas for refrigeration of medicines, etc.
- Lighting: In remote areas PV systems are used to power lights in homes. This benefits remote communities in terms of better education.

9.3.2.1 *Approach to designing an off-grid PV system with battery as storage*

The design of stand-alone PV system with battery as storage is pretty simple. Initial steps in the design process include:

- Determining the load (energy (kWh) not power (kW)). In this the nominal system voltage, range of voltages that would be tolerated by load must be considered, average load per day must be determined and load profile throughout the year must be determined. Table 9.1 can be used to find the load.

Table 9.1. Calculating total load demand per day.

				d.c.		
Appliances	Quantity	V	I	$P_{d.c.}$	Hrs used	Daily Energy (Wh)
				a.c.		
Appliances	Quantity	V	I	$P_{a.c.}$	Hrs used	Daily Energy (Wh)
						Total daily energy (a.c. + d.c.)

It is suggested by Ref. 7 that in designing PV systems for homes, the electrical appliances, where large heating or cooling is required, should not be used since they consume large amounts of energy and this will consequently increase the size of PV systems and cost. Furthermore, in solar power systems the most efficient types of lighting and appliances should be used in order to reduce the energy consumption for carrying out a certain job.

At this point it should also be noted that even though d.c. appliances are expensive and using this would require larger wires to cope with the larger currents, in the end it would consume the least amount of energy and thus the PV modules size and cost would decrease. The load analysis described above has to be carried out carefully to see which d.c. and a.c. loads would be used in order to decrease the total energy consumption.

$$\text{Total daily load demand (Wh)} = \text{total daily energy} + 20\% \text{ of total daily energy.}$$
(9.6)

Additional 20% is taken to cater for loss

- Calculating the battery size. In this step the number of days in a week where there is no sun is multiplied with the total daily load, and the battery losses is also catered for.

$$\text{Battery storage capacity (Wh)} = \frac{\text{total daily load demand} \times \text{No. of no sun days}}{\eta_{\text{Coulomb}}}.$$
(9.7)

The coulomb efficiency η_{Coulomb} is usually taken as 80%.

This battery storage capacity is then divided with the nominal voltage of the system to get the battery size in ampere-hours (Ah). In stand-alone systems the nominal system voltage for d.c. is 12 V, 24 V or 48 V and a.c. is 120 V. The system voltage is usually chosen to ensure that the total discharge current of the battery does not exceed 100 A. According to Ref. 11 it is noted that for d.c. systems, the system voltage should be that required by the largest loads. Most d.c. PV systems smaller than 1 kW operate at 12 V d.c. (The maximum current would be $1000 \div 12 = 83.3$ A.)

Also according to Ref. 11 inverter specifications have to be studied since it will provide the total and instantaneous a.c. power required. An inverter will be selected that will meet the load and keep the d.c. current below 100 A. For example, if the a.c. load is 3600 W and the a.c. voltage is 240 V, the a.c. current will be 15 A. Excluding losses in the inverter, the d.c. power must be the same; 3600 W. If a 12 V inverter is selected the d.c. current would be 300 A — not recommended. Use a 48 V inverter to make the input current to 75 A.

The voltages of the inverter, battery charger, PV array and any other equipment connected to the d.c. system will be determined by the nominal voltage of the system.

- Calculating the number of PV modules required

By analyzing the local solar radiation data (daily and monthly radiation) and by considering total daily load demand, the appropriate size of solar modules will be selected. The number of parallel and series combination of solar modules will be selected based on the maximum current and voltage that will be supplied to the load.

Example 9.1:

The a.c. load for a remote home is 2200 Wh/day. A PV system with battery storage is considered for powering this home. If the inverter efficiency is taken as 85%, coulomb efficiency is taken as 80%, PV de-rating is 90% (10% losses due to dirt and temperature) and system voltage is 24 V.

(a) Calculate the size of batteries for maximum 5 days of storage if a 12 V battery with 100 Ah is considered.
(b) Determine the PV size if for the site on average there are 3 hours of full sun and a 12 V PV module is considered which has a rated current of 6.99 A.

Solution:

(a) d.c. load input to inverter

$$\frac{2200 \; Wh}{0.85} = 2588 \; Wh \; d.c.$$

$$Battery \; storage \; capacity \; (Wh) = \frac{total \; daily \; load \; demand \times No. \; of \; no \; sun \; days}{\eta_{Coulomb}}$$

$$= \frac{2588 \; (5)}{0.80} = 16175 \; Wh.$$

Since 12 V battery @ 100 Ah is considered and the system voltage is 24 V, 2 batteries will be in series.

Ah for battery is

$$Ah\ for\ battery = \frac{16175\ Wh}{24\ V} = 674\ Ah.$$

Number of parallel strings is

$$\#\ of\ parallel\ strings = \frac{674\ Ah}{100\ Ah} = 6.74 = 7\ parallel\ strings.$$

Altogether there are 2 batteries in series with 7 parallel strings = 14 batteries in total.

(b) PV sizing

Number of PV modules in series is

$$\#\ of\ PV\ modules\ in\ series = \frac{System\ voltage}{V_{PV\ module}}$$
$$= \frac{24\ V}{12\ V} = 2\ modules\ in\ series.$$

Since a.c. load is 2200 Wh/day and d.c. load for input to inverter is 2588 Wh/day, the amphere-hours (Ah) needed at the inverter is

$$Ah\ needed\ @\ inverter = \frac{2588\ Wh}{24\ V} = 107.83\ Ah/day\ @\ 24\ V.$$

One string has 2 modules, therefore, Ah to inverter is

$$6.99\ A \times 3\ hrs\ full\ sun \times 0.90 \times 0.90 = 16.99\ Ah/day.$$

Hence, the number of parallel strings is

$$\#\ of\ parallel\ strings = \frac{107.83\ Ah}{6.99\ Ah} = 6.35 = 7\ parallel\ strings.$$

Therefore the PV array will have 2 modules in series with 7 parallel strings.

The tilt angle of the PV array must also be determined. If the location is in the Southern hemisphere then the solar module must be facing north and if in the Northern hemisphere, then facing south. If there is no solar tracking system for the module, then the optimum tilt angle for the solar module must be determined. This is usually done by adding the geographical latitude to 20° in the southern hemisphere.

- Assessing the need for any back-up energy of flexibility for load growth

Electrical energy demand from past years must be studied to assess the need for a back-up power source.

9.3.2.2 *Stand-alone PV system with battery and diesel generator as back-up*

In stand-alone systems there may be times when the battery bank is incapable of providing increase in load demand or peak demand. In this case the battery may over discharge and thus have lower battery life. To cater for load growth some back-up needs to be thought of to make the system reliable and efficient. Since having an oversized PV system would increase the overall cost of the system and the PV system would not be efficiently utilized if it was sized for peak demands. Back-up diesel generators provide a good solution to this problem.

The size of the load and the seasonal insolation variability at a site are two main indicators that work together to alert the designer to a possible hybrid application.[11] Plotting the load versus the array/load ratio (Fig. 9.8) gives an indication of whether a hybrid system should be considered. If the point falls between the two curves or above the top curve, then sizing a PV-generator hybrid system is recommended so that cost comparisons with the PV-only design can be made. Muselli *et al.*[12] proposed the optimal configuration for hybrid systems by minimizing the kilowatt-hour (kWh) cost. Researchers[13–15] have used the Hybrid Optimization Model for Electric Renewables (HOMER)[16] software to find optimum sizing and minimizing cost for hybrid power system with specific load demand.

A conventional diesel generator system design simply involves selecting a locally available unit that is closest to the peak load requirements of the application. However, designing a hybrid system, Fig. 9.9 (solar/diesel with battery storage), involves matching load demand with expected energy to be generated. In a hybrid

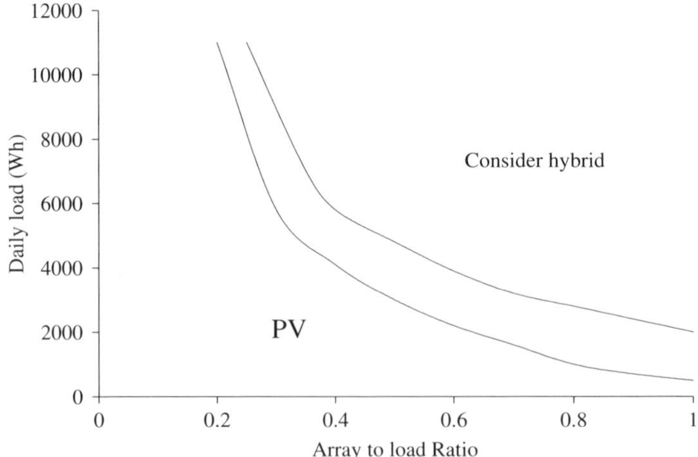

Fig. 9.8.   Hybrid design determination.[11]

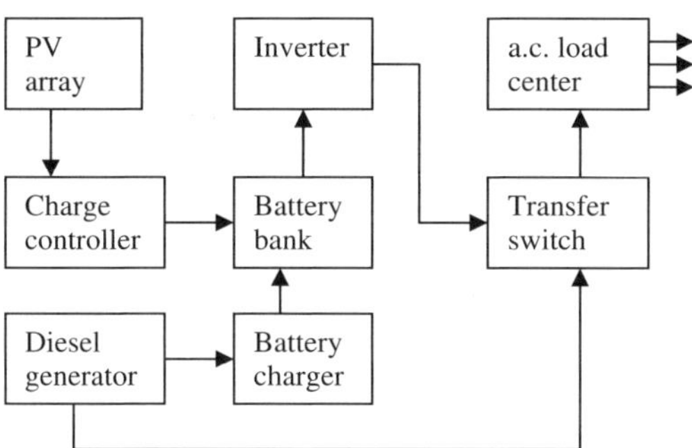

Fig. 9.9.   Schematic of a stand-alone PV with battery with diesel generator as back up.[17]

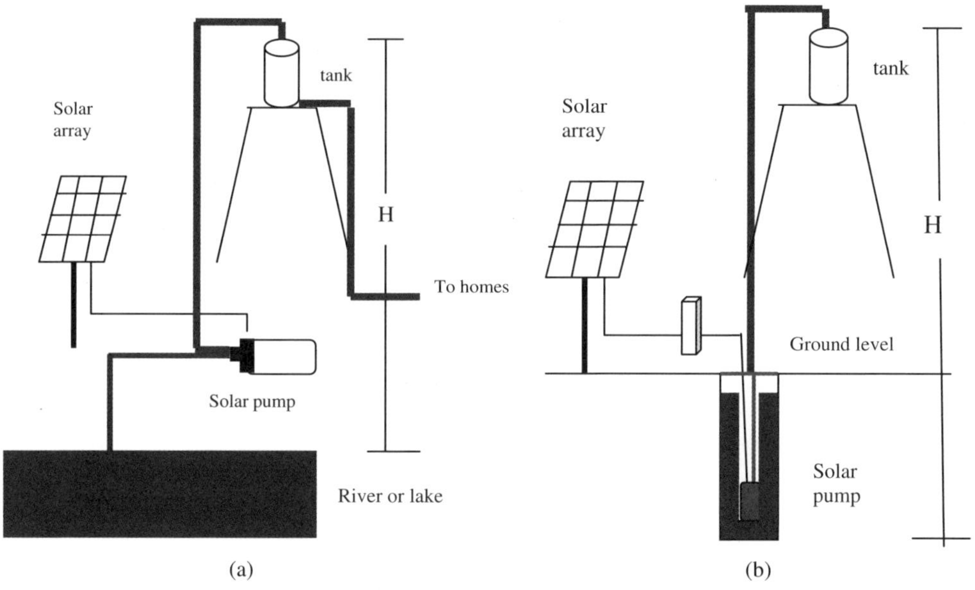

Fig. 9.10.   Surface pumps and submersible pumps.[18]

system when the battery bank cannot supply the load center then the generator will power the a.c. loads as well as a battery charger to help to recharge the batteries.

9.3.2.3   *Stand-alone solar water pumping system*

A solar powered water pumping system is essentially comparable to any other water pumping system. It basically uses the PV module to power the electric pump to pump water from the ground to some kind of storage tank (Fig. 9.10).

All solar powered pumping systems have a PV array, holding structure, a motor-pump, water tank (pressurized or unpressurized), control panel and protection. The array can be coupled directly to a d.c. motor or, through an inverter, to an a.c. motor. The motor is connected to any one of a variety of variable-speed pumps (surface or submersible pump). The use of batteries should be avoided in solar water pumping system because batteries decrease the system efficiency and increases the overall cost of the system.

9.3.2.3.1   Sizing/designing PV water pumping system

A solar powered water pumping system is designed to provide a given flow of water in gallons per minute (GPM) for a given pressure or lift (head). Some of the steps that are followed are as follows:

1. The first step is to estimate the amount of water (gallons) that will be required per day.
2. Distances (vertical and horizontal) that the water will be pumped has to be deter-mined. The total dynamic head (TDH) has to be calculated which is the sum of static lift of the water (static lift is measured from ground level to the low water level in the well), the static height of the storage tank (static height is measured from ground level to the top of the tank) and the losses from friction (friction losses in terms of equivalent height are determined by pumping rate and the size of the pipe).
3. Determine the size of the pump that one would need and the amount of power needed from the PV.

   The pumping rate of the pump (Q) GPM would be determined from the amount of water required per day (Eq. (9.8))

$$Q = \frac{\text{gallons per day}}{\text{peak sun hours per day}} \times \frac{\text{hour}}{60 \text{ minutes}}. \tag{9.8}$$

The power (W) that is delivered by the pump to the fluid is given by

$$P_{\text{out}_{\text{pump}}} = \rho g H Q, \tag{9.9}$$

where ρ is the fluid density (kgm^{-3}), g is the gravitational acceleration (9.81 ms^{-2}), H is the head in m and Q is the pump rate (m^3s^{-1}).

Pump input power can also be determined from Eq. (9.10)[3]

$$P_{\text{in}_{\text{pump}}} = \frac{0.1885 \times H \, (feet) \times Q \, (GPM)}{\eta_{\text{pump}}}, \tag{9.10}$$

where η_{pump} is the pump efficiency, the efficiency of the pump can be 25% for suction pumps and 35% for submersible pumps.

The number of PV modules connected in series that would be required would be found by

$$\text{Number of PV modules in series} = \frac{\text{Voltage of solar pump}}{\text{Voltage of one solar module}}. \tag{9.11}$$

Since the power out from the PV system is input for the pump, one would get Eq. (9.12)

$$P_{in_{pump}} = P_{out_{PV}}$$

$$P_{in_{pump}} = \eta_{due\ to\ dirt/temp\ loss} \times V_{one\ module} \times No.\ of\ modules\ in\ series$$
$$\times I_R \times No.\ of\ parallel\ strings$$

$$\therefore No.\ of\ parallel\ strings$$

$$= \frac{P_{in_{pump}}}{\eta_{due\ to\ dirt/temp\ loss} \times V_{one\ module} \times No.\ of\ modules\ in\ series \times I_R},$$
$$(9.12)$$

where I_R is the rated current of the PV module.

Equation (9.12) is for the PV module connected directly to the solar pump. However, if an inverter is placed between the PV module and the pump then the inverter efficiency needs to be considered in Eq. (9.12); it would be multiplied with $P_{out_{PV}}$.

4. Size of the storage tank. A typical storage tank is sized so that it can store water for certain number of "no sun" days. A float switch is usually installed inside the tank to control the pump according to the water level.

Example 9.2:

If you require 2000 gallons per day and the site has 4.8 peak sun hours per day, (i) determine the pumping rate and (ii) calculate the friction loss in terms of vertical height if you are using 200 feet of 3/4 inch pipe. Consider friction loss 10.6 feet per 100 feet.

Solution:

(1)

$$\text{Pumping rate, } Q = \frac{\text{gallons per day}}{\text{peak sun hours per day}} \times \frac{\text{hour}}{60\,\text{minutes}}$$
$$= \frac{2000\,\text{gallons per day}}{4.8\,\text{hours}} \times \frac{1\,\text{hours}}{60\,\text{minutes}}$$
$$= 6.94\,\text{GPM}.$$

(ii) As friction loss is 10.6 feet per 100 feet of pipe used at a pump rate of $\sim$7 GPM. Hence for the 200 feet of pipe used the frictional loss in terms of height is $(10.6 \times 2) = 21.2$ feet.

Example 9.3:

For solar water pumping system sum of static lift and static height comes out to be 200 feet. For this system the pump is a 24 V d.c. surface pump with 30% efficiency. Find the size of PV array if the solar module chosen has a voltage of 15 V which has rated current of 5.99 A and water demand, pipe and solar data information are taken from Example 9.2.

Solution:
The pumping rate, Q comes out to be 6.94 GPM.

The frictional loss in terms of height comes out to be 21.2 feet. Hence the TDH is calculated to be $(200\,0\text{ft} + 21.2\,\text{ft}) = 212.2\,\text{ft}$.

The power input the solar pump is

$$P_{\text{in}_{\text{pump}}} = \frac{0.1885 \times H\,(feet) \times Q\,(GPM)}{\eta_{\text{pump}}}$$

$$= \frac{0.1885 \times 221.2 \times 6.94}{0.3}$$

$$= 964.57\,W.$$

The number of PV modules in series would be

$$\text{Number of PV modules in series} = \frac{\text{Voltage of solar pump}}{\text{Voltage of one solar module}}$$

$$= \frac{24\,\text{V}}{15\,\text{V}} = 1.6 \cong 2\,\text{modules}.$$

Number of parallel strings is

$$No.\ of\ parallel\ strings$$

$$= \frac{P_{\text{in}_{\text{pump}}}}{\eta_{\text{due to dirt/temp loss}} \times V_{\text{one module}} \times No.\ of\ modules\ in\ series \times I_R}$$

$$= \frac{964.57\,W}{0.8 \times 15\,V \times 2 \times 5.99} = 6.7 \cong 7\ strings.$$

9.4   Concluding Remarks

PV systems have a wide range of applications both in grid systems and stand-alone systems. In grid connected PV systems matching output power from PV array to the load is not significant due to the use of net meters. These meters run in both directions thereby when homes have excess electricity it can sell it to the grid and the meters runs backwards and if there is deficit power from PV systems then power can be bought from the grid and the meter would run forward. However, in stand-alone systems load must be matched with the PV system. Before PV systems are installed vigilant design of PV systems are required. This chapter has attempted to provide guidelines to designing a PV system for electrification of remote homes and solar water pumping systems.

References

1. "Design and installation of solar power systems: Training manual," Training Productive Authority of Fiji (TPAF), Suva, Fiji.
2. T. Markvart (ed.), *Solar Electricity* (John Wiley & Sons Ltd., UK, 1994).
3. G.M. Masters, *Renewable and Efficient Electric Power Systems* (John Wiley & Sons, Hoboken, New Jersey, 2004).
4. NYSERDA, "Guide to solar-powered water pumping systems in New York state," www.nyserda.org (2009).
5. Solarbuzz, "Solar cell technologies," http://www.solarbuzz.com/technologies.htm (2010).
6. R. Menzies, "Designing a solar power systems," *Proc. Solar Photovoltaic Energy Workshop'98*, Monash University, Caulfield (1998).
7. Bright Green Energy, "What types of solar PV panels are available?" http://www.wirefree direct.com/Types_Solar_PV_Panels.asp (2009).
8. Southwestern Solar, "Solar panel systems," http://sw-solar.com/solar-basics/about-solar/ (2009).
9. imexSOLAR, "Controllers," http://www.imex-solar.com/enlisch/solarchargecontroller.htm (2009).
10. "Solar power inverter," http://www.homesolarandwindinfo.com/ solar-power-inverter/ (2009).
11. Sandia National Laboratories, "Stand-alone photovoltaic systems: A handbook of recommended design practices," http://www.sandia.gov/pv/docs/PDF/Stand%20Alone.pdf (1995).
12. M. Muselli, G. Notton, P. Poggi and A. Louche, "PV-hybrid power systems sizing incorporating battery storage: An analysis via simulation calculations," *Renewable Energy* **20** (2000) 1–7.
13. S. Kamel and C. Dahl, "The economics of hybrid power systems for sustainable desert agriculture in Egypt," *Energy* **30** (2005) 1271-1281.
14. M.J. Khan and M.T. Iqbal, "Pre-feasibility study of stand-alone hybrid energy systems for applications in Newfoundland," *Renewable Energy* **30** (2005) 835–854.
15. R.D. Prasad, R.C. Bansal and M. Sauturaga, "A case study of wind-diesel hybrid configuration with battery as a storage device for a typical village in Pacific Island Country," *Int. J. Agile Systems and Management* **4** (2009) 60–75.
16. National Renewable Energy Laboratory, "HOMER Getting Started Guide," Version 2.1, NREL (2005).
17. "Solar system design," http://www.ehvacdesign.com/solar_system_design.htm (2010).
18. "Solar electrification systems," http://www.windgen.biz/Solar%20Electrification%20Systems/ SolarSystem.htm (2009).

Chapter 10

Solar Thermal Electric Power Plants

Keith E. Holbert

School of Electrical Computer and Energy Engineering, Arizona State University,
P.O. Box 875706, Tempe, AZ 85287-5706

keith.holbert@asu.edu

Concentrating solar power appears to be the method of choice for large capacity, utility-scale electric generation in the near term. In particular, distributed trough systems represent a reasonably mature approach, and the power tower configuration is also a viable candidate. Both of these technologies have the possibility of energy storage and auxiliary heat production during sunlight unavailability. Other systems examined in this chapter include dish-Stirling engines, solar chimneys and solar ponds. Economics inevitably drives the choice of technology.

10.1 Introduction

Heliothermal is the process of conversion of solar energy to thermal (heat) energy. Solar thermal can be used directly for heating purposes (e.g., domestic hot water) — and this was the approach taken by many early residential solar installations. In terms of large-scale electric utility power generation, solar energy can be utilized to heat a fluid and eventually produce steam to drive a traditional turbine-generator set, which produces AC power (in contrast to solar cells that have DC output). The approach known as concentrating solar power (CSP) is the major subject of this chapter. The solar thermal domestic implementation remains important since some electric utilities provide monetary incentives to customers who install such solar systems, and that subject is addressed in Chap. 12 of this handbook.

10.2 Solar Thermal Systems

The leading contenders for exploiting solar assets for electric power generation are photovoltaic (PV) and concentrating solar power (CSP) facilities. Other schemes

225

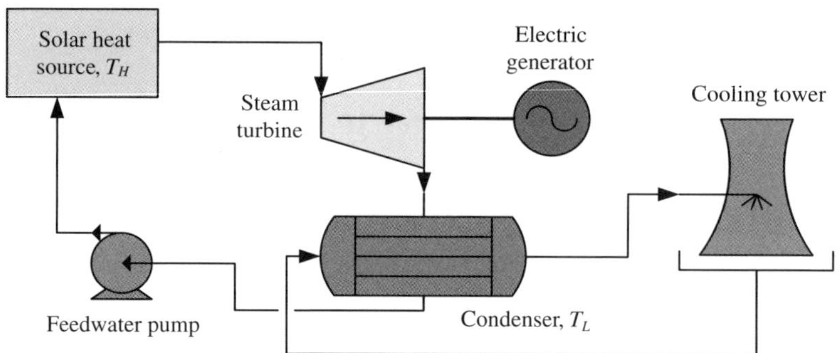

Fig. 10.1.   Solar thermal plant employing the Rankine cycle.

for converting sunlight to power include solar salt ponds and solar chimneys. In the near-term, CSP appears to be leading PV for utility-scale power generation due in part to its maturity and relative cost. In addition, there are two other advantages of solar thermal over PV. First, most CSP units can be constructed with an integral thermal energy storage (TES) system thereby providing the capability of generating electricity into the evening hours. Second, solar thermal plants can be equipped with auxiliary burners in order to generate electricity when sunlight is unavailable (e.g., at night time and during inclement weather).

Most CSP plants employ a thermal cycle, as depicted in Fig. 10.1, similar to those of coal and nuclear power plants except that the heat source is from sunlight. The thermal-to-mechanical energy conversion efficiency of a solar thermal electric generating station is Carnot cycle limited, that is, the maximum theoretical thermal efficiency is

$$\eta = 1 - T_L/T_H, \tag{10.1}$$

where T_L and T_H are the minimum and maximum absolute temperatures within the cycle. Since T_L is approximately equal to the ambient environmental temperature at the power plant site, this relation motivates the design of a thermal cycle that adds heat at the highest possible temperature (T_H); and hence, it promotes the utilization of sunlight concentrators to achieve high temperature operation.

Solar thermal technologies can be classified in terms of the temperature reached by the working fluid, as exhibited in Fig. 10.2. Flat plate collectors are employed in domestic hot water heating applications. Since lower thermal efficiency typically results in higher capital costs, electric utilities will tend to adopt concentration methods that achieve higher temperatures and efficiencies. The utility-scale CSP systems may be categorized as

- *Large point focus*: employing heliostats to concentrate light at a central receiver (power tower systems), with plants sizes of 100 kWe to 100 MWe;

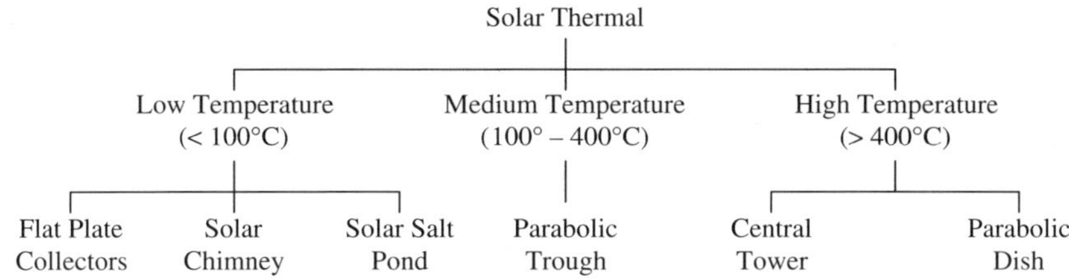

Fig. 10.2.   Categorization of solar thermal technologies according to temperature.

Table 10.1.   Representative solar thermal collector operating parameters.[1-4]

Thermal collector type	Concentration ratio	Working fluid temperature
Flat plate and solar salt pond	1	40°–100°C
Line focus (parabolic trough)	10–50	150°–350°C
Point focus (parabolic dish and central receiver)	100–1500	250°–1500°C

- *Small point focus*: using parabolic hemispherical dishes to reflect light to a focal point for each individual dish, (5–25 kWe); and
- *Line focus systems*: utilizing parabolic shaped troughs or linear Fresnel reflectors but having lower efficiency than the above, with plant output ranging from a few to hundreds of MWe.

In each of these three systems, a concentrator redirects (reflects) sunlight to a receiver (absorber). As can be noted in Table 10.1, working fluid temperatures for concentrating collectors are noticeably above ambient; hence, receivers may utilize vacuum barriers to prevent absorber heat loss due to convection to the surroundings.

10.2.1   *Solar thermal storage*

A key advantage of solar thermal compared to photovoltaics is the capability of integrating thermal energy storage (TES) within a CSP plant. Not only can such a TES system allow the plant to continue generating electricity during brief periods of sunlight loss (i.e., cloudiness), more importantly, electricity production can be continued after sunset and into the evening, which generally corresponds to a peak utility load period. Similarly, solar thermal plants can incorporate auxiliary burners to produce heat at night and during inclement weather. In the near term, TES is expected to provide capacity factors approaching 40%, and in the long term up to 70%.[5]

Figure 10.3 illustrates how a solar thermal power plant can attain base load generation by diverting thermal energy to storage during the mid-day hours while

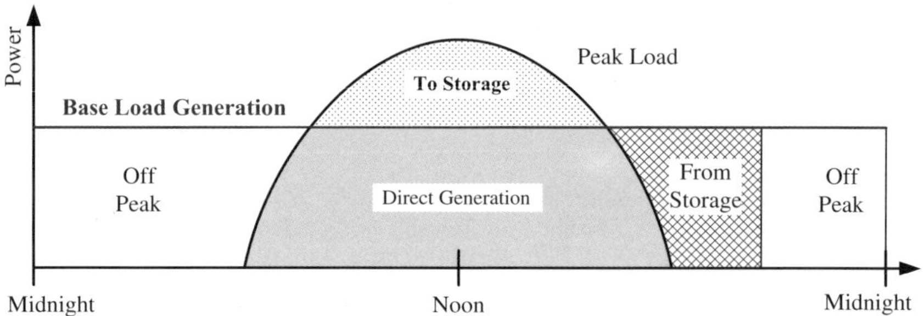

Fig. 10.3.   Storage and withdrawal of solar thermal energy.

also directly supplying electric power to the grid. In the early evening hours, the stored thermal energy is withdrawn and converted to electricity. Once the stored energy is depleted, auxiliary burners utilizing biomass, hydrogen or fossil fuels could be employed to produce electricity during off-peak times. In this manner CSP facilities can accomplish grid dispatchable power generation, that is, firm power delivery.

Two types of TES systems have been demonstrated on a significant scale. The first, called a thermocline, utilizes a single storage tank with a thermal gradient of hot fluid on top and cooler fluid at the bottom. The second employs dual storage tanks with one serving to hold the cold fluid while the other contains the hot medium. The latter dual tank system permits a greater fraction of its stored energy to be extracted.

Because the solar field must be oversized for a CSP plant with thermal storage, the facility may be described in terms of its solar-to-electric capacity ratio (β), which is the ratio of the solar field thermal output capacity to the plant electric power output.

10.2.2   *Hybrid facilities*

There are also hybrid solar thermal plants in which fossil or biomass fuels, for instance, can be used as a backup heat source. As depicted in Fig. 10.4, the Solar Electric Generating Systems (SEGS) units built in the 1980s employ natural gas to provide up to 25% of the thermal energy for steam.[6] Hybridization can be accomplished in trough, power tower, and dish engine systems, although implementation in the latter is more difficult.[7] It is noteworthy that present hybrid CSP facilities have only moderate thermal efficiency, leading to greater emissions on a per kWh basis from fossil fuel burning than would occur in a modern fossil power plant; however, designs exist that utilize an integrated solar combined cycle system (ISCCS). An ISCCS could employ a Brayton top cycle using a fossil-fired gas turbine and a Rankine bottom cycle that receives its heat from both the gas turbine waste heat

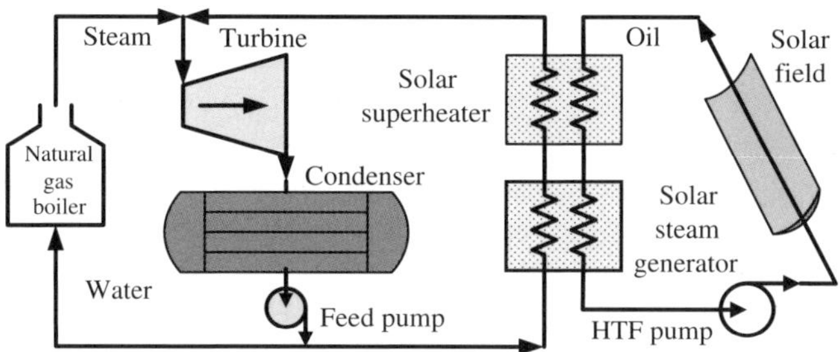

Fig. 10.4.   Schematic of SEGS hybrid power plant using oil as heat transfer fluid (HTF); after Refs. 4 and 9. Reheat is implemented in later SEGS units.[6]

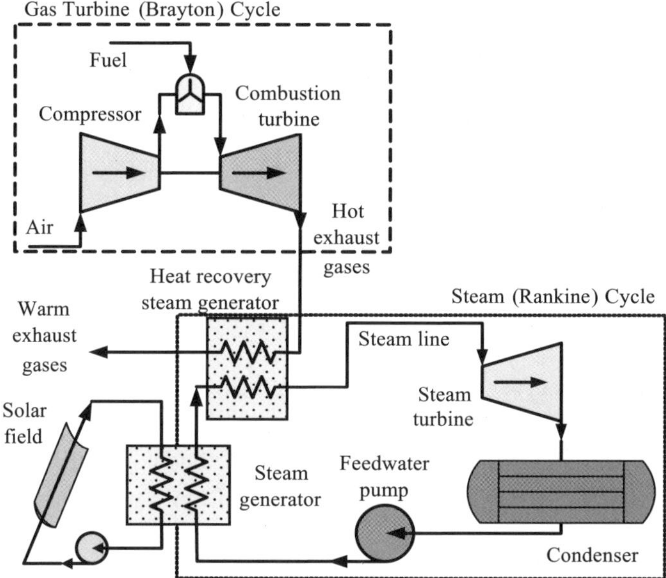

Fig. 10.5.   An integrated solar combined cycle system (ISCCS).

and solar heat supplication, as illustrated in Fig. 10.5. Various permutations of the ISCCS exist, including those designed to save fuel and others intended to boost the power plant electric output.[8]

10.2.3 *Economic considerations*

Table 10.2 compares projected costs for central receiver solar thermal and single-axis flat-plate photovoltaic power plants versus two other leading green energy sources. The table shows that the capital and the operating and maintenance (O&M) costs for solar thermal plants are high. The high O&M expenses are attributed to the

Table 10.2. Projected costs of renewable energy based electric generating stations.[14]

Technology	Unit Size	Capital Cost	Fixed O&M	Generation Cost*
Solar thermal	100 MW	$4693/kW	$56.78/kW	19.7 ¢/kWh
Photovoltaic	5 MW	$5750/kW	$11.68/kW	21.5 ¢/kWh
Wind (onshore)	50 MW	$1797/kW	$30.30/kW	7.9 ¢/kWh
Wind (offshore)	100 MW	$3416/kW	$89.48/kW	16.6 ¢/kWh

*Based on a capacity factor of 25% and an annual fixed charge rate of 8%.

costs associated with servicing and replacing plant components.[10] Assuming a pure (i.e., non-hybrid) solar plant, there are no fuel costs, and the levelized electricity generation cost is

$$e = \frac{IF_B + OM}{P_{e,\text{rate}}\, CF\ (8760\,\text{hrs/yr})},\tag{10.2}$$

where I is the levelized annual fixed-charge rate; F_B is the construction cost; OM is the annual operating and maintenance costs; $P_{e,\text{rate}}$ is the net plant electric output; and CF is the capacity factor. From the above equation, the importance of maximizing the capacity factor using techniques such as sensible heat storage becomes obvious. Capacity factor estimates vary from 25% to 29% without storage and from 40% to 56% with storage.[11–13]

Because central receiver and trough plants utilize conventional steam turbine-generator and balance-of-plant equipment, these plants benefit from economy of scale leading to unit sizes on the order of 100 MW. Dish systems, because of their smaller physical size, benefit from the possibility of mass production at a dedicated fabrication facility. In the case of the SEGS plants, the levelized cost of electricity was reduced from 24 ¢/kWh for the first unit to about 8 ¢/kWh for the ninth (and last) system.[15] Other estimates indicate overall CSP generation costs from power tower, parabolic trough and dish-Stirling engines to be 9.0, 13.4 and 16.7 ¢/kWh, respectively.[7] Besides improving marketability, CSP plants with thermal storage decrease capital costs on a per kWh basis since smaller capacity turbine-generator and balance of plant equipment is sufficient.

10.3   Concentrating Solar Power Systems

Within this section, three CSP plants are examined

- parabolic trough collectors (PTC),
- central receiver systems (power towers), and
- dish-Stirling engine systems.

Table 10.3. Comparison of CSP Technologies.[7,16]

Parameter	Parabolic trough	Power tower	Dish-Stirling engine
Peak efficiency	21%–24%	22%–23%	29%–30%
Annual capacity factor without thermal storage	23%–25%	25%–29%	25%
Annual capacity factor with thermal storage for $\beta = 1.8x$	33% (4 hrs)	48% (8 hrs)	n/a
Net annual solar-to-electric efficiency	13%	13%	15%
Acres/MW of collectors	5	8	4

An overall comparison of the three technologies is provided in Table 10.3. The gross solar-to-electric conversion efficiency can be described by

$$\eta_{\text{gross}} = \eta_{\text{con}}\eta_{\text{rec}}\eta_{\text{cyc}}\eta_{\text{gen}}, \tag{10.3}$$

where η_{con} is the concentrator efficiency; η_{rec} is the receiver efficiency; η_{cyc} is the thermal cycle efficiency; and η_{gen} is the electric generator efficiency. The net efficiency would also account for in-house (parasitic) use of electricity by the power plant itself.

Modern solar concentrators are of two main configurations: (1) heliostats, which are mirrors that focus sunlight at a single point as shown in Fig. 10.6(a), and (2) parabolic reflectors which redirect sunlight to single or multiple points. The parabolic concentrators include trough geometries that focus light along a line, and dish arrangements that concentrate at a single focal point as depicted in Figs. 10.6(b) and 10.6(c), respectively. Dish systems and heliostats require dual-axis solar tracking, whereas troughs need only use single-axis rotation.

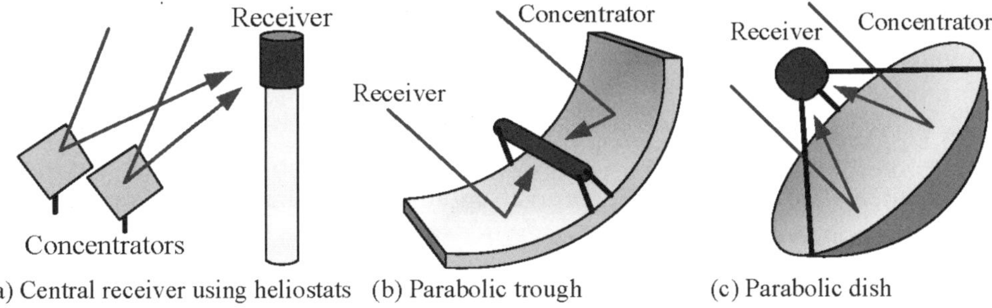

Fig. 10.6. Three primary solar concentrating systems.

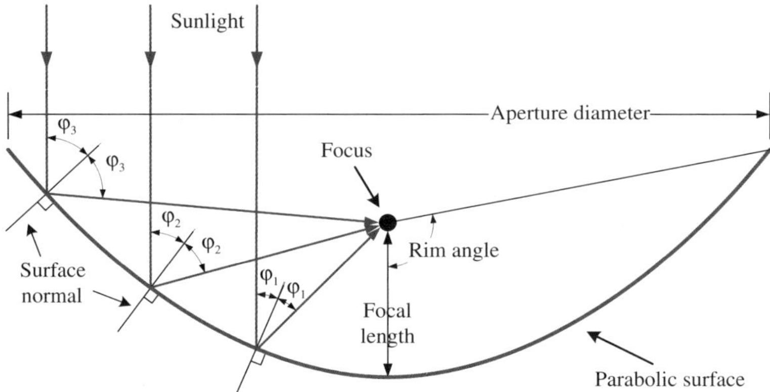

Fig. 10.7. Sunlight concentration by a parabolic surface. The focus is a focal point in the case of dish systems and a focal line for trough concentrators.

Carnot efficiency considerations of Eq. (10.1) demonstrate the need for heat addition at high temperature, which is accomplished by concentrating sunlight. The concentrating ratio is defined as

$$C = A_{ap}/A_{\text{rec}}, \tag{10.4}$$

where A_{ap} is the aperture (projected) area of the concentrator, and A_{rec} is the receiver area. Figure 10.7 illustrates the important dimensions associated with a parabolic reflector. The maximum concentrations for linear and dish concentrators are 212 and 45000, respectively.[17]

A listing of major CSP facilities is given in Table 10.4. Luz International Limited constructed the series of Solar Electric Generating Systems (SEGS) in the Mojave Desert of California (CA).[9] The SEGS plants benefitted from both favorable power contracts with an electric utility, and federal and state tax incentives.[10] It is noteworthy that Luz went bankrupt in 1991 when tax incentives (which had been as high as 35% in 1984–1986) and mandatory purchase contracts were withdrawn.[18] Presently, parabolic trough plants represent the most mature CSP technology,[16] as might be surmised from the number of operating plants in Table 10.4. Specific plants from this table are further discussed later in this chapter.

10.3.1 *Parabolic trough collectors*

As an alternative to the central receiver approach, distributed collectors, such as parabolic trough concentrators (PTCs), may be employed. A PTC is a line focus system in that a long mirrored concentrator in the shape of a parabolic trough is directed to heat the fluid within a conduit located on the focal line of the trough. The level of concentration within the PTC is less than that of a power tower, resulting in lower peak temperatures. In addition, a power plant requires many PTCs. At the

Table 10.4. Major concentrating solar power (CSP) plants.

Facility	Unit size (MW)	Location	Operating date(s)	Solar field area (m^2)
Central Receiver (Power Tower)				
Solar One	10	Barstow, CA	1982–1988	72650
Solar Two	10	Barstow, CA	1995–1998	82750
PS10	11	Seville, Spain	2007	74880
PS20	20	Seville, Spain	2009	150520
Line Focus (Parabolic Trough)				
SEGS I	13.8	Daggett, CA	1985	82960
SEGS II	30	Daggett, CA	1986	190338
SEGS III	30	Kramer Junction, CA	1987	230300
SEGS IV	30	Kramer Junction, CA	1987	230300
SEGS V	30	Kramer Junction, CA	1988	250500
SEGS VI	30	Kramer Junction, CA	1989	188000
SEGS VII	30	Kramer Junction, CA	1989	194280
SEGS VIII	80	Harper Lake, CA	1990	464340
SEGS IX	80	Harper Lake, CA	1991	483960
NV Solar One	64	Nevada, USA	2007	357200
Solnova 1	50	Seville, Spain	2010	300000
Solnova 3	50	Seville, Spain	2010	300000

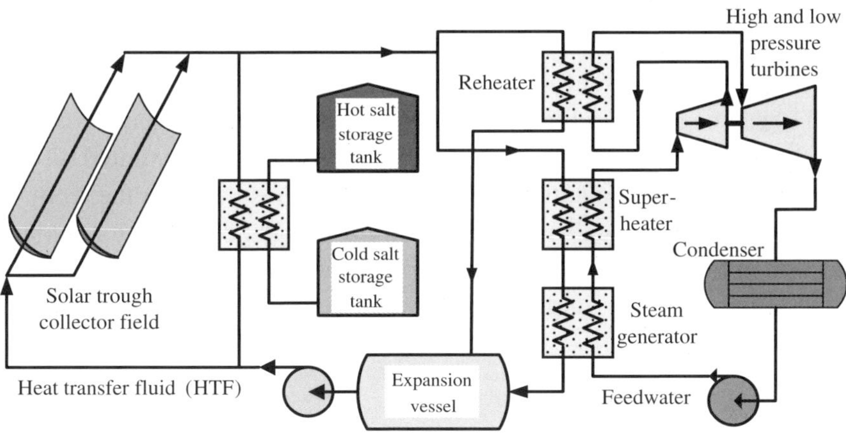

Fig. 10.8.   Trough type CSP plant with thermal storage, after Ref. 19.

present time, new solar thermal construction appears to favor PTCs, such as that presented in Fig. 10.8, over the power tower configuration.

PTCs are typically aligned with the focal line along the north-south axis. A horizontal north-south trough generally collects slightly more energy than an east-west

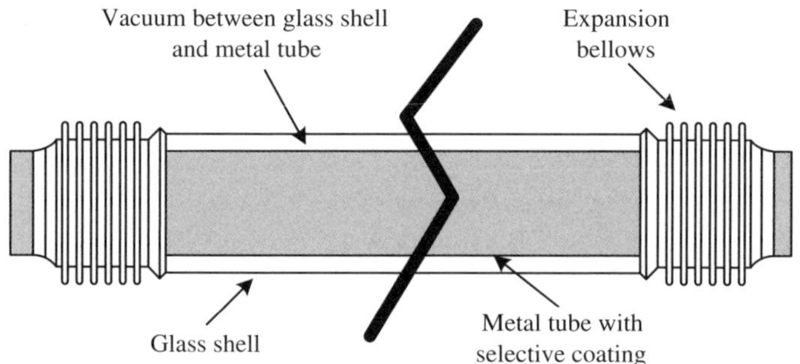

Fig. 10.9.   Parabolic trough receiver tube; adapted from Refs. 20 and 21.

alignment on an annual basis, although an E–W orientation typically provides a more constant annual output.[4] Single-axis tracking is sufficient for PTCs.

The receiver tube in the PTC is constructed using two concentric tubes placed on the focal line of the trough. As shown in Fig. 10.9, the inner metallic tube, which transports the working fluid, is separated from the outer transparent tube by vacuum. The outer glass tube and vacuum permit sunlight entry while preventing the convection heat loss, although there is transmittance loss through the glass.

Various working fluids may be utilized. For low temperature applications ($<200°C$), demineralized water with an ethylene-glycol mixture can be used. However, to achieve efficient high-temperature operation, a synthetic oil, such as biphenyl-diphenyl oxide, is employed.[21] More recently, approaches using high-pressure steam and molten salts have been investigated. If water–steam can be used directly, then the need for and cost of the heat exchangers and heat transfer fluid (HTF) can be avoided; however, heat storage may become more complicated. The DISS (Direct Solar Steam) Project has successfully demonstrated direct solar steam generation in parabolic trough collectors[22] and results favor the recirculation approach over a once-through method.[23] Multiple trough field collector layouts exist as illustrated in Fig. 10.10. The inverse return layout results in an even pressure drop (Δp) within the collector field, whereas the direct return and center feed require

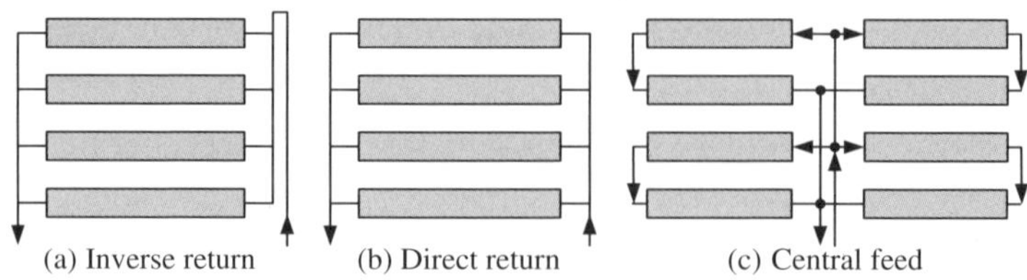

Fig. 10.10.   Layouts for a trough collector solar field, after Ref. 24.

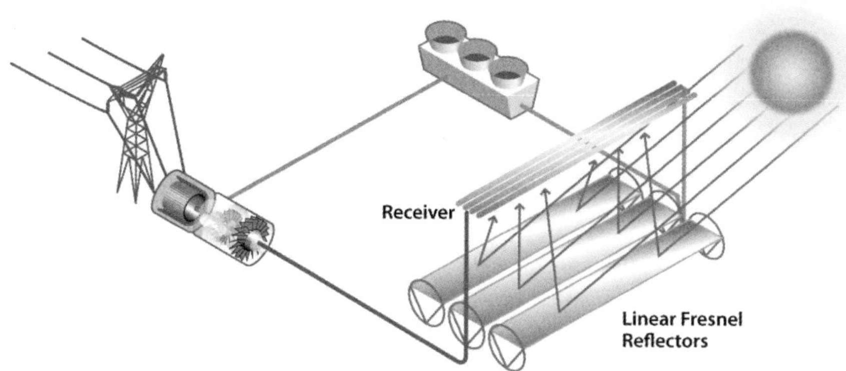

Fig. 10.11.   Linear Fresnel CSP (courtesy: U.S. Department of Energy).

valving to ensure a uniform Δp; however, the central feed requires less piping than the other two options.

The linear Fresnel system, as depicted in Fig. 10.11, is a cross of sorts between the parabolic trough and power tower systems. Each receiver is placed about 10 m above the ground, and multiple reflectors are oriented near ground level to redirect sunlight to the fixed receiver. A secondary reflector above the receiver prevents loss of light. In a variation, known as the compact linear Fresnel reflector (CLFR), the reflectors can be directed toward an alternative receiver that would make most effective use of the sunlight depending on the time of day. The interleaving of closely packed reflectors in the CLFR should increase the utilization of the available solar irradiance.[25]

10.3.1.1   *SEGS, NV Solar One and Solnova 1 and 3*

The 30-MW SEGS units use concentrators employing silvered low-iron glass reflectors (see Fig. 10.12) with the LS-2 and LS-3 collector assemblies achieving concentration ratios of 71 and 82, respectively.[26] The first SEGS plant employed a 3-hr thermal storage unit using mineral oil (Therminol VP-1). The SEGS I dual-tank storage system cost $25/kWt with the oil constituting 42% of the cost.[27] However, in 1999, an accidental fire destroyed the storage unit.[7]

The 64-MWe Nevada Solar One was completed in 2007 at a cost of $266 million. This is the largest CSP trough facility to be built in the world since 1991. Using a concentration ratio of 71, the 760 parabolic troughs heat the working fluid (therminol) from a field inlet temperature of 300°C to a field outlet of 390°C. Nevada Solar One does not have a TES unit and is allowed to use only 2% natural gas to steady temperature fluctuations in the HTF and for freeze protection.[28]

Fig. 10.12.   Solar trough collectors at Kramer Junction (courtesy of DOE/NREL).

The Solnova 1 and 3 trough systems heat a synthetic oil to 395°C. Each collector field is composed of 90 rows of four trough modules, which are each 12.5 m long by 5.76 m wide, yielding a mirrored surface area of 3×10^5 m^2. The heat is transferred to produce 390°C, 100 bar steam that feeds a 50 MWe turbine–generator set. The plant includes a natural gas fired boiler capable of supplying 12–15% of its capacity during low solar radiation conditions. The overall plant efficiency is to be near 19%.

10.3.2   *Central receiver systems*

Central receiver systems employ a large field of mirrors known as heliostats. The heliostats are individually oriented to redirect sunlight toward the top of the power tower. The heat transfer media of the thermal power cycle may be pumped to the top of the central receiver or a mirror can be located at the top of the tower to reflect the light to a ground-level receiver.[29–31] Configurations using multiple towers have also been proposed.[32] The heat transfer fluid (HTF), typically a molten salt mixture, is circulated to transport its received heat to a secondary fluid such as water, as depicted in Fig. 10.13. The fact that the HTF can also be used as a thermal energy storage medium is an advantage, but the high freezing point of a salt requires deployment of heat tracing and its concomitant parasitic energy use. A representative molten salt is a mixture of 40% potassium nitrate (KNO_3) and 60% sodium nitrate ($NaNO_3$) as was employed in Solar Two. This salt melts at 220°C and is thermally stable to ~600°C.[27,33]

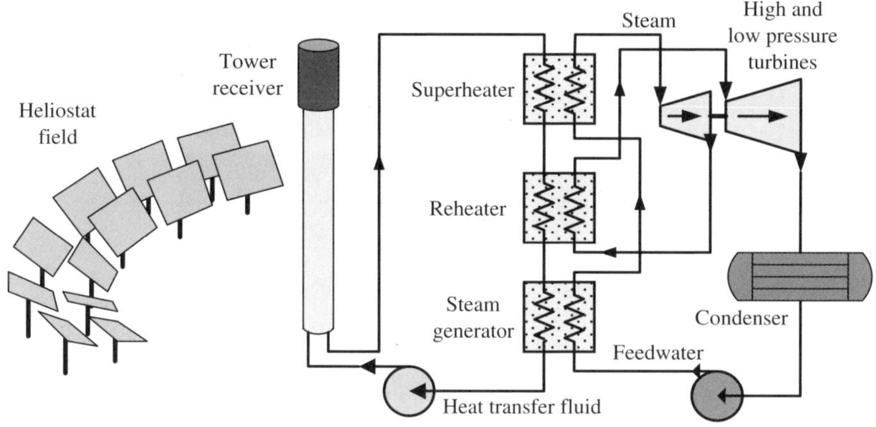

Fig. 10.13.   Power tower CSP plant without thermal storage.

The heliostats are a significant component of the plant capital costs. For example, in the case of the Solar One plant, heliostats composed 45% of the total cost.[34] In addition to an individual dual-axis positioning mechanism, a heliostat requires structural support that can withstand high winds. Analysis has determined that excessive shading of heliostats is avoided with reflector-to-land area ratios from 0.1 to 0.4, and averaging around 0.25.[35] Optimal heliostat field shapes have been determined based on plant power rating, as shown in Fig. 10.14.[36]

10.3.2.1   *Solar One and Solar Two*

The Solar One central receiver was constructed near Barstow, California, at a cost of $140 million, and operated from 1982 to 1988. The receiver in the 94.5-m tall

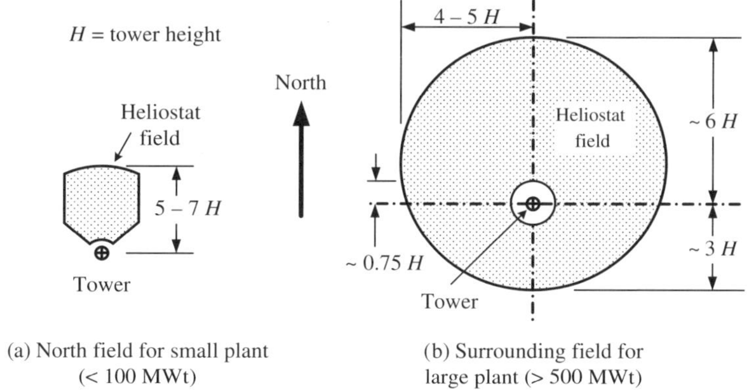

Fig. 10.14.   Heliostat field shapes in the Northern Hemisphere, after Refs. 34 and 37. For small thermal output, the heliostats are strictly north of the tower.

Fig. 10.15.   Solar One central receiver CSP facility (courtesy of DOE/NREL).

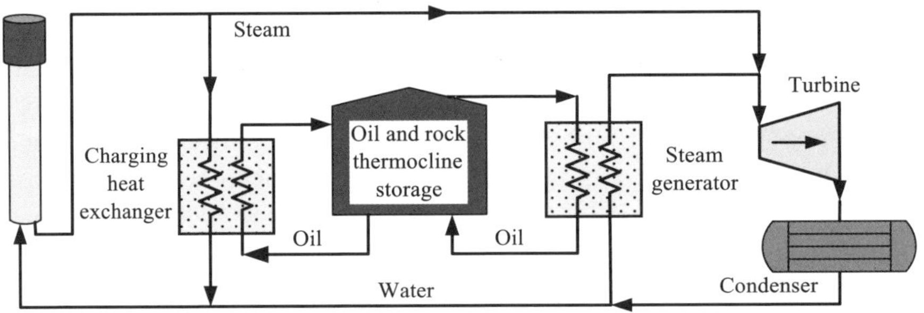

Fig. 10.16.   Simplified schematic of Solar One process flow; after Refs. 37 and 38.

tower heated the feedwater to generate steam for a direct Rankine cycle. The 1818 heliostats of 39.1 m^2 each were arranged in a surround field configuration,[38] as seen in Fig. 10.15. As depicted in Fig. 10.16, the Solar One unit employed a thermocline storage system composed of oil within a gravel and sand mixture. The storage medium had an energy density of 0.05 MWh/m^3 within a temperature range of 220° to 300°C.[34]

Solar One was upgraded to Solar Two by utilizing molten salt and increasing the reflector field, and it operated from 1996 to 1999.[39] With a total reflective surface area of 81,400 m^2, the heliostats provided an average solar flux of 430 kW/m^2 and a peak of 800 kW/m^2.[33] The molten nitrate salt was used both as a heat transfer medium from the receiver to a steam generator and for 3 hours of TES using a dual tank system, as shown in Fig. 10.17. The molten salt was heated from 290°C to 565°C from its passage through the central tower receiver. Sited in Spain, Solar Tres is to be a 15-MWe power plant utilizing Solar Two technology.

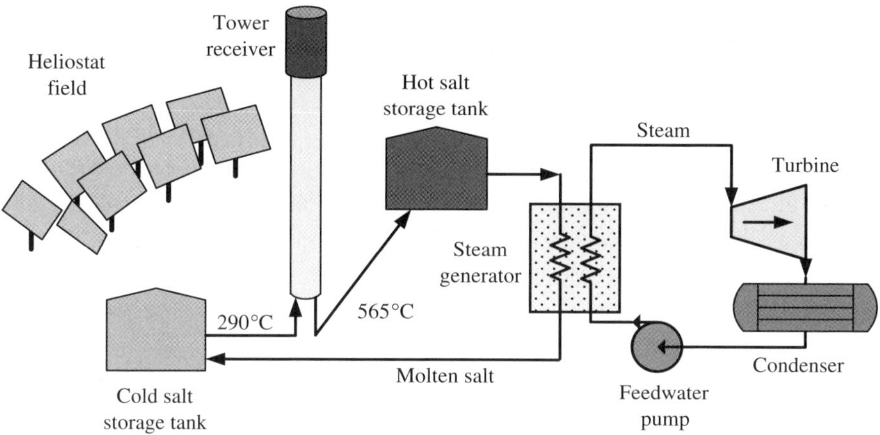

Fig. 10.17.   Simplified schematic of Solar Two process flow; adapted from Refs. 33 and 39.

10.3.2.2 *Planta Solar 10 and 20*

The 11-MWe Planta Solar (PS10) solar tower power plant began operation in 2007 on 150 acres near Seville, Spain. The facility utilizes 624 heliostats that each has an area of $120\,m^2$ in a north field configuration, yielding a reflective area of $7.5 \times 10^4\,m^2$. This unit differs from Solar One and Two in that 257°C, 40 bar steam is directly made in the 115-m tower cavity receiver. As back-up, the plant has the capability of storing an hour's worth of steam, and natural gas firing can supply 12–15% of its capacity. The steam cycle efficiency is 27%; and the overall solar-to-electric energy conversion efficiency is near 17%. In April 2009, Abengoa Solar began operation of a larger (20 MW) power tower plant, designated PS20, which uses a 160-m tall tower and doubles the heliostat field size of the PS10.

10.3.3 *Stirling engine*

A Stirling cycle that employs a piston engine can be employed in combination with a dish collector that uses dual-axis tracking. The engine shaft is connected to an electric generator. Such dish-based Stirling engine systems can achieve high solar-to-electric efficiencies, but an individual unit tends to provide smaller electric outputs — on the order of 25 kW.[40] Assuming an efficiency of 25% and using an irradiance of $1\,kW/m^2$, one quickly realizes that the dish for a 25 kW system would have a diameter greater than 11 m. To provide large-scale electric generation, the dishes can be installed as arrays of individual dishes. In addition, the modular nature of dish systems makes them appropriate for distributed generation and remote generation (e.g., end of the power line installations). However, dish-Stirling systems are estimated to have capital costs of approximately \$10,000 per kWe.[41]

The Stirling system consists of four major components: (1) the dish concentrator, (2) a cavity receiver, (3) the Stirling engine, and (4) an electric generator or alternator. The combination of the latter two items forms what is referred to as the power conversion unit (PCU), which is annotated in Fig. 10.18. The dish concentrator was addressed earlier and the electric generator employed is not novel, therefore, the cavity receiver and Stirling engine and cavity are emphasized herein. However, as seen in Fig. 10.19, one feature of many dish-Stirling systems is the use of multiple surfaces (mirror facets) to concentrate the sunlight.

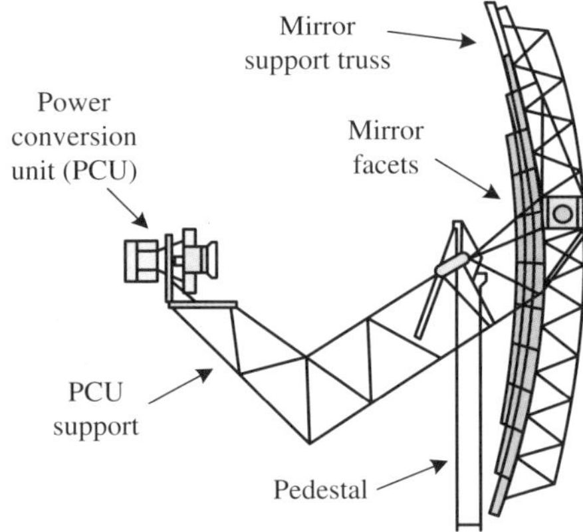

Fig. 10.18.	Dual-axis tracking dish concentrator with truss support, after Ref. 40.

Fig. 10.19.	Solar dish Stirling system (courtesy of Stirling Energy Systems, Inc.).

The cavity receiver, which is located at the focal point of the dish, absorbs the concentrated heat energy and transfers the heat to the cycle working fluid. The working fluid, which is typically helium or hydrogen, is sealed within the Stirling engine. The Stirling cycle uses a four stage process of heating, expansion, cooling, and contraction of the sealed working gas (fluid),[42] and there are no valves. The alternating heating and cooling of the gas produces more power during the heating-expansion stages than is consumed during the cooling–contraction phases, thereby yielding a net power output. Although a Stirling engine has the potential to be very efficient in converting heat into mechanical work, dish-Stirling systems are not suitable for thermal energy storage.

10.4   Low Temperature Solar Thermal Approaches

In addition to the mainstream CSP technologies, other solar thermal electricity generation schemes that the reader should be cognizant of include solar chimneys and salt ponds, which are addressed below.

10.4.1   *Solar salt pond*

In the case of a solar salt pond, a shallow ($\sim$2 m deep) pool with salt forms the solar collector, as shown in Fig. 10.20. Unlike a fresh body of water in which warm water rises to the surface and its heat is lost by evaporation, a solar salt pond traps the warm water near the bottom of the pool. Two different implementations of solar salt ponds exist[43]: (1) the saturated salt pond and (2) the non-convective gradient pond. In either case, piping carrying a low boiling point working fluid (e.g., ammonia) can be placed near the bottom of the pond to extract the heat ($\sim$100°C) for use in a thermal power conversion cycle. It is noteworthy that the solar salt pond itself forms an integral TES mechanism.

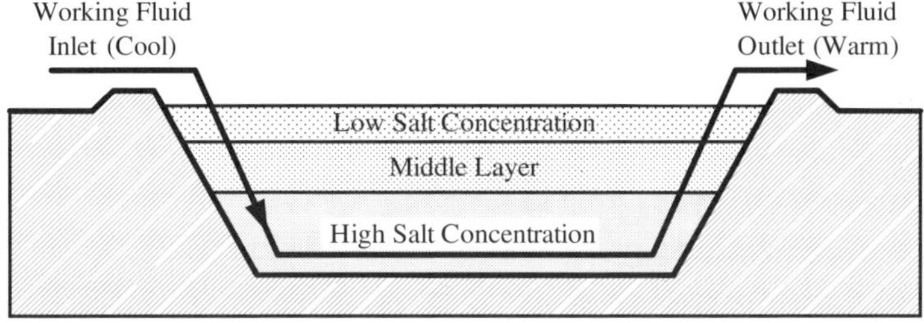

Fig. 10.20.   Solar salt pond side view.

Solar salt ponds have been demonstrated on a small scale. In particular, the largest constructed was located beside the Dead Sea in Israel, and was initially sized for a 5 MWe output using a 1 km^2 pond. The actual unit operated with 1% solar-to-electric efficiency[44] until it was shut down in 1989. There are several issues with solar salt ponds:

- water loss due to evaporation,
- the need to maintain a critical water-salt chemistry balance to form the required salt concentration gradient,
- inefficiency due to the small temperature difference which, via Carnot, means that the thermal efficiency is very low, and
- the surface area of the pond needs to be large for a reasonable power output, which leads to high capital cost.

10.4.2 *Solar chimney*

A solar chimney plant is a hybrid plant that utilizes sunlight-induced temperature differences to create a natural draft within a tall tower, as shown in Fig. 10.21. A greenhouse of sorts is established within the outer solar collector. The draft created within the tower is then harnessed using wind turbines, thereby making the solar chimney a hybrid solar–wind system. Thermal mass elements can be placed within the collector region of the facility so that heat storage occurs during daylight, and then during nighttime or inclement conditions the heat is released and the draft continues at a reduced level for some limited time period.

A pilot solar chimney plant was constructed at Manzanares, Spain and operated from 1982 to 1988. The 50 kWe plant had a tower of 194.6 m height and 10.16 m diameter, with collector area covering 46,800 m^2 at average height of 1.85 m.[45] The pilot plant produced a temperature rise of 20°C yielding an upward air flow of 9 m/s

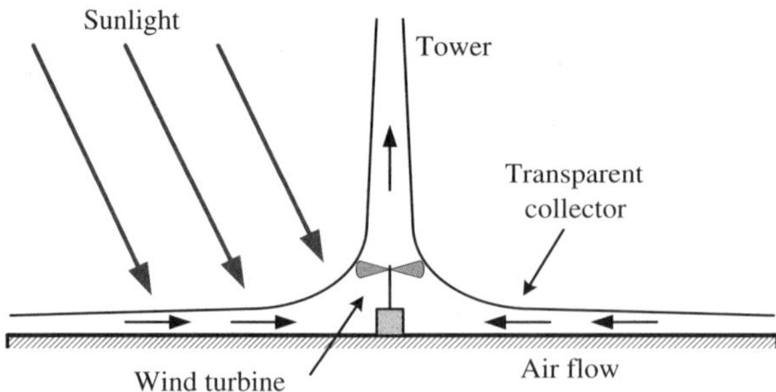

Fig. 10.21.   Solar chimney utilizing greenhouse effect to create air draft.

under load conditions. The plant achieved a solar-to-electricity efficiency of only 0.53%.[46]

10.5   Environmental Impact

All electricity generation schemes have some environmental impact. As a thermal cycle, CSP has the potential to generate thermal pollution. Moreover, the solar thermal approaches that employ a steam turbine cycle require condenser cooling water — such clean water is generally at a premium in the arid regions of the world in which sunlight is correspondingly so plentiful.[47] Straining water resources is not the only impact of solar thermal facilities; for example, McCrary *et al.* found an avian mortality rate of $\sim$2 birds/week at the Solar One plant.[48] Other possible environmental impacts include land use, accidental chemical releases, aesthetics, and effects on the local ecosystem, including loss of habitat.[49–51] With respect to land use, CSP facilities require about 2 ha (5 acres) per MW of installed capacity. However, use of CSP has the potential to offset emissions, such as CO_2, SO_2 and NO_x, from other electric generation technologies.

Conversely, the environment also has an effect on solar plants. Interestingly, the June 1991 eruption of Mount Pinatubo in the Philippines led to a 15–20% reduction in the direct beam radiation at mid-latitudes.[52] The SEGS units recorded a direct normal insolation of 6.32 kWh/m^2/day in 1992 as compared to an average of 7.94 kWh/m^2/day for 1988–1990;[6] the operators attributed the 20% reduction to a combination of Mt. Pinatubo and the weather phenomenon known as El Nino, which brings above normal rainfall to California.

10.6   Concluding Remarks

With concerns over climate change (global warming) due to carbon dioxide emissions from fossil fuel burning, green electricity generation technologies, such as solar, will see increasing interest and investment. It is also noteworthy that besides electricity generation, CSP facilities have been proposed for other uses including desalination,[53,54] coal gasification,[55] water splitting to yield hydrogen,[56] and destruction of hazardous chemicals.[57]

References

1. J. Weisman and R. Eckart, *Modern Power Plant Engineering* (Prentice-Hall, 1985).
2. R.L. San Martin, "Solar energy," AccessScience [online] (McGraw-Hill, 2005).
3. G.W. Crabtree and N.S. Lewis, "Solar energy conversion," *Physics Today* **60** (2007) 37–42.

4. S.A. Kalogirou, "Solar thermal collectors and applications," *Progress in Energy and Combustion Science* **30** (2004) 231–295.

5. M. Mehos, "Concentrating solar power," in *Physics of Sustainable Energy, Using Energy Efficiently and Producing It Renewably*, eds. D. Hafemeister *et al.* (American Institute of Physics, 2008), pp. 331–339.

6. G.J. Kolb, "Evaluation of power production from the solar electric generating systems at Kramer Junction: 1988 to 1993," *ASME Int. Solar Energy Conf.*, Lahaina, HI, 19–24 March 1995.

7. A. Leitner, "Fuel from the sky: Solar power's potential for western energy supply," National Renewable Energy Laboratory, NREL/SR-550-32160 (2002).

8. G.J. Kolb, "Economic evaluation of solar-only and hybrid power towers using molten salt technology," Sandia National Laboratories, SAND96-2454C (1996).

9. "Solar electric generating stations (SEGS)," *IEEE Power Engineering Review* **9** (1989) 4–8.

10. R. Ramakumar, N.G. Butler, A.P. Rodriguez and S.S. Venkata, "Economic aspects of advanced energy technologies," *Proc. IEEE* **81** (1993) 318–332.

11. H.W. Price and S. Carpenter, "The potential for low-cost concentrating solar power systems," National Renewable Energy Laboratory, NREL/CP-550-26649 (1999).

12. L. Stoddard, J. Abiecunas and R. O'Connell, "Economic, energy, and environmental benefits of concentrating solar power in California," National Renewable Energy Laboratory, NREL/SR-550-39291 (2006).

13. H. Price and D. Kearney, "Reducing the cost of energy from parabolic trough solar power plants," National Renewable Energy Laboratory, NREL/CP-550-33208 (2003).

14. Energy Information Administration, "Assumptions to the annual energy outlook," U.S. Department of Energy, Report DOE/EIA-0554 (2009), p. 89.

15. M. Lotker, "Barriers to commercialization of large-scale solar electricity: Lessons learned from the Luz experience," Sandia National Laboratories, SAND91-7014 (1991).

16. International Energy Agency, "Renewables for power generation 2003: Status and prospects," Organization for Economic Co-operation and Development (2003).

17. S.K. Roy and H. Liu, "Solar heating," in *Encyclopedia of Electrical and Electronics Engineering*, ed. J. G. Webster, vol. 2 (Wiley & Sons, 1999), pp. 649–658.

18. C. Philibert, "The present and future use of solar thermal energy as a primary source of energy," The InterAcademy Council (2005).

19. U. Herrmann and D.W. Kearney, "Survey of thermal energy storage for parabolic trough power plants," *J. Solar Energy Engineering* **124** (2002) 145–152.

20. E. Zarza, L. Valenzuela and J. León, "Solar thermal power plants with parabolic-trough collectors," *Proc. 4th Int. Conf. Solar Power from Space*, Granada, Spain, ESA SP-567 (2004), pp. 91–98.

21. H. Price *et al.*, "Advances in parabolic trough solar power technology," *J. Solar Energy Engineering* **124** (2002) 109–125.

22. E. Zarza, "The DISS Project: Direct steam generation in parabolic trough systems. Operation and maintenance experience and update on project status," *J. Solar Energy Engineering* **124** (2002) 126–133.

23. M. Eck and W.-D. Steinmann, "Direct steam generation in parabolic troughs: First results of the DISS Project," *J. Solar Energy Engineering* **124** (2002) 134–139.

24. M. Romero-Alvarez and E. Zarza, "Concentrating solar thermal power," in *Handbook of Energy Efficiency and Renewable Energy*, eds. F. Kreith and D.Y. Goswami (CRC Press, 2007).

25. D.R. Mills and G.L. Morrison, "Compact linear Fresnel reflector solar thermal powerplants," *Solar Energy* **68** (2000) 263–283.

26. G.E. Cohen, D.W. Kearney and G.J. Kolb, "Final report on the operation and maintenance improvement program for concentrating solar power plants," Sandia National Laboratories, SAND99-1290 (1999).

27. Pilkington Solar International GmbH, "Survey of thermal storage for parabolic trough power plants," National Renewable Energy Laboratory, NREL/SR-550-27925 (2000).

28. R. Peltier, "Nevada Solar One, Boulder City, Nevada," *Power* **151** (2007) 40–43.

29. M. Romero, R. Buck and J.E. Pacheco, "An update on solar central receiver systems, projects, and technologies," *J. Solar Energy Engineering* **124** (2002) 98–108.

30. A. Yogev, A. Kribus, M. Epstein and A. Kogan, "Solar 'tower reflector' systems: A new approach for high-temperature solar plants," *Int. J. Hydrogen Energy* **23** (1998) 239–245.

31. A. Rabl, "Tower reflector for solar power plant," *Solar Energy* **18** (1976) 269–271.

32. P. Schramek and D.R. Mills, "Multi-tower solar array," *Solar Energy* **75** (2003) 249–260.

33. C. Tyner *et al.*, "Solar power tower development: Recent experiences," Sandia National Laboratories, SAND96-2662C (1996).

34. M.M. El-Wakil, *Powerplant Technology* (McGraw-Hill, 1984).

35. A.F. Hildebrandt and L.L. Vant-Hull, "Power with heliostats," *Science* **197** (1977) 1139–1146.

36. K.W. Battleson, "Solar power tower design guide: Solar thermal central receiver power systems, a source of electricity and/or process heat," Sandia National Laboratories, SAND81-8005 (1981).

37. W.B. Stine and R.W. Harrigan, *Solar Energy System and Design* (John Wiley, 1985).

38. L.G. Radosevich and A.C. Skinrood, "The power production operation of Solar One, the 10 MWe solar thermal central receiver pilot plant," *J. Solar Energy Engineering* **111** (1989) 144–151.

39. J.E. Pacheco, H.E. Reilly, G.J. Kolb and C.E. Tyner, "Summary of the Solar Two test and evaluation program," Renewable Energy for the New Millennium Conference, Sydney (2000).

40. W.B. Stine and R.B. Diver, "A compendium of solar dish/stirling technology," Sandia National Laboratories, SAND93-7026 UC-236 (1994).

41. T. Mancini *et al.*, "Dish-Stirling systems: An overview of development and status," *J. Solar Energy Engineering* **125** (2003) 135–151.

42. B. Kongtragool and S. Wongwises, "A review of solar-powered Stirling engines and low temperature differential Stirling engines," *Renewable and Sustainable Energy Reviews* **7** (2003) 131–154.

43. H. Tabor, "Solar ponds," *Solar Energy* **27** (1981) 181–194. Also see article correction in *Solar Energy* **30** (1983), 85.

44. H.Z. Tabor and B. Doron, "The Beith Ha'arava 5 MW(e) solar pond power plant (SPPP) — progress report," *Solar Energy* **45** (1990) 247–253.

45. W. Haaf, K. Friedrich, G. Mayr and J. Schlaich, "Solar chimneys. Part I: principle and construction of the pilot plant in Manzanares," *Int. J. Solar Energy* **2** (1983) 3–20.

46. D. Mills, "Advances in solar thermal electricity technology," *Solar Energy* **76** (2004) 19–31.

47. K.E. Holbert and J. Haverkamp, "Impact of solar thermal power plants on water resources and electricity costs in the Southwest," *Proc. North American Power Symp.*, Starkville, Mississippi, 4–6 October 2009.

48. M.D. McCrary *et al.*, "Avian mortality at a solar energy power plant," *J. Field Ornithology* **57** (1986) 135–141.

49. T. Tsoutsos, N. Frantzeskaki and V. Gekas, "Environmental impacts from the solar energy technologies," *Energy Policy* **33** (2005) 289–296.

50. S.A. Abbasi and N. Abbasi, "The likely adverse environmental impacts of renewable energy sources," *Applied Energy* **65** (2000) 121–144.

51. D. Pimentel *et al.*, "Renewable energy: Current and potential issues," *BioScience* **52** (2002) 1111–1120.

52. J.J. Michalsky, R. Perez, R. Seals and P. Ineichen, "Degradation of solar concentrator performance in the aftermath of Mount Pinatubo," *Solar Energy* **52** (1994) 205–213.

53. F. Trieb *et al.*, "Combined solar power and desalination plants for the Mediterranean region — sustainable energy supply using large-scale solar thermal power plants," *Desalination* **153** (2003) 39–46.

54. F. Trieb and H. Müller-Steinhagen, "Concentrating solar power for seawater desalination in the Middle East and North Africa," *Desalination* **220** (2008) 165–183.

55. P.V. Zedtwitz and A. Steinfeld, "The solar thermal gasification of coal — energy conversion efficiency and CO_2 mitigation potential," *Energy* **28** (2003) 441–456.

56. A. Kogan, "Direct solar thermal splitting of water and on site separation of the products I. Theoretical evaluation of hydrogen yield," *Int. J. Hydrogen Energy* **22** (1997) 481–486.

57. G.C. Glatzmaier, R.G. Nix and M.S. Mehos, "Solar destruction of hazardous chemicals," *J. Environmental Science and Health* **A25** (1990) 571–581.

Chapter 11

Maximum Power Point Tracking
Charge Controllers

Ashish Pandey

SunUrja Renewable Energy (P) Ltd,
Unit 1, TBIU Block, IIT Delhi,
New Delhi 110016, India
ashish@sunurja.com

Nivedita Thakur

The Energy and Resources Institute (TERI),
New Delhi, India

Ashok Kumar Mukerjee

CES, IIT Delhi, New Delhi 110016, India

This chapter starts with an overview of charging technologies. It then touches upon various sources of losses such as power conversion loss and charge–discharge loss in batteries. Thereafter a detailed discussion on tracking loss is provided to develop the basis for Maximum Power Point Tracking (MPPT). Detailed mathematical modeling is then provided to build a relationship between various parameters which is then used to explain the basic algorithm. Advance topics associated with MPPT systems complete with pseudo-code and experimental results and in-depth explanations are presented.

11.1 Solar Battery Charging

The humanity has since long understood that fossil fuel driven development is not sustainable. Eventually alternatives have to be explored to continue with the current pace of economic growth. Imported from some of the most unstable regions of the world, the fossil fuel also compromises a nation's energy security. Only recently, artificially driven fuel prices triggered economic recession all over the

247

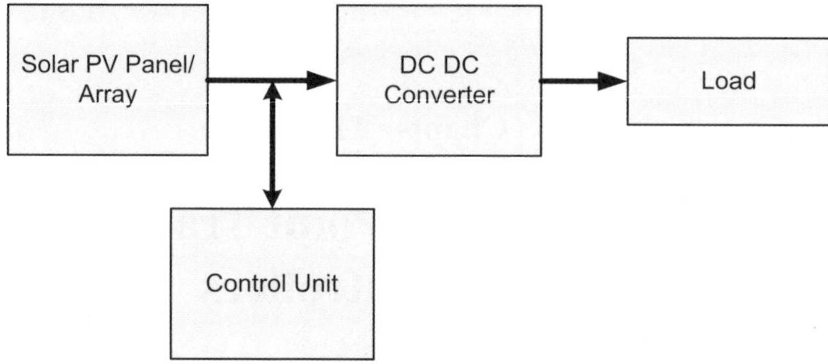

Fig. 11.1.   Solar PV battery charging system.

world. While these concerns gave tremendous impetus to adaptation of Solar Technologies as a "Green" alternative, high initial cost remains a stumbling block in its wider adaptation.

Battery charging is one area where Solar Photo Voltaic (PV) systems can eminently replace diesel-generators. They also find use in applications such as solar lighting especially in areas where the grid is not present. In developed countries it is mostly limited to camping sites and log cabins but in the developing world these can be primary source of energy to billions who have yet to see a functional grid reaching their homes.

The charge controller can be a simple on-off arrangement to protect deep discharge overcharging of the battery, or something more involving such as a Pulse Width Modulated (PWM) or MPPT charge controller, as shown in Fig. 11.1.

In these systems the solar PV array sources the current that is used to charge the batteries when the sun is available. The load uses this energy as and when required. Solar battery charging can be simple and straightforward. The charge controller may only ensure that the battery must never be overcharged or under-discharged. This ensures that life of the battery is not unduly shortened. While meeting this condition is simple, the real challenges in solar PV battery charging are several. These challenges emerge from complex physics of the Solar PV system and the even more complex electro-chemistry of chemical storage batteries. The primary challenge is to know and minimize the various losses that incur in the system. The real sources of losses that must be optionally addressed by the controller stem from this complexity are discussed next.

11.2   Various Sources of Losses

The electricity from the grid extracted from burning coal is priced at about US\$0.06 in India. On the other hand the electricity from a fully battery backed solar PV system will be 10 times more expensive. Clearly the solar photo-voltaic resource is an expensive alternative. Extensive government subsidies are keeping the cost somewhat bearable but it is important, nevertheless, to closely look at the various sources of losses and consider minimizing them when designing or selecting charge controllers.

11.2.1 *Power conversion loss*

This is the loss that occurs in the power electronics. Some energy is dissipated in magnetic components such as transformers and inductors while considerable energy is lost during switching of power devices. Substantial conduction or ohmic losses also take place inside the power electronics devices.

The charge controller inherently dissipates power through these three sources. A power electronics equipment operates at efficiency levels as low as 80%. Typically several such equipment are in series. One example is a charger in series with an inverter, and in such a situation efficiency is as low as 60%. Under special situation when no switching is taking place and only conduction losses are in picture, the efficiency can be around 95%.

11.2.2 *Charge/discharge loss*

While power conversion losses is a fairly well understood phenomenon, losses in charging and discharging chemical batteries is more complex. Due to underlying electrochemistry of a battery, a fraction of the power made available to the battery is dissipated as heat. Similarly the entire power that is stored in a battery cannot be used by the load. Overall only 80% of the power fed to the battery becomes available to the load. This aspect has to be considered when looking for fast charging systems or fixing the discharge rate of the batteries. Complex electro-chemistry of these batteries ensures that much of the charge–discharge process is understood only by empirical means and results have shown that PWM charging can improve the efficiency considerably.

11.2.3 *Tracking loss*

Physics of solar cells present an equally daunting challenge. It turns out that the maximum power available at the output of a solar PV cell (and therefore module or array as well) is a function of its operating voltage and temperature. The value of this voltage is in turn a function of instantaneous incidental solar insolation and ambient temperature as shown in Fig. 11.2. Typically about 30% of the power is

　　　　　　　A. Pandey, N. Thakur and A. K. Mukerjee

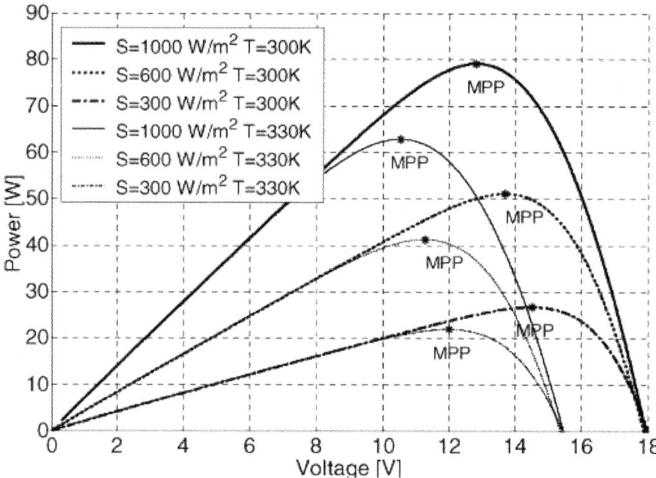

Fig. 11.2.　　IV Curve of solar PV cell as a function of solar insolation and cell temperature.

lost to tracking which would have been otherwise been available to the load. In the next two sub-sections we delve deeper into understanding this loss.

11.2.3.1　*Illustrating the tracking loss*

Let us look closely at Fig. 11.2. Now let us assume that in the morning a battery in a deep discharged state is connected to the panel. What will happen when the battery starts charging from 11 V from a panel with IV characteristics shown in Fig. 11.2?

Let us say that when solar insolation is 1000 W/m² and temperature of the cell is 330 K, the panel will source 62 W of power to the battery. The situation will change when a passing cloud reduces the insolation to 600 W/m² and temperature to 300 K. While the total power available was 50, only 45 W will be sourced.

A loss of 5 W in a system is prohibitively expensive!

This process of tracking loss resulting from that situation is elaborated in Fig. 11.3. The battery voltage is shown along the dotted line. When the solar insolation is 1000 W/m² and cell temperature is 300 K the panel is charging the battery at close to full power and is transferring 79 W as shown by the red band. However if the cell temperature increases to 330 K at the same insolation then this system will give only about 53 W even though the power available was 63 W as depicted by the large pink band. Over the day, both insolation and temperature are continually changing and these bands add up to a significant overall loss. As we see in the next sub section, this adds up to 30% in a typical system.

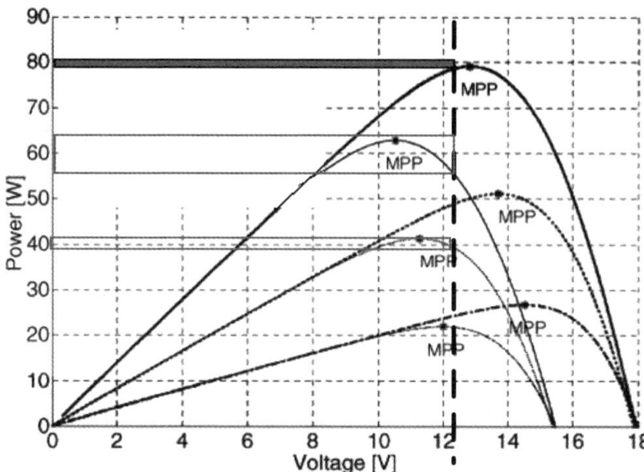

Fig. 11.3.   Illustrating loss in charging a battery at 12.5 V under varying insolation and temperature condition.

11.2.3.2   *Calculating the tracking loss*

Interesting insight into this loss can be obtained by calculating the Wp requirement of the panel and AH for the battery. Let us run a typical design flow for a PV power plant which will reveal the tracking loss, and design a PV power plant to meet the load condition defined in Table 11.1.

Sizing the battery bank and PV array

The load will be powered by an inverter which will convert the DC voltage at the battery terminals to the required AC voltage. Like any power electronics system we must consider 20% losses in the inverter system to which the load will be connected.

Table 11.1.   Example load condition.

Calculating the load	Watts/Unit	No	Watts Total	Hrs/Day =	Wh/Day
Fan	100	8	800	9 =	7200
Tube Light	60	8	480	9 =	4320
				Total Wh/Day	11520
	Enter Backup required for no. of days		1		
	Enter DC Bus Voltage at the Battery Bank		24		
	Enter Average Sun per day		6		
	Enter Depth of Discharge (Not more than 80%)		50%		

To compensate for this loss, the total Whr that is to be stored in the batteries should be 20% more than what is demanded by the load. This means that a total of 13824 Whr of energy needs to be stored in the batteries to provide 11520 Whr demanded by the load.

Since the DC bus voltage is 24 V, the the battery must provide 576 AH to meet the energy requirement of the load. The battery bank is designed to provide the desired terminal voltage and AH capacity.

Battery sizing

Amp Hour from Battery per day: 576 AH

The entire energy stored in the battery should not be used up. This is essential as a deep discharge will shorten the life of the battery. Therefore we assume only 50% discharge and hence the AH of the battery bank must accordingly be 1152 AH.

We take a Tubular Lead Acid 12 V, 150 AH battery and connect 8 such battery in parallel to provide 1200 AH. Two such bank are connected in series to get 24 V. The total batteries required will be 16.

Array sizing

The charge–discharge process of the battery bank is also lossy. This loss of about 20% power must also be compensated by the panels. This means we need 691.2 AH from the panel every day to replenish the batteries. The panel will start charging the battery bank in the morning and will continue to do so till sunset. Now let us assume that in the location where the PV system is located the average sunshine per day is 6 hours. This means that the total energy harvested by the panel over the day will be equivalent to operating the panel at its peak capacity for 6 hours. Therefore the panel must supply only a sixth of the required AH viz 115.2 A can be sourced from the array.

This requirement will be met by taking an 119 Wp panel with operating values 17 V and 7 A, we connect 17 in parallel and 2 such bank in series. Total Panel will be 34.

To meet the energy requirement of 11520 Whr, investment is made in panels that can source 24276 Whr of energy. Of this only 47% is utilized.

We have assumed power conversion loss and charge control loss to be 80%. This leads to the tracking efficiency of 75%.

11.3 Charge Control in Battery Backed PV Systems

The battery bank has the highest lifecycle ownership cost in a battery backed PV system. The bank also dissipates a sizeable amount of energy during its charging and discharging. Therefore it is imperative to understand the charge control process while designing an optimal PV system. In selecting a charge controller a deep cost-benefit evaluation must be carried out taking into account both charge–discharge efficiency and life of battery bank. Application specific features such as fast charging also need to be analyzed before converging on a specific charge controller.

An in-depth study of charge control and controller is outside the scope of this chapter but an outline is necessary for sake of completeness. The panels cannot be connected directly to the battery bank in most systems because deep discharging and overcharging of the batteries can significantly reduce the life and overall efficiency of the battery bank. Charge controllers with a whole range of features are commercially available. The basic controller only protects the battery from deep discharge and overcharge. The advance designs incorporate PWM charging for more efficient and fast charging. The MPPT controllers reduce the tracking losses in the panel. Some controllers incorporate algorithms that enhance battery life and reduce the charge imbalance in systems with many cells. Commercially available systems are essentially three types. The on-off system or linear charge controller, the PWM controller and MPPT. They are explained in the next section.

11.3.1 *Linear charge control*

The most basic of any charge control strategy — linear charge controllers — are also the cheapest controllers available. These controllers only ensure that the batteries are never deep-discharged or overcharged. This can be done by series controllers which disconnect the battery from the panels or shunt controllers which shorten the panels and thereby effectively disconnecting the batteries from the panel. In either case the battery life is shortened in the long run by deposition of sulfur on the surface of the lead due to exposure to continuous charging currents. Such controllers can only be recommended for very small systems.

11.3.2 *PWM charge controller*

Empirical understanding of charging chemical batteries has shown that the charging efficiency can be increased by using pulsed current. This enhances the efficiency to about 90%; a substantial improvement from a typical 80% charge discharge efficiency achievable by linear control methods.

Use of active power electronics allows implementation of many more features. The charge controller can implement a process for removing sulfur from the surface of the lead, reduce charge imbalance in individual cells common in large multi-cell banks. Some application may demand fast battery charging and the controller can implement algorithms for the same. More expensive than linear controllers, PWM chargers are recommended for larger systems using lead-acid batteries.

11.3.3 *MPPT controller*

MPPT charge controllers use specialized algorithms that continually adjust the operating voltage of the panel such that the maximum available energy is harvested. More expensive, these controllers are currently available in ratings of few amps to a hundred amps. As technology matures and newer techniques reduce the cost of implementation these systems are finding their way into smaller systems such as solar street lights and other erstwhile areas of linear charge controllers.

The MPPT controllers ensures maximum utilization of PV resource and PWM controllers ensure maximum utilization of the energy storage system. Controllers incorporating basic features of both MPPT and PWM charge controllers will significantly improve the performance of battery back PV systems in the near future.

11.4 Maximum Power Point Tracking (MPPT)

MPPT systems adjust the duty cycle of the converter in such a way that the operating voltage of the panel is consistently maintained at its maximum operating point. Mathematically this process can be evaluated by modeling various components of solar PV system.

The mathematical model establishes the theoretical relationship between the duty cycle and various parameter of the PV system that helps us design algorithms that enable maximum power transfer.

11.4.1 *Solar cell*

The modeling starts with the solar cell. Essentially the panel or even array displays the same behavior defined by the basic building block — the PV cell. A cell has a non-linear relationship between its output voltage and current. The values of these parameters depend upon solar irradiance and cell-temperature as given in the following equation.

$$I = I_L - I_{OS}\left\{\exp\left[\frac{q(V + IR_S)}{AkT}\right] - 1\right\} - \frac{V}{R_{SH}}. \tag{11.1}$$

In the above equation, I is the output current of the solar cell, I_L is the current across the *p-n* junction (light generated current; this parameter depends upon the solar insolation), I_{OS} is the reverse saturation current of cell, q is the electronic charge, V is the output voltage of the cell, R_S stands for the series resistance (ideally zero),

A is the ideality factor, k denotes the Boltzmann's constant and T is the absolute operating temperature. R_{SH} denotes the shunt resistance, which is ideally infinity, and therefore the last term in Eq. (11.1) is generally dropped.

11.4.2 *DC-DC converter*

In most applications the converter used for MPPT is a buck converter. In these converters the Duty Cycle D that is the ratio of "on" time to total time period is given as

$$D = V_O/V, \tag{11.2}$$

where V and V_O are the input and output voltage of the buck converter. The buck converter and the load (Fig. 11.1) can be viewed together as a load on the solar array system. According to the maximum power transfer theorem, this equivalent resistance has to be same as the series resistance R_S of solar cell to achieve maximum power transfer. The equivalent load resistance of the converter can be derived from the model of the buck converter and can be given as

$$R_{\text{eq}} = \frac{\eta R_L}{D^2} \tag{11.3}$$

and hence,

$$V = \left(\frac{\eta R_L}{D^2}\right) I, \tag{11.4}$$

where η and R_L are the converter efficiency and load resistance, respectively.

The duty cycle D is adjusted to match load impedance R_{eq} to source impedance R_S, according to Eq. (11.4), R_{eq} is inversely proportional to the square of the duty cycle. The experimental curve verifying this relation is shown in Fig. 11.4(a).

According to the solar cell characteristic equation, the dependency of panel current and power upon the converter duty cycle are given by the following equations.

$$I = I_L - I_O \left\{ \exp\left[\frac{q\eta R_L^l}{AkTD^2}\right] - 1 \right\} \tag{11.5}$$

and

$$P = I_L \left(\frac{\eta R_{\text{LOAD}}}{D^2}\right) - I_O \left(\frac{\eta R_{\text{LOAD}}}{D^2}\right) \left\{ \exp\left[\frac{q\eta R_{\text{LOAD}}^l}{AkTD^2}\right] - 1 \right\}. \tag{11.6}$$

11.4.3 *MPPT algorithm*

Equations (11.3) and (11.4) establish the relationship between the power available from the panel and the duty cycle. It can also be seen in Figs. 11.4(a) and 11.4(b). Once it becomes evident that voltage of the panel is a function of duty cycle of the

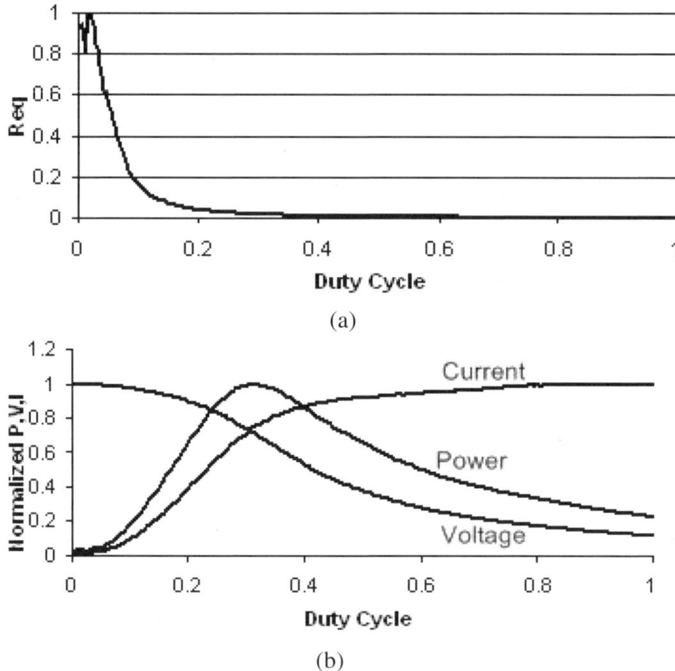

Fig. 11.4.　(a) Relation between equivalent resistance and duty cycle, (b) relation between panel current, voltage and power with duty cycle of converter.

converter, or alternatively, the equivalent resistance is a function of duty cycle, these parameters can be used for tracking the maximum power point.

The most effective way of finding this duty cycle is by a search algorithm commonly referred as the Perturb & Observe (P&O) algorithm (Fig. 11.5). This algorithm is one of the standards in industry and has the advantage of simple software and hardware realization. The Maximum Power Point (MPP) is searched by changing the duty cycle according to the following equation.

$$D(k) = D(k-1) \pm C. \qquad (11.7)$$

Here $D(k)$ and $D(k-1)$ are duty cycle at kth and $k-1$th sampling instant. C is a constant search step. The sign of this step C defines the direction of search as detailed in the pseudo code.

11.5　Advance Issues and Algorithms

In Sec. 11.2.3 the cost saving potential of MPPT charge control is discussed. A theoretical potential of saving almost 30% cost exists by using a MPPT charge controller or implementing P&O algorithm as discussed in the preceding section.

Initialize Duty Cycle, $D = Dk - 1$
Initial Voltage and Current (V and I)
Calculate initial Power ($Pk - 1$)
$C = +C$

LOOP
$D(k) = D(k - 1) + C$
Calculate P_k
If $(P_k > P_{k-1})$
$C = +C$ ELSE
$C = -C$
$D(K) = D(K - 1)$

Goto LOOP

Fig. 11.5.   Pseudo-code for P&O algorithm.

This cost reduction however remains only a theoretical potential due to a number of factors that makes true operation at maximum power point a difficult preposition.

The search algorithms that are used for tracking the MPP are essentially variants of gradient following method — a well known search algorithm. All implementations of the P&O algorithm or algorithms based on instantaneous evaluation of power including the basic implementation discussed in Sec. 11.4 will inherit all the limitations of the gradient-following method. The system never rests at the MPP but continually oscillates around this point due to a finite step size C or ΔD. This constant step size ΔD is also responsible for the most prominent and well known of the limitation of the algorithm that is the dynamics versus tracking trade-off which relates to finding the optimal search step. Another is a problem specific to the solar PV system — the drifting of the operating point away from MPP during changing atmospheric conditions.

11.5.1 *Search steps*

The step size ΔD in the search algorithm is constant. A choice of large step size in P&O algorithm will improve the dynamic response but in turn will cause oscillations during the steady state. Since the atmospheric condition is fairly dynamic the large step size offers distinct advantage but results in continuous power loss during the steady state operation. A choice of small step size impairs the dynamics of the converter. It also leads to poor utilization of the solar cell since the system is unable to track the MPP quickly under changing atmospheric conditions and during startups.

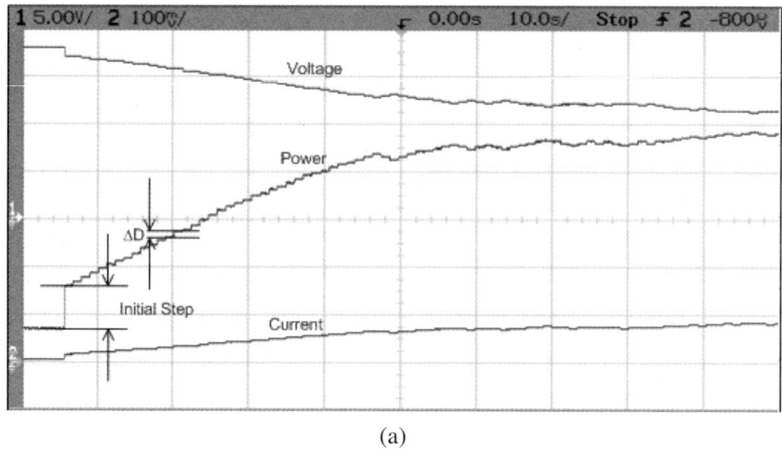

(a)

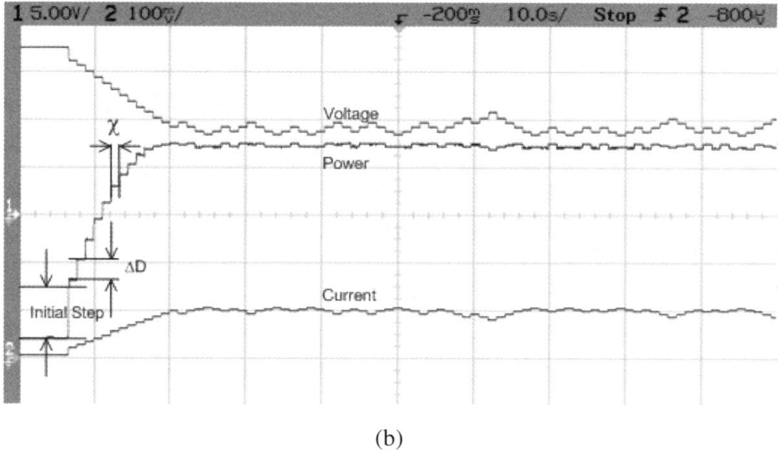

(b)

Fig. 11.6.   Start up and steady state performance of P&O algorithm (a) at $\Delta D = 0.5\%$ and (b) $\Delta D = 2.5\%$.

This trade-off is evident from experimental results documented in Figs. 11.6(a) and 11.6(b). In both cases the system is operated from an initial duty cycle which is shown as initial step in the Fig. 11.6. This is done as the duty cycle typically corresponding to MPP point is rarely small and so starting from zero duty cycle is not required. In the figures the step-size is indicated by ΔD.

Analyzing the results in Fig. 11.6(a) where the step size is 0.5% we clearly see that the time taken to reach a steady state near the MPP is very long. The system is also more susceptible to noise but steady state oscillations are minimized. Looking at the results in Fig. 11.6(b) where the step size is increased to 2.5% the system latches up to MPP very quickly and is also less susceptible to noise but there is considerable amount of steady state power loss due to continuous and significant oscillations around the MPP.

Variable step size algorithm

An immediate conclusion of empirical observations documented in Figs. 11.6(a) and 11.6(b) is to introduce a variable step size mechanism in the algorithm. A variable step size would require calculating a parameter that is large when the system is far away from the MPP and monotonically decreases as the MPP is reached.

A study of an experimental P-V curve and its derivatives indicates that the derivatives are uniquely suitable for step size after proper scaling. The derivative is large when the operating point is away from MPP, and monotonically decrease as the MPP is approached.

From the results shown in Fig. 11.7, it is clear that the derivative of power with regard to voltage can be used as the scaling parameter for the step change in duty cycle for the next sampling cycle while directly manipulating the duty cycle to locate the MPP. This is given by the following equation

$$D(k + 1) = D(k) \pm M \frac{|P(k) - P(k - 1)|}{|V(k) - V(k - 1)|}. \tag{11.8}$$

Here M is a suitable scaling parameter, $D(k)$, $P(k)$ and $V(k)$ are Duty cycle, Power and Voltage at Kth instant. $D(k + 1)$, $P(k + 1)$, $V(k + 1)$ are duty cycle, Power and Voltage at $(k + 1)$th instant, respectively.

Use of the variable step size can significantly improve the system dynamics as can be seen in the experimental test results given in Fig. 11.8. A large step size during

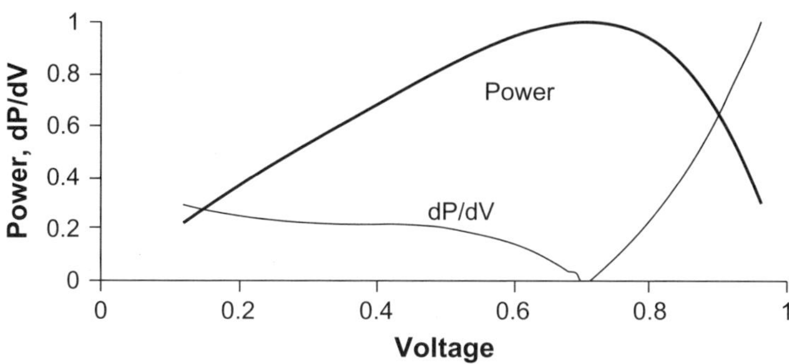

Fig. 11.7.   Experimental PV curve from solar PV panel and its derivative.

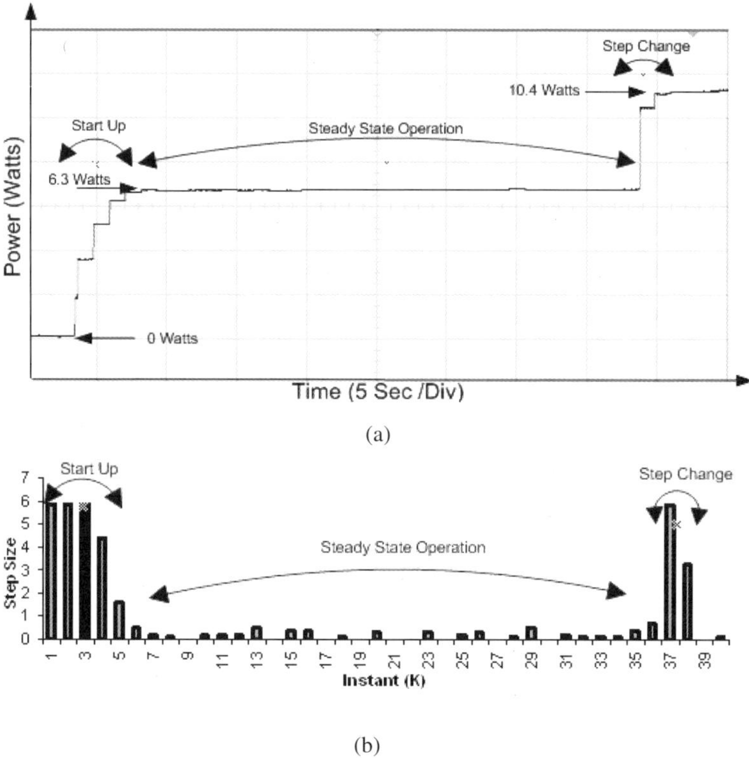

Fig. 11.8.   (a) Tracking performance during startup and a step change in the power level, (b) graph showing changes in step size at the start up and during and after a step change in power level.

startup takes the system quickly to maximum power point. A very small step size during steady state operation eliminates energy loss due to steady state oscillations. Susceptibility to noise is considerably reduced as well.

Analyzing the empirical observation in Fig. 11.8 clearly indicates that the step size is increased when the operating point is away from the MPP. As shown during the startup the step size is large which rapidly reduces as the system approaches the MPP. During the steady state operation the step size reduces to a naught as the derivative becomes negligible at the peak. When a step change is introduced in the power level the peak shifts and hence the derivative term builds up. A corresponding increase in step size immediately takes the system to its next steady state.

11.5.2   *Drift*

Drift is a situation when an MPPT algorithm drifts away from the MPP during changing atmospheric condition. In most parts of the year the atmospheric conditions

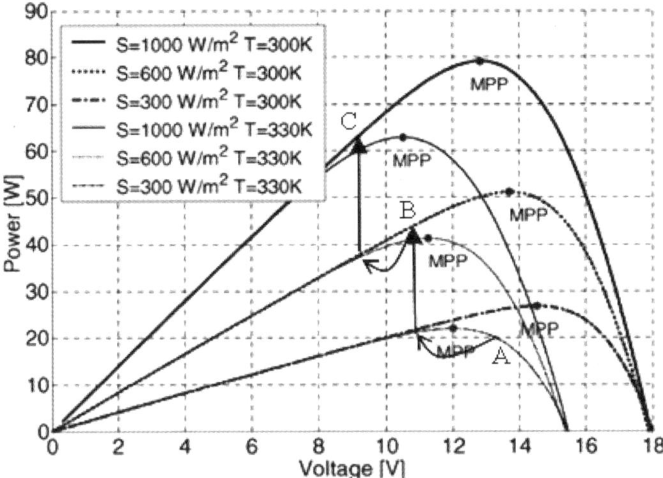

Fig. 11.9.   Mechanism of drift in MPPT algorithms.

are highly dynamic. Movement of clouds and wind ensure that the operating point is continually in a dynamic state. Therefore drift considerably impairs the energy harvest capability of the MPPT system.

The genesis of the drift in standard algorithms stems from the limitation arising out of considering instantaneous power values in deciding the search direction. The mechanism of the drift is explained using a typical situation shown graphically in Fig. 11.9.

Let us assume that the system was under a steady state oscillation about MPP in a stable cloudy condition with insolation of $300\,\text{W/m}^2$. At a time when the operating point was oscillating towards the MPP from A by reducing the duty cycle, the cloud begins to scatter and the solar insolation begins to increase to $600\,\text{W/m}^2$. Due to this change the operating point shifts to B.

The search algorithm is tricked into believing that by reducing the duty cycle the power level is increased as it has no way to cross check that points A and B lie on different *I-V* curve. Now as the cloud cover is eliminated and the insolation is $1000\,\text{W/m}^2$ the search algorithm continually reduces the duty cycle, effectively shifting the system farther and farther away from MPP to point C. It is only when a steady state is reached that the algorithm will correct itself and start increasing the duty cycle to take the operating point to MPP.

Avoiding drift: The full curve evaluation (FullCurvE) algorithm

Drift is a limitation of algorithms that chooses a search direction by evaluating changes in instantaneous values of power. This way the algorithm cannot

                                    A. Pandey, N. Thakur and A. K. Mukerjee

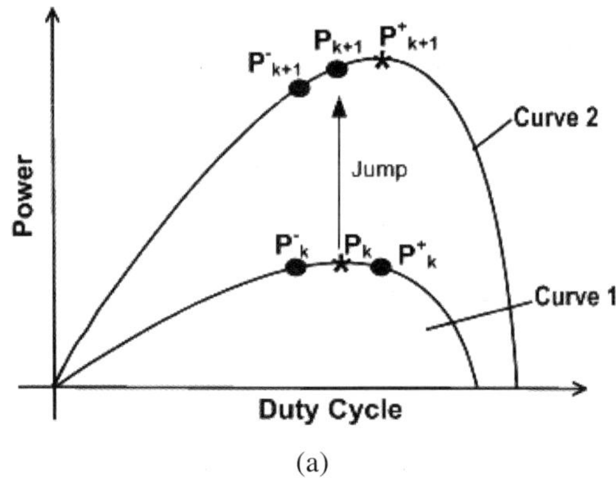

(a)

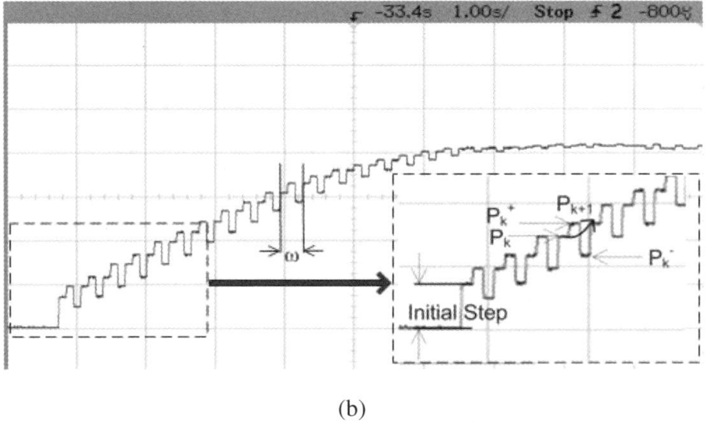

(b)

Fig. 11.10.   (a) Full curve evaluation algorithm, (b) start up with FullCurvE algorithm.

discriminate whether the power value changes due to duty cycle correction by the algorithm or due externally to changing atmospheric condition.

One of the ways of overcoming it would be an algorithm that evaluates the complete *I-V* curve. This is done by the Full Curve Evaluation (FullCurvE) Algorithm. FullCurvE algorithm, which aims to eliminate this deficiency by evaluating the entire power curve at each sampling cycle. Working of the algorithm is shown in Fig. 11.10(a).

At sampling instant k, the power is evaluated at the duty cycle D_k. Subsequently the duty cycle is perturbed at $D_k + \Delta D$ and $D_k - \Delta D$ to get the corresponding power levels P^+ and P^-, respectively. In case of change in the insolation level between successive sampling runs, the operating point automatically jumps to the corresponding point in the new curve as shown in Fig. 11.10(a), since the duty cycle is not changed. In the next sampling run all the three points are identified to evaluate the shape of the curve at the particular operating point as seen experimentally in Fig. 11.10(b). The performance of the algorithm is defined by the sub-sampling frequency depicted as ω in the figure.

The FullCurvE algorithm is given as below.

Sample P_k, P_{k+}, P_{k-}
If ($P_{k-} < P_k$ AND $P_k <= P_{k+}$)
$D = D + \Delta D$
OR
If ($P_{k-} >= P_k$ AND $P_k > P_{k+}$)
$D = D - \Delta D$
ELSE
No Change

11.6 Conclusion

A battery backed PV system will play an important role in the assimilation of PV technology. Their range of application will be enormous. Due to high initial cost of these systems and also substantial lifecycle ownership cost of the battery bank it is important to closely understand the features in the charge controller that can simultaneously reduce the various losses and also enhance life of the battery bank.

Of the various losses in these systems tracking losses are sizable and need to be contained. This can only be done by employing MPPT charge control. Typically 30% more energy is harvested by such a system. In a battery backed system the cost of a PV component is more than 50% of the overall system cost and this increase in efficiency has a significant impact on the price and therefore on the viability itself of battery backed systems.

The MPPT systems are as efficient in containing tracking losses as the MPPT algorithm that runs inside them. The standard algorithm which is the P&O algorithm has two serious limitations. The trade-off in its dynamic and steady state performance can lead to significant losses as ability to handle one of the two aspect is always limited. Drift is another source of loss.

Market forces and technology innovation are driving down the price of these system constantly. The introduction of MPPT chargers in smaller applications aimed at wider assimilation in developing economies such as Solar Lanterns and Street Light will help reduce cost of these systems and bring them within reach for common people.

11.7 Further Readings

(a) MPPT algorithms[1,2]

- The MPPT system must proliferate small systems such as solar lights and lanterns. To enable this it is required to use simple algorithms and fewer sensed variables. These papers present the fundamental approach towards evaluating MPPT algorithms. A whole range of algorithms are presented and compared.

(b) Fast tracking and low-oscillation (Variable step size)[3,4]

- The trade-off in P&O algorithm is identified in these pioneering works.

(c) Drift avoidance, fast tracking and low cost solution[5–9]

- Several advance algorithms for drift avoidance and fast tracking is presented and compared in these papers. It is shown, by fast tracking and avoidance of drift simultaneously a significant energy harvest is possible.

(d) Partial shading (local maxima) and hybrid methods[10–15]

- PV systems are designed taking shadowing into account. However, for small systems this may not always be possible. Moreover in vehicles and other applications where the panel is moving, it becomes even more difficult. Partial shading of the panels will cause local maxima in the I-V curve.

References

1. D.P. Hohm and M.E. Ropp, "Comparative study of maximum power point tracking algorithms," *Progress in Photovoltaics: Research and Applications* **11** (2003) 47–62.
2. E. Koutroulis, K. Kalaitzakis and N.C. Voulgaris, "Development of a microcontroller-based, photovoltaic maximum power point tracking control system," *IEEE Trans. Power Electronics* **16** (2001) 46–54.
3. P. Huynh and B.H. Cho, "Design and analysis of a microprocessor-controlled peak-power-tracking system for solar cell arrays," *IEEE Trans. Aerospace and Electronic Systems* **32** (1996) 182–190.
4. N. Femia, G. Petrone, G. Spagnuolo and M. Vitelli, "Optimization of perturb and observe maximum power point tracking method," *IEEE Trans. Power Electronics* **20** (2005) 963–973.
5. A. Pandey, N. Dasgupta and A.K. Mukerjee, "High performance algorithms for drift avoidance and fast tracking in solar MPPT system," *IEEE Trans. Energy Conversion* **23** (2008) 681–689.
6. K.H. Hussein, I. Muta, T. Hoshino and M. Osakada, "Maximum photovoltaic power tracking: An algorithm for rapidly changing atmospheric conditions", *Generation, Transmission and Distribution, IEE Proc.* **142** (1995) 59–64.
7. N. Dasgupta, A. Pandey and A.K. Mukerjee, "Voltage-sensing-based photovoltaic MPPT with improved tracking and drift avoidance capabilities," *Solar Energy Materials and Solar Cells* **92** (2008) 1552–1558.
8. A. Pandey, N. Dasgupta and A.K. Mukerjee, "A simple single sensor MPPT solution," *IEEE Trans. Power Electronics* **22** (2007) 698–700.

9. N. Dasgupta, A. Pandey and A.K. Mukerjee, "Current-sensor-based photovoltaic MPP tracking algorithm with efficient dynamic response," *IEEE Int. Conf. on Industrial Technology*, ICIT '08, China.

10. K. Kenji, T. Ichiro and S. Yoshio "A study of a two stage maximum power point tracking control of a photovoltaic system under partially shaded insolation conditions," *Solar Energy Materials and Solar Cells* **90** (2006) 2975–2988.

11. A. Kovach, "Effects of inhomogeneous irradiation distribution on a PV array in an urban environment," *Photovoltaic Energy Conversion, 1994, Conf. Record of the 24th; IEEE Photovoltaic Specialists Conf.* **1** (1994) 994–997.

12. C.R. Sullivan and M.J. Powers, "A high-efficiency maximum power point tracker for photovoltaic arrays in a solar-powered race vehicle," *24th Annual Power Electronics Specialists Conference, IEEE 1993. PESC '93*, 20–24 June 1993, pp. 574–580.

13. W. Wenkai, N. Pongratananukul, Q. Weihong, K. Rustom, T. Kasparis and I. Batarseh, "DSP-based multiple peak power tracking for expandable power system," *18th Annual IEEE Applied Power Electronics Conf. and Exposition, 2003. APEC '03* (2003), pp. 525–530.

14. S. Jain and V. Agarwal, "A new algorithm for rapid tracking of approximate maximum power point in photovoltaic systems," *IEEE Power Electronics Letters* **2** (2004) 16–19.

15. C. Hua and J. Lin, "A modified tracking algorithm for maximum power point tracking of solar array," *Energy Conversion and Management* **45** (2004) 911–925.

Chapter 12

Non-grid Solar Thermal Technologies

Mahendra S. Seveda

Department of Farm Power and Machinery,
College of Agricultural Engineering and post Harvest Technology,
(Central Agricultural University),
Ranipool, Gangtok, Sikkim, India-737135
sevda_mahendra@rediffmail.com

Narendra S. Rathore

Department of Renewable Energy Sources,
College of Technology and Engineering, Maharana Pratap,
University of Agriculture and Technology,
Udaipur, Rajasthan, India-313001
rathoren@rediffmail.com

Vinod Kumar

Department of Electrical Engineering,
College of Technology and Engineering, Maharana Pratap,
University of Agriculture and Technology,
Udaipur, Rajasthan, India-313001
vinodcte@yahoo.co.in

Solar energy is a very large, inexhaustible source of energy. The solar thermal technologies use the sun's heat. Non-grid solar thermal technologies like water heating systems, solar cookers, solar drying applications and solar thermal building designs are simple and can be readily adopted. These technologies help to conserve energy in heating and cooling applications. Solar thermal devices use direct heat from the sun, concentrating it to produce heat at useful temperatures.

12.1 Introduction

Solar energy is the most promising of the renewable energy sources in view of its apparent limitless potential. The sun radiates its energy at the rate of about 3.8×1023 kW per second. Most of this energy is transmitted radially as electromagnetic radiation which comes to about 1.5 kW/m^2 at the boundary of the atmosphere.[1] Solar energy provides an inexhaustible source of energy to mankind. Earlier times, solar energy was used for drying and other purposes but to a very low efficiency. Solar radiation is available at any location on earth. The total world average power at the earth's surface in the form of solar radiation exceeds the total current energy consumption by 15000 times, but its low density and geographical and time variations pose major challenges to its efficient utilization. Utilization of solar energy can change the energy consumption pattern of rural masses in particular for various energy needs provided solar devices are simple and require low maintenance and economical.[1]

Solar energy can be utilized by two routes namely solar thermal and solar photovoltaic. Solar thermal energy devices convert radiant energy of the sun into thermal energy for different productive works. It is a promising way of utilizing solar energy by converting it into thermal energy which is largely required by industrial and domestic consumers. Solar thermal energy devices convert radiant energy of the sun into thermal energy for different productive works. The conversion of sunlight into thermal energy is easily or conveniently achieved by means of a metallic or plastic cover painted black with ordinary black board paint or having selected coating over it and covered with one or double glazing cover for transmitting solar radiation inside it.[2] In the terms of the amount of energy available for utilization the solar energy option is the largest energy resource with the added advantage of being environment friendly and renewable.

12.2 Solar Collectors

The heart of any solar thermal device is the solar collector. Solar radiation is absorbed in a solar collector and the absorbed solar energy is converted into heat to raise the temperature of the working fluid.[3] The two types of collectors are the flat-plate collector and focusing collector.

12.2.1 *Flat plate collector*

Flat-plate collectors are the more commonly used type of collector today. They are arrays of solar panels arranged in a simple plane. They use both beam and diffuse

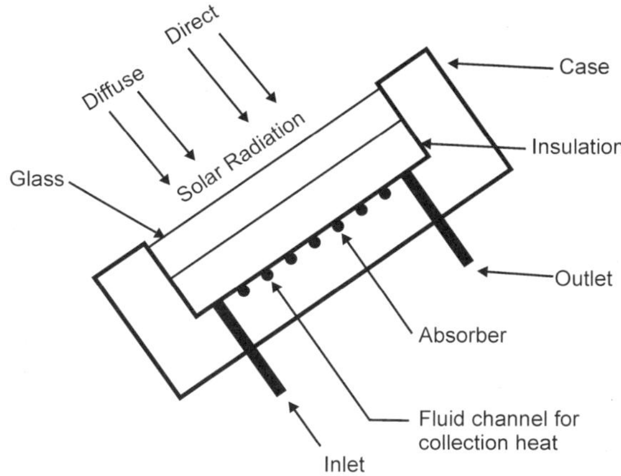

Fig. 12.1.   Schematic view of a flat plate collector.

radiation of the sun. They can be of nearly any size and have an output that is directly related to a few variables including size, facing and cleanliness. These variables all affect the amount of radiation that falls on the collector. Often these collector panels have automated machinery that keeps them facing the sun. The additional energy they take in due to the correction of facing more than compensates for the energy needed to drive the extra machinery. Flat-plate collectors are in wide use for domestic household hot-water heating and for space heating, where the demand temperature is low. The schematic view of flat plate collector is shown in Fig. 12.1.

The basic parts noted are an absorber, transparent cover sheets and an insulated box. The absorber is usually a sheet of high thermal conductivity metal with tubes or ducts either integral or attached. Its surface is painted or coated to maximize radiant energy absorption and in some cases to minimize radiant emission.[1,3] The cover sheets called glazing, let sunlight pass through to the absorber but insulate the space above the absorber to prohibit cool air from flowing into this space. The insulated box provides structure and sealing and reduces heat loss from the back or sides of the collector.

12.2.2 *Concentrating (focusing) solar collector*

Focusing collectors are essentially flat-plate collectors with optical devices arranged to maximize the radiation falling on the focus of the collector. These are currently used only in a few scattered areas. Solar furnaces are examples of this type of collector. The schematic diagram of focusing collector is shown in Fig. 12.2. Although they can produce far greater amounts of energy at a single point than the flat-plate

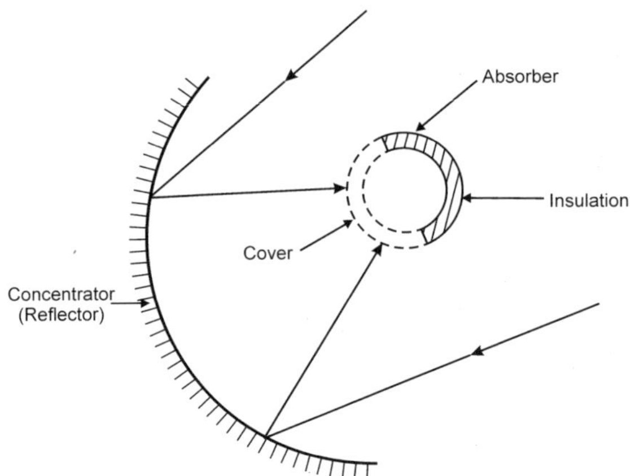

Fig. 12.2.   Schematic diagram of a focusing collector.

collectors can, they lose some of the radiation that the flat-plate panels do not. Radiation reflected off the ground will be used by flat-plate panels but usually will be ignored by focusing collectors (in snow covered regions, this reflected radiation can be significant). One other problem with focusing collectors in general is due to temperature. The fragile silicon components that absorb the incoming radiation lose efficiency at high temperatures and if they get too hot they can even be permanently damaged.[1] The focusing collectors by their very nature can create much higher temperatures and need more safeguards to protect their silicon components.

12.3   Solar Drying

Drying means removal of moisture from a substance involving simultaneous transfer of heat to the substance which is known as a unit operation. The solar dryer is a device which uses solar energy for drying. There are several processes by which heat can be transferred to a substance like conduction, convention and radiation. Traditionally, drying of agricultural products is done on the open ground directly under the sun. This method is known as open sun drying. This leads to losses due to uncontrolled drying, besides causing contamination of the product. Various types of solar dryers have been developed as an alternative to open-air sun drying and other conventional drying methods.[4] These include cabinet type solar dryer suitable for small scale use, roof integrated solar heating systems and solar dryer based on flat plate collectors. Solar heated air can suitably be used for space heating during winter in cold regions and can meet process heat requirement in industries. There are different type solar dryer used for drying application worldwide. Broadly, the solar dryer can be classified into two categories (a) passive solar dryer (natural circulation) and (b) active solar dryer (forced circulation).

12.3.1 *Passive solar dryer*

A passive solar dryer is one in which the drying product is directly exposed to the sun's rays. Direct passive dryers are best for drying small batches of fruits and vegetables such as bananas, pineapples, mangoes, potatoes, carrots and French beans.[5] This type of dryer comprises of a drying chamber that is covered by a transparent cover made of glass or plastic. The drying chamber is usually a shallow, insulated box with air-holes in it to allow air to enter and exit the box. The food samples are placed on a perforated tray that allows the air to flow through it and the food.

The experiment conducted on an indirect maize dryer.[6] The dryer consisted of a single-glazed passive solar air heater with a 1 m^2 single flat-plate absorber and an air gap of 5 cm from the glazing. The air heater was connected to an insulated drying bin equipped with a chimney as illustrated in Fig. 12.3. The entire dryer assembly was made from hardboard. To improve efficiency, the air heater was modified with a wider air gap (15 cm) to accommodate three layers of wire-mesh absorber between the glazing and the flat-plate absorber. The dryer was capable of drying 90 kg of wet maize from a moisture content of about 20 per cent wet basis to 12 per cent within 3 days on a bright day.

Punjab Agricultural University, Ludhiana, India developed domestic size solar dryer for drying chillies, garlic, ginger, mango powder, coriander, onion, fenugreek leaves, etc.[7] The main parts of the dryer are the hot box, base frame, trays and shading plates. The frame of the hot box is made of angle iron (19 $\times$ 19 $\times$ 1.6 mm thick).

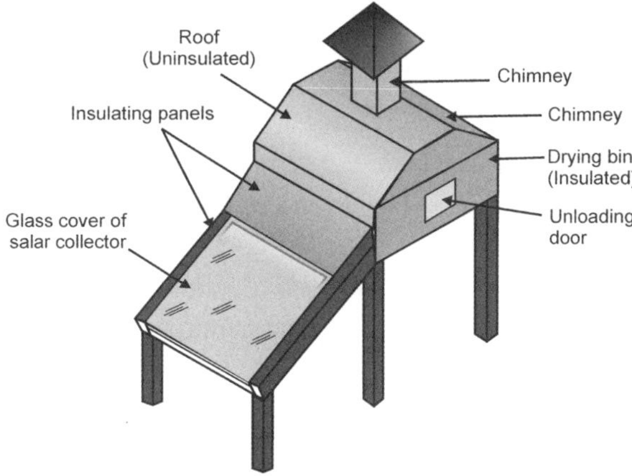

Fig. 12.3.   Natural circulation solar maize dryer.[6]

Fig. 12.4.   Domestic solar dryer.[7]

The back, top and one side of the box are insulated with 5 cm thick thermocole sandwiched between two GI sheets. On the other side of the dryer, an insulated door is provided for loading and unloading the perforated trays. The door is made of 2.5 cm thick thermocole insulation sandwiched between two 26 gauge GI sheets. For air flow, 40 holes of 8 mm diameter with total area of 2×10^{-3} m^2 were provided. The interior of the dryer is painted with dull black paint for absorption of solar radiation. A 4 mm thick transparent window glass sheet is fixed as glazing on the front of the hot box for solar energy interception. The glazing is fixed to the hot box with the help of an aluminum angle (Fig. 12.4). The aperture area of this dryer has been kept at 0.36 m^2 such that it is capable of drying about 1 kg of fresh product per day.

A distributed natural-circulation type solar-energy dryer was designed and developed[8] as shown in Fig. 12.5. The dryer comprised of the following basic units: (a) an air-heating solar-energy collector, (b) appropriately insulated ducting, (c) a drying chamber and (d) a chimney. It was found suitable for drying most of the agricultural produces. The results from the experimental facility have demonstrated the superior drying characteristics of the integral type, natural-circulation solar-energy dryers over traditional open sun drying.

The solar greenhouse dryer based on natural circulation principal was reported by the Brace Research Institute glass-roof solar dryer.[9-11] The dryer (Fig. 12.6) consisted of two parallel rows of drying platforms (along the long side) of galvanized

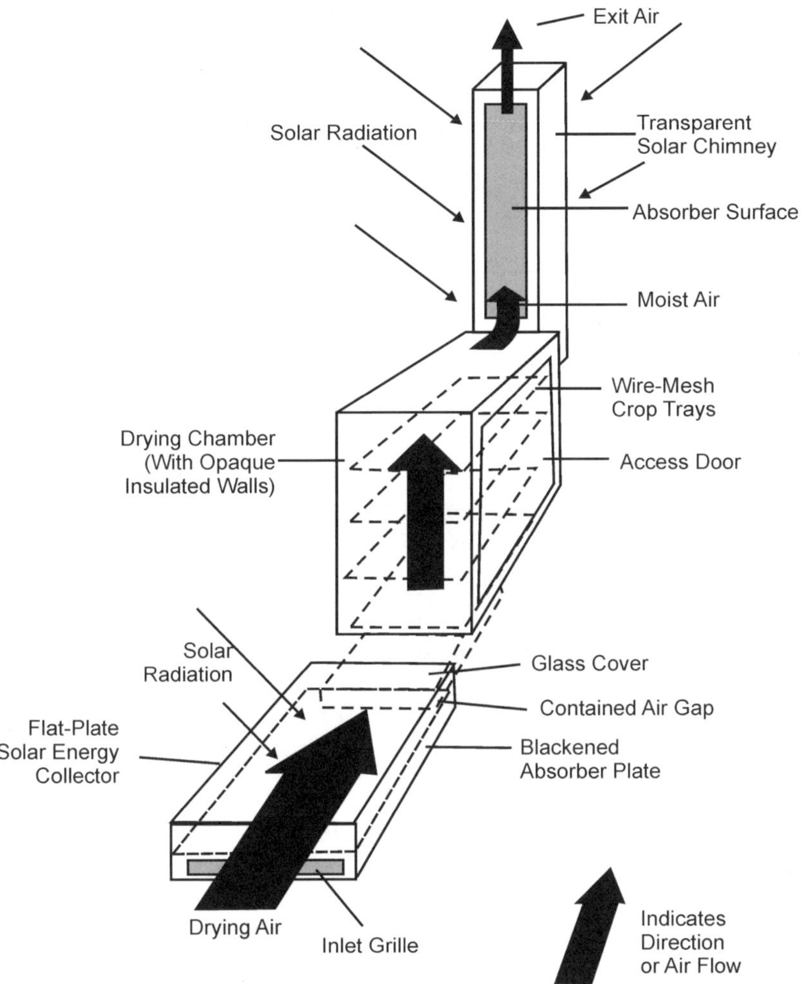

Fig. 12.5.   Distributed-type (indirect) natural-circulation solar energy dryer.[8]

iron wire mesh surface laid over wooden beams. A fixed slanted glass roof over the platform allowed solar radiation over the product. The dryer, aligned lengthwise in the north-south axis, had black coated internal walls for improved absorption of solar radiation. A ridge cap made of folded zinc sheet over the roof provides an air exit vent. Shutters at the outer sides of the platforms regulated the air inlet.

12.3.2   *Active solar dryer*

In the active solar drying system, hot air was generated outside the drying chamber. In an active dryer, the solar heated air flows through the solar drying chamber in such

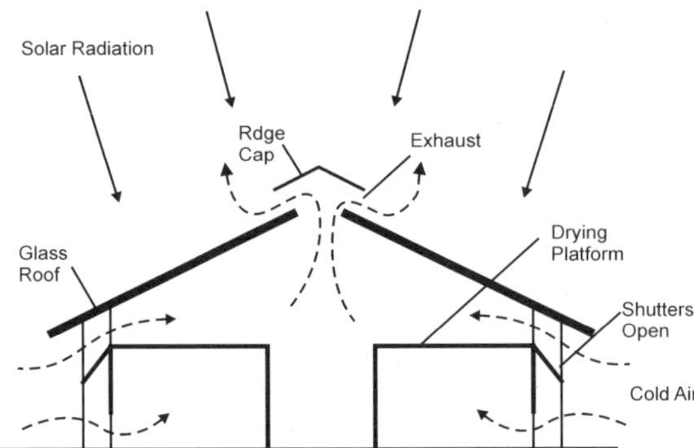

Fig. 12.6. Natural-circulation glass-roof solar-energy dryer.[10]

Fig. 12.7. (a) Experimental setup of open sun drying and greenhouse drying under natural convection,[13] (b) Experimental setup of greenhouse drying under forced convection.

a manner as to contact as much surface area of the food as possible. Thinly sliced foods are placed on drying racks, or trays, made of a screen or other material that allows drying air to flow to all sides of the food. Once inside the drying chamber, the warmed air will flow up through the stacked food trays. The drying trays must fit snugly into the chamber so that the drying air is forced through the mesh and food.[12] Active solar dryers are known to be suitable for drying higher moisture content foodstuffs such as papayas, kiwi fruits, brinjals, cabbages and cauliflower slices.

The experiment conducted on the effect of the greenhouse on the convective heat and mass transfer under forced modes[13] for cabbages and peas by using the data of crop drying (Fig. 12.7). It was concluded that (a) the convective mass transfer coefficient inside greenhouse drying under natural mode at initial stage is lower then for open sun drying, (b) the convective mass transfer coefficient in greenhouse drying under forced mode is double that of natural convection in the initial stage of drying, (c) the maximum rate of moisture evaporation took place in the beginning

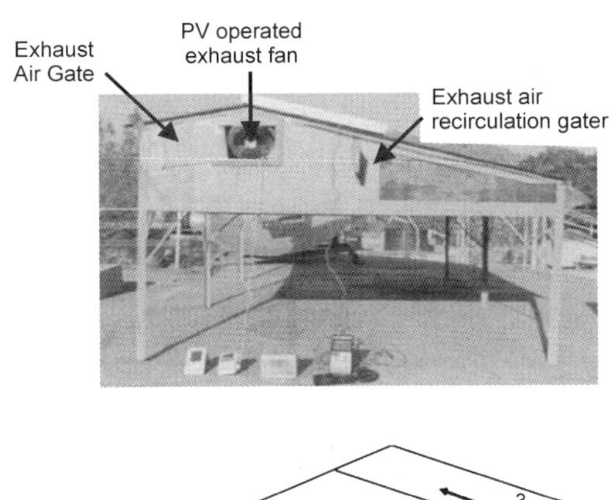

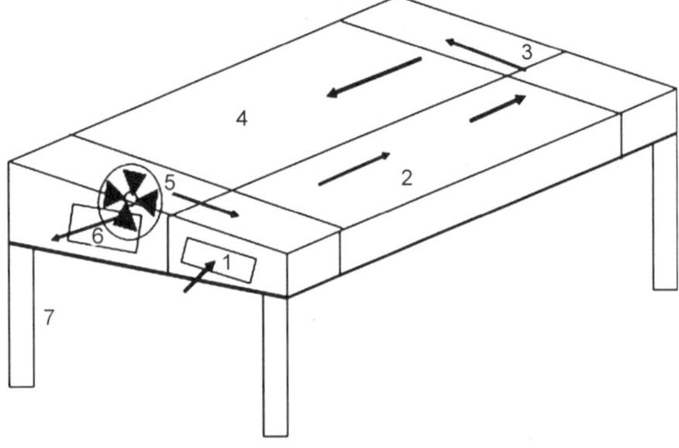

1. Air inlet 2. Air heater 3. Packed bed storage 4. Drying unit 5. Exhaust fan
6. Exhaust air out 7. Supporting leg

Fig. 12.8. Exhaust air recirculation arrangement in dryer.[14]

of the drying time (5–6 h). The mass transfer rate became essentially constant after 20 h of drying time, (d) the behavior of the convective mass transfer coefficient in the beginning of drying was like that of a wetted surface and at the end of the drying was like that of a dry surface, (e) the convective mass transfer coefficient as a function of drying time has been established with the help of a two term exponential curve model.

The performance and evaluation of the mixed mode type solar dryer[14] for drying onion flakes (Fig. 12.8). The drying and thermal efficiencies and heat utilization factors were recorded as 21, 74 and 31 per cent respectively, more compared to the recirculation of exhaust air test. The quality of dried onion flakes without recirculation of exhaust air test was superior. The recirculation of exhaust air was founded feasible only with use of desiccant material.

A batch type solar dryer was designed and simulated[15] for agriculture produce. The dryer was tested during periods of low sunshine; a heater was used to boost up the temperature (Fig. 12.9). The onion was chosen as the dried product because

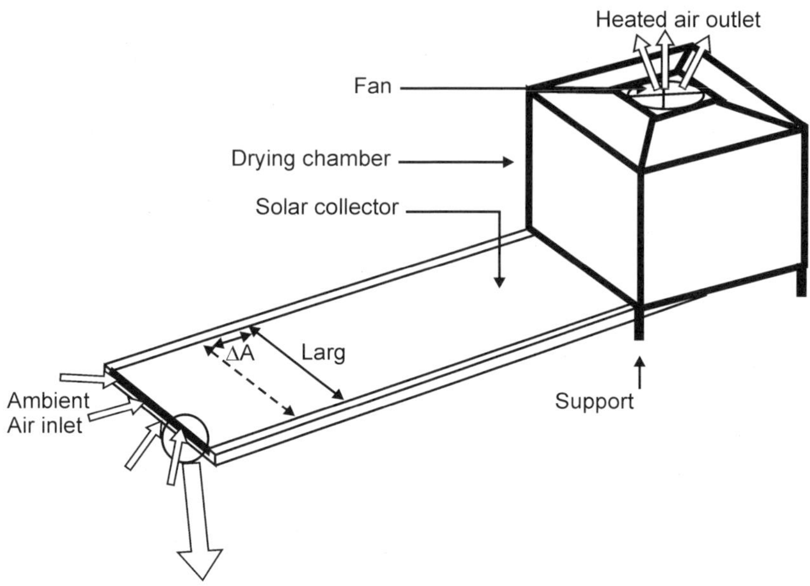

Fig. 12.9.   Batch type solar dryer.[15]

of its swift deterioration property. The shrinking effect has been taken into consideration. The results showed that drying is affected by the surface of the collector, the air temperature and the product characteristics. Significant improvements were registered in the results, after the heater is added.

12.4 Solar Cooking

Cooking is the most energy intensive activity in domestic sector. Cooking is the major energy consuming operation for rural household women. Solar energy being renewable and free source of energy can be used for thermal applications viz., drying, household cooking, etc. Cooking food with solar energy is a simple application solar thermal technology.[16] The solar cooker works on the principle of solar energy absorption and requires no fuel, emits no smoke, does not spoil the cooking utensils nor stain the kitchen walls with soot, and keeps the environment clean. The principal methods of cooking food are boiling, frying, roasting and baking. In conventional food habit the boiling technique is employed for preparation of pulse, vegetables and rice and during this process the temperature of food being cooked is about 100°C. For other methods of cooking, higher temperature is required. For frying and boiling, heat is supplied from all the sides of the material being cooked and heat is transferred to the food through convention and radiation. There are three broad categories of solar cookers available for cooking of food, namely, focusing type, box type and advanced type.

12.4.1 *Focusing type solar cooker*

In this type of solar cooker some kind of solar energy concentrator is used, which ultimately reflects the sun's rays to a common point on which a cooking pot is placed. In this type of solar cooker, the cooker utilizes the direct radiation only.

12.4.2 *Box type solar cooker*

This type of solar cooker, commonly known as a hot box, consists of a well insulated box, the inside of which is painted dull black and is covered by one or more transparent covers. The purpose of these transparent covers is to trap heat inside the solar cooker. These covers allow the radiation from the sun to come inside but do not allow the heat from the hot black absorbing plate to come out of the box. Because of this the temperature of the blackened plate inside the box increases and can heat up the space inside to temperature up to 140°C which is adequate for cooking. The box type solar cooker is illustrated in Fig. 12.10. The box type solar cooker has

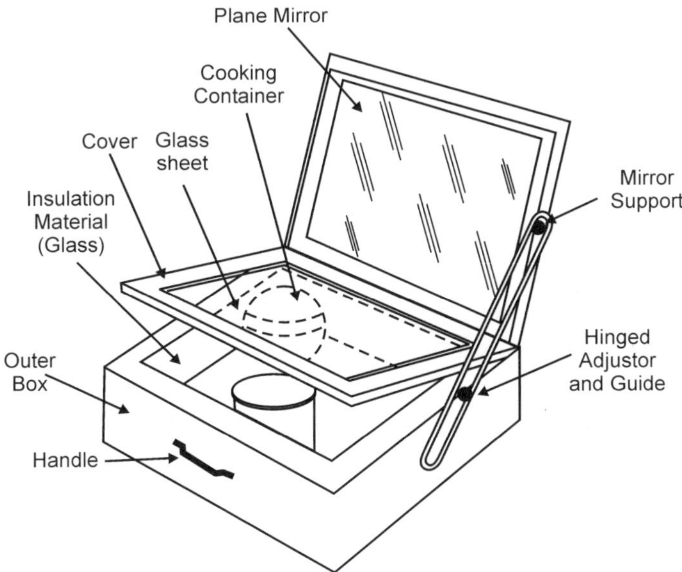

Fig. 12.10.   Schematic view of box type solar cooker.

reached the commercialization stage. The important parts of a simple box type solar cooker are:

(1) *Outer box*
The outer box of a solar cooker may be made of wood, iron sheet or fiber reinforced plastic with suitable dimensions.

(2) *Inner box*
The inner box can be made from galvanized iron or mild steel or aluminum sheet. All four sides and the bottom of the inner box which are exposed to the sun are coated with blackboard paint which absorbs solar radiation.

(3) *Thermal insulation*
The space between the outer box and the inner box is fitted with insulating material such as glass wool, thermocole, etc. This prevents heat losses from the cookers.

(4) *Double glazing*
Generally a double glass cover is provided for a solar cooker. These covers have length and breadth slightly greater than the inner box and can be fixed in a wooden frame maintaining a small spacing between the two glasses. This space contains air which acts as an insulator and prevents heat escaping from inside. The wooden frame is attached to the outer box by means of hinges. A rubber strip is affixed all around on the edges of this frame to prevent any heat leakage.

(5) *Mirror*
A mirror is used in a solar cooker to increase the radiation input on the absorbing surface. Sunlight incident on the mirror gets reflected from it and enters the box after passing through the glass covers. This radiation is in addition to the radiation entering the box directly and helps to quicken the cooking process by raising the inside temperature of the cooker. The use of a mirror can enhance the solar radiation input to the cooker by about 50 per cent.

(6) *Cooking containers*
The cooking containers are generally made of aluminum or stainless steel. The containers are painted dull black on the outer surface so that they also absorb radiation directly.

12.4.3 *Advanced type solar cooker*

It is a separate collector and cooking chamber type solar cooker where solar energy is collected at a separate place with the help of either a flat plate or focusing collector and then this stored heat is transferred to the cooking vessel placed either at a separate place. Further, here the cooking in some cases can either be done with stored heat or the solar heat is directly transferred to the cooking vessel in the kitchen.

12.4.4 *Advantages of solar cooker*

The solar cooker has a number of advantages over the traditional cooking devices. These are:

1. It preserves the nutrition value of the food (because the cooking is done at low temperature).
2. It does not require constant attention.
3. It saves time for the housewife.
4. It is pollution free.
5. It is safe and simple.
6. It saves money and fuel.
7. It helps in preserving our environment.
8. It keeps the food hot for a long time.

12.5 Solar Water Heating

The radiation from the sun is collected by the solar collector which converts the radiation into heat energy. The device through which heat energy is transported through water is called a solar water heating system. The most common way of using solar energy is through hot water by solar water heaters. Hot water is required for domestic and industrial uses such as houses, hotels, hospitals and mass-production and service industries.[16] Solar water heating systems are basically of two types, namely, collector coupled to storage tank and collector-cum-storage system.

12.5.1 *Collector coupled to storage tank*

Collector coupled to storage tank type of water heaters are used as domestic solar water heaters, where the maximum temperature required is not more than 70°C. In this type of solar water heating system there can be two possible ways of extracting heat from the collector, namely, through natural convention, generally known as the thermosyphon effect and through a forced flow of water using an electrically operated pump. The basic parts of solar water heating systems are:

1. Collector consisting of front glazing, metallic absorber, black insulated and collector box.
2. Insulated storage tank with or without heat exchanger.
3. Piping.
4. Controls and pumps.

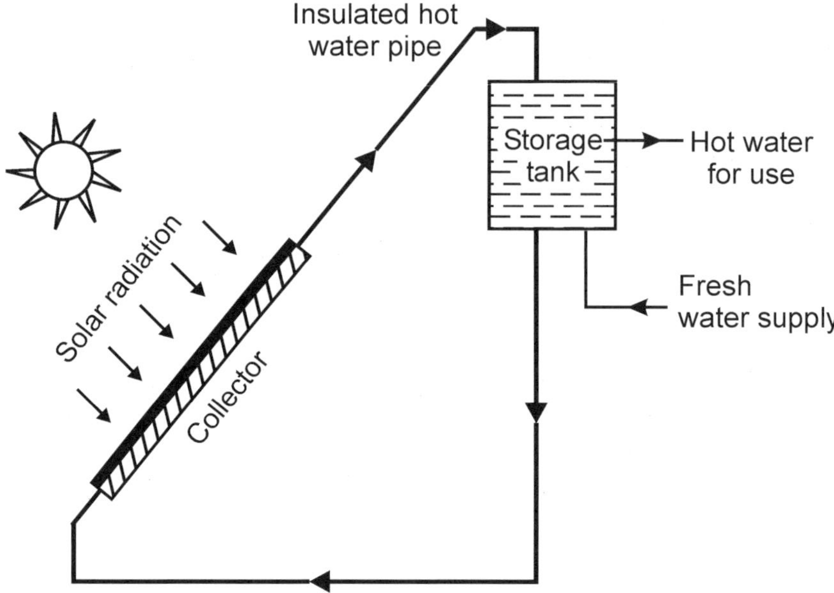

Fig. 12.11.　　Schematic of a thermosyphon solar water heater.

12.5.1.1　*Thermosyphon system*

The most commonly used solar water heating system for domestic needs is through a natural circulation type that consists of a flat plate solar collector connected to an insulated storage tank. The collector is usually placed below the insulated storage tank. Cold water from the storage tank flows down to the inlet of the collector, gets heated in the collector and rises to the storage tank by the thermosyphon effect.[17] The schematic of a thermosyphon solar water heater is given in Fig. 12.11. A density difference created by the temperature gradient causes the fluid to flow up in the collector by the thermosyphon effect. The collectors are usually oriented south with an inclination angle to latitude of the place.

12.5.1.2　*Forced circulation system*

The only additional component in a forced circulation system is a pump, which circulates water in the collector bank. This is due to higher flow rates, leading to higher efficiency. The system may additionally have some control instruments like temperature indicators, differential thermostats, flow meters, etc. In very cold climates, water alone cannot be circulated through the collector loop. In these climates it is necessary to mix anti-freeze materials in the heat extracting fluid circulating through the collector channels. In this case a heat exchanger may also be used which prevents antifreeze fluid to flow directly through the collectors.

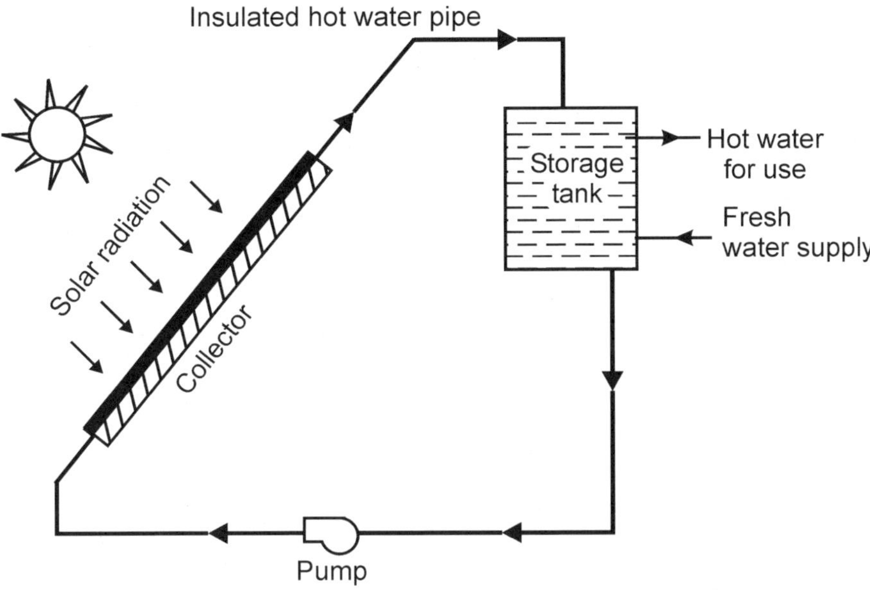

Fig. 12.12.   Schematic of a forced flow solar water heater.

The schematic diagram of a forced flow solar water heater is given in Fig. 12.12. In very cold climates water alone cannot be circulated through the collector loop. In these climates it is necessary to mix anti-freeze material in the extracting fluid circulating through the collector channel. In such systems therefore, a heat exchanger is introduced in the circuit. A heat exchanger is also required if two different types of fluids are used in the collector system and the utility circuit or the quantity of water is not suitable due to the presence of impurities.

12.5.2 *Collector-cum-storage system*

The collector-cum-storage type water heating system is extremely simple. It consists of essentially three main components namely front glazing, absorber sheet and insulated storage tank.[16] In such a water heater, sunlight passes through the front glazing and gets absorbed by the absorber which further heats the water. A schematic diagram of system is shown in Fig. 12.13. The storage tank acts as a collector cum storage for hot water.

12.6 Solar Distillation

Solar distillation is based on flat plate collector technology and is used to convert salty brackish or sea water into potable water. The device used for this purpose is also

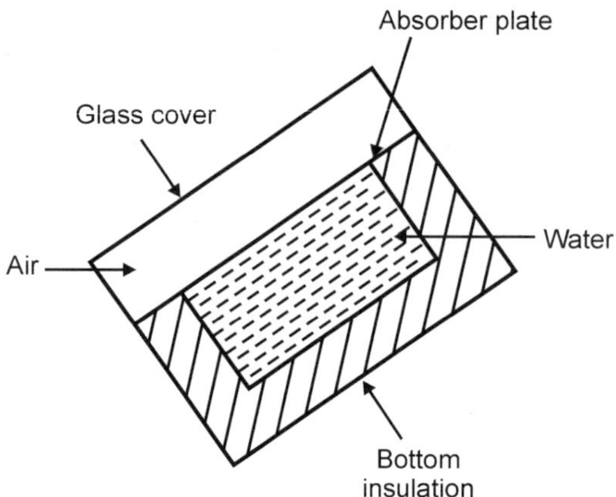

Fig. 12.13.   Collector cum storage solar water heater.

known as a solar still. In solar distillation technology water is being evaporated by solar energy and is being condensed naturally, thus getting potable water. A number of solar still designs have been developed and are currently used in India.[18] However, the most simple and popular design is known as the basin type solar still. It has a rectangular or square basin made of aluminum/fibre reinforced plastic concrete. The basin is blackened. The top is covered with a transparent material like glass with proper sealing to allow solar energy to enter the basin and channels at the lower ends of the sloping roof to collect pure water. When the sun rays pass through the top transparent cover, they are absorbed by the water and the blackened basin. As the water heats up, it changes into water vapor, which rises and condenses on the inside surface of the transparent cover into the channels, which are collected in a container. The silt or dirt remains in the bottom of the blackened basin.

This technology is suitable for supplying drinking water in coastal areas and to large pockets of different parts where water is not suitable for drinking purposes. The distilled water from solar stills can also be used for feeding batteries. Also in rural schools of any developing country, solar stills can find applications for supplying distilled water to school laboratories as well as demonstrate the utility of solar energy to the students. Modular solar stills, which can deliver 2–4 litres of distilled water every day for each square metre area, could be ideally used in large numbers for providing potable water in remote areas for battery charging, for small health centres, etc. Since these stills work on the principle of vapor pressure difference between the atmosphere and on the surface of water, they deliver more output in dry climates. A simple horizontal basin type solar still is shown in Fig. 12.14.

The operation and maintenance of the basin type solar still is simple. For efficient use, it is to be ensured that water to be distilled is present in the basin in the required

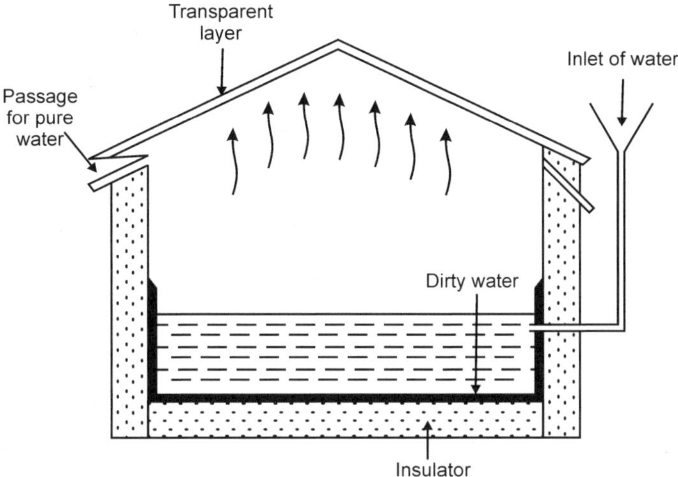

Fig. 12.14.   Schematic diagram of a basin type solar still.

quantity whenever in use. For removing salt and dirt, solar stills need to be cleaned every week by flushing the basin with water; the glazing should be cleaned at least once each day by sprinkling water or wiping it with a soft duster to improve transmitivity of solar insolation in the basin.

12.7   Solar Heating of Buildings

There are two basic types of solar heating of buildings:

1. Active solar heating of buildings.
2. Passive solar heating of buildings.

12.7.1   *Active solar heating of buildings*

In the active solar heating systems, separate solar collectors are needed for heating the space as per requirements.[19] The heating of buildings in winter is one of the major requirements for all types of building. Basically there are three ways for active heating of buildings: solar air heaters, solar liquid systems and solar heat pump systems. The simplest configuration of active solar heating of a building is shown in Fig. 12.15.

The active solar heating of buildings consists of four major parts:

(1) *Solar collectors*
Solar collectors may be air collectors, liquid collectors and heat pump type. In the air collectors the air is heated and afterward the heat is transferred to living space

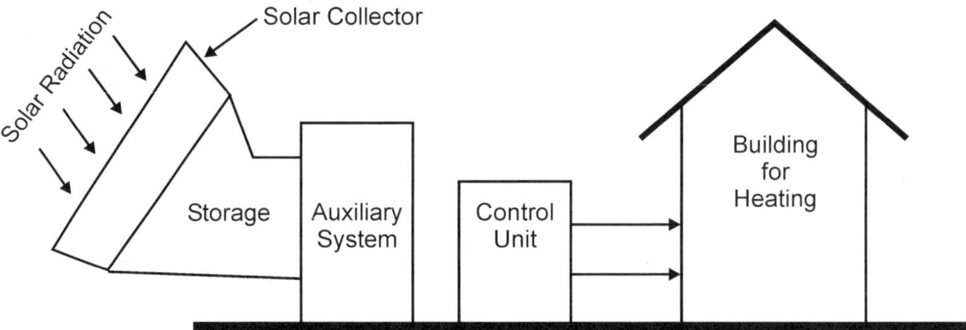

Fig. 12.15.   Simplest configuration of active solar heating of building.[8]

from storage unit by means of air or liquid. Whereas in liquid collectors, any suitable liquid, preferably water, is used for heating purpose.

(2) *Storage system*

Storage devices are used to store heat for use at night and on intermittent days.

(3) *Auxiliary heating system*

It is provided to supply heat when required of solar heating devices.

(4) *Controls*

Electrical or mechanical or electronically operated controls are used for supplying stored energy into space for heating purposes.

12.7.2   *Passive solar heating of buildings*

The concept of solar passive heating as well as day lighting involve many parameters like building envelope design, orientation, fenestration, ventilation, shading, materials like transparent insulation and low heat loss glazing. A complete passive solar heating system includes collection of energy through south facing glass, storage of energy through the use of thermal mass (usually in the form of concrete, brick, water or phase change material), regulation of energy through provision of overhangs, shades or other insulating material for windows or providing a door to close off a sun space at night and distribution of energy, which usually consists of vents, dampers, duct work and small fans.[20]

Therefore the design of houses according to the passive approach requires a detailed understanding of the complex inter relationship between architectural textures, human behavior and climatic factors. Solar passive architecture incorporates several features such as shape and orientation of building, shading devices and use of appropriate building materials for conserving energy used in heating, cooling and interior lighting of building considering prevailing solar radiation conditions in the

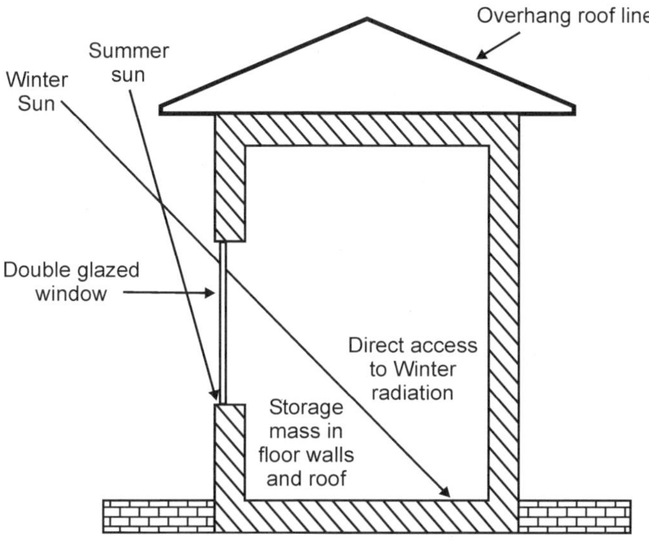

Fig. 12.16.   Direct solar gain type passive solar heating of building.

area. There are three primary solar passive heating configurations, namely, direct solar gain, indirect solar gain and isolated solar gain.

12.7.2.1   *Direct solar gain*

In this system, the sun shines directly through south facing windows into the space to be heated. Thermal storage is provided to absorb and store the energy and to prevent overheating on sunny days.[21] The schematic diagram of direct gain type passive solar heating of buildings is shown in Fig. 12.16.

12.7.2.2   *Indirect solar gain*

Indirect gain attempts to control solar radiation reaching an area adjacent but not part of the living space. Heat enters the building through windows and is captured and stored in thermal mass (e.g., water tank, masonry wall) and slowly transmitted indirectly to the building through conduction and convection. Efficiency can suffer from slow response (thermal lag) and heat losses at night. Other issues include the cost of insulated glazing and developing effective systems to redistribute heat throughout the living area. Examples: Trombe walls, water walls and roof ponds. The schematic diagram of an indirect gain type passive solar building is shown in Fig. 12.17.

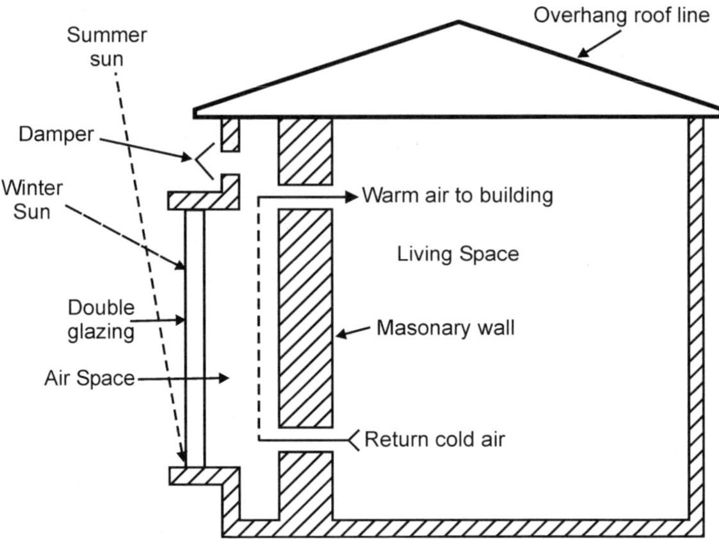

Fig. 12.17. Indirect solar gain type passive solar heating of building.

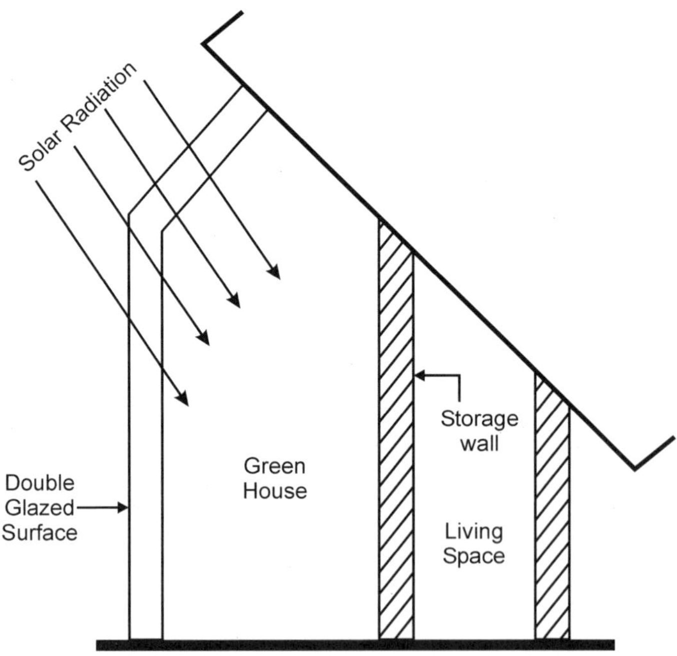

Fig. 12.18. Isolated solar gain type passive solar heating of building.

12.7.2.3 *Isolated solar gain*

Isolated gain involves utilizing solar energy to passively move heat from or to the living space using a fluid, such as water or air by natural convection or forced convection. Heat gain can occur through a sunspace, solarium or solar closet. These areas may also be employed usefully as a greenhouse or drying cabinet. An equator-side sun room may have its exterior windows higher than the windows between the sun room and the interior living space, to allow the low winter sun to penetrate to the cold side of adjacent rooms. Glass placement and overhangs prevent solar gain during the summer. Earth cooling tubes or other passive cooling techniques can keep a solarium cool in the summer. Measures should be taken to reduce heat loss at night, e.g., window coverings or movable window insulation. The schematic diagram of an isolated gain type passive solar building is shown in Fig. 12.18.

12.8 Conclusions

This chapter has focused exclusively on solar water heaters, solar dryers, solar cookers, solar stills and solar heating of buildings. Solar water heaters are used for heating water in residential or commercial premises, solar dryers are mainly used for drying agricultural and industrial products and solar cookers are used in households and institutions for cooking.

References

1. J.A. Duffie and W.A. Beckman, *Solar Engineering of Thermal Processes* (John Wiley and Sons, Inc. New York, 1991).
2. J.L. Threlkeld, *Thermal Environmental Engineering* (Prentice-Hall, New Jersey, 1970).
3. G.N. Tiwari, *Solar Energy* (Narosa Publishing House, New Delhi, India, 2002).
4. J.P. Hartnett, "The case of alternative energy sources," *Alternative Energy Sources* (ed.) J.P. Hartnett (Academic Press, 1976), pp. 19–38.
5. K.S. Jayaraman, D.K.D. Gupta and N.B. Rao, Solar drying of vegetables, *Developments in Drying* (Kasetsart University Press, Bangkok, 2000), pp. 179–186.
6. B.D. Shukla and G. Singh, *Drying and Dryers (Food and Agricultural Crops)* (Jain Brothers, New Delhi, 2003).
7. P.P. Singh, S. Singh and S.S. Dhaliwal, "Multi-shelf domestic solar dryer," *Energy Conversion and Management* **47** (2006) 1799–1815.
8. O.V. Ekechukwu and B. Norton, "Experimental studies of integral-type natural circulation solar-energy tropical crop dryers," *Energy Conversion and Management* **38** (1997) 1483–1490.
9. Saulnier, "Survey of solar agricultural dryers," *Joint Conf. Am. Section ISES and Solar Energy Society Canada Inc.* **17** (1976), 7–21.
10. T.A. Lawand, "The potential of solar agricultural dryers in developing areas," *UNIDO Conf. 5, Tech. for Solar Energy Utilization* (1977) pp. 125–132.

11. B. Kilkis, "Solar-energy assisted crop and fruit drying systems: Theory and applications," *Proc. Sem Energy Conservation and Use of Solar and other Renewable Energies in Bio-industries* (1981), pp. 307–333.

12. L. Imre, "Solar drying," *Handbook of Industrial Drying* (New York, Marcel Dekker, 1995), pp. 373–452.

13. D. Jain and G.N. Tiwari, "Effect of greenhouse on crop drying under natural and forced convection I: Evaluation of convective mass transfer coefficient," *Energy Conversion and Management* **45** (2004) 765–783.

14. S. Kothari, N.L. Panwar and S. Chaudhari, "Performance evaluation of exhaust air recirculation system of mixed mode solar dryer for drying of onion flakes," *Int. J. Renewable Energy Technology* **1** (2009) 29–41.

15. L. Bennamoun and A. Belhamri, "Design and simulation of a solar dryer for agriculture products," *J. Food Engineering* **59** (2003) 259–266.

16. A.N. Mathur and N.S. Rathore, *New and Renewable Energy Sources* (Bohra Ganesh Publications, Udaipur, 1996).

17. E.E. Anderson, *Fundamentals of Solar Energy Conversion* (Addision Wesley Publishing Company, Reading, Masschusetts, 1983).

18. G.N. Tiwari and M.A. Noor, "Characterization of solar still," *Int. J. Solar Energy* **18** (1996) 147.

19. N.S. Rathore, N.L. Panwar and A.K. Kurchania, *Non-conventional Energy Sources* (Himanshu Publication, New Delhi, 2008).

20. F.C. McQuiston, J.D. Parker and J.D. Spitler, *Heating, Ventilation and Air Conditioning: Analysis and Design* (John Wiley and Sons, 2000).

21. F. Kreith and J.F. Kreider, *Principles of Solar Engineering* (Mc Graw-Hill Book Company, 1978).

Chapter 13

Solar Tunnel Dryer — A Promising Option for Solar Drying

Mahendra S. Seveda

Department of Farm Power and Machinery,
College of Agricultural Engineering and post Harvest Technology
(Central Agricultural University), Ranipool,
Gangtok, Sikkim, India-737135
sevda_mahendra@rediffmail.com

Narendra S. Rathore

Department of Renewable Energy Sources,
College of Technology and Engineering,
Maharana Pratap University of Agriculture and Technology,
Udaipur, Rajasthan, India-313001
rathoren@rediffmail.com

Vinod Kumar

Department of Electrical Engineering,
College of Technology and Engineering,
Maharana Pratap University of Agriculture and Technology,
Udaipur, Rajasthan, India-313001
vinodcte@yahoo.co.in

The solar tunnel dryer is one of the promising options for drying various agricultural and industrial products on a large scale. In this type of dryer, the loading and unloading is easy and more quantity can be dried at lower cost. The solar tunnel dryer is essentially a poly house made of a hemi cylindrical frame cover, over which a high density polythene sheet is wrapped, where temperature, air flow rate and relative humidity required for drying the products can be maintained by placing the axis of tunnel in such a way that maximum sunlight can be trapped during the day hours. The solar tunnel dryer can be an efficient system for drying agricultural and industrial products with reduced drying time and cost of drying operations.

289

13.1 Introduction

Energy is a critical input for economic growth, social development and human welfare. Indeed, a direct correlation exists between the level of economic development and the consumption of energy. The standard of living is found to rise with increase in the per capita consumption of energy. Further, energy is considered as a prerequisite for the development. Life cannot be in existence without energy. Energy is not only meant for comforts of modern life, but it is also required for essential necessities of human beings.

The energy received from the sun is called solar energy. All forms of energy on the earth are derived from the sun. The sun provides us heat and light energy free of cost. Solar energy, which is the ultimate source of most forms of energy used now, is clean, safe and exists in viable quantities in many countries. The drawbacks in using solar radiation as energy are that it cannot be stored and it is a dilute source of energy. Also, availability of solar energy varies with time. It varies on a day-night cycle due to the rotation of the earth on its axis. It also varies on an annual cycle due to the revaluation of earth around the sun in an elliptical path. Solar energy has been used since prehistoric times, but in a most primitive manner. Before 1970, some research and development was carried out in a few countries to exploit solar energy more efficiently, but most of this work remained mainly academic. After the dramatic rise in oil prices in the 1970s, several countries began to formulate extensive research and development programmes to exploit solar energy.[1]

Solar energy is one of the most promising renewable energy sources in the world compared to non-renewable sources for the purpose of drying of agriculture and industrial products. The concept of a dryer powered by solar energy is becoming increasingly feasible because of the gradual reduction in price of solar collectors coupled with the increasing concern about atmospheric pollution caused by conventional fossil fuels used for drying crops.[2]

Solar energy is used for drying of the various industrial and agricultural products in open sun and in a scientifically designed solar dryer. Solar drying is a continuous process where moisture content, air and product temperature change simultaneously along with the two basic inputs to the system, i.e., the solar insolation and the ambient temperature. The drying rate is affected by ambient climatic conditions. This includes: temperature, relative humidity, sunshine hours, available solar insolation, wind velocity, frequency and duration of rain showers during the drying period.

Open sun drying of various industrial and agricultural products has been practiced since the dawn of time. Open sun drying is slow and exposes the products to various losses and deterioration in quality. A number of industries have, therefore, accepted mechanical drying of the produce. Fuel wood, petroleum fuel, coal or electricity is used for air heating in the mechanical dryers. Solar air dryers have great potential for replacement of industrial scale drying of industrial and agricultural products. Besides, contributing saving of precious fossil fuel, fuel wood, electricity, the solar drying may also be cost effective.

The solar tunnel dryer, a new technology for use at agricultural and industrial level for large scale drying has been developed. The solar tunnel dryer is essentially a walk-in type poly house, which can accommodate larger quantities of fruits and vegetables for drying at reasonable cost. The loading and unloading of material in this dryer is quite easy. Further, the cost of dryer installation is less compared to other commercial dryers. Solar tunnel drying is a quite promising technology in the field of value addition, which not only maintains quality of product for use but also saves tremendous amounts of conventional energy sources. The added advantage of the solar tunnel dryer is its relatively lower cost of construction and it making use of locally available material for construction.[3] However, the drying efficiency seems too low. Therefore, there is a need to minimize heat losses in the existing set up and integrate various options for perfection in design, and simultaneously more drying efficiency can be obtained.

13.2 Principle of Drying

The drying of any material involves the migration of water from the interior of the material to its surface and the removal of the water from the surface. The rate of movement differs from one substance to another. The rate of drying is dependent on the volume, temperature and moisture content of the air passing over the material.[4] The usual practice is to heat ambient air which lowers relative humidity and increases its capacity to absorb water. This warm dry air is then passed over the material to be dried. The warm air absorbs the moisture and dries the produce and then the moisture laden air is exhausted. The energy requirement for drying different products can be determined from the initial and final moisture content of each product. Products have different drying rates and maximum allowable temperatures. In many cases, only a small temperature rise in the air is necessary to achieve proper drying conditions.

13.3 Open Sun Drying

Drying is a universal method for preserving food and thereby helps in storage and easy transportation, because food becomes lighter on account of moisture removal. Traditionally food products are dried by spreading them in the open sun in thin layers. This method is known as open sun drying. The solar radiation falling on the drying material surface is partly reflected and partly absorbed. The absorbed radiation and surrounding heated air heat up the drying material surface. A part of this heat is utilized to evaporate the moisture from the drying material surface to the surrounding air. Part of this heat is lost through radiation (long wavelength) to the atmosphere and through conduction to the ground surface.[5] The process of heat and mass transfer occurs simultaneously in open sun drying. The rate of drying depends on a number of external parameters (solar radiation, ambient temperature, wind velocity, and relative humidity) and internal parameters (initial moisture content, type of drying material, drying material absorptive and mass of product per unit exposed area).

Open sun drying is economical and is a simple method of drying. Sun drying of fruits and vegetables is still practiced largely unchanged from ancient times. The working principle of open sun drying is shown in Fig. 13.1. The solar radiation falling on the crop surface is partly reflected and partly absorbed. The absorbed radiation and surrounding heated air heat up the crop surface. Part of this heat is utilized to evaporate the moisture from the crop surface to the surrounding air. This part of heat is lost through radiation (long wavelength) to the atmosphere and through conduction to the ground surface.[6]

Traditional sun drying takes place by storing the product under direct sunlight. Sun drying is only possible in areas where, in an average year, the weather allows foods to be dried immediately after harvest. The main advantages of sun drying are low capital and operating costs and the fact that little expertise is required.

The main disadvantages of this method are as follows: contamination, theft or damage by birds, rats or insects; slow or intermittent drying and no protection from

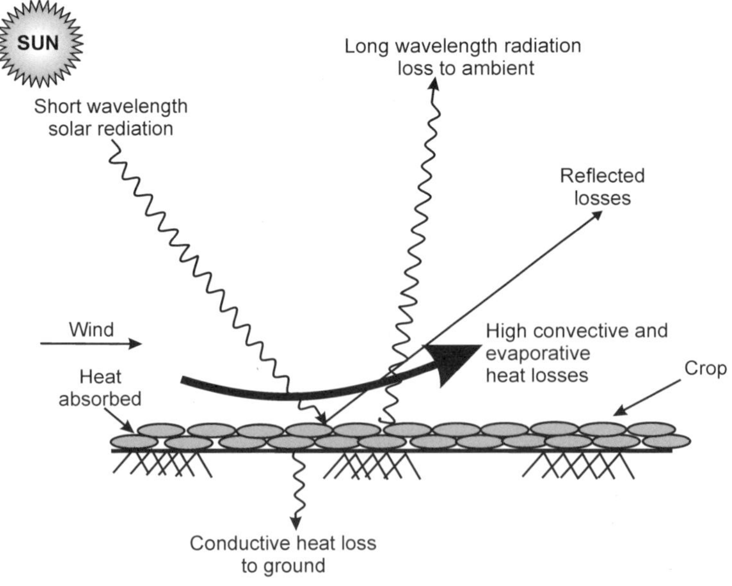

Fig. 13.1.   Working principle of open sun drying.[6]

rain or dew that wets the product, encourages mould growth and may result in a relatively high final moisture content; low and variable quality of products due to over or under drying; large areas of land needed for the shallow layers of food; laborious since the crop must be turned, moved if it rains; direct exposure to sunlight reduces the quality (color and vitamin content) of some fruits and vegetables. Moreover, since sun drying depends on uncontrolled factors, production of uniform and standard products is not expected. The quality of sun dried foods can be improved by reducing the size of pieces to achieve faster drying and by drying on raised platforms, covered with cloth or netting to protect against insects and animals.

13.4　Types of Solar Dryers

Solar dryers have some advantages over sun drying when correctly designed. They give faster drying rates by heating the air to 10–30°C above ambient, which causes the air to move faster through the dryer, reduces its humidity and deters insects.[1] The faster drying reduces the risk of spoilage, improves quality of the product and gives a higher throughput, so reducing the drying area that is needed. However, care is needed when drying fruits to prevent rapid drying, which will prevent complete drying and would result in case hardening and subsequent mould growth. Solar dryers also protect food from dust, insects, birds and animals. They can be constructed from locally available materials at a relatively low capital cost and there are no fuel costs. Thus, they can be useful in areas where fuel or electricity are expensive, land for sun drying is in short supply or expensive, or sunshine is plentiful but the air humidity is high. Typical drying times in solar dryers range from 1 to 3 days depending on the sun, air movement, humidity and the type of food to be dried.[7]

There are mainly two types of solar dryers: passive and active solar dryers.[8] Passive solar dryers can further be classified as direct, indirect and direct cum indirect type. In direct type passive solar dryers the solar radiation is transmitted through a transparent cover and absorbed by blackened interior surface and produce kept for drying. Due to accumulation of energy the temperature inside the dryer increases, as a result there is a continuous flow of air on the drying material.[9] Such dryers are most suited for a limited quantity of fruits and vegetables at domestic level. A schematic view of a direct type passive solar dryer is shown in Fig. 13.2.

Indirect dryers are suitable for color sensitive produce, as the produce is not exposed directly to sunlight. The passive direct cum indirect dryer consists of an air heating collector and a drying unit with a transparent cover on the top and the system is connected in series.[1] The solar energy that passes through the transparent cover in the collector heats the absorber which transfers heat to the air. A schematic view of direct cum indirect type passive solar dryer is shown in Fig. 13.3.

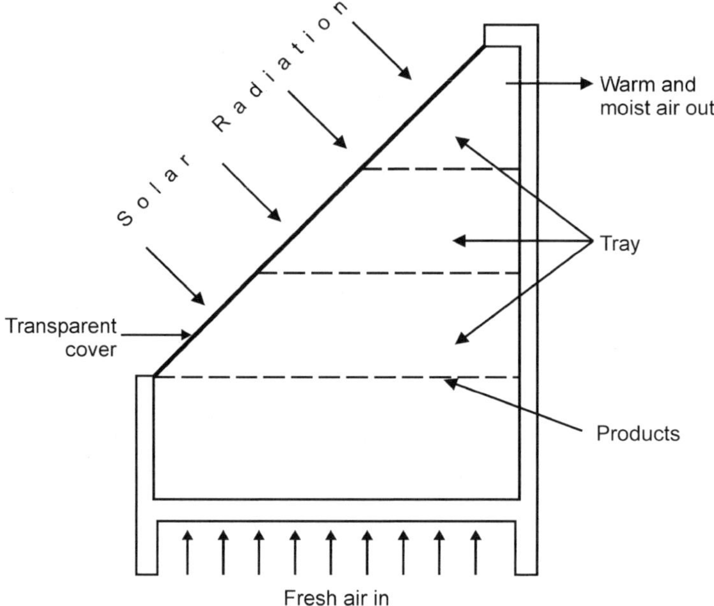

Fig. 13.2.　Schematic view of a direct type passive solar dryer.

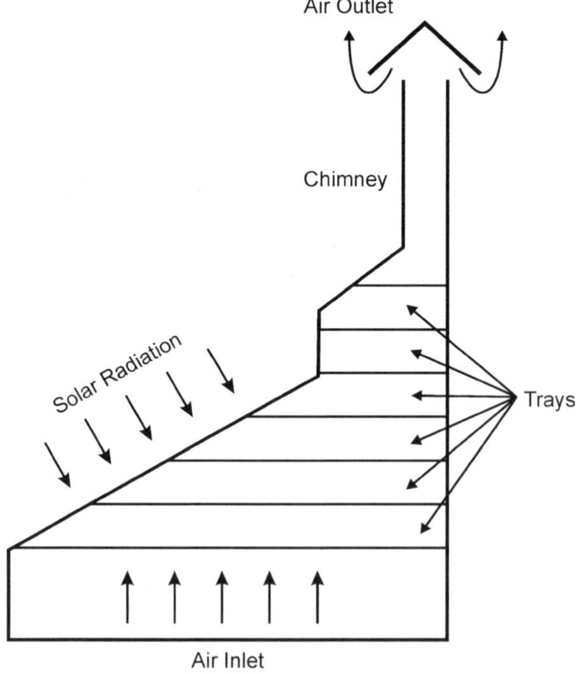

Fig. 13.3.　Schematic diagram of a direct cum indirect type passive solar dryer.

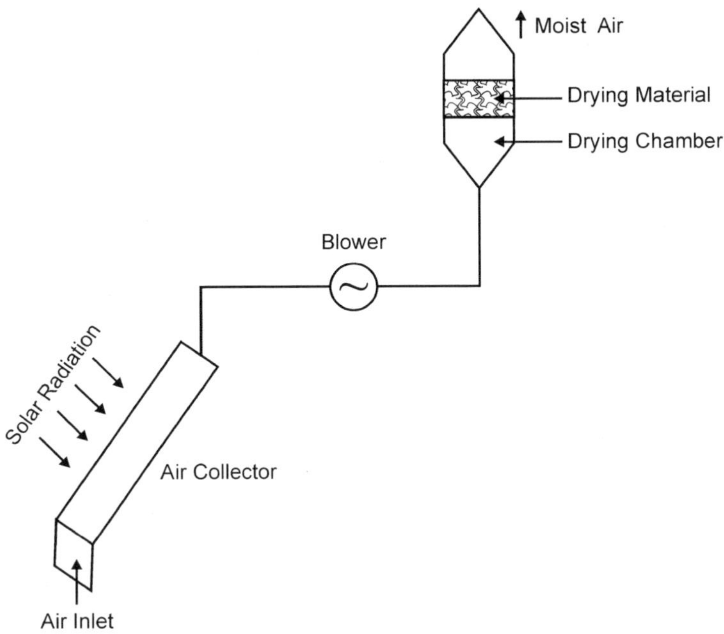

Fig. 13.4.   Schematic view of a forced convection solar dryer.

The forced convection tunnel solar dryer is an example of a mixed mode passive solar dryer where the power required to operate the fan is supplied by a solar module. The active type dryers are suited for large scale application where blowers provide force circulation of heated air.[1] A schematic view of a forced convection solar dryer is shown in Fig. 13.4.

13.5   Factors Affecting Solar Drying

Solar drying is a continuous process where moisture content, air and product temperature change simultaneously along with the two basic inputs to the system, i.e., the solar insolation and the ambient temperature.[10] The drying rate is affected by ambient climatic conditions.

13.5.1   *Temperature*

The drying rate of drying materials depends on the air temperature. In direct solar dryers the drying material absorbs solar radiation and transfers a portion of the heat to the drying air entering the dryer. This air in turn carries away evaporated moisture from the drying material while in this case of an indirectly heated solar dryer, the air is preheated in the collector and then enters the drying unit.

13.5.2 *Solar radiation*

The availability of solar radiation depends on the climatic and geographic locations of solar drying. The radiation intensity is determined by the latitude of the location, position of the sun, the day of the year and the clearness of the sky.

13.5.3 *Relative humidity*

The drying potential depends on the difference between the moisture content of the product and the equilibrium moisture content of the drying material determined by the temperature and relative humidity of the drying air. If the relative humidity is higher, the drying potential is lower and hence a lower drying rate.

13.5.4 *Moisture content*

The selection of the design of solar dryers is guided by initial moisture content and the amount of moisture to be removed. The amount of moisture of the product to be removed is dictated by the expected storage period of the dried product. The lower the final moisture content, the higher is the expected storage life of the product. Again, the drying rate of the product is also influenced by the moisture content of the product.

13.6 Selection of Solar Dryers

The solar dryer is an improved form of sun drying in which drying is accomplished in a closed structure under relatively controlled conditions utilizing the thermal energy of sun.[11] Important factors to be considered in selection for a type of solar dryer for a particular product are: (i) the amount of product to be dried, (ii) the recommended temperature for intended use and (iii) amount of moisture to be removed for an expected storage life. In addition to this intensity of solar radiation, air temperature, relative humidity and moisture content of the product are the main factors that affect the drying process.

In view of the enormous choices of dryer types one could possibly deploy for most products, selection of the best type is a challenging task that should not be taken lightly nor should it be left entirely to dryer vendors who typically specialize in only a few types of dryers. The user must take a proactive role and employ vendors' experience and bench-scale or pilot-scale facilities to obtain data, which can be assessed for a comparative evaluation of several options. A wrong dryer for a given application is still a poor dryer, regardless of how well it is designed. Note that minor changes in composition or physical properties of a given product can influence its drying characteristics, handling properties, etc., leading to a different product and in some cases severe blockages in the dryer itself. As a minimum, the following quantitative information is necessary to arrive at a suitable dryer:

- Dryer throughput; mode of feedstock production (batch/continuous).
- Physical, chemical and biochemical properties of the wet feed as well as desired product specifications; expected variability in feed characteristics.
- Upstream and downstream processing operations.
- Moisture content of the feed and product.
- Quality parameters (physical, chemical, biochemical).
- Value of the product.
- Need for automatic control.
- Toxicological properties of the product.
- Type and cost of fuel, cost of electricity.
- Environmental regulations.
- Space in plant.

For high value products like pharmaceuticals, certain foods and advanced materials, quality considerations override other considerations since the cost of drying is unimportant. In some cases, the feed may be conditioned (e.g., size reduction, flaking, palletizing, extrusion, back mixing with dry product) prior to drying, which affects the choice of dryers. As a rule, in the interest of energy savings and reduction of dryer size, it is desirable to reduce the feed liquid content by less expensive operations such as filtration, centrifugation and evaporation.

13.7　Solar Tunnel Dryer

A solar tunnel dryer is a hemi cylindrical shaped tunnel having metallic frame structure covered with ultraviolet stabilized polythene sheet of 200 micron, where agricultural and industrial products on a large scale could be dried under a at least partially controlled environment and which is large enough to permit a person to enter and carry out operations such loading/unloading materials and inspecting the condition of products during drying. The solar radiation is transmitted through a plastic sheet of the tunnel dryer.[12] This plastic cover has the property to transfer short-wave length radiation. The solar radiation is converted into long wave length radiation inside the solar tunnel dryer, as the plastic sheet is opaque to long wave length radiation. The radiation is trapped inside the dryer, raising its temperature. The cement concrete floor has been painted black for better absorption of solar radiation. This particular effect is called the greenhouse effect.[13]

The axis of the solar tunnel dryer is kept in an east-west direction so that maximum exposure of southern radiation can be obtained. Inlets of fresh air are provided along the periphery of the tunnel near ground level towards the south side. Chimneys have been provided on the top of the curved tunnel surface to exhaust hot moist air. A north wall is provided to reduce heat loss which is more significant during winter months. Approximately 68 per cent of the solar tunnel dryer area is receiving sunlight whereas the remaining area does not boost the temperature inside it. Therefore this insulation wall serves the purpose of minimizing heat loss. The solar tunnel dryer consists of a solar air collector cum drying chamber. The drying chamber is divided into multiple trays. The heated air passing over the product in each tray evaporates moisture. The hot moist air exhausts through a chimney. The solar tunnel dryer is simple, efficient and convenient in operation for drying purposes, which can be easily integrated for large scale drying. It is very useful in all types of climatic conditions.

The solar tunnel dryer was designed and developed for accommodating wide varieties of materials such as agricultural produce, industrial chemical products and industrial non-chemical products. Perfection in the design has been made keeping in view manufacturing aspects, choice of construction material, specification to accommodate large quantity, ease of operation and least cost of installation. Suitable chimney size for optimum air flow rate and north wall size for prevention of heat losses especially for winter season and maintaining adequate temperature for drying has been designed as per standard method of thermodynamics and heat transfer. The schematic illustration of the solar tunnel dryer is given in Fig. 13.5.

The solar energy trapped inside the solar tunnel dryer raises temperature of the air and accelerates evaporation of moisture in the product kept inside the solar tunnel dryer. The moist air is removed through a natural convention current. The micro-climate inside the solar tunnel dryer may be controlled by regulating air flow rate through the dryer using an exhaust fan. The salient features of solar tunnel dryer are as follows:

1. It is a modular walk-in-type hemi-cylindrical poly house type design.
2. The metallic frame structure of the tunnel dryer was covered with ultra-violet stabilized semi-transparent polyethylene sheet of at least 200-micron thickness. It has a long life and does not allow the trapped radiation to escape.
3. The axis of the solar tunnel dryer is east-west direction so that maximum exposure of southern radiation can be obtained.
4. A gradient of 2–3° is provided along the length of the tunnel to induce natural convection airflow.
5. The cement concrete floor was painted black for better absorption of solar radiation. Five-cm thick glass wool insulation was provided to reduce heat loss through the floor.

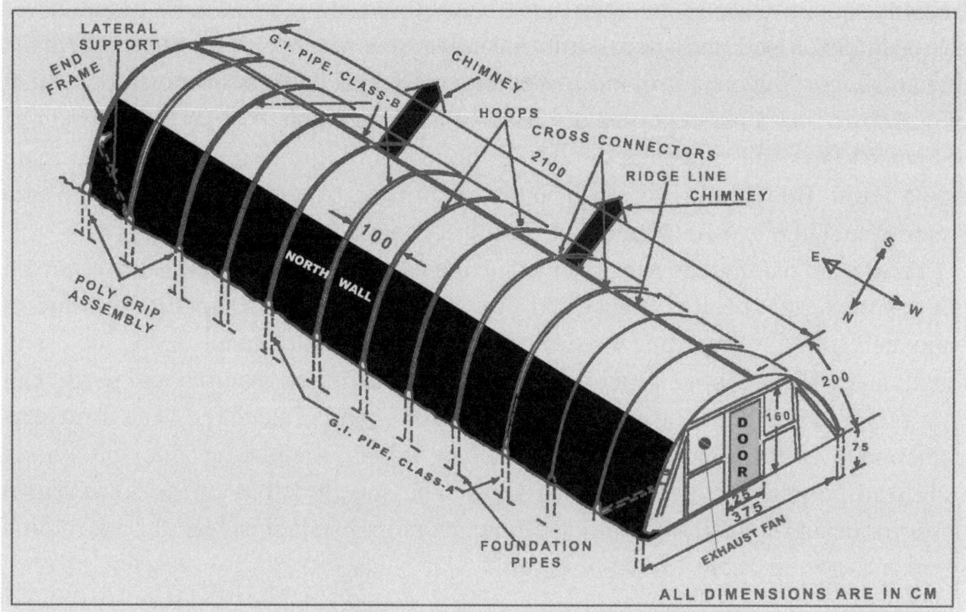

Fig. 13.5. Schematic diagram of solar tunnel dryer.

6. A black body was installed to reduce heat losses from the northern side of the tunnel.

7. Inlets for fresh air were provided all along the periphery of the tunnel near ground level.

8. The upper end of the tunnel was provided with a steel door of at least 1.6 m × 0.75 m size for loading and unloading of the material.

9. Exhaust fan of 1000–1200 m^3/h air flow rate capacity and 0.75 kW power rating (as per requirement of the products). The exhaust fan operation was automatic to maintain average relative humidity of the inside air between 30–40 per cent.

10. The number and size of chimney was decided on the basis of amount of moisture removed on a day and the drying rate required.

13.8 Case Studies on Solar Tunnel Dryer for Drying Agricultural Product (Embilica Officinalis Pulp)

Emblica officinalis (Aonla) is an important dry land fruit crop. Emblica officinalis fruit rich in Vitamin C and pectin has an important medicinal value in Ayurvedics.[14] As the crop is becoming popular among the farmers its area under cultivation is increasing at a steady rate. However, farmers are not achieving adequate returns.

Keeping this in view, a new approach for Aonla drying using a solar tunnel dryer has been developed and successfully installed at a number of emblica officinalis orchards for drying one ton or more of emblica officinalis pulp from moisture content of 424.93 to 10.08 per cent on a dry basis in a single batch. A solar tunnel dryer of 3.75 m × 17 m size has been designed, developed and commissioned at M/s Shrinath Aonla Farm, Banswara, Rajasthan, India, for drying of a one-ton Aonla pulp in a batch from 424.93 to 10.08 per cent moisture content (db).

The design parameters were decided on the basis of amount of moisture removed per day, material specific gravity and flow rate required for removing moisture in stipulated time.[15] According to specification of the solar tunnel dryer, including length and width of tunnel, number and height of chimney required and north wall area required for boosting required temperature, were calculated. The structural components of the solar tunnel dryer include hoops, foundation, floor ultraviolet stabilized polythene film and drying trays. The specifications of the solar tunnel dryer for drying 1.0 ton of embilica officinalis pulp installed at M/s Shrinath Aonla Farm is shown in Table 13.1.

Table 13.1.   Design details of solar tunnel dryer for drying of embilica officinalis pulp.

Components/Particulars	Specifications
Product dried	Embilica officinalis pulp
Loading capacity	1000 kg per batch
Initial moisture content	424.93% (db)
Final moisture content	10.08% (db)
Drying period required	16 h
Collector material (cover)	UV polythene 200 micron sheet
Orientation	E-W direction
North wall	Glass wool sandwich in metallic cover
Size of north wall	17 m × 1.61 m
No. of chimneys	5
Size of chimney (diameter and length)	0.20 m and 0.75 m
Tray size	800 mm × 400 mm × 40 mm
Number of trays	2 trays per trolley
Number of trolleys	15
Door size	1.6 m × 0.75 m
Length of solar tunnel dryer	17.0 m
Diameter of solar tunnel dryer	3.75 m
Floor area of solar tunnel dryer	63.75 m^2
Area of hemi cylindrical shape of solar tunnel dryer	111.18 m^2

The dryer is large enough to permit a person to enter it and carry out operations such as loading/unloading materials for drying. The cement concrete floor has been painted black for better absorption of solar radiation. A black body made up of metallic sheet has been provided on the northern side of the tunnel for preventing heat losses from the north side of the tunnel. Inlets for fresh air have been provided along the periphery of the tunnel near ground level, five chimneys of 20 cm diameter and 75 cm height have been provided on top of the curved surface to allow exhaust of hot air. The upper end of the tunnel has been provided with a door of size 1.60 m × 0.75 m to facilitate loading/unloading of the material. The closed view of the solar tunnel dryer installed at M/s Shrinath Aonla Farm, Banswara, Rajasthan, India, is shown in Fig. 13.6.

13.8.1　*Experimental study*

Experiments were conducted on the solar tunnel dryer installed at M/s Shrinath Aonla Farm. Under full load conditions embilica officinalis pulp was spread in thin layers of approximately 4 cm thickness in the trays of size 0.4 m × 0.8 m. A total of ten trays were placed inside the tunnel in position to accommodate one ton of embilica officinalis pulp. Drying tests were started at 9:00 hours and stopped at 17:00 hours in the months of December and May. For studying the microclimatic variation inside a solar tunnel dryer, the following parameters were selected for monitoring:

1. Temperature.
2. Solar insolation.

Fig. 13.6.　Closed view of solar tunnel dryer installed at M/s Shrinath Aonla's Farm.

3. Moisture content variations.
4. Relative humidity.
5. Air flow rate.

(1) *Temperature*

Full load testing of the dryer was conducted for evaluating the performance in actual loaded condition. This test was conducted by loading the dryer at its designed capacity. The tunnel dryer was evaluated for finding the performance in actual loaded condition with drying product. To measure the temperature of air at various locations of the dryer, three sensors inside and one sensor outside the solar tunnel dryer were kept. Sensors number one, two and three were kept inside the solar tunnel dryer at mid bottom point, at 1 m above mid bottom point, and at the north wall, respectively, and sensor number zero was kept outside the solar tunnel dryer. All temperature data were registered at an interval of an hour by a data logger.

(2) *Solar insolation*

Solar insolation is an important parameter in the energy balance of atmosphere and earth surface. All bodies emit energy in the form of electromagnetic waves, when they are at a temperature above absolute zero. The source of this thermal radiation or temperature radiation is in the molecular motion. During collision (or more generally as a result of interaction between molecules), part of their energy is transformed into radiation. The emission and the absorption of thermal radiation are governed by the temperature and nature of emitting and absorbing substance. It is expressed in W/m^2.

(3) *Moisture content variations*

(a) *Determination of moisture content*

The moisture content of the product was determined by the oven drying method. Samples of the product were taken and weighed. Then they were placed in moisture boxes which were placed in oven at temperature of 105°C for 24 hours. Then the samples were again weighed and their moisture content was determined by two methods (i) wet basis (wb) and (ii) dry basis (db). The moisture content on the wet basis method was computed using Eq. (13.1).

$$M_{\text{wb}} = \frac{W_o - W_d}{W_o} \times 100. \tag{13.1}$$

The moisture content using the dry basis method was computed using Eq. (13.2).

$$M_{db} = \frac{W_o - W_d}{W_d} \times 100, \qquad (13.2)$$

where
W_o = Initial weight of sample,
W_d = Bone dry weight of sample.

(b) *Determination of moisture ratio*

The moisture ratio was calculated by using Eq. (13.3).

$$\text{MR} = \frac{M - M_e}{M_o - M_e} = e^{-k\theta}, \qquad (13.3)$$

where
M = Moisture content at any time θ,
M_e = Equilibrium moisture content,
M_o = Initial moisture content,
k = Drying constant,
θ = time.

(c) *Determination of drying rate*

The drying rate was calculated from Eq. (13.4).

$$k = \frac{W_L}{T}, \qquad (13.4)$$

where
W_L = Amount of moisture evaporated,
T = Time taken.

(4) *Relative humidity*

The ratio of actual vapor pressure in the air-water mixture and the saturated water vapour at same temperature is known as the relative humidity of air and expressed in percentage. It is largely dependent on atmospheric temperature. During experiments the relative humidity was measured and recorded by a digital hygrometer.

(5) *Determination of air flow rate*

The air flow rate was calculated using Eq. (13.5).

$$\text{AR} = V_a \times A_i, \qquad (13.5)$$

where
V_a = Air velocity,
A_i = Area of inlet opening.

13.8.1.1 *Performance evaluation of solar collector*

The performance of the solar collector was evaluated in terms of collector efficiency. The hourly collector efficiency was calculated by following the formula for each day.

(a) Determination of mass flow rate

The mass flow rate of air was calculated using Eq. (13.6).

$$m = \text{AR} \times D_a, \tag{13.6}$$

where
D_a = Density of air.

(b) Determination of heat gained by drying air

Heat gained by air in the solar tunnel dryer was calculated using Eq. (13.7).

$$Q_g = mC_a(T_2 - T_1), \tag{13.7}$$

where
C_a = Specific heat of air,
T_1 = Ambient temperature,
T_2 = Dryer air temperature.

(c) Heat received by the collector

The heat received by the collector was calculated using Eq. (13.8).

$$Q_c = I \times 0.86 \times A_c, \tag{13.8}$$

where
I = Solar insolation,
A_c = Area of collector.

(d) Solar collector efficiency

Solar collector efficiency was calculated using Eq. (13.9).

$$\eta_c = \frac{Q_g}{Q_c} \times 100. \tag{13.9}$$

13.8.1.2 *Economic analysis*

For the success and commercialization of any new technology, it is essential to know whether the technology is economically viable or not. Therefore, an attempt was made to determine the economics of the solar tunnel dryer. Different economic indicators such as net present worth (NPW), benefit cost ratio (B/C ratio) and pay back period were used for economic analysis of the solar system under this study.

13.8.2 *Performance evaluation of solar tunnel dryer for drying emblica officinalis (aonla) pulp*

The performance of the solar tunnel dryer was evaluated in actual loaded condition during the months of December and May for drying Aonla pulp at M/s Shrinath Aonla Farm. First, Aonla was shredded. In the full load testing, one ton of Aonla pulp was placed in the solar tunnel dryer and 50 kg of pulp was placed in the open air for drying for comparison purpose. An interior view of the solar tunnel dryer under full load condition is given in Fig. 13.7.

The performance evaluation test includes measuring solar insolation, ambient temperature, ambient relative humidity, wind velocity, air flow rate inside the dryer, air temperature and relative humidity inside the solar tunnel dryer. During the full load testing, two days are required for drying the total amount of emblica officinalis pulp in the solar tunnel dryer from moisture content of 424.93 to 10.08 per cent (db). The testing on full load was conducted for two consecutive days in the months of December and May.

Fig. 13.7.   Inside view of solar tunnel dryer for drying foremblica officinalis pulp.

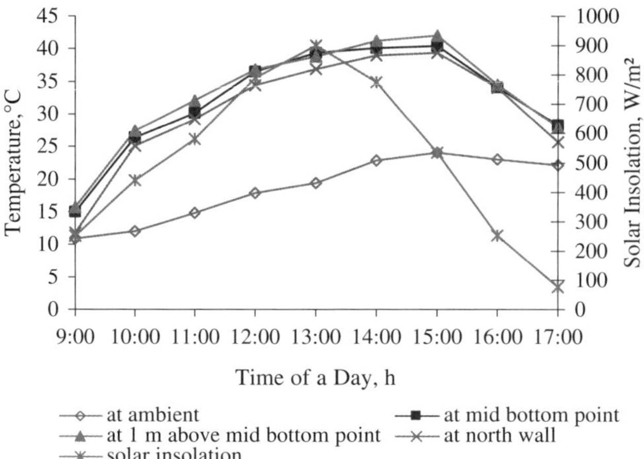

Fig. 13.8. Variation in solar radiation, ambient temperature and temperature at different locations in the solar tunnel dryer with time at full load condition for foremblica officinalis pulp drying in the winter month (December).

13.8.2.1 *Solar flux density and temperature*

As shown in Fig. 13.8 the maximum temperature inside the solar tunnel dryer was 42°C at 15:00 hours while the minimum temperature inside the solar tunnel dryer was 11.8°C at 9:00 hours in a typical day in the month of December against the maximum and minimum ambient temperatures of 24.1 and 10.9°C, respectively. It was also observed that the maximum and minimum ambient solar insulation in this month was 901 W/m² at 13:00 hours and 76 W/m² at 17:00 hours, respectively.

Similarly, in a typical day in the month of May the maximum and minimum temperature inside the solar tunnel dryer was observed as 61.7°C at 15:00 hours and 32.8°C at 9:00 hours, respectively, which is shown in Fig. 13.9. Correspondingly, the maximum and minimum ambient temperature was 40.8°C at 15:00 hours and 29.4°C at 9:00 hours, respectively. It was also observed that the maximum and minimum solar insolation in this month was 1009 W/m² at 13:00 hours and 596 W/m² at 9:00 hours, respectively.

13.8.2.2 *Variation of relative humidity*

As shown in Fig. 13.10 the maximum relative humidity inside the tunnel was 54 per cent at 9:00 hours, while the minimum relative humidity was 29 per cent at 15:00 hours in the month of December. The maximum ambient relative humidity in this month was 82 per cent at 9:00 hours while the minimum ambient relative humidity was 35 per cent at 15:00 hrs, respectively.

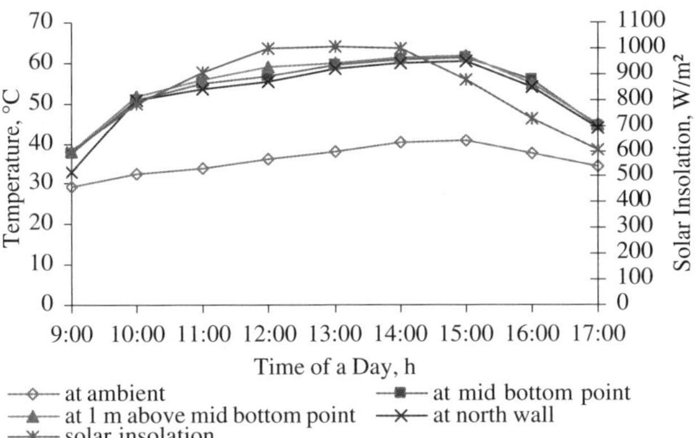

Fig. 13.9. Variation in solar radiation, ambient temperature and temperature at different locations in the solar tunnel dryer with time at full load condition for foremblica officinalis pulp drying in the summer month (May).

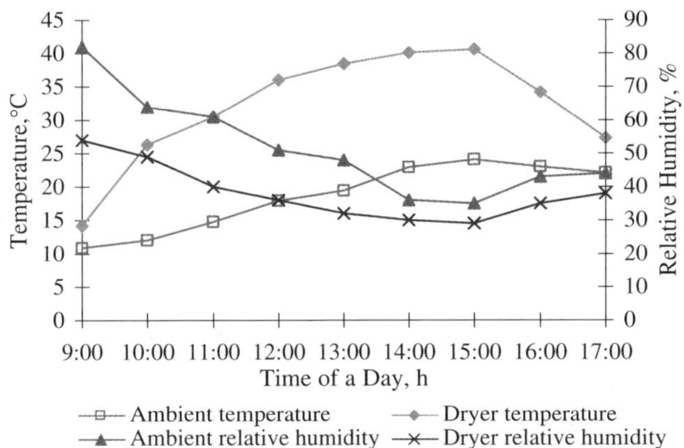

Fig. 13.10. Ambient and average dryer temperature and relative humidity at full load condition for foremblica officinalis pulp drying in the winter month (December).

Figure 13.11 indicates that the maximum relative humidity inside the tunnel was 50 per cent at 9:00 hours, while the minimum relative humidity was 26 per cent at 15:00 hours in the month of May. The maximum ambient relative humidity in this month was 62 per cent at 9:00 hours while the minimum ambient relative humidity was 29 per cent at 15:00 hours, respectively.

13.8.2.3 *Variation of air flow rate*

The maximum air flow rate inside the tunnel was 4192.21 m^3/h at 13:00 hours, while the minimum air flow rate was 2968.2 m^3/h at 17:00 hours in the month of December.

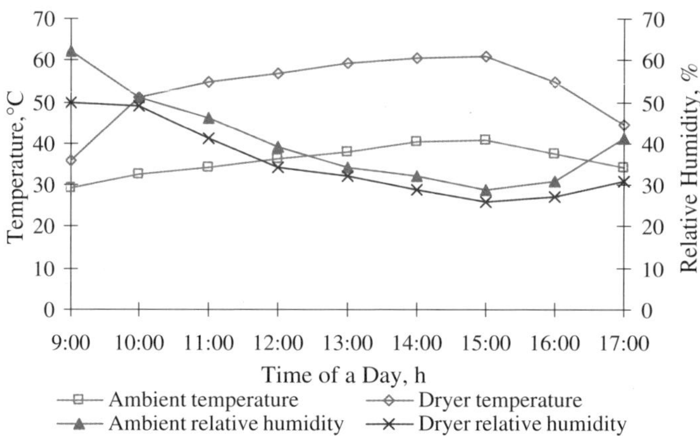

Fig. 13.11.   Ambient and average dryer temperature and relative humidity at full load condition for foremblica officinalis pulp drying in the summer month (May).

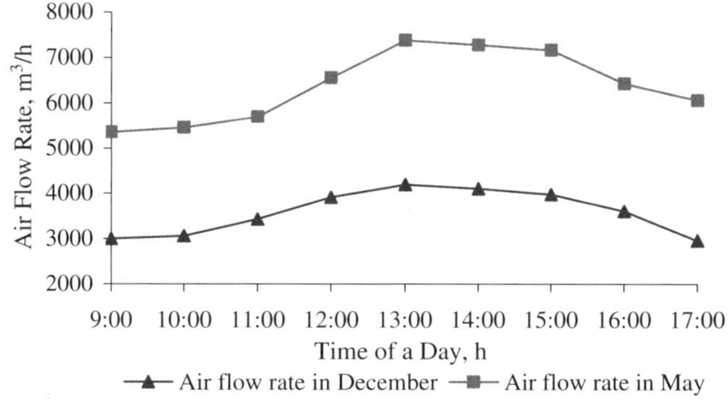

Fig. 13.12.   Variation of air flow rate at full load condition for foremblica officinalis pulp drying in the months of December and May.

The results trend is given in Fig. 13.12. It is observed that the maximum air flow rate inside the tunnel was $7374.6\,\mathrm{m^3/h}$ at 13:00 hours, while the minimum air flow rate was $5355\,\mathrm{m^3/h}$ at 09:00 hours in the month of May (Fig. 13.12).

13.8.2.4   *Variation of moisture content*

Comparison of the moisture content of Aonla pulp in the solar tunnel dryer with conventional sun drying for a typical experimental run was conducted in the months of December and May. Foremblica officinalis pulp was dried from a moisture content of 424.93 per cent (db) from 10.08 per cent (db) in 16 hours of drying in the solar tunnel dryer as compared to 40 hours of drying in the open sun.

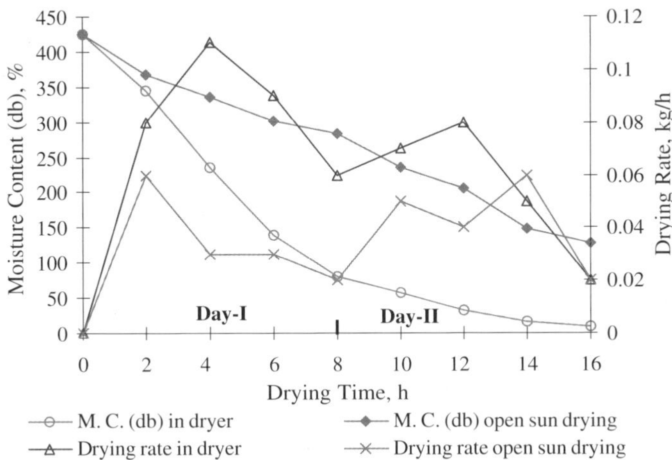

Fig. 13.13.   Relationship between moisture reduction and drying time in the solar tunnel dryer and open sun drying for foremblica officinalis pulp drying in the winter month (December).

Figure 13.13 indicates that the moisture content of the pulp was reduced from 424.93 to 80.38 per cent on dry-basis on the first day of drying for the month of December. At the end of the drying, the pulp moisture content was 10.08 per cent on dry-basis, while in the open air it was 128.4 per cent. It was also noted that about 44.95 per cent moisture was removed on the first day of drying while 43.73 per cent moisture was removed on the second day of drying.

The variation of the drying rate with drying time is shown in Fig. 13.13 for the solar tunnel dryer and open sun drying for a typical day of the month of December. It was observed that the drying rate for the first two hours was 0.08 kg/h and at the end of the trial the drying rate was 0.02 kg/h. It was observed that the drying rate follows the pattern of radiation. The drying rate increased up to 13 hours and then decreases on each day of drying.

The variation of moisture content, drying rate with drying time is shown in Fig. 13.14 for the solar tunnel dryer and open sun drying for a typical day of the month of May. It was observed that the initial moisture content of the Aonla pulp was 424.93 per cent on dry-basis at the starting of drying in the month of May.

It was found that the moisture content of the Aonla pulp decreases with drying time and finally reaches 10.33 per cent on dry-basis after 16 sunshine hours, while in the open air it was 93.12 per cent. It was observed that the drying rate for the first two hours was 0.07 kg/h and at the end of the trial the drying rate was 0.02 kg/h. The drying rate was found to be comparatively higher in the solar tunnel dryer than open sun drying. It was further observed that the drying rate varies with cumulative drying hours (Fig. 13.14).

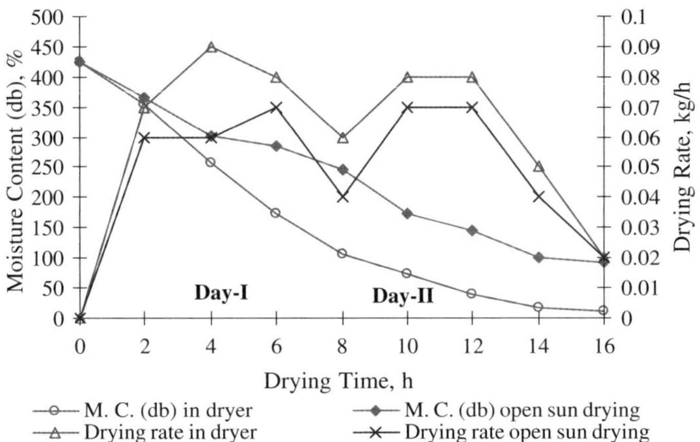

Fig. 13.14.　Relationship between moisture reduction and drying time in the solar tunnel dryer and open sun drying for foremblica officinalis pulp drying in the summer month (May).

13.8.2.5　*Variation of solar collector efficiency*

The efficiency of the collector ranged from 12.24 to 60.5 per cent in the month of December. The maximum collector efficiency of 60.5 per cent was attained at 17:00 hours. The average collector efficiency was 31.99 per cent. The results trend is illustrated in Fig. 13.15. The efficiency of the collector ranged from 17.07 to 44.30 per cent in the month of May. The maximum collector efficiency of 44.30 per cent was attained at 15:00 hours. The average collector efficiency was 35.61 per cent (Fig. 13.15).

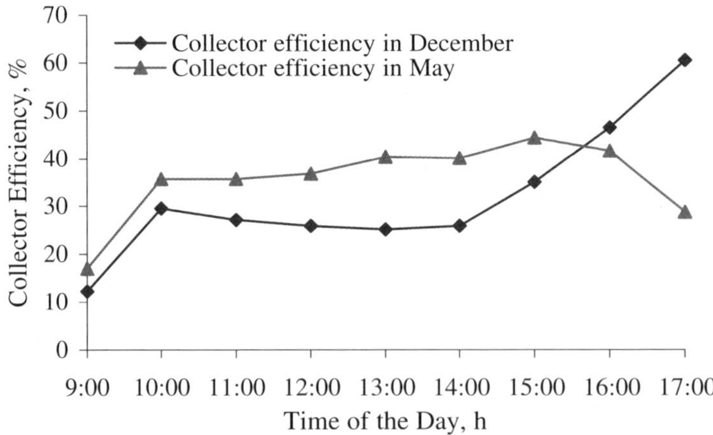

Fig. 13.15.　Collector efficiency of the solar tunnel dryer at full load condition for foremblica officinalis pulp drying in the months of December and May.

Table 13.2. Economic Indicators for drying foremblica officinalis pulp in the solar tunnel dryer.

Economic indicators	Value
Net present worth (Rs.)	7730844
B/C ratio	3.01
Pay back period (Years)	2 years 5 months

13.8.3 *Techno-economics analysis*

The cost of construction of the solar tunnel dryer for drying foremblica officinalis pulp is Rs. 74543. Techno-economic analysis of the solar tunnel dryer for drying foremblica officinalis pulp was carried out using three different economic indicators namely net present worth (NPW), benefit cost ratio (B/C ratio) and pay back period shown in Table 13.2.

13.9 Case Studies on Solar Tunnel Dryer for Drying Industrial Product (Di-basic Calcium Phosphate)

Di-basic calcium phosphate ($CaHPO_4 \cdot 2H_2O$) is an odorless, tasteless, white powder mineral based inorganic compound widely used for supplementing phosphorous and calcium to animals along with feed. It is also used in pharmaceutical grade and dentifrice grade. The initial moisture content of the di-basic calcium phosphate (DCP) is about 62.87 per cent (db) and this is to be reduced to around 10.62 per cent (db) for further usages. Presently, mechanical methods are used for drying the material. It consists of a tray drying system which equipped with a 5.6 kW blower, a 7.5 kW suction motor and 3.7 kW hammer mill. The air is heated to a temperature of around 60°C in a diesel-fired hot air generator. Its drying is costly, time consuming and labor intensive.

Looking to the power requirement and high cost of the existing mechanical dryer, a solar tunnel dryer has been designed, developed and commissioned at M/s Phosphate India Pvt. Ltd, Udaipur, Rajasthan, India, for drying 1.5 tons di-basic calcium phosphate (DCP) from 62.87 to 10.62 per cent moisture content (db) of in a batch. It was a hemi-cylindrical shaped walk-in type and had a base area of 3.75 m × 21.00 m for drying 1.5 tons per batch. Five-cm thick glass wool insulation had been provided to reduce heat loss through the floor. It is based on the theoretical calculation for critical insulation thickness for this dryer.[16] The orientation of the solar tunnel dryer is in a east-west direction. The structural components of the solar tunnel drier include hoops foundation, floor, ultraviolet stabilized polythene film

Fig. 13.16.   Closed view of the solar tunnel dryer installed at the factory site for drying of di-basic calcium phosphate.

and drying trays. A closed view of the solar tunnel dryer installed at the factory site is shown in Fig. 13.16. The specifications of the solar tunnel dryer for drying 1.5 tons of di-basic calcium phosphate are shown in Table 13.3.

13.9.1   *Experimental study*

The experiments were conducted on the solar tunnel dryer installed at M/s Phosphate India Pvt. Ltd. The performance of the solar tunnel dryer was evaluated by loading di-basic calcium phosphate, and measuring the following parameters: (a) radiation incident on the dryer, (b) air temperatures at various locations in the dryer, (c) Moisture content variation, (d) relative humidity and (e) air flow rate.

To measure the temperature of air at various locations of the dryer, three sensors inside and one sensor outside the solar tunnel dryer were kept for recording. Sensors number one, two and three were kept inside the solar tunnel dryer at the mid bottom point, at 1 m above the mid bottom point and at the north wall, respectively, and sensor number zero was kept outside the solar tunnel dryer. All temperature data was registered at an interval of one hour by a data logger. The solar insolation was measured by a solar insolometer. Relative humidity was measured by a digital hygrometer. Air flow was calculated by multiplying the air velocity with the area of flow.

Keeping view, the months for the winter and summer season (January and June) were adopted for conducting the experiments. Drying tests were started at 8:00 hours

Table 13.3. Design details of solar tunnel dryer for drying of di-basic calcium phosphate.

Components/Particulars	Specifications
Product dried	Di-basic calcium phosphate
Loading capacity	1500 kg per batch
Initial moisture content	62.87% (db)
Final moisture content	10.62% (db)
Drying period required	18 h
Collector material (cover)	UV polythene 200 micron sheet
Orientation	E-W direction
North wall	Glass wool sandwich in metallic cover
Size of north wall	21 m × 1.57 m
No. of Chimneys	5
Size of chimney (diameter and length)	0.20 m and 0.75 m
Tray size	800 mm × 400 mm × 40 mm
Number of trays	24 trays per trolley
Number of trolleys	10
Door size	1.6 m × 0.75 m
Length of solar tunnel dryer (STD)	21.0 m
Diameter of solar tunnel dryer	3.75 m
Floor area of solar tunnel dryer	78.75 m^2
Area of semi-cylindrical shape of STD	134.74 m^2

and stopped at 17:00 hours in the month of January and while in the month of June tests were started at 8:00 hours and stopped at 18:00 hours. The economic evaluation was done for drying di-basic calcium phosphate in the solar tunnel dryer.

13.9.2 *Performance evaluation of the solar tunnel dryer for drying di-basic calcium phosphate*

Full load testing of the solar tunnel dryer was made for evaluating the performance in an actual loaded condition of 1.5 tons of materials during the months of June and January for drying di-basic calcium phosphate. Di-basic calcium phosphate with 62.87 per cent initial moisture content (db) was taken loaded in the trays of the solar tunnel dryer. Wet di-basic calcium phosphate is spread in a thin layer of approximately 4 cm thickness in the trays of size 40 cm × 80 cm. Twenty-four trays are loaded onto a trolley. Each tray carries an approximate of 6.25 kg material. Ten trolleys containing 1.5 tons of di-basic calcium phosphate was moved into the tunnel dryer in the morning and 50 kg of di-basic calcium phosphate was placed in the open air for drying for comparison purpose. The inside view of the solar tunnel

Fig. 13.17.   Inside view of solar tunnel dryer under full load condition for drying of di-basic calcium phosphate.

dryer under the full load condition is given in Fig. 13.17. The testing on full load was made for consecutive days in the months of winter (January) and summer (June).

13.9.2.1 *Solar insolation and temperature*

The variation of solar insolation, ambient air temperature and air temperature at different locations inside the solar tunnel dryer for the month of January during the drying of di-basic calcium phosphate are shown in Fig. 13.18. It was observed

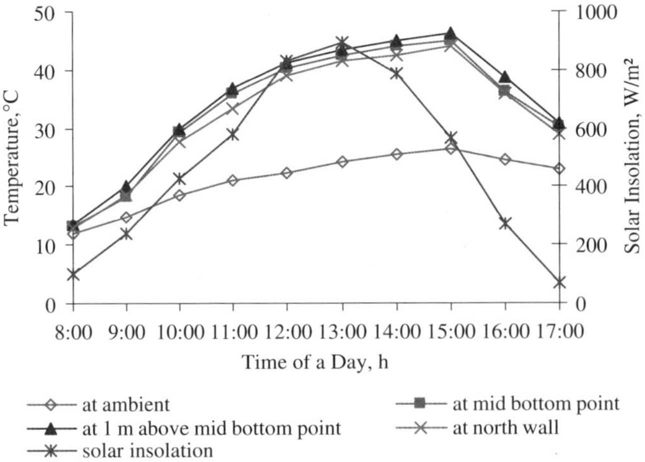

Fig. 13.18.   Variation of solar radiation, ambient temperature and temperature at different locations in the solar tunnel dryer with time at the full load condition for di-basic calcium phosphate drying in the winter month (January).

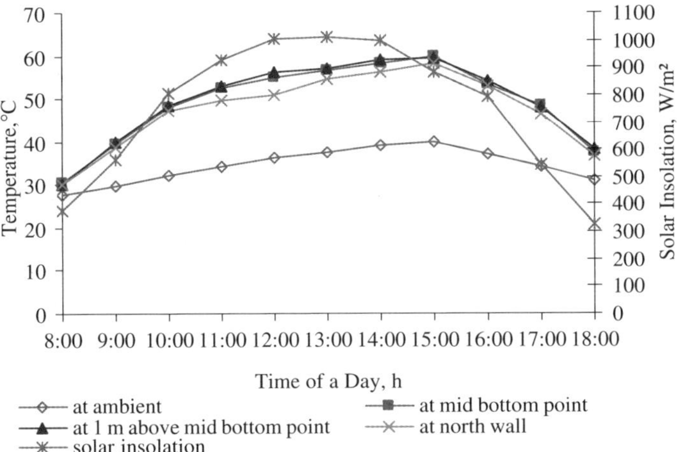

Fig. 13.19.   Variation of solar radiation, ambient temperature and temperature at different locations in the solar tunnel dryer with time at the full load condition for di-basic calcium phosphate drying in the summer month (June).

that the maximum temperature inside the solar tunnel dryer was 46.1°C at 15:00 hrs while the minimum temperature inside the solar tunnel dryer was 13°C at 8:00 hrs in a typical day in the month of January against the maximum and minimum ambient temperature of 26.3°C and 12°C, respectively. During drying of the di-basic calcium phosphate the solar insolation varied from 68 to 891 W/m².

Figure 13.19 shows that the maximum temperature attended inside the tunnel was 60.0°C at 15:00 hours while the minimum inside temperature was 30.1°C at 8:00 hours in a typical day in month of June. Corresponding, the maximum ambient temperature was 40.0°C at 15:00 hours while minimum ambient temperature was 27.6°C at 8:00 hours. It was observed that the maximum and minimum solar insulation in this month was 1010 W/m² at 13:00 hours and 323 W/m² at 18:00 hrs, respectively.

13.9.2.2   *Variation of relative humidity*

Figure 13.20 indicates that the maximum relative humidity inside the tunnel was 71 per cent at 8:00 hours, while the minimum relative humidity was 35 per cent at 15:00 hours in the month of January. The maximum ambient relative humidity in this month was 86 per cent at 8:00 hours while the minimum ambient relative humidity was 39 per cent at 15:00 hours, respectively.

It was observed from Fig. 13.21 that the maximum relative humidity inside the tunnel was 50 per cent at 8:00 hours, while the minimum relative humidity was 29 per cent at 15:00 hours in the month of June. The maximum ambient relative

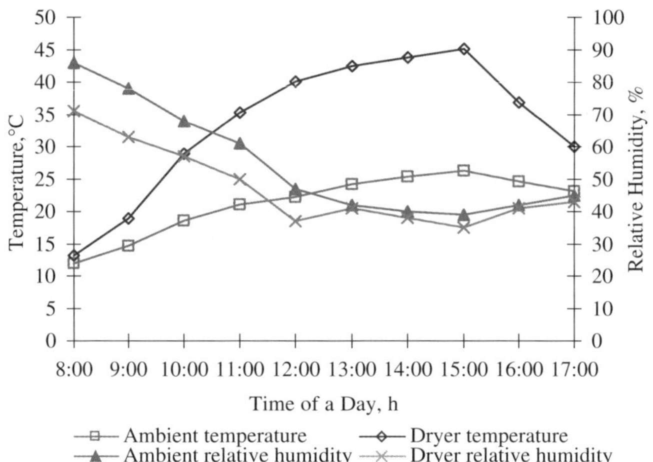

Fig. 13.20. Ambient and average dryer temperature and relative humidity at the full load condition for di-basic calcium phosphate drying in the winter month (January).

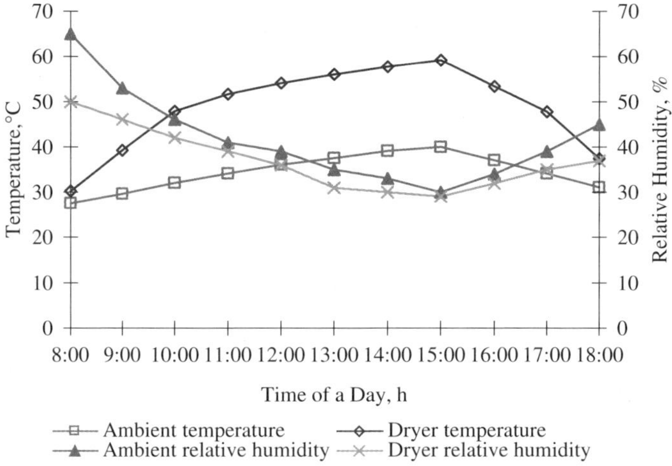

Fig. 13.21. Ambient and average dryer temperature and relative humidity at the full load condition for di-basic calcium phosphate drying in the summer month (June).

humidity in this month was 65 per cent at 8:00 hours while the minimum ambient relative humidity was 30 per cent at 15:00 hours, respectively.

13.9.2.3 *Variation of air flow rate*

The maximum air flow rate inside the tunnel was 5253.12 m³/h at 13:00 hours, while the minimum air flow rate was 2591.9 m³/h at 17:00 hours in the month of January. The results trend is given in Fig. 13.22. It is observed that the maximum air flow

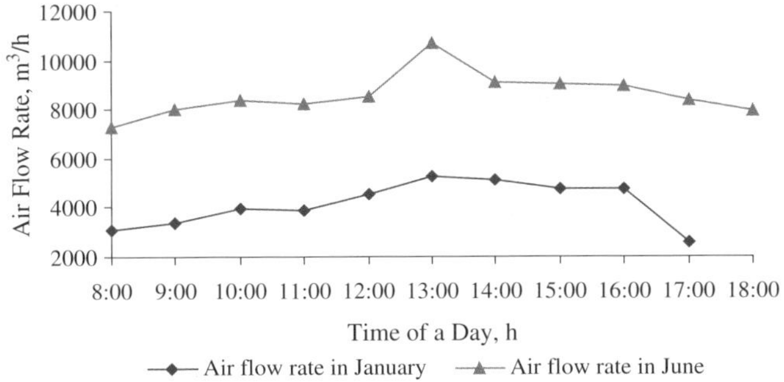

Fig. 13.22. Variation of air flow rate at the full load condition for di-basic calcium phosphate drying in the months of January and June.

rate inside the tunnel was $10713.6\,\text{m}^3/\text{h}$ at 13:00 hours, while the minimum air flow rate was $8491.37\,\text{m}^3/\text{h}$ at 08:00 hours in the month of June (Fig. 13.22).

13.9.2.4 *Variation of moisture content*

It is observed from Fig. 13.23 that the moisture content was reduced from 62.87 to 20.13 per cent (db), while in the open air it was 45.62 per cent (db) during the first day of drying in a typical day of month of January. It was further reduced to 10.62 per cent (db), while in the open air it was 31.94 per cent (db) at the end of second day.

The variation of drying rate with drying time is shown in Fig. 13.23 for the solar tunnel dryer and open sun drying for a typical day of the month of January. It was observed that the drying rate for the first two hours was 0.02 kg/h and at the end of the trial the drying rate was 0.01 kg/h. It was observed that the drying rate

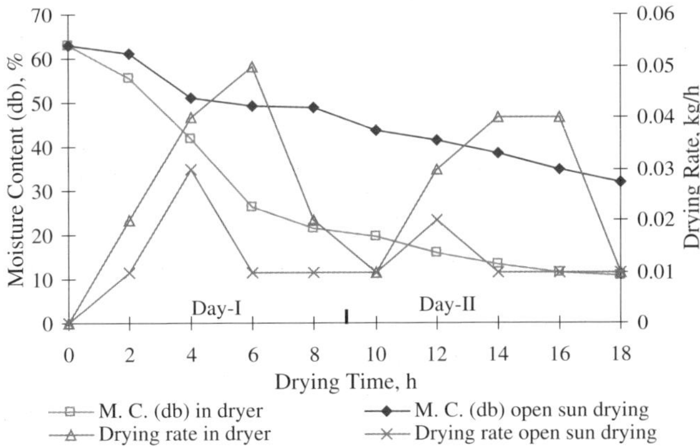

Fig. 13.23. Relationship between moisture reduction and drying time in the solar tunnel dryer and open sun drying for di-basic calcium phosphate drying in the winter month (January).

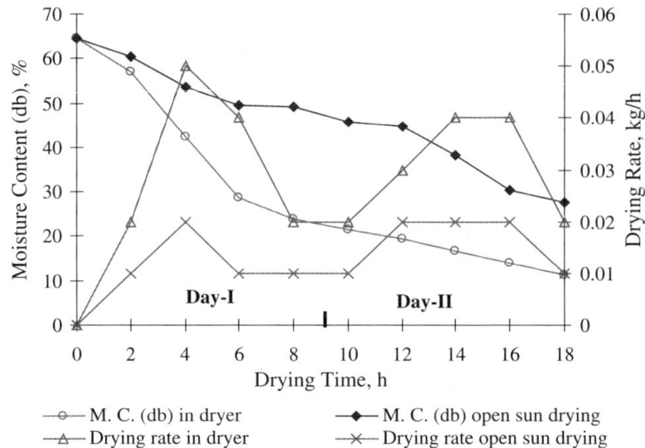

Fig. 13.24.   Relationship between moisture reduction and drying time in the solar tunnel dryer and open sun drying for di-basic calcium phosphate drying in the winter month (June).

follows the pattern of radiation. The drying rate increased up to 13 hours and then decreases on each day of drying. It was further observed that the drying rate varies with cumulative drying hours. It follows the trend of solar radiation, i.e., increasing up to 13 hours and then decreasing, on both days of drying.

The variation of moisture content and drying rate with drying time is shown in Fig. 13.24 for the solar tunnel dryer and open sun drying for a typical day of the month of June. The initial moisture content of di-basic calcium phosphate was 64.5 per cent (db). It was observed that the moisture content of di-basic calcium phosphate decreases with drying time and reduced to 11.40 per cent (db) in 18 sunshine hours, while in the open air it was 27.52 per cent (db). It was observed that the drying rate for the first two hours was 0.02 kg/h and at the end of the trial the drying rate was 0.02 kg/h. The drying rate was found to be comparatively higher in the solar tunnel dryer than open sun drying. It further observed that the drying rate varies with cumulative drying hours. It follows the trend of solar radiation, i.e., increasing up to 13 hours and then decreasing on both day of drying (Fig. 13.24).

13.9.2.5 *Variation of solar collector efficiency*

The efficiency of the collector ranged from 13.78 to 64.10 per cent in the month of January. The maximum collector efficiency of 64.10 per cent was attained at 17:00 hours. The average collector efficiency was 31.04 per cent. The results trend is illustrated in Fig. 13.25. The efficiency of the collector ranged from 12.16 to 47.82 per cent in the month of June. The maximum collector efficiency of 47.82 per cent was attained at 17:00 hours. The average collector efficiency was 36.53 per cent (Fig. 13.25).

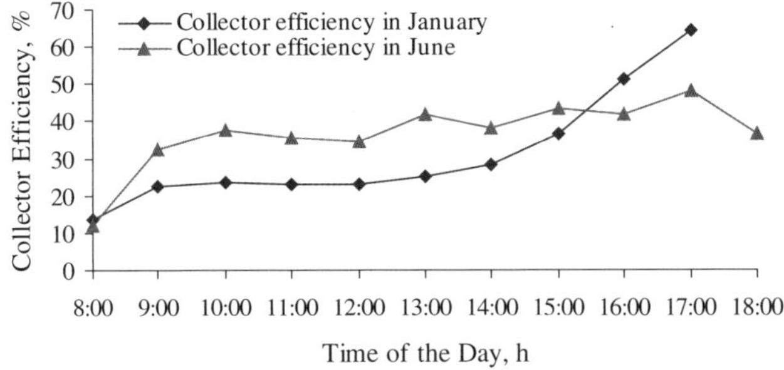

Fig. 13.25.   Collector efficiency of the solar tunnel dryer at full load condition for di-basic calcium phosphate drying in the months of January and June.

Table 13.4.   Economic indicators for drying DCP in the solar tunnel dryer.

Economic indicators	Value
Net present worth (Rs.)	6207592
B/C ratio	3.06
Pay back period (Years)	2 years 5 months

13.9.3 *Techno-economics analysis*

The cost of construction of the solar tunnel dryer for drying di-basic calcium phosphate is Rs. 85353. Techno-economic analysis of the solar tunnel dryer for drying di-basic calcium phosphate was carried out using three different economic indicators shown in Table 13.4.

13.10 Conclusions

This chapter has described the comparison between traditional sun drying and other solar drying techniques to show that the use of the solar tunnel dryer leads to a considerable reduction of the drying time and to a significant improvement of the product quality in terms of color, texture and taste.

Case studies on solar tunnel drying in this chapter present the field level performance of different products. This study deals with the critical design specifications and field performance of the solar tunnel dryer for drying one ton of foremblica officinalis pulp and 1.5 tons of di-basic calcium phosphate at two different locations.

Earlier drying of all the products were mainly carried out by the traditional method of sun drying, which was not only time consuming but had no control over the quality of products. However the use of the improved solar tunnel dryer has led to a considerable reduction in drying time and improvement quality. The results were shown in terms of the variation of air temperature, solar flux density, relative humidity, air flow rate, moisture content, etc., which indicate that the performance of the solar tunnel dryer was better.

Techno-economic analysis of the solar tunnel dryers was also carried out by using different economic indicators. It was observed that the solar tunnel dryer is an economical technology for drying all kinds of agriculture, horticulture and agro-industrial materials.

References

1. P. Singh, N.S. Rathore, A.K. Kurchania and A.N. Mathur, *Sustainable Development with Renewable Energy Sources* (Yash Publishing House, Bikaner, India, 2005).
2. S. Lawrence, A. Pole and G.N. Tiwari, "Performance of a solar crop dryer under PNG climatic conditions," *Energy Conversion and Management* **30** (1990) 333–342.
3. C.M. Joy, P.P. George and K.P. Jose, "Solar tunnel drying of red chillis (Capsicum annum. L.)," *J. Food Science Technology* **38** (2001) 213–216.
4. M.A. Karim and M.N.A. Hawlader, "Development of solar air collectors for drying applications," *Energy Conversion and Management* **45** (2004) 329–344.
5. B.K. Bala and J.L. Woods, "Simulation of the indirect natural convection solar drying of rough rice," *Solar Energy* **53** (1994) 259–266.
6. D. Jain and G.N. Tiwari, "Thermal aspects of open sun drying of various crops," *Energy* **28** (2003) 37–54.
7. M.S.A. Al-Amri, "Thermal performance tests of solar dryer under hot and humid climatic conditions," *AMA-Agricultural Mechanization in Asia, Africa and Latin America* **28** (1997) 56–60.
8. E.A. Arinze, S. Sokhansanj, G.J. Schoenau, B. Crerar and A. Opoku, "Design experimental and economic evaluation of a commercial-type solar dryer for production of high-quality hay," *Drying Technology* **16** (1998) 597–626.
9. M.K. Malaiya and R.S.R. Gupta, "Cabinet type natural convection dryer with chimney," *Agricultural Engineering Today* **11** (1987) 37–39.
10. C. Palaniappan and S.V. Subramanian, "Performance of $500\,m^2$ area solar air heater for drying of spices — A case study," *IREDA News Souvenir* (1997).
11. I.N. Simate, "Optimization of mixed mode and indirect mode natural convection solar dryers," *Renewable Energy* **28** (2003) 435–453.
12. H.P. Garg and R. Kumar, "Studies on semi-cylindrical solar tunnel dryers: Estimation of solar irradiance," *Renewable Energy* **13** (1998) 393–400.
13. K.R. Manohar and P. Chandra, "Drying of agricultural produce in a greenhouse type solar dryer," *Int. Agricultural Engineering J.* **9** (2000) 139–150.
14. R.C. Verma and A. Gupta, "Effect of pre-treatments on quality of solar-dried amla," *J. Food Engineering* **65** (2004) 397–402.
15. O.V. Ekechukwu and B. Norton, "Design and measured performance of a solar chimney for natural circulation solar energy dryers," *Renewable Energy* **10** (1997) 81–90.
16. B.K. Bala and M.R.A. Mondol, "Experimental investigation on solar drying of fish using solar tunnel dryer," *Drying Technology* **19** (2001) 427–436.

Section 3

Bio Fuels

Chapter 14

Biomass as a Source of Energy

Mahendra S. Seveda

Department of Farm Power and Machinery,
College of Agricultural Engineering and post Harvest Technology
(Central Agricultural University),
Ranipool, Gangtok, Sikkim,
India-737135
sevda_mahendra@rediffmail.com

Narendra S. Rathore

Department of Renewable Energy Sources,
College of Technology and Engineering,
Maharana Pratap University of Agriculture and Technology,
Udaipur, Rajasthan, India-313001
rathoren@rediffmail.com

Vinod Kumar

Department of Electrical Engineering,
College of Technology and Engineering,
Maharana Pratap University of Agriculture
and Technology, Udaipur,
Rajasthan, India-313001
vinodcte@yahoo.co.in

Biomass is a renewable energy resource derived from the carbonaceous waste of various human and natural activities. Biomass is the storage of solar energy in chemical form in plant and animal materials. It is one of the commonly used, precious and versatile resources on earth. Biomass has been used for energy purposes ever since man discovered fire. Biomass energy can be a sustainable, environmentally benign and economically sound source.

14.1 Introduction

Energy demand is increasing continuously due to rapid growth in population and industrialization development. The development of energy sources is not keeping pace with spiraling consumption. Even developed countries are not able to compensate even after increasing the energy production multifold.[1] The major energy demand is compensated from conventional energy sources such as coal, oil, natural gas, etc.[1] Two major problems, which every country is facing with these conventional fuels, are as follows:

(1) These energy sources are on the verge of becoming extinct. The world's oil reserves are estimated to be depleted by 2050.

(2) Energy extraction from these conventional fuels causes pollution. It is well known that SO_2 emission produced by burning fossil fuels is the major cause of acid rain. Globally, an increase in the emissions rates of greenhouse gases, i.e., CO_2 presents a threat to the world climate. As an estimate in the year 2000, over 20 million metric tons of CO_2 are expected to be released in the atmosphere each year.[2,3] If this trend continues, some extreme natural calamities are expected, such as excessive rainfall and consequent floods, droughts and local imbalances.

Presently, there is an utmost need of alternative energy resources which are cheap, renewable and do not cause pollution. Therefore, attention is being given to alternate and renewable energy sources such as solar, wind, thermal, hydroelectric, biomass, etc. Biomass is a carbon neutral resource in its life cycle and the primary contributor of greenhouse effect. Biomass is the fourth largest source of energy in the world after coal, petroleum and natural gas, providing about 14 per cent of the world's primary energy consumption. Renewable biomass is being considered as an important energy resource all over the world. Biomass is used to meet a variety of energy needs, including generating electricity, fueling vehicles and providing process heat for industries.[4,5] Among all the renewable sources of energy, biomass is unique as it effectively stores solar energy. It is the only renewable source of carbon that can be converted into convenient solid, liquid and gaseous fuels through different conversion processes.[6]

Biomass is the term used for all organic material originating from plants (including algae), trees and crops and is essentially the collection and storage of the sun's energy through photosynthesis. Biomass energy is the conversion of biomass into useful forms of energy such as heat, electricity and liquid fuels. Biomass for bio energy comes either directly from the land, as dedicated energy crops, or from residues generated in the processing of crops for food or other products such as pulp and paper from the wood industry.[7] Another important contribution is from post consumer residue streams such as construction and demolition wood, pallets used in transportation, and the clean fraction of municipal solid waste. The biomass to bio energy system can be considered as the management of flow of solar generated materials, food, and fibre in our society.[8] Not all biomass are directly used to produce energy but rather it can be converted into intermediate energy carriers called bio fuels. This includes charcoal (higher energy density solid fuel), ethanol (liquid fuel), or producer-gas (from gasification of biomass).

Biomass energy is by far the largest renewable energy source, representing 10.4 per cent of the world's total primary energy supply or 77.4 per cent of global renewable energy supply. However, biomass energy represents only 1 per cent of the total fuel used for electricity production. In Southeast Asia, it provides 26 per cent of the total primary energy supply and accounts for 87 per cent of the renewable energy supply. It is mostly in the form of traditional biomass that meets the energy needs of the rural population and small scale industries.[9]

Biomass is usually not considered a modern energy source, given the role that it has played, and continues to play, in most developing countries.[10] In developing countries it still accounts for an estimated one-third of primary energy use while in the poorest up to 90 per cent of all energy is supplied by biomass. Over two billion people cook by direct combustion of biomass, and such traditional uses typically involve the inefficient use of biomass fuels, largely from low cost sources such as natural forests, which can further contribute to deforestation and environmental degradation. The direct combustion of biomass fuels, as used in developing countries today for domestic cooking and heating, has been called "the poor man's oil" ranking at the bottom of the ladder of preferred energy carriers where gas and electricity are at the top.[11]

Biomass usage as a source of energy is of interest due to the following envisaged benefits:

(1) Biomass is a renewable, potentially sustainable and relatively environmentally friendly source of energy.
(2) A huge array of diverse materials, frequently stereo-chemically defined, are available from the biomass giving the user many new structural features to exploit.[12]
(3) Increased use of biomass would extend the lifetime of diminishing crude oil supplies.
(4) Biomass fuels have negligible sulfur content and, therefore, do not contribute to sulfur dioxide emissions that cause acid rain.
(5) The combustion of biomass produces less ash than coal combustion and the ash produced can be used as a soil additive on farms, etc.

(6) The combustion of agricultural and forestry residues and municipal solid wastes for energy production is an effective use of waste products that reduces the significant problem of waste disposal, particularly in municipal areas.

(7) Biomass is a domestic resource which is not subject to world price fluctuations or the supply uncertainties as of imported fuels.

(8) Biomass provides a clean, renewable energy source that could improve our environment, economy and energy securities.[13,14]

(9) Biomass usage could be a way to prevent more carbon dioxide production in the atmosphere as it does not increase the atmospheric carbon dioxide level.

14.2 Types of Biomass

The six types of biomass are agricultural biomass, forest biomass, energy plantation, marine biomass, biomass from animal waste and municipal waste.

14.2.1 *Agricultural biomass*

Agricultural biomass which could be used for energy production is defined as biomass residues from field agricultural crops (stalks, branches, leaves, straw, waste from pruning, etc.) and biomass from the byproducts of the processing of agricultural products (residue from cotton ginning, olive pits, fruit pits, etc.).

14.2.2 *Biomass from animal waste*

The potential biomass from animal waste includes primarily waste from intensive livestock operations, from poultry farms, pig farms, cattle farms and slaughterhouses. The animal waste is a rich source of fuel. The dung cakes prepared with animal wastes can be used for meeting cooking energy requirement in rural and semi-urban areas.

14.2.3 *Forest biomass*

Forest biomass, which is used or can be used for energy purposes consists of firewood, forestry residues (from thinning and logging), material cleared from forests to protect them from forest fires, as well as byproducts from wood industries.

14.2.4 *Municipal waste*

The municipal waste consists of solid wastes as human excreta, garbage, city wastes and commercial wastes. It also includes liquid form domestic sewage and effluent from community institutional activities. There are a number of ways these wastes can be recycled and resources recovered in terms of fuel gas, manure and liquid fuel, etc.

14.2.5 *Energy plantation*

Energy plantation means growing selected species of trees and shrubs which are harvestable in a comparatively shorter time and are specific means for fuel. The sources of energy plantation depend on the availability of land and water and careful management of the plants. The fuel wood may be used either directly into wood burning stoves and boilers or processed into methanol, ethanol and producer gas. There are many species suitable for energy plantation. Few of them are: Acacia nilotica, Acacia auriculiformis, Dalbargia sissoo, Eucalyptus comaldulusis, Prosopis juliflora, Leucaena lencocephala, Albizzia lebbek, Casuarina, etc.

14.2.6 *Marine biomass*

Floating water plants (e.g., water hyacinths) are pest plants in many rivers, lakes and ponds in tropical and semitropical areas of the world. The growth rate of water hyacinths is very high and it is nutrient rich with net productivity of up to 25 tons of dry product per acre per year.

14.3   Energy Content of Biomass

Biomass refers to all forms of plant-derived material that can be used for energy viz.: wood, herbaceous plants, crop and forest residues, animal wastes, etc. Because biomass is a solid fuel it can be compared to coal. On a dry-weight basis, heating values range from 17.5 GJ per ton for various herbaceous crops like wheat straw, sugarcane bagasse to about 20 GJ per ton for wood. The corresponding values for bituminous coal and lignite are 30 GJ per ton and 20 GJ per ton respectively. At the time of its harvest biomass contains considerable amount of moisture, ranging from 8 to 20 per cent for wheat straw, to 30 to 60 per cent for woods, to 75 to 90 per cent for animal manure, and 90 to 95 per cent for water hyacinth. In contrast the moisture content of the most bituminous coals ranges from 2 to 12 per cent.[15] Thus the energy densities for the biomass at the point of production are lower than those for coal. On the other side chemical attributes make it superior in many ways. The ash content of biomass is much lower than for coals, and the ash is generally free of the toxic metals and other contaminants and can be used as soil fertilizer.

Biomass is generally regarded as a low-status fuel, and in many countries rarely finds its way into statistics. It offers considerable flexibility of fuel supply due to the range and diversity of fuels which can be produced. Biomass energy can be used to generate heat and electricity through direct combustion in modern devices, ranging from very-small-scale domestic boilers to multi-megawatt size power plants electricity (e.g., via steam turbines), or liquid fuels for motor vehicles such as ethanol, or other alcohol fuels. Biomass-energy systems can increase economic development without contributing to the greenhouse effect since biomass is not a net emitter of CO_2 to the atmosphere when it is produced and used sustainably. It also has other benign environmental attributes such as lower sulphur and NO_x emissions and can help rehabilitate degraded lands. There is a growing recognition that the use of biomass in larger commercial systems based on sustainable, already accumulated resources and residues can help in improving natural resource management.

14.4   Harvesting Methods of Biomass

These are several different harvesting methods that allow the plant to re-generate through sprouting. These are as follows

14.4.1   *Coppicing*

It is one of the most widely used harvesting methods in which the tree is cut at the base, usually between 15–75 cm above the ground level. A new shoot is developed from the stump or root. These shoots are sometimes referred to as sucker or sprouts. Management of sprouts should be carried out according to use. For fuel wood, in order for the number of sprouts to container to grow, depends on the desired sizes of fuel wood. If many sprouts are allowed to grow for a long period, the weights of the sprouts will become heavy and the sprouts may tear away from the main trunk.[16] Several rotations of coppicing are usually possible with many species. The length of the rotation period depends on the required tree products from the stand. It is a suitable method for production of fuel wood (Fig. 14.1(a)).

14.4.2   *Pollarding*

It is the harvesting system in which the branches, including the top of the tree, are cut at a height of about 2 m above the ground and the main trunk is allowed to remain standing. The new shoots sprout or emerge from the main stem to develop a new crown. This results in continuous increase in the diameter of the main stem although not in height. Finally when the tree loses its sprouting vigor, the main stem is also cut for use as large diameter poles. An advantage of this method over coppicing is that the new shoots are high enough off the ground that they are out of reach of most grazing animals.[16] The neem tree (*Azadirachata indica*) is usually harvested in this manner. The branches may be used for poles and fuel wood (Fig. 14.1(b)).

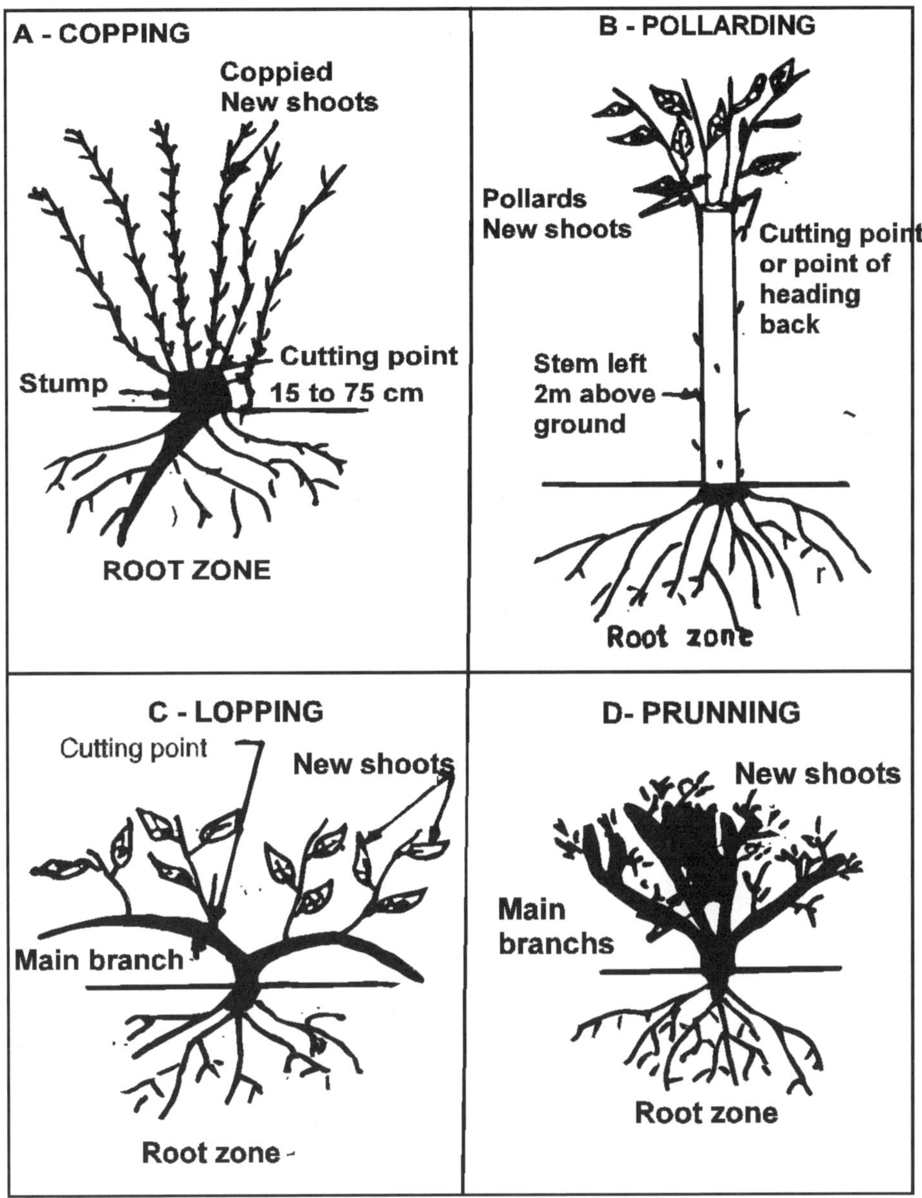

Fig. 14.1.   Different harvesting methods.[16]

14.4.3 *Lopping*

In this method most of the branches of the tree are cut. The fresh foliage starts sprouting from the bottom to the top of the denuded stem in spite of severe defoliation, surprisingly quickly.[16] The crown also regrows and after a few years, the tree is lopped again. The lopped trunk continues to grow and increase in height, unless this is deliberately prevented by pruning it at the top (Fig. 14.1(c)).

14.4.4 *Pruning*

It is a very common harvesting method. It involves the cutting of smaller branches and stems. The clipped materials constitute a major source of biomass for fuel and other purposes, such as fodder mulching between tree sows.[16] It is also often required for the maintenance of fruits and forages trees, alley cropping and live fences. The process of pruning also increases the business of trees and shrubs for bio fencing. Root pruning at a distance of 2–3 meters from the hole is effective to reduce border tree competition for water and nutrient with the crops (Fig. 14.1(d)).

14.4.5 *Thinning*

It is an important traditional forestry practice in fuel wood plantation. The primary objectives of the thinning are to enhance diametric growth of some specific trees through early removal of poor and diseased tree to improve the plantation by reducing the competition for light and nutrients.[16] Depending on initial plant density, initial thinning can be used for fuel wood or pole production. If maximum biomass production is the main objective of the plantation, regardless of quality, thinning may not be needed.

14.5 Conversion of Biomass

Biomass can be converted to a number of secondary energy carriers (electricity, gaseous, liquid and solid fuels, and heat) using a wide range of conversion routes. The conversion routes to fuels and electricity can be distinguished in thermal, chemical, and biochemical conversion routes.[17] Thermo chemical processes involve the pyrolysis, liquefaction, gasification, and supercritical fluid extraction methods. The products of the thermo chemical processes are divided into a volatile fraction consisting of gases, vapors, and tar components and a carbon-rich solid residue.[18]

Thermal conversion of biomass has received special attention since it leads to useful products and simultaneously contributes to solving pollution problems arising from biomass accumulation.[19] Biochemical processes are essentially microbic digestion and fermentation. A biochemical process converts biomass to ethanol and methanol. Transportation fuels from biomass are at present mainly derived from sugar- or starch-containing crops (e.g., ethanol from sugar cane or maize).

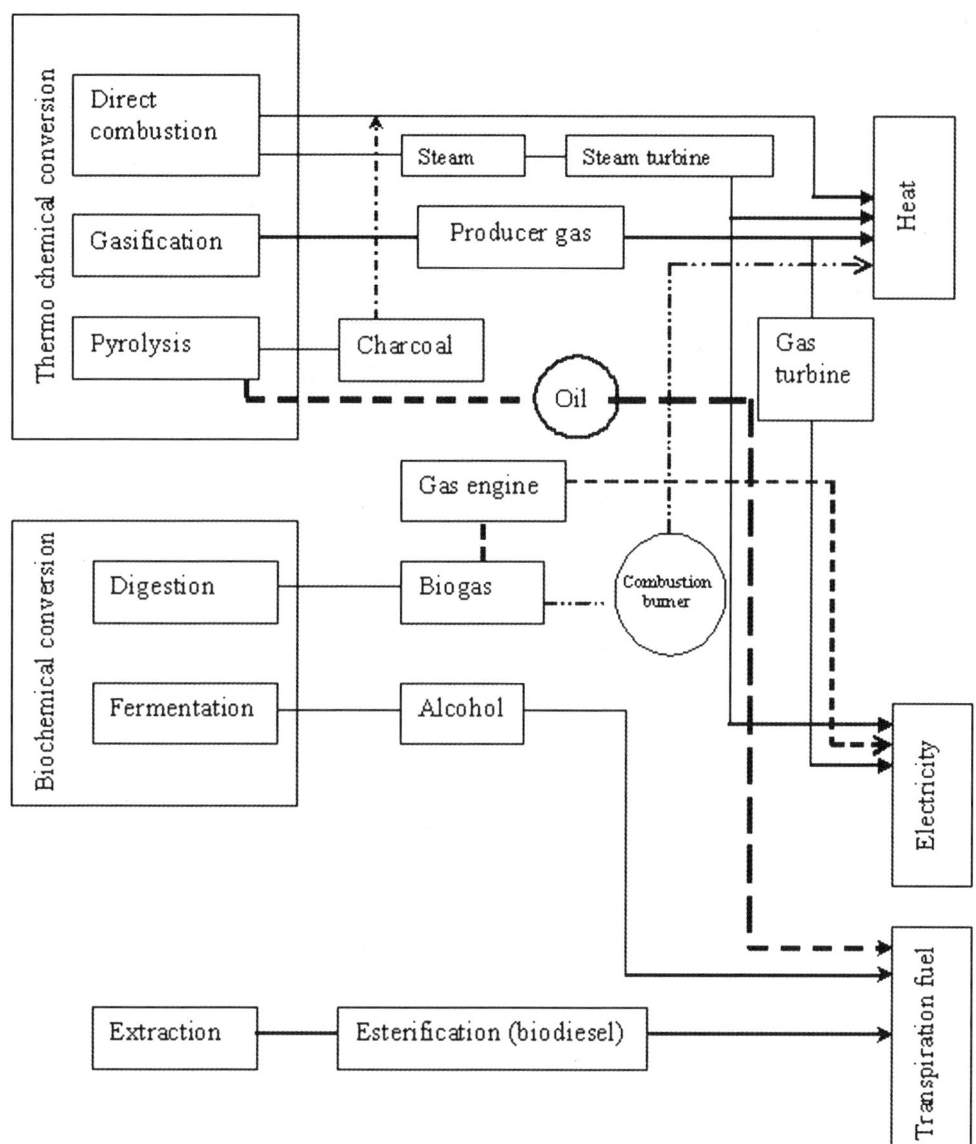

Fig. 14.2.   Biomass conversion routes.

From lignocellulosic crops, advanced technologies are the conversion via gasification to methanol and hydrogen, the conversion to ethanol using a hydrolysis and fermentation step, and the conversion to long-chain hydrocarbon fuel as Fischer–Tropsch.[17] Conversion of biomass into fuel through different routes is illustrated in Fig. 14.2.

14.6 Thermo-Chemical Conversion of Biomass

Thermo-chemical conversion of biomass is one of the most common and convenient routes for conversion into energy. This includes combustion, gasification, liquefaction and carbonization. In all these processes, pyrolysis plays a key role in the reaction kinetics and hence in reactor design and determining product distribution, composition and properties.

14.6.1 *Direct combustion*

The direct combustion of biomass is the simplest and earliest method of converting the chemically stored energy into heat. In direct combustion, the hydrogen and carbon in a fuel combined with oxygen, a process which releases heat. This is the most common thermo chemical method of converting biomass to energy. Biomass burns in three successive but overlapping stages, first, the contained water evaporates and then there is distillation and probably combustion of volatiles and finally the high temperature reaction of fixed carbon with oxygen. Only the last two steps release heat and provide energy.

Direct combustion of biomass provides the majority of rural population of the world with energy for cooking and heating. Wood is burnt daily in a variety of ways all over the world in open fires, simple cooking stoves, Dutch ovens, and sophisticated package boilers and fluidized bed units. Open fires, widely used in India, deliver only 5 to 10 per cent of the available energy for cooking. The efficiency of wood burning stoves can be increased to an overall efficiency between 20 to 30 per cent. This will reduce not only requirement of fuel wood by 5 to 10 fold but also the time, energy and money that the Third World population now has to spend on collecting fuel. Production of charcoal from wood, as well as briquettes, pellets and chips of wood and of agricultural and other plant residues can further save energy.

Besides these traditional uses, direct combustion is industrially used for generating steam and electricity.[20] However, direct combustion of biomass often requires carefully tailored equipment to satisfy the constraints imposed by the properties of biomass feed stocks. High ash fuels for instance can lead to slagging in boilers. The heating value of the biomass may also be low due to high moisture and ash contents. Variation in feedstock composition and size distribution also creates difficulties in controlling combustion. Finally, biomass causing high corrosion or erosion will require special materials of construction.

Everyday, municipal, commercial and light industrial wastes are converted to energy in about 450 mass burning facilities throughout the world by direct combustion. Over 99 per cent of the world's solid waste is processed into energy by a mass burning, refuse-fired energy system. Japan has many such systems to process waste.

14.6.2 *Pyrolysis*

Pyrolysis, another process of thermo chemical conversion, is broadly a system which thermally decomposes carbonaceous materials. It has, unlike direct combustion, at least one zone in which the thermal decomposition takes place in the absence of oxygen. Pyrolysis can be based on either coal or biomass. All three phases of products are generated, solid char, a non condensable gas and an organic liquid-oil. Oxygen is only required in small and measured amounts to support combustion which releases just sufficient heat to initiate the endothermic (energy-consuming) pyrolysis process. Some examples of pyrolysis are the destructive distillation of wood and other agricultural products to produce methanol, charcoal and gas with a low to medium joule value.[21]

Pyrolysis is usually carried out at near atmospheric pressures and temperatures of up to about 1100°C. Several factors, such as the final temperature, the type of biomass, the rate of heating, the proportion of oxygen present, and the equipment design all affect the yield and the product make up. Generally charcoal yields are 30 to 40 per cent of the dry biomass feed. Gaseous product yields based on the weight of the dry mass fed, range from 5 to 20 per cent depending on temperature.[21] The main constituents of the gas are usually carbon monoxide, carbon dioxide, hydrogen and methane with higher proportions of hydrogen at elevated temperatures.

The liquid product, which is low in the troublesome sulphur, ash and nitrogen, can be as much as 25 weight per cent of the dry feed. It is composed of phenolic tar and an aqueous solution of pyroligneous liquor (mainly acetic acid, methanol and acetone). After removal or neutralization of the acid, the liquid can be used as fuel. Tar can be separated from the aqueous phase by distillation or extraction. The charcoal produced is also low in sulphur and nitrogen, and much easier to store, transport and distribute than the biomass feedstock itself. Of course the energy value of the char is lower than of the biomass and gas is also of low to medium joule value.

14.6.3 *Charcoal production*

Charcoal production is a form of pyrolysis with very limited available oxygen, where the vapors and gases are driven off. Modern charcoal furnaces operating at about 600°C produce 25–35 per cent of the dry biomass feed as charcoal and the gases can be used for kiln drying. The charcoal produced is 75–85 per cent carbon and is useful as a compact, controllable fuel. It can be burnt to provide heat on a large or small scale. Charcoal is a better fuel for cooking compared with firewood.

14.6.4   *Formation of producer gas (gasification)*

Gasification is a process of turning solid biomass into a combustible gas. They contain carbon, hydrogen, and oxygen along with some moisture. Under controlled conditions, characterized by low oxygen supply and high temperatures, most biomass materials can be converted into a gaseous fuel known as producer gas, which consists of carbon monoxide, hydrogen, carbon dioxide, methane and nitrogen. This thermo-chemical conversion of solid biomass into gaseous fuel is called gasification. Producer gas is the mixture of combustible and non-combustible gases. The quantity of gases constituents of producer gas depends upon the type of fuel and operating condition. The heating value of producer gas varies from 4.5 to 6 MJ/m^3 depending upon the quantity of its constituents.[21] In energy terms, the conversion efficiency of the gasification process is in the range of 60–70 per cent. Typical properties of producer gas from gasifier are shown in Table 14.1.

14.6.4.1   *Types of gasifiers*

A gasifier is a chemical reactor where various complex physical and chemical processes take place during burning of biomass in limited air supply. A biomass gasifier can be classified based on the direction of the gas flow and capacity of the gasifier. There are three types of gasifiers: Updraft, Downdraft and Crossdraft. Table 14.2 lists therefore, the advantages and disadvantages generally found for the various classes of gasifiers.

Table 14.1.   Constituent of producer gas.

Compound	Symbol	Gas (volume %)	Dry gas (volume %)
Carbon monoxide	CO	21.0	22.1
Carbon dioxide	CO_2	9.7	10.2
Hydrogen	H_2	14.5	15.2
Water vapor	H_2O	4.8	—
Methane	CH_4	1.6	1.7
Nitrogen	N_2	48.4	50.8

Gas High heating value:

Generator gas (wet basis)	5506 kJ Nm^{-3}
Generator gas (dry basis)	5800 kJ/Nm^{-3}
Air ratio required for gasification	2.38 kg wood (kg air)$^{-1}$
Air ratio required for gas combustion	1.15 kg wood (kg air)$^{-1}$

Table 14.2.   Advantages and disadvantages of various gasifiers.

Gasifier type	Advantage	Disadvantages
Updraft	- Small pressure drop - Good thermal efficiency - Little tendency towards slag formation	- Great sensitivity to tar and moisture and moisture content of fuel - Relatively long time required for start up of IC engine - Poor reaction capability with heavy gas load
Downdraft	- Flexible adaptation of gas production to load - Low sensitivity to charcoal dust and tar content of fuel	- Design tends to be tall - Not feasible for very small particle size of fuel
Crossdraft	- Short design height - Very fast response time to load - Flexible gas production	- Very high sensitivity to slag formation - High pressure drop

(a) *Updraft gasifier*

An updraft gasifier has air passing through the biomass from bottom and the combustible gases come out from the top of the gasifier (Fig. 14.3). This type of gasifier is easy to build and operate. The gas produced has less ash but contains tar and water vapor because of the passing of gas through the unburnt fuel. Hence, updraft gasifiers are suitable for tar free fuels like charcoal. An updraft gasifier tends to a high thermal efficiency because the gases from the combustion zone passes upwards through the in-coming fuel which preheats it. The sensible heat given by the gas is used to preheat and dry fuel. The disadvantages of updraft gasifiers are the excessive amount of tar in raw gas and poor loading capability.

(b) *Downdraft gasifier*

In a downdraft gasifier, air is introduced into a downward flowing packed bed and solid fuels and gas is drawn off at the bottom (Fig. 14.4). In this type of gasifiers, the volatiles and the tar produced from the pyrolysis zone have to pass through the combustion and reduction zone where they are mostly cracked and gasified. The gas produced contains less of tar and more of ash. These gasifiers are suitable for fuels like wood and agricultural wastes.

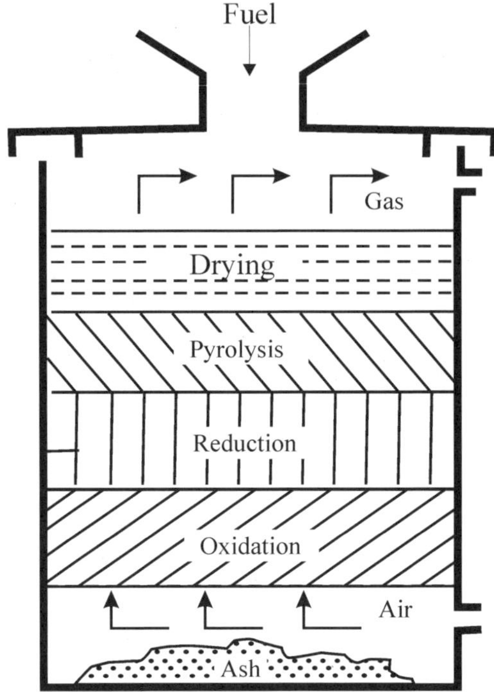

Fig. 14.3.   Updraft gasifier.

A lower overall efficiency and difficulties in handling higher moisture and ash content are common problems in small downdraft gas producers. The time (20–30 minutes) needed to ignite and bring plant to working temperature with good gas quality is shorter than updraft gasifier. This gasifier is preferred to an updraft gasifier for internal combustion engines. They can be used for engine application and they may be used for power generation. They are cheap and easy to make.

(c) *Crossdraft gasifier*

In the crossdraft gasifiers, air is introduced on one side of the gasifier and the gas outlet is on opposite side. Normally an air inlet nozzle is extended at the centre of the combustion zone as shown in Fig. 14.5. The main advantage of the crossdraft gasifiers are its rapid response of change in load, simple construction and light weight. These gasifiers are selected for clean fuel like charcoal. This design characteristic limits the type of fuel for operation to low ash fuels such as wood, charcoal, and coke. The load following ability of crossdraft gasifier is quite good due to concentrated partial zones which at temperatures up to 2000°C. Start up time (5–10 minutes) is much faster than that of downdraft and updraft units. The relatively higher temperature in a crossdraft gasifier has an obvious effect on gas composition such as high

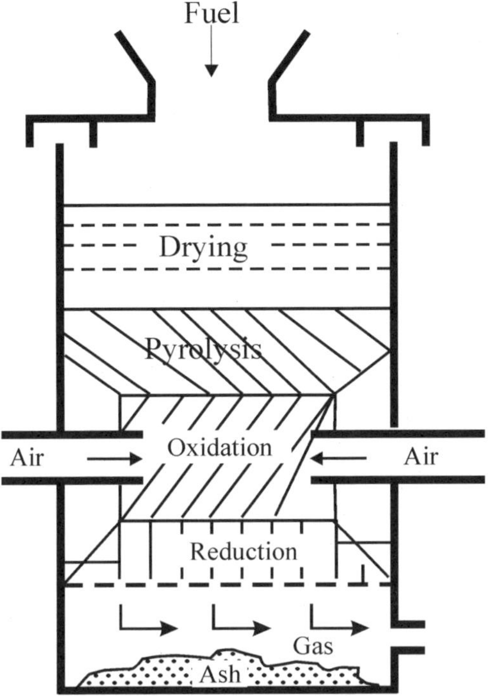

Fig. 14.4. Downdraft gasifier.

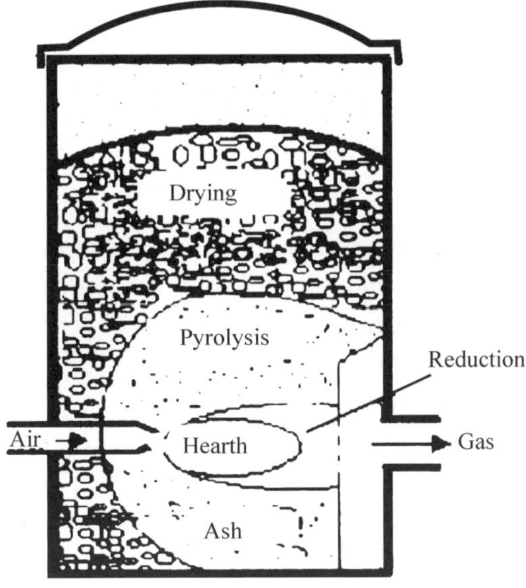

Fig. 14.5. Crossdraft gasifier.

carbon monoxide, and low hydrogen and methane content when dry fuel such as charcoal is used. Crossdraft gasifier operates well on dry blast and dry fuel.

14.6.4.2 *Process zones*

Four distinct processes take place in a gasifier as the fuel makes its way to gasification. They are: Drying of fuel, Pyrolysis, Combustion and Reduction.

(a) *Drying of fuel*

Biomass is fed at the top of the gasifier hopper and it moves down as the process proceeds. Usually, the air dried biomass contains moisture in the range of 13–15 per cent, and the heat radiation from the oxidation zone evaporates the moisture of biomass in the upper most layers. The temperature of this zone remains less than 120°C.

(b) *Pyrolysis*

As the dried mass moves down, it is subjected to stronger heating from the oxidation zone. At a temperature above 200°C the material starts losing its volatiles. No air is admitted until then. Once the temperature reaches 400°C, a self sustaining exothermic reaction takes place, in which the material structure of wood or other organic solids breaks down. This zone in which biomass, in all its volatility, gets converted into two parts.

 (i) **Primary pyrolysis zone**: The normal temperature in this zone is 200°C–600°C.
(ii) **Secondary pyrolysis zone**: The temperature range is 300°C–800°C.

(c) *Combustion*

The combustible substance of a solid fuel is usually composed of elements carbon, hydrogen and oxygen. In complete combustion carbon dioxide is obtained from carbon in fuel and water is obtained from the hydrogen, usually as steam. The combustion reaction is exothermic and yields a theoretical oxidation temperature of 1450°C. The main reactions, therefore, are:

$$C + O_2 = CO_2 \ (+393 \, \text{MJ/kg mole}), \tag{14.1}$$

$$2H_2 + O_2 = 2H_2O \ (-242 \, \text{MJ/kg mole}). \tag{14.2}$$

(d) *Reduction*

The products of partial combustion (water, carbon dioxide and uncombusted partially cracked pyrolysis products) now pass through a red-hot charcoal bed where the following reduction reactions take place.

$$C + CO_2 = 2CO \ (-164.9 \, MJ/kg \, mole), \tag{14.3}$$

$$C + H_2O = CO + H_2 \ (-122.6 \, MJ/kg \, mole), \tag{14.4}$$

$$CO + H_2O = CO_2 + H_2 \ (+42 \, MJ/kg \, mole), \tag{14.5}$$

$$C + 2H_2 = CH_4 \ (+75 \, MJ/kg \, mole), \tag{14.6}$$

$$CO_2 + H_2 = CO + H_2O \ (-42.3 \, MJ/kg \, mole). \tag{14.7}$$

Reactions (3) and (4) are main reduction reactions and being endothermic have the capability of reducing gas temperature. Consequently the temperatures in the reduction zone are normally 800–1000°C. The lower the reduction zone temperature ($\sim$700–800°C) is, the lower is the calorific value of gas.

14.6.5 *Liquefaction*

Another technique that uses high temperatures to convert biomass is liquefaction. It can be both a direct and indirect process of thermo chemical conversion. The former is usually catalytic, the feed being first converted into a gaseous intermediate from which liquid fuels are then synthesized. In one of the direct liquefaction processes, the biomass slurry containing the catalyst is fed to a high pressure (but not more than 280 kg per sq. cm) and medium temperature (340°C) reactor where a lique-faction reaction takes place in a reducing gas atmosphere (with carbon monoxide and/or hydrogen also being fed to the reactor).[22] A series of complex reactions occur simultaneously — pyrolysis, gasification and volatilization. Hydrogasification by pyrolysis in the presence of the hydrogen produces mainly methane and water. The gas, containing the combustibles, is removed and can be used as a heat source to pre-dry the wood feedstock. The liquid-solid mixture is then separated by distillation. The bottoms containing residual solids can be recycled together with a portion of the oil which is not removed as product. After the oil product is separated, vacuum dis-tillation would allow a product as desired. The product composition will, of course, depend on the biomass feed.

In the direct liquefaction process, there are two stages. In the first, the biomass is gasified to an intermediate species which in the second is converted to methanol, gasoline or polymerized liquid hydrocarbons. The intermediate mixture could be synthesis gas (carbon monoxide and hydrogen), light olefins (ethylene and propylene), etc. Methanol synthesis has been in wide commercial production for many years. Most methanol production today is based on natural gas as the feed source.[23] This generally requires a higher hydrogen to carbon monoxide ratio than a biomass gasification unit.

Only methanol is being considered as a prospective alcohol to be derived from biomass by thermo chemical means. At present, methanol from wood would be more expensive than that from coal, partly due to the smaller scale of biomass units. Nonetheless, there are opportunities to decrease the costs. Clustering the gasification units near biomass sources and transferring the gas to a central methanol unit could greatly reduce cost. The Mobil Corporation has also developed a process for converting methanol and other aggregates into high octane gasoline:

$$n(CH_3OH)n \xrightarrow[\text{catalyst}]{} (-CH_2-)n + nH_2O. \qquad (14.8)$$

With methanol feedstock a zeolite catalyst allows a yield of about 85 per cent of hydrocarbons (gasoline) containing virtually no component heavier than C_{10}.

14.7 Biodiesel Production

Bio-diesel is an eco-friendly, alternative diesel fuel prepared from domestic renewable resources, i.e., vegetable oils and animal fats. These natural oils and fats are made up mainly of triglycerides. These triglycerides when reacted chemically with lower alcohols in presence of a catalyst result in fatty acid esters. These esters show striking similarity to petroleum derived diesel and are called "bio-diesel".

The major application of bio-diesel is in the transportation sector as an alternate to mineral diesel. Many automobiles builders like Ford, John Deere, Massey–Ferguson, Mercedes, BMW, Volkswagen, Volvo, etc., have accepted bio-diesel as the fuel suitable for their vehicles in existing diesel engines. However, mostly bio-diesel is used in 10 per cent or 20 per cent blends rather than as neat bio-diesel. This blending approach also avoids the need to build a separate and costly infrastructure for storing bio-diesel. Though bio-diesel is recommended for use in almost all diesels run vehicles, the fuel must meet the ASTM/DIN specifications.[24]

There are four primary ways to make biodiesel, direct use and blending, microemulsions, thermal cracking (pyrolysis) and transesterification. The most commonly used method is transesterification of vegetable oils and animal fats. The transesterification reaction is affected by a molar ratio of glycerides to alcohol, catalysts, reaction temperature, reaction time and free fatty acids and water content of oils or fats. The mechanism and kinetics of the transesterification show how the reaction occurs and progresses.[25]

Trans-esterification, also called alcoholysis, is the displacement of alcohol from an ester by another alcohol in a process similar to hydrolysis. This process has been widely used to reduce the viscosity of triglycerides. The transesterification reaction is represented by the general equation, which is the key reaction for bio-diesel production.

$$RCOOR' + R'' \quad RCOOR'' + R'OH. \tag{14.9}$$

If methanol is used in the above reaction, it is termed as methanolysis. The reaction of triglyceride with methanol is represented by the general equation.

$$
\begin{array}{c}
\left[\begin{array}{l} OCOR_1 \\ OCOR_2 \\ OCOR_3 \end{array}\right. + 3\,ROH \xrightleftharpoons[Catalyst]{Catalyst}
\begin{array}{c} R_1COOR \\ + \\ R_2COOR \\ + \\ R_3COOR \end{array} \;+\;
\left[\begin{array}{l} OH \\ OH \\ OH \end{array}\right.
\end{array}
$$

Triglyceride   Alcohol   Alkyl ester   Glycerol

Transeterification of triglyceride with alcohol

$$\tag{14.10}$$

Triglycerides are readily trans-esterified in the presence of alkaline catalyst (Lye) at atmospheric pressure and temperature of approximately 60–70°C with an excess of methanol. The mixture at the end of reaction is allowed to settle. The excess methanol is recovered by distillation and sent to a rectifying column for purification and recycled. The lower glycerol layer is drawn off while the upper methyl ester layer is washed with water to remove entrained glycerol. Methyl esters of fatty acids are termed as bio-diesel.

14.8 Bioethanol Production

The principle fuel used as a petroleum substitute is bioethanol. Bioethanol is produced through natural fermentation of the starch and sugars present in different forms of biomass by biological organisms.[26] Ethanol can be produced from different kinds of raw materials. Ethanol pathways from different raw materials are given in Table 14.3. The raw materials are classified into three categories of agricultural raw materials: simple sugars, starch, and lignocellulose.[27]

The components of the lignocellosic biomass include cellulose, hemicelluloses, lignin, extractives, ash, and other compounds. Ethanol can be produced from cellulose feedstocks such as corn stalks, rice straw, sugar cane bagasse, pulpwood, switchgrass, and municipal solid waste.[29] Fermentation involves the conversion of feedstock by microorganisms in a process similar to digestion, but the products are alcohols or organic acids instead of methane. Feedstock energy conversion by fermentation is in wide-scale use, with a global production of about 20 GW ethanol.

Table 14.3.   Ethanol pathways from different raw materials.[28]

Raw material	Processing
Wood	Acid hydrolysis + fermentation
Wood	Enzymatic hydrolysis + fermentation
Straw	Acid hydrolysis + fermentation
Straw	Enzymatic hydrolysis + fermentation
Wheat	Malting + fermentation
Sugar cane	Fermentation
Sugar beet	Fermentation
Corn grain	Fermentation
Corn stalk	Acid hydrolysis + fermentation
Sweet sorghum	Fermentation

Bioethanol is mainly produced by the sugar fermentation process, although it can also be produced by the chemical process of reacting ethylene with steam.

The main source of sugar required to produce ethanol comes from fuel or energy crops. Bioethanol, or rather ethanol, itself belongs to the chemical family of alcohols and has a structure of C_2H_5OH. It is a colorless liquid and has a strong odor. Brazil and the United States account for over 70 per cent of all ethanol production in the world today with the USA producing an estimated 6,500 million gallons a year. The basic steps for large scale production of ethanol are: fermentation of sugars, distillation, dehydration and denaturing (optional).

Ethanol can be produced from a large variety of carbohydrates with a general formula of $(CH_2O)_n$. Fermentation of sucrose is performed using commercial yeast such as Saccharomyces ceveresiae. The chemical reaction is composed of enzymatic hydrolysis of sucrose followed by fermentation of simple sugars. First, invertase enzyme in the yeast catalyzes the hydrolysis of sucrose to convert it into glucose and fructose:

$$C_{12}H_{22}O_{11} \rightarrow C_6H_{12}O_6 + C_6H_{12}O_6. \qquad (14.11)$$

$$\underset{\text{Sucrose}}{} \qquad \underset{\text{Glucose}}{} \quad \underset{\text{Fructose}}{}$$

Second, zymase, another enzyme also present in the yeast, converts the glucose and the fructose into ethanol:

$$C_6H_{12}O_6 \rightarrow 2C_2H_5OH + 2CO_2. \qquad (14.12)$$

Gluco-amylase enzyme converts the starch into D-glucose. The enzymatic hydrolysis is then followed by fermentation, distillation, and dehydration to yield anhydrous bioethanol. In contrast to fossil fuel, ethanol is a renewable raw material that can be obtained from biomass and address the issue of the greenhouse effect.

Furthermore, the infrastructure needed for ethanol production and distribution is already established in countries like Brazil and the USA since ethanol is currently distributed and used as an octane enhancer or oxygenate blended with gasoline.[30]

14.9 Conclusions

Biomass is the most important source to increase the production of energy based on renewable energy sources. In this chapter, various biomass conversion routes such as thermo chemical, bio chemical and chemical conversion have been covered, which includes biomass gasifiers. There are many routes to convert biomass to useable forms of fuel. The various present technologies for both conversion and utilization of biomass energy have shown that they have promising future improvement potential, especially if sustainably sourced with little or no negative impact on the environment and human well being.

References

1. R.C. Saxena, D.K. Adhikari and H.B. Goyal, "Biomass-based energy fuel through biochemical routes: A review," *Renewable and Sustainable Energy Reviews* **13** (2009) 167–178.
2. A.E. Putun, A. Ozcan, H.F. Gercel and E. Putun, "Production of biocrudes from biomass in a fixed bed tubular reactor; product yields and compositions," *Fuel* **80** (2001) 1371–78.
3. A.V. Bridgewater, "Renewable fuels and chemicals by thermal processing of biomass," *Chemical Engineering Journal* **91** (2003) 87–102.
4. A.V. Bridgewater, D. Meier and D. Radlein, "An overview of fast pyrolysis of biomass," *Organic Geochemistry* **30** (1999) 1479–1493.
5. A.V. Bridgewater, "Principles and practice of biomass fast pyrolysis processes for liquids," *J. Analytical and Applied Pyrolysis* **51** (1999) 3–22.
6. N. Ozbay, A.E. Putun, B.B. Uzun and E. Putun, "Biocrude from biomass: Pyrolysis of cotton seed cake," *Renewable Energy* **24** (2001) 615–625.
7. N.S. Rathore and N.L. Panwar, *Renewable Energy Sources for Sustainable Development* (New India Publishing), 2007.
8. M. Balat, "Use of biomass sources for energy in Turkey and a view to biomass potential," *Biomass and Bioenergy* **29** (2005) 32–41.
9. M.C. Romel and D.B. Khang, "Characterization of biomass energy projects in Southeast Asia," *Biomass and Bioenergy* **32** (2008) 525–532.
10. E. Erdogdu, "An expose of bioenergy and its potential and utilization in Turkey," *Energy Policy* **36** (2008) 2182–2190.
11. M. Parikka, "Global biomass fuel resources," *Biomass and Bioenergy* **27** (2004) 613–620.
12. J.J. Bozell, "Renewable feed-stocks for the production of chemicals," *Proc. 217th ACS National Meeting* (1999), pp. 204–209.
13. K. Othmer, *Encyclopedia of Chemical Technology*, vol. 11, 3rd ed., (1980), pp. 347–360.
14. L.P. White and L.G. Plasket, *Biomass as Fuel* (Academic Press, 1981).
15. N.S. Rathore, N.L. Panwar and S. Kothari, *Biomass Production and Utilization Technology* (Himanshu Publication, Udaipur, India, 2007).

16. A.N. Mathur and N.S. Rathore, *New and Renewable Energy Sources* (Bohra Ganesh Publications, Udaipur, India, 1996).

17. M. Hoogwijk, A. Faaij, B. Eickhout, B. Vries and W. Turkenburg, "Potential of biomass energy out to 2100," *Biomass and Bioenergy* **29** (2005) 225–257.

18. S. Yaman, "Pyrolysis of biomass to produce fuels and chemical feedstocks," *Energy Conversion and Management* **45** (2004) 651–671.

19. P.A.D. Rocha, E.G. Cerrella, B.R. Bonelli, and A.L. Cukierman, "Pyrolysis of hardwood residues: On kinetics and chars characterization," *Biomass and Bioenergy* **16** (1999) 79–88.

20. M. Balat, "Turkey's hydropower potential and electricity generation policy overview beginning in the twenty first century," *Energy Sources* **27** (2005) 949–962.

21. A. Faaji, "Modern biomass conversion technologies," *Mitigation and Adaptation Strategies for Global Change* **11** (2006) 343–375.

22. A. Demirbas, "Mechanisms of liquefaction and pyrolysis reactions of biomass," *Energy Conversion and Management* **41** (2000) 633–646.

23. A. Demirbas, "Biomass resource facilities and biomass conversion processing for fuels and chemicals," *Energy Conversion and Management* **42** (2001) 1357–1378.

24. F. Ma and M.A. Hanna, "Biodiesel production: A review," *Bioresource Technology* **70** (1999) 1–15.

25. A. Murugesana, C. Umaranib, T.R. Chinnusamya, M. Krishnana, R. Subramanianc and N. Neduzchezhaind, "Production and analysis of bio-diesel from non-edible oils: A review," *Renewable and Sustainable Energy Reviews* **13** (2009) 825–834.

26. L.A. Lucia, D.S. Argyropoulos, L. Adamopoulos and A.R. Gaspar, "Chemicals and energy from biomass," *Canadian J. Chemistry* **84** (2006) 960–970.

27. A. Demirbas, "Bioethanol from cellulosic materials: A renewable motor fuel from biomass," *Energy Sources* **27** (2005) 327–337.

28. M. Balat, "Global status of biomass energy use," *Energy Sources* **31** (2009) 1160–1173.

29. M. Balat, "Current Alternative Engine Fuels," *Energy Sources* **27** (2005) 569–577.

30. L.V. Mattos and F.B. Noronha, "Partial oxidation of ethanol on supported Pt catalysts," *J. Power Sources* **145** (2005) 10–15.

Chapter 15

Forest Biomass Production

Severiano Pérez[*], Carlos J. Renedo, Alfredo Ortiz and Mario Mañana

*Department of Electric and Energy Engineering,
University of Cantabria, 39005, Santander, Spain*
[*]perezrs@unican.es

Carlos Tejedor

*Bosques 2000 S.L. Grupo Sniace,
Ganzo 39300 Torrelavega, Spain*
bosques2000@sniace.com

This chapter presents the resource known as forest biomass. An analysis of its potential energy, associated to its two sources, forest residues and energy crops, is carried out. It offers energy density values for various tree species. Also, it discusses the collection and transportation systems and their performance. Finally, it shows the environmental implications of the exploitation of this resource, both in carbon sinks, and in the impact generated by the nutrients extraction from the soil.

15.1   Introduction

Forest biomass is defined as the biodegradable part of products and waste generated in the forest. In the mountains of Europe, the primary forest biomass is made up of vegetable material from silvicultural operations including: pruning, selection of shoots, phytosanitary felling, etc. The waste from the exploitation of forest wood is also included, this being from final cuts or intermediate cuts. Vegetable material from energy crops, both woody and herbaceous, installed on forest land can be also used. Usually, it is measured in tons per hectare of green or dry weight, and if it is studied for energy purposes in MJ per ha and year. The interesting part, from the energy and operating points of view, is the flight biomass, i.e., which is what lies on top of the soil, as the parts belonging to the land lack interest (roots).

Since the beginning of human history, biomass has been an essential energy source for humans. With the advent of fossil fuels, this energy source has lost importance in the industrial world, with its main uses currently being domestic.

In Europe, France is the country with the greatest amount of biomass consumed (more than 9 million tonnes of equivalent oil (toe)), followed by Sweden.

The factors influencing the consumption of biomass in Europe are:

- Availability of the resource: this is the factor to be considered first in order to determine the access and seasonality of the resource.
- Geographical factors: weather conditions indicate the temperature and the water availability in each zone, and whether it can be covered by biomass.
- Energy factors: the biomass profitability as an energy resource will depend on market prices at every moment.

In the country members of the Organization for Economic Co-operation and Development (OECD) the electricity consumption has risen from 100 to 149 kWh per capita in the last 20 years. This growth in energy consumption requires the development of new primary energy sources such as forest biomass. Forest biomass has important advantages, since it does not compete with any food consumption as do agricultural food crops.

From the economic point of view, the external costs of electricity generation from biomass, in comparison with other fuels, are clearly favorable to biomass as shown in Table 15.1.[1]

In addition to issues related to energy, forest biomass is associated to a number of effects that can be classified into:

- Environmental effects.
- Socio-economic effects.

Table 15.1.   External costs of electricity generation from biomass compared with fossil fuels.

Country	Coal	Fuel	Biomass
Denmark	4–7		1
Finland	2–4	2–5	1
France	7–1	8–11	1
Greece	5–8	3–5	0–0.8
Germany	3–6	5–8	3
Netherlands	3–4		0.5
Portugal	4–7		1–2
Sweden	2–4		0.3
United Kingdom	4–7	3–5	1

As regards the environmental effects, some of them are identified as positive:

- A falling concentrations of CO_2 in the atmosphere.
- Fewer problems associated with soil degradation (erosion, desertification,...).
- Lowered fire risk, facilitating also their extinction.
- Improved plant health of the forests.
- Improvement in mountain forest management.

The negative environmental impacts are listed as follows:

- Loss of soil nutrients. This may question energy sustainability.
- Negative effects caused by the use of heavy machinery in exploitation operations (noise, pollution, changes in the topography of the land due to construction of tracks...).
- Local pollution in the vicinity of the power plants.

Within the group of socio-economic impacts, the following can be highlighted:

- It helps to guarantee energy supply while respecting the environment.
- It helps to diversify energy sources.
- It gives greater price security.
- It lowers transport electricity costs due to the location of the decentralized power plants.
- Employment: this is a new source of income, since it is based on an asset that has not been previously marketed.
- Its favorable impact on the industry as a whole.
- Sustainable development.
- Rural development and improvement of inter-regional equity.

The depletion of oil as time goes by and the impact that derived energies generate in the environment lead towards clean and sustainable energies being developed. Forest biomass belongs to this set of energies and should be developed in regions where the soil and climatic characteristics allow the development of forest species, which can be exploited for energy purposes with acceptable profitability levels.

15.2 Bioclimatic Potential

In order to study the potential to generate vegetation in a given area, a phytoclimatic tool known as bioclimatic diagram is used, Fig. 15.1. The method for preparing it is that proposed by Ref. 2.

These diagrams are based on variables strongly linked to the vegetative capacity of regions, such as average temperature, rainfall, evapotranspiration, etc. Through

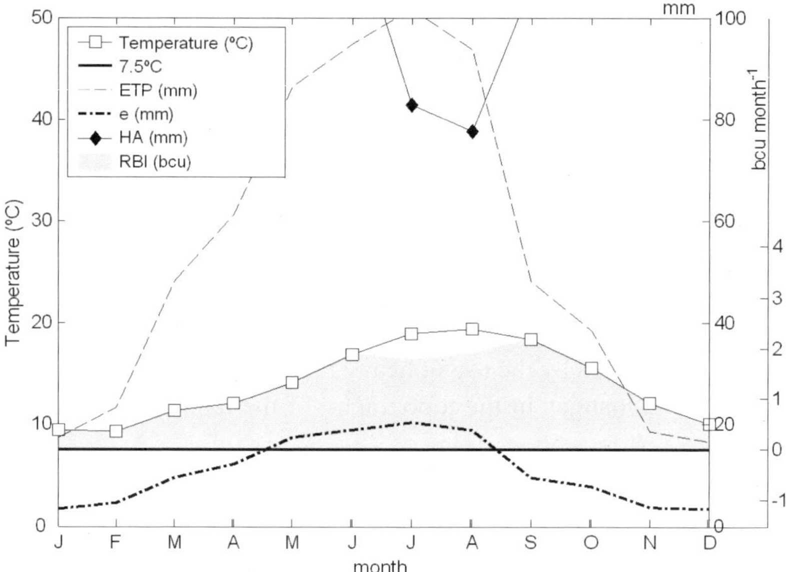

Fig. 15.1.   Bioclimatic diagram of northern Spain.

these diagrams one can quantify the diverse climatic conditions affecting plants (phytological). This quantification is represented on charts where there appear areas that show the vegetative activity in plants, which is proportional to those areas.

Through the analysis of these diagrams, the capability of a region to generate forest biomass can be determined. For the preparation of these diagrams, a number of variables are employed, which are described below.

15.2.1 *Potential bioclimatic intensity (PBI)*

Each forest species is associated with a temperature which is optimal, provided that the other requirements of the plant are met, such as moisture, nutrients, etc. This optimal situation results in maximum stable vegetative activity. For all forest species, one can say that there is a degree of proportionality between the growth curve and the increase in the corresponding temperature over 7.5°C. Thus, the area between the curve of monthly average temperatures and the line matching to 7.5°C, is called potential bioclimatic intensity. The vegetative activity of the plant is proportional to the bioclimatic intensity (if they have sufficient water).

15.2.2 *Potential evapotranspiration (E)*

Potential evapotranspiration is defined as the amount of water lost by the soil when this is covered with vegetation actively growing, and taking into consideration the fact that the soil contains the water required by plants. As a result, those periods in which there is no water limitation and its average temperature exceeds 7.5°C, the potential bioclimatic intensity coincides with the real one, while in periods where the water availability is not sufficient, the real bioclimatic intensity (RBI) will be lower than the potential one.

15.2.3 *Hydric availability (HA)*

This is the amount of water that plants have available. This parameter does not coincide with rainfall, as this would give soil a non-retaining capacity. When HA is lower than E, comes to a halt growth.

15.2.4 *Residual evapotranspiration (e)*

This is defined as the potential evapotranspiration when sap is stopped. This concept is introduced, in an approximate way, to quantify the effect generated in the ideal development of the plant by an insufficient level of water.

Figure 15.1 presents the bioclimatic diagram of the northern coast of Spain. It corresponds to an Atlantic climate characterized by mild winters and summers, influenced by the proximity of the sea. It shows a coincidence between the RBI and PBI, except during summer months in which the effect of drought makes the RBI lower than the PBI. This diagram shows the great capability for biomass generation in this area. As noted here, the growth is not stopped in winter since the average temperature curve does not go down below the minimum value for the existence of vegetative activity (7.5°C).

15.3 Forest Species

One can distinguish two types of forest biomass depending on its origin. On one hand, there is forest biomass waste from species whose primary objective is to generate wood, and on the other hand, the biomass generated by forest energy crops. Thus, there is forest biomass from a wide range of forest genera. A large proportion of these species, due to their ecological value and their location in strategic and/or protected areas, would not be subject to use either for wood or for energy purposes. In this way, the impact that would be generated by its exploitation will seriously deteriorate these ecosystems. However, they can provide the forest waste generated by the removal of dead trees, pruning, etc.

The group of species that provide the greatest amount of forest biomass are all called species of rapid growth. Within this group, the most representative members belong to the genera: Pinus, Populus, Salix, Eucalyptus. In each region, the forest genus best suited to the climate and soil characteristics must be selected.

The genera Populus and Salix are used in Nordic countries. Research results have led to the development of specific clones for some species, which maximize the quality and quantity of the biomass generated.[3–6]

The species belonging to the genera Pinus and Eucalyptus have been traditionally selected to maximize wood production in quality and quantity. The selection of trees to generate biomass with energy production purposes is still under research.[7,8] One of the main barriers to overcome is the notorious difficulty involved in cloning certain species of these genera, because of their low reproduction capacity by cuttings.

In the future, research should seek to improve the production of the current species in both quantity and quality, under the lens of environmental sustainability. New tree species should also be investigated, and these should be at the stage of forest biomass production in those places where the traditional species do not provide guarantee of development.

15.4   Evaluation of Forest Biomass

The assessment of forest biomass is made from two perspectives:

- The energy point of view.
- The productivity point of view.

Both points are discussed for two sources of forest biomass, that is, residual biomass from waste produced in conventional forest exploitation and biomass from forest energy crops.

15.4.1   *Residual forest biomass from traditional exploitation*

15.4.1.1   *Theoretical quantification*

The information on the amount of tree biomass is a valuable tool for energy management of this resource. The studies conducted so far, are for fast-growing species, namely, Eucalyptus globulus and Pinus radiate, which are the most interesting from the industrial point of view.

The equation that relates the amount of dry biomass of an evergreen tree is given by the expression[18]:

$$W = \alpha D^\beta H^\gamma, \tag{15.1}$$

where:
W is the weight in grams of dry biomass on the ground,
D is the diameter of the tree in millimeters,

H is the tree height in centimeters,

α, γ and β are coefficients specific to each species.

Given that the total biomass is the sum of its individual parts, it can be written that:

$$W = W_m + W_h + W_o, \tag{15.2}$$

where:

W_m is the fraction for wood and bark,

W_h is the fraction for the foliage,

W_o is the remaining biomass.

Taking logarithms and for each portion of the tree, we obtain:

$$\ln W_i = \alpha + \beta \ln D_i + \gamma \ln H_i. \tag{15.3}$$

Then,

$$W_i = \exp^{(\alpha + \beta \ln D_i + \gamma \ln H_i)}, \tag{15.4}$$

where the subscript i indicates the corresponding fraction of the tree.

In different countries, to estimate the biomass generated by the tree and the percentage of the parts that make this up,[7,9–11] trials with irrigation and fertilization have been carried out in experimental stands. The forest waste is divided into its component parts, i.e., leaves, seeds, fine branches, large branches and bark. Table 15.2 shows the equations that allow to be estimated the amount of biomass in a stand of Pinus radiata in northern Spain for each of its fractions.[7]

Table 15.3 shows the amount of biomass produced per area unit for Pinus radiata[7] with an average age of 21 years.

Table 15.2. Allometric equations for estimating Pinus radiata biomass.[7]

Total biomass	$-3.38 + 1.82\,\mathrm{LnDn} + 0.97\,\mathrm{LnH}$
Wood	$-4.83 + 1.72\,\mathrm{LnDn} + 1.43\,\mathrm{LnH}$
Bark	$1.11 - 0.80\,\mathrm{LnDn}$
Thick branches	$-3.75 + 2.01\,\mathrm{LnDn}$
Fine branches	$-3.99 + 1.89\,\mathrm{LnDn}$
Branchlets	$-4.94 + 1.96\,\mathrm{LnDn}$
Needles	$-4.05 + 1.97\,\mathrm{LnDn}$
Remains	$-2.47 + 1.95\,\mathrm{LnDn}$

where:

D_n is the normal diameter in centimetres

H is the height in meters.

Table 15.3. Amount of dry biomass divided into the different components in a stand of Pinus radiata.[7]

Component	Ton/Ha	% Total
Wood	111.2	69.9
Bark	13.1	8.2
Large branches	11.0	6.9
Fine branches	6.2	3.9
Branchlets	3.0	1.9
Leaves	6.7	4.2
Cone	7.6	4.8

Table 15.4. Allometric equations for estimating the biomass of Eucalyptus globulus.[12]

Component	Adjusted equation
Wood	$0.062\ D_n^{2.35}H^{1.001}$
Bark	$0.0093\ D_n^{2.46}$
Large branches	$0.0076\ D_n^{3.39}G^{-0.83}$
Fine branches	$0.0264\ D_n^{2.63}G^{-0.81}$
Branchlets	$0.0451\ D_n^{3.08}G^{-1.59}$
Needles	$0.0043\ D_n^{3.69}G^{-1.22}$
Shoots (<5 cm)	$0{,}2536\ D_n^{2.4579}$

where:
D_n is the normal diameter in centimeters.
H is the height in meters.
G is the base area in m^2/ha.

It should be stressed that these equations and data are obtained from experimental measures on stands, which give an idea of the order of magnitude. They have a worse fit in those areas with different characteristics to those of the stands where the data was taken. These deviations may be due to several factors: soil, forestry, climatics, geographic, genetics, and so on.

In the same way, for the species Eucalyptus globulus, the equations that quantify its biomass in Galicia (Spain) are presented in Table 15.4.[12]

For the species Eucalyptus globulus, Table 15.5 shows the average amount of biomass per hectare at an age between 14 and 18 years.[13] These results come from stands of first cut down operation, without fertilization, with a non-altered soil profile at the moment of planting.

Table 15.5. Amount of biomass in a stand of *Eucalyptus globulus*[13] around 15 years old.

Fraction	tonnes/ha	% Total
Bark	33.1	9.74
Branches	22.0	6.47
Branchlets	5.6	1.65
Leaves	20.8	6.12
Wood	258.4	76.02

Table 15.6. Average weight of each fraction of the total waste.[13,14]

Species	Leaves	Branches + branchlets	Bark
Eucalyptus sp.	25.56	33.92	40.51
Fagus sylvatica	4.50	66.71	25.00
Quercus robur	6.10	49.58	42.61
Pinus radiata	8.80	24.17	67.03

As for Pinus radiata, these equations and data are the result of trials in experimental stations. So, the extrapolation to other scenarios depends on the deviation of these with respect to the base case, which in many cases may be relevant. What it does is to give an idea of the percentages of each fraction that makes up the tree.

Table 15.6 shows the approximate weight percentages of each fraction over the total waste, for the different species. For the species Eucalyptus the percentages of each fraction that make up the waste were taken from data from Refs. 13 and 14.

These theoretical methods should be compared with real results, in order to refine the equations and thus allow a better fit between the theoretical and the real results.

15.4.1.2 *Energy characterization*

The calorific value is the amount of heat, per mass unit, that a fuel gives off when it is burned.

Two calorific values are defined:

- The net calorific value (NCV) is the heat that is normally used. It is defined as the heat given off when burning a fuel, while the moisture content in the smoke does not condense.
- The gross calorific value (GCV) is that given off in the combustion when smoke moisture is collected condensated. It is greater or equal than the NCV, as the condensation latent heat of the water contained in the smoke is taken.

Table 15.7. Calorific value of waste from several forest species at 5% moisture.[15]

	GCV (kJ/kg)	NCV (kJ/kg)
Fagus sylvatica	18,161	16,680
Quercus robur	17,484	15,972
Pinus radiata	18,096	16,557
Eucalyptus globulus	17,682	16,144
Eucalyptus nitens	18,708	17,169

The relation between the GCV and the NCV is given by the following equation:

$$NCV = GCV\ 2,442\ 0.01\ (H_b + H_a) - 2,442\ 0.09\ H_d, \qquad (15.5)$$

where:

H_d is the weight percentage of hydrogen in the dry sample.

H_b is the weight percentage of moisture in the sample.

H_a is the moisture percentage in the combustion air.

The method for the energy characterization of forest waste is based on determining the calorific value through a calorimeter unit.

Once the calorific value of each part has been determined, and using their percentage in the total waste, the average calorific value of the waste can be determined. Table 15.7 shows the GCV and the NCV from the waste of several forest species. The fraction corresponding to the leaves has the highest calorific value, reaching values higher than 20000 kJ/kg at 5% moisture. Note that the moisture content exerts a great influence on the calorific value of forest biomass, this dropping as the moisture increases.

Energy density is defined as the amount of energy in the waste of one species per area unit and year, generally the hectare. This parameter represents forestry and energy variables. Figure 15.2 shows the energy densities of several forest species. The fast-growing species are those which produce more biomass per ha. The number of hectares dedicated to the cultivation of a species for feeding a power plant can be calculated with the energy density.

15.4.2 *Forest biomass from energy crops*

Forest energy crops are those formed by forest species only intended for energy purposes. The potential of the different forest species as energy crops is an issue that is currently being studied by comparing the different species, both from the energy and the forestry points of view.

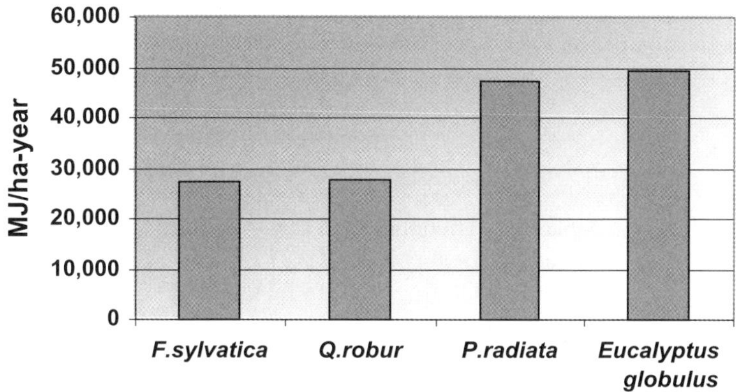

Fig. 15.2.   Energy density of waste from various forest species.[15]

In order to exploit a species as an energy crop, it must meet two main premises:

- Adaptation to the territory in which it is to be grown, which means resistance to pathogens, climate.
- A good energy density, to optimize both productive and energetic aspects.

15.4.2.1 *Biomass production*

The stands of forest species to produce biomass for energy purposes are characterized by high planting density and short rotation harvesting (3–5 years). This means that the youth stage is the one of greatest relevance.

As regards other species belonging to the Willow and Poplar genera, there are field experiments with specific clones for the generation of biomass in countries of northern Europe. Thus, Table 15.8 presents the production of Poplar and Willow clones in two different locations.[16]

Concerning the genus Eucalyptus, the experiences on biomass generation are not only recent but few and far between in Europe, due to the climatic requirements of the genus. However, in other places of the world, there are studies that have estimated the potential of the Eucalyptus species as a biomass generator.[17]

In order to assess the amount of biomass in a stand without cutting it down, some procedures using heights and diameters can obtain an estimate of the amount of biomass in a stand of a particular species. These methods make it possible to calculate the amount of the fractions that make up the biomass (leaves, branches, bark...).

Table 15.8. Comparison of Poplar and Willow yields in oven dry ton per ha.[16]

	Species	Variety	Standing biomass (odt ha^{-1})		
			year 1	year 2	year 3
Site 1	Poplar	Trichobel	11.91	23.23	41.71
	Poplar	Beaupré	15.87	24.7	36.54
	Poplar	Ghoy	14.62	22.89	35.89
	Willow	Jorunn	14.98	22.39	33.33
	Willow	Germany	7.71	12.52	22.01
	Willow	Q83	19.16	22.97	34.06
Site 2	Poplar	Trichobel	4.9	11.6	23.68
	Poplar	Beaupré	7.29	12.24	21.31
	Poplar	Ghoy	7.48	13.91	23.76
	Willow	Jorunn	21.15	26.51	36.61
	Willow	Germany	9.5	20.12	30.47
	Willow	Q83	11.16	20.34	30.85

Table 15.9. Adjusted parameters for Eucalyptus globulus stands as energy crops.[18]

Fraction	α	β	γ
Total biomass	−2.8982	0.1984	1.7425
Leaves	0.7897	0.2921	0.8769
Wood + Bark	−6.8579	0.2474	2.2294
Rest	−2.5669	0.3346	1.3349

Table 15.9 shows the values of α, β and γ, given by Ref. 18 for the juvenile stage of the Eucalyptus globulus species.

From the field data (height and diameter), one can estimate the amount of dry biomass per tree at a certain age. Figure 15.3 shows the dry biomass of several species of Eucalyptus at two different ages.[15] In this figure, significant differences appear between the species of Eucalyptus with respect to the amount of biomass generated. However, this does not mean you have to discard some species over others because, depending on location and climate, one species may be more suitable than another. For example, Eucalyptus gunni is the most suitable, out of all those of its type, in places with very cold climates.

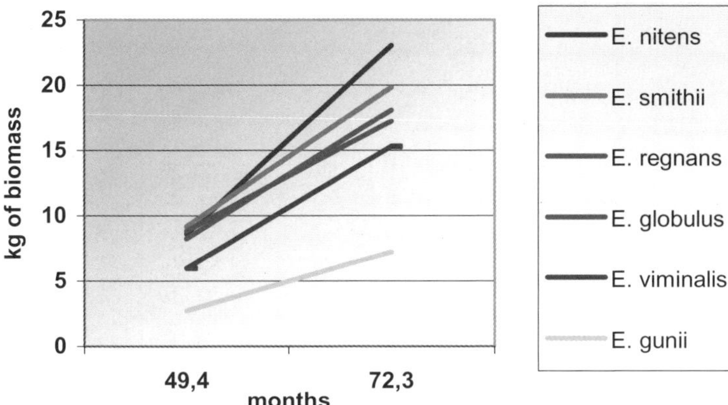

Fig. 15.3.   Amount of dry biomass of several species of Eucalyptus in their youth.[15]

15.4.2.2   *Energy characterization of species as energy crops*

The method used in energy characterization is similar to that discussed in the case of forest waste. In both forms of biomass, the moisture content exerts a major influence on the calorific value.

The energy characterization of forest biomass generated by the Populus genus offers higher calorific values than 18596 kJ/kg,[19] for the wood on a dry basis. In the case of the Willow genus, this value is around 19220 kJ/kg.[20]

For the Eucalyptus genus, a more detailed energy characterization is presented. Table 15.10 shows the gross calorific value of two species of Eucalyptus, at youth and with varying degrees of moisture. As in the case of waste, as the moisture increases, the calorific value decreases. The leaves are the fraction with the highest calorific values, around 20000 kJ/kg. This value is linked to the components in the leaves that present higher calorific values such as aromatic oils and alcohols.

With the calorimetric and productive data, the energy stored by these species at different ages is obtained. Thus, Fig. 15.4 shows the amount of energy stored by several species of Eucalyptus at two different ages.[15] Note that in the energy density the silvicultural variables are more significant than the energy ones.

For forest energy crops of the Salix genus, the energy densities obtained in stands of Canada are presented in Table 15.11.[20]

One can see that the energy density of the species of the Eucalyptus genus is larger than that of the Salix genus. Note that the planting density in a Eucalyptus stand is around 2000 stems per ha, while that of the genus Salix is greater than 20000 stems per ha. This has a great effect on the establishment cost of stands and consequently in the final cost of the biomass.

Table 15.10.   Average, standard deviation and error of the gross calorific value in kJ/kg for the Eucalyptus globulus and Eucalyptus nitens species at youth with varying degrees of moisture.[15]

Moisture (%)	Leaves	Wood	Branches	Bark
		E. globulus		
50–40	13583 ± 189 (1.4%)	8048 ± 105 (1.3%)	7618 ± 208 (2.7%)	6432 ± 104 (1.6%)
20–15	20892 ± 245 (1.2%)	16839 ± 248 (1.5%)	16848 ± 199 (1.5%)	14979 ± 158 (1.0%)
5–0	20589 ± 201 (1.0%)	18684 ± 340 (1.8%)	18746 ± 136 (0.7%)	15728 ± 201 (1.3%)
		E. nitens		
50–40	14305 ± 125 (0.5%)	9128 ± 149 (1.6%)	12716 ± 214 (1.7%)	7477 ± 127 (1.7%)
20–15	21339 ± 99 (1.0%)	16730 ± 215 (1.3%)	16849 ± 154 (0.9%)	14527 ± 351 (2.4%)
5–0	22731 ± 231 (1.1%)	18599 ± 179 (1.0%)	18788 ± 204 (1.1%)	16541 ± 257 (1.6%)

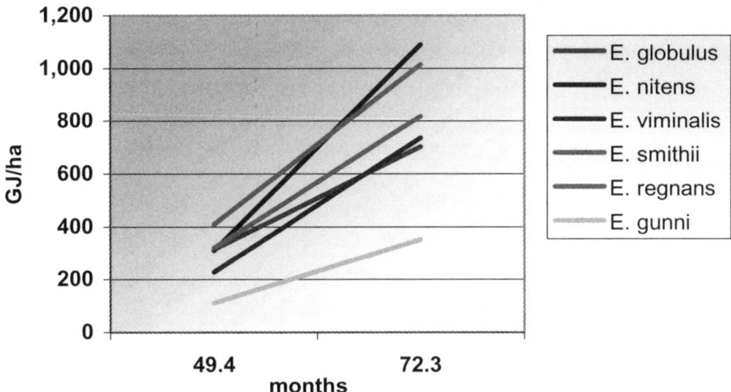

Fig. 15.4. Energy density of several Eucalyptus species at two ages.[15]

Table 15.11.   Energy density in GJ/ha of three varieties of Salix used as energy crops at 2 years of age.[20]

	Site 1	Site 2
Salix discolor	480	344
Salix petiolaris	197	185
Salix viminalis	580	383

15.5   Collection Systems for Forest Biomass

How the waste is distributed in a stand is of crucial importance for the performance and optimization of the collection processes. It is essential that the operations prior to waste collection be integrated as part of the overall process, i.e., from the cutting down of the tree to the use of the waste as fuel.

There are, generally speaking three situations,[21] Fig. 15.5 represents these cases in graphical form:

- Case 1: in this case, the biomass will be extended all over the plot, as a result of the methodology used for forestry operations, since it does not take further use into account. It is the most common situation because the waste has no market at present. From the point of view of collection efficiency, it is the worst alternative.
- Case 2: the remains of wood or biomass (energy crops) extraction from the forest, are stacked in small piles along the surface of the plot. This is the case in which the work is done manually or through peeling machinery, as are the head processors.

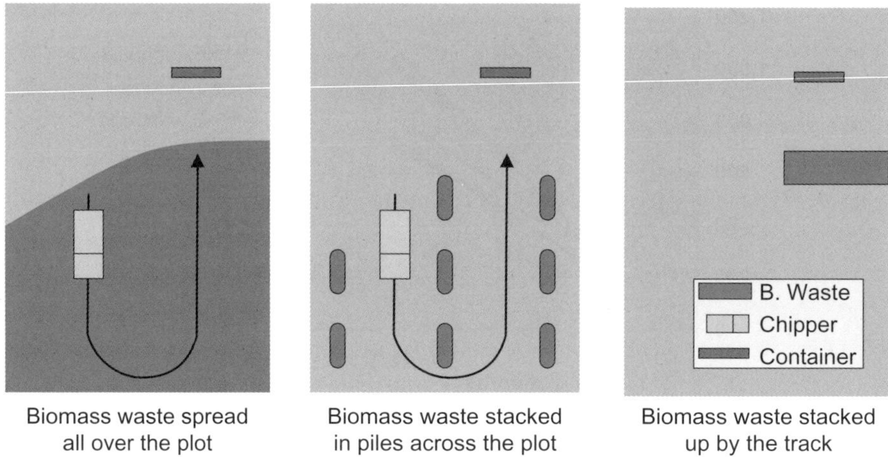

Fig. 15.5.   Collection scenarios.[21]

- Case 3: the remains are stacked next to the track for vehicles. A forest tractor is responsible for collecting the parcel and transporting the waste to a place where it can be processed. This situation is all the more desirable because of the need to reduce the moisture content of the biomass, for which previous stacking is required.

15.5.1   *Waste collection equipment*

From the source of production to the power plant, forest biomass passes through a series of stages which can have a great effect on the profitability of this fuel. Figure 15.6 shows these stages.

Forest biomass has a low density and high moisture content. These characteristics make collection and transportation of such fuel difficult.

This section briefly describes the equipment designed to increase the density of biomass in order to reduce transportation costs. Note that very high moisture also makes transporting it more expensive. An increase in the amount of water decreases the dry matter transported.

The technologies used to increase the density are chipping and compacting. Choosing either of these technologies is subject to variables such as orography, distance to the power station, type of biomass, and so on; all of them influencing the cost of taking biomass to the power plant.

15.5.1.1   *Chipping equipment*

The working process of chipping equipment consists of the following steps:

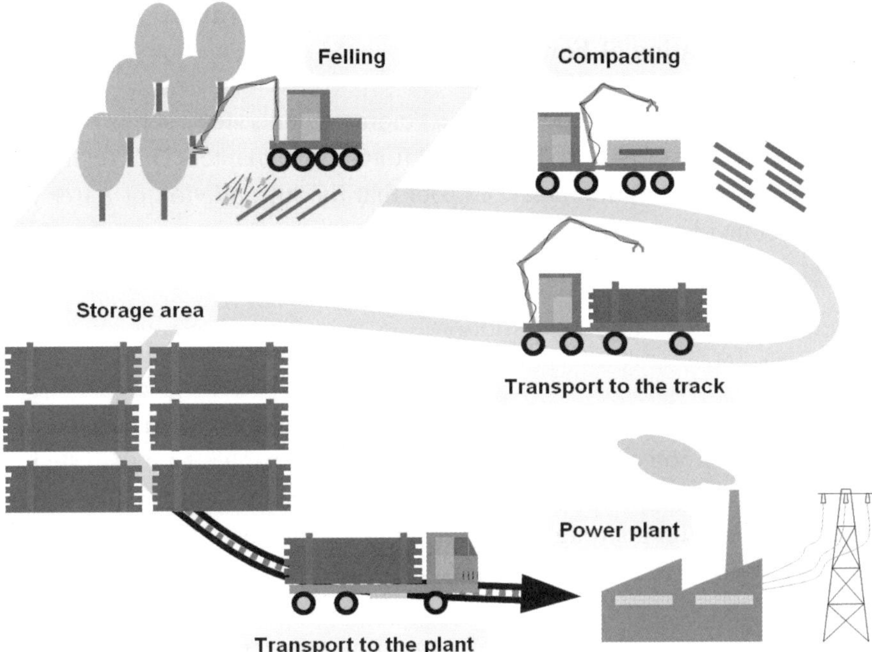

Fig. 15.6. Biomass logistics.

- Feeding.
- Chipping.
- Shipment of chips to the main container.
- Downloading chips on the container.
- Collection of containers by conventional truck.

These devices, rather than splintering the waste, increase its density by compacting it in cylindrical bales, of approximately 70 cm in diameter and of varying length. Because it does not require auxiliary equipment for its operation, it has the advantage of being highly flexible in daily work. It is a Finnish machine and therefore adapted to the specific characteristics of this country, both forestry and orographic. The equipment is mounted on a forest self-loader of the same brand, thus forming a fully integrated vehicle.

The cost difference between compacters and chippers depends on each situation, although as a general rule, the cost per green tonne is lower in compacting technology.[21]

15.6 Environmental Impact Resulting from the Generation and Exploitation of Forest Biomass

The development and exploitation of forest biomass generates a series of social and economic impacts discussed in the introduction to this chapter. This section delves more deeply into the environmental impact that the introduction of a tree species for use as fuel involves.

The section is divided into two parts:

- Consequences for carbon sequestration.
- Impact on soils.

The promotion of the global production policies and use of renewable energy have two crucial problems, the fight against climate change, as manifested in the Kyoto Protocol, and the reduction in fossil fuel use, which is a particularly sensitive issue to countries with a strong external energy dependence. These environmental and economic demands are also cross-linked by the need to converge to a sustainable development.

In the Kyoto Protocol, a series of commitments were established which the developed countries agreed to meet in terms of CO_2 emissions. Forest stands, whatever their purpose is, are CO_2 sinks that help to mitigate the greenhouse effect.

This section analyzes the amount of CO_2 stored in a forest of Eucalyptus globulus at two ages. In Table 15.12, the data on the weight percentage of each part of the species Eucalyptus globulus and their percentage in C are shown.[22]

15.6.1 *Forest waste*

Based on the percentages of C for each fraction, the amount of CO_2 accumulated can be estimated by knowing the biomass weight. This value is variable and depends

Table 15.12. Weight percentage of each component of a tree and their percentage in C.[22]

Component	% on the total	% of C	
		Adult stage	Young stage
Wood	76	46.9	46.2
Bark	9.7	43.5	42.8
Branches	6.47	46.8	46.4
Branchlets	1.65	48.2	46.1
Leaves	6.12	53.4	55.7

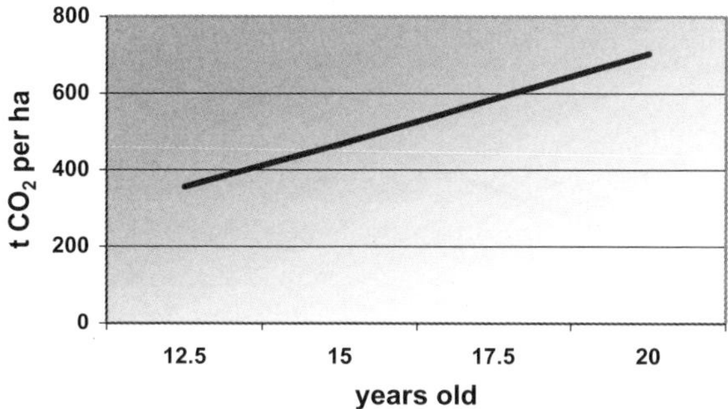

Fig. 15.7.   Amount of CO_2 stored per hectare in an adult stand of Eucalyptus globulus.[15]

Table 15.13.   Carbon accumulated in the soil by stands of E. globulus.[22]

Age	6–9 years (youth)	15–18 years (adult)
Amount of C (t/ha)	119.9	124.7

on the forest characteristics, i.e., the more biomass contained by the forest, the more CO_2 it accumulates. Figure 15.7 shows the cumulative amount of CO_2 in an adult forest in northern Spain.

In addition to capturing carbon in the tree, it is also sequestered in the soil. Table 15.13 shows the amount of C fixed in the case of Eucalyptus globulus stands.[22]

Both the removal of forest waste and forest exploitation for energy purposes generate a gradual impoverishment of the soil. To maintain the sustainability of the resource, preventive programs must be implemented to prevent this loss; either through artificial fertilization or restocking of the ash resulting from combustion of biomass.

In the case of conventional forest exploitation the waste is abandoned in the soil, and it provides the nutrients removed during the growth of the trees. This is mainly due to two causes:

- Harvesting periods exceeding 15 years, implying possible natural regeneration (atmospheric inputs, weathering of rock, . . .).[23]
- The highest concentration of nutrients being the leaves and bark.[22]

This section describes the case of the nutrient nitrogen, but the same is true for other nutrients including P, K, Ca and Mg. Thus, Fig. 15.8 shows the amount of

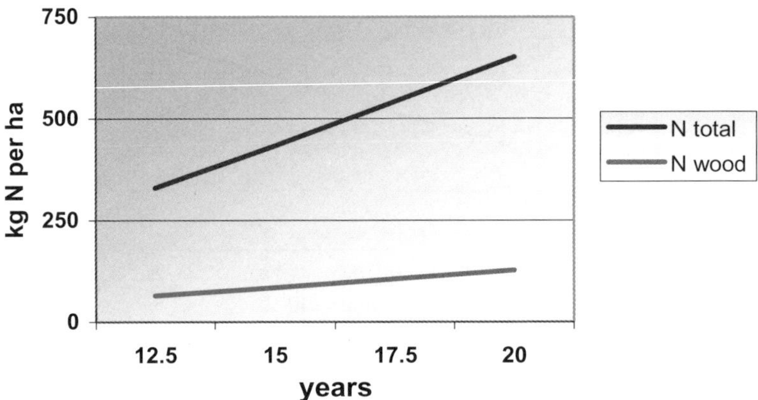

Fig. 15.8.   Amount of nitrogen extracted per hectare in a stand of Eucalyptus globulus at different ages.[15]

nitrogen corresponding to a stand of Eucalyptus globulus, per kg and hectare, when the entire tree is used and also when only the wood is extracted.

Figure 15.8 shows that N is an element that is found in low proportions in the wood. The difference between the amount of N extracted in wood exploitation and that extracted in whole tree exploitation, for an average age of 15 years, is approximately 348 kg of N per ha. The amount of N that is extracted with waste removal is much greater than if only the wood is removed. Taking into account natural inputs, Fig. 15.9 shows the amount of N needed by the soil, in the case of removing waste. It is to be noted that if only wood is removed, then it is not necessary to provide nitrogen.

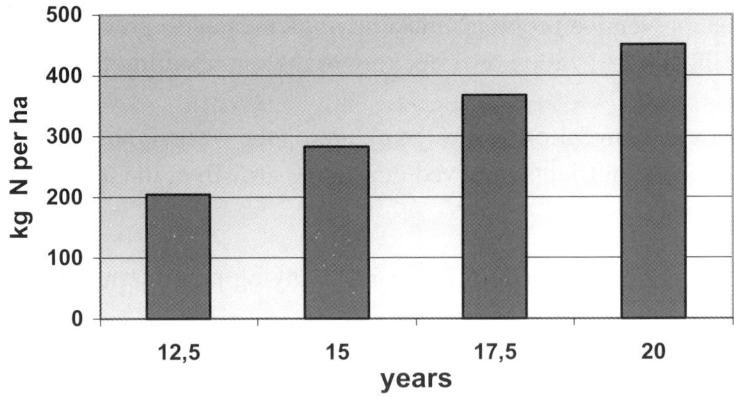

Fig. 15.9.   Amount of N to be replaced externally depending on age.

15.6.2 *Energy crops*

In the case of forest energy crops, the input from external fertilization is essential to ensure an acceptable yield. The difference between energy crops and forest wood exploitation is that the first one presents higher planting densities and much shorter harvesting periods. Several studies show that for short harvesting periods, taking advantage of the entire tree, the nutrients removal is much higher than in cases where only the wood is removed.[24,25] Three harvests of Eucalyptus globulus every seven years remove twice the amount of K, Ca and Mg than one every 21 years.[26] Figure 15.10 shows the amount of nitrogen extracted per hectare in Eucalypus stands for energy purposes.[15] It should be noted that the Eucalyptus nitens and Eucalyptus smithii species are those which extract most nitrogen. The same is true for other nutrients (P, K, Ca and Mg).

In the case of clones belonging to the Populus genus, the amount of nitrogen extracted by each tree is presented in Fig. 15.11.[27]

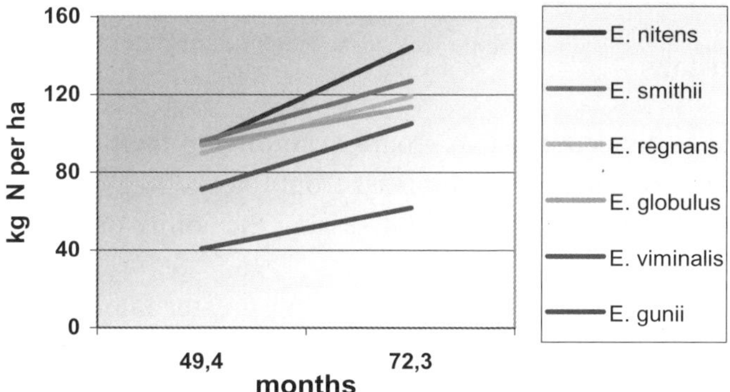

Fig. 15.10.   Amount of N harvested per ha and species depending on age.

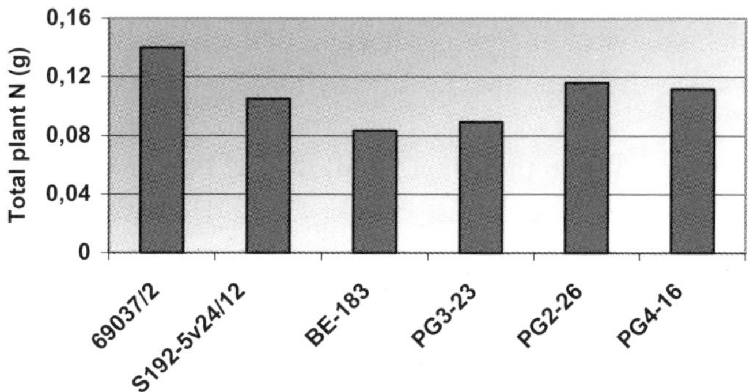

Fig. 15.11.   Amount of N extracted by six clones of Populus.[27]

Table 15.14.   Amount of nutrients removed per hectare and year in forestry and agricultural crops.[15,28]

	N	P	K	Ca	Mg
Wheat	56.4	10.8	10.8	1.1	4.6
Corn	87.1	18	23.9	1.3	5.3
Sunflower	45	7.5	13.4	2.7	5.9
Soybeans	142.8	16.1	46.4	7.1	6.5
E. nitens	7.50	1.25	2.23	0.45	0.98

From the fertilization point of view, energy crops require a lower amount than that needed in agricultural uses. Table 15.14 shows the amount of nutrients extracted by agricultural crops compared to the Eucalyptus nitens in kg per ha and year.

15.7   Conclusions

Forest biomass is divided into two groups according to their origin: biomass from traditional forest exploitation and biomass from forest energy crops.

Bioclimatic diagrams are tools for assessing the ability of a region to generate forest biomass.

Fast growing species are those which generate greater amount of biomass. There are equations that allow the biomass estimation for the different fractions that compose a tree, according to their height, diameter and base area. The leaves are the fraction that has the highest calorific value, and the bark is the one that presents the lower calorific value. Both fractions are those that more ashes generate and that greater concentration of nutrients accumulates. Combining productivity with specific energy, the amount of energy per hectare of each species (energy density), is obtained. The wastes from the species Eucalyptus are those that accumulate higher energy density.

The species belonging to the genera Willow and Poplar are commonly used in Europe and North America as forest energy crops. However, in temperate zones where drought is a limiting factor for growth in the months of spring and summer, the genus Eucalyptus is best suited to generate biomass in short rotations. Within this genus, the Eucalyptus nitens is the one that greatest amount of biomass generates at 72 months of age.

Forest biomass is a fuel that must be compacted or crushed to increase its density. This is necessary to reduce the cost per tonne transported. The way in which biomass is distributed has a major impact on the performance of the collection. There are two main technologies to increase the density: compacting and crushing.

From the environmental point of view, the exploitation of forest biomass as a fuel has two great effects. On the one hand, it helps to mitigate the greenhouse effect because the plants during their growth absorb CO_2 from the atmosphere, but the removal of forest waste and energy crops stands generate a gradual impoverishment of the soil if the nutrients are not replenished again. This is very important because it may question the sustainability of the system itself.

References

1. European commision, "External costs. Research results on socio environmental damages due to electricity and transport," Luxembourg: Office for Official Publicatios of the European Communities (2003).
2. J.L. Montero de Burgos and J.L. González Rebollar, "Diagramas bioclimáticos," ICONA, Ministerio de Agricultura, Pesca y Alimentación del Gobierno de España.
3. R.S. Zalesny, A.H. Wiese, E.O. Bauer and D.E. Riemenschneider, "Ex situ growth and biomass of Populus bioenergy crops irrigated and fertilized with landfill leachate," *Biomass and Bioenergy* **33** (2009) 62–69.
4. M. Stolarski, S. Szczukowski, J. Tworkowski and A. Klasa, "Productivity of seven clones of willow coppice in annual and quadrennial cutting cycles," *Biomass and Bioenergy* **32** (2008) 1227–1234.
5. I. Vande Walle, N. Van Camp, L. Van de Casteele, K. Verheyen and R. Lemeur, "Short-rotation forestry of birch, maple, poplar and willow in Flanders (Belgium) I — Biomass production after 4 years of tree growth," *Biomass and Bioenergy* **31** (2007) 267–275.
6. H. Sixto, M.J. Hernández, M. Barrio, J. Carrasco and I. Canellas, "Plantaciones del género Populus para la producción de biomasa con fines energéticos: Revisión," *Investigaciones agrarias: Sistemas y recursos forestales* **16** (2007) 1131–7965.
7. A. Merino, C. Rey, J. Brañas and R. Rodriguez, "Biomasa arbórea y acumulación de nutrientes en plantaciones de Pinus radiata," *Investigaciones agrarias: Sistemas y recursos forestales* **12** (2003) 85–98.
8. R.E.H. Sims, K. Senelwa, T. Maiava and B.T. Bullock, "Eucalyptus species for biomass energy in New Zealand," *Biomass and Bioenergy* **16** (1999) 199–205.
9. E.M. Birk and J. Turner, "Response of flooded gum (eucalyptus grandis) to intensive cultural treatments: Biomass and nutrient content of Eucalyptus plantations and native forests," *Forest Ecology and Management* **47** (1992) 1–28.
10. R.N. Cromer and D. Cameron, "Response to nutrients in Eucalyptus grandis: 1. Biomass accumulation. 2. Nitrogen accumulation," *Forest Ecology and Management* **62** (1993) 211–243.
11. D. Reed and M. Tomé, "Total aboveground biomass and net drymatter accumulation by plant component in young Eucalyptus globulus in response to irrigation," *Forest Ecology and Management* **103** (1998) 21–32.
12. F. Sanz Infante and V.G. Piñeiro, "Aprovechamiento de la biomasa forestal en la cadena monte-industria," *Revista Cismadera* **10** (2003) 6–27.
13. J. Brañas, F. González-Río, "Estimación de la biomasa arbórea en plantaciones de Eucalipto en el norte de Galicia y Asturias," Universidad de Santiago de Compostela.

14. C. Gracia, S. Sabaté, J. Vaireda and J. Ibañez, "Aboveground biomass expansion factors and biomass equations of forest in Catolonia," CREAF: Universidad de Barcelona (2002).
15. S. Pérez, "Potencial energético de la biomasa forestal en Cantabria y sus implicaciones medioambientales," Ph.d. Thesis, Universidad de Cantabria, Spain (2008).
16. I. Tubby and A. Armstrong, "Establishment and management of short rotation coppice," *Forestry Comission*, Practice Notes (2002).
17. K. Senelwa and R.E.H. Sims, "Fuel characteristics of short rotation forest biomass," *Biomass and Bioenergy* **17** (1999) 127–140.
18. D. Reed and M. Tomé, "Total above ground biomass and net dry matter accumulation by plant component in young Eucalyptus globulus in response to irrigation," *Forest Ecology and Management* **103** (1998) 21–32.
19. P.C. Gimeno, "Estudio de los poderes caloríficos de las especies forestales españolas del género Quercus en España," Tesis doctoral, Universidad de León, Spain (1990).
20. M. Labrecque, T.I. Teodorescu and S. Daigle, "Biomass productivity and wood energy of salix species after 2 years growth in sric fertilized with wastewater sludge," *Biomass and Bioenergy* **12** (1997) 409–417.
21. F. Sanz Infante and G. Piñeiro Veiras, "Aprovechamiento de la biomasa forestal producida por la cadena monte-industria," *Area de innovación y tecnología del CIS-madera* **10** (2003) 6–37.
22. J. Brañas, F. González-Río and A. Merino, "Contenido y distribución de nutrientes en plantaciones de Eucalyptus globulus del noroeste de la Península Ibérica," *Investigaciones agrarias* **9** (2002) 317–335.
23. E. Dambrine, J.A. Vega, T. Taboada and L. Rodriguez, "Bilans d'eléments minéraux dans de petits bassins versants forestiers de Galice (NW Espagne)," *Annals of Forest Science* **57** (2000) 23–28.
24. C.T. Smith, W.J. Dyck, P.N. Beets, P.D. Hodgkiss and A.T. Lowe, "Nutrition and productivity of Pinus radiata following harvest disturbance and fertilization of coastal sand dunes," *Forest Ecology and Management* **66** (1998) 5–38.
25. A.E. Tiarks and J.D. Haywood, "Site preparation and fertilization effects on growth of slash pine for two rotations," *Soil Science Society American J.* **60** (1996) 1654–1663.
26. H. Folster and P.K. Khanna, "Dynamic of nutrient supply in stand soils, 339–378 en E.K.S.," ed. Nambiar y A. G. Brown, Management of soil, nutrients and water in tropical plantation forest, ACIAR monograph 43, Caberra.
27. A. Karacic and M. Weih, "Variation in growth and resource utilisation among eight poplar clones grown under different irrigation and fertilisation regimes in Sweden," *Biomass and Bioenergy* **30** (2006) 115–124.
28. L. Ventimiglia, H.G. Carta and S.N. Rillo, "Exportación de nutrientes en campos agrícolas," *Informaciones agronómicas para el Cono sur No. 7*, INPOFOS Cono Sur, Buenos Aires Acassuso (2000).

Chapter 16

Bioethanol

Alfredo Ortiz[*], Severiano Pérez, Carlos J. Renedo,
Mario Mañana and Fernando Delgado

*Department of Electric and Energy Engineering,
University of Cantabria, 39005, Santander, Spain*
*ortizfa@unican.es

This chapter discusses different aspects of the production and utilization of bioethanol. It starts by reviewing the technical fundamentals of the various manufacturing systems, depending on the raw material used. It then presents statistical aspects of global production to continue establishing the strengths and weaknesses of their use compared to other conventional energy sources. Moreover, the environmental impacts of their manufacture and utilization are considered. It also sets the cost of investment and development in plants for manufacturing. Finally, it discusses the possible uses of bioethanol.

16.1 Technical Fundamentals

Ethyl alcohol or ethanol (C_2H_5OH) is an alcohol whose molecule has two carbon atoms, Fig. 16.1. It is a colorless liquid with a strong odor, which burns easily producing a small bright blue flame. It is obtained by distilling products from the fermentation of sugars or starch crops. Those are found in living matter such as cereals, sugar beet, sugar cane or biomass.

Besides it being part of many drinks such as wine, liquor, beer, etc., and having industrial applications, it is a good fuel that can be used alone or mixed with gasoline. This last use, and its vegetable origin, makes it a renewable energy source.

One can distinguish three groups of uses: biofuel, industrial and food (alcoholic beverages). Table 16.1 shows the percentage of these uses.[1]

Plants grow based on the process of photosynthesis, where sunlight, carbon dioxide from the atmosphere, water and nutrients from the soil form complex organic molecules that are concentrated in the fibrous part of the plant. Their composition

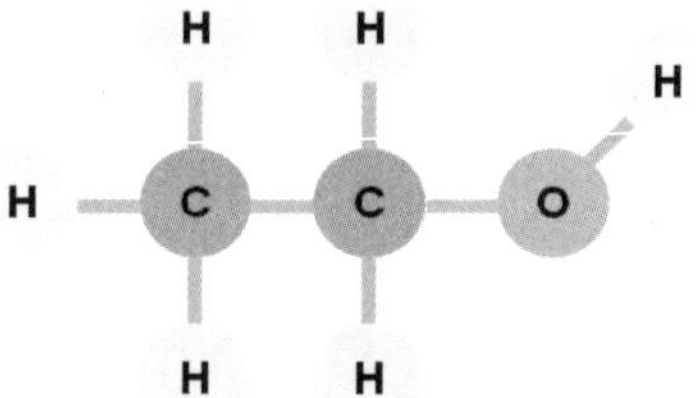

Fig. 16.1.　Ethanol molecule.

Table 16.1.　Global use of ethanol.[1]

Fuel	Industrial	Food (beverages)
70%	12%	18%

depends on their origin, but the highest percentages are made mainly up of three polymers:[2]

- Cellulose (35% to 50%), consists of long chains of glucose molecules (simple 6-carbon sugars) built on a solid three-dimensional crystal structure.
- Hemicellulose (20% to 35%), is a branched polymer composed primarily of xylose molecules (simple 5-carbon sugars) and other sugars.
- Lignin (10% to 25%), is not a carbohydrate but a rigid aromatic polymer.

The plant components most suitable for ethanol production are those that have high percentages of cellulose and hemicellulose, which are carbohydrates and can be transformed into simple sugars by their rupture. These are simple sugars which are converted into alcohol by yeasts and bacteria. The lignocellulosic biomass requires a lot of treatment in order to break the protective structure and split the cellulose into sugars, for which reason this process is not currently competitive. Thus, treatments for converting biomass into ethanol depend on its origin and composition. Some processes are commonly used, while others are still being developed.

Production potential is limited by the supply of raw materials. Therefore, it is linked to climatic factors, such as the amount of land dedicated to energy crops. This conflict disappears, or at least is minimized, if non-cultivated land or waste from agriculture, forestry, industrial or urban origin are used for ethanol production. Some examples of common waste are cereal straw, cereal or rice hulls, waste from cut down forests, or municipal solid waste (MSW). The waste has the advantage of its low cost, since it is not part of products or processes, except when used in cattle feeding. Due to the diverse origin of MSW, it can contain other materials whose separation preprocess could mean and excessive increase in the price of the ethanol obtained.

The use of ethanol as fuel has gone through several stages. In the origins of the automobile industry it was the main fuel for Otto cycle engines. It was later that the oil based industry was developed. With the oil crisis of the 70s new investments were made in order to develop the bioethanol economy, Brazil being the first country that began to mix gasoline and ethanol at a ratio of 22 to 78.[3] In 1979 Brazil produced the first cars that could run with hydrated alcohol, and in the 80s most of the cars were designed to run exclusively on ethanol.

Until the 80s the main motivation for ethanol production was its use as an alternative fuel to petroleum in automotive industry. It was in the mid-80s when the environmental interest appeared, mainly associated to the reduction of gaseous emissions. The growing interest in climate change caused by emissions of "greenhouse effect gases", has been behind the search for more environmentally friendly fuels. The combustion of bioethanol produces the same amount of CO_2 as the plant absorbed during its growth, with the exception of what is emitted due to the activity required in the process of its production, so the balance is CO_2 neutral.

As fuel, ethanol is used in two ways:

- In blends with gasoline (mainly in concentrations of 5 or 10%, E5 and E10 respectively).
- As a gasoline additive, as Ethyl Tertiary Butyl Ether (ETBE).

16.1.1 *Transformation of biomass into bioethanol*

For the production of bioethanol, three families of products are mainly used.

- Sugar, typically from cane or beet.
- Grains, through the fermentation of sugars from starch.
- Lignocellulosic biomass, by fermentation of the sugars contained in cellulose and hemicellulose.

Table 16.2[4] classifies the different materials used in ethanol production, showing the corresponding source and the typical uses. Then, Fig. 16.2 presents the basic process of ethanol production; whatever the initial raw material is, it consists of three phases:

- Hydrolysis. It consists in the addition of water to the raw material to obtain a sugar solution. With the respect to grain, it allows the extraction of oil. For cellulosic biomass, the addition of water can divide their complex molecular structure and turn them into simple sugars.

Table 16.2.   Biomass materials considered for bioethanol production.[4]

Material	Source	Uses
	Agriculture	
Grain straw, cobs, stalks husks	Grain harvesting	Animal feed, burning as fuel, composting, soil conditioning
Grain bran	Grain processing	Animal feed
Seeds, peels, stones, reject fruit	Fruit and vegetable harvesting	Animal feed, fish feed, seeds for oil extraction
Bagasse	Sugar cane industries	Burning as fuel
Sheels, husks, fiver, presscake	Oils and oilseed plants	Animal feed, fertilizer, burning as fuel
	Forestry	
Wood residues, bark, leaves	Logging	Soil conditioning and mulching, burning as fuel
Woodchips, shavings, sawdust	Milling	Pulp and paper, chip and fiber board
Fiber waste, sulphite liquor	Pulping	Use in pulp and board industries as fuel
Paper, cardboard, furniture	Municipal solid waste	Soil conditioning and mulching, burning as fuel

- Fermentation of sugars. There it consists of using yeast or bacteria to produce ethanol and CO_2 from the glucose, the fructose ($C_6H_{12}O_6$ both) and the sucrose ($C_{12}H_{22}O_{11}$) contained or released by the raw material.

$$C_{12}H_{22}O_{11} \text{(sucrose)} + H_2O \Rightarrow C_6H_{12}O_6 \text{ (glucose)}$$
$$+ C_6H_{12}O_6 \text{ (fructose)}, \quad (16.1)$$
$$C_6H_{12}O_6 \text{(glucose/fructose)} \Rightarrow 2\,C_2H_5OH \text{ (ethanol)} + 2\,CO_2. \quad (16.2)$$

Thus, in terms of molecular weight conversion, 180 kg of glucose can theoretically produce 92 kg of ethanol and 88 kg of carbon dioxide.

- Distillation of ethanol from the fermentation. The alcohol produced by fermentation contains a significant amount of water that must be removed before the fuel can be used. As the ethanol boiling point (78.3°C) is lower than that of water (100°C), the mixture is heated until the alcohol evaporates, condensing the bioethanol.

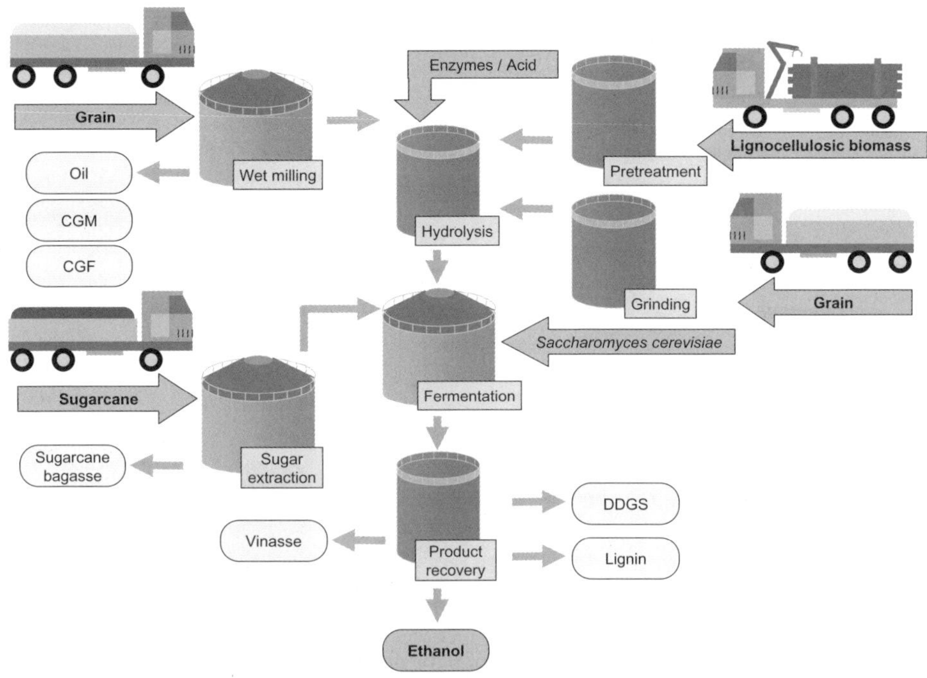

Fig. 16.2. Schematic overview of the process to produce bioethanol,[5] where CGM is corn gluten meal, CGF is corn gluten feed, and DDGS is dried distiller grains of solubles.

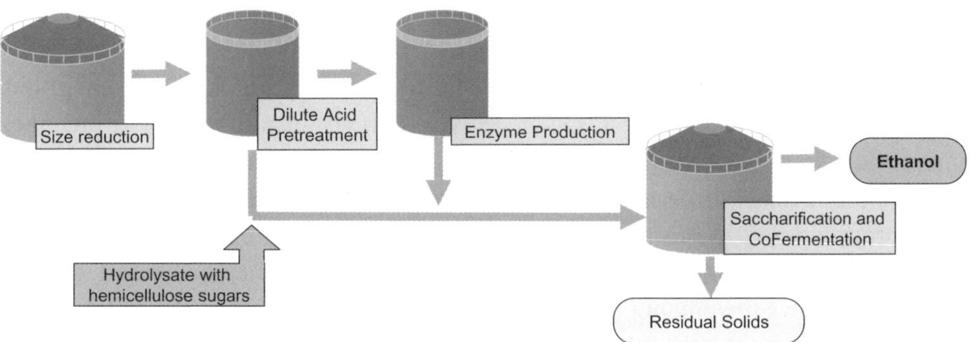

Fig. 16.3. Schematic overview of the process of simultaneous saccharification and distillation.

16.1.1.1 *Production of bioethanol from sugar*

More than half the current world production of bioethanol is based mainly on sugar crops.[1] Taking the production model based on sugar cane, saccharification would not be necessary, since the raw material contains enough sugar, Fig. 16.3. The process

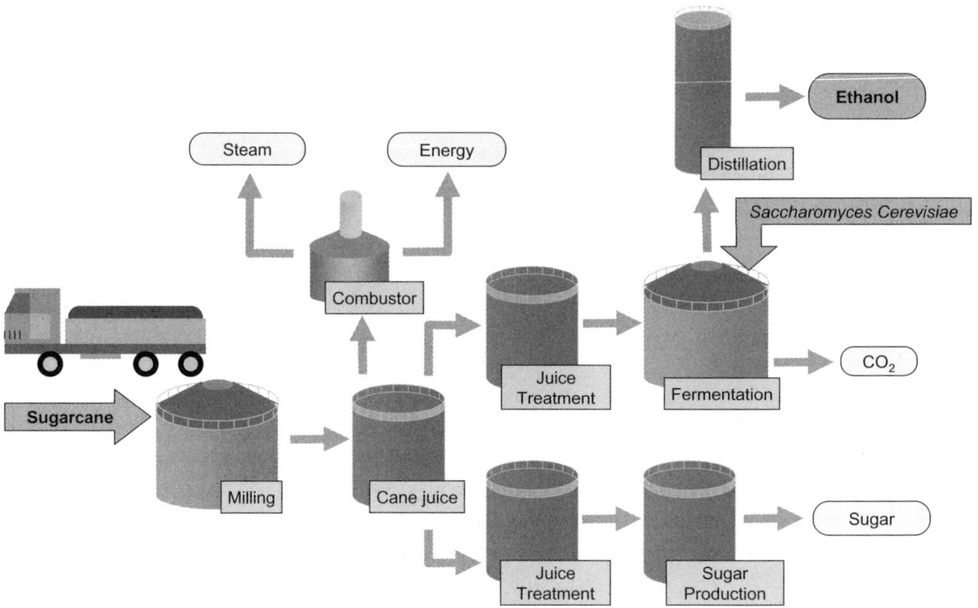

Fig. 16.4. Schematic overview of the process of producing bioethanol from sugar cane.

would start by grinding the cane to extract the juice of the sugar, and then performing a fermentation process, which, if it is carried out with all its products, only ethanol is produced.[6] Another option could be to take part or all of the juice in order to produce sugar, Fig. 16.4.

The microorganism mainly used for the fermentation of starch sugars and grains in industrial processes is a yeast, saccharomyces cerevisiae, although there are other microorganisms, bacteria and fungi, which can also ferment glucose to ethanol. The economic analysis on the fermentation of xylose, has identified higher ethanol concentrations and yields as the most important factors influencing production costs, increased volumetric productivity this being an important secondary target.[7] Table 16.3 shows the ethanol production by yeasts on different carbon sources.

After fermentation, the ethanol is separated from the resulting solution by distillation, a hydrated ethanol being obtained. Depending on it future use, a later stage of dehydration would be required in order to increase its concentration up to 99.8%.[6]

16.1.1.2 *Production of bioethanol from starch (grain)*

After the sugar products, most of the bioethanol is produced from grain, mainly corn.[1] It is of note that corn is used in the U.S., and barley and wheat in Spain.[9] The use of these raw materials for ethanol production reduces their availability in the

Table 16.3.   Ethanol productions by yeasts on different carbon sources.[8]

Yeast	Carbon source	Temperature (°C)	Fermentation time (h)	Maximum ethanol (g/l)
Saccharomyces cerevisiae	Glucose 200 g/l	30	94	91.8
Saccharomyces cerevisiae	Sucrose 220 g/l	28	96	96.7
Saccharomyces cerevisiae	Galactose 20–150 g/l	30	60	40.0
Saccharomyces cerevisiae	Molases 1.6–1.5 g/l	30	24	18.4
Saccharomyces pastorianus	Glucose 50 g/l	30	30	21.7
Saccharomyces Bayanus	Glucose 50 g/l	30	60	23.0
Kluyveromyces fragilis	Glucose 120 g/l	30	192	49.0
Kluyveromyces marxianus	Glucose 50 g/l	30	40	24.2
Candida utilis	Glucose 50 g/l	30	80	22.7

food market. Using this raw material, there are two possibilities for the preparation and production of sugars: wet milling and dry milling.[3,5,6]

Wet milling

This system is applied when using grain as the raw material, where other products are obtained, such as corn oil, gluten for feeding, corn steep liquor, syrup, fructose, dextrose, etc., in addition to producing bioethanol. It is used by approximately two-thirds of producers in the U.S.; it requires greater energy expenditure, is more expensive, and produces less ethanol, but it is more profitable by generating subproducts with economic value.

It is a complex process,[3,5,6] since many steps are necessary in the pretreatment of the grain and in the separation into its various components. The corn is soaked in large tanks in a solution containing small amounts of sulphur dioxide and lactic acid. These two chemicals in water at a temperature of about 50°C, help to soften the grains in a process that can last from one to two days. During this time the corn swells and softens. Slightly acidic conditions of the dissolution help to break down the proteins and to release the starch in the grain.

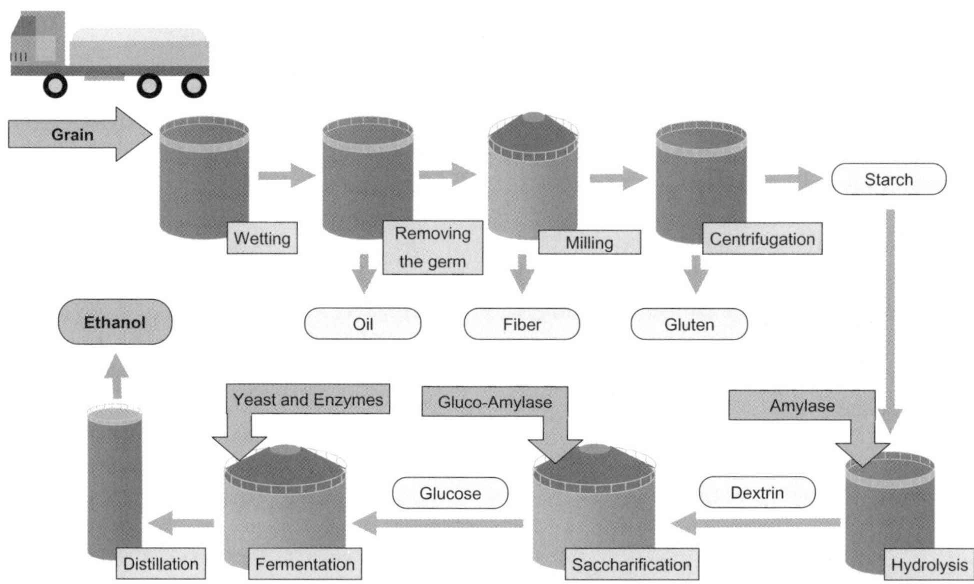

Fig. 16.5. Schematic overview of the bioethanol production process by wet milling of corn.

The next part of the process is to extract the germ. This is carried out by passing it through a separator where the germ of the grain floats (due to its oil content) and it can be collected easily. The softening of the grain facilitates the milling, which separates the fiber. A centrifuge separates protein and starch, of which ethanol is obtained by fermentation and subsequent distillation.[3,5,6] An outline of the process is shown in Fig. 16.5.

Dry milling

This process does not require a prior cereal pretreatment, but once cleaned, the grain is milled until broken down into fine particles. A flour is produced with the germ, fiber and starch of corn. To produce a "sweet" solution, the flour is hydrolyzed or converted into sucrose using enzymes or an acid dissolution. The mixture is cooled and yeast added before the fermentation and distillation processes can be started.[3,5,6]

It is possible to obtain syrup by centrifugation. From the resulting mass, after obtaining alcohol, a subproduct called Dried Destiller Grains of Solubles (DDGS) is obtained, which are distributed in the form of pellets and can be used as feed for livestock.[3,5,6] This technology is used in plants of small and medium size. Figure 16.6 shows the outline of the process.

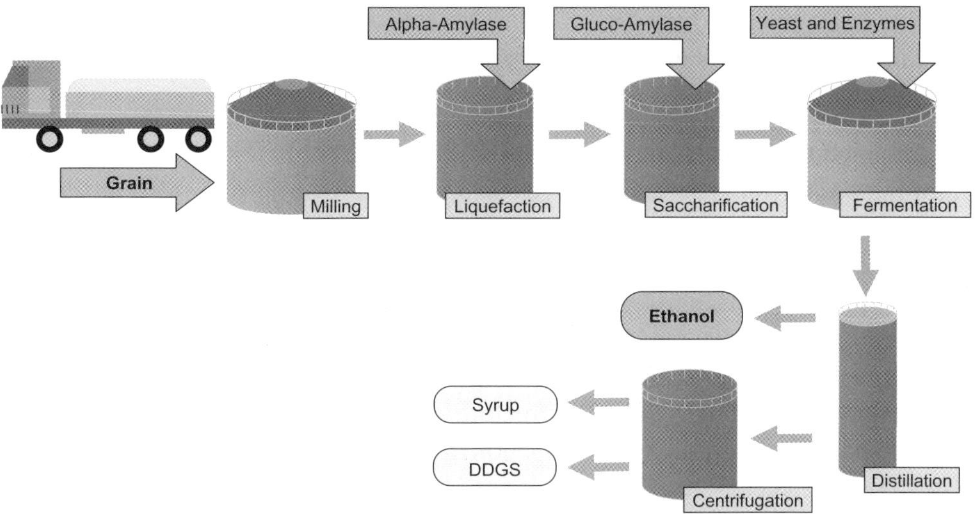

Fig. 16.6.   Schematic overview of the bioethanol production process by dry milling of corn

16.1.1.3   *Production of bioethanol from lignocellulosic materials*

There are many sources of lignocellulosic material such as forestry, horticultural and agricultural waste, industrial waste, the organic part of municipal solid waste. The first two are the main sources in the ethanol production.[5]

Those of forest origin are composed of waste from forest harvesting, from paper and wood industries and from forest energy crops. Within the forest genera producing this type of biomass, it can be distinguished the species Eucalyptus, Populus and Salix.

Among the species of herbaceous crops that are used for ethanol production are:

- *Cynara cardunculus* is more suitable for places where the growth-limiting factor is water. Their productions are around 14.5 tonnes of dry matter per ha.[10]
- *Miscanthus* is a perennial, rhizomatous, C4 grass, of the *Poaceae* Family, originates from Asia.[11] *Miscanthus* cropping is therefore attractive to renewable energy producers due to annual, high yields and the relative ease of husbandry using conventional equipment. It is also a crop with a low nitrogen requirement showing little biomass yield impact from fertilization.[12]
- Switchgrass (*panicum virgatum L.*) is a native warm-season, perennial grass indigenous to the Central and North Americantall-grass prairie into Canada. The plant is an immense biomass producer that can reach heights of 10 feet or more. Its high cellulosic content makes switchgrass a candidate for ethanol production as well as a combustion fuel source for power production. Their yields range

between 26 and 36.7 t per ha.[13,14] It has advantages as feedstock due to low levels of lignin, which reduces the severity and cost of pre-treatment versus forestry materials.[15]

Although the process may vary depending on the feedstock, obtaining ethanol from lignocellulosic materials requires an enzymatic hydrolysis, which breaks the links in the cellulose chains. This process is now very expensive because of the high cost of enzymes production.

As a result of the complex structure of lignocellulosic materials, these require a pretreatment to make the sugar substrate of the hydrolytic enzymes accessible. The pretreatments are: physical (increasing surface activity), chemical (solubilizing the lignin part and modifying the structure of the cellulose chains) or biological (attacking the lignin through the production of hydrolytic enzymes). Finally, through a process of fermentation, ethanol is synthesized from a microorganism culture in the presence of suitable nutrients.[3,5,6] Figure 16.7 shows the schematic overview of the bioethanol production process from lignocelullosic biomass.

To optimize the performance of the fermentation of lignocellulosic biomass, the microorganism needs to be capable not only of fermenting hexoses (glucose) but also pentose. Saccharomyces cerevisiae is not able to ferment pentose to ethanol. Obtaining ethanol from pentose is more complex, mainly due to the aeration factor and the need for fermenting organisms.[16,17]

Currently, there are many microorganisms capable of fermenting pentose or mixtures of pentose and hexoses but these have not yet reached an optimal level. This is the main reason why the production of ethanol from lignocellulosic biomass

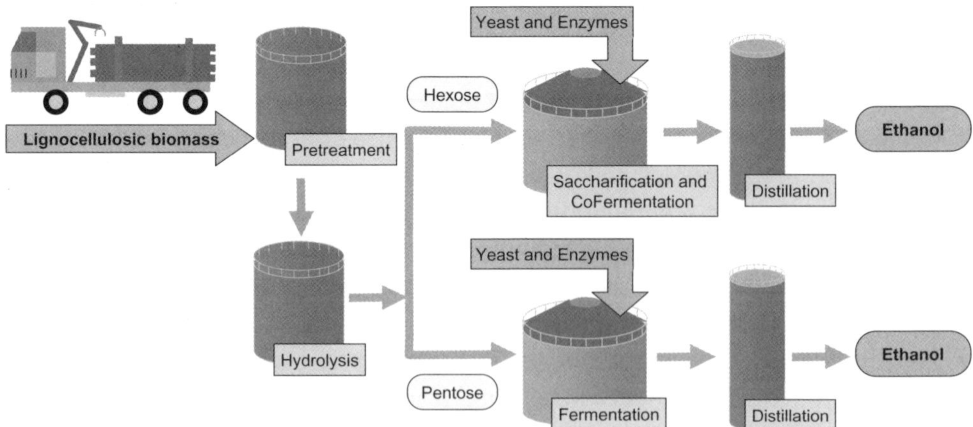

Fig. 16.7.   Schematic overview of the bioethanol production process from lignocelullosic biomass.

has not been commercially developed yet. However, the fact that the type of waste used as raw material has no economic value in the context where it is generated, in addition to the fact that its elimination may cause environmental problems, and that it those not in competition with the food market, show the great potential there is for lignocellulosic materials to be used as feedstock in the production of bioethanol.[18,19]

16.1.2 *Use of bioethanol*

The primary energy use of ethanol is in the automotive sector, and is combined with gasoline in two ways:[20]

- As an additive. In Europe, in order to eliminate leaded gasoline, it was decided to oxygenate gasoline by products derived from alcohols. The compounds derived from ethanol show some benefits over those derived from methanol, for instance, gross calorific value (PCS), higher octane index and lower vapor pressure. The use of bioethanol allows the use of this renewable resource as an additive in gasoline.
- As a pure fuel or blended with gasoline. The mix of up to 5% ethanol with gasoline, E-5 (5% bioethanol), need not to be differentiated from unblend gasoline. It is generally accepted that one can use the E-10 mixture (10% bioethanol) in internal combustion engines without the need for the engine to be modified in any way.

The use of mixtures in proportions greater than 10% requires the adaptation of the engine. Today, there are vehicles on the market called flexible fuel vehicle (FFV) that can use any mixture of up to 85% ethanol (E-85). The use of pure bioethanol (E-100) requires a special engine.

16.2 Level of Development

World production of ethanol has undergone a sustained growth in recent years, going from nearly 30000 million liters in 2001 to nearly 66000 million liters in 2008, Table 16.4.[21]

According to international statistics, Brazil has historically been the largest world producer, but production in the U.S. has experienced greater increases, and in 2007 U.S. was placed as a world leader. The Eurobserver web offers detailed statistics on European Union biofuel production.[22]

The challenge for ethanol is to increase the profitability of the production process if a real industrial development is to be reached. This requires action in four ways:

- Improvement of process technologies for both the raw materials currently used (sugar and starch), and for other alternative raw materials (lignocellulose, agricultural waste, ...).

Table 16.4. World production of ethanol in millions of litters.[21]

Country	2004	2005	2006	2007	2008
Brazil	15100	16001	17000	18999	24500
U.S.	13381	16141	18378	24601	34070
China	3649	3801	3850	1840	1900
European Union	2249	2521	2983	2158	2777
Canada	231	231	579	799	900
Others	6159	7298	5932	1196	1500
Total	40769	45993	48722	49593	65647

- Revaluation of the products obtained during the production process to improve the economic viability of the process. The products are different depending on the type of material used.
- Advancement of other energy uses for bioethanol. Currently, there are developing processes for the production of electricity in fuel cells.
- Development of policies to encourage production and consumption.

In this way, the experience of bioethanol production from the two world leaders, Brazil and USA, is briefly discussed.

Brazil is a pioneer in production and export of ethanol. This industry has generated 2.5 million jobs and saves annually the country nearly two billion dollars in gasoline imports. As sugarcane is the raw material, and it is collected on a seasonal basis, the Government is trying to create a major stock that will provide a supply of ethanol throughout the year.[23]

The experience of using ethanol as fuel in Brazil started in the 30s with a mixture of 5% in gasoline. In the 70s, again due to higher oil prices, it was decided to increase the percentage of mixture gradually, reaching 20% in the early 80s. At present, most vehicles use only ethanol as fuel. Currently the government is not pushing for vehicles using solely ethanol, but for those of flexible combustion, which allow the use of any portion of hydrated alcohol, depending on the relative prices at service stations. The flexible combustion engine has been developed by the largest automaker in Brazil.

With regard to the Brazilian experience, what should be highlighted is the government's political wish to promote sustainable growth in ethanol consumption. This implies the creation of incentives in all the stages, namely, production, marketing and use of bioethanol.

In the U.S. ethanol is primarily produced from the starch contained in grains such as corn, grain sorghum, and wheat. Despite being the world's largest producer, it has needed to import 2.1 billion liters. Strong growth in demand for ethanol is expected thanks to the banning of MTBE and the requirement of a minimum level of renewable fuels.

Ethanol industry in the U.S. continues to set records for production. In 2007 they had 139 refineries with a capacity of 30 billion liters per year. These numbers are increasing substantially and there are currently 196 refineries with a capacity of 47.5 billion liters per year. There is also another 7100 million liters per year corresponding to facilities in construction.[21]

16.3 Strengths and Weaknesses

In this section, strengths and weaknesses of bioethanol are analyzed, compared to other traditional energy sources.

16.3.1 *Advantages compared to traditional sources*

The bioethanol uses presents advantages for both the system and the users.

16.3.1.1 *For the system*

The advantages for the system are the following ones:

- It contributes to the diversification and security of energy supply of car fuels, which reduces dependence on a foreign energy supply to the transport sector.
- It helps to develop the rural economy, creating jobs, fixing the population in rural areas, reducing the amount of abandoned farmland, etc.
- The added value given to the raw materials traditionally used in the food market. It prevents a drop in price caused by surplus production. The production of bioethanol may lead to a real alternative for farmers.
- It reduces exhaust emissions. Ethanol contains 35% oxygen, which improves the combustion of fuel and therefore reduces emissions of carbon monoxide. In addition, bioethanol contains no sulphur, so it produces no sulphur dioxide.

16.3.1.2 *For the users*

The advantages for the users are the following ones:

- A spill is biodegradable, and presents no greater risks in its use than fossil fuels.
- It provides a better and smooth functioning of the engine as it is the highest octane fuel of the market. Also, ethanol in blends promotes a cleaner burning since it leaves no residue in the fuel circuit and acts as an antifreeze.

- It has a higher vaporization heat than gasoline, resulting in a reduction in the maximum temperature of combustion, which increases engine efficiency.

16.3.2 *Disadvantages compared to traditional sources*

The bioethanol uses presents disadvantages for both the system and the users.

16.3.2.1 *For the system*

The disadvantages for the system are the following ones:

- Currently, obtaining bioethanol has a cost greater than the purchase of a litre of gasoline in the oil markets. In order to eliminate this additional cost for the client the state reduces taxes on biofuels, with the consequent loss of revenues.
- The oil companies are opposed to the addition of an ethanol blend in gasoline, since it reduces their involvement in the business. In today's market the entry of ethanol blends with gasoline is premature: they require additional logistics and there is in certain markets an overproduction of gasoline (the result of having to satisfy the growing market for diesel).
- There is an increase in evaporative emissions caused by use of ethanol blends-fuel. Although ethanol has a relatively low vapor pressure, when used as a gasoline additive, its effective vapor pressure is high. By increasing the vapor pressure, it increases its volatility, and hence emissions in the filling of fuel tanks.

Another effect that increases evaporative emissions is the increase in fuel permeability through the plastic and rubber components of the circuits feeding the engines.

- Ethanol shows an affinity for water, and the presence of small amounts of water in the ethanol-gasoline blends can cause separation into two phases, which reduces engine efficiency. This requires an alignment of the storage and transport systems so as to be completely free of water.

16.3.2.2 *For the users*

The disadvantages for the users are the following ones:

- The mixtures above 10% (E-15, E-25) typically require small adjustments to the engine.
- The use of E-85 and above requires adapted engines (flexible fuel vehicle, FFV).

16.3.3 *How to overcome the disadvantages*

In order to overcome the disadvantages of bioethanol, it is necessary to take into account the following elements:

- The price of bioethanol can be reduced via improvements in the productive process, and developing processes that enable the processing of lignocellulosic raw materials and industrial waste.
- To ensure compliance with the specifications for ethanol — fuel blends as to the limits of vapor pressure, one solution is the use of gasoline with lower vapor pressures so that the mixture meets these parameters.

This requires the elimination, or at least reducing the amount of butanes and pentanes in gasoline, making the process of producing gasoline more expensive.

- Storage and transportation facilities, as well as the operation modes to prevent contamination with water, can be adjusted. Generally, it is not advisable to carry these mixtures in pipes, using trucks to carry bioethanol up to the distribution points. Additives can also be used to increase ethanol's water receiving capacity without it involving a separation of gasoline.
- Engines to be used with mixtures should be adjusted. Some manufacturers are already doing this for the use of mixtures in low proportion.
- Establishment of measures to promote and support both the production and use of bioethanol and their blends (subsidies, information campaigns, etc.).

16.3.4 *Horizons to overcome barriers*

In this section, the next barriers to be overcome by ethanol technology are analyzed.

16.3.4.1 *Production costs*

Over the past 20 years the construction costs of a bioethanol production plant using the dry milling process have fallen between 25 and 30%. Production costs have been reduced to almost 50%.[20]

There is a growing trend in the levels of agricultural and industrial efficiency in the production of bioethanol from sugar cane in recent years. Studies of the International Energy Agency for the United States and European Union show a decreasing trend in the cost of bioethanol production between 10 and 15% for 2010. If one takes into account the upward trend in oil prices, ethanol could well achieve a similar production cost per equivalent liter of gasoline.[20]

In the case of bioethanol production from lignocellulosic materials, the IEA expects that it will have decreased by 30% in 2010, compared to the estimated cost in

2002. Although the cost would still be above the cost per equivalent liter gasoline, this would significantly expand the availability of raw material for producing ethanol.[20]

16.3.4.2 *Evaporative emissions*

This topic is currently under review by a study commissioned by the European Commission, but it can be generally stated that the control of evaporative emissions is based on:

- Control of the mixtures vapor pressure through the use of gasoline with lower vapor pressure.
- Use of effective systems to reduce emissions in vehicles. In this regard, in recent years automobile manufacturers have implemented emission control systems which are proving very effective: for example, active carbon canisters, perfectly prepared to absorb evaporative emissions of ethanol and other compounds.

16.3.4.3 *Logistical problems*

Cases like those of Brazil, US and Sweden indicate that the adaptation of systems for storage and transportation of ethanol and their mixtures can be carried out without major problems, it requiring only a willingness to do so.

16.3.4.4 *Vehicle adaptation*

Programs to promote the use of ethanol in Brazil or the United States have shown that it is entirely feasible to adapt the engines of vehicles for ethanol blends. The development of FFV or adapted vehicles for use of E-95 and E-100 can be overcome with political determination and social awareness.

16.4 Environmental Impact

For ethanol three types of impacts are considered: engine emissions in the exhaust after combustion, accidental discharges of fuel and evaporative emissions.

16.4.1 *Engine emissions*

Applying the Life Cycle Assessment (LCA) to the fuel (production, transport and fuel consumption), the emissions from an engine are lower if ethanol is used instead of gasoline.[20] This is because it is produced from a renewable source. The analysis depends on the raw materials, the process and products obtained from the latter.

How the different levels of emissions will be affected by the addition of bioethanol to petrol is directly related to the engine model used and its regulation. This means that the comments made below are general and are not always valid.

- Carbon dioxide (CO_2): the balance of CO_2 emitted is considered neutral because it was captured during the growth of plants used as feedstock.

The use of fossil fuels releases CO_2 into the atmosphere, fixed over millions of years, resulting in a net increase in emissions. If the vegetation is renewed in the same proportion as it degrades, the use of biofuels is part of a cycle of continuous movement, and it does not increase CO_2 emissions.

- Carbon monoxide (CO): oxygenated gasoline, such as bioethanol blends, emit lower levels of CO, because the presence of oxygen allows a more complete combustion process.
- Particulates: bioethanol reduces emissions of particulates, especially fine ones, those that have a greater influence on the health problems of children, the elderly and those people with respiratory problems.
- Sulphur dioxide (SO_2): as bioethanol contains no sulphur, a reduction in these emissions in proportion to the presence of bioethanol in the mixture will be achieved.
- Acetaldehyde and Formaldehyde: the use of bioethanol slightly increases levels of acetaldehyde and peroxiacetil nitrates. However, the emission of these compounds is by compensating the reductions in emissions of formaldehyde, since the toxicity of the latter is higher than that of acetaldehyde.
- Nitrogen Oxides (NO_x): The use of bioethanol represents only a small increase in NO_x emissions. Modern vehicle engines automatically regulate the air/fuel ratio depending on the level of oxygen in the fuel, so it is not a change from conventional fuels.

16.4.2 *Accidental discharges*

Bioethanol is currently more secure than petrol when accidental discharges occur, since it degrades rapidly in surface, subterranean and soil water.

16.4.3 *Evaporative emissions*

Evaporative emissions come from the vehicle volatile organic compounds (VOC) and they do not come from the exhaust pipe of the engine. The largest sources of evaporative emissions are given off by fuel storage and transport systems.

The use of bioethanol-fuel blends involves an increase in these emissions due to the combination of a higher vapor pressure, and an increased permeability of the fuel through the plastic and rubber components.

16.4.4 Environmental assessment: "Ecotest"

The "Ecotest" has been promoted by the International Automobile Federation, and is assessed by the German Automobile Club (ADAC).[24] The environmental impact of a vehicle in urban and interurban operation is assessed using two indexes, one for CO_2, and one for the other pollutants (CO, HC, NO_x and PM). The final classification is given with a growing number of stars in terms of lower emissions. In general, gasoline vehicles are better considered by this method than the diesel ones. Those are only improved by the hybrids and those using natural gas as fuel.

The use of pure ethanol or mixed with gasoline in different percentages reduces emissions of CO_2, CO, particulates and toxic organic compounds. The potential increase of NO_x will have a negligible result with respect to the effect of the other reductions. The results are dependent on the percentage of ethanol in petrol. The use of increasing rates will push these vehicles from the 3-star to the 4-star category.[24]

It is difficult at this time to make an overall assessment of the impacts because there is not sufficient data regarding emissions not associated with combustion. In any case, there does exist potential for converting average impact vehicles in low impact vehicles.

16.4.5 Emissions: Comparison E85/petrol

A recent LCA of biofuels in Spain, carried out by CIEMAT[25] on behalf of the Ministry of Environment has concluded that the production, distribution and use of biofuels allow the following reductions in greenhouse gas emissions:

- A mixture of gasoline with 85% bioethanol (E85) allows a 70% reduction in emissions of greenhouse gas emissions (CO_2eq) per kilometer compared to gasoline.
- A mixture of gasoline with only a 5% bioethanol allows a reduction in emissions 3%.
- The pure biodiesel (B100) reduces greenhouse gas emissions (CO_2eq) between 57% (biodiesel from raw vegetable oils) and 88% (biodiesel from used vegetable oils) per kilometer compared to fossil diesel. A mixture of diesel with a 10% biodiesel (B10) allows a reduction in emissions of between 6% and 9% respectively.

16.5 Economics

From the economic point of view, biofuels are more expensive than fossil fuels.[20] This is mainly due to the raw material cost, which represents a high percentage of the biofuel end price.

The biofuels' market is not conventional, as their development is closely linked to the total or partial reduction of the tax rate on oil products.

16.5.1 *Investment levels*

Generally, the investment in ethanol production plants is relatively high, representing the second largest cost, after the raw materials. Investment costs have been steadily declining in recent years due to the standardization of equipment and this trend is expected to continue in the medium term.

Investment costs can vary depending on the selected raw material, the combination of the plant and the existing facilities, or the development of the country. For reference, the capital costs of the major plants in operation in Spain are provided in Table 16.5.[26]

16.5.2 *Production costs*

In order to calculate the production costs of bioethanol it is necessary to take into account the following elements:

- Amortization of the investment costs.
- Operating and maintenance costs: the main element being the cost of raw material (in this case approximately 60% of total) with a yield of 379 liters of ethanol per tonne of wheat grain.

Table 16.5. Investment levels of bioethanol production in Spain.[26]

Plant	Operation starting date	Investment cost (M€)	Production capacity (M liters/year)
Ecocarburantes españoles Cartagena (Spain)	2000	70	100 (cereal) 50 (wine alcohol)
Bioetanol Galicia A Coruña (Spain)	2002	85	126 (cereal) 50 (wine alcohol)
Biocarburantes de Castilla y León Salamanca (Spain)	2006	150	195 (cereal) 5 (lignocellulose biomass)

Table 16.6. Annual average prices of raw materials for ethanol production in Spain (€100 kg).

Products	2003	2004	2005	2006	2007	2008
Wheat	13.80	14.15	13.96	13.93	21.03	21.89
Corn	14.79	14.67	13.50	15.19	20.45	18.20
Barley	12.15	12.63	13.28	12.57	18.36	16.97
Sugar beet	5.88	6.08	5.50	4.37	3.51	3.71

- The rest of operating costs: labor, energy consumption, consumables, maintenance, etc.
- Additional incomes from sale of subproducts such as the Destilled Dried Grain Soluble (DDGS). The performance of DDGS from wheat is 1.35 kg/liter of bietanol produced.

The sharp rise experienced by oil prices in recent times is reducing the differential cost of bioethanol production. However, in order to ensure the economic viability of the bioethanol production, governmental grants are still needed.

Regarding the stability of the prices of raw materials, there are significant fluctuations depending on the campaign, changes in crops grants, and so on. For reference, the annual average prices received by farmers in the last six years in Spain are presented in Table 16.6.[27]

Global prospects for increased production of biofuels and bioethanol in particular, point to tensions between the use of raw materials for food production and its use for energy purposes. So, fluctuations and the high price of cereals and sugar beet has led to the use of other raw materials which are not regulated by the food market as lignocellulosic materials (agricultural waste, yard pruning, etc.), and other crops rich in sugar or starch that does not compete in the food market.

This means that cost optimization and increased production of bioethanol is related to the development of new processes capable of producing bioethanol from alternative and low-cost raw materials, such as lignocellulosic materials, organic waste of different types, and so on. Table 16.7 presents the costs distribution and its percentage in a US ethanol plant.[28]

Global prospects for increased production of biofuels and bioethanol in particular, point to tensions between the use of raw materials for food production and its use for energy purposes. So, fluctuations and the high price of cereals and sugar beet has led to the use of other raw materials which are not regulated by the food market as lignocellulosic materials (agricultural waste, yard pruning, etc.), and other crops rich in sugar or starch that does not compete in the food market.

Table 16.7. Operating costs for yellow poplar sawdust-ethanol in the United States.[28]

Input	Production cost (cent euro/liter)	Production cost (% of total)
Biomass feedstocks	6.98	60
Chemicals	1.51	13
Nutrients	1.17	10.1
Diesel	0.17	1.5
Makeup water	0.17	1.5
Utility chemicals	0.23	1.9
Solid waste disposal	0.23	1.9
Electricity credit	−1.36	−11.7
Fixed costs	2.55	21.9
Total	11.64	

This means that cost optimization and increased production of bioethanol is related to the development of new processes capable of producing bioethanol from alternative and low-cost raw materials, such as lignocellulosic materials, organic waste of different types, and so on. Table 16.7 presents the costs distribution and its percentage in a US ethanol plant.[28]

16.5.3 *Support system*

The support measures generally applied are based on the following points:

- Grants to pilot or industrial plants.
- Partial or complete exemption from taxes.
- Promoting the use: applying mandatory percentages or banning MTBE in favor of ETBE.
- Agricultural policy to reduce the cost of raw materials for energy use.

It is worth remembering that the tax on oil is an important source of funding for states, and many reservations have been raised by finance ministers about the reduction or exemption of the tax on biofuels. Without a reduction in taxes, the price of biofuels is still higher than the price of fossil fuels.

16.6 Combination with Conventional Sources

Currently, the energy purposes of ethanol are based on its use as a fuel (either alone or mixed with gasoline), as an additive (both petrol and diesel) and as a source of hydrogen for fuel cells.

16.6.1   *Bioethanol as fuel*

The bioethanol can be used as fuel or as a mixture with gasoline. This use is analyzed in this section.

16.6.1.1   *Ethanol fuel only*

The first vehicle that was designed to use ethanol was built to be used on farms, so that their owners could produce alcohol from the fermentation of corn. Later models were developed to use both ethanol and gasoline.

Currently, in order to make explosion gasoline vehicles run on ethanol, a number of amendments are required; these including the tank, fuel pipes, injectors and engine management system. In recent years, a number of vehicles able to run either on gasoline or on ethanol, or a mixture of both, have come on to the market. FFV adapt the carburetion of the engine based on the ethanol/gasoline rate of the fuel. They adjust the amount of air introduced to optimize engine operation.

Ethanol as the only fuel is used mainly in Brazil and Argentina. Its use in temperatures below 15°C could lead to ignition problems, so for this not to happen the most common solution is to add a small amount of gasoline. The mixture most widely used is the E85, which is composed of 85% ethanol and 15% gasoline.

Unfortunately, ethanol contains less energy per liter than gasoline, so, in order to provide a similar performance to petrol, the compression ratio should be increased.

Ethanol has a much higher octane than gasoline, around 110, which does not burn so efficiently in conventional engines. Continued use of fuels with a high proportion of ethanol, such as E85, produces corrosion on metal parts and damages rubber components.

16.6.1.2   *Direct mixture of ethanol and gasoline*

Ignition engines can operate on blends up to 10% dehydrated alcohol, without modifications to the engine, being necessary. Mixtures up to 25% may require minor modifications.

If there are no modifications to the engines, the vehicle behavior varies with respect to conventional fuel in the following aspects:

- Reducing power and torque (about 2% for mixtures with 15%).
- Increased consumption (4% for mixtures of 15%).
- Increased corrosion of metal parts and rubber components.

However, if the engine is adjusted by increasing the compression ratio, and adapting carburetion to the new stoichiometric relation, the following is achieved:

- Increased horsepower and torque (9% with a mixture of 20%).
- Reduced consumption (7% compared to gasoline only).
- Improved combustion with a lower rate of carbonization and emission reduction of polluting gases (CO and HC).

16.6.2 *Bioethanol as an additive in gasoline: ETBE*

Another alternative application of bioethanol is its use as a gasoline additive.[29,30] With the introduction of unleaded petrol, for which the use of catalytic converters is necessary to reduce harmful emissions, the octane number of gasoline is reduced.

Gasoline can explode for two reasons: the appearance of a spark or high temperature and pressure conditions. In engines, the explosion should be avoided until the spark plug creates the spark. The octane number is the antiexplosion capacity of a gasoline. The higher the number, the greater the pressure and temperature needed to detonate the fuel without a spark.

After the elimination of lead from gasoline, and to recover the octane, oxygenate additives such as methanol, ethanol, tertiary butyl alcohol (TBA) or tertiary butyl-methyl ether (MTBE), were added.

Initially MTBE obtained from petroleum was used. In recent years, ETBE, which is produced by a catalytic reaction of isobutene and ethanol at a ratio of 1:0.8, is becoming a better option because it produces less toxic emissions.

ETBE is a compound with superior qualities to both MTBE and bioethanol. Its advantages are:

- Low solubility in water, lower than MTBE, and therefore greater resistance to the "phase separation".
- Lower oxygen content (15.7%) than MTBE (18.2%), so no need to modify the carburetor.
- Lower vapor pressure (0.27 bar) than MTBE (0.54 bar) and ethanol (1.22 bar).
- Emissions reduction of carbon monoxide and non-burned hydrocarbons.
- Lower corrosive power than alcohols.
- Increased calorific value.
- Improved manufacturing performance respect to the MTBE.

For these reasons its use is gaining ground in Europe, prevailing over the mix with gasoline. In Spain all the ethanol dedicated to the automotive sector is converted to ETBE, and it is being used in blends up to 15% of the gasoline volume (ETBE15).

16.6.3 *Diesel additive to improve the ignition*

For diesel engines the diesel-ethanol blend, known as E-diesel, contains up to 15% ethanol. Compared with regular diesel, it significantly reduces emissions of particulates and other pollutants, and it improves the characteristics of the cold start.[31]

This product is currently in the development stage and is not marketed. Some companies such as Abengoa Bioenergy R&D are working to eliminate major technical and regulatory barriers to its commercialization.

16.6.4 *Fuel cells*

Electric vehicles powered by fuel cells are considered by many experts as the most promising technology for road transport in the medium term. Although currently they still require further development, they are expected to have high energy efficiency. In a fuel cell, chemical energy, which comes from the combination of hydrogen and oxygen, is converted directly into electricity without going through a combustion process, the only emission into the atmosphere being steam.[32]

Fuel cells for vehicles need to operate under certain restrictions on weight, size, temperature and low maintenance, and will be demanded a rapid response.

Hydrogen is the best fuel for these devices, however the security measures and the installation cost associated with production, storage, transport, adaptation of service stations and depots of vehicles opens the door to the use of alternative fuels.

Ethanol may be one of these alternatives, since it can be obtained from renewable sources and it is much easier to handle than hydrogen. Currently, research on this application works primarily in two ways:

- The reforming of ethanol to obtain hydrogen, both in the vehicle and in the service station.
- The direct use of ethanol in fuel cells.

16.7 Conclusions

Different aspects of the production and utilization of bioethanol have been discussed in this chapter. The technical fundamentals of the various manufacturing systems, depending on the raw material have been studied. The ethanol production technologies should be further developed to achieve efficiency levels that make competitive this fuel.

Statistical aspects of global production have been presented. In recent years the growth of ethanol production has been very important, U.S. having overtaken Brazil as the leading producer.

Strengths and weaknesses of their use compared to other conventional energy sources have been established. For the full development of this energy source it is necessary to establish a regulatory framework and an efficient marketing and distribution system.

Moreover, the environmental impacts of their manufacture and utilization have been considered. The use of ethanol and other renewable energy resources will contribute to the reduction of fossil fuels consumption and of the greenhouse gases emissions, palliating climate change.

References

1. FO-Licht 2004 World Ethanol and biofuels report.
2. Sudo *et al.*, *Biomass Handbook*, Chap 5.3 (Gordon and Breach, New York, 1989).
3. J.M. García and J.A. García, "Biocarburantes líquidos: Biodiesel y bioetanol, Madi+d," http://www.madrimasd.org/informacionidi/biblioteca/publicacion/doc/VT/vt4_Biocarburantes_liquidos_biodiesel_y_bioetanol.pdf.
4. R.L. Howard *et al.*, "Lignocellulose biotechnology: Issues of bioconversion and enzyme production," *African Journal Biotechnology* **2** (2003) 603.
5. C. Wyman, *Encyclopedia of Energy*, vol. 2, (Elsevier, 2004).
6. Las Energías Renovables en España, CENER, Ed Fundación Gas Natural (2006).
7. N.D. Hinman, *et al.* "Xylose Fermentation: An economic analysis," *Applied Biochemical Biotechnology* **20** (1989) 391–410.
8. D. Mousdale, *Biofuels: Biotechnology, Chemistry and Sustainable Development* (Taylor and Francis Group, 2008).
9. Y. Lechón, H. Cabal and R. Saez, "Life cycle analysis of wheat and barley crops for bioethanol production in Spain," *Int. J. Agriculture Resources* **4** (2005) 113–122.
10. L.G. Angelini *et al.*, "Long-term evaluation of biomass production and quality of two cardoon (Cynara cardunculus L.) cultivars for energy use," *Biomass and Bioenergy* **33** (2009) 810–816.
11. M. Matamura and T. Yukimura, "Fundamental studies on artificial propagation by seeding useful wild grasses in Japan. VI. Germination behaviour of three native species of genus Miscanthus," *Research Bulletin of Faculty of Agriculture Gifu University* **38** (1995) 339–349.
12. M. Himken *et al.*, "Cultivation of Miscanthus under west European conditions: Seasonal changes in dry matter production, nutrient uptake and remobilization," *Plant and Soil* **189** (1997) 117–126.
13. D.R. Keshwani and J.J. Cheng, "Switchgrass for bioethanol and other value-added applications: A review," *Bioresource Technology* **100** (2009) 1515–1523.
14. W.E. Thomason *et al.*, "Switchgrass response to harvest frequency and time and rate of applied nitrogen," *J. Plant Nutrition* **27** (2004) 1199–1226.
15. M.A. Sanderson *et al.*, "Switchgrass as a sustainable bioenergy crop," *Bioresource Technology* **56** (1996) 83–93.
16. I.S. Goldstein, *Organic Chemicals for Biomass* (CRC Press, Boca Raton FL, 1981).
17. N.W. Lutzen *et al.*, "Cellulose and their application in the conversion of lignocellulose to fermentable sugars," *Philosophical Trans. Royal Society London* **300** (1983) 283–284.
18. P. Mishra and A. Singh, "Microbial pentose utilization," *Advances in Applied Microbiology* **39** (1993) 91–152.
19. T.W. Jeffries and Y.S. Jin, "Ethanol and Thermotolerance in the bioconversion of xylose by yeast," *Advances in Applied Microbiology* **47** (2000) 221–268.

20. IEA (International Energy Agency), "Biofuels for transport: An international perspective," May 2004, http://www.iea.org/textbase/nppdf/free/2004/biofuels2004.pdf.
21. Renewable Fuel Association, http://www.ethanolrfa.org.
22. Biofuels barometer, eurobserver, http://www.eurobserv-er.org/.
23. J. Goldemberg, "The Brazilian biofuels industry," http://www.biotechnologyforbiofuels.com/content/1/1/6.
24. German Automobile Club (ADAC), http://www.adac.de/Tests/Autotest/Ecotest/default.asp? Quereinstieg=EcoTest.
25. Centro de Investigaciones Energéticas Medioambientales y Tecnológicas (CIEMAT), Análisis del ciclo de vida de combustibles alternativos para el transporte (2005), http://cindoc.ciemat.es/adjuntos_documentos/BioetanolCiemat2005.pdf.
26. Abengoa web, http://www.abengoabioenergy.com/sites/bioenergy/en/.
27. Spanish Environment Ministry 2008, http://www.marm.es/ http://www.mapa.es/es/estadistica/pags/PreciosPercibidos/indicadores/indicadores_precios.htm.
28. R. Wooley *et al.*, "Procces desing and costing of bioethanol technology: A tool for determining the status and direction of research and development," *Biotechnology Progress* **15** (1999) 794.
29. Y. ChunDe *et al.*, "Experimental investigation of effects of bio-additives on fuel economy of the gasoline engine," *Sci. China Ser. E-Tech Sci.* **51** (2008) 1177–1185.
30. E. Weber *et al*, "Addition of an azeotropic ETBE/ethanol mixture in eurosuper-type gasolinas," *Fuel* **85** (2006) 2567–2577.
31. E. Weber *et al.*, "Effect of ethers and ether/ethanol additives on the physicochemical properties of diesel fuel and on engine tests," *Fuel* (2006) 815–822.
32. E. Orucu *et al.*, "Investigation of ethanol conversion for hydrogen fuel cells using computer simulations," *J. Chemical Technology and Biotechnology* **80** (2005) 1103–1110.

Chapter 17

Biodiesel

Carlos J. Renedo[*], Alfredo Ortiz,
Severiano Pérez, Mario Mañana and Inmaculada Fernández

*Department of Electric and Energy Engineering,
University of Cantabria, 39005, Santander, Spain*
[*]*renedoc@unican.es*

This chapter deals with biodiesel which represents one of the alternatives to diesel for the automotive and industrial sector. Firstly, the main characteristics that make biodiesel an attractive biofuel are explained. The use of this fuel could help to reduce the world's dependence on petrol. Secondly, the chapter details the raw materials which are used to obtain biodiesel and their principal advantages and disadvantages. Moreover, the processes of biodiesel industrial production are showed. Thirdly, the level of utilization of biodiesel in the world and specifically in the European Union, which is the leader in biodiesel production, are commented on. Once the situation of biodiesel development is detailed, both the advantages and disadvantages, and the crucial challenges of biodiesel production are explained. Finally, the chapter shows some issues of production costs of biodiesel and its conclusions.

17.1 Technical Fundamentals

Biodiesel is a liquid fuel produced from vegetable oils and animal fats. The properties of biodiesel are virtually the same as for diesel fuel in terms of density and cetane number.[1] It does, however, present a higher flashpoint and can be used in blends with fossil diesel, and individually if engines are suitably adapted. Thus, biodiesel is an alternative to diesel for both the automotive and industrial sector. With respect to its nomenclature, the letter B is typically used followed by the percentage of mixture (so, for pure biodiesel it is B100, and B20 for a diesel with a 20% addition of biodiesel).

At first, the term biodiesel is used as a generic name that brings together various products used as an alternative to diesel fuel in diesel engines: vegetable oils, mixtures of diesel and vegetable oils, ethyl or methyl esters of oils and/or vegetable or animal fats and its blends with diesel, and so on.

The definition of biodiesel specifications proposed by the ASTM (American Society for Testing and Materials Standard) describes it as mono alkyl esters from long chain fatty acids derived from renewable lipids such as vegetable oils or animal fats, and used in compression spark ignition engines.

Table 17.1 shows a comparison of the properties of diesel fuel, vegetable oils and their esters. It can be seen here that the characteristics of the esters are more similar to that of diesel than to that of unchanged vegetable oil.[2]

Vegetable oils also have a high viscosity, containing fatty acids, phospholipids and other impurities that hinder their direct use as fuel. In order to avoid the engine modifications that are required for use, and substantially improve their characteristics as fuel, they are transformed into methyl or ethyl esters, which have similar properties to diesel.[3]

The conversion of oil to methyl or ethyl ester can be carried out by means of different processes, including transesterification, pyrolysis and emulsification. The process for ester formation from any oils or animal fats is transesterification, and esterification if it is based on fatty acids.

Table 17.1. Differences between the properties of diesel, common vegetable oils and biodiesel.

Property	Diesel oil	Sunflower oil	Sunflower methyl ester	Rapeseed oil	Rapeseed methyl ester
Density at 15°C (kg/l)	0.84	0.92	0.89	0.9	0.883
Flash point (°C)	63	215	183	200	153
Kinematic viscosity at 37.8°C (mm²/s)	3.2	35	4.2	39	4.8
Cetane Number	45–50	33	47–51	35–40	52
Combustion heat (MJ/kg)	44	39.5	40	—	40
Turbidity point (°C)	0–3	−6.6	3	—	−3
Sulfur (% weight)	0.3	<0.01	<0.01	<0.01	<0.01
Carbon residue (% weight)	0.2	0.42	0.05	—	—
On set of volatilitation (°C)	70	280	70	280	70
End of volatilitation (°C)	260	520	260	520	250

The methyl esters derived from vegetable oils have some physical characteristics and properties similar to diesel, which allows them to be mixed in any proportion and used in conventional diesel vehicles, without making major changes to the basic design of the engine However, in proportions of more than 5% it is necessary to replace the material used in the fuel circuit by another more resistant one, as the solvent power of biodiesel can cause its deterioration.[3]

17.1.1 *A renewable resource*

Fuels derived from crops and other organic matter, called biofuels, have a number of advantages. Firstly, they can help to reduce the growth in CO_2 emissions, making it possible to fulfil Kyoto Protocol commitments. The CO_2 balance of biofuels is low (it is not zero as the energy consumed in the production process has to be taken into account). On the other hand, they reduce dependence on oil by diversifying and improving security in the supply of fuel. Also, these fuels can be alternative sources of income for rural areas.

The production of vegetable oils is possible from more than 300 different species. Although crops like olive, coconut,, etc., have the highest yields of oil, the complexity of their harvesting operations makes them less interesting than arable crops.[3,4] In this case, the soil and climatic conditions, the agricultural yields, the oil content of seeds and the mechanization possibilities limit the species to a few profitable oilseeds.

The raw materials currently used for obtaining vegetable oils in biodiesel production are the seeds of conventional oleaginous plants. These are followed, in terms of volume, by palm and coconut oils, animal fats and used oils.

As biofuel production is based on vegetable products, the characteristics of agricultural markets should be taken into account. At the same time, the energy market will exert a definitive influence on the viability of the product. In this sense, it is noteworthy that the biofuels industry depends not only on the local availability of raw material, but also on the existence of sufficient demand.

The development of the biofuel market can help to promote other agriculture policies, fostering job creation in the primary sector, establishing population in rural areas, developing agricultural and industrial activities, and, at the same time, reducing the effects of desertification by planting energy crops.

17.1.1.1 *Conventional oleaginous crops*

The crops used to produce biodiesel in each country are different due to climate reasons. For example, in the south of Europe sunflower (Helianthus annuus) is the crop used conventionally whereas in the north use rapeseed (Brassica napus).[5] Other countries farm other crops such as soybean (United States, Brazil or Argentina),[6] coconut (Philippines) or palm (Malaysia and Indonesia).

Oils differ mainly in their fatty acids content. Those with high proportions of unsaturated fatty acids such as sunflower oil improve the viability of biodiesel at low temperatures, but decrease its resistance to oxidation, resulting in a high iodine value. For this reason, the possibility of genetically modified species should be considered, so as to optimize the properties of the biodiesel produced, as is the case of the high oleic sunflower oil.[7,8]

The agricultural yield and its energy balance depends on if land is irrigated or not. In the last case the weather conditions play a crucial role in crop production mainly when there are droughts. Table 17.2 shows the production of sunflower, rapeseed and soybean in Spain.[9]

At this time, the increased production of vegetable oils is used in the food sector (of the total output, less than a quarter goes to other sectors: production of glycerin, hydraulic oils, paints, industrial lubricants, etc.).

The use of these oils to produce biodiesel in Europe has been associated with the regulations established by European agricultural policy, which allows the cultivation of oil seeds at reasonable prices. However, the dedication of lands exclusively to the production of energy raw materials poses a risk, as these areas vary in time because the set-aside scheme depends on supply and demand for food grains.

So, the raw materials for biodiesel compete in the food market, suffering instability in prices. In these circumstances, it seems logical that research should be directed at the development of other raw materials. These materials ought to provide

Table 17.2.   Average area under cultivation and production (Spain).[9]

		Sunflower	Rapeseed	Soybean
Area under cultivation (ha)	1992 to 2002	1175012	42349	4667
	2002	763082	4911	536
	2008	724488	11397	590
Irrigated/rainfed (%)	1992 to 2002	21 / 79	75 / 25	96 / 4
	2002	5 / 95	14 / 86	97 / 3
	2008	8 / 92	10 / 90	1 / 99
kg/ha Irrigated	1992 to 2002	1731	1430	2120
	2002	2007	n.a.	n.a.
	2008	2073	2209	n.a.
kg/ha Rainfed	1992 to 2002	705	1022	1082
	2002	1092	n.a.	n.a.
	2008	1200	1903	n.a.

more advantages in the production process and also in biofuel functionality in order to obtain higher agricultural yields. These characteristics will get competitive prices outside the food sector.

17.1.1.2 *Alternative oleaginous crops*

Independently of conventional vegetable oils, a large number of oleaginous species non-oriented towards the food sector are being tested and their potential use as energy crops evaluated.[10] Species are better adapted to the conditions of the country where they grow. The crops are aimed at maximizing the return on oil production and reduce the costs of seed production.

The main energy crops oriented towards biodiesel production at present are primarily the following: Camelina sativa, Crambe abyssinica, Jatropha curcas, Pogianus, and in Spain Brassica carinata and Cynara cardunculus.

The use of wild species to produce biodiesel needs to solve some significant problems which include: irregular and staggered ripening, spontaneous opening of the fruits, etc. For this reason, these wild species should fulfil a set of requirements in order to provide competitive advantages:

- Are not subject to prices set by the food market.
- Adapt to the soil and climatic conditions of the location.
- Show a high agronomic potential and oil yield.
- Are in the development stage, with some problems already overcome, for instance: elimination of dehiscence problems, regulated maturation, reduced levels of glucosinolates, etc.
- Provide an adequate fatty acid profile for the production process or introduce a biofuel functionality.

17.1.1.3 *Animal fats*

Although its use is not widespread, animal fat, particularly that of the cow or buffalo, can also be used as feedstock for the transesterification of biodiesel. Previously, the lower quality tallow was used in the production of feed for animals. However, after the problems experienced with mad cow disease, its use in cattle feed has been limited.[11] The legal requirement to incinerate bovine remains with specified risk means that, given the high costs of transport and incineration, processing into biodiesel is an attractive alternative.

17.1.1.4 *Used frying oils*

There are other materials that also contain triglycerides to obtain biodiesel. The oil used for frying is one of the alternatives with the best prospects in the production of biodiesel. Not only is it the cheapest raw material, easily available in comparison with the new vegetable oils, but its use renders waste treatment costs unnecessary.[12] Countries of southern Europe such as Spain or Italy are big consumers of vegetable oils, focusing on the consumption of olive and sunflower oils. On the other hand, these oils have a low level of reuse and therefore do not suffer major changes and are good candidates for use as biofuel.

This oil, despite its vegetal origin, is not biodegradable, so it becomes a dangerous pollutant that causes difficulties in sewage treatment plants. There is a big logistical problem for the collection of oil. So far, only a small percentage of these oils is included as controlled release (in Spain about 20%) and used as feedstock in the manufacture of biodiesel and soap. Spain is trying to create a system for collecting frying oil, olein and fat in three stages: industrial, catering and domestic.

Although the use of frying oil has economic and environmental benefits it presents some challenges that must be taken into account:

- Poor collection system, currently hampering the supply of the plants (improvements in the collection system imply, in general, an increase in the cost).
- Low yields of the process due to difficulties in the optimization of biodiesel production, which are mainly caused by the inherent heterogeneity of the waste.

17.1.1.5 *Olein*

Olein (or soap pastes or acid oils) are also used as a raw material for obtaining biodiesel. The olein is a waste product from the refining process of vegetable oils. They consist of a mixture of triglycerides and free fatty acids, the latter exceeding 50%. Traditionally, its use has been as an ingredient in animal feed.[13]

Being a waste, its availability and composition homogeneity fluctuates. Its use presents the advantage of upgrading a low cost subproduct and the inconvenience of a process that requires a more aggressive purification treatment and decoloration of the resulting biodiesel to ensure compliance with its quality standard.

17.1.1.6 *Oils from other sources*

It is interesting to highlight the production of lipid, with similar compositions to vegetable oils, by microbial processes from algae, bacteria and fungi, as well as from microalgae.[14]

Finally, it should be noted that the use of bioethanol and biomethanol in the process of esterification of these oils for biodiesel production is being studied, as is the development of crops for energy purposes, and not for food.

17.1.2 *Transformation mechanisms of the resource into energy*

17.1.2.1 *Reactions in biodiesel production*

The production of biodiesel is a well known chemical process. Although based on the free fatty acid content of raw material there can be different production processes. Although the esterification process is possible, the method used commercially to obtain biodiesel is transesterification. This consists of the reaction, in the presence of a suitable catalyst, of the oil or fat with a low molecular weight alcohol, typically methanol or ethanol. In this process, depending on the alcohol used, ethyl or methyl esters of fatty acids (biodiesel) are obtained, with glycerine as a subproduct.[15,16]

Transesterification reactions of triglycerides

This is based on the reaction of triglyceride molecules, which are the primary component of vegetable oil or animal fat, with low molecular weight alcohols (methanol, ethanol, propanol, butanol) to produce esters and glycerine.[17]

The transesterification reaction, Fig. 17.1, develops in a molar ratio of alcohol to triglyceride, 3 to 1. As it is a two-way reaction, a quantity of alcohol is added to displace the reaction towards the formation of methyl ester. In addition, the formation of the basis of the glycerine, immiscible with methyl esters, plays an important role in shifting the reaction towards the right, reaching conversions close to 100%.

Chemically, the reaction actually consists of three consecutive reversible reactions where the triglyceride is converted sequentially into diglyceride, monoglyceride, and finally into glycerin, Fig. 17.2 In each of the three reactions a mol of methionine is generated. These are reversible reactions, so, to move the reaction towards the production of methyl esters, there must be an excess of methanol for this to be carried out.[3]

Depending on the raw material, the basic process of biodiesel production can undergo differences[1,3,4]:

- From conventional vegetable oils: the crude vegetable oil must undergo a refining process to remove free fatty acids, waxes and gums that can interfere with the

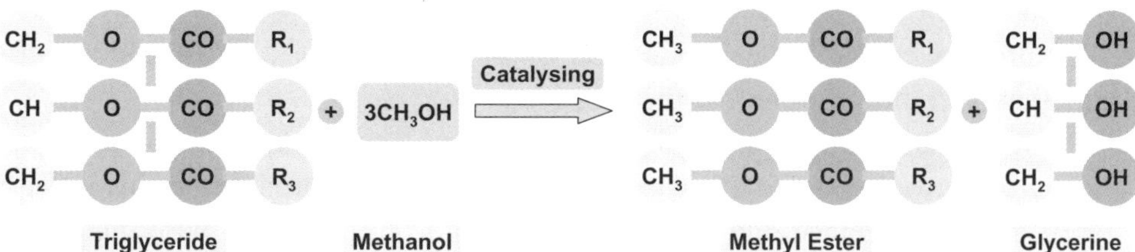

Fig. 17.1. Transesterification reaction.

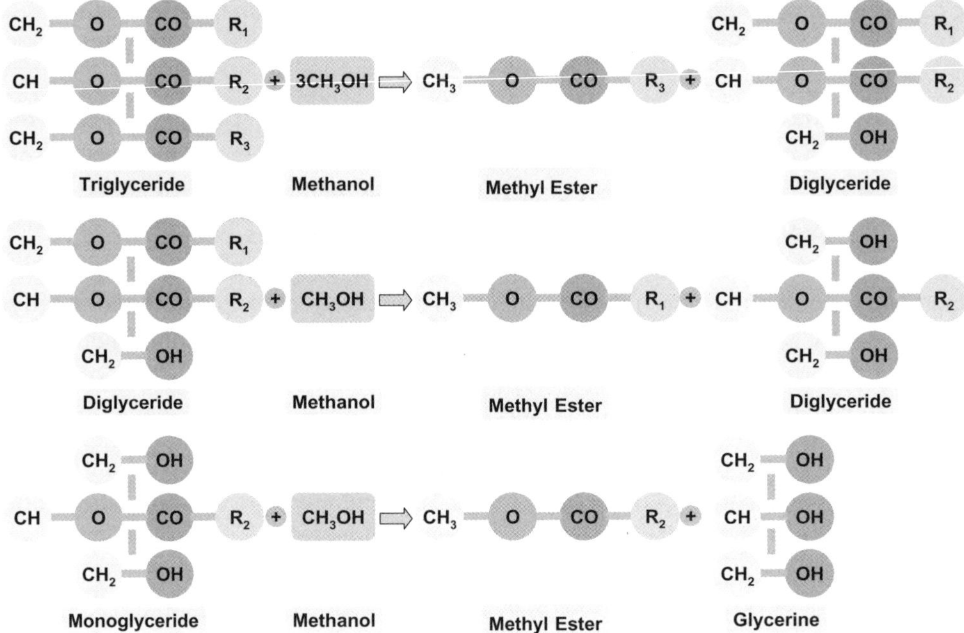

Fig. 17.2.   Reactions involved in the transesterification.

process. This supplies the raw material that will power the transesterification unit.

- From frying oils: frying oils should be, first, degummed, dried and cleaned of impurities by centrifugation and filtration processes. This stage is critical to ensure the smooth operation of the post-process. Depending on the free acids, a pre-esterification with methanol is required (in the presence of an acid, usually sulfuric), which can improve the transesterification yield by reducing acidity and minimizing the formation of soaps.

From here on the processes are similar, although the stages of purification and decoloration of both biodiesel and glycerine tend to be more complex in the case of used oils, since, despite the pre-cleaning, there are always impurities left.[1,3,4]

- The process requires the preparation of the mixture of catalyst and alcohol (methanol in this case) before the reaction.
- The transesterification reaction is brought about by raising the temperature (typically to between 60–70°C). The analysis of the matter transfer in this process is of great importance due to the inmiscibility of triglycerides with methanol,, and to the inmiscibility of methyl esters with glycerine, so that there are phase separations through in the process.
- In a first decantation, the crude biodiesel is separated from the glycerin.

- Crude biodiesel is subjected to a washing process to separate one part of the non-reacting alcohol and to remove impurities. Subsequent drying makes it possible to obtain biodiesel. Other operations of purification, such as discoloration, can be necessary to ensure compliance with marketing standards. The recovered methanol is reused in the process.
- The transesterification process produces as a subproduct about 10% glycerin. This crude glycerin contains impurities from crude oil, fractions of the catalyst, mono and diglycerides and traces of methanol. In order to sell it on the market it must be purified.
- The methanol recovered at this stage is reused in the process.
- In the glycerine purification fatty acids are separated. Those can be esterified once again to produce more biodiesel, or be used as raw material for soap production or for animal feed.
- The glycerine obtained after the washing process has technical quality. If, later on, it is distilled, the glycerine obtained is of pharmaceutical grade (99.8%). The glycerine can be channeled into traditional uses (cosmetics, pharmaceutical, food) or modern ones (animal feed, plastics industry, etc.).

The diagram shown in Fig. 17.3 is an overview of biodiesel production.[4]

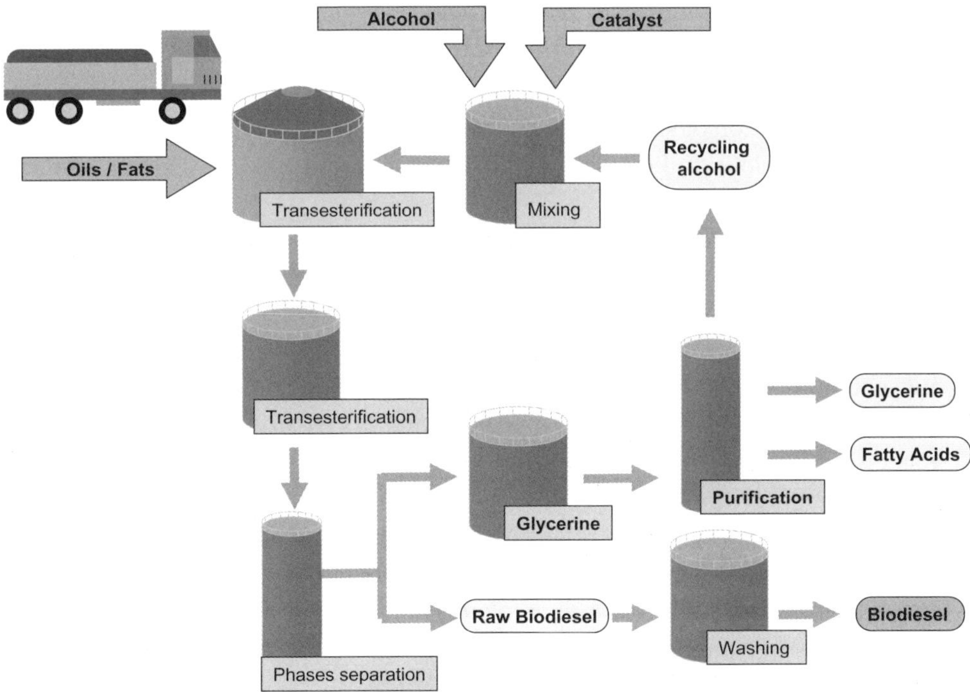

Fig. 17.3.   Scheme for biodiesel production.[4]

Apart from the general scheme for diesel production, on the market there are different types of technologies that propose industrial processes with several variants: continuous or batch, with recovery of fatty acids or not, with recovery of the catalyst in the form of salts that can be reused as fertilizer; with previous esterification phases to minimize the free acidity of the oils, etc.[1]

Plants between 500–10000 t/year usually work with batch systems in two stages. The advantage of this type of process is its low investment cost and flexibility. In contrast, they have trouble in achieving consistency in product quality and the process is not very reliable.

For over 10000 t/year, continuous reactors are normally used, which provide greater consistency, greater security and better design options for the reactor and subprocesses such as separation and glycerine purification.

In addition to this processing type, alternative optimization is being developed, such as the use of heterogeneous catalysts that facilitate the refining of the glycerine obtained or the use of enzymatic pathways for the production of biodiesel.

Variables affecting the transesterification reaction

Among the most important variables affecting the transesterification reaction are the following ones:

- *Fatty acids and moisture content.* The complete reaction requires a value of free fatty acids (FFA) of less than 3%. The higher the oil's acidity is, the lower the conversion is. Moreover, both catalytic excess and catalytic deficiency can cause the formation of soap,[18] and in addition, as noted, the presence of moisture decreases the yield of the reaction, since water reacts with the catalyst to form soaps.
- *Type and concentration of catalyst.* The bases, in particular sodium and potassium hydroxides, are the most widely used, especially if the oil has a high degree of fatty acids and high moisture.
- Although the process of transesterification with alkaline catalysts has a very high conversion in a shorter period of time, it presents some drawbacks: the catalyst must be removed from the final product, recovery of glycerol may be hard, alkaline water resulting from the process should be treated, and, fatty acids and water affect the reaction, Figs. 17.1 and 17.2.
- Enzyme catalysts can obtain relevant results in both aqueous and non-aqueous systems.[19] In this way, the glycerol can be separated easily, and the fatty acids contained in the reused oil can be completely converted into alkyl esters. However, the use of these catalysts has a higher cost than the alkaline ones.

- *Alcohol/oil molar ratio and alcohol type.* The stoichiometric relation requires three moles of alcohol and one mole of triglyceride to produce three moles of esters and one mole of glycerol. Transesterification requires an excess of alcohol to drive the reaction to the right side. For maximum conversion a 6:1 molar ratio should be used.[3] However, a high value of the alcohol molar ratio affects the separation of glycerine because of increased solubility. When glycerine is retained in the solution, the reverse reaction goes to the left, lowering the ester yield.

- The formation of ethyl ester is more difficult than that of methyl ester. Ethanol and methanol do not dissolve with triglycerides at room temperature and the mixture should be agitated mechanically to allow mass transfer. During the reaction, an emulsion is usually formed, which quickly drops in the metanolysis forming a layer rich in glycerol, while another area rich in methyl ester stays on top. In contrast, in the etanolysis this emulsion is not stable and the separation and purification of the ethyl esters is much more complicated.

- *Effect of reaction time and temperature.* Increasing these parameters increases the conversion. Table 17.3 shows different oil types that have been the subject of optimization studies.[12,20–24]

17.1.2.2 *Processes for industrial production of biodiesel*

As has been mentioned in the previous section, transesterification — the industrial process used in biodiesel production — consists of three consecutive reversible reactions. The triglyceride is sequentially converted into diglycerides, monoglycerides and glycerine. In each reaction one mole of methyl ester is released. The whole

Table 17.3. Transesterification optimization for the production of biodiesel from different oils.

Plant oil (country)	Alcohol	Temp. (°C)	Reaction time (hours)	Molar ratio alcohol:oil	Catalyst
Soybean and castor (Brazil)	Ethyl	70	3	9	NaOH
Pongamia pinnata (India)	Methyl	60	—	10	KOH
Waste frying oils (Portugal)	Methyl	—	1	4.8	NaOH
Rapeseed (Korea)	Methyl	60	0.33	10	KOH
Sunflower (Spain)	Methyl	25	—	6	KOH

process is conducted in a reactor, where the reactions occur in the later stages of separation, purification and stabilization.

This section describes the different processes for the production of biodiesel including transesterification and esterification (usually based on the fatty acids produced as a subproduct of the transesterification). Finally, it ends with the supercritical conditions, in which the addition of catalysts is not necessary.

The transesterification process

It is the most common process for producing biodiesel, and presents different variants. Many of these technologies can be combined in different ways by varying process conditions and feeding. The choice of technology should be made depending on the desired capacity of production, supply, quality and recovery of alcohol and catalyst.

The *batch process* is the simplest method for the production of biodiesel, and is used in the plants of lower capacity and different quality of feeding, Fig. 17.4.[3]

The alcohol/triglyceride ratio commonly used is 4 to 1. The most common catalyst is NaOH, although KOH is also used, ranging from 0.3% to 1.5% (depending

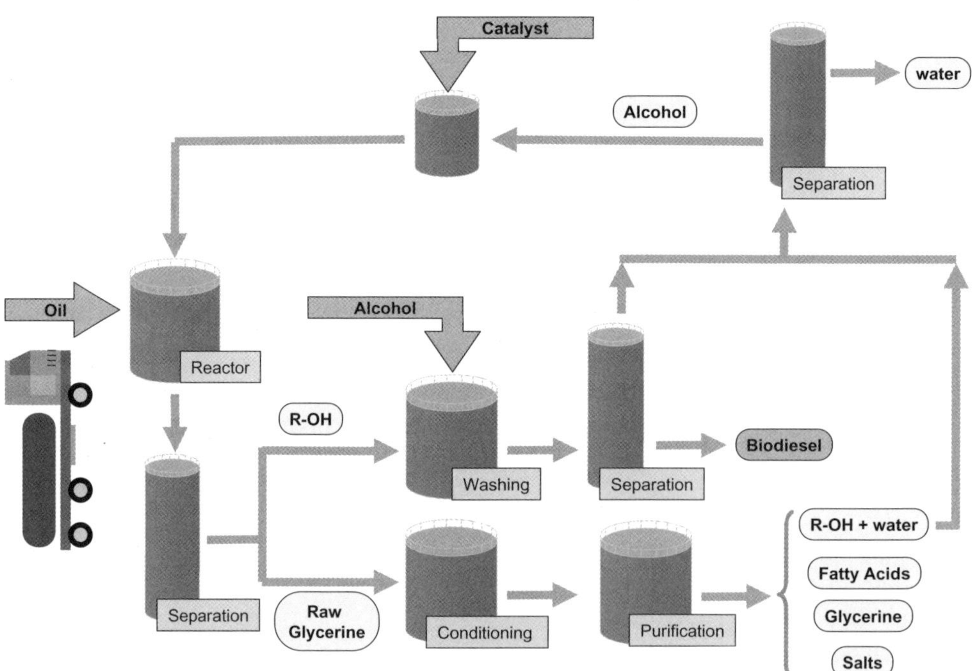

Fig. 17.4.   Transesterification process.[3]

on whether the catalyst used is KOH or NaOH). If acid catalysts are used, it is necessary to increase the temperatures and reaction times.

These reactors are sealed or fitted with a reflux condenser. To achieve a proper mixture of oil, catalyst and alcohol, a rapid agitation is required. Towards the end of the reaction, and to allow the separation of glycerol from the ester phase, the agitation should be reduced. The reaction usually lasts between 20 minutes and 1 hour at temperatures around 65°C, although it can range between 25 and 85°C. Results of between 85% and 94% are obtained. Higher temperatures and higher ratios of alcohol/oil can increase the yield of the reaction.

Some operating plants use two-step reactions, the glycerol being removed between them, which increase the final yield to greater percentages, as high as 95%.

The continuous process is the most suitable for large capacity plants, which require more staff since more uniform feeding is required. Continuous Stirred Tank Reactors (CSTR) are used and these can vary in volume to allow greater residence time and to improve the outcome of the reaction.[25] Thus, after the decantation of glycerol, the reaction in a second CSTR is much faster, with 98% of reaction product.

An essential element in the operation of the CSTR reactors is the assurance that the composition in the reactor is virtually constant, so the mix must be controlled at all times. This has a growing effect on glycerol dispersion in the ester phase, the result of which is that the time required for the separation of phases increases.

There are various processes to facilitate the esterification reaction using an intense mixture. They use tubular reactors, where the mix moves longitudinally, with little mixing in the axial direction. This Plug Flow Reactor (PFR), behaves like small CSTR reactors working one after the other. Figure 17.5 shows a block diagram for a transesterification process by means of plug flow reactors.[3]

The result is a continuous system that requires lower residence times, between 6 and 10 minutes, the reactors needed being smaller. This type of reactor can operate at high temperature and pressure, which increases the conversion rate.

The most common *Esterification Process* involves heating a mixture of alcohol and acid catalyst, typically sulfuric acid, using the cheaper reagent in excess to increase performance and shift the balance towards the right.[3,4,26] In Fig. 17.6, the diagram shows a plant with an esterification process. The sulfuric acid, in addition to being a catalyst, serves as a hygroscopic substance which absorbs water formed in the reaction. Sometimes it is replaced by concentrated phosphoric acid.

The *Combined Esterification-Transesterification Process* refines fatty acids coming from the supply system or through a different treatment in the esterification unit. Caustic catalysts are added and the reaction product is separated by centrifugation (a process called Caustic Stripping). Refined oils are dried and sent to the transesterification unit for subsequent processing. In this way, fatty acids can be transformed into methyl esters using an acid esterification process.[3,4,26]

  C. J. Renedo et al.

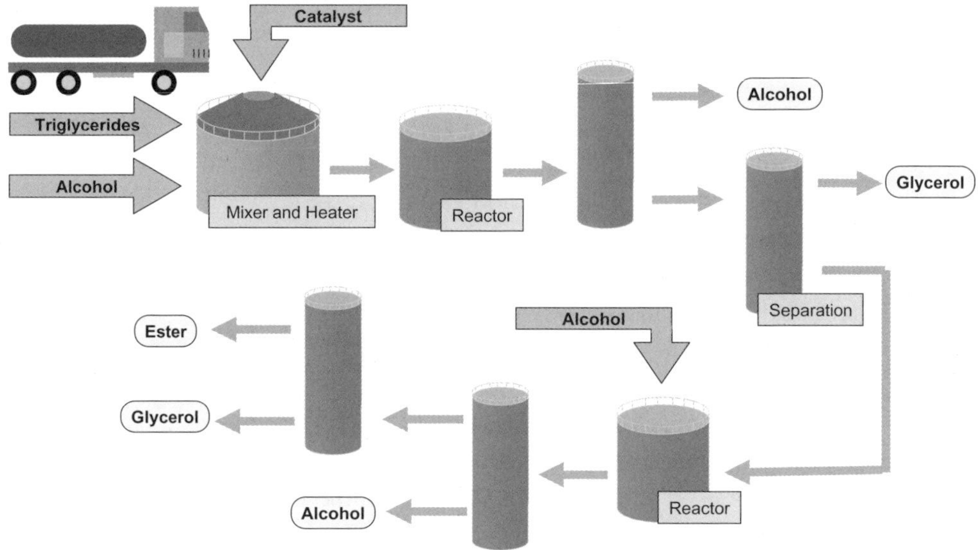

Fig. 17.5.   Biodiesel production process by plug flow reactors.[3]

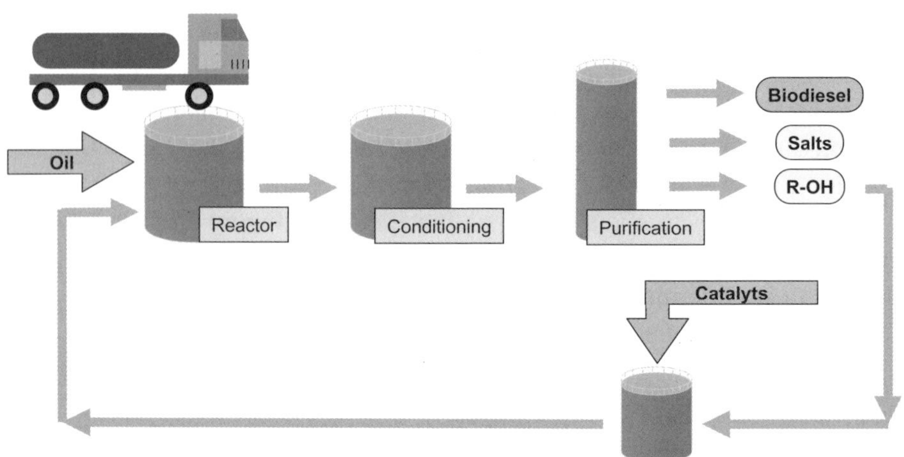

Fig. 17.6.   Esterification process.[3]

The processes of acid catalysis can be used for direct esterification of free fatty acids (FFA). An alternative would be to use a base catalyst to deliberately form soap in the FFA. The soap is recovered, the oil dried and subsequently used in a conventional system using base catalysts.

In the esterification-transesterification reaction, the fatty acids, subproducts of the reaction, can be used to later feed the esterification reactor. The process diagram of esterification/transesterification is reproduced in Fig. 17.7.[3]

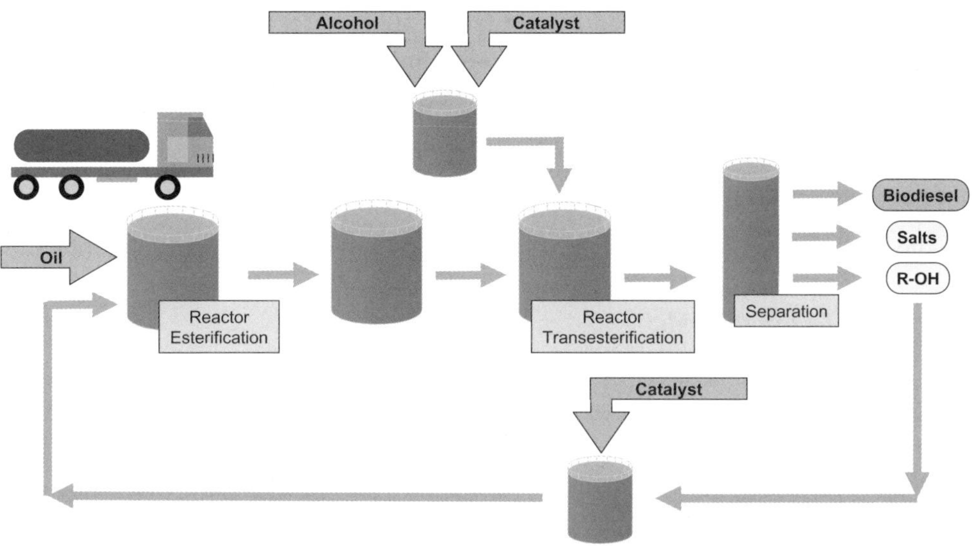

Fig. 17.7. Process of biodiesel production by esterification/transesterification. Acid catalysis process.[3]

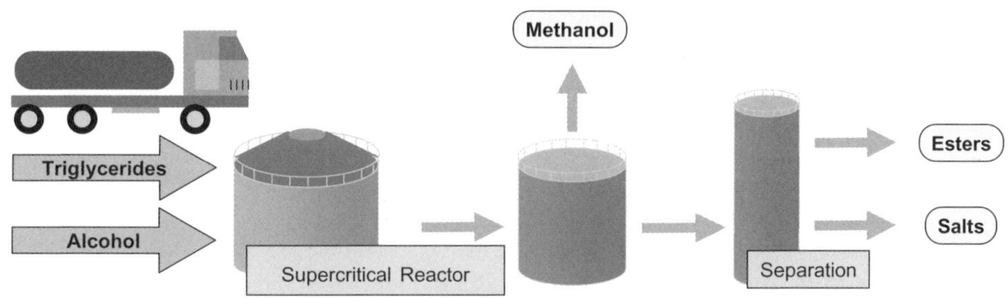

Fig. 17.8. Biodiesel production by supercritical process.[3]

The process can be performed in *supercritical conditions* if the fluid is subjected to temperatures and pressures beyond their critical point, which causes a number of unusual properties.[27] The difference between the liquid and vapor phases is removed, with only a fluid phase present. In addition, solvents containing OH groups, such as water or primary alcohols, take on the properties of superacids, which makes it no longer necessary to add catalytic converters. Figure 17.8 shows the block diagram of a supercritical process.[3]

These processes typically use alcohol-oil ratios above 40/1, the temperature tends to exceed 350°C and pressures reach 80 atm, and the reaction is complete in about 4 minutes. The installing and operating costs are higher and the amount of energy required is also higher, so, although the results from this process are very interesting, the scaling of these facilities at industrial level can be difficult.

In the synthesis of biodiesel glycerine is a subproduct.[25] Glycerine is used in the manufacture, storage, wetting and softening of many products. They can be alkyd resins, cellophane, tobacco, explosives (nitroglycerin), drugs and cosmetics, urethane foam, food and beverages, etc.

However, the increasing availability of glycerine is already causing a reduction in sale prices with the consequent diminishing of returns for those involved in the biodiesel sector. In the future, the glycerine market could suffer saturation problems.

17.1.2.3 *The use of biodiesel*

The idea of using vegetable oil as fuel has been around since the invention of the diesel engine. Rudolph Diesel experimented with a wide range of fuels (from coal to peanut oil). However, at the beginning of the 20th century, diesel engines were adapted to the use of petroleum distillate, inexpensive and available. The rising cost of petroleum experienced in recent decades and the support for renewable biofuels has revived interest in biodiesel.

The main use of biodiesel is in the automotive industry, although it is perfectly valid for use as a substitute for the diesel used in heating boilers. Moreover, biodiesel dissolves hydrocarbons, which make it useful to clean up oil spills (e.g., the Erika accident on the French coast).

The ways of using vegetable oils or methyl esters in diesel engines can be[3]:

- Diesel engine adaptation for oils.
- Use of specific engines (Elsbett).
- Using mixtures of diesel and oil.
- Use of mixtures of diesel with methyl esters.
- Use of mixtures of diesel with alcohol.

None of the options: the diesel engine adjustment, the use of a mix of biodiesel with vegetable oil and the employ of the Elsbett engine have the level of implementation and development that the mixes comprised of diesel with esters (especially methyl) and with alcohol have, which appear as the most commercially viable.

Direct use of vegetable oils

The nature of vegetable oils poses a number of disadvantages that makes its direct use in engines unadvisable, as these require a preparation consisting of a previous degumming and filtration. The seeds are pressed to separate the oil from the cake, usually by subjecting the seeds to warm up and to the action of a solvent to achieve oil extraction yields close to 100%. The cake produced as a subproduct has a high protein content so it is possible to sell it on the cattle feeding market, which lowers

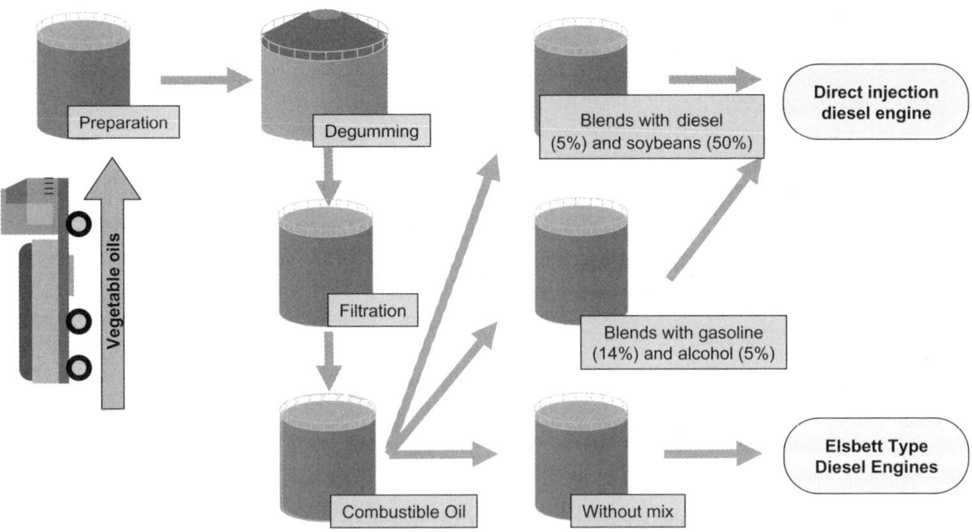

Fig. 17.9.   Vegetable oils as automotive fuel.[3]

the costs for the extraction process.[28] Figure 17.9 shows an outline of the potential use of vegetable oils as car fuels.[3]

In order to use vegetable oils in direct injection engines it is necessary to use mixtures of vegetable oil with diesel. This allows the characteristics of the fuel to be maintained, by changing the ratio of the components.[29] In any case, these mixtures of vegetable oil and diesel continue to be part of the various problems associated with vegetable oils, which are:

- Polysaturated fatty acids have a high reactivity which makes them very susceptible to polymerization and formation of gums.
- They do not burn completely, with carbonaceous deposits resulting from this.
- Thickening of the lubricating oil.
- High viscosity, making difficult to pump it, especially in cold weather.
- A low cetane number, which means poor combustion in fast engines.
- Dirty fuel injectors with a deteriorating performance of the engine.

One solution to these problems would be the following modifications to the diesel engine:

- Preheating of the fuel.
- Injection in pre-chamber. Self-cleaning injectors.
- More adiabatic engines: an indented combustion chamber, requiring less cooling.
- An auxiliary diesel starting system.

It has been demonstrated that blends of up to 1/2 (biodiesel/diesel) can be used in unmodified diesel engines, for example in the case of soybean oil.[30] Another alternative is to use blends made of raw vegetable oil, petrol (14%) and alcohol (5%), which can be used directly as fuel for diesel engines.[3]

Elsbett engines

As mentioned above, the use of unprocessed oils require that either a series of changes be carried out on the engines or that Elsbett engines be used, which can be applied to engines up to twelve cylinders.

The Elsbett engine is an adiabatic engine, that is, producing very little heat exchange with the environment and experiencing between 25 and 50% energy loss through the cooling system. It does not have the conventional cooling system and this makes it possible to work at higher temperature and therefore with greater thermodynamic efficiency. In addition, it totally burns the fuel and therefore it can be considered a virtually clean engine.[31] Moreover, the fact that it burns vegetable oil means it releases no sulfur dioxide. It is an engine developed for the combustion of crude vegetable oil, unrefined and non etherified; it neither carbonizes nor leaves residual substances; it has better thermal efficiency than diesel or gasoline, which provides more useful mechanical energy.

The mechanical components that differ from a conventional diesel engine are:

- An articulated piston with the upper part thermally and acoustically isolated, located inside a combustion chamber of spheroidal shape.
- One or two injectors per cylinder, single-hole and self-cleaning, injecting vegetable oil into the combustion chamber tangentially. This allows a perfect nebulization, i.e., the air-fuel mixture is very thin and this prevents it from making carbonaceous deposits.
- The lid of the cylinder has a small annular chamber through which the lubricating oil used as coolant circulates. Because the cooling system is not water, the lid does not seal the cylinder. A small oil radiator will close the circuit of the lubricating oil cooler.

The fact that you do not need water for cooling saves pieces, weight and volume on the engine. Another important aspect to highlight is the spherical combustion chamber, which allows an excess of air in the combustion of vegetable oil and the engine temperature to stratify. Thus, while the core of the combustion can reach 1300°C, the contact area of the piston does not exceed the typical 650°C of any engine. The final temperature of the exhaust gas is only slightly higher than for conventional diesel engines. Also, the combustion is made with less air and therefore reduces the emission of nitrogen oxides. The Elsbett engine modifications let diesel cars run either on diesel or on vegetable oil, with a good thermodynamic yield and without problems affecting the smooth operation of the engine.

The Elsbett engine is not the only one known as poly-fuel and semi-adiabatic. However, it is the only system that can be applied in any diesel engine with minimal intervention and at a reasonable cost, 2500 to 3200 Euros. The intervention consists basically in eliminating the block water chamber, changing the top of cylinders and pistons, and adding a small radiator for oil cooling. The only condition is that the engine must not have ceramic elements.

Using biodiesel

Although biodiesel can be used (in pure form or mixed in different proportions with conventional diesel) in diesel engines, the following functional differences should be considered[3,4,32–34]:

- A higher cetane number than conventional diesel, so it has greater autoignition ability.
- As a result of the delay time in the combustion, higher temperatures and pressures are reached, so that the emissions of nitrogen oxides generally increase.
- It is slightly denser than conventional diesel, which should be taken into account when creating blends.
- The behavior at low temperature varies depending on the fatty acids profile of the oil, but, in general, biodiesel has a higher solidification point than conventional diesel. This implies that cold temperatures (below 10°C) can cause problems such as starting the engine, and sealing the filter and the feeding system.
- Biodiesel provides a solvent effect for sediment accumulated in a vehicle that uses diesel. Typically, the filter system should be capable of acting as a barrier, but in severe cases, this could lead to vehicles becoming immobilized. Depending on the condition of the car, and on the use of pure biodiesel or blending in high proportions, it may be advisable to clean the tank and the piping of the feeding system to prevent filter obstruction.
- Like conventional diesel, biodiesel can corrode certain materials such as brass, bronze, copper, lead, tin and zinc. Such materials are usually replaced by aluminum or stainless steel. Additionally, biodiesel can also affect joints, seals, hoses and pipes, particularly those manufactured in natural rubber or nitrile.
- Since it is impossible to control whether or not vehicles fueling at service stations have been prepared to use 100% biodiesel, the mixture BDP-10 (10% biodiesel + 90% diesel) or BDP-5 are used, considering in this case the biofuel as an additive.

- Since the 90s, several car manufacturers (primarily German makes) have already changed the rubber hoses for others made of plastic materials which do not dissolve in biodiesel.
- Combustion engines require lubrication to prevent friction in the moving parts. Reducing levels of sulfur and aromatics in conventional fuels has limited the traditional diesel lubrication. In this way, the addition of small amounts of biodiesel (1–2%) improves the lubricant characteristics of low sulfur diesel, making it possible to eliminate other lubricant additives. However, when using 100% biodiesel, the lubricant oil is contaminated because of the low viscosity of the ester.

Biodiesel has a calorific value slightly lower than diesel (about 11%), which implies a slight increase in consumption and a decrease in engine power. These two effects depend on the ratio of the mixture, and can reach 15% and 7% for pure biodiesel.

- The accidental loss of fuel into the environment causes a contamination of soils and water resources. For this reason, it is important to reduce the time that the environment needs to break the pollutant down into other harmless substances. Biodiesel degrades almost completely (98.3%) in just 21 days.
- Generally, for biodiesel blends under 20% no modifications to the engines are required. For the use of pure biodiesel or blends over 20%, minor modifications such as hoses, gaskets, etc., are necessary.

Regulation on the use of biodiesel as fuel

Biodiesel and its blends can be used in compression-ignition engines designed, according to the standard ASTM-D975, to operate with fossil diesel. Regardless of the raw material's origin, it must comply with the European quality standard EN-14214.

17.2   Level of Development

17.2.1 *Use and industrial development*

The global production of biodiesel has reached 14273560 t in 2008.[35] The main producers are represented in Fig. 17.10 and their productions are detailed in Table 17.4.

The European Union is the world leader in the development of the biodiesel sector. It produces roughly the 54% of biodiesel global production. The growth of biodiesel in Europe, as seen in the past 16 years, has recently accelerated, rising from 55000 t in 1992 to 7750000 t in 2008.[36]

Fig. 17.10.   Worldwide distribution of biodiesel production in 2008 (Ml).

Table 17.4.   Main countries where biodiesel is produced.

Country	Production [Ml]	Porcentage [%]
Germany	3175	20
United States	2650	16
France	2044	13
Argentina	1205	7
Brazil	1100	7
Indonesia	684	4
Italy	670	4
Malaysia	541	3
Thailand	394	2
China	338	2
Belgium	312	2
Poland	310	2
Portugal	302	2
Austria	240	1
Spain	233	1
Others	1888	12

It is clear that behind this sector's development are policies to promote production and consumption.

In the case of European Union the Directive 2003/30/EC proposes the replacement of 2 and 5.75% of the total energy content of petrol and diesel consumed in the transport sector by the end of 2005 and 2010, respectively.

It refers to total figures and does not give specific goals for each biofuel. However, these figures may vary depending on private sector initiatives for the different biofuels. If these objectives are transposed to the estimated consumption, in the case of biodiesel the obtained figures are collected in Table 17.5.

In late 2005, the average replacement in the EU remained below the proposed 2% (it was 1.4%). However, the trend and the latest measures taken by the European Commission and member countries in recent years indicate that the 2010 target will be met.

Specifically in Spain, the development experienced in last years and the short-term perspective for the biodiesel related industrial sector, has provided a dramatic increase in levels of production and consumption, and a significant approach to the Directive's goal. In July 2005 there was a revision of the Plan of Development of Renewable Energy through the Renewable Energy Plan 2005–2010 whose targets were revised upwards, allowing compliance with the European directive for biodiesel, Table 17.6.[38]

Table 17.5.   Biodiesel objectives for Europe.[37]

	2005	2010
Estimated consumption of diesel	23336 kTep	29768 kTep
Directive targets	2%	5.75%
Predicted consumption of biodiesel	467 kTep	1712 kTep

Note: kToe: Thousand tonnes of oil equivalent. 1 t biodiesel provides 0.9 Toe.

Table 17.6.   Aims for primary energy during the period 2005–2010.[38]

Resources	Tep
Pure vegetable oils	1021800
Used vegetable oils	200000
Applications	
Biodiesel	1221800

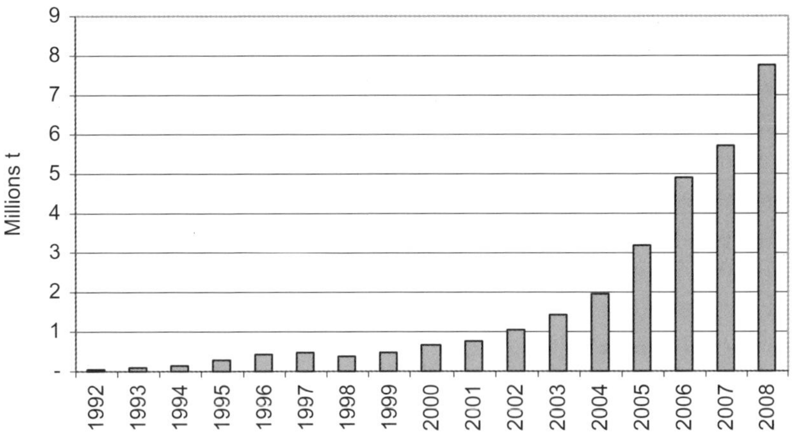

Note: The production includes the incorporation of new members of European Union.

Fig. 17.11.   Production of biodiesel in the European Union since 1992 (t).[36]

The growing trend in the production of biodiesel in the European Union is reflected in the Fig. 17.11.[36]

Germany is the country with the greatest development, with a production of 2817000 tonnes of biodiesel in 2008.[36] This is aided by higher prices and taxes imposed on diesel and by many government support programmes which are quite favourable to the development of biofuels (with a total tax exemption for pure and mixed biofuels, from January 1, 2004).

In the case of other countries such as USA, Canada, Brazil, India, Malaysia, etc., active biodiesel programmes have been developed in order to provide legislative support and draw up national strategies on biodiesel development.

For example, the centerpiece of Indian biodiesel development and commercialization is the National Biodiesel Mission (NBM), supported by the Planning Commission of the Government of India. NBM was planned for two phases. Phase I was termed as demonstration phase and has been carried out from year 2003 to 2007. The work done during this phase was the setting up of nurseries for Jatropha seed collection, the cultivation of 400000 ha of waste land, and the installation of transesterification plants to produce biodiesel (80000 Mt/year). Phase II was designed for a self sustaining expansion of the programme leading to the production of biodiesel to meet 20% of the diesel requirements of the country by 2011–2012. Moreover, recent research and development on microalgae has exposed a new road ahead that can be promising sustainable feedstock for biodiesel production.[39] In China, its government is planning to establish a vehicular biodiesel production capacity of 2Mt by 2020. This target is clearly stated in the Long-term Development Strategies for Renewable Energy.[40]

Since the early 80s, the Malaysian government had realized the importance of developing biofuel and in particular, biodiesel in the long term. This was achieved by the Malaysian Palm Oil Board (MPOB) when the project of developing palm biodiesel was initiated at laboratory scale in 1982. However, it was not until the Fifth Fuel Policy under the Eighth Malaysia Plan (2001–2005) that huge efforts were undertaken to encourage utilization of renewable resources and renewed interest in biodiesel was developed. In this way, the Government of Malaysia drafted the National Biofuel Policy in 2005 to carve out a comprehensive framework and specific initiatives for use of biofuels in Malaysia.[41]

In the First National Energy Conference (1998), the Taiwanese government decided to seek new clean energies, and biodiesel development was listed as one of the planned programs. In 2001, the government promulgated the "Administrative Regulations on the Production and Sales of Renewable Energies such as Ethanol, Biodiesel, or Oil from Recycled Waste". These administrative regulations listed ethanol, biodiesel, or oil recycled from waste under the scope of application of the "Petroleum Administration Act". To promote the utilization of biodiesel, Article 38 of the "Petroleum Administration Act" explicitly prescribes that producers of renewable energies are exempted from the petroleum stockpiling obligations and petroleum fund payments.[42]

In Brazil, in response to the petroleum shortage during the decades of 1970 and 1980, the Government created in the 1980s a program called PROALCOOL, which implemented and regulated the use of ethanol as fuel. At the end of 20th century, the Federal Government restarted the discussion about the use of biodiesel, and many studies were carried out by inter-ministerial commissions in partnership with universities and research centers. Furthermore, Brazil's Government created an Inter-ministerial Work Group in 2003, this Group considered that biodiesel should be introduced immediately in the Brazilian energy matrix. In 2004, the National Program of Production and Use of Biodiesel (PNPB) was launched, its main objective was to guarantee the economically viable production of biodiesel, and the social inclusion as well as regional development. The most important action from PNPB was the introduction of biofuels derived from oils and fats in the Brazilian energy matrix by means of Law No. 11097 dated January 13, 2005. In this law, the optional use of B2 until the beginning of 2008 is foreseen; after that year, B2 will be mandatory. Between 2008 and 2013, it will be possible to use blends up to 5% of biodiesel, and after this period B5 will be mandatory. The Brazilian Government plans to use the PNPB also for developing familiar and sustainable agriculture in underdeveloped areas. Biodiesel producers who acquire raw material in productive arrangements that include familiar agriculture with a purchase guarantee, receive the social fuel label. This label, regulated by the Agrarian Development Ministry, guarantees not only fiscal exemptions for industries, but also better conditions for financing. Another action of PNPB has been the creation of a research network involving scientists from university and research institutes from all Brazilian regions. The aim of this network is to develop science and technology for all the biodiesel production chain.[43]

It is remarkable that in Argentina, despite the increasing interest in biodiesel, the issuing of new regulations has suffered from a lack of inter-disciplinary coordination and communication between different ministries. Following two successive "National Programs on Biofuels", respectively issued by the Department for Environment and Sustainable Development and the Department for Agro-food Activities, and due to the lobbying of several stakeholders aiming to give a strong push to the development of biofuels in the country, a new law was promoted in 2007. This law provides a mix of fiscal incentives and blending quotas to act as stimulus for the biofuels industry. In addition, the new law established a coordinating body, the National Biofuels Commission, designed to harmonize the approaches taken by agribusiness-focused and energy/industrial-focused ministries.[44]

These are some examples of different initiatives implemented in biodiesel-producing countries whose objective is to increase biodiesel production in order to get a development more sustainable.

17.2.2 *Challenges of biodiesel sector*

The main challenges faced by this sector are:

- Optimizing costs of biodiesel production. In the process of biodiesel production from conventional vegetable oils, the majority today, between 70–80% of the cost is due to the impact of raw material. The main research lines focus on obtaining oil from crops that allow cost to be reduced and the optimization of the process and the functional characteristics of the biodiesel obtained. Although a significant part of the cost comes from the raw material, there is a margin for improvement in yields and production costs. For the recovery of oil from frying the pre-treatment still requires optimization to improve process performance. Finally, it is important to consider returns from subproducts production. In particular, what is of great importance is the search for alternative markets for glycerine. During recent years, the price of pure glycerine has fallen consistently, mainly due to increased supply. So, the income that previously involved the sale of this subproduct is losing importance in the results. The search for alternative applications that help to enhance this product may contribute significantly to improving the competitiveness of the cost of biodiesel.

- Biodiversity protection. The rapid expansion of raw-materials plantations to support the extra demand of biodiesel is threatening the biodiversity of the natural ecosystems. More proactive actions from governments are needed in order to prevent the extinction of species. Multi-feedstock biodiesel production could play a huge role in ensuring security and sustainability for the future. The support of the necessary research to find new ways of producing biodiesel is another basic action.

- Development of specific additives for biodiesel. So far, the largest consumption of biodiesel has been made at rates below 20%. The trend has been to use conventional diesel additives, without any specific developments to optimize the functionality of the biodiesel. The use of specific additives could help to overcome some disadvantages and/or to ensure compliance with the specifications in terms of cost optimization of raw material/process (running at low temperatures, stability, etc.).
- Searching for biodiesel applications with high added value. Some research promotes the use of specific types of biodiesel as an additive in the automotive sector (lubricants and others).
- Public support. In some countries, specifically in developing countries, the public awareness still remains low. The advocates of biodiesel development involve only the government, related industry players and some environmentalists. Most of society is ignorant or has limited knowledge on biodiesel issues. They need to be informed or educated through proper channels because they are vital to attract more investments and efforts to develop biodiesel industry. Therefore, it will be an important task to encourage greater public involvement to develop biodiesel industry.

17.3 Strengths and Weaknesses

In this section, strengths and weaknesses of biodiesel are analyzed, compared to other traditional energy sources.

17.3.1 *Advantages compared to traditional sources*

The biodiesel uses present advantages in common with other biofuels:

- Reduced dependence on energy supplies from the transportation sector (currently 98% dependent on oil in the European Union) and cooperation in diversifying and securing an energy supply to the automotive sector.
- Development of rural economy: creating jobs, establishing population in rural areas, a reduction of abandoned farmland, and so on.
- Environmental benefits: reduction of CO_2 emissions in the transport sector in order to meet the Kyoto Protocol. Emissions of carbon monoxide, sulfur oxides, non-burned particulate, methane and aromatic hydrocarbons are cut down.

- More biodegradable than conventional diesel (in 21 days 98.3%, compared to 50%), this means that any spills are much less harmful to the environment.
- Secure management and transport because it is not very toxic and has a flashpoint at approximately 150°C (the flashpoint for conventional diesel is 50°C).
- In countries like Spain the current refining system presents a shortfall in diesel, as a result of the "dieselisation" process in the automotive market. The introduction of biodiesel can reduce this deficit.
- The conversion of diesel vehicles (generally more polluting than those based on gasoline) into more environmentally friendly vehicles.
- It provides greater lubricant properties, which extends engine life and reduces noise.
- The ester provides a certain amount of solvent power, which acts as an additive anti-dirt and does not produce char. This keeps the engine, the fuel injectors and the ducts cleaner, and the effect increases with the proportion of ester in the mixture.

17.3.2 *Disadvantages compared to conventional sources*

The biodiesel uses presents disadvantages for both the system and the users.

17.3.2.1 *For the system*

- Biodiesel production is more costly than buying a litre of diesel on the market. To compensate the cost difference and to avoid the effect on the client, States must reduce taxes on fuels, with the consequent income loss.
- Several studies indicate that to achieve the 5.75% substitution of diesel with biofuels it would require an occupancy of between 4–13% of farmland in the EU-25 (depending on the raw material and the technological development).
- The massive deployment of energy crops could have a negative impact on biodiversity.[45]
- Biodiesel causes a slight increase in emissions of nitrogen oxides (NO_x).
- It requires an adaptation of the storage systems and taxes associated with it.
- The need to maintain strategic reserves of fuel hinders the expansion of distribution networks.
- To maintain the technical specifications of fossil fuels for biodiesel makes the sale of blends higher than 15 or 20% impossible.
- The widespread use of biodiesel results in large volumes of aqueous wastes with high biological demand. Although, there are some researches which have used biodiesel wastes containing glycerol to produce hydrogen by fermentation through a Klebsiella pneumoniae strain,[46] this process has not been developed

at industrial scale yet. Biodiesel wastes containing glycerol can be used by a Klebsiella pneumoniae strain to produce hydrogen by fermentation as a source of locally generated combustible gas for local heating or on-site use for biomass drying.[46]

17.3.2.2   *For the user*

- Acceptance by the general public and awareness of the benefits of bioenergy are limited.
- When one first starts using biodiesel in an engine, the main difference appears with the filter change, since it should be carried out sooner than normal. This is because the solvent feature causes the fuel to pick up the existing dirt on the engine. The lack of small precautions to be taken into account, can lead to misunderstandings about the correct performance of the engine.
- Possible problems starting at low temperatures. The pure biodiesel has a freezing point between 0 and $-5°C$. In the case of mixtures with rates lower than B30, the freezing temperature decreases. This problem only occurs in cold climates.
- Reduction of engine performance. With mixtures up to B30, a slight increase in consumption and a slight loss of power can be noted. For B100 the reduction in power is 7%, and the increased consumption 16%.
- A different smell with respect to conventional diesel.

17.3.3   *How to overcome the disadvantages*

In order to overcome the disadvantages of biodiesel, it is necessary to take into account the following elements:

- The improvement in production cost requires the reduction in the cost of raw material (70–80% of total). This in turn requires the development of alternative raw materials (energy crops) not linked to the food market. With regard to technological improvements in production, since it is already a mature technology, no major reductions in cost can be expected.
- The use of land in retreat should be promoted, and crops with improved yields developed.
- The expected negative impact on biodiversity from a massive occupation of crops could be overcome locally through the use of indigenous and varied crops. The objective should be to develop small economies which help to create local market opportunities although the size of manufacturing plants is restricted.
- In order to reduce NO_x emissions some solutions have been proposed: the delay of the injection point in the engine, and the use of an oxidation catalyst that filters

the soluble fraction of the fuel. The latter system would further reduce emissions of hydrocarbons, particulates and carbon monoxide.

- To solve the problem of starting at low temperatures, the following should be taken into account: the use of suitable additives for the oil, the percentage of the mix, and the seasonal climatic conditions of the country or region.
- To allow the stocks of raw material to be considered as part of the strategic fuel reserves.
- The development of technical specifications for biodiesel and the different percentages of the fuel blend.
- Public communication of the benefits of using biodiesel and its environmental consequences, as well as the minor precautions to be taken into account in cars.

17.3.4 *Horizon to overcome the barriers or disadvantages*

Although cost reductions are expected in 2010, the reduction of the differential cost with respect to fossil fuels will come more easily from the increasing cost of the oil than from the reduction in the production cost of biodiesel. New technologies are being considered for industrial implementation in the medium term (enzyme catalysis, catalysis in chemical fixed support, etc.). These are not expected to produce significant improvements in costs, but in a reduction of process problems and subproducts management.

When comparing the use of biodiesel with the use of conventional diesel, there is seen to be a slight increase in NO_x emissions (around 6%). The introduction of catalysts can make biodiesel engines pollute less than without catalysts and than vehicles using conventional diesel. However, all possible efforts should be made to improve engine performance with biodiesel.

17.4 Environmental Impact

The use of biodiesel as a substitute for diesel has environmental benefits by reducing the emission of greenhouse gases and other pollutants. These reductions will depend on the engine and its operation, and are proportional to the mixture used (B100, B20, etc.).

The quantification of these benefits is usually undertaken through the Life Cycle Analysis (LCA), measuring the net aggregate of emissions over the entire production process and the consumption of fuel. However, the results of these tests may vary in each case, depending on the raw materials, the processes and the use of the subproducts considered.

Fig. 17.12. CO_2 cycle.

In a LCA, for every tonne of fossil diesel, 2.8 tonnes of CO_2 are emitted. The carbon content of biodiesel is lower, and emits approximately 2.4 t. It can be also assumed that the carbon content will be completely recaptured in the next crop to produce vegetable oil as feedstock. Thus, taking into account the complete cycle, the CO_2 net emissions can be considered zero. Figure 17.12 shows the CO_2 cycle.

When burning diesel, the sulfur (in the European Union it contains a maximum of 350 parts per million of sulfur) is released into the atmosphere as SO_x, contributing to the formation of acid rain. Biodiesel is virtually free of sulfur (less than 0.0024 ppm).[47]

NO_x emissions from B100 biodiesel increase by 6% compared to the diesel ones. However, the absence of sulfur in biodiesel enables the use of techniques for controlling NO_x emissions, which cannot be used in conventional diesel. Thus, the use of biodiesel could limit such emissions.

Biodiesel contains oxygen, which improves the combustion process and the emission profile, which significantly reduces CO emissions (by at least 20%).

Particulate emissions from biodiesel are lower with respect to the average particulate emissions from conventional diesel (at least 40%).

Table 17.7.   Comparative table of biodiesel versus fossil diesel emissions.[49]

Emission type	B100	B20
CO	−43.2%	−12.6%
Hydrocarbons	−56.3%	−11.0%
Particulates	−55.4%	−18.0%
NO_x	+5.8%	+1.2%
Toxic emissions	−60%/−90%	−12%/−20%
Mutagenicity	−80%/−90%	−20%
CO_2 (LCA)	−78.3%	−15.7%

Note: For CO_2, the Life Cycle Assessment is reviewed.

Table 17.8.   Contaminants depending on the biodiesel origin.[50–53]

Biodiesel source	Pollutants investigated
Neem oil (Bangladesh)	CO ↓, NO_x ↑, smoke ↓
Soybean (Turkey)	CO↓, NO_x ↑, particulates ↑, hydrocarbons ↓
Rapeseed oil (Korea)	CO↓, NO_x ↑, smoke ↓, CO_2 ↑
Soybean oil (U.S.)	Particulares ↓
Waste cooking oil (Spain)	Particulares ↓, smoke ↓
Soybean, Rapeseed oil (Germany)	Mutagenicity of particulares ↑
Palm oil (China)	CO ↓, polyaromatics ↓, particulates ↓, hydrocarbons ↓
Brassica carinata (Italy)	NO_x ↑, particulates ↓

Another environmental advantage presented by biodiesel is its characteristic biodegradability, as it degrades to 98% in the first 21 days (50% in conventional diesel).

In addition to the above data, B100 reduces the risk of cancer by 94% and, consequently, B20 does by 27%.[48]

In Table 17.7, exhaust emissions from engines with biodiesel with respect to fossil diesel are compared.[49]

Since 2000, the number of studies exploring pollutants has increased. Table 17.8 shows the contaminants recently studied depending on the biodiesel origin.[50–53]

17.4.1   *Environmental assessment "Ecotest"*

The "Ecotest" assessment has been promoted by the International Automobile Federation, and is assessed by the German Automobile Club (ADAC). This index assesses the environmental impact of a vehicle in urban and interurban operation using two indexes, first for CO_2, and second for the other pollutants (CO, HC, NO_x and PM). At the end, a classification with an increasing number of stars, based on decreasing emissions, is given. In general, gasoline vehicles are better considered by this method than the diesel ones, these being only surpassed by the hybrids and those using natural gas as fuel.[54]

The "Ecotest" assessment for different vehicles indicates that the use of biodiesel modifies the results, so, a car can go from being a "three star" vehicle to a "five star" vehicle by changing from diesel to biodiesel. This is due to the emission reduction of CO_2 and particulates.

A recent LCA on biofuels in Spain, carried out by CIEMAT and commissioned by the Environment Spanish Ministry[55] has concluded that the replacement of diesel by pure biodiesel (B100) enables there to be a reduction of 57% (biodiesel from raw vegetable oils) and of 88% (biodiesel from used vegetable oils) in emissions of greenhouse gases (CO_2eq) per kilometre, compared to fossil diesel. A mixture of diesel with a 10% biodiesel (B10) makes a reduction in emissions of between 6% and 9% respectively possible.

17.5 Economics

In this section, economic aspects of the biodiesel production: the investment levels, the production costs and the prices of automotive diesel, are analyzed.

17.5.1 *Investment levels and production costs*

It should be noted that the limiting factors that have marked the development of the biodiesel industry in recent years have not been technical, but economic. These factors include the cost of raw material, the production cost of biodiesel, the price of fossil fuels and the taxation of energy products.

Regardless of the type of technology, most of the cost of the final product is the raw material. In the case of vegetable oils it is between 70 and 80%. In the case of waste oil/animal fat, the impact on the cost of biodiesel is reduced, but generally more investment is necessary, and due to its quality, it must be mixed with diesel to comply with specifications for sale.

The investment cost of a typical plant producing biodiesel from vegetable oils is in an average range of 250 to 270 €/t, for a plant of 50000 t/year. Of course, this value is subject to the peculiarities of the process, raw material used, etc. In the short term, it does not expect significant reductions in this parameter.

The biodiesel production costs in Spain for a 50000 t/year plant from sunflower oil, are approximately: 7.6 c€/l of fixed costs and 7.8 c€/l of variable costs (including

Table 17.9. Comparative production costs of biodiesel (2002).[56]

US	€/ litre	European Union	€/litre
Biodiesel from used oil	0.23–0.42	Biodiesel from used oil	0.23–0.42
Biodiesel from soybean	0.37–0.70	Biodiesel from rape	0.37–0.74

the sale of subproduct glycerine and the distribution of biodiesel, and excluding the cost of raw materials).[38]

The production and sale of subproducts is an added income that can support the profitability of a biodiesel plant. The main product that is obtained in any technology is glycerine, and even though in recent years the price has gone down, it remains as a major source of income. Its current price is around 500 €/t. Depending on different technologies, other subproducts can be recovered: fatty acids (for animal feeding) and minerals (such as agricultural fertilizer), although they are of less value.

In the case of conventional vegetable oils, the raw material competes in the food market, so, it is necessary to consider the prices of this sector and its high variability depending on market conditions. Given the enormous fluctuations in the cost of oil, and that these represent the largest percentage of the final cost, it is difficult to establish a standard cost for biodiesel.

The Spanish Renewable Energy Plan provides a price for feedstock (sunflower oil) of 59 c€/litre, a biodiesel production cost of 74.5 c€/litre. Table 17.9 shows the ranges of biodiesel production costs from different raw materials in the United States and in the European Union.[56]

Figure 17.13 presents the production total costs per biodiesel litre compared to the partial costs, in 2005 and 2030. The figure shows the high cost of feedstocks on the total. In United States the cost is more competitive than in EU.[57]

17.5.2 *Prices of automotive diesel*

Since biodiesel competes in some way with automotive diesel prices, its retail price should be taken into account to calculate the biodiesel production profitability.

Since the selling price of biodiesel uses the diesel price as a reference, the upward trend of oil prices in recent times is improving the profit margins in biodiesel.

17.6 **Combination with Conventional Sources**

The 2003/30/EC directive of the European Parliament, 8 May 2003, on the promotion of biofuels or other renewable fuels for the transport sector, details the various ways

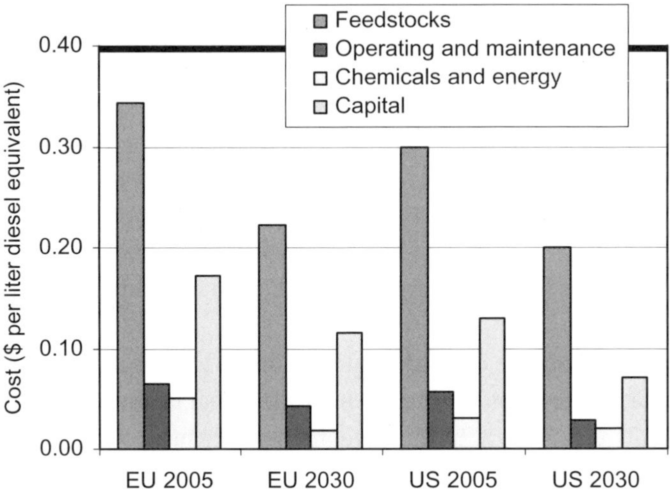

Fig. 17.13. Production costs of biodiesel in the European Union and United States (including subsidies crops production) from 2005 and 2030.[57]

in which biofuels could be made available:

- In its pure state or in high concentration in petroleum products, according to specific quality standards for applications in the transport sector. In the case of biodiesel the standard is EN-14214.
- Mixed with petroleum products, in accordance with the relevant European standards where the technical specifications for transport fuels are established (EN 228 and EN 590).

As a result of technological advances, most vehicles currently in circulation in the European Union can use a low biofuel blend without problems. Recent technological advances make it possible to use higher percentages of biofuel in the blend. Some countries are already using biofuel blends of 10% or higher. In certain European cities there are municipal fleets of vehicles that allow the use of mixtures with higher concentrations, even on pure biofuels.

17.7 Conclusions

The continued use of petroleum-based fuels is now widely recognized as unsustainable because of depleting supplies and the contribution of these fuels to the environment pollution. However, the hydrocarbons sector plays a vital role in the economic growth of a country.

The challenge, therefore, is to secure suitable energy supplies at the least possible cost. This challenge has stimulated the search for other liquid fuels as an alternative to diesel. One of these new fuels is biodiesel.

Biodiesel is a well established fuel and will reach greater market share in the coming decades. Its benefits include greenhouse gas reductions which will contribute to domestic and international targets, potential air quality benefits, diversification of the fuel sector and contribution to the creation of new jobs and development of rural areas.

There are several sources to obtain biodiesel: conventional oleaginous crops (sunflower, rapeseed, soybean...), alternative oleaginous crops (Camelina sativa, Crambe abyssinica, Jatropha curcas, Pogianus, Brassica carinata, Cynara cardunculus...), animal fats, used frying oils, olein and oils from other sources (microbial processes from algae, bacteria and fungi, as well as from microalgae). The method used commercially to obtain biodiesel is transesterification which consists of the reaction of the oil or fat with a low molecular weight alcohol (methanol or ethanol) and in presence of a suitable catalyst.

Although biodiesel can be used (in pure form or mixed in different proportions with conventional diesel) in diesel engines, there are some functional differences which should be considered in order to optimize its use.

The biodiesel sector has undergone a crucial development in last decades thanks the initiatives implemented by biodiesel-producing countries whose objective is to reduce environmental impact. However, there are some barriers such as high cost of raw materials, biodiversity loss or lack of society awareness that hinder the complete development of biodiesel sector. For this reason, it is necessary to go on working in the development of knowledge which help to create cheaper and more productive raw materials as well as technologies which enable the use of biodiesel in different applications not only in automobile sector. Finally, the public ought to be prepared to sacrifice their money for a cleaner environment and be willing to pay the extra cost of biodiesel.

References

1. F. Kreith and D. Yogi, *Handbook of Energy Efficiency and Renewable Energy* (CRC Press, 2007).
2. K. McDonnell *et al.*, "Properties of rapeseed oil for use as a diesel fuel extender," *J. American Oil Chemistry Society* **76** (1999) 539.
3. J.M. García and J.A. García, "Biocarburantes líquidos: Biodiesel y bioetanol," Madrid, http://www.madrimasd.org/informacionidi/biblioteca/publicacion/doc/VT/vt4_Biocarburantes_liquidos_biodiesel_y_bioetanol.pdf.
4. "Las energías renovables en España," CENER, Ed Fundación Gas Natural (2006).

5. F. Culshaw and C. Butler, "A review of the potential of biodiesel as a transport fuel," Department of trade and industry energy technology support unit report ETSU-R-71, Her Majesty's Stationery Office, London (1993).

6. "Commercial biodiesel production plants," (2007) and "Biodiesel production plants in pre-construction," (2006), National Biodiesel Board, www.nbb.org.

7. G. Vicente, M. Martínez and J. Aracil, "Ésteres metílicos como combustible," Materias primas y propiedades, *Tecno-Ambiente* **85** (1998) 9–12.

8. G. Vicente, M. Martínez and J. Aracil, "Biodiésel: Una alternativa real al gasóleo mineral," *Ingeniería Química* **377** (2001) 135–145.

9. Ministry of Environment, "Rural and Marine Medium in Spain (MAPA)," Survey about crops' area and productivity ESYRCE, http://www.mapa.es/es/estadistica/pags/encuesta cultivos/anteriores.htm (2009).

10. D. Cocco, "Comparative study on energy sustainability of biofuel production chains," *Proc. Institution of Mechanical Engineers, Part A: Journal of Power and Energy*, vol. 221 (2007).

11. J.M. Dias *et al.*, "Mixtures of vegetable oils and animal fat for biodiesel production: Influence on product composition and quality," *Energy Fuels* **22** (2008) 3889–3893.

12. P. Felizardo *et al.*, "Production of biodiesel from waste frying oils," *Waste Management* **26** (2006) 487–494.

13. Y. Watanabe *et al.*, "Conversion of acid oil by-produced in vegetable oil refining to biodiesel fuel by immobilized Candida antarctica lipase," *J. Molecular Catalysis B: Enzymatic* **44** (2007) 99–105.

14. A.B.M. Sharif Hossain *et al.*, "Biodiesel fuel production from algae as renewable energy," *Am. J. Biochemistry and Biotechnology* **4** (2008) 250–254.

15. F. Ma and M.A. Hanna, "Biodiesel production: A review," *Bioresource Technology* **70** (1999) 1–15.

16. J.M. Marchetti *et al.*, "Possible methods for biodiesel production," *Renewable and Sustainable Energy Review* **11** (2007) 1300–1311.

17. L.C. Meher, S.D. Vidya and S.N. Naik, "Technical aspects of biodiesel production by transesterification — A review," *Renewable and Sustainable Energy Review*, vol. 10, pp. 248–268.

18. M.P. Dorado *et al.*, "An alkalicatalyzed transesterification process for high free fatty acid oils," *Trans. ASAE*, vol. 45, pp. 525–529.

19. H. Fuduka, A. Kondo and H. Noda, "Biodiésel fuel production by transterifications of oils," *J. Bioscience and Bioengineering* **92** (2002) 405–416.

20. D. De Oliveira *et al.*, "Optimization of alkaline trans esterification of soybean and castor oil for diesel production," *Applied Biochemical Biotechnology* **553** (2005) 121–124.

21. S.K. Karmee and A. Chadha, "Preparation of biodiesel from crude oil of Pongamia Pinn," *Bioresource Technology* **96** (2005) 1425.

22. G.T. Jeong and D.H. Park, "Batch production of biodiesel fuel from rapeseed oil," *Applied Biochemical Biotechnology* **668** (2006) 129–132.

23. G. Vicente, M. Martinez and J. Aracil, "Optimization of integrated biodiesel production I. A study of the biodiesel purity and yield," *Bioresource Technology* **98** (2007) 1724.

24. G. Vicente, M. Martinez and J. Aracil, "Optimisation of integrated biodiesel production II. A study of the material balance," *Bioresource Technology* **98** (2007) 1754.

25. L.G. Schumache, J. Van Gerpen and B. Adams, *Encyclopedia of Energy, Vol 2, Biodiesel Fuel* (Elsevier, 2004).

26. Y.C. Sharma, B. Singh and S.N. Upadhyay, "Advancements in development and characterization of biodiesel: A review," *Fuel* **87** (2008) 2355–2373.

27. S. Glisic and D. Skala, "The problems in design and detailed analyses of energy consumption for biodiesel synthesis at supercritical conditions," *J. Supercritical Fluids*, vol. 49, pp. 293–301.

28. D. Agarwal *et al.*, "Performance evaluation of a vegetable oil fuelled compression ignition engine," *Renewable Energy* **33** (2008) 1147–1156.

29. P. Mondal, M. Basu and N. Balasubramanian, "Direct use of vegetable oil and animal fat as alternative fuel in internal combustion engine," *Biofuels, Bioproducts and Biorefining*, vol. 2, pp. 155–174.
30. F. Ma, L.D. Clements and M.A. Hanna, "The effect of catalyst, free fatty acids, and water on transeterification of beef tallow," *Trans. ASAE* **41** (1998) 1261–1264.
31. http://www.elsbett.com/.
32. A.K. Agarwal and L.M. Das, "Biodiesel development and characterization for use as a fuel in compression ignition engines," *J. Engineering for Gas Turbines and Power*, vol. 123, pp. 440–447.
33. National Biodiesel Board, "Biodisel usage checklist and National lubricity benefits," http://www.biodiesel.org/f.
34. National Renewable Energy Laboratory, "Biodiesel-clean, Green Diesel Fuel," http://www.nrel.gov/.
35. Biofuels Platform, http://www.biofuels-platform.ch.
36. EurObserv'ER, "Biofuels Barometer," http://www.eurobser-er.org/default.asp.
37. APPA, "A Strategy for Biofuels in Spain (2005–2010)," http://www.appa.es/.
38. IDAE, "Plan de energías renovables en España (2005–2010)," http://www.idae.es/.
39. S.A. Khan, H.M.Z. Rashmi, S. Prasad and U.C. Banerjee, "Prospects of biodiesel production from microalgae in India," *Renewable and Sustainable Energy Reviews* **13** (2009) 2361–2372.
40. X. Yan and R.J. Crookes, "Reduction potentials of energy demand and GHG emissions in China's road transport sector," *Energy Policy* **37** (2009) 658–668.
41. S. Lim and L.K. Teong, "Recent trends, opportunities and challenges of biodiesel in Malaysia: An overview," *Renewable and Sustainable Energy Reviews* (2009), doi: 10.1016/j.rser.2009.10.027.
42. Y.H. Huang and J.H. Wu, "Analysis of biodiesel promotion in Taiwan," *Renewable and Sustainable Energy Reviews* **12** (2008) 1176–1186.
43. G.P.A.G. Pousa, A.L.F. Santos and P.A.Z. Suarez, "History and policy of biodiesel in Brazil," *Energy Policy* **35** (2007) 5393–5398.
44. J.A. Mathews and H. Goldsztein, "Capturing latecomer advantages in the adoption of biofuels: The case of Argentina," *Energy Policy* **37** (2009) 326–337.
45. European Energy Agency, http://www.eea.europa.eu/.
46. F. Liu and B. Fang, "Optimization of bio-hydrogen production from biodiesel wastes by Klebsiella pneumoniae," *Biotechnology J.* **2** (2007) 374.
47. U.S. Department of Energy, "Biodiesel-clean green fuel," http://www.energy.gov/.
48. M. Engidanos, A. Soria, B. Kavalov and P. Jensen, "Techno-economical analysis of biodiesel production in the EU: A short summary for decision-makers," European Commission Joint Research Center, May 2002.
49. U.S. Department of Energy, "Biodiesel: handling and use guidelines," http://www.energy.gov/.
50. M.N. Nabi, M.S. Akhter and M.M. Zaglul Shahadat, "Improvement of engine emissions with convencional diesel fuel and diesel-biodiesel blends," *Bioresource Technology* **97** (2006) 372.
51. M. Canakci, "Combustion characteristics of a turbocharged DI compression ignition engine fueled with petroleum diesel fuels and biodiesel," *Bioresource Technology* **98** (2007) 1167.
52. M. Cardone, "Brassica carinata as alternative oil crop for the production of biodiesel in Italy: Engine performance and regulated and unregulated exhaust emissions," *Environmental Science Technology* **36** (2002) 4656.
53. M. Lapuerta, J. Rodriguez-Fernandez and J.R. Agudelo, "Diesel particulate emissions from used cooking oil biodiesel," *Bioresource Technology* **99** (2008) 731.
54. German Automobile Club (ADAC), http://www.adac.de.
55. Abengoa Bioenergy, http://www.abengoabioenergy.com/sites/bioenergy/en/.
56. IEA, "Biofuels for transport: An international perspective," May 2004.
57. D. Moudsale, *Biofuels: Biotechnology, Chemistry and Sustainable Development* (Taylor & Francis Group, LLC, 2008).

Section 4

Ocean and Small Hydro Energy Systems

Chapter 18

Technologies and Methods used in Marine Energy and Farm System Model

V. Patel Kiranben

A.D.I.T, New V. V. Nagar, India
kichipatel@rediffmail.com

M. Patel Suvin

Arcogul, Chikhodra, India
suvin_m_patel@yahoo.co.in

"We must and we can start the world development process that leads to an environmentally sustainable world habitat for humanity there is no alternative . . . there is none."

Dr. John P. Craven

This chapter focuses on some of the key challenges to be met in the development of marine energy, it presents a prototype form to being a widely deployed contributor to a future energy supply of the world. Large-scale wave and tidal current prototypes have been demonstrated around the world, but marine renewable energy technology is still 10–15 years behind that of wind energy. However, having started later, the developing technology can make use of more advanced science and engineering, and it is therefore reasonable to expect rapid progress. Many scientific advances are required to meet these challenges, and their likelihood is explored based on current and future capabilities. This chapter incorporates aspects of technology, power production effects and capital cost factor implications. The aim is to give grounding in the nature of the industry, the current state of the industry and the key factors, which will potentially shape and limit the growth of the industry. This is achieved by evaluating tidal power from technological, environmental and socioeconomic viewpoints.

18.1 Introduction

The development of offshore renewable energy can be seen as environmentally desirable, especially for the amelioration of climate change, including meeting international carbon dioxide reduction targets. Whilst renewable energy sources are seen as clean and inexhaustible, there are nevertheless some adverse impacts from any marine energy development and these deserve further consideration. The renewable energy industry is in some ways still in its infancy and, as such, not all of its impacts are clear or fully assessed. It is important that all possible impacts are identified and resulting research and monitoring is carried out in a timely manner during the development of all marine renewable projects, including wind, wave and tidal power. In order for any marine renewable project to be seen as part of a move towards sustainable development it is important that any adverse impacts are considered, costs and benefits fully evaluated and adverse environmental impacts minimized through careful project design and implementation. Such investigation will help enable appropriate placement of renewable energy sites and careful monitoring will feed into adaptive management. A thoroughly precautionary approach is necessary in all developments. This chapter describes some of the key challenges to be met in the development of marine renewable energy technology, from its present prototype form to being a widely deployed contributor to future energy supply. Since 2000, a number of large-scale wave and tidal current prototypes have been demonstrated around the world, but marine renewable energy technology is still 10–15 years behind that of wind energy. However, having started later, the developing technology can make use of more advanced science and engineering, and it is therefore reasonable to expect rapid progress. Although progress is underway through deployment and testing, there are still key scientific challenges to be addressed in areas including resource assessment and predictability, engineering design and manufacturability, installation, operation and maintenance, survivability, reliability and cost reduction. Many scientific advances are required to meet these challenges, and their likelihood is explored based on current and future capabilities.

18.2   Marine Energy: How Much Development Potential is There?

How much marine energy is available for development? Information on the amount of electrical capacity available and extractable from two forms of marine energy: wave and kinetic stream. Of all the large natural resources available for generating electricity, ocean energy may be one of the last investigated for its potential. To fill this knowledge gap, we performed an assessment of the available and extractable ocean resources. When discussing marine resources, we refer to two forms: wave and kinetic stream. Although there are other marine energy sources — the thermal energy resulting from the large temperature differences between deep and cold ocean waters and sun-warmed surface waters, the chemical energy in ocean salinity gradients, and marine biomass — we only discuss wave and kinetic stream energy. There are two forms of tidal energy: potential (i.e., harnessing the potential energy changes associated with the tidal rise and fall of sea level) and kinetic (i.e., harnessing the kinetic energy associated with the motion of the tidal stream). In this part, we will discuss the kinetic form, which can be tapped without building barrages or dams. There are three types of kinetic energy from water: tidal, ocean current, and river streams. These are renewable energy resources that can be converted to electricity without greenhouse gas emissions. The technology to convert these resources to electricity, albeit in its infancy, has been deployed in demonstration projects. Commercial projects are expected in the next five to ten years. Given proper care in design, deployment, operation, and maintenance, ocean wave and kinetic stream energy could be two of the most environmentally benign electricity generation technologies yet developed.

18.3   Understanding the Power of Marine Energy

To understand the generating potential of a given site, there are two relationships that are important to know. The first is the factors that affect the power contained in a wave. The power fluctuation of a wave is a factor of two variables: the significant wave height (in meters squared) and the mean wave period (in seconds). The annual average wave power fluctuation in deep water that is required for a site to have commercial interest is about 20 to 50 kilowatts per meter. The second is the kinetic power density of a stream of water. This relationship depends on both the density of the seawater (in kilograms per cubic meter) and the instantaneous speed or velocity of the stream (in meters per second). Kinetic power density varies considerably over a tidal cycle and can vary with depth. To make a relevant comparison between sites, values for kinetic power density usually are averaged over the year and over total water depth (e.g., annual depth-averaged power densities for sites with commercial interest in the U.S. are about 2 to 5 kilowatts per meter squared). The energy from ocean waves and tidal streams, along with ocean-based wind energy, make the world's oceans a source of renewable energy that may in the next few decades greatly outstrip solar energy as the economical alternative of choice. Options for exploiting the energy available from the world's oceans include offshore wind, wave and tidal stream energy.

Extracting clean and economically viable energy from the world's oceans has fascinated researchers and engineers for centuries. The first patent on a wave energy device was taken out in 1799, and more than 300 such devices have been patented since. Commercial application has been limited to a small number of devices that use wave energy to power navigation buoys. However, concerns over climate change may fuel progress. All of the marines renewable offer energy with low environmental impact and near-zero emissions. Tidal power has received great attention because of its high energy density, high predictability and low environmental impacts. World-widely, the estimated tidal energy potential reaches $500\sim1000\,T$ Wh/yr.[2] However, this energy largely remains untapped. Early tidal power technology was based on barrages which extract potential energy from tides functioning like a traditional hydro dam (i.e., harness energy by utilizing the elevation difference between high tides and low tides), and this technology has been abandoned since the 1980s due to its significant ecological footprint.[3] The subsequent development of tidal power had been slow in the 1980s, and has been reinvigorated in the past decade as tidal current turbines are experimentally employed to extract kinetic energy from tidal currents.[5,8] Due to the accessibility and geographical limitations, recent tidal current turbines are quite often projected to be constructed in a confine channel, which are called in-stream tidal current turbines. The working principle of an in-stream tidal turbine is similar to that of a typical wind turbine. An in-stream tidal turbine farm (i.e., a group of tidal turbines located at one site) is similar to an offshore or on-land wind farm. However, those experiences from wind study cannot be readily transferred to tidal energy output estimation for the reasons that (1) although offshore wind farms have been commercialized for a while, the related research primarily focuses on the operation and maintenance of the farms[9–11]; (2) wind farms are often located in open areas where geological conditions do not restrict design options, while in-stream tidal turbine farms are generally restricted by the geological conditions; and (3) the physical characteristics of tidal turbines, such as the Reynolds Number is different from that of wind turbines. The development of in-stream tidal current turbine farm model is still in the infancy stage. Studies have been conducted to estimate energy potentials in a number of oceanic countries, such as Brazil,[12] Canada,[13] India,[15] Spain,[4] the United Kingdom,[16] and the United States.[17] These studies have been used by governmental agencies to determine whether and how to build in-stream turbine farms. In estimating energy, all of these studies assume that the efficiency of each turbine in a farm is equal to the maximum efficiency of a stand-alone turbine. Thus, the predicted energy outputs from these studies are inaccurate. The inaccuracy in energy output estimation is regarded as a significant barrier to the industrialization of tidal power since the results are not convincing to investors.[18] In order to develop an accurate and responsive approach to estimate energy output from an in-stream turbine farm, one should derive a relationship between the power output of an individual turbine and the farm.

18.4　Global Development of Marine Energy

There has been a considerable growth in world wind turbine capacity. Northern Europe has significant capacity for offshore wind, and testimonies to this are the large-scale investments in developments in Denmark, Germany, the Netherlands, Norway, Sweden and the U.K. (Pelc and Fujita, 2002). Plans are underway in Nantucket Sound, Massachusetts, U.S. for one of the largest wind developments of its kind in the world, consisting of 130 generators over a 28 square mile area (Santora *et al.*, 2004). Such a development would produce an average of 170 mega watts (MW) per day, with a maximum of 420 MW, sufficient energy to power three-quarters of the electrical power needs of Cape Cod (Kempton *et al.*, 2005). There are other developers investigating sites along the east coast of the U.S. (Jarvis, 2005). The total installed wind generating capacity of the U.S. is 11603 MW (AWEA, 2007). The U.S. has also developed a hybrid system that incorporates tide, wave and wind power (Pelc and Fujita, 2002). Innovative use of renewable includes a development of offshore wind and wave power to contribute towards self-sustainability of a French Island community (Babarit *et al.*, 2006). Europe accounts for almost half of the wave and tidal projects that have been proposed or installed around the world and over half of the potential projects that are proposed or planned for the future (Figs. 18.3 and 18.4). Scotland in particular has set an ambitious goal for renewable energy generation, aiming for a target of 40% of all energy generation by 2020 (Scottish Executive, 2003). A marine renewable test facility located off the Orkney Islands in Scotland provides an experimental facility for developers of marine energy technologies investigating wave and tidal power (BWEA, 2007) and has the goal of becoming the leading international marine renewable centre (Winskel *et al.*, 2006). The first commercial wave plant in the U.K. was installed in 2000 and has since been providing power to the national grid via the island of Islay, Scotland (Pelc and Fujita, 2002). Efforts to minimize possible negative impacts of marine wind turbines on migratory species. It also calls upon the Parties to:

- Identify areas where migratory species are vulnerable to wind turbines;
- Apply and strengthen comprehensive strategic environmental impact assessment to identify appropriate sites;
- Evaluate possible negative ecological impacts prior to decision making;
- Assess cumulative environmental impacts;
- Take full account of the precautionary principle in development; and
- Take account of impact and monitoring data as they emerge.

Potential impacts can also be expected for non-migratory species. This may be particularly true for a resident or semi resident population where their home range coincides with a renewable energy development.

18.5 Possible Impacts

Several desktop studies have been conducted to investigate the potential for impacts on cetaceans relating to the noise associated with various aspects of renewable energy developments. Results have included the potential for injury at short range due to pile-driving and behavioral impacts to considerable distances, including exclusion from habitat. Case-by-case differences are acknowledged, depending on substrate type and other environmental parameters. As far as the authors are aware, there have not been any field studies that have looked at the impacts of turbine farms, or other sources of renewable energy, on any baleen species of cetacean, white-beaked dolphins or any odontocetes other than harbor porpoises to date.

18.5.1 *Potential environmental impacts*

The potential environmental impacts of marine renewable energy developments may be long, medium and short-term and each stage of a development has associated impacts. Generally speaking, the greatest concern has been raised about the operational phase because of its perceived potential for long term impacts. However, construction activities, such as pile driving, also warrant considerable concern and, in fact, each phase in the "life-cycle" of the development requires thorough investigation. Activities associated with turbine farm development that are of particular importance to cetaceans are listed (Table 18.1). The results of some studies on impacts of wind developments are currently providing seemingly conflicting results which probably reflects the fact that we are still in the early stages of a "new industry" and there is little information available about the potential impacts of tidal and wave power. In common with wind power, the infrastructure associated with generation

Table 18.1. Activities associated with Marine farm development of particular importance to cetaceans.

Activities likely to cause *longer term* impacts:	<ul><li>The physical presence of structures (e.g., wind farm towers, tidal and wave generators and associated anchoring and artificial reef effects) and associated effects on habitat;</li><li>Continual operational noise and vibrations emanating from turbines and potentially other marine energy generators once in place;</li><li>Electromagnetic impacts due to cabling that may impact navigation and affect food sources, particularly in the case of elasmobranches;</li><li>Increased vessel traffic, for instance from maintenance operations; and,</li><li>Collision with turbine blades on tidal generators.</li></ul>
Activities likely to cause at least *short* and *medium* term impacts:	<ul><li>Seismic exploration;</li><li>Intense noise during construction due to ramming/piling, drilling and dredging operations;</li><li>General construction noise and disturbance;</li><li>Increased vessel activities during exploration and construction;</li><li>Increased turbidity due to construction and cable laying; and, later,</li><li>Decommissioning (which may involve the use of explosives).</li></ul>

of tidal power will involve major construction work and including pile driving in some cases. Consideration must be given to the proximity to tidal races, where productivity is likely to be increased (and these are thus likely to be feeding grounds), as well as the submerged nature of the structures, which may bring moving parts into contact with wildlife. Studies of impacts from offshore industries including the oil and gas industry may provide useful guidance as a starting point. The effects of renewable energy generation may combine with other stressors also affecting individuals and populations, such as other sources of chemical and noise pollution, and thereby produce an impact on marine life that may be greater than predicted for any one source or indeed the predicted sum of the impact of the stressors. The fact that developments are typically only evaluated on their own without taking into account what else is affecting a region or population deserves to be challenged.

Renewable energy should fully and transparently consider environmental implications, and commit to appropriate research programs. In European waters, this is in theory provided for by Strategic Environmental Assessments, as required by the SEA European Directive (2001/42/EC). In consideration of cumulative impacts, the positive potential impacts of a reduction in effort relating to traditional energy sources might be taken into account.

18.5.2 *Indirect impacts*

In addition, indirect impacts may result from adverse impacts to fish and shellfish stocks. Marine farm construction and development may impact fish spawning, over wintering, nursery and feeding grounds, and migratory pathways. Whilst this is focused on cetaceans, investigations should also consider basking sharks, fish and other marine life as well as potential aerial impacts on bats and seabirds. As many new and, increasingly, large-scale renewable energy developments are planned all around the world it is imperative in each case that robust baseline data are available before any development site is determined and construction commences, to allow evaluation of conditions prior to monitoring possible effects post construction. The development of guidelines for the protection of marine mammals is clearly appropriate, where choice of site, and scale, are key considerations. Independent and high quality Environmental Impact Assessments and Strategic Environmental Assessments should be undertaken. Results from any work should be made available as soon as possible to inform the design of further projects and allow adaptive management. Should developments proceed in areas in which cetaceans are known to exist, mitigation measures that are put in place to reduce impacts below significant levels must be sufficiently researched and effective. Offshore renewable energy developments are but one of the many pressures on cetacean populations, and it is important that any adverse effects are identified and minimized as soon as possible. Research and monitoring needs should be built into national programmes for the development of offshore renewable industries. Lessons should be learned from other offshore industries. Whilst the renewable industry is in its infancy we are in a good position to lead in the development of best practice in the generation of renewable and environmental responsible energy.

18.6 Ocean Wave Energy

Power generation using wave energy is at a much earlier stage of development. Wave energy offers more predictable outputs than wind, but in early 2003 there was only around one megawatt of generating capacity installed worldwide, all of it essentially with demonstration prototypes. Proposed projects are likely to take this to about 6 MW over the next few years. The wave industry is characterized by a wide variety of novel devices and a large number of small firms. Devices can be classified by generic technology type, though there is some overlap. Methods of extraction of marine energy are indicated in Fig. 18.1.

18.6.1   *Types of wave energy devices*

— Pneumatic devices, such as the oscillating water column (OWC), use wave motion to compress and decompress air, and drive a turbine.
— Float-based devices utilize a buoyant float moving with the waves, reacting against a seabed anchor in order to harness energy.
— Spillover devices utilize wave height to replenish a reservoir of seawater, which runs a turbine.
— Raft-type devices use the relative motion of adjacent rafts or pontoons to harness wave energy. Moving-body devices articulate in the water, inducing motion, which may be used to drive a hydraulic motor.

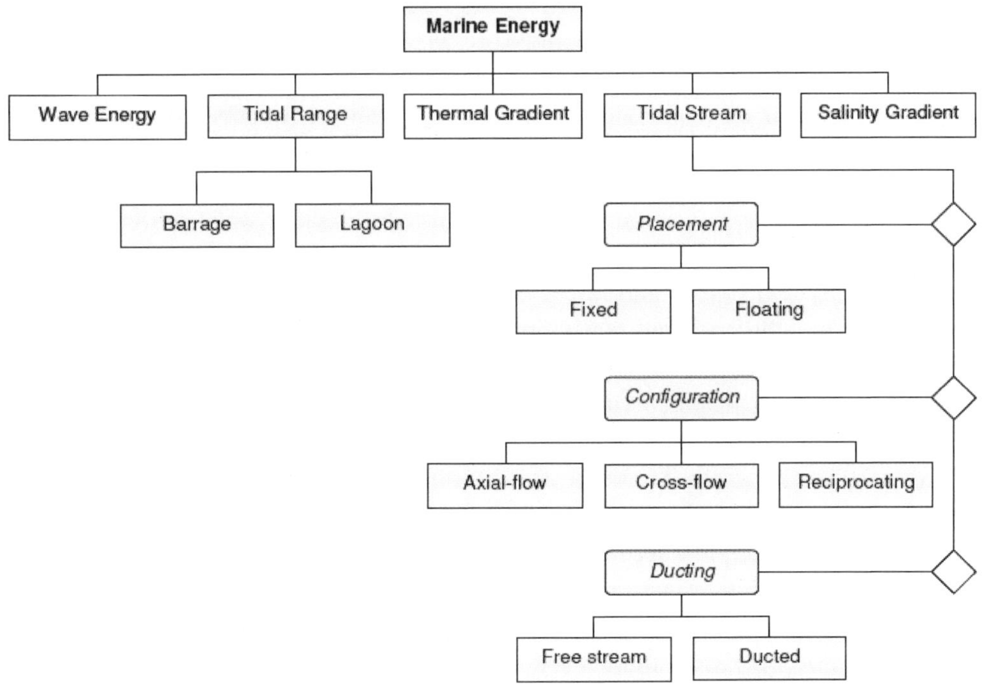

Fig. 18.1.   Methods of extracting marine energy.

Commercial-scale wave energy is yet to become a reality and as such empirical evidence on costs is limited. Of those devices that have been deployed (for the most part near-shore and shoreline OWC devices), costs are in the region of 6–8 U.S. cents per kWh. Three designs — the Limpet, Osprey and Pelamis — have secured support from the Irish and Scottish renewable schemes — though supplementary investment has also been required (for example, E.U. grants). The other devices are still at the research stage, though some are much closer to commercial deployment than others. (Float based devices are already in use for niche applications such as navigation buoys.) The Osprey is designed to provide a mounting platform for wind turbines and hence offers the prospect of the first hybrid wind-wave device. Hybrids have the potential to improve the utilization of sub-sea power connections and to raise the ratio of output to construction cost. Waves are generated by wind passing over the sea: as long as the waves propagate slower than the wind speed just above the waves, there is an energy transfer (refer to Fig. 18.2) from the wind to the most energetic waves. Both air pressure differences between the upwind and the lee side of a wave crest, as well as friction on the water surface by the wind shear stress causes the growth of the waves.[4] The wave height increases with increases in (see Ocean surface wave):

— wind speed,
— time duration of the wind blowing,
— fetch — the distance of open water that the wind has blown over, and
— water depth (in case of shallow water effects, for water depths less than half the wavelength).[5]

In general, large waves are more powerful. Specifically, wave power is determined by wave height, wave speed, wavelength, and water density. Wave size is determined by wind speed and fetch (the distance over which the wind excites the waves) and by the depth and topography of the seafloor (which can focus or disperse the energy of the waves). A given wind speed has a matching practical limit over which time or distance will not produce larger waves. This limit is called a "fully developed sea". Oscillatory motion is highest at the surface and diminishes exponentially with depth. However, for standing waves (clapotis) near a reflecting coast, wave energy is also present as pressure oscillations at great depth, producing microseisms.[4] These pressure fluctuations at greater depth are too small to be interesting from the point of view of wave power. The waves propagate on the ocean surface, and the wave energy is also transported horizontally with the group velocity. The mean transport rate of the wave energy through a vertical plane of unit width, parallel to a wave crest, is called the wave energy flux (or wave power, which must not be confused with the actual power generated by a wave power device).

Power Generation Cycle

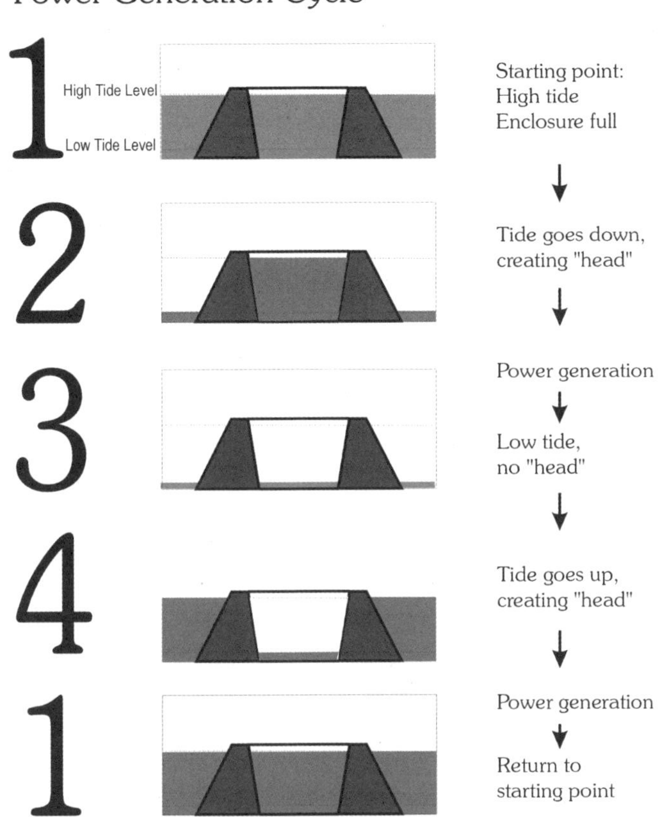

Fig. 18.2.   Power generation cycle.

18.6.2 *Wave power formula*

In deep water, if the water depth is larger than half the wavelength, the wave energy flux is

$$P = \frac{(0.5\,\text{kW})}{\text{m}^3 \cdot \text{s}} H_{\text{mo}}^2 T, \tag{18.1}$$

where
P the wave energy flux per unit wave crest length (kW/m);
H_{mo} is the significant wave height (meter), as measured by wave buoys and predicted by wave forecast models. By definition, H_{mo} is four times the standard deviation of the water surface elevation;
T is the wave period (second);
ρ is the mass density of the water (1025 kg/m^3), and
g is the acceleration by gravity (9.81 m/s^2).

The above formula states that wave power is proportional to the wave period and to the square of the wave height. When the significant wave height is given in meters, and the wave period in seconds, the result is the wave power in kilowatts (kW) per meter wavefront length.

Example: Consider moderate ocean swells, in deep water, a few kilometers off a coastline, with a wave height of 3 meters and a wave period of 8 seconds. Using the formula to solve for power, we get meaning there are 36 kilowatts of power potential per meter of coastline.

$$P \approx 36 \, \frac{\text{kW}}{\text{m}}. \tag{18.2}$$

In major storms, the largest waves offshore are about 15 meters high and have a period of about 15 seconds. According to the above formula, such waves carry about 1.7 MW/m of power across each meter of wavefront. An effective wave power device captures as much as possible of the wave energy flux. As a result the waves will be of lower height in the region behind the wave power device.

18.6.3   *Wave energy and wave energy flux*

In a sea state, the average energy density per unit area of gravity waves on the water surface is proportional to the wave height squared, according to the linear wave theory:[4,6]

$$E = \frac{1}{16} \rho g H_{\text{mo}}^2, \tag{18.3}$$

where E is the mean wave energy density per unit horizontal area (J/m^2), the sum of kinetic and potential energy density per unit horizontal area. The potential energy density is equal to the kinetic energy,[4] both contributing half to the wave energy density E, as can be expected from the equipartition theorem. In ocean waves, surface tension effects are negligible for wavelengths above a few decimetres.

As the waves propagate, their energy is transported. The energy transport velocity is the group velocity. As a result, the wave energy flux, through a vertical plane of unit width perpendicular to the wave propagation direction, is equal to:

$$P = E c_{\text{g}} \tag{18.4}$$

with c_{g} the group velocity (m/s). Due to the dispersion relation for water waves under the action of gravity, the group velocity depends on the wavelength λ, or equivalently, on the wave period T. Further, the dispersion relation is a function of the water depth h. As a result, the group velocity behaves differently in the limits of deep and shallow water, and at intermediate depths[4,6]:

Deep water corresponds with a water depth larger than half the wavelength, which is the common situation in the sea and ocean. In deep water, longer period waves propagate faster and transport their energy faster. The deep-water group velocity is half the phase velocity. In shallow water, for wavelengths larger than twenty times the water depth, as found quite often near the coast, the group velocity is equal to the phase velocity.[5]

18.6.4 *Modern technology*

Wave power devices are generally categorized by the method used to capture the energy of the waves. They can also be categorized by location and power take-off system. Table 18.2 gives idea about electricity network required before starting the process of extraction. Method types are point absorber or buoy; surfacing following or attenuator; terminator, lining perpendicular to wave propagation; oscillating water column; and overtopping. Locations are shoreline, nearshore and offshore. Types of power take-off include: hydraulic ram, elastomeric hose pump, pump-to-shore, hydroelectric turbine, air turbine,[11] and linear electrical generator. Some of these designs incorporate parabolic reflectors as a means of increasing the wave energy at the point of capture.

In the United States, the Pacific Northwest Generating Cooperative is funding the building of a commercial wave-power park at Reedsport, Oregon.[12] The project will utilize the PowerBuoy technology Ocean Power Technologies (Fig. 18.3) which consists of modular, ocean-going buoys. The rising and falling of the waves moves the buoy-like structure creating mechanical energy which is converted into electricity and transmitted to shore over a submerged transmission line. A 40 kW buoy has a diameter of 12 feet (4 m) and is 52 feet (16 m) long, with approximately 13 feet of the unit rising above the ocean surface. Using the three-point mooring system, they are designed to be installed one to five miles (8 km) offshore in water 100 to 200 feet (60 m) deep.

An example of a surface following device is the Pelamis Wave Energy Converter. The sections of the device articulate with the movement of the waves, each resisting motion between it and the next section, creating pressurized oil to drive a hydraulic

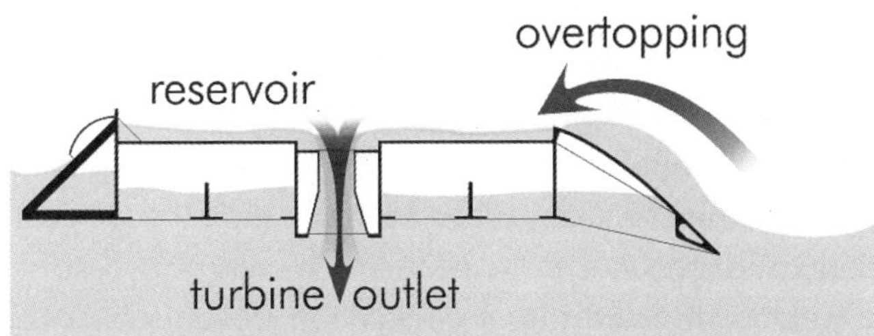

Fig. 18.3.　The front of the Pelamis m/c bursting a wave at the Agucadoura Wave Park.

Table 18.2.   Electricity network connections for Marine energy.

	Tidal range		Tidal stream	
Connection voltage (kV)	Barrage	Lagoon	Deep	Shallow
	275–400	132	11–132	11–132
Required network upgrades for connection	Transmission network upgrades	Some local distribution and transmission network upgrades may be required	Local distribution network upgrades will be required and Transmission network upgrades as capacity increases	Some local distribution and transmission network upgrades may be required
Constraints caused by existing system	Tidal range resource is close to consumer demand and strong transmission network	Tidal range resource is close to consumer demand & strong transmission network	Current distribution and transmission network constraints mean there is only capacity available for very small schemes	Most of the shallow resource is close to consumer Demand & strong distribution and transmission network

ram which drives a hydraulic motor.[13] The machine is long and narrow (snake-like) and points into the waves; it attenuates the waves, gathering more energy than its narrow profile suggests. Its articulating sections drive internal hydraulic generators (through the use of pumps and accumulators). With the Wave Dragon (Fig. 18.4), wave energy converter large "arms" focus waves up a ramp into an offshore reservoir. The water returns to the ocean by the force of gravity via hydro-electric generators. The AquaBuOY, made by Finavera Renewables Inc., is a wave energy device: Energy transfer takes place by converting the vertical component of wave kinetic energy into pressurized seawater by means of two-stroke hose pumps. Pressurized seawater is directed into a conversion system consisting of a turbine driving an electrical generator. The power is transmitted to shore by means of a secure, undersea transmission line. A commercial wave power production facility

Fig. 18.4.   Wave Dragon seen from reflector, prototype 1:4½.

utilizing the AquaBuOY technology is beginning initial construction in Portugal.[14] The company has 250 MW of projects planned or under development on the west coast of North America.[15] The SeaRaser, build by Alvin Smith; which uses a entirely new technique (pumping) for gathering the wave energy.[16] A device called CETO, currently being tested off Fremantle, Western Australia,[17] consists of a single piston pump attached to the sea floor, with a float tethered to the piston. Waves cause the float to rise and fall, generating pressurized water, which is piped to an onshore facility to drive hydraulic generators or run reverse osmosis water desalination.[18] Another type of wave buoys, using special polymeres, is being developed by SRI.[19] Wavebob is a leading wave energy converter which has been conducting research and development over a ten-year period. With a track record of testing from small scale to sea-going trials in the Atlantic Ocean near Galway in Ireland, Wavebob is now considered one of the leading technologies to emerge in this new sector.

18.6.5   *World wide challenges*

Large storm waves pose a challenge to wave power developers. These are some of the challenges to deploying wave power devices: Efficiently converting wave motion into electricity; generally speaking, wave power is available in low-speed, high forces, and the motion of forces is not in a single direction. Most readily-available electric generators operate at higher speeds, and most readily-available turbines require a constant, steady flow.

Constructing devices that can survive storm damage and saltwater corrosion; likely sources of failure include seized bearings, broken welds, and snapped mooring lines. Knowing this, designers may create prototypes that are so overbuilt that materials costs prohibit affordable production.

High total cost of electricity; wave power will only be competitive when the total cost of generation is reduced. The total cost includes the primary converter, the power takeoff system, the mooring system, installation and maintenance cost, and electricity delivery costs.

Impacts on the marine environment, such as noise pollution, could have negative impact if not monitored, although the noise and visible impact of each design varies greatly.[8]

In terms of socio-economic challenges, wave farms can result in displacement of commercial and recreational fishermen from productive fishing grounds, can change the pattern of beach sand nourishment, and may represent hazards to safe navigation.[20]

In the U.S., development of wave farms is currently hindered by a maze of state and federal regulatory hurdles and limited research and development funding.

18.6.6 *Wave farms*

The world's first commercial wave farm opened in 2008 at the Aguçadora Wave Park near Póvoa de Varzim in Portugal. It uses three Pelamis P-750 machines with a total installed capacity of 2.25 MW.[3,21] A second phase of the project is now planned to increase the installed capacity to 21 MW using a further 25 Pelamis machines.[22] Funding for a 3 MW wave farm in Scotland was announced on February 20, 2007 by the Scottish Executive, at a cost of over 4 million pounds, as part of a £13 million funding packages for marine power in Scotland. The farm will be the world's largest with a capacity of 3 MW generated by four Pelamis machines.[23] Funding has also been announced for the development of a Wave hub off the north coast of Cornwall, England. The Wave hub will act as giant extension cable, allowing arrays of wave energy generating devices to be connected to the electricity grid. The Wave hub will initially allow 20 MW of capacity to be connected with potential expansion to 40 MW. Four device manufacturers have so far expressed interest in connecting to the Wave hub.[24,25]

The scientists have calculated that wave energy gathered at Wave Hub will be enough to power up to 7500 households. Savings that the Cornwall wave power generator will bring are significant: about 300000 tons of carbon dioxide in the next 25 years.[26] A CETO wave farm of the coast of Western Australia has been operating to prove commercial viability and after preliminary environmental approval is poised for further development.

18.7 Ocean Tide Energy

Tidal energy is generated by the relative motion of the Earth, Sun and the Moon, which interact via gravitational forces. Periodic changes of water levels, and associated tidal currents, are due to the gravitational attraction by the Sun and Moon. The magnitude of the tide at a location is the result of the changing positions of the Moon and Sun relative to the Earth, the effects of Earth rotation, and the local shape of the sea floor and coastlines. Because the Earth's tides are caused by the tidal forces due to gravitational interaction with the Moon and Sun, and the Earth's rotation, tidal power is practically inexhaustible and classified as a renewable energy source. A tidal energy generator uses this phenomenon to generate energy. The stronger the tide, either in water level height or tidal current velocities, the greater the potential for tidal energy generation. Tidal movement causes a continual loss of mechanical energy in the Earth–Moon system due to pumping of water through the natural restrictions around coastlines, and due to viscous dissipation at the seabed and in turbulence. This loss of energy has caused the rotation of the Earth to slow in the 4.5 billion years since formation. During the last 620 million years the period of rotation has increased from 21.9 hours to the 24 hours[3] we see now; in this period the Earth has lost 17% of its rotational energy. While tidal power may take additional energy from the system, increasing the rate of slowdown, the effect would be noticeable over millions of years only, thus being negligible.

18.7.1 *Properties of tidal turbines*

Tidal stream devices extract energy from the diurnal flow of tidal currents (caused by the gravitational pull of the moon). Unlike wind and wave power, tidal streams offer entirely predictable output. However, as the lunar cycle is of around 25 hours' duration, the timing of peak outputs differs by around an hour each day and tidal energy cannot be guaranteed at times of peak demand. Typically, tidal turbines, similar in appearance to wind turbines, are mounted on the seabed. They are designed to exploit the higher energy density, but lower velocity, of tidal flows compared to wind. Tidal stream differs from established technology for exploiting tidal energy (e.g., estuarine tidal barrages, such as the 240 MW barrage operating in France) in that tidal flows are not captured and controlled by means of a large dam-like structure. Rather, tidal turbines operate in the free flow of the tides, meaning that large construction costs and disruption of estuarine ecosystems associated with barrages may be avoided. However, as tidal streams are a diffuse form of energy and the purpose of the barrage is to concentrate tidal flow, this also means that large numbers of turbines, spread over relatively large areas of seabed, are required if significant amounts of energy are to be extracted. Until recently, the diffuse nature of the resource, combined with the relatively high costs of engineering and installing turbines able to withstand the rigors of the sea, confined tidal stream to university laboratories. However, several large grid-connected demonstration projects are expected to enter the water in the near future. Tidal stream is thus a few years behind wave energy.

18.7.2 *Categories of tidal power*

Tidal power can be classified into three main types:

— Tidal stream systems make use of the kinetic energy of moving water to power turbines, in a similar way to windmills that use moving air. This method is gaining in popularity because of the lower cost and lower ecological impact compared to barrages.

— Barrages make use of the potential energy in the difference in height (or head) between high and low tides. Barrages are essentially dams across the full width of a tidal estuary, and suffer from very high civil infrastructure costs, a worldwide shortage of viable sites, and environmental issues.

— Tidal lagoons, are similar to barrages, but can be constructed as self contained structures, not fully across an estuary, and are claimed to incur much lower cost and impact overall. Furthermore they can be configured to generate continuously which is not the case with barrages. Modern advances in turbine technology may eventually see large amounts of power generated from the ocean, especially tidal currents using the tidal stream designs but also from the major thermal current systems such as the Gulf Stream, which is covered by the more general term marine current power. Tidal stream turbines may be arrayed in high-velocity areas where natural tidal current flows are concentrated such as the west and east coasts of Canada, the Strait of Gibraltar, the Bosporus, and numerous sites in Southeast Asia and Australia. Such flows occur almost anywhere where there are entrances to bays and rivers, or between land masses where water currents are concentrated.

18.7.3 *Farm system model*

The measure of power plant is usually developed based on a cost-effectiveness analysis that can be also applied in the in-stream turbine farm. It is defined as the ratio of the total cost (sum of operational and maintenance cost and capital cost) to the total energy output, given as follows,

$$Energy\ cost = \frac{omc + cap}{Energy}, \tag{18.5}$$

where *omc* denotes levelized operation and maintenance cost, *cap* denotes levelized capital cost, and *Energy* denotes lifetime energy output of the entire farm. The objective of a cost-effectiveness analysis is to minimize the energy cost by minimizing the total cost and maximizing the energy output. The total cost incurring in producing tidal energy can be minimized by carefully selecting the operation and maintenance strategies, which is detailed in Ref. 20. In this chapter, we focus on the energy output.

18.7.3.1 *Energy output*

The energy output here refers to the amount of energy in the load center, which is ready to be delivered to the existing electricity grid, which can be calculated as follows,

$$Energy = \int P_{\text{out}} dt, \tag{18.6}$$

where P_{out} denotes the electrical power output from the in stream tidal turbine farm (Fig. 18.5), which is ready to be delivered to the electricity grid, and t and T denote the time increment and the lifetime of the in-stream tidal turbine farm, respectively. The tidal current flow is highly predictable and its velocity is almost constant during one turbine revolution. Because the tidal energy output is an integral of the power output with respect to time, we can use the power output to represent the total energy output when calculating the energy cost. The process of the power output has two elements: (1) the electrical power system; and (2) the mechanical system.

18.7.3.2 *Electrical power*

For a given in-stream tidal turbine farm, the electrical power can be expressed as follows,

$$P_e = f(P_m) \approx f_e P_m, \tag{18.7}$$

$$P_{\text{out}} = f(P_e) \approx f_t P_e, \tag{18.8}$$

where P_e, P_m, f_e, and f_t denote the electrical power, the mechanical power, the electrical efficiency coefficient, and the transmission efficiency coefficient of the farm, respectively. By combining Eqs. (18.7) and (18.8), we can rewrite the electric power output as follows,

$$P_{\text{out}} \approx f_t f_e P_m. \tag{18.9}$$

Farm configurations have very little influence on f_e and f_t; therefore, the focus of electric power output can be maintained by focusing on mechanical power output P_m.

18.7.3.3 *Mechanical power*

The ideal mechanical power from a continuous flow can be written as follows,

$$P_{\text{ideal}-m} = 1/2 \rho A U_\infty^3, \tag{18.10}$$

where ρ, A, and U_α denote the density of seawater, turbine frontal area, and the incoming flow velocity. The ideal mechanical power is a function of turbine frontal area and incoming flow velocity.

The real mechanical power from the ocean flow is much less than the ideal power due to hydrodynamic energy losses. The ratio of the real power to the ideal mechanical power is defined as the mechanical power efficiency, given as follows,

$$\eta = \frac{P_m}{1/2\rho AU^3}. \tag{18.11}$$

The mechanical efficiency can be predicted precisely using Reynolds Averaged Navier–Stokes methods and potential methods.[21]

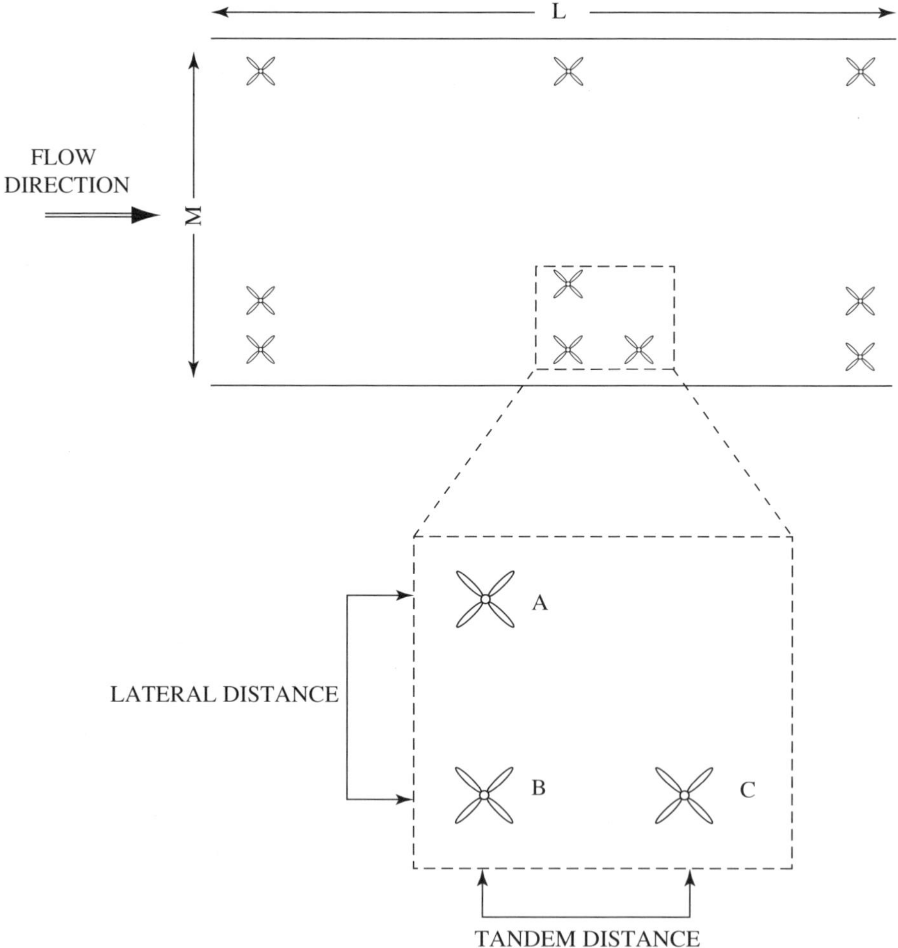

Fig. 18.5.   Illustration of front view of an ideal turbine farm.

18.7.3.4 *Nondimensionlization*

According to the discussion above, we can use power output to represent energy output and use power efficiency to represent power output; therefore, we can use efficiency to represent the energy output. Thus the objective here (i.e., to maximize the energy output) can be considered as to maximize the total mechanical efficiency of the turbine farm, i.e., the sum of the efficiency of the turbines in the entire tidal turbine farm. For a turbine farm, given the number of turbines of a specified size, the total mechanical power efficiency can be written as follows,

$$\eta = \Sigma \eta_{m,i}, \tag{18.12}$$

where $\eta_{m,i}$ denotes the power efficiency from turbine number i, and N denotes the total number of turbines in the farm.

18.7.4 *Tidal stream generators*

A relatively new technology, tidal stream generators draw energy from currents in much the same way as wind turbines. The higher density of water, 832 times the density of air, means that a single generator can provide significant power at low tidal flow velocities (compared with wind speed). Given that power varies with the density of medium and the cube of velocity, it is simple to see that water speeds of nearly one-tenth of the speed of wind provide the same power for the same size of turbine system. However this limits the application in practice to places where the tide moves at speeds of at least 2 knots (1 m/s) even close to neap tides. Since tidal stream generators are an immature technology (no commercial scale production facilities are yet routinely supplying power), no standard technology has yet emerged as the clear winner, but a large variety of designs are being experimented with, some very close to large scale deployment. Several prototypes have shown promise with many companies making bold claims, some of which are yet to be independently verified, but they have not operated commercially for extended periods to establish performances and rates of return on investments.

Engineering approaches

The European Marine Energy Centre[4] categorizes them under four heads although a number of other approaches are also being tried.

Axial turbines

These are close in concept to traditional windmills operating under the sea and have the most prototypes currently operating. These include: Kvalsund, south of Hammerfest, Norway.[5] Although still a prototype, a turbine with a reported capacity of 300 kW was connected to the grid on 13 November 2003. A 300 kW Periodflow marine current propeller type turbine — Seaflow — was installed by Marine Current Turbines off the coast of Lynmouth, Devon, England, in 2003.[6] The 11 m diameter turbine generator was fitted to a steel pile which was driven into the seabed. As a prototype, it was connected to a dump load, not to the grid. Since April 2007 Verdant Power[7] has been running a prototype project in the East River between Queens and Roosevelt Island in New York City; it was the first major tidal-power project in the United States.[8] The strong currents pose challenges to the design: the blades of the 2006 and 2007 prototypes broke off, and new reinforced turbines were installed in September 2008.[9,10]

Following the Seaflow trial, a fullsize prototype, called SeaGen, was installed by *Marine Current Turbines* in Strangford Lough in Northern Ireland in April 2008. The turbine began to generate at full power of just over 1.2 MW in December 2008[11] and was reported to have fed 150 kW into the grid for the first time on 17 July 2008.[12] It is currently the only commercial scale device to have been installed anywhere in the world.[13]

OpenHydro,[14] an Irish company exploiting the Open-Centre Turbine developed in the U.S., has a prototype being tested at the European Marine Energy Centre (EMEC), in Orkney, Scotland.

Vertical and horizontal axis turbines

Invented by Geroges Darrius in 1923 and Patented in 1929 these can be deployed either vertically or horizontally.

The Gorlov turbine[15] is a variant of the Darrieus design featuring a helical design which is being commercially piloted on a large scale in S. Korea.[16] Neptune Renewable Energy has developed Proteus[17] which uses a barrage of vertical axis (Fig. 18.6) crossflow turbines for use mainly in estuaries. In late April 2008, Ocean Renewable Power Company, LLC (ORPC)[3] successfully completed the testing of its proprietary turbine-generator unit (TGU) prototype at ORPC's Cobscook Bay and Western Passage tidal sites near Eastport, Maine.[18] The TGU is the core of the OCGen technology and utilizes advanced design cross-flow (ADCF) turbines to drive a permanent magnet generator located between the turbines and mounted on the same shaft. ORPC has developed TGU designs that can be used for generating power from river, tidal and deep water ocean currents.

Oscillating devices

These do not use rotary devices at all but rather aerofoil sections which are pushed sideways by the flow. Oscillating stream power extraction was proven with the

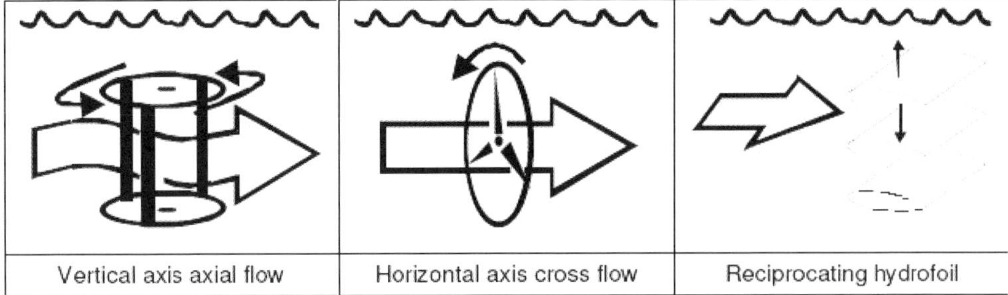

Fig. 18.6.   Classification of trubines based on axis direction.

omni or bi-directional Wing'd pump windmill.[19] During 2003 a 150 kW oscillating hydroplane device, the Stingray, was tested off the Scottish coast.[20]

Venturi effect

This uses a shroud to increase the flow rate through the turbine, as shown in Fig. 18.7. These can be mounted horizontally or vertically. The Australian company Tidal Energy Pty Ltd undertook successful commercial trials of highly efficient shrouded tidal turbines on the Gold Coast, Queensland in 2002. Tidal Energy has commenced a rollout of their shrouded turbine for a remote Australian community in northern Australia where there are some of the fastest flows ever recorded (11 m/s, 21 knots) — two small turbines will provide 3.5 MW. Another larger 5 meter diameter turbine, capable of 800 kW in 4 m/s of flow, is planned for deployment as a tidal powered desalination showcase near Brisbane Australia in October 2008. Another device, the Hydro Venturi, is to be tested in San Francisco Bay.[21] Trials in the Strait of Messina, Italy, started in 2001 of the Kobold concept.[22]

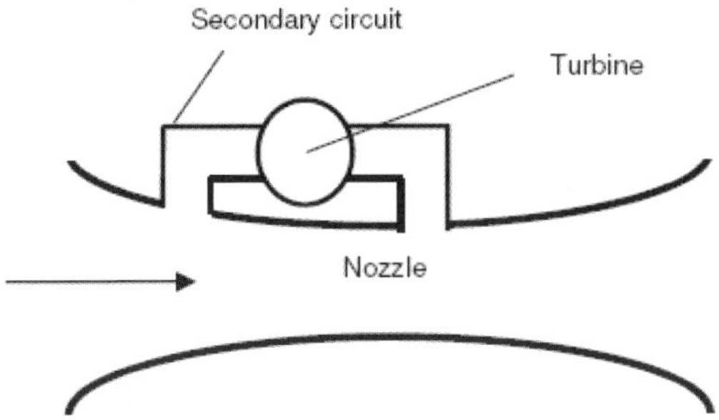

Fig. 18.7.   Schematic of a Venturi system.

18.7.4.1 *Energy calculations*

Various turbine designs have varying efficiencies and therefore varying power output. If the efficiency of the turbine "C_p" is known the equation below can be used to determine the power output. The energy available from these kinetic systems can be expressed as:

$$P = 0.5\,C_p \rho A V^3, \tag{18.13}$$

where

C_p is the turbine coefficient of performance,
$P =$ the power generated (in watts),
$\rho =$ the density of the water (seawater is 1025 kg/m^3),
$A =$ the sweep area of the turbine (in m^2),
$V^3 =$ the velocity of the flow cubed (i.e., $V*V*V$).

Relative to an open turbine in free stream, shrouded turbines are capable of as much as 3 to 4 times the power of the same rotor in open flow, depending on the width of the shroud.[26] However, to measure the efficiency, one must compare the benefits of a larger rotor with the benefits of the shroud.

Potential sites

As with wind power, selection of location is critical for the tidal turbine. Tidal stream systems need to be located in areas with fast currents where natural flows are concentrated between obstructions, for example at the entrances to bays and rivers, around rocky points, headlands, or between islands or other land masses.

The following potential sites are under serious consideration:

- Pembrokeshire in Wales.[27]
- River Severn between Wales and England.[28]
- Cook Strait in New Zealand.[29]
- Kaipara Harbour in New Zealand.[30]
- Bay of Fundy[31] in Canada.
- East River[32,33] in New York City.
- Golden Gate in the San Francisco Bay.[32]
- Piscataqua River in New Hampshire.[30]
- The Race of Alderney and The Swinge in the Channel Islands.[25]

18.7.4.2 *Barrage tidal power*

An artistic impression of a tidal barrage, including embankments, a ship lock and caissons housing a sluice and two turbines. With only a few operating plants globally

Fig. 18.8.   Layout of barrage.

(a large 240 MW plant on the Rance River, and two small plants, one on the Bay of Fundy and the other across a tiny inlet in Kislaya Guba, Russia) and a suggested barrage across the Severn river (from Brean Down to Lavernock Point), England and Wales (near Cardiff) the barrage method of extracting tidal energy involves building a barrage across a bay or river as in the case of the Rance tidal power plant in France. Turbines installed in the barrage wall (Fig. 18.8) generate power as water flows in and out of the estuary basin, bay, or river.

These systems are similar to a hydro dam that produces Static Head or pressure head (a height of water pressure). When the water level outside of the basin or lagoon changes relative to the water level inside, the turbines are able to produce power. The largest such installation has been working on the Rance river, France, since 1966 with an installed (peak) power of 240 MW, and an annual production of 600 GWh (about 68 MW average power). The basic elements of a barrage are caissons, embankments, sluices, turbines, and ship locks. Sluices, turbines, and ship locks are housed in caissons (very large concrete blocks). Embankments seal a basin where it is not sealed by caissons. The sluice gates applicable to tidal power are the flap gate, vertical rising gate, radial gate, and rising sector. Barrage systems are affected by problems of high civil infrastructure costs associated with what is in effect a dam being placed across estuarine systems, and the environmental problems associated with changing a large ecosystem.

18.7.4.3   *Ebb generation*

The basin is filled through the sluices until high tide. Then the sluice gates are closed (at this stage there may be "Pumping" to raise the level further). The turbine gates are kept closed until the sea level falls to create sufficient head across the barrage, and then are opened so that the turbines generate until the head is again low. Then the sluices are opened, turbines disconnected and the basin is filled again. The cycle repeats itself. Ebb generation (also known as outflow generation) takes its name because generation occurs as the tide changes tidal direction.

Flood generation

The basin is filled through the turbines, which generate at tide flood. This is generally much less efficient than ebb generation, because the volume contained in the upper half of the basin (which is where ebb generation operates) is greater than the volume of the lower half (filled first during flood generation). Therefore the available level difference important for the turbine power produced — between the basin side and the sea side of the barrage, reduces more quickly than it would in ebb generation. Rivers flowing into the basin may further reduce the energy potential, instead of enhancing it as in ebb generation. Which of course is not a problem with the "lagoon" model, without river inflow.

Pumping

Turbines are able to be powered in reverse by excess energy in the grid to increase the water level in the basin at high tide (for ebb generation). This energy is more than returned during generation, because power output is strongly related to the head. If water is raised 2 ft (61 cm) by pumping on a high tide of 10 ft (3 m), this will have been raised by 12 ft (3.7 m) at low tide. The cost of a 2 ft rise is returned by the benefits of a 12 ft rise. This is since the correlation between the potential energy is not a linear relationship, rather, is related by the square of the tidal height variation.

Two-basin schemes

Another form of energy barrage configuration is that of the dual basin type. With two basins, one is filled at high tide and the other is emptied at low tide. Turbines are placed between the basins. Two-basin schemes offer advantages over normal schemes in that generation time can be adjusted with high flexibility and it is also possible to generate almost continuously. In normal estuarine situations, however, two-basin schemes are very expensive to construct due to the cost of the extra length of barrage. There are some favourable geographies, however, which are well suited to this type of scheme.

Environmental impact

The placement of a barrage into an estuary has a considerable effect on the water inside the basin and on the ecosystem. Many governments have been reluctant in recent times to grant approval for tidal barrages. Through research conducted on tidal plants, it has been found that tidal barrages constructed at the mouths of estuaries pose similar environmental threats as large dams. The construction of large tidal plants alters the flow of saltwater in and out of estuaries, which changes the hydrology and salinity and possibly negatively affects the marine mammals that use the estuaries as their habitat.[30] The La Rance plant, off the Brittany coast of northern France, was the first and largest tidal barrage plant in the world. It is also the only site where a full-scale evaluation of the ecological impact of a tidal power system, operating for 20 years, has been made.[30] French researchers found that the isolation of the estuary during the construction phases of the tidal barrage was detrimental to flora and fauna, however; after ten years, there has been a "variable degree of biological adjustment to the new environmental conditions".[19] Some species lost their habitat due to La Rance's construction, but other species colonized the abandoned space, which caused a shift in diversity. Also as a result of the construction, sandbanks disappeared, the beach of St. Servan was badly damaged and high-speed currents have developed near sluices, which are water channels controlled by gates.[20]

Turbidity

Turbidity (the amount of matter in suspension in the water) decreases as a result of smaller volume of water being exchanged between the basin and the sea. This lets light from the Sun to penetrate the water further, improving conditions for the phytoplankton. The changes propagate up the food chain, causing a general change in the ecosystem.

Tidal fences and turbines

Tidal fences and turbines can have varying environmental impacts depending on whether or not fences and turbines are constructed with regard to the environment. The main environmental impact of turbines is their impact on fish. If the turbines are moving slowly enough, such as low velocities of 25–50 rpm, fish kill is minimalized and silt and other nutrients are able to flow through the structures.[30] For example, a 20 kW tidal turbine prototype built in the St. Lawrence Seaway in 1983 reported no fish kills.[30] Tidal fences block off channels, which makes it difficult for fish and wildlife to migrate through those channels. In order to reduce fish kill, fences could be engineered so that the spaces between the caisson wall and the rotor foil are large enough to allow fish to pass through.[30] Larger marine mammals such as seals or dolphins can be protected from the turbines by fences or a sonar sensor auto-breaking system that automatically shuts the turbines down when marine mammals are detected.[29] Overall, many researches have argued that while tidal barrages pose environmental threats, tidal fences and tidal turbines, if constructed properly, are likely to be more environmentally benign. Unlike barrages, tidal fences and turbines do not block channels or estuarine mouths, interrupt fish migration or alter hydrology, thus, these options offer energy generating capacity without dire environmental impacts.[30]

Salinity

As a result of less water exchange with the sea, the average salinity inside the basin decreases, also affecting the ecosystem. "Tidal Lagoons", as shown in Fig. 18.9, do not suffer from this problem.

Sediment movements

Estuaries often have high volume of sediments moving through them, from the rivers to the sea. The introduction of a barrage into an estuary may result in sediment accumulation within the barrage, affecting the ecosystem and also the operation of the barrage.

Fish

Fish may move through sluices safely, but when these are closed, fish will seek out turbines and attempt to swim through them. Also, some fish will be unable to escape the water speed near a turbine and will be sucked through. Even with the most fish-friendly turbine design, fish mortality per pass is approximately 15%. (From pressure drop, contact with blades, cavitation, etc.). Alternative passage technologies (fish ladders, fish lifts, fish escalators, etc.) have so far failed to solve this problem for tidal barrages, either offering extremely expensive solutions, or ones which are used by a small fraction of fish only. Research in sonic guidance of fish is ongoing. The Open-Centre turbine reduces this problem allowing fish to pass through the open centre of the turbine. Recently a run of the river type turbine has been developed in France. This is a very large slow rotating Kaplan type turbine mounted on an angle.

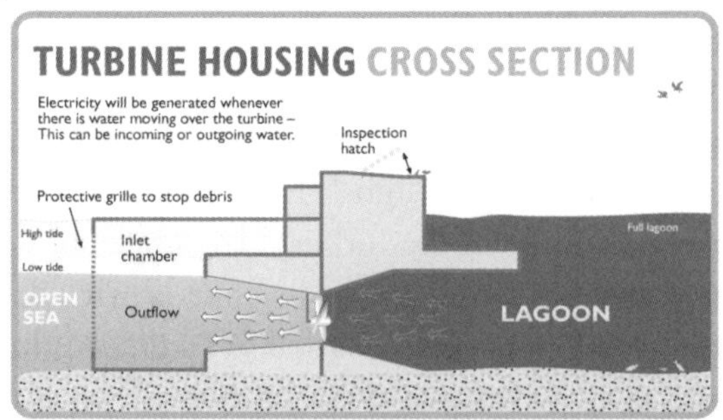

Fig. 18.9. Diagram of the tidal lagoon.

Testing for fish mortality has indicated fish mortality figures to be less than 5%. This concept also seems very suitable for adaption to marine current/tidal turbines.[15]

Energy calculations

The energy available from barrage is dependent on the volume of water. The potential energy contained in a volume of water is:[14]

$$E = 1/2 A\rho g h^2, \tag{18.14}$$

where

h is the vertical tidal range,

A is the horizontal area of the barrage basin,

ρ is the density of water $= 1025$ kg per cubic meter (seawater varies between 1021 and 1030 kg per cubic meter) and

g is the acceleration due to the Earth's gravity $= 9.81$ meters per second squared.

The factor half is due to the fact, that as the basin flows empty through the turbines, the hydraulic head over the dam reduces. The maximum head is only available at the moment of low water, assuming the high water level is still present in the basin.

Below is an calculation example of tidal power generation.

Assumptions:

Let us assume that the tidal range of tide at a particular place is 32 feet $= 10\,\text{m}$ (approx.)

The surface of the tidal energy harnessing plant is 9 km^2 (3 km × 3 km)

$$= 3000\,\text{m} \times 3000\,\text{m} = 9 \times 10^6\,\text{m}^2.$$

Specific density of sea water $= 1025.18\,\text{kg/m}^3.$

Mass of the water $=$ volume of water $\times$ specific gravity

$$= (\text{area} \times \text{tidal range}) \text{ of water} \times \text{mass density}$$

$$= (9 \times 10^6\,\text{m}^2 \times 10\,\text{m}) \times 1025.18\,\text{kg/m}^3$$

$$= 92 \times 10^9\,\text{kg (approx)}.$$

Potential energy content of the water in the basin at high tide

$$= 1/2 \times \text{area} \times \text{density} \times \text{gravitational acceleration} \times \text{tidal range squared}$$

$$= 1/2 \times 9 \times 10^6\,\text{m}^2 \times 1025\,\text{kg/m}^3 \times 9.81\,\text{m/s}^2 \times (10\,\text{m})^2$$

$$= 4.5 * 10^{12}\,\text{J (approx.)}.$$

Now we have 2 high tides and 2 low tides every day. At low tide the potential energy is zero.

Therefore the total energy potential per day = Energy for a single high tide $\times$ 2

$$= 4.5 \times 10^{12} \, \text{J} \times 2$$

$$= 9 \times 10^{12} \, \text{J}.$$

Therefore, the mean power generation potential

$$= \text{Energy generation potential/time in 1 day}$$

$$= 9 \times 10^{12} \, \text{J}/86400 \, \text{s}$$

$$= 104 \, \text{MW}.$$

Assuming the power conversion efficiency to be 30%:

The daily-average power generated $= 104 \, \text{MW} \times 30\%/100\%$

$$= 31 \, \text{MW} \, (\text{approx}).$$

A barrage is best placed in a location with very high-amplitude tides. Suitable locations are found in Russia, U.S.A, Canada, Australia, Korea, and the U.K. Amplitudes of up to 17 m (56 ft) occur for example in the Bay of Fundy, where tidal resonance amplifies the tidal range.

Economics

Tidal barrage power schemes have a high capital cost and a very low running cost. As a result, a tidal power scheme may not produce returns for many years, and investors may be reluctant to participate in such projects. Governments may be able to finance tidal barrage power, but many are unwilling to do so also due to the lag time before investment return and the high irreversible commitment. For example the energy policy of the United Kingdom[14] recognizes the role of tidal energy and expresses the need for local councils to understand the broader national goals of renewable energy in approving tidal projects. The U.K. government itself appreciates the technical viability and siting options available, but has failed to provide meaningful incentives to move these goals forward.

18.8 Mathematical Modeling of Tidal Schemes

In mathematical modeling of a scheme design, the basin is broken into segments, each maintaining its own set of variables. Time is advanced in steps. Every step, neighbouring segments influence each other and variables are updated. The simplest type of model is the *flat estuary* model, in which the whole basin is represented by one segment. The surface of the basin is assumed to be flat, hence the name. This model gives rough results and is used to compare many designs at the start of the design process. In these models, the basin is broken into large segments (1D), squares (2D) or cubes (3D). The complexity and accuracy increases with dimension. Mathematical modeling produces quantitative information for a range of parameters, including:

- Water levels (during operation, construction, extreme conditions, etc.).
- Currents.
- Waves.
- Power output.
- Turbidity.
- Salinity.
- Sediment movements.

18.9   Global Environmental Impact

A tidal power scheme is a long-term source of electricity. A proposal for the Severn Barrage, if built, has been projected to save 18 million tonnes of coal per year of operation. This decreases the output of greenhouse gases into the atmosphere. If fossil fuel resources decline during the 21st century, as predicted by the Hubbert peak theory, tidal power is one of the alternative sources of energy that will need to be developed to satisfy the human demand for energy.

18.10   Operating Tidal Power Schemes

The first tidal power station was the Rance tidal power plant built over a period of 6 years from 1960 to 1966 at La Rance, France.[26] It has a 240 MW installed capacity. The first tidal power site in North America is the Annapolis Royal Generating Station, Annapolis Royal, Nova Scotia, which opened in 1984 on an inlet of the Bay of Fundy.[25] It has a 18 MW installed capacity. The first in-stream tidal current generator in North America (Race Rocks Tidal Power Demonstration Project) was installed at Race Rocks on southern Vancouver Island in September 2006.[14,25] The next phase in the development of this tidal current generator will be in Nova Scotia.[16] A small project was built by the Soviet Union at Kislaya Guba on the Barents Sea. It has a 0.5 MW installed capacity. In 2006 it was upgraded with a 1.2 MW experimental advanced orthogonal turbine. Jindo Uldolmok Tidal Power Plant in South Korea is a tidal stream generation scheme planned to be expanded progressively to 90 MW of capacity by 2013. The first 1 MW was installed in May 2009.[17] A 1.2 MW SeaGen system became operational in late 2008 on Strangford Lough in Northern Ireland.[18]

18.11 Conclusions

Generation and distribution of the tidal power is a very complicated problem because of the lack of detail analysis of turbine working principle and the complexity of the ocean natural environment. Given the present knowledge of tides, principle of turbine and computational ability, this study developed a systematic framework of tidal farm modeling method. In general, some conclusions are drawn here, the approach can estimate the energy output based on the given turbine design, turbine farm distribution, and local condition. The approach allows users to test the sensitivity of individual component, although complete sensitivity analysis is not conducted in this study. Characteristics of every component are given as inputs in the methodology.

References

1. A.C .Baker, *Tidal Power* (Peter Peregrinus Ltd., London, 1991).
2. G.C. Baker, E.M. Wilson, H. Miller, R.A. Gibson and M. Ball, "The Annapolis tidal power pilot project," *Waterpower '79 Proc.* (1980), pp. 550–559.
3. T.J. Hammons, *Tidal Power, Proc. IEEE* **81** (2004) 419–433.
4. George E. Williams, "Geological constraints on the Precambrian history of Earth's rotation and the Moon's orbit," *Reviews of Geophysics* **38** (2000) 37–60.
5. Kate Galbraith, "Power from the restless sea stirs the imagination," *Times*, 22 September 2008.
6. http://www.nytimes.com/2008/09/23/business/23tidal.html?em (2009).
7. http://www.marineturbines.com/3/news/.
8. H. Lamb, *Hydrodynamics*, 6th edition (Cambridge University Press, UK, 1994).
9. O.M. Phillips, *The Dynamics of the Upper Ocean*, 2nd edition (Cambridge University Press, UK, 1977).
10. R.G. Dean and R.A. Dalrymple, *Water Wave Mechanics for Engineers and Scientists* (World Scientific, Singapore, 1991).
11. Y. Goda, *Random Seas and Design of Maritime Structures* (World Scientific, Singapore, 2000).
12. http://www.esru.strath.ac.uk/EandE/Web_sites/012/RE_info/wave%20power.htm (2008).
13. L.H. Holthuijsen, *Waves in Oceanic and Coastal Waters* (Cambridge University Press, Cambridge, 2002).
14. O. Reynolds, "On the rate of progression of groups of waves and the rate at which energy is transmitted by waves," *Nature* **16** (1877) 343–344.
15. L. Rayleigh (J.W. Strutt), "On progressive waves," *Proc. London Mathematical Society*, Vol. 9, pp. 21–26. doi:10.1112/plms/s1-9.1.21. Reprinted as Appendix in: *Theory of Sound* 1, MacMillan, 2nd revised edition (1894).
16. J. Cruz, *Ocean Wave Energy — Current Status and Future Prospects* (Springer, 2008).
17. M. McCormick, *Ocean Wave Energy Conversion* (Dover, 2007).
18. J. Twidell, A.D. Weir and T. Weir, *Renewable Energy Resources* (Taylor & Francis, 2006).

19. "Engineering committee on oceanic resources — Working group on wave energy conversion," John Brooke, ed., Wave Energy Conversion, Elsevier, pp. 7.

20. Y. Li, B.J. Lence and S.M. Calisal, "Modeling a tidal turbine farm system with vertical axis tidal turbine," *IEEE Int. Conf. Systems, Man, and Cybernetics (SMC 2007)*, Montréal, QC, Canada, 7–10 October 2007.

21. T.J. Hammons, "Tidal Power," *Proc. IEEE* **89** (1993) 419.

22. E.Van Walsum, "Barrier to tidal power: Environmental effects," *International Water Power and Dam Construction* (2003), pp. 38–42.

23. L.1. Tejedor, A. Izquierdo, D.V. Sein and B.A. Kagan, "Tides and tidal energetics of the Strait of Gibraltar (Spain): A modeling approach," *Tectonophysics*, Vol. 294, (1998).

24. R. Charlier, "Reinvention or aggorniamento tidal power at 30 years," *Renewable and Sustainable Energy Reviews* (1997).

25. C. Lang, "Harnessing tidal energy takes new turn," *IEEE Spectrum*, Vol. 13, September 2003.

26. G. Marsh, "Tidal turbine harness power of the sea," *Reinforced Plastics* **48** (2004) 44–47.

27. R. Pelc and R. Fujita, "Renewable energy from the ocean," *Marine Policy* **26** (2002) 471–479.

28. L. Rademakers, H. Braam, M.B. Zaaijer and G.J.W. van Bussel, "Assessment and optimization of operation and maintenance of offshore wind turbines," *Proc. European Wind Energy Conference* Madrid, Spain (2003).

29. G.J.W. van Bussel and W. Bierbooms, "Analysis of different means of transport in the operation and maintenance strategy for the reference DOWEC offshore wind farm," *Proc. OW EMES*, Naples, April 2003.

30. G.J.W. van Bussel and W. Bierbooms, "The DOWEC Offshore Reference Wind farm: Analysis of transportation for operation and maintenance," *Wind Engineering* **27** (2003) 381–392.

31. S. Anderson, "Tide-generated energy at the Amazon estuary: The use of traditional technology to support modern development," *Renewable Energy* **3** (1993) 271–278.

32. D. Deokar, "Modeling tidal power plant at Saphale," *IEEE Region 10 Annual Int. Conf. Proc./TENCON* **2** (1999) 544–547.

Chapter 19

Operational Challenges of Low Power Hydro Plants

Arulampalam Atputharajah

Department of Electrical and Electronic Engineering,
University of Peradeniya,
Peradeniya, 20400, Sri Lanka
atpu@ee.pdn.ac.lk

This chapter discusses the electrical circuits and operations of low power hydro plants. Mainly small hydro and micro hydro power plants are covered. Both the grid connected and islanded operations are summarized. Grid connection issues and power quality problems are explained with some examples. Especially in the case of small hydro power plants, operational problems and solutions through strengthening its grid connection codes are enlightened. Finally in the case of micro hydro power plants, their electrical circuits are illustrated in details. Circuits used to obtain single phase output from a three-phase machine with machine optimized operation, electronic load controllers to maintain voltage and frequency are discussed.

19.1 Introduction

Before the industrial revolution, firewood and coal were the main sources of energy. As late as in 1890 oil constituted a mere figure of the energy market. However, with the implementation of oil based industrial applications, oil consumption grew up.

In the case of hydro power plants, the increasing power demand was compensated by building large hydro plants. However, during the last decade it was found that mostly no more potential sites were available for large hydro power plants. Further increasing environmental constrains and rapid development of Renewable Energy (RE) and Distributed Generation (DG) technologies,[1] the low power hydro plants were potentially identified as good solutions.

The severe energy crisis in the last decade forced the world to pay attention on RE Sources. Today many countries' energy policies are stressing to increase the renewable energy penetration from about 10% to 30% of its generation. Some countries like UK, USA, Spain, Denmark, Germany, Sweden, India and Sri Lanka, made policies on renewable energy as mandatory with a penetration target. Also they are supported with increased tariff and relaxing power purchase agreements with Government subsidies.

Figure 19.1 shows one unit of a 2×1 MW mini hydro power plant. For example in Sri Lanka when considered only small hydro power plants,[2] it has increased from 0.12 MW in 1996 to 100 MW in 2006 and it increased further to 125 MW in June 2008 (total country power generation is about 2 GW). However, effective and efficient operation of small hydro plants is challenging when tripping occurred at the low voltage networks. Also the plant utilization is restricted when the grid voltage violates its limits.

Off-grid power generations are one of the major technologies used in small scale RE power systems. This is very important to develop the rural areas in an economical manner with a properly coordinated electrification plan. Initially the rural areas, where less population occurs and the grid expansion is very expensive, are electrified with the off-grid. The off-grid micro hydropower generation is the

Fig. 19.1.   Small hydro power plant (one unit 1250 kVA).

Fig. 19.2.   Micro hydropower plant (25 kW).

economical solution to the rural electrification of the countries, which have highly hydro resources spread by streams.

Figure 19.2 shows a 25 kW micro hydro plant operated islanded mode. For example, in Sri Lanka there are about 200 micro hydropower projects that can be seen especially in the tea plantation lands.[3] The off-grid technology has already electrified about 1.2 MW proving power to about 4500 households in rural areas till June 2008. In the case of off-grid technology, a cheap and automatic voltage and frequency control of the plant is required. Even though this control circuits are available in the market, it needs to be improved further more with research and development. This will allow to run the system smoothly with unman operations.

This chapter presents an overview of the hydro plants, and discusses challenging technical problems related to plant operations, network integration and their solutions in low power hydro power plants. Mainly it is focused on: (1) network operations on voltage violation and reactive power control of the grid connected small hydro power plants and (2) single phase supply from off-grid micro hydro power plants and its power control.

19.2   Low Power Hydro Plants

19.2.1   *Categories*

Categorizing the hydro plants vary in different countries based on their total power generation. In small countries where 2 GW is the total power generation, the low power hydro plant schemes are classified into (1) small hydropower projects, which are more than 10 MW, (2) mini hydropower projects, which are from 300 kW to 10 MW and (3) micro hydropower projects, which are less than 300 kW.

In addition based on their operational features, there are two categories of hydropower schemes according to the use of water:

(1) Storage scheme: this blocks the river flow and makes a reservoir using a dam. Water will release when power generation is needed.
(2) Run-of-river scheme: here the river flow is diverted through a channel and a penstock line to the turbine for power generation.

There are many advantages on the run-of-river scheme such as (i) low cost, as the bigger dams are not involved, (ii) simplicity gives a long-term reliability and (iii) environmental friendly, as river flow patterns of downstream is not violated and no flooding of the valley upstream of the project. However, only one disadvantage is that water does not reserve extra water from rainy seasons to the dry season.[4,5]

19.2.2 *Important components*

Figure 19.3 shows the important components of a small hydro power plant. The main components of a low power hydro plant are a weir and intake structure such as desilting and forebay tank, penstock, powerhouse, tailrace, control structure with spill way, generating plant with a set of turbine-generator unit, a step-up transformer and a transmission line if it is required to connect to the National Grid.

Designs of these components are very well explained in several books and technical papers.[6,7] However, the interconnection issues and technical information related to optimizing the small hydro power plants are not well addressed in past literatures. Commonly the small hydro plants are connected to the distribution systems therefore in terms technical connection requirements, they fall under the category of Distributed Generations (DG).

19.2.3 *Interconnection issues*

These low power hydro plants are normally taken as DG types. There are many possibilities for interconnecting the DG type power plants with the power system grid. The complexity of the DG operation generally depends on the level of interaction with the existing network, especially on voltage control issues. Mostly due to over voltage problems, the DGs are unable to utilize to get their maximum power. When DG pumps active power through distribution lines it increases the terminal voltage due to more resistive line impedance. Therefore voltage control through reactive power control is mostly recommended. Few DG interconnections are explained, which are related to the low hydro power plants.[8–10]

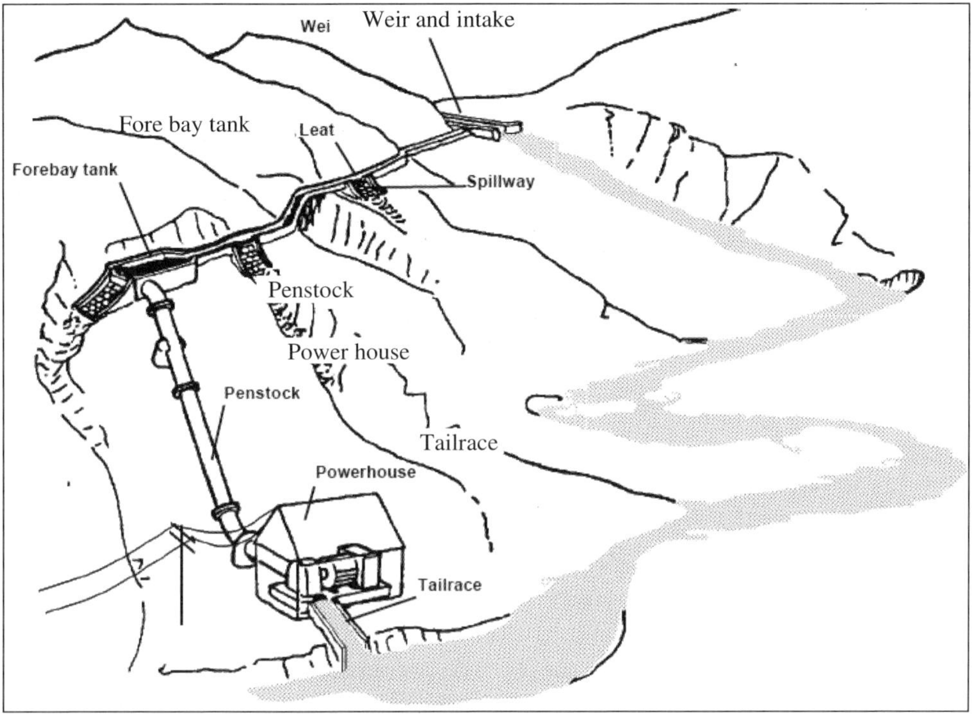

Fig. 19.3.   Important components of a small hydropower plant.

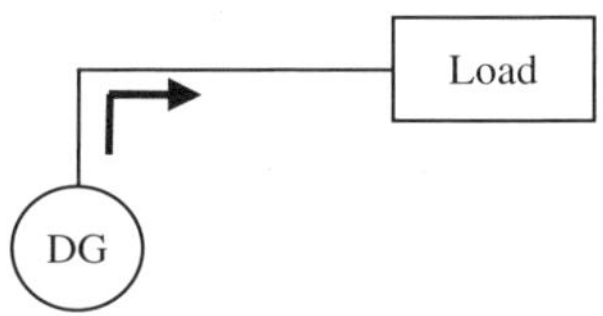

Fig. 19.4.   Stand alone DG connection.

(a) *Isolated standalone power plant*

Figure 19.4 shows the stand alone DG connection. In this case the load is met by DG only and no network connection with the grid network. This is more suitable for micro hydro level of power plants. Here the voltage and frequency control must be engaged at the DG. This is more suitable where the national grid extension is very expensive and the required area is very far from the grid network.

(b) *Network support is used as back up*

Figure 19.5 shows the DG connection where the network support is used a back up supply. DG provides power to Load 2 (part of the total load). The network covers

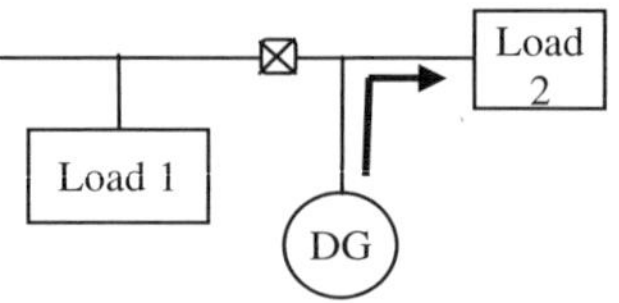

Fig. 19.5.   Isolated system with automatic transfer.

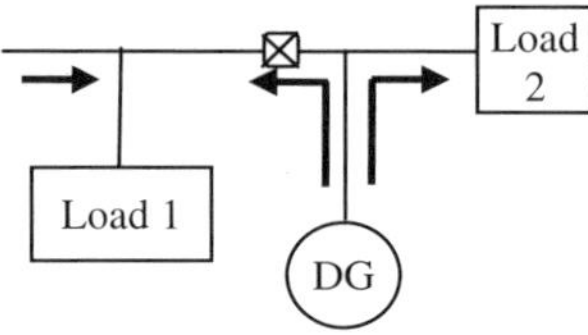

Fig. 19.6.   DG connected to the network with no power export.

Load 1 and 2 when needed. DG does not need to be operated with the network in parallel. Here the DG must be engaged with frequency and voltage control. Mostly the loads in the "Load 2" section are based on passive loads, which will not be very sensitive to the frequency of the supply. The automatic or manual transfer switch can be used based on the facility required and type of DG.

(c) *Connected to the network without exporting power to the network*

Figure 19.6 shows the DG connection where it is connected to the network but not exporting power to the network. DG provides power to Load 2 and partially to the Load 1. The network covers balance of the Load 1. Here the DG may be engaged for voltage control based on the grid code requirements. DG need not to contribute to frequency control. The maximum load in the "Load 2" section is less than the DG's maximum power while the minimum of the total load is higher than DG ratings. Therefore here the DG will not pump any excess power to the grid so that there would not be any reverse power flow in the grid network. This minimizes changes in the grid protection system.

(d) *Connected to the network with exporting power to the network*

Figure 19.7 shows the DG connection where its agreement supports for the export of power to the network. Therefore here the DG provides power to Load while exporting extra power to the grid network. There are two types: (i) always exports power to the network, where DG rating is higher than the load and (ii) exports power only when excess power is available. DG may be engaged for voltage control based on the grid code requirements. But it is recommended to go with voltage control wherever required so that the DG can be properly utilized as its power export needs

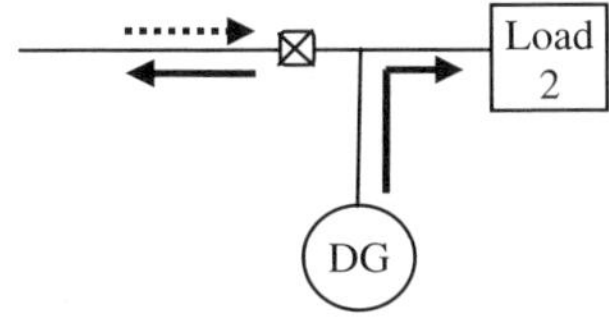

Fig. 19.7.   DG connected to the network with power export.

not to be constrained by network voltage limits. This kind of DG connection is beneficial to the utilities. For example, if the system peak load is from domestic loads, if this DG is operated with an industry which is not operated during the peak time (closing down by 5 pm), then the DG supports level the system load by double factor, so that the transmission network utilization becomes very high and thus the payback is reduced. The attractive payback catches the attention of transmission line infrastructure development.

(e) *Stand by operation or peak load operation*

Figure 19.8 shows the DG connection where it is used for peak load operation or back up generation. This type of DG connection is mainly used to the sensitive industrial loads to avoid disturbance during any power failures. Some industries use DG to supply their demand during peak load period. This is mainly dependant on the utility electricity tariff. If the tariff is too high during the peak load period then industries go for this arrangement. In this category, the industrial load is connected to DG or utility at once. DG will not operate in parallel with utility supply. The DG operating time will be minimal.

(f) *Connected to the network exporting power to the utility*

Figure 19.9 shows the normal DG connection. Here the DG is connected to the network all the time and exports power to the utility. DG may be engaged for voltage control based on the grid code requirements. This concept does not include

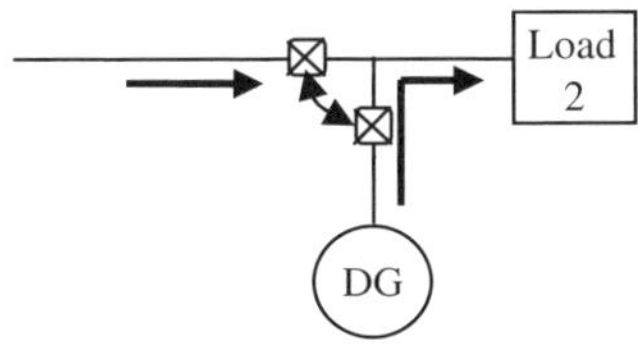

Fig. 19.8.   DG operated as stand by or during peak load period.

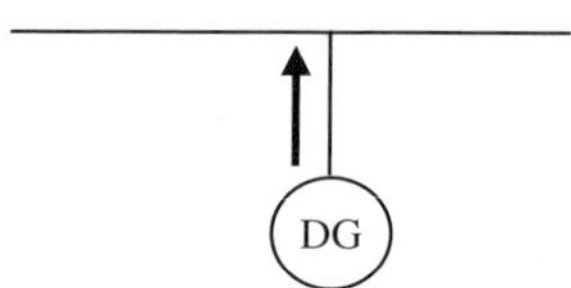

Fig. 19.9.   DG connected to the utility network with power export.

any industrial loads. This is mainly dealt with Utilities through Independent Power Producers (IPP) agreements. This kind of DG connection is also beneficial to the utilities by minimizing the transmission losses. Here the DGs are allowed to install closer to the load centers and operate in parallel to the grid network. The DG operation is fully determined by the utility rather than industries.

19.2.4   *Operational challenges and grid code requirements*

The DGs are connected to the low voltage networks, where the load varies frequently and rapidly. Further, the Small Hydropower Plants (SHP) are located in rural areas, where the grid network is very weak, far in distance and mostly depend on low voltage networks. It will be experiencing several tripping happening in the network mainly due to tree touching. Therefore additional care should be taken on voltage control at steady state operations. The transient operation must be considered especially on fast load rejection due to frequent tripping of the low voltage grid network.[7]

The SHPs are usually operated according to DG concept in a voltage-following mode. Therefore the DG will not attempt to control the network voltage. However the power injection into the network changes the voltage as a result of current flows through the system impedance. These changes create interactive operations with the existing utility voltage control equipment such as Load Tap Changers (LTCs) and feeder regulators.

Figures 19.10 and 19.11 show a scenario, where a DG at downstream of a regulator can cause low voltage at the end of the circuit due to the use of line drop compensation on the regulator. Here the load along the feeder is also assumed as gradually decreasing due to power distribution. The power injected by the generator

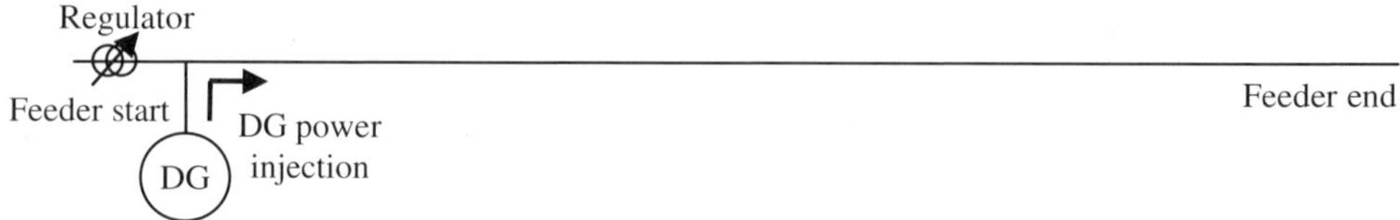

Fig. 19.10.   Circuit configuration with the DG at downstream of a regulator used to analyze the voltage along the feeder.

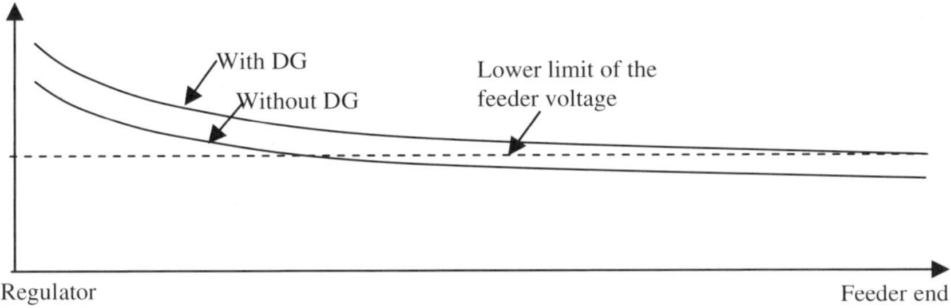

Fig. 19.11.   Voltage distribution along the feeder with and without the DG.

will not allow the regulator to properly act for the voltage changes. In the case of reverse power flow the regulator can change all the way to the lowest tap, which would cause very low voltage at the downstream feeder end.

To overcome the voltage problem the SHPs can be operated together with the voltage control devices in the distribution system through proper coordination. This will allow the DG to export power without violating the feeder voltages. To achieve this some options are listed: (i) To reduce the line drop, raise the compensator voltage set point slightly. This moves the regulator constant voltage point closer to the regulator and reduces the impact of the DG, (ii) if the reverse power flow causes the voltage violation then advance regulator controllers can be used to change the operating mode, (iii) moving either the generator or the regulator can also be considered at the planning stage, so that the regulator is placed at the downstream of the generator, (iv) installation of a new regulator can also be an option, especially if the DG is large relative to the feeder capacity, (v) another consideration is widening the regulator bandwidth setting, which reduces the line drop compensation. Increasing the time delay can also help to prevent excessive regulator tap changes, depending on the rate of change of the DG's energy source, (vi) if possible, integrate the DG closer to the feeder end. This will solve any problem due to low voltage at the feeder end. However a proper study must be made at the planning stage to find the best option for a case.

19.3   Micro Hydro Plants

19.3.1   *Overview*

Boosting the economy of a country is mainly achieved by providing electricity to all nations. A major challenge to achieving this target is providing the electricity in an economical way to the rural areas, which are far away from the national grid and the houses are spread widely. Further fossil fuel operated generations are also not wanted according to the environmental constraints. Therefore the stand alone operated micro hydro plant concept was used as a potential energy source especially when there are small waterfalls or rivers. Sri Lanka is one of very good example country for the off-grid electrification. According to Renewable Energy for Rural Economic Development (RERED) project reports,[11] the off-grid technology has already electrified about 1.2 MW, providing power to about 4500 households in rural areas till June 2008. In the case of off-grid technology, a cheap automatic voltage and frequency control of the plant is one of the important section that needs to be developed. This will allow to run the system unman and smoothly.

19.3.2 *Micro hydro generation technology*

In micro hydro schemes, either synchronous or induction machines can be used as the generator. Due to the simple construction, robust operation, wider availability and cheap cost, induction generator sets are much popular in micro-hydro industries. Further single phase network is more economical than a three phase network in rural areas. Therefore generally three phase induction generators are used with well known C-2C arrangement to obtain a single phase output while maintaining the balance operation in the machine.[12]

Figures 19.12 and 19.13 show the C-2C circuit configuration and phasor diagram of the voltages and currents in the circuit. Here the single phase load is connected across the terminals *a* and *b*. Therefore the total output power is same as the single

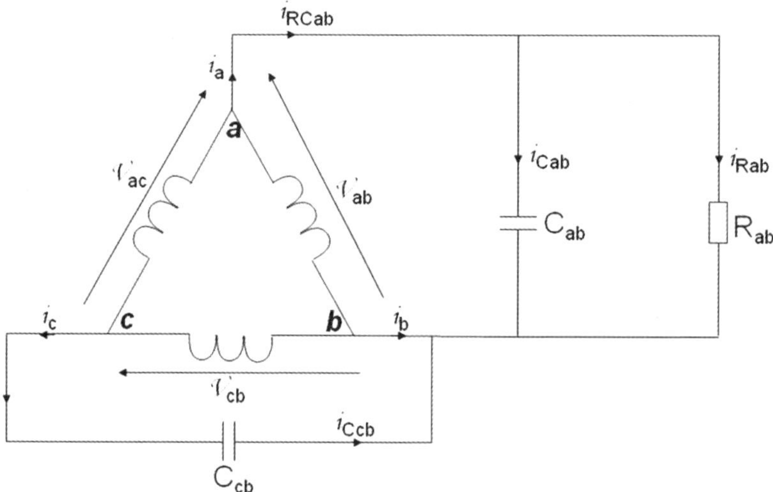

Fig. 19.12. C-2C circuit configuration to obtain the single phase supply from the three phase induction generator.

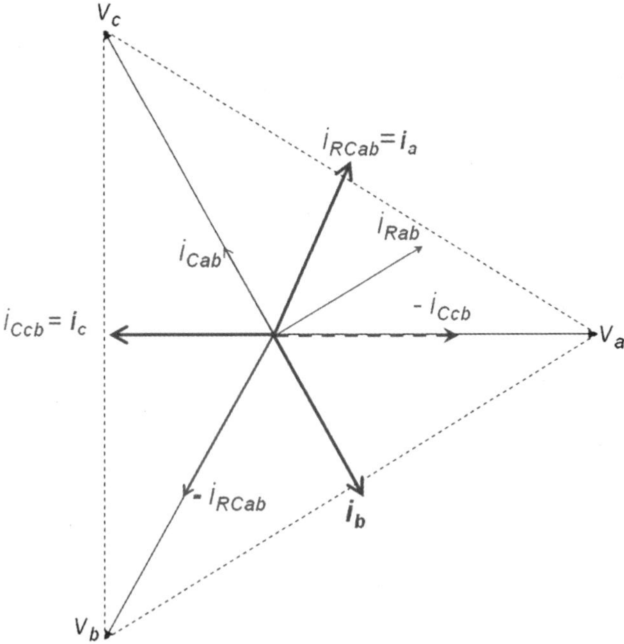

Fig. 19.13.   Phase diagram of the current and voltages in the C-2C circuit configuration.

phase power that can be written as:

$$P_{total\_1\phi} = v_{ab} \times i_{Rab} = V_{LL} \times I_{LL} \times \cos 30°. \qquad (19.1)$$

Further, it can be proved that in order to make the load on the machine to be balanced, the angle between the i_a and the i_{Cab} should be 60°. Then the angle between the i_a and i_{Rab} should be 30°. Then i_{Rab} can be written in terms of i_a. Therefore the total power can be written as

$$P_{total\_1\phi} = v_{ab} \times i_a \times \cos 30°. \qquad (19.2)$$

Hence, the power delivered by the machine becomes 50% of its ratings, as shown below, when C-2C configuration is used.

$$P_{total\_1\phi} = \frac{1}{2} P_{total\_3\phi}. \qquad (19.3)$$

This is the only drawback of the C-2C configuration. However, it simplifies the required number of electronic controllers to be one with a single phase arrangement. Therefore this configuration is mostly used in many micro hydro scale generating schemes especially with islanded operation.

19.3.3 *Load controllers used in micro hydro generation technology*

Normally, a mechanical governor system is used to match the input power to the turbine according to the load demand. However due to its complexity, a new mechanism of Electronic Load controller (ELC) came into the industry and became very popular.[13] In ELC, the load is controlled using electronic devices. This maintains a constant electrical load on the generator in spite of change in user loads. ELC uses a ballast load to damp the extra power that is not required by the users, so that it maintains the constant load on the generator. This permits the use of a turbine without governor control system if it is supplied roughly with constant head and water flow. Hence the ELC maintains the machine speed. However, it has been reported that in micro hydro systems, ELC fails very frequently and requires frequent maintenance. Hence, in order to develop a robust ELC system, a thorough understanding of the effects of the variations of the various parameters on the system is a necessity.

19.3.4 *Recent developments on micro hydro generation technology*

Failures of electronic ballast were suspected due to risk on the control techniques used. The traditional control technique uses only the load current and tries to keep the load constant. Studies have been done to accommodate the generator speed and terminal voltage to this traditional control technique. The addition of speed and terminal voltage parameters makes more stable control with solid determination on the system status. However, this also makes the control technique complicated with many input parameters.

In case of C-2C configuration, it was proven that its utilization is only 50% of the machine rating. Recent researches are focused on increasing its utilization. Therefore a new technique to generate two phase power from a three phase generator using C-2R-R arrangement is proposed.[14] This becomes more attractive as it utilizes the machine's rating up to 86.6%. As shown in Figs. 19.14 and 19.15, and according to Eqs. (19.4) to (19.7), it can be shown that with two phase output configuration, it is capable of delivering more power than that with the traditional C-2C method. Two phase loads can be connected with common delta point as neutral terminal and due to two phase configuration the total load power is increased.

$$P_{total\_2\phi} = v_{ab} \times i_{Rab} + v_{cb} \times i_{Rcb}, \tag{19.4}$$

$$P_{total\_2\phi} = V_{LL} \times I_{LL} + V_{LL} \times I_{LL} \times \cos 60°, \tag{19.5}$$

$$P_{total\_3\phi} = \sqrt{3} \times V_{LL} \times I_{LL}, \tag{19.6}$$

$$P_{total\_2\phi} = \frac{\sqrt{3}}{2} P_{total\_3\phi}. \tag{19.7}$$

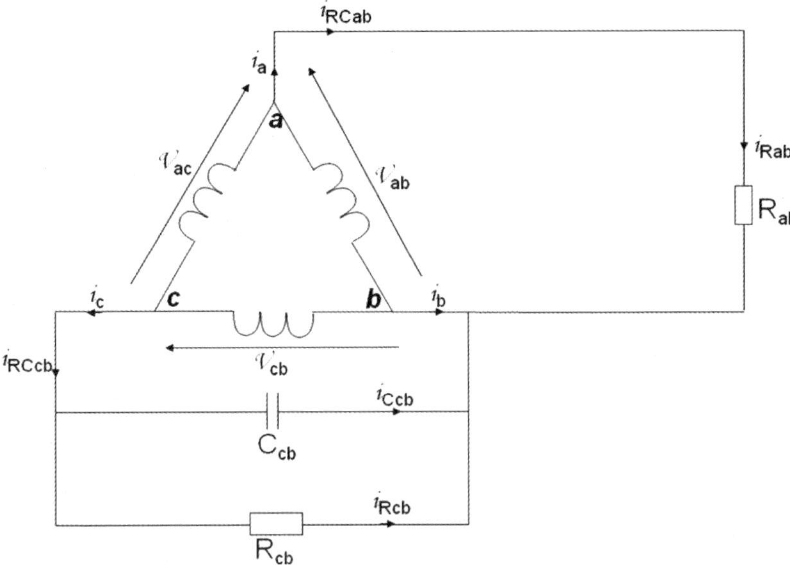

Fig. 19.14.   C-2R-R circuit configuration to obtain two phases supplies from a three phase induction generator.

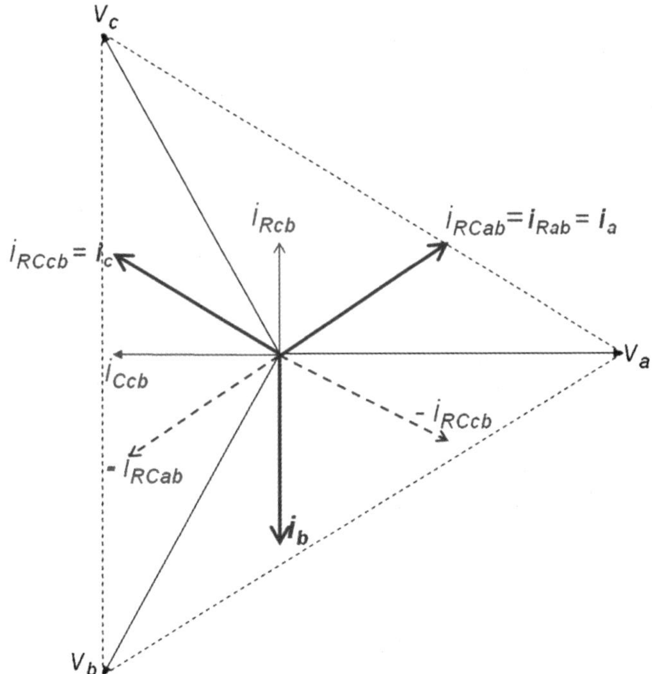

Fig. 19.15.   Phase diagram of the current and voltages in the C-2C circuit configuration.

Hence, the power delivered by the machine becomes 86.6% of its ratings, as shown in Eq. (19.7), when C-2R-R configuration is used. However, this technique is still in the research and development stage as it has minimum amount of capacitors for exciting the machine. Therefore if the machine is not having enough residual flux then the starting up procedure of this configuration still seems to be complicated.

19.4 Concluding Remarks

This chapter has summarized the mini hydro and micro hydro generations. The mini hydro generation technology has been very well developed. Its connection methods are discussed in detail for the benefit of operators of both plant and the national grid. The micro hydro technology is mainly used in islanded connections. In rural areas single phase connections are more economical when compared with three phase connections. Therefore the traditionally used C-2C configuration is explained to get a single phase output from the three phase machine. Further recent research and development on this micro hydro technology is explained with a two phase output from the three phase machine.

References

1. N. Jenkins, R. Allan, P. Crossley, D. Kirschen and G. Strbac, "Embedded Generation," *IEE Power and Energy Series*, vol. 31, (2000).
2. Siemens Power Technologies International Manchester UK, "Final report on overview of technical requirements, connection and management of embedded generation," Project No: 108763-0001, Renewable Energy for Rural Economic Development Project, http://www.energyservices.lk/pdf/finalreport_27_07_05v3p.pdf.
3. AC Nielsen Lanka (Pvt) Ltd, Administrative Unit of the (RERED) Project, "Final report on the consultancy for conducting a consumer satisfaction survey for the village hydro schemes and solar home systems," Renewable Energy for Rural Economic Development Project, December 2006, http://www.energyservices.lk/pdf/ consumer_satis_surv_06.pdf.
4. G.R.C.B. Gamlath, A. Atputharajah and I.H.D. Sumanaratne, "Electrical systems of a grid interconnected 2.0 MW mini hydropower project at siripagama," *Int. Conf. on Small Hydro Power, Conf. Proc. Kandy*, Sri Lanka, 22nd–24th October 2007, pp. 85–93.
5. G.R.C.B. Gamlath, "Design of a grid connected mini hydropower plant at Siripagama," Thesis paper, Department of Electrical and Electronic Engineering, University of Peradeniya, Sri Lanka, October 2007.
6. A. Harvey, A. Brown, P. Hettiarachchi and A. Inversin, "Micro-hydro design manual — A guide to small-scale water power schemes," *Intermediate Technology Publications* (1993), pp. 153–304, 321–348.
7. T.S. Bhatti, R.C. Bansal and D.P. Kothari, *Small Hydro Power Systems* (Dhanpat Rai & Co. Delhi).

8. A.I. Weerasekera, A. Arulampalam and J.B. Ekanayake, "Limitation on connecting mini hydro power plants to the Sri Lankan power system network: A case study at Balangoda Grid Sub-station," *Peradeniya University Research Sessions 2007 Proc.*, University of Peradeniya, Sri Lanka, 30 November 2007, pp. 263–265.

9. A.I. Weerasekera, "Impact identification of distributed generation on Sri Lankan power system," Thesis Paper, Department of Electrical and Electronic Engineering, University of Peradeniya, Sri Lanka, January 2010.

10. Ceylon Electricity Board, "Guide for grid interconnection of embedded generators-part 2," Head office, CEB, Sri Lanka, December 2000.

11. Renewable Energy for Rural Economic Development Project, Sri Lanka, http://www.energyservices.lk/theproject/introduction.htm.

12. J.B. Ekanayake, "Induction generator for small hydro schemes," *IEE Power Engineering J.* **16** (2002) 61–67.

13. B. Singh and G.K. Kasal, "Analysis and design of voltage and frequency controllers for isolated asynchronous generators in constant power applications," *Int. Conf. on Power Electronics Drives and Energy Systems* (PEDES'06), 12–15 December 2006, pp. 1–7.

14. T.S. Weerakoon, R.P.S. Chandrasena and A. Arulampalam, "Novel C2R-R configuration for micro-hydro plants used in islanded systems," *J. Energy and Environment* (2009), pp. 51–58.

Chapter 20

Frequency Control in Isolated Small Hydro Power Plant

Suryanarayana Doolla

*Department of Energy Science and Engineering,
Indian Institute of Technology Bombay, 400076, India
suryad@iitb.ac.in*

In this chapter, the importance of power system parameters (frequency and voltage) and various techniques to maintain them within limits are discussed. In the case of an isolated small hydro power plant, frequency is generally maintained constant either by dump load/load management or by input flow control. A huge amount of energy is wasted in the former technique while the latter was found to be expensive. Frequency control by using a combination of dump load and input flow control is discussed in this chapter.

Mathematical modeling of frequency control techniques for an isolated small hydro power plant along with the transfer function block diagram are covered in detail. For more understanding of the concept of frequency control, case studies using Simulink tool box of MATLAB are presented. Finally the trade-offs of conventional, combined flow control and reduced dump load techniques are summarized.

20.1 Introduction

Frequency and voltage are two important parameters of an electric power system. In a conventional power system it is assumed that the load frequency control and excitation control are non-interactive, as small changes in active power are mainly dependent on changes in generator speed and are almost independent of changes in terminal voltage, while small changes in terminal voltage are mainly dependent on machine excitation and are almost independent of changes in generator speed. Furthermore, the excitation control is fast acting whereas the power frequency control is slow acting. Therefore the load frequency control loop and the load-voltage control loop are assumed to be decoupled. The frequency control loop is slow acting as the major time constant is contributed by the turbine and moment of inertia of the generator. In a small hydro power (SHP) dump load is generally used for frequency control.

The consumer equipment operates at a high efficiency with minimum wear and tear if frequency and voltage of the supply are maintained at nominal values. The primary aim of frequency control is to maintain balance between power demand and power generation. In the case of isolated SHP plants, the input hydro flow rate is held constant so that constant power is produced. Therefore the sum of the actual load and dump load is constant and equal to the power generated. The dump load controller based on the control signal adjusts the dump load so that the deviation in frequency reduces to zero. The load can vary from no load to full load, therefore, the minimum size of the dump load selected is equal to the full load.

In some cases the SHP generation is not sufficient to meet the demand of the local community. The other available local sources like wind, photo voltaic (PV) along with diesel power generation are put in parallel so that generation can be enhanced. Such systems are called isolated or stand alone hybrid power systems.[1–4]

The advantage of such a hybrid system is the reduced size of the dump load (or priority switched load) and additional peak demand can be met by using short duration storage devices such as battery energy storage system (BESS),[5–8] aqua electrolyzer hydrogen storage system,[9–11] superconducting magnetic energy storage system,[12–15] flywheel,[16,17] etc. In this section various frequency controls, using load management and using dump load are discussed in detail.

20.1.1 *Frequency control using dump load*

In a stand alone small-hydro generation system due to a non-availability of storage facility for water, the total input has to be converted into electrical energy. Since the input to the generator is essentially constant, the excess power due to a decrement in load is dissipated into the dump load.

What is dump load?
Dump load is generally a fixed resistive load whose power consumption is controlled by generally semiconductor switches.[18] Dump load can be fed either by a 6-pulse rectifier or a chopper and the power consumed by the resistive bank is determined by the firing angle or by a duty cycle of the intermediate circuit.[2,4,19–21]

An impedance controller (a combination if resistor and inductor) can also be used as a dump load[22] where both real and reactive powers can be controlled. A capacitor bank is often connected in parallel to such a system, to which it acts as a harmonic filter for higher order harmonics. Detailed mathematical modeling of such systems is given in Refs. 3, 23 to 25. Such systems are also called electronic load governors or electronic load controllers. Detailed modeling of such a controller is given in the next section.

Distributed load control systems can be more robust than centralized systems because if one of the load controller fails the system can still continue to function. A fuzzy logic based distributed load controller was designed and tested on a micro-hydro power system[5,26] for controlling frequency. The results were encouraging with the fuzzy controller compared to the conventional load controller.

The concept of using an electric water heater as a dump load to meet the load balance has been proposed in Refs. 18, 27 and 28. This kind of system has proved to be cost effective when there is a hot water requirement. The maximum size of the dump load available is limited and it varies from country to country.

The only drawback of frequency control using dump load/load management is that the excess power is wasted in heaters and all the water available at the high head in the forebay tank is transferred through the tail-race. But the water head in the penstock is almost held constant by the arrangement of the forebay tank and spillway. Therefore the power output of the hydro scheme can be controlled by controlling the flow rate of the water in the penstock.[29–32] The flow rate can be regulated by diverting water through a small section of multi-pipes fitted with control valves and by placing the arrangement in series with the penstock. This arrangement is preferably placed close to the forebay tank so that surges due to sudden closing of the valves can be avoided.

Using a combination of control valve and dump load there are two best schemes.

- On/off control valve plus dump load (reduced in size).
- On/off control valve plus variable control valve (using servo/stepper motor).

The functional block diagram of an isolated SHP plant equipped with dump load as well as flow control mechanism is shown in Fig. 20.1.

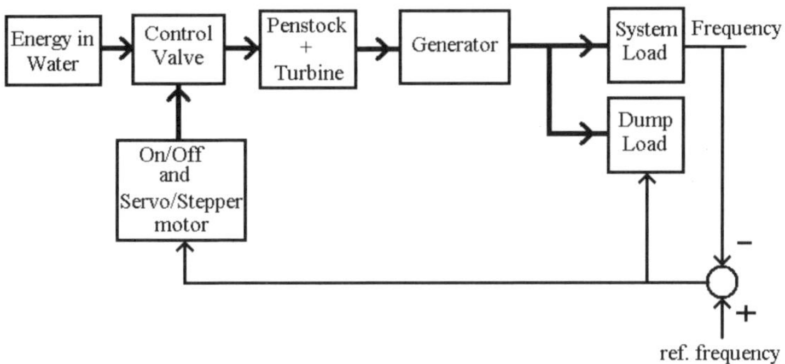

Fig. 20.1.   Functional block diagram for frequency control of an isolated SHP plant.

Mathematical modeling, simulation (using Simulink tool box of MATLAB) and analysis of conventional frequency control and using above two schemes is discussed in the subsequent sections.

20.2   Mathematical Modeling of an Isolated SHP Plant

20.2.1   *System modeling with conventional control*

Figure 20.2 shows a generalized block diagram of an isolated SHP plant with dump load and electronic load controller (ELC).

20.2.1.1   *Electronic Load Controller (ELC)*

The dump load along with ELC is capable of maintaining the frequency constant and there is no control over the input side of the hydro turbine.[18,20–29,33,34] The increase in frequency due to a decrement of system load is compensated by an addition of the dump load in parallel to the system load by the ELC. This kind of frequency control techniques using dump load and ELC are capable of compensating wide fluctuations in load from nominal.

The change in frequency of the system due to a disturbance of load is sensed and the appropriate firing signal is generated for the gate circuit of the thyristor bridge connected to the dump load. By controlling the firing angle of the thyristor bridge, the voltage across the dump load is controlled, which has a direct control over power and hence the frequency of the system. The sensing of disturbance in the load and generation of the firing angle for the thyristor bridge in a closed loop system is done by the ELC. The whole process involves measurement devices, and firing angle circuits, and hence the transfer function block diagram of the measurement circuit

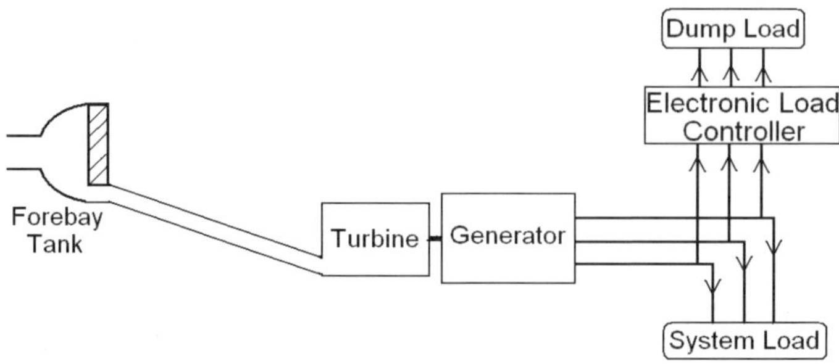

Fig. 20.2.   Schematic diagram of an isolated SHP plant with dump load and its controller.

along with firing angle generation is given by[30]:

$$\Delta Y_1(s) = \frac{K_M}{1 + sT_M} \Delta Y(s), \tag{20.1}$$

where K_M is gain constant of the transfer function and T_M is the time taken by the measurement circuit to generate the firing angle to the thyristor bridge from the instant of disturbance in load. If the measuring circuit is designed accurately then it should not take more than a cycle (0.02 sec) to perform this task.

20.2.1.2 *Dump load*

As discussed earlier the dump load is a resistive bank connected with the thyristor bridge. The modeling of the dump load is the same as the normal resistive load[35] and hence its transfer function is given by:

$$\Delta P_{DL}(s) = \frac{K_{DL}}{1 + sT_{DL}} \Delta Y_1(s). \tag{20.2}$$

20.2.1.3 *Proportional plus Integral (PI) controller*

The most important component of ELC is its controller. The PI controller is generally used to reduce peak deviation and eliminates steady state error in frequency. The transfer function equation is given by:

$$\Delta Y(s) = \left(K_{PDL} + \frac{K_{IDL}}{s} \right) \Delta F(s). \tag{20.3}$$

The sign of K_{PDL} and K_{IDL} is taken Positive as the dump load has to increase with the increase in frequency.

20.2.1.4 *Generator and load*

The power system load comprises of resistive loads, motors, generators, transformers, etc. For the sake of small signal analysis, the transfer function of the generation system along with load[35,36] is considered as,

$$\Delta F(s) = \frac{K_P}{1 + sT_P} [\Delta P_G(s) - \Delta P_L(s) - \Delta P_{DL}(s)], \tag{20.4}$$

where K_P and T_P are power system gain and time constants.

20.2.1.5 *Hydro turbine*

The transfer function of a hydraulic turbine for load frequency studies is given as[35–40]:

$$\Delta P_G(s) = \frac{1 - sT_W}{1 + \dfrac{sT_W}{2}} \Delta P_E(s), \tag{20.5}$$

where T_W is the water starting time in the penstock and represents the delay of water in the penstock in this transfer function and is proportional to t_P, therefore

$$T_W = kt_P. \tag{20.6}$$

Also, the time taken by water to cross the penstock, t_P is given by:

$$t_P = \frac{l}{v} = \frac{l}{\sqrt{2gh}}, \tag{20.7}$$

where

l = length of penstock, m
v = velocity of water in penstock, m/sec
g = acceleration due to gravity, m/sec^2
h = hydraulic head at the gate, m.

The range of T_W is generally taken as 0.5 to 4.0 sec based on the head of water. The transfer function block diagram of an isolated SHP plant equipped with dump load circuit along with ELC is shown in Fig. 20.3.

20.2.2 *System modeling of an isolated SHP plant*

In the conventional method, frequency is maintained constant by keeping the total load (actual + dump load) constant. The addition or reduction of the dump load depends upon the variation in frequency of the system. Water for irrigation is of prime importance for survival of the local population at a large number of locations of SHP plants. Moreover the plant load factors are in the range of 40 to 60%, therefore, water can be saved by employing proper control techniques. Modeling of various components of isolated SHP are given in this section. Detailed analysis and simulation of such a system is given in the next section.

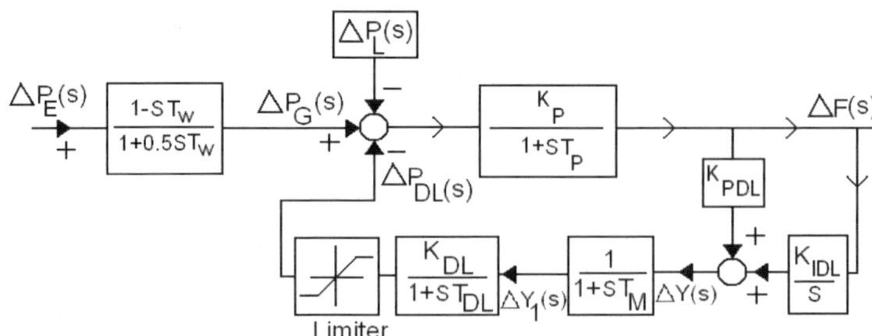

Fig. 20.3. Transfer function block diagram of conventional isolated SHP plant equipped with dump load.

20.2.2.1 *Modeling of the on/off control valve*

The on/off control valve introduced in the pipes of the multiple pipe flow control (MPFC) system plays a vital role in the reduction of size of the dump load. The on/off control valve is opened or closed slowly to avoid turbulence. The decision of the time for "on/off" the control valve, when the load increases or decreases, plays a vital role in the system dynamics. Once the decision circuit decides to toggle the generation, it is passed through an integrator to ensure linear opening or closing of the valve. A limiter is added at the output of the integrator so as to ensure that the on/off control valve operates within its limits. The transfer function block diagram of the on/off control valve circuit used in the study is shown in Fig. 20.4.

20.2.2.2 *Modeling of the servo motor controlled valve*

Servo motor control systems are finding widespread applications, mostly in automatic production process. This includes nearly all types of packaging machinery, material handling, assembly and other applications including robotics.[41,42] The programmability feature of servo systems increases its adaptability to different control applications. A very common method of position control using a servo motor is by controlling armature voltage and keeping the field constant.[36,43–45] A schematic diagram of a dc servo motor with armature control for controlling the water flow rate in the penstock is shown in Fig. 20.5.

The flow of water is controlled by controlling the position of the valve. The transfer function of dc servo motor is given as[36,46]:

$$\frac{\theta(s)}{E_a(s)} = \frac{K}{s[L_a J s^2 + (L_a b + R_a J)s + R_a b + KK_b]}, \quad (20.8)$$

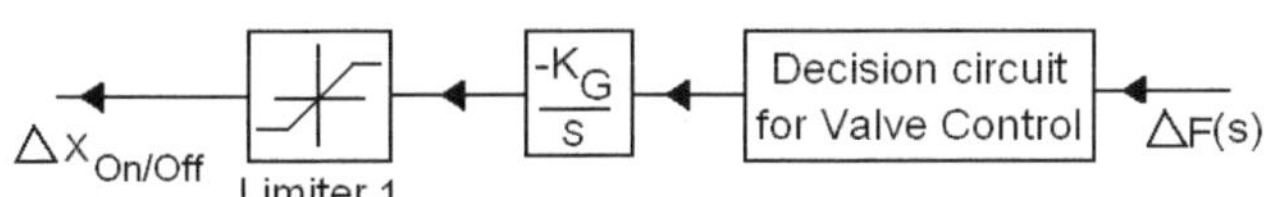

Fig. 20.4. Transfer function block diagram for the on/off control valve circuit.

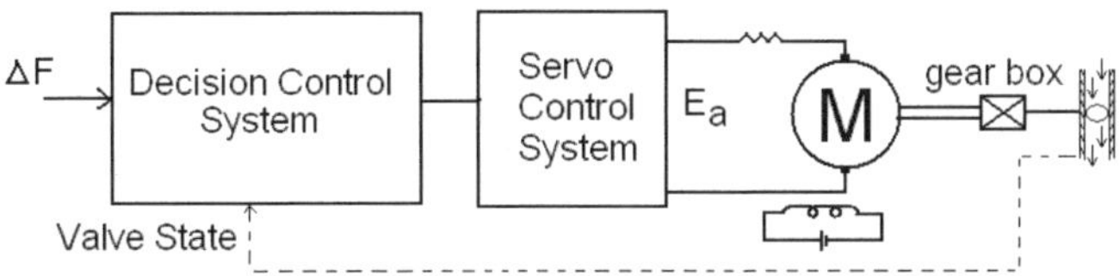

Fig. 20.5. Servo motor controlled valve.

where

R_a = armature resistance, ohm
L_a = armature inductance, henry
θ = angular displacement of the motor shaft, radian
J = equivalent moment of inertia of the motor and load, kgm^2
b = viscous-friction coefficient of the motor and load, N-m/rad/sec
K_b = constant
K = angular velocity and voltage constant in rpm/voltage.

The inductance L_a in the armature circuit is usually small and may be neglected. If L_a is neglected, then the transfer function of Eq. (20.8) is reduced to:

$$\frac{\theta(s)}{E_a(s)} = \frac{K_{\text{Servo}}}{s(sT_{\text{Servo}} + 1)}, \qquad (20.9)$$

where

$K_{\text{Servo}} = K/(R_a b + K K_b)$ is motor gain constant
$T_{\text{Servo}} = R_a J/(R_a b + K K_b)$ is motor time constant.

20.3 Frequency Control using On/Off Control Valve with Reduced Size of Dump Load[29–31,47]

The functional block diagram of the scheme consisting of a multi pipe flow control system (MPFC) with reduced dump load is shown in Fig. 20.6(a).

In this scheme the penstock flow is controlled through two pipes as shown in Fig. 20.6, which results in a reduction of the dump load size and also of the gate valve. The dump loads are readily available upto certain rating. Therefore this scheme can be used to control higher rating SHP plants by employing available dump loads which otherwise is not possible. Detailed simulation and analysis of the MPFC scheme is limited to two pipes in this section. However, a detailed analysis of several such schemes is given in Refs. 29–31 and 47.

20.3.1 *Two pipe control*

In this scheme, a small section of penstock employs two pipes of equal diameter, one fitted with an "on/off control valve", as shown in Fig. 20.6(b). The water flow rate is equal in both the pipes when the control valve is fully open. The on/off control valve is either fully open or closed depending upon the loading condition. If the valve is fully closed, the flow rate is reduced so that the power produced is reduced to half. The water head is maintained constant by an overflow of excess water through the

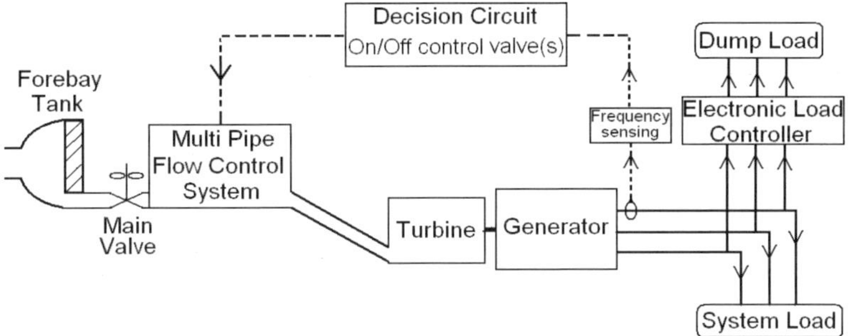

(a) Multi Pipe Flow Control (MPFC) System

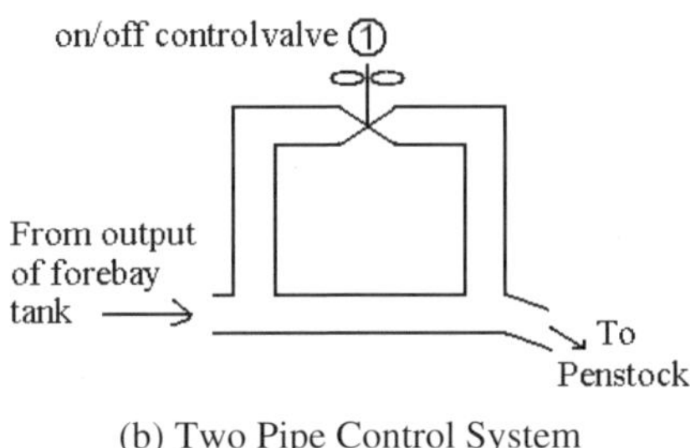

(b) Two Pipe Control System

Fig. 20.6. Block diagram of isolated SHP plant with MPFC and reduced size dump load.

Table 20.1. Various classification of cases for the two pipe control scheme.

Case	Initial state	After disturbance	Valve action
1A	$P_L^0 > 0.5P_{L,\text{Max}}$	$P_L^0 + \Delta P_L > 0.5P_{L,\text{Max}}$	No change
1B	$P_L^0 > 0.5P_{L,\text{Max}}$	$P_L^0 + \Delta P_L \leq 0.5P_{L,\text{Max}}$	From Open to Close
1C	$P_L^0 \leq 0.5P_{L,\text{Max}}$	$P_L^0 + \Delta P_L > 0.5P_{L,\text{Max}}$	From Close to Open

spillway to the diverting channel. The water from the diverting channel is given to fields at higher altitudes. This method therefore reduces the size of the dump load to 50% of the normally required.

Based on the combination of the state of the on/off control valve and initial load, the two pipe control can further be divided into three cases. The action of the on/off control valve for disturbances in the system load for all the three cases of a two pipe control system is tabulated in Table 20.1. Where, (P_L^0) is the initial load on the system and (ΔP_L) is the disturbance in load.

The transfer function block diagram of an isolated SHP plant employing the MPFC scheme for the reduction of a dump load is shown in Fig. 20.7. As the

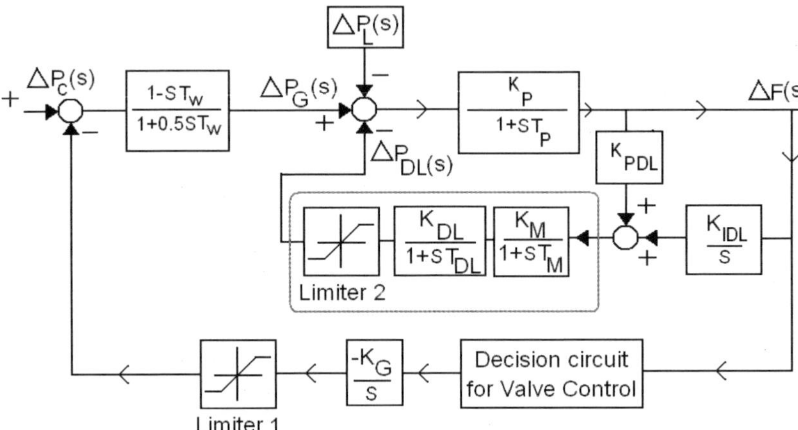

Fig. 20.7. Transfer function block diagram of an isolated SHP plant with the on/off control valves and reduced size dump load.

current study involves only small disturbances, it is expected that only one valve of the MPFC will change its state at any point of time. Therefore the transfer function block diagram will remain the same for two or more pipes. The values of the limiters and some of the parameters may differ based on the system considered.

20.3.1.1 *Typical example of an isolated SHP plant with 50% dump load*

Example 1. Let us consider a 1200 kW isolated SHP plant (rating of the plant P_R) for simulation with a nominal load of 1000 kW, inertia constant of 5 seconds and system frequency of 50 Hz. The water time constant is assumed to be 1.0, 2.2 and 4.0 seconds for low, medium and high heads, respectively. Calculation of power system constants for Case 1A for given system is as follows:

With two pipes of equal flow rate in MPFC system, a dump load of 50% is sufficient to control the frequency. Let us assume that the initial load on the system (P_L^0) is 900 kW while the maximum load on the system could be 1000 kW. In order to meet the gap between generation and load, the dump load will contribute a load (P_{DL}^0) of 100 kW. The power system parameters for this condition are given as:

$$D = \frac{(P_L^0 + P_{DL}^0)/P_R}{F^0} = \frac{(900 + 100)/1200}{50} = 0.01667$$

$$K_P = \frac{1}{D} = \frac{1}{0.01667} = 60$$

$$T_P = \frac{2H}{F^0 \times D} = \frac{2 \times 5}{50 \times 0.01667} = 12.$$

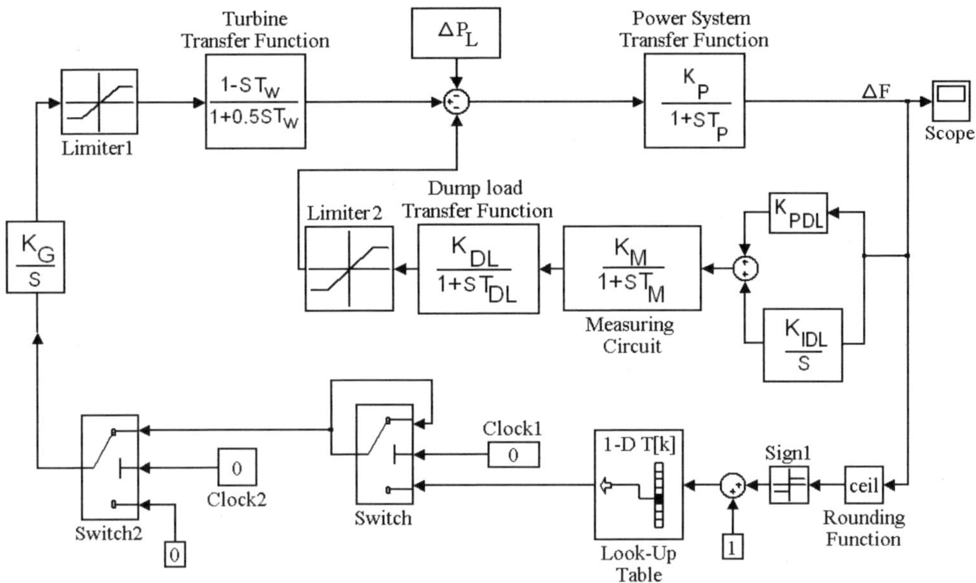

Fig. 20.8.   MATLAB/SIMULINK block diagram for Example 1.

The transfer function block diagram of an isolated SHP plant with reduced dump load size and an on/off valve control used for simulation is shown in Fig. 20.8. The power system constants depend on the initial loading and generation conditions. The values of power system constants for various cases of the two pipe control scheme are given in Table 20.2. The parameters of the dump load controller and on/off control valve used in the simulation are:

$$K_{DL} = 0.41667 \text{ pu}, \quad T_{DL} = 0.05 \text{ sec}, \quad K_{PDL} = 1, \quad K_{IDL} = 0.1, \quad K_M = 1,$$
$$T_M = 0.02 \text{ sec}, \quad K_G = 0.0075.$$

The disturbance conditions and the state of the valve for this example are given in Table 20.3. The values of limiter-1 (on/off control valve) and limiter-2 (dump load) of Fig. 20.4 are given in Table 20.4.

Table 20.2.   Power system constants for two pipe control.

Case	Operating load (kW)	$P_L^0 + \Delta P_D$ (kW)	D	K_P	T_P
1A	>500	1000	0.01667	60	12
1B	>500	1000	0.01667	60	12
1C	≤500	500	0.00833	120	24

Table 20.3. Disturbance conditions and valve state for Example 1.

Case	Initial load (kW)	Load disturbance (kW)	Valve state
1A	900	$\pm 15, \pm 21, \pm 24$	No Change (NC)
1B	520	$-15, -21, -24$	NC, O2C, O2C
1C	480	$+15, +21, +24$	NC, C2O, C2O

Valve Status: O2C (open to close), C2O (close to open).

Table 20.4. Limiter values based on initial conditions.

Case	Nominal load (pu)	Dump load (pu)	Limiter 1 (pu)	Limiter 2 (pu)
1A	0.75	0.0833	-0.41667 to 0	-0.833 to 0.333
1B	0.433	0.4	-0.41667 to 0	-0.4 to 0.01667
1C	0.4	0.01667	0 to 0.41667	-0.01667 to 0.4

20.3.1.2 *Case 1A*

The initial state of the on/off control valve is open/close and there will not be any action of the on/off control valve in this case for small disturbances in load. If the load disturbance on the system varies such that $0 \leq P_L^0 + \Delta P_L \leq 0.5 P_{L,\text{Max}}$ or $0.5 P_{L,\text{Max}} < P_L^0 + \Delta P_L \leq P_{L,\text{Max}}$, then the dump load varies between the minimum and maximum value, so as to maintain the frequency constant. The transient responses of the system for a step disturbance of ± 21 kW in load are shown in Fig. 20.9. The dump load varies its value in accordance with a change in load such that $\Delta P_{DL} + \Delta P_L = 0$, thereby bringing frequency back to its normal value i.e., ΔF to zero in about 60 sec.

20.3.1.3 *Case 1B*

The initial state of the on/off control valve is "open". When the nominal load $P_L^0 > 0.5 P_{L,\text{Max}}$ and the load disturbance occurs such that $P_L^0 + \Delta P_L \leq 0.5 P_{L,\text{Max}}$, then the on/off control valve closes to reduce the generation by 50%. Once the on/off control valve is completely closed and a load disturbance occurs in the system in such a way that $P_L^0 + \Delta P_L \leq 0.5 P_{L,\text{Max}}$, then only the dump load will vary between its limits so as to maintain the frequency constant. The transient responses of frequency and dump load deviations with a change in the on/off control valve position of the

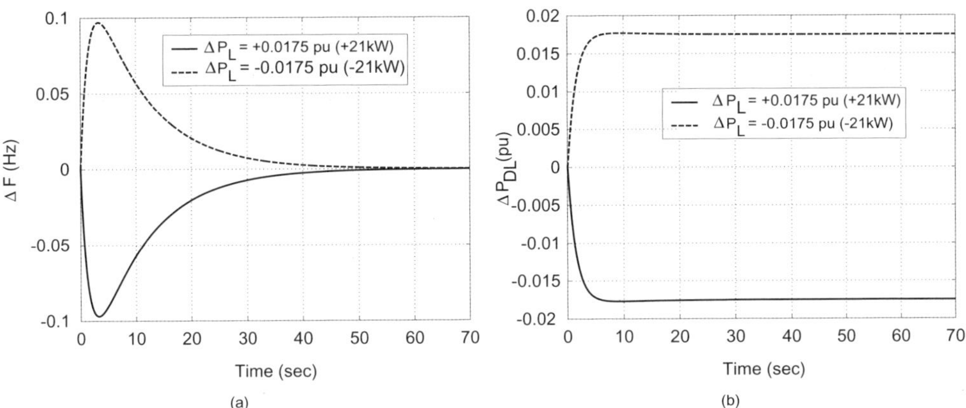

Fig. 20.9.   Transient responses of the system, Case 1A, for step changes in load, showing deviations in (a) ΔF (b) ΔP_{DL}.

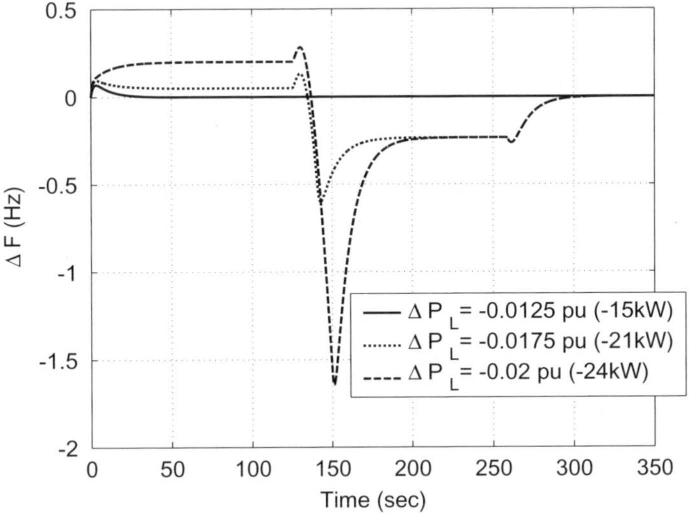

Fig. 20.10.   Transient frequency responses of the system, Case 1B, for step changes in load, ΔP_L.

system for different step disturbances in load are shown in Figs. 20.10 and 20.11, respectively.

It can be observed from Figs. 20.10 and 20.11 that for a 15 kW decrease in load, ΔF initially increases and then slowly decreases to zero as the dump load adjusts itself to vanish the frequency deviation.

As the load decreases further, ΔF increases and it attains a steady state value in about 50 sec as the dump load has already reached its positive maximum. The

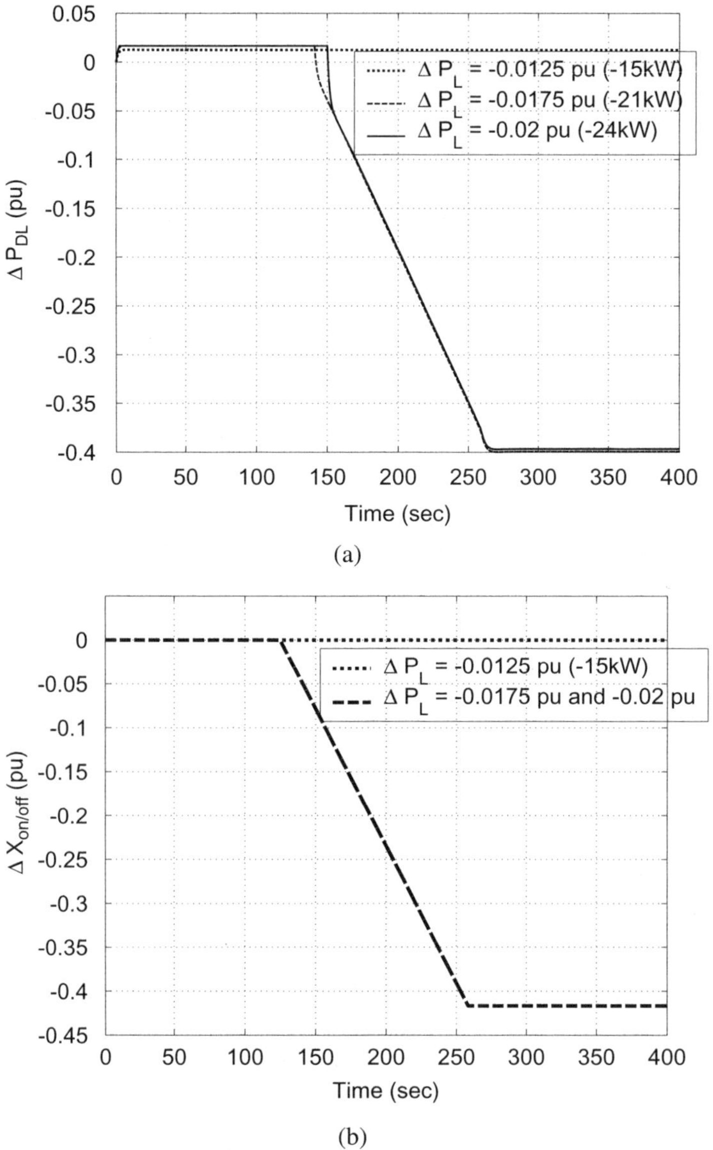

Fig. 20.11.   Transient responses of the dump load and the on/off control valve position of the system, Case 1B, for step changes in load ΔP_L.

control logic starts closing the on/off control valve at about 125 sec as shown in Fig. 20.11(b), therefore, ΔF momentarily decreases and then increases sharply to a new steady state value and remains constant from 200 to 260 sec as shown in Fig. 20.10. This steady state error in ΔF is due to change in the on/off control valve

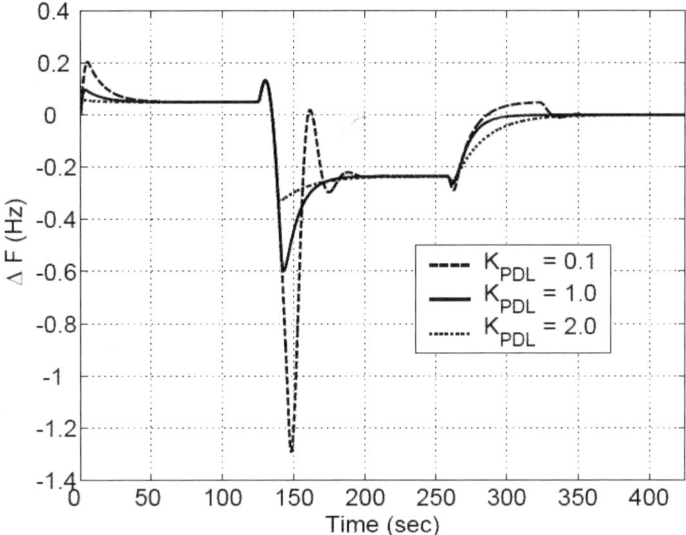

Fig. 20.12.   Transient frequency responses of the system, Case 1B, for a step change in load, $\Delta P_L = -21\,\text{kW}$ for various values of K_{PDL}.

state and it exists till the valve reaches its final position. Once the on/off control valve reaches its final position at 260 sec, then only the dump load controller will be acting and reduces ΔF to zero at around 300 sec.

Effect of PI controller gains

The transient responses for various values of K_{PDL} are shown in Fig. 20.12. It can be observed that with an increase in proportional gain the peak deviation in ΔF is decreased and the settling time has increased. Also when K_{PDL} is decreased, the peak value of ΔF has increased and ΔF starts oscillating during the transition of the on/off control valve. The steady state error and settling time decreases with an increase in K_{IDL}, but the peak deviation of ΔF increases. Upon further increasing the value of K_{IDL}, the system will go into oscillations, with larger peak deviation in ΔF as shown in Fig. 20.13.

Effect of gain of on/off control valve

The rate at which the on/off control valve is closed has a significant influence on the transient performance of the system. The transient responses for the various rates of closing the on/off control valve, K_G, are shown in Fig. 20.14. The higher the slope of the on/off control valve (i.e., fast closing) is, the more is the peak deviation in ΔF and less is the settling time of the frequency response. Therefore, a moderate slope

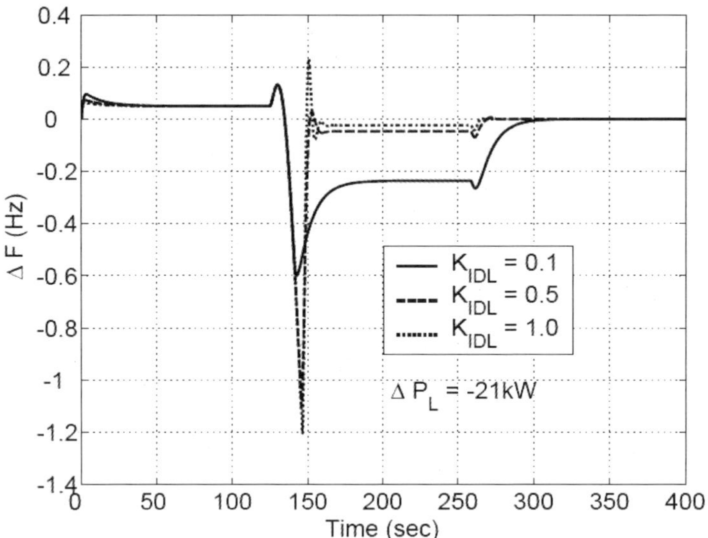

Fig. 20.13.   Transient frequency responses of the system, Case 1B, for a step change in load, $\Delta P_L = -21\,\text{kW}$ for various values of K_{IDL}.

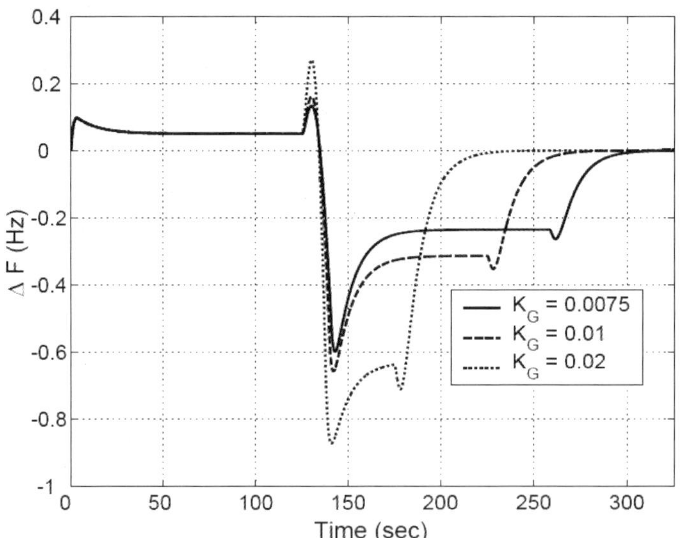

Fig. 20.14.   Transient responses of the system, Case 1B, for a step change in load, $\Delta P_L = -21\,\text{kW}$, for different rate(s) of closing of the on/off control valve K_G, showing deviations in ΔF.

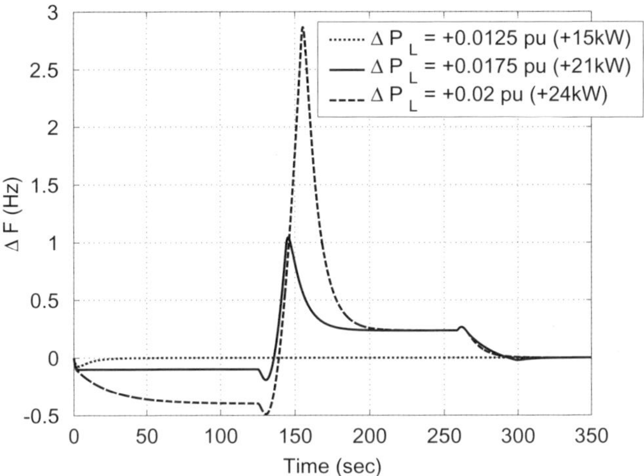

Fig. 20.15. Transient frequency responses of the system, Case 1C, for step changes in load, ΔP_L.

for the on/off control valve has to be selected, keeping peak deviation and settling time in view.

20.3.1.4 *Case 1C*

The initial state of the on/off control valve is "close". When the nominal load $P_L^0 \leq 0.5 P_{L,\text{Max}}$ and the load disturbance occurs such that $P_L^0 + \Delta P_L > 0.5 P_{L,\text{Max}}$, then the on/off control valve opens to increase the generation by 50%. Once the on/off control valve is completely open and a load disturbance occurs in the system in such a way that $P_L^0 + \Delta P_L > 0.5 P_{L,\text{Max}}$, then only the dump load will vary between its limits so as to maintain the frequency constant. The transient responses of such a system for unit step disturbances in load are shown in Fig. 20.15.

For a 15 kW increase in load, ΔF initially decreases and then slowly increases to zero as the dump load adjusts itself to vanish deviation in frequency. There will be no change in the state of the on/off control valve for this value of load disturbance as shown in Fig. 20.16. For higher step disturbances in load, ΔF initially decreases and attains a steady state value (as the dump load attains its minimum limit) in about 50 sec. The control logic starts opening the valve at about 125 sec as shown in Fig. 20.16(b). The frequency momentarily increases and then decreases sharply to a new steady state value and remains constant from 200 to 260 sec as shown in Fig. 20.15. The integral action of the dump load controller will further decrease the steady state error in ΔF to zero at around 300 sec. The observations made for the effect of gains of both PI and on/off controller in the Case 1B also hold good for this case.

                                            S. Doolla

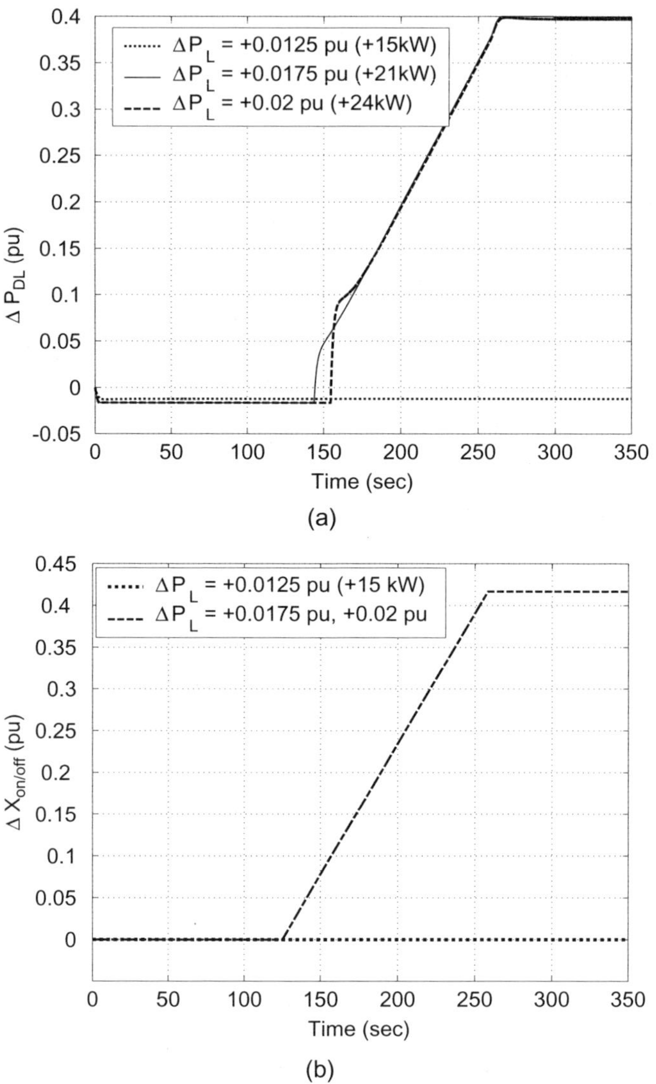

(a)

(b)

Fig. 20.16.   Transient responses of the dump load and the on/off control valve position of the system, Case 1C, for step changes in load, ΔP_L.

20.4   Frequency Control using Servo Motor Along with On/Off Control Valve[29−31,47]

The functional block diagram of the scheme using a combination of the servo motor controlled valve and the on/off control valve is shown in Fig. 20.17. The frequency control in this scheme is accomplished by using the servo motor controlled valve

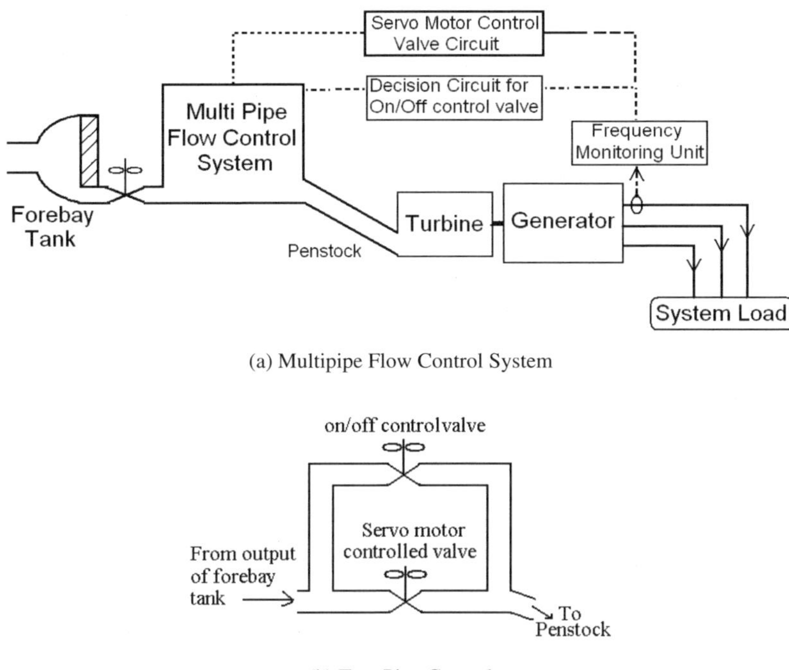

(a) Multipipe Flow Control System

(b) Two Pipe Control

Fig. 20.17.   Block diagram of an isolated SHP plant with a servo motor controlled valve and on/off control valves.[29]

along with the on/off control valve. A detailed study of such a control system using two pipe control is discussed in detail in this section.

20.4.1 *Two pipe control*

In this scheme, the penstock flow is regulated through two longitudinal small sections of pipes as shown in Fig. 20.17(b). One pipe is fitted with an on/off control valve with a flow rate so that 50% power of the maximum rated load is produced in the "on" state. The second pipe is fitted with a valve which is controlled by a servo motor. The flow rate in the second pipe is continuously controlled by controlling the input signal to the servo motor. The diameters of both the pipes are selected in such a way that the flow rate in either pipes is same when both valves are opened. The on/off control valve is either fully open or closed depending upon the loading condition. When the load is less than 50% of the rated maximum load, the on/off control valve will remain closed and the servo motor controlled valve will take care of the deviation in frequency due to a disturbance in load. Whenever there is a disturbance in load, the servo motor controlled valve changes the flow rate so as to maintain the system frequency constant. The water head is maintained constant by

Table 20.5. Various classification of cases for the two pipe control scheme.

Case	Initial state	After disturbance	Valve action
2A	$P_L^0 > 0.5 P_{L,\text{Max}}$	$P_L^0 + \Delta P_L > 0.5 P_{L,\text{Max}}$	Continuously vary
2B	$P_L^0 > 0.5 P_{L,\text{Max}}$	$P_L^0 + \Delta P_L \leq 0.5 P_{L,\text{Max}}$	From Open to Close
2C	$P_L^0 \leq 0.5 P_{L,\text{Max}}$	$P_L^0 + \Delta P_L > 0.5 P_{L,\text{Max}}$	From Close to Open

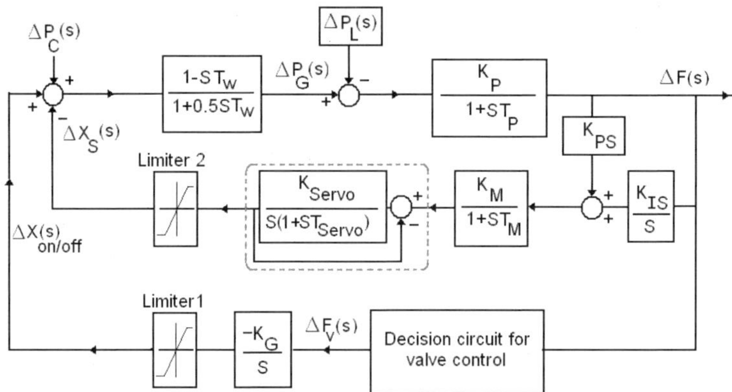

Fig. 20.18. Transfer function block diagram of an isolated SHP plant with servo motor and on/off control valves.

an overflow of excess water through the spillway and diverting to the fields through a channel for irrigation.

Based on the combination of state of the on/off control valve, and initial load, the two pipe control can further be divided into three cases. The action of the on/off control valve for disturbances in the system load for all the three cases of a two pipe control system is tabulated in Table 20.5.

The transfer function block diagram of an isolated SHP plant employing a MPFC scheme for elimination of the dump load is shown in Fig. 20.18.

As the current study involves only small disturbances, it is expected that only one valve of the MPFC will change its state at any point of time. Therefore the transfer function block diagram will remain the same for two or more pipes. The values of the limiters and some of the parameters may differ based on the system considered.

20.4.1.1 *Typical example of isolated SHP plant without dump load*

Example 2. Let us consider a 1200 kW isolated SHP plant (rating of the plant P_R) for simulation with a nominal load of 1000 kW, inertia constant of 5 seconds and system frequency of 50 Hz. The water time constant is assumed to be 1.0, 2.2 and

4.0 seconds for low, medium and high heads, respectively. Calculation of power system constants for Case 2A for the given system is as follows:

The initial load on the system is assumed to be 900 kW while the maximum load on the system could be 1000 kW. The contribution of various valves in order to meet the present demand are as follows:

On/off control valve (fully open)   : 500 kW
Servo motor control valve (partially open)   : 400 kW.

The power system parameters for such a condition are given by:

$$D = \frac{P_L^0/P_R}{F^0} = \frac{(900)/1200}{50} = 0.015$$

$$K_P = \frac{1}{D} = \frac{1}{0.015} = 66.6667$$

$$T_P = \frac{2H}{F^0 \times D} = \frac{2 \times 5}{50 \times 0.015} = 13.33334.$$

The transfer function block diagram of an isolated SHP plant with a servo motor controlled valve and an on/off valve control used for simulation is shown in Fig. 20.19. The power system constants depend on the initial loading and generation

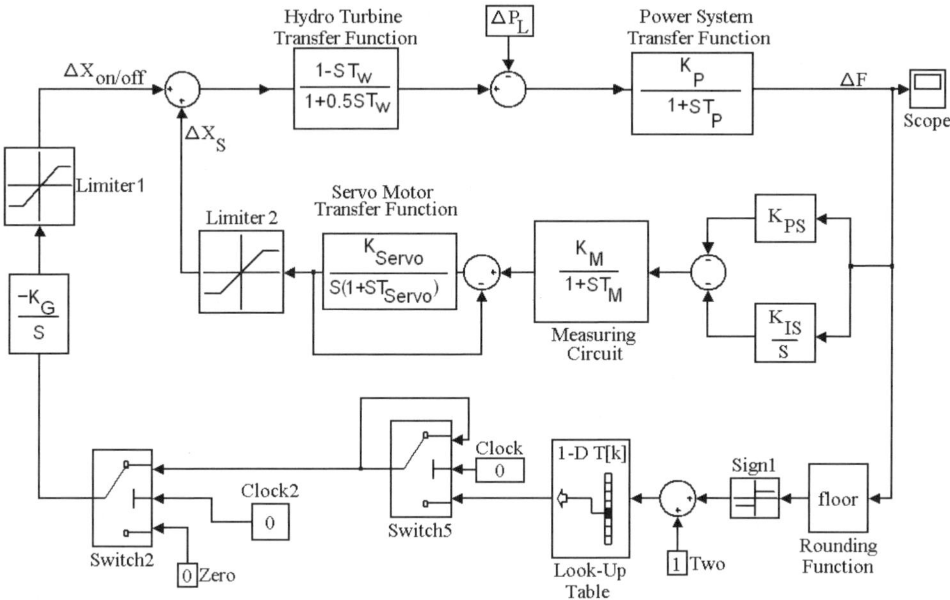

Fig. 20.19. A generalized simulink block diagram of an isolated SHP plant with servo motor and on/off control valves.

Table 20.6.   Power system constants for two pipe control.

Case	Initial operating condition (kW)	Initial load P_L^0	D	K_P	T_P
2A	$P_L^0 > 500$	900	0.015	66.67	13.33
2B	$P_L^0 > 500$	520	0.008667	115.3846	23.0769
2C	$P_L^0 \leq 500$	480	0.008	125	25

conditions. The values of the power system constants for various cases of the two pipe control scheme of Fig. 20.19 are given in Table 20.6. The parameters of the servo motor and the on/off control valve used for simulation studies are:

$$K_{\text{Servo}} = 2.5 \text{ pu}, \quad T_{\text{Servo}} = 0.1 \text{ sec}, \quad K_{PS} = 8.52, \quad K_{IS} = 0.4, \quad K_M = 0.004,$$
$$T_M = 0.02 \text{ sec}, \quad K_G = 0.004.$$

The disturbance conditions and the state of the valve for this example are given in Table 20.7. The values of limiter-1 (on/off control valve) and limiter-2 (servo motor control valve) of Fig. 20.19 are given in Table 20.8.

Table 20.7.   Disturbance conditions and valve state for Example 2.

Case	Initial load (kW)	Load disturbance (kW)	Valve state
2A	900	$\pm 15, \pm 21, \pm 24$	No Change (NC)
2B	520	$-15, -21, -24$	NC, O2C, O2C
2C	480	$+15, +21, +24$	NC, C2O, C2O

Valve Status: O2C (open to close), C2O (close to open).

Table 20.8.   Limiter values based on initial conditions.

Case	Nominal load (pu)	Servo motor control valve (pu)	Limiter 1 (pu)	Limiter 2 (pu)
2A	0.75	0.333	-0.41667 to 0	-0.333 to 0.0833
2B	0.433	0.01667	-0.41667 to 0	-0.01667 to 0.4
2C	0.4	0.4	0 to 0.41667	-0.4 to 0.01667

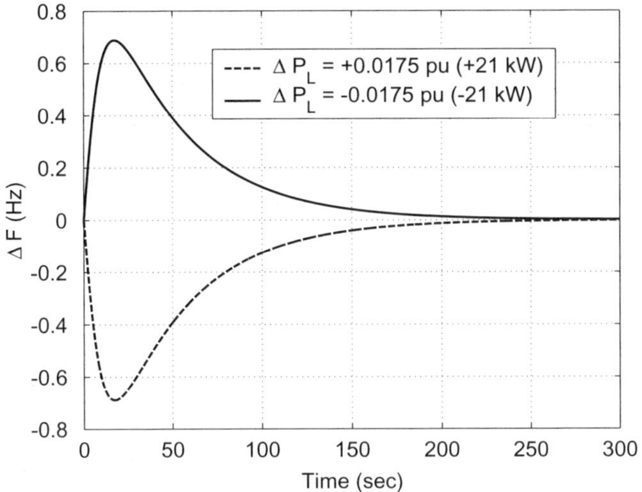

Fig. 20.20.   Transient frequency response of system, Case 2A, for step changes in load.

20.4.1.2 *Case 2A*

The initial state of the on/off control valve is "open/close" and there will not be any action of the on/off control valve in this case for small disturbances in the load. If the load disturbance on the system varies such that $0 \leq P_L^0 + \Delta P_L \leq 0.5 P_{L,\text{Max}}$ or $0.5 P_{L,\text{Max}} < P_L^0 + \Delta P_L \leq P_{L,\text{Max}}$, then the servo motor controlled valve of the MPFC system varies between its limits so as to maintain the frequency constant. The transient responses of the system for a step disturbance in the load are shown in Figs. 20.20 and 20.21.

The servo motor controlled valve position ΔX_S, slowly, opens/closes in response to the disturbance in the load and reaches a corresponding steady state error in about 250 sec as shown in Fig. 20.21(a). Initially when the servo motor controlled valve opens to increase the power generation, ΔP_G, it decreases first and then increases and vice-versa as shown in Fig. 20.21(b). It has been observed that the settling time of the transient responses is about 250 sec. The dynamic response is quite sluggish as compared to the case of dump load in Sec. 1.3. The reason is that the dump load is a completely static electrical device with small time constant as compared to the servo motor operated valve controlling the flow of input water in the penstock. Hence it takes more time to settle down in the present case.

20.4.1.3 *Case 2B*

For a step change in load the transient responses of the system are shown in Figs. 20.22 and 20.23. It can be seen that for a step decrement in a load of 15 kW,

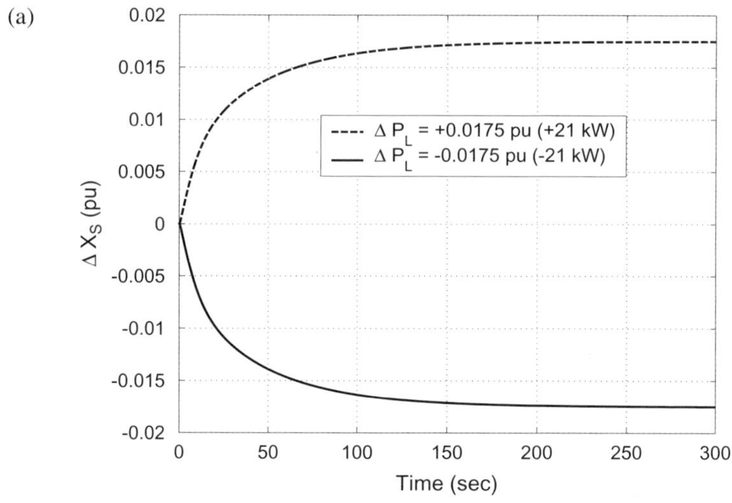

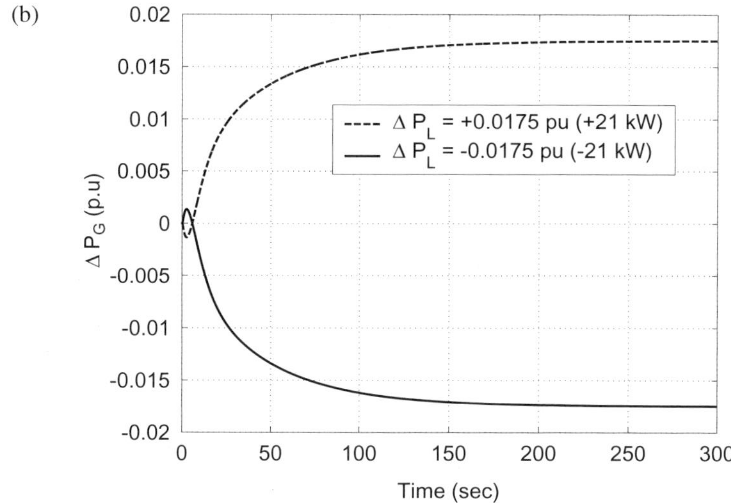

Fig. 20.21.   Transient response of the system, Case 2A, for step changes in load, showing deviations in (a) ΔX_S (b) ΔP_G.

there is an initial increase in a ΔF, and then it slowly reduces to zero, as the servo motor controlled valve decreases the generation.

For load disturbances of $-21\,\text{kW}$ and $-24\,\text{kW}$, it can be observed that ΔF increases initially and then settles at some positive value, this is because the servo motor has already reached its minimum position, and no further corrective action can be taken by the servo motor. Now, the on/off control valve begins to "close" after 125 sec, and reduces the generation. When the on/off control valve starts closing, there is

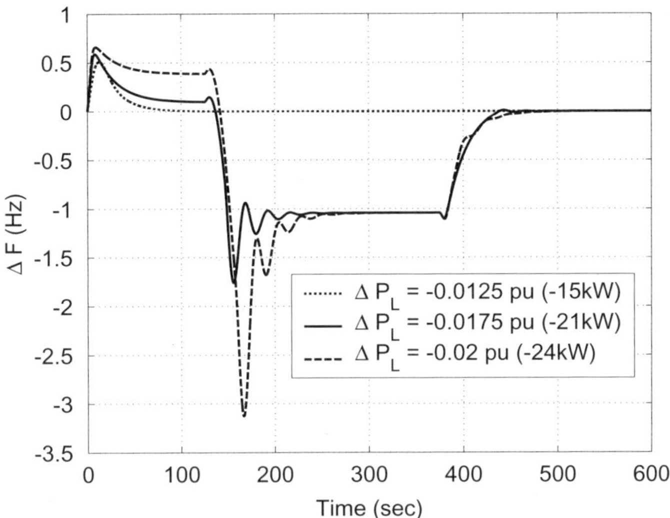

Fig. 20.22.   Transient frequency responses of the system, Case 2B for step changes in load, ΔP_L.

a sudden decrease in ΔF, while the servo motor starts increasing the generation, as a result it slowly starts rising and it settles down at -1.04 Hz, from 250 to 350 sec. The reason that ΔF remains constant during this interval is that the decrease in generation by the on/off control valve is partially compensated by the increase in generation of the servo motor controlled valve with no oscillations as shown in Fig. 20.23.

After 360 sec, the deviation in frequency starts diminishing because of the integral action of the servo motor controlled valve and finally it settles to zero around 450 sec. The steady state error in ΔF, when the on/off control valve is in transition and the settling time remains the same but the peak value depends upon the magnitude of ΔP_L.

The rate at which the on/off control valve is closed has significant influence on the peak deviation in ΔF, therefore, two different methods for closing the on/off control valve are suggested.

In the first method, the rate at which the on/off control valve is closed is initially kept "high" and after sometime the slope is "decreased", while it is exactly opposite in the second method. Both these methods can be accomplished by incorporating the small transfer function gain block shown as dotted lines in Fig. 20.24, in the on/off control valve loop. The switching in/out time for the new gain block is controlled by incorporating a clock device externally. Using the new technique of the on/off valve control transient responses of the system for small signal disturbances are investigated for both the methods.

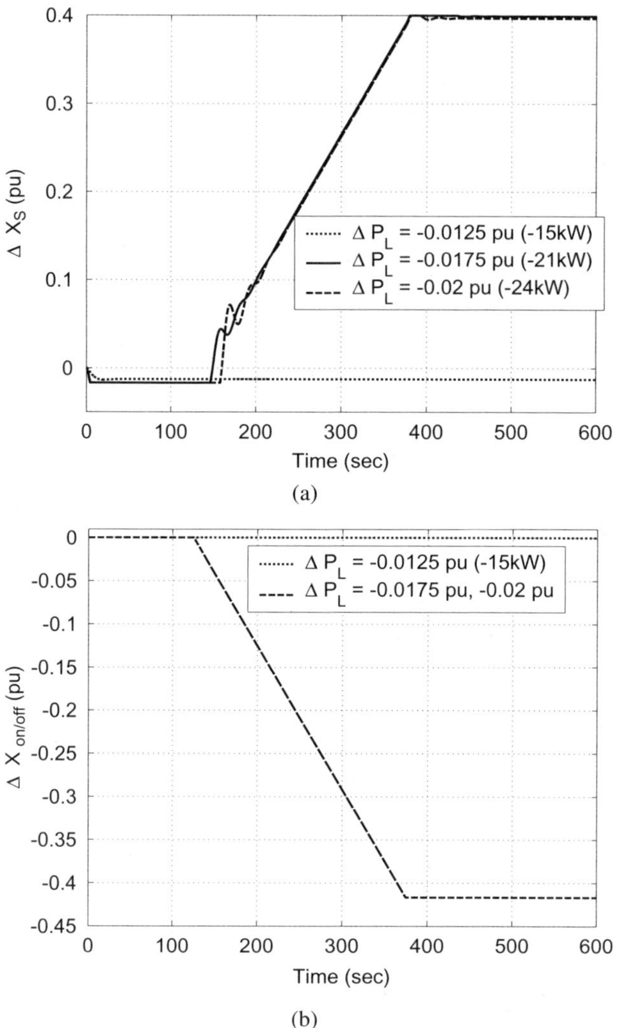

Fig. 20.23. Transient responses for the system, case 2B, for step changes in load, showing deviations in (a) ΔX_S (b) $\Delta X_{\text{on/off}}$.

20.4.1.4 *Case 2C*

The initial state of the on/off control valve is "close". When the nominal load $P_L^0 \leq 0.5 P_{L,\text{Max}}$ and the load disturbance occurs such that $P_L^0 + \Delta P_L > 0.5 P_{L,\text{Max}}$, then the on/off control valve opens to increase the generation by 50%. Once the on/off control valve is completely open and a load disturbance occurs in the system in such a way that $P_L^0 + \Delta P_L > 0.5 P_{L,\text{Max}}$, then only the servo motor controlled valve will vary between its limits so as to maintain the frequency constant. The transient frequency responses of the system for various load disturbances are shown in Fig. 20.25.

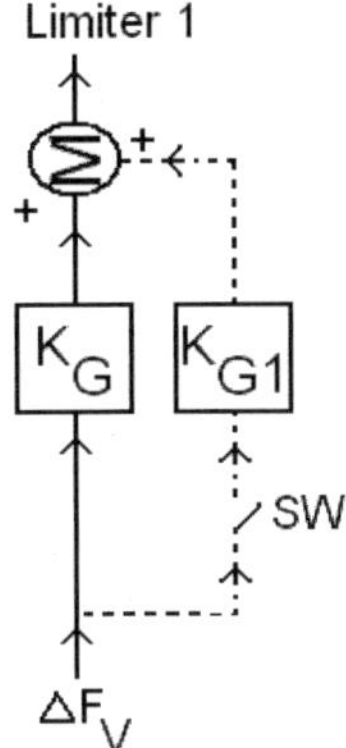

Fig. 20.24.   Dual slope controller for on/off control valve.

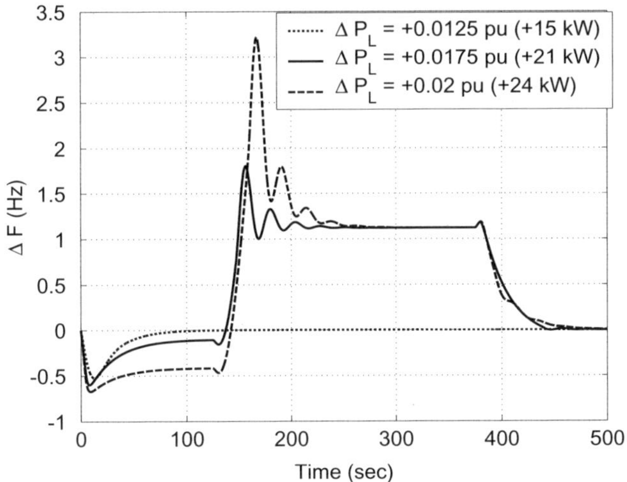

Fig. 20.25.   Transient frequency responses of the system, Case 2C, for step changes in load, ΔP_L.

For a 15 kW step increment in load, ΔF initially decreases and then it reduces to zero as the servo motor controlled valve adjusts itself to increase the generation as shown in Fig. 20.26. It can be observed from Fig. 20.26(b) that there is no action of the on/off control valve as the disturbance is not sufficient to cause transition.

For a 21 kW and 24 kW step increment in load, ΔF initially decreases and then attains a constant value for a certain time. This is because the servo motor controlled valve has reached its positive maximum limit as shown in Fig. 20.26(a). If this steady state error persists till 125 sec, then the on/off control valve will open to increase the generation by 50%, resulting in the simultaneous closing operation of the servo motor controlled valve to slowly reduce the generation. In this process ΔF has a large swing and then settles down at +1.1 Hz from 250 to 375 sec.

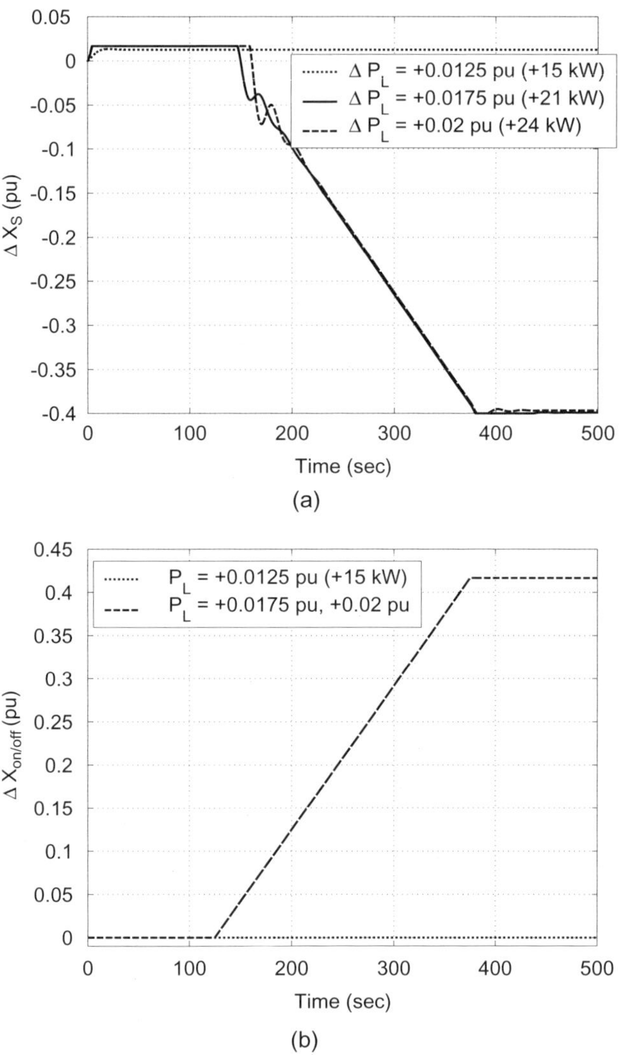

Fig. 20.26.   Transient responses of the system, Case 2C, for step changes in load, showing deviations in (a) ΔX_S (b) $\Delta X_{\text{on/off}}$.

The reason that ΔF remains constant for some time is that the increase in generation by the on/off control valve is partially compensated by the decrease in generation of the servo motor controlled valve with no oscillations as shown in Fig. 20.25.

Once the on/off control valve is fully opened then, ΔF will reduce to vanish at around 450 sec due to the integral action of the servo motor controlled valve. The peak deviation in ΔF is more in Case 2C as compared to Case 2B. This is because the servo motor controlled valve in Case 2C moves from an initial position to maximum open state and then partially closes to produce very small power along with the 50% power produced by the on/off control valve.

The performance deteriorates if the initial rate of closing of the on/off control valve is high. Also, the peak deviation in ΔF rises drastically if the initial slope is higher. So, by using the initial low rate of closing and subsequent high rate of closing the on/off control valve, the transient performance of the system improves considerably. The transient responses of the system with an initial low and subsequent high rate of opening of the on/off control valve are shown in Fig. 20.27. It

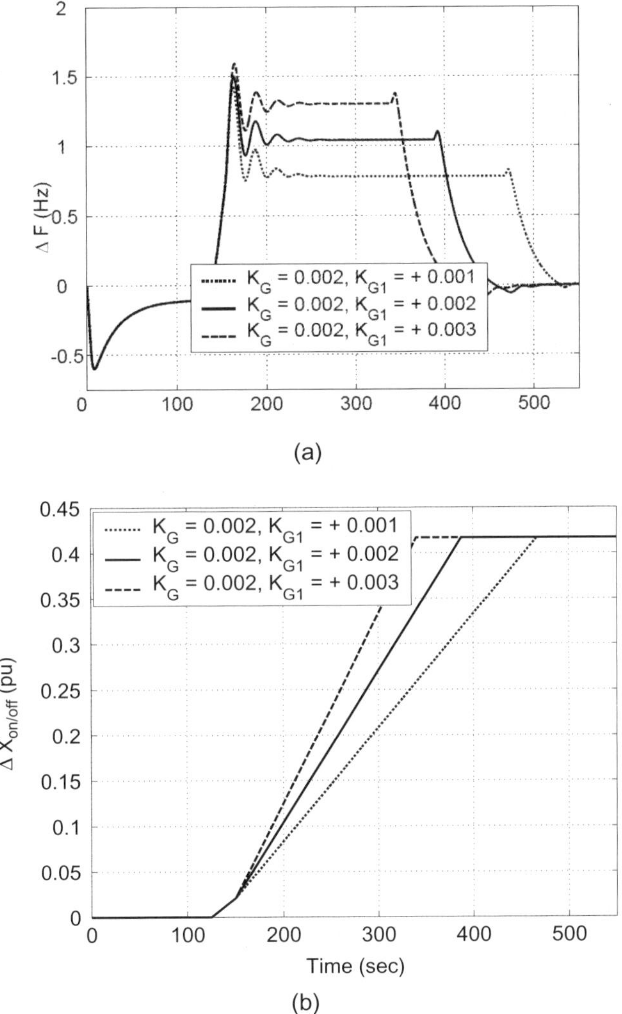

Fig. 20.27. Transient responses of the system, Case 2C, for a step change in load, ΔP_L for initial LOW and subsequent HIGH rate of opening of the on/off control valve showing deviations in (a) ΔF (b) $\Delta X_{on/off}$.

is very clear from Fig. 20.27 that the system performance increases if the rate of closing/opening of the valve is initially low and subsequently high.

20.5 Conclusions

A balance between generation and load is essential in order to maintain frequency constant in any electrical power system. This is generally accomplished by using a load management technique in the case of isolated SHP plants. But it is essential to save water for irrigation as in most of the cases, survival of local communities depends upon it and hence frequency control techniques employing saving of water are discussed in detail in this chapter. A complete mathematical model of an isolated SHP plant for frequency control with the combination of a dump load and/or servo motor controlled valve along with on/off control valves. It is observed that the system response with a servo motor controlled valve is sluggish due to the involvement of mechanical time constant. Various case studies using Simulink tool box of MATLAB were presented for more clarity.

References

1. T.S. Bhatti, A.A.F. Al-Ademi and N.K. Bansal, "Load-frequency control of isolated wind-diesel-microhydro hybrid power systems (WDMHPS)," *Energy* **22** (1997) 461–470.
2. A. Tomilson, J. Quaicoe, R. Gosine, M. Hinchey and N. Bose, "Modelling an autonomous wind-diesel system using SIMULINK," *IEEE Canadian Conf. Electrical and Computer Engineering* (St. Johns, Nfld, May, 1997), pp. 35–38.
3. R. Gagnon, B. Saulnier, G. Sybille and P. Giroux, "Modeling of a generic high-penetration no-storage wind-diesel system using matlab/power system blockset," *Global Windpower Conf.*, Paris, France (2002), pp. 1–6.
4. A.S. Neris, N.A. Vovos and G.B. Giannakopoulos, "Dynamics and control system design of an autonomous wind turbine," *Third IEEE Int. Conf. Electronics, Circuits, and Systems*, Rodos, 13–16 October 1996, pp. 1017–1020.
5. R. Sebastian and J. Quesada, "Distributed control system for frequency control in a isolated wind system," *Renewable Energy* **31** (2006) 285–305.
6. D. Kottick, M. Balu and D. Edelstein, "Battery energy storage for frequency regulation in an island power system," *IEEE Trans. Energy Conversion* **8** (1993) 455–459.
7. M.J. Khan and M.T. Iqbal, "Pre-feasibility study of stand-alone hybrid energy systems for applications in newfoundland," *Renewable Energy* **30** (2005) 835–854.
8. H. Kunisch, K. Kramer and H. Dominik, "Battery energy storage: Another option for load frequency control and instantaneous reserve," *IEEE Trans. Energy Conversion* **1** (1986) 41–46.
9. D.S. Tarnay, "Hydrogen production at hydro-power plants," *Int. J. Hydrogen Energy* **10** (1985) 577–584.
10. M. Santarelli and S. Macagno, "Hydrogen as an energy carrier in stand-alone applications based on PV and PV micro-hydro systems," *Energy* **29** (2004) 1159–1182.
11. M. Santarelli, M. Cali and S. Macagno, "Design and analysis of stand-alone hydrogen energy systems with different renewable sources," *Int. J. Hydrogen Energy* **29** (2004) 1571–1586.

12. A. Demiroren, H.L. Zeynelgil and N.S. Sengor, "Automatic generation control for power system with SMES by using neural network controller," *Electrical Power Components and Systems* **31** (2003) 1–25.

13. S. Banerjee, J.K. Chatterjee and S.C. Tripathy, "Application of magnetic energy storage unit as load-frequency stabilizer," *IEEE Trans. Energy Conversion* **5** (1990) 46–51.

14. I. Ngamroo, Y. Mitani and K. Tsuji, "LFC and stabilisation of multi-area interconnected power system by decentralized control of a SMES and solid-state phase shifters," *Proc. EMPD, Energy Management and Power Delivery* (1998), pp. 85–90.

15. H. Ohsaki, S. Taniguchi, S. Nagaya, S. Akita, S. Koso and M. Tatsuta, "Development of SMES for power system control: Present status and perspective," *Physica C.* **412–414** (2004) 1198–1205.

16. R. Sebastian, M. Castro, F. Yeves and J. Peire, "Control of the diesel-flywheel-synchronous set(s) in the hybrid wind-diesel system of Jandia plant in Fuerteventura (Spain)," *Proc. ISES Solar World Congress*, Adelaide, Australia (2001), pp. 2038–2042.

17. D.E. Hampton, C.E. Tindall and J.M. McArdle, "Emergency control of power system frequency using flywheel energy injection," *Int. Conf. Advances in Power System Control, Operation and Management APSCOM* (Hong Kong, November, 1991), pp. 662–667.

18. P. Freere, "Electronic load/excitation controller for a self-excited squirrel cage generator micro-hydro scheme," *5th IEEE Int. Conf. Electrical Machines and Drives* (London, UK, June, 1991), pp. 266–270.

19. N. Jaleeli, L.S. VanSlyck, D.N. Ewart, L.H. Fink and A.G. Hoffmann, "Understanding automatic generation control," *IEEE Trans. Power Systems* **7** (1992) 1106–1122.

20. R. Widmer and A. Arter, "Harnessing water power on a small scale — village electrification," Technical report, SKAT, Switzerland (1992), pp. 23–20.

21. A. Harvey, A. Brown, P. Hettiararchi and A. Inversin, *Micro Hydro Design Manual: A Guide to Small-Scale Water Power Schemes* (Intermediate Technology Publications, London, 1993).

22. R. Bonert and G. Hoops, "Stand alone induction generator with terminal impedance controller and no turbine controls," *IEEE/PES Summer Meeting*, Long Beach, California, 9–14 July 1989, pp. 28–31.

23. R. Bonert and S. Rajakaruna, "Self-excited induction generator with excellent voltage and frequency control," *IEE Proc. Generation, Transmission and Distribution* **145** (1998) 33–39.

24. B. Singh, S.S. Murthy and S. Gupta, "Analysis and implementation of an electronic load controller for a self-excited induction generator," *IEE Proc. Generation, Transmission and Distribution* **151** (2004) 51–60.

25. D. Henderson, "An advanced electronic load governor for control of micro hydroelectric generation," *IEEE Trans. Energy Conversion* **13** (1998) 300–304.

26. K. Pandiaraj, P. Taylor, N. Jenkins and C. Robb, "Distributed load control of autonomous renewable energy systems," *IEEE Trans. Energy Conversion* **16** (2001) 14–19.

27. M.H. Nehrir, B.J. Lameres, G. Venkataramanan, V. Gerez and L.A. Alvarado, "An approach to evaluate the general performance of stand-alone wind/photovoltaic generating systems," *IEEE Trans. Energy Conversion* **15** (2000) 433–439.

28. S.S. Murthy, R. Jose and B. Singh, "A practical load controller for stand alone small hydro systems using self excited induction generator," *Int. Conf. Power Electronic Drives and Energy Systems for Industrial Growth* (New Delhi, India, December, 1998), pp. 359–364.

29. S. Doolla and T.S. Bhatti, "Automatic generation control of an isolated small hydro power plant," *Electric Power Systems Research* **76** (2006) 889–896.

30. S. Doolla and T.S. Bhatti, "Frequency control of an isolated small hydro power plant," *Int. Energy J.* **7** (2006) 25–41.

31. S. Doolla and T.S. Bhatti, "Load frequency control of an isolated small hydro power plant," *IEEE Trans. Power Systems* **21** (2006) 1912–1919.

32. S. Doolla and T.S. Bhatti, "A new load frequency control technique for an isolated small hydro power plant," *Int. J. Power and Energy* **221** (2007) 51–57.

33. H. Miland, R. Glockner, P. Taylor, R.J. Aaberg and G. Hagen, "Load control of a wind-hydrogen stand-alone power system," *Int. J. Hydrogen Energy* **31** (2006) 1215–1235.

34. S.S. Murthy, B. Singh, A. Kulkarni, R. Sivarajan and S. Gupta, "Field experience on a novel picohydel system using self excited induction generator and electronic load controller," *5th IEEE Int. Conf. Power Electronics and Drive Systems*, London, UK, November, 2003, pp. 842–847.

35. P. Kundur, *Power System Stability and Control* (McGraw-Hill, New Delhi, 1993).

36. O.I. Elgerd, *Electric Energy System Theory: An Introduction* (Tata McGraw-Hill Publishing Company Ltd, New Delhi, 1982).

37. R. Oldenburger and J.J. Donelson, "Dynamic response of a hydroelectric plant," *AIEE Trans. Power Apparatus Syst.* **81** (1962) 403–419.

38. D.G. Ramey and J.W. Skooglund, "Detailed hydrogovernor representation for system stability studies," *IEEE Trans. Energy Conversion* **PAS-89** (1970) 106–112.

39. IEEE committee report on dynamic models for steam and hydro turbines in power system studies (1973).

40. Hydraulic turbine and turbine control models for system dynamic studies (1992).

41. T.E. Kissel, *Servomotor Applications: Industrial Electronics*, 2nd edn (Prentice Hall, PTR, 1992).

42. L.C.T. Dugler and S. Uyan, "Modelling, simulation and control of a four-bar mechanism with a brushless servo motor," *Mechatronics* **7** (1997) 369–383.

43. K. Ogata, *Modern Control Engineering*, 2nd edn. (Prentice Hall International, India, 1995).

44. I.J. Nagrath and M. Gopal, *Control Systems Engineering* (New Age International Publishers Ltd, India, 1998).

45. M. Hitoshi, "Compact servo driver for torque control of DC-servo motor based on voltage control," *IEEE Int. Conf. Advanced Intelligent Mechatronics*, Atlanta, USA (1999), pp. 19–23.

46. N. Khongkoom, A. Kanchanathep, S. Nopnakeepong, S. Tanuthong, S. Tunyasrirut and R. Kagawa, "Control of the position DC servo motor by fuzzy logic," *TENCON*, Kaula Lumpur, Malaysia (2000), pp. 354–357.

47. S. Doolla, "Automatic frequency control of an isolated small hydro power plant," PhD thesis, I I T Bombay, India (2007).

Section 5

Simulation Tools, Distributed Generation and Grid Integration

Chapter 21

Simulation Tools for Feasibility Studies of Renewable Energy Sources

Juan A. Martinez-Velasco

Universitat Politècnica de Catalunya, Barcelona, Spain

martinez@ee.upc.edu

Jacinto Martin-Arnedo

EiPE, Manresa, Spain

jacinto.martin@eipe.cat

This chapter reviews the main capabilities of the most common software packages for feasibility studies of renewable energy installations. The chapter details the models implemented in these tools for representing loads, resources, generators and dispatch strategies, and summarizes the approaches used to obtain the life-cycle cost of a project. A short description of a methodology for estimating greenhouse gas (GHG) emission reductions is also included. Although some of the reviewed tools may analyze systems that combine heat and electric power from renewable and non-renewable sources, only renewable technologies for generation of electricity are considered. Two detailed examples illustrate the scope of feasibility studies.

21.1 Introduction

Renewable energy resources are those whose future availability is not affected by their use, although some resources can cease if harvesting is not performed in a sustainable manner. Renewable energy technologies transform a renewable energy resource into heat, cooling, electricity or mechanical energy. Their use has steadily increased over the last years, and they provide now cost-effective alternatives to fossil fuel-based technologies.

To benefit from these technologies, it is important to assess accurately that their implementation is profitable. The selection of optimum technologies and sizes must be based on several aspects (technical, economic, environmental, and social), and consider load characteristics as well as potential energy storage, which can be particularly important when non-dispatchable intermittent energy resources (e.g., wind, photovoltaic) are involved. Obtaining a feasible solution is not easy, since there can be many technology alternatives; in addition, this task may require the integration of different forecasting and simulation techniques. In general, renewable energy technologies have higher initial costs but lower operating costs than conventional technologies, because the marginal cost of renewable energy resources may be assumed zero. To determine the life-cycle cost of a project, all costs over its lifetime must be added taking into account the time value of money. This value includes initial costs, annual costs for operation and maintenance, costs for replacement of equipment, costs for project decommissioning, and financial costs. Fuel costs, and even greenhouse gas (GHG) emission penalties, must be also included in case of hybrid systems.

According to *Clean Energy Project Analysis. RETScreen Engineering and Case Textbook* (RETScreen International, Clean Energy Decision Support Centre, 2005), the complete implementation of a renewable energy project may consist of the following four steps: (1) pre-feasibility analysis; (2) feasibility analysis; (3) engineering and development; (4) construction and commissioning. A decision is made after the completion of each of the above steps to whether to stop or to continue with the next step. A pre-feasibility analysis is a quick examination, based on simple cost calculations and judgements, and aimed at determining whether a project may deserve a more serious investment of time and resources. A feasibility analysis is a more in-depth analysis, which requires a detailed collection of resource, cost and equipment data. A feasibility analysis will usually involve computer simulations, which must help to decide whether or not to proceed with the project, and provide information about design, economical viability, and (environmental and social) impact of the project.

This chapter deals with renewable energy technologies for generation of electricity. Systems studied in this chapter will consist of at least one renewable energy source for electricity generation and at least one destination of the produced energy. A system may be autonomous (off-grid) or grid-connected (on-grid), can include, in addition to renewable energy input, conventional non-renewable on-site generation and electric energy input from the grid, and may take advantage of energy storage technologies.

Figure 21.1 depicts the schematic diagram of an energy system for electricity generation. The figure does not provide detailed information about the components and about the system behavior; it is just aimed at showing the energy flow between the different parts of the system. Converters are needed in systems with a DC bus. From the definitions presented by T. Lambert, P. Gilman, and P. Lilienthal ("Micropower System Modeling with HOMER", Chap. 15 of *Integration of Alternative Sources of Energy*, by F.A. Farret and M. Godoy Simões, John Wiley, 2006), a load is a demand of energy, a resource is anything coming from outside the system that is

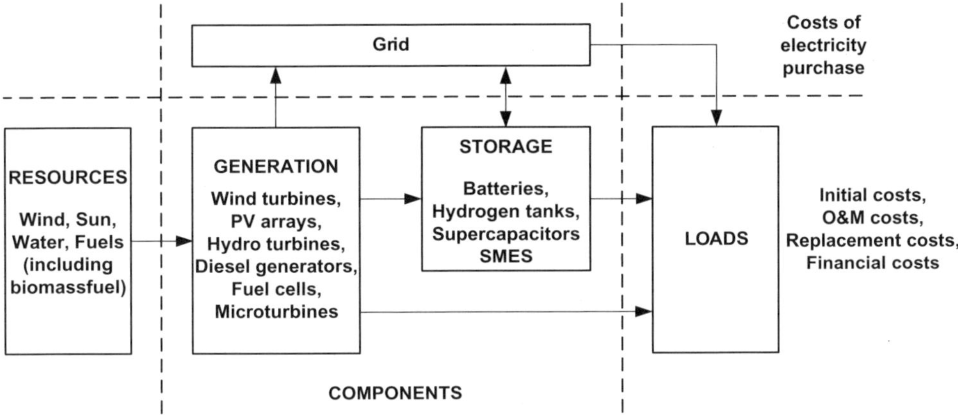

Fig. 21.1.   Schematic diagram of a renewable energy system for electricity production.

used to produce energy, and a component is any part that generates, delivers, converts or stores energy. Aspects to be considered are the great variety of generating and energy storage devices, the different behavior of the various energy technologies, and the fact that some generating units are connected to the utility network via a static converter, while others can be connected directly. All these issues complicate modeling, analysis and simulation of systems.

The organization of the chapter is as follows. Section 21.2 summarizes the models used in current software packages for representing generation systems and dispatch strategies. Section 21.3 presents some approaches used to obtain the life-cycle cost of a project and to rank the economical merit of the different alternatives. A methodology for estimating GHG emission reductions is summarized in Sec. 21.4. The capabilities of the most common software packages for feasibility studies are analyzed in Sec. 21.5, which also includes a summary of simulation tools for economic operation of distributed energy systems. Section 21.6 presents two illustrative examples.

21.2   Modeling for Feasibility Studies

The capabilities of present software packages for feasibility studies are very different; however, there are some common aspects to be considered when modeling the different parts (resources, generators, converters, energy storage devices, loads, control strategies) of a renewable energy system. The following subsections describe models for loads, energy resources and components, as well the dispatch strategies that can be used to operate the different parts.

The simulation tools analyzed in this chapter are used for long-term performance predictions and can be classified into several categories depending on the way in which intermittent variables (e.g., wind speed, solar insulation, and load) are specified and/or obtained (*Hybrid2 — A Hybrid System Simulation Model, Theory Manual*):

- *Time series*: The model uses time series for intermittent variables.
- *Probabilistic*: The model uses statistical techniques and requires long-term data that is specified monthly, seasonally or yearly.
- *Time series/probabilistic*: The model combines time series and statistical approaches; i.e., a time series approach is used to account for variations over intervals typically ranging from a few minutes to one hour; fluctuations within a time step are obtained by means of statistical techniques.

21.2.1 *Loads*

Software packages distinguish between several types of loads: primary, deferrable, dump (see "Micropower System Modeling with HOMER", Chapter 15 of *Integration of Alternative Sources of Energy* and *Hybrid2 — A Hybrid System Simulation Model, Theory Manual*).

(1) A **primary load** is an electrical demand that the system must meet at any time according to a given schedule. Primary load is specified in kW for each time step during the period of study. In general, users can either import a file containing data or take advantage of code capabilities to synthesize load data. Some packages allow users to specify different profiles for weekdays and weekends, and to include randomness when synthesizing data to avoid repetitive patterns.

(2) A **deferrable load** is an electrical demand that can be met within a defined time interval, so there is some flexibility as to when it must be supplied. Water pumps and battery-charging stations are examples of deferrable loads. This load offers an important advantage for systems with intermittent renewable energy input, because it can be postponed until there is excess of energy from renewable sources. A deferrable load may be specified by means of the characteristics of an associated storage tank and its average value during a certain interval; e.g., a month. The user specifies the storage tank capacity, in kWh, and the rate limits at which the power system can put energy into the tank. When simulating a system serving a deferrable load, the program tracks the level in the tank and puts any excess of renewable energy into the tank. As long as the tank level remains above zero, the system does not use a dispatchable power source (e.g., a generator, a battery bank, the grid) to put energy into the tank. If the level in the tank drops to zero, the deferrable load is temporarily treated as primary load, and any available power source will put energy into the tank.

(3) A **dump load** is used to absorb any energy excess that cannot be used by deferrable loads. A dump load is normally an actual (storage) device, which may be used to control the grid frequency where there is no other form of

frequency regulation (e.g., in a diesel system in which all the diesels are shut off). It may be also used when there is more power produced on the grid than can be used or stored (e.g., where there is a low load, but a large amount of renewable generation and the storage is full). The power rating and the excess power that must be dumped during each time step characterize a dump load. This type of load is not honored by all packages.

21.2.2 *Resources*

The term resource includes renewable resources (solar, wind, hydro, and biomass) and any fuel used by some generators. The characterization of renewable energy resources requires data on the available resource plus information on the measurement of the resource data, variability of the data, and geographic factors that affect the determination of the actual available renewable resource. The solar resource depends strongly on latitude and climate, the wind resource on large-scale atmospheric circulation patterns and geographic influences, the hydro resource on local rainfall patterns and topography, and the biomass resource on local biological productivity. Renewable resources may vary enormously by location, and may exhibit strong seasonal and hour-to-hour variability. The nature of the available renewable resources affects the behavior and economics of renewable power systems, since they determine the quantity and the timing of renewable power production.

Wind resource — It is characterized by a number of parameters that depend on the simulation package. In general, the main input is the wind speed the turbine will experience in a typical year. The user can provide measured wind speed data, or take advantage of the code capability to generate synthetic data from average wind speeds. Several packages calculate the wind speed distribution as a Weibull probability density function, which gives the probability to have a certain wind speed from the average (mean) wind speed. The Weibull function is characterized by the shape factor (greater than 1) and the scale factor (typically ranging from 1 to 3). A lower shape factor indicates a relatively wide distribution of wind speeds around the average, while a higher shape factor indicates a relatively narrow distribution. A lower shape factor will normally lead to a higher energy production for a given average wind speed.

Other characteristics parameters are the autocorrelation factor, the diurnal pattern strength, and the hour of peak wind speed. The autocorrelation factor is a measure of how strongly the wind speed in a time step (e.g., one hour) tends to depend on the wind speed in the preceding time step. The diurnal pattern strength and the hour of peak wind speed indicate the magnitude and the phase, respectively, of the average daily pattern in the wind speed.

Another parameter to be specified is the anemometer height. If the wind turbine's hub height is different from the anemometer height, the code calculates the wind speed at the turbine hub height using either the logarithmic law, which assumes that the wind speed is proportional to the logarithm of the height above ground, or the power law, which assumes that the wind speed varies exponentially with height. To use the logarithmic law, the user enters the surface roughness length, which is a parameter characterizing the roughness of the surrounding terrain. To use the power law, the user enters the power law exponent.

Factors that affect the local wind resource are the air density and the nature of the turbulence. The local air density, needed to calculate the output of the wind turbine, can be specified as input or determined from the local temperature and pressure using the ideal gas law. The turbulence length scale, the reference wind velocity for the turbulence length scale, and the turbulence intensity may be also used to determine the wind power at a given site, but they are not honored by some packages.

Solar resource — It is the amount of solar radiation that strikes ground surface. It is characterized by measured insolation data and parameters related to the site, which may include, depending on the simulation package, geographic information (site latitude and longitude, ground reflectance) and environmental information (the ambient temperature or a time series of ambient temperature). Input data can be expressed as either average total solar radiation on the horizontal surface (kW/m^2) or the average clearness index. The clearness index is the ratio of the total horizontal radiation at ground surface to the extraterrestrial radiation, which is the total radiation that strikes the top of the atmosphere.

The prediction of the power supplied by a photovoltaic (PV) panel requires the determination of the insolation on the panel surface. This insolation is composed of direct and diffuse radiation, each of which depends on the clearness index. The direct radiation is also a function of the position of the sun and the physical orientation of the solar panels. During simulation, the code obtains the sun location at each time step and the corresponding insolation on a horizontal surface. Some codes may synthesize the hourly radiation from monthly average data and latitude.

Hydro resource — It is characterized by the amount of water that is available during a certain period of time (e.g., one year) for electricity production. This information can be provided by entering a flow-duration curve, the measured hourly stream flow data, or monthly averages, under the assumption that the flow rate remains constant within each month. A flow-duration curve is specified by a certain number of values that represent the flow equaled or exceeded a certain percentage of the time. Another parameter to be specified is the residual flow, which is the minimum flow that must be left in the river or bypass the hydro turbine for environmental reasons. The net stream flow available to the turbine is the stream flow minus the residual flow. Some package calculates the firm flow, which is defined as the flow being available a given percentage of the time, usually equal to 95%.

Fuels — They are used by generators to produce electricity. The physical properties of a fuel include its density, lower heating value, carbon and sulphur content. The economic properties of a fuel are the price and the annual consumption limit, if any.

Biomass — It takes various forms (e.g., wood waste, agricultural residue, animal waste) and its availability depends in part on human effort for harvesting, transportation and storage. Biomass is not intermittent, although it may be seasonal and may have a cost. Some packages allow users to specify the biomass availability in the same way that other intermittent variables (wind speed, solar radiation); that is, the user can indicate the availability of biomass by importing a data file or specifying average values during a certain interval, e.g., a month. Additional parameters to define the biomass resource are price, energy content of the biomass fuel, carbon content, and gasification ratio. The energy content is used to calculate the efficiency of the generator that consumes this fuel. The carbon content value is needed to obtain the net amount of carbon released to the atmosphere by harvesting, processing, and consumption of the biomass feedstock. When biomass feedstock may be converted to a fuel that can be consumed by a conventional generator, the user has to specify the gasification ratio, which is the ratio of the mass of generator-ready fuel resulting from the mass of biomass feedstock.

Variables and factors needed to characterize resources, as well as the output derived from each type of resource, are summarized in Table 21.1.

Table 21.1. Description of energy resources.

Resource	Main characteristics	Main outputs
Wind	Average wind speed, Anemometer height, Wind speed variation law (logarithmic/exponential), Air density/Altitude above sea, Autocorrelation, Diurnal pattern strength, Hour of peak wind speed, Weibull distribution factors	Wind speed time series
Sun	Average insolation, Latitude and longitude, Ground reflectance	Solar radiation time series, Average clearness index
Water	Average stream flow, Residual flow	Net stream flow
Fuel	Type (gas, diesel, ethanol, hydrogen,...), Price, Lower heating value, Density, Carbon and sulphur content	
Biomass	Average availability (tonnes/day), Price, Lower heating value, Gasification ratio, Carbon content	Biomass availability

21.2.3 *Components*

A component is any part of a micropower system that generates, delivers, converts, or stores energy. Components that generate electricity can be divided into two groups: (1) those which use intermittent renewable resources (wind turbines, photovoltaic modules, hydro turbines), and (2) dispatchable energy sources (generators, the grid). Some components (e.g., converters, electrolyzers) convert electrical energy into another form. Finally, some components (e.g., batteries) store energy. A summary of the models implemented in current software packages for all these types of components is presented below.

Wind energy system — It typically consists of: (1) a wind turbine, which converts the energy in the wind into mechanical energy; (2) an electric generator, which converts the mechanical energy into electricity; (3) a tower, which supports the turbine-generator set above the ground to capture higher wind speeds; (4) a control system used to start and stop the wind turbine, and to monitor the proper operation of the machinery.

The main specifications that define a wind turbine model are the power curve and the response factor. The power curve is a graph of power output versus wind speed at hub height; it is a function of the turbine design and is specified by the turbine manufacturer. The turbine response factor is a measure of the relationship between the variability of the wind and the variability of the resulting electrical power. A wind turbine will have a cut-in wind speed at which the turbine starts to generate power, a rated wind speed, at which it starts to generate rated power, and a high-wind cut-out wind speed at which it is shut down for safety.

Several factors affect the power output, and some adjustments may be required if the wind speed is not measured at the turbine hub height, or if the turbine is to be operated under non-standard atmospheric conditions; that is, the calculations are affected by differences in wind turbine and anemometer heights, and by air density. In general, the wind turbine model assumes that the power curve applies at a standard air density of $1.225 \, \text{kg/m}^3$, which corresponds to standard temperature and pressure conditions.

Each time step, the wind turbine model calculates the power output in a four-step process: (1) determines the average wind speed each time step at the anemometer height by using the wind resource data; (2) calculates the wind speed at the turbine hub height (using either the logarithmic law or the power law); (3) uses the turbine power curve to calculate its power output assuming standard air density; (4) multiplies the power output value by the air density ratio, which is the ratio of the actual air density to the standard air density. The air density ratio is assumed constant throughout the year.

The total power generated by multiple wind turbines depends on the individual power curve of each type of turbine and the number of turbines of each type. The variability of the total power will depend on the variability of power from each turbine, their relative spacing, and characteristics of the site. Wind speed varies not only with time, but also spatially. Thus, multiple wind turbines may not experience exactly the same wind. If the power from multiple wind turbines is combined and they all experience the same wind regime, the resulting wind power would be the same as that produced by one large wind turbine. However, when they are so far from each other that they experience different winds, the result may be a net reduction in the variability of the total power. When the package allows the simulation of multiple turbines with different characteristics at different locations, the specifications are then the total number of turbines, the characteristics of each wind turbine, the spacing between wind turbines, and the wind power scale factor.

Photovoltaic Array — It is a device that produces DC electricity in direct proportion to the total solar radiation incident upon it. The nominal/rated capacity of a PV array is the power produced under standard conditions of 1 kW/m^2 of sunlight and a cell temperature of 25°C.

Although the output of a PV array depends strongly and non-linearly on the voltage to which it is exposed, and its maximum power point (the voltage at which the power output is maximized) depends on the solar radiation and the ambient temperature, simplified but accurate enough algorithms may be used. A simple model represents a PV array as a device whose behavior is independent of its temperature and the voltage to which it is exposed, being the power output obtained by means of the following equation:

$$P_{PV} = P_{PVr} f_{PV} I_T, \tag{21.1}$$

where P_{PVr} is the rated capacity of the PV array (kW), f_{PV} is the derating factor, and I_T is the global solar radiation (direct plus diffuse) incident on the surface of the PV array (kW/m^2).

The rated capacity in this model accounts for both the area and the efficiency of the PV module. The derating factor is a scaling factor to account for any effect that can cause the output of the PV array to deviate from that expected under ideal conditions; e.g., dust, wire losses, an elevated temperature. The total solar radiation incident on the array surface is obtained taking into account the orientation of the PV array, which may be fixed or vary according to a tracking scheme, the location, the time of year, and the time of day. The incident radiation on the panel surface is usually determined by means of the following procedure: (1) the code first calculates the extraterrestrial radiation based on the day of the year, and the site latitude and longitude; (2) next, it obtains the clearness index; (3) this index is used to determine the direct and diffuse components of the total radiation via empirical correlations; (4) finally, the radiation on the tilted surface of the PV panel is determined, based on the incident direction of the direct and diffuse solar radiation components and the ground reflectance.

If the PV array is connected directly to a DC load or a battery bank, it will often be exposed to a voltage different from the maximum power point, and its performance will vary. A maximum power point tracker (MPPT) is a solid-state device placed between the PV array and the rest of the DC components of the system that decouples the array voltage from that of the rest of the system, and ensures that the array voltage is always equal to the maximum power point. The effect of the PV array voltage is ignored when the model assumes that a MPPT is present.

A more sophisticated model must include the effect of temperature, the dependence of the temperature with the clearness index, and a correction factor to account for any angle deviation from the optimum tilt angle.

Another PV array representation deduces the power output from the current-voltage relationship. A one-diode model forms the basic circuit model used to establish the current-voltage curve, and includes the effects of radiation level and cell temperature on the output power. A PV panel is composed of individual cells connected in series and parallel, and mounted on a single module. The characteristic parameters of the array model are the parameters that define the basic current-voltage relationship of a cell. Inputs during simulation are the insolation data, the terminal DC voltage and the ambient temperature. The outputs at each time step are voltage, current and generated power.

Hydro turbine — It converts the power of falling water into electricity at a constant efficiency. The viability of a hydro project is site specific: (1) the power output depends on the stream flow and the head, which is the vertical distance through which the water falls; (2) the amount of energy that can be generated depends on the quantity of water available and the variability of water flow. Small hydro projects can generally be categorized into two groups:

1. *Run-of-river developments*: The hydro plant uses only the available water in the natural flow of the river, there is no water storage, and the power output fluctuates with the stream flow. In general, these projects do not provide much firm capacity, often require supplemental power, and are best suited to provide energy to a larger electricity system.

2. *Water storage developments*: A hydro plant can provide power on demand, either to meet a fluctuating load or to provide peak power, when some volume of water can be stored in one or more reservoirs. Pumped storage is a type of storage development where water is recycled between downstream and upstream reservoirs: water is used to generate power during peak periods, and pumped

back to the upper reservoir during off-peak periods. The viability of pumped storage projects depends on the difference between the values of peak and off-peak power. Recycling of water results in net energy consumption, so energy used to pump water must be generated by other sources.

A hydro turbine model must provide a means to assess the available energy at a small hydro site. As for many other components, hydro turbine models available in current packages have different capabilities. A complete model should address both run-of-river and reservoir developments; in reality, most models do correctly represent run-of-river developments but they are limited for water storage developments.

A simple hydro turbine model is characterized by the available head, the head loss that occurs in the intake pipe due to friction, the design flow rate of the turbine and its acceptable range of flow rates. The power output of the hydro turbine is approximated by the following formula:

$$P_{\text{hyd}} = \eta_{\text{hyd}} \rho_{\text{wat}} g h (1 - f_{\text{hyd}}) Q_{\text{tur}}, \tag{21.2}$$

where η_{hyd} is the turbine efficiency, ρ_{wat} is the density of water, g is the gravitational acceleration, h is the available head, f_{hyd} is the pipe head loss, and Q_{tur} is the flow rate through the turbine.

The turbine does not operate if the stream flow is below the minimum; on the other hand, the flow rate through the turbine cannot exceed the maximum. The stream flow available to the hydro turbine at each time step comes from the hydro resource data.

The turbine efficiency depends on a number of factors such as the net head, the runner diameter, or turbine specific speed and design coefficient. Depending on the package, it can be entered manually or calculated by the tool.

Generator — It consumes fuel to produce electricity. A simple generator module can model a wide variety of generators, such as internal combustion engine generators, microturbines, thermoelectric generators, or fuel cells. The physical properties of a generator are its maximum and minimum electrical power output, its expected lifetime in operating hours, the type of fuel it consumes, and its fuel curve, which relates the quantity of fuel consumed to the electrical power produced. In general, the fuel curve is approximated by a straight line with the following equation:

$$F = F_0 P_{\text{genr}} + F_1 P_{\text{gen}}, \tag{21.3}$$

where F_0 is the fuel curve intercept coefficient, F_1 is the fuel curve slope, P_{genr} is the rated capacity of the generator (kW), and P_{gen} is the electrical output of the generator (kW).

Most packages allow users to enter emission coefficients, which specify the emissions of pollutants emitted per quantity of fuel consumed. An additional parameter for a generator that provides also heat is the heat recovery ratio, which is the fraction of the waste heat that can be captured to serve the thermal load.

The user can schedule the operation of the generator to force it on or off at certain times. During intervals that the generator is not forced on neither off, the code decides whether it should operate based on the needs of the system and the relative costs of the other power sources. During intervals that the generator is forced on, the code decides at what power output level it operates. Results produced by the generator model are output power, fuel consumption, operating hours, number of starts, and pollutant emissions.

Grid — It is a component from which the system can purchase electricity, and to which the system can sell electricity. The complexity of the grid model depends on the software package. A general grid model includes a grid power price, a sellback rate and a demand charge based on the peak demand within the billing period. The grid power price is the price that the utility charges for energy purchased from the grid, the sellback rate is the price that the utility pays for power sold to the grid, the demand rate is the price the utility charges for the peak grid demand. The user can define and schedule several rates, each of which can have different values of grid power price, sellback rate, and demand rate. The grid model may also consider net metering, with which the utility charges the customer based on the net grid purchases (purchases minus sales) over the billing period.

The grid capacity may be characterized by two variables: the maximum grid demand and the maximum power sale. The first variable is the maximum amount of power that can be drawn from the grid, while the second one is the maximum rate at which the power system can sell power to the grid. This value is zero when the utility does not allow sellback. A grid-connected generator is turned on whenever the load exceeds the maximum grid demand, which acts as a control parameter that affects the operation and economics of the system.

Some grid models allow users to enter grid emission coefficients, which are used to calculate the emissions of pollutants associated with buying power from the grid, as well as the avoided emissions resulting from the sale of power to the grid. Emission coefficients are usually specified in grams of pollutant emitted per kWh consumed.

Battery bank — A battery is a device capable of storing a certain amount of electricity at fixed round-trip energy efficiency, with limits as to how quickly it can be charged or discharged, how deeply it can be discharged without causing damage, and how much energy can cycle through it before being replaced. A battery bank is a collection of one or more individual batteries, whose properties are assumed to remain constant throughout its lifetime and are not affected by external factors such as temperature.

The battery model implemented in some packages is based on the kinetic model presented by J. F. Manwell and J. G. McGowann ("Lead acid battery storage model for hybrid energy systems", *Solar Energy*, vol. 50, pp. 399–405, 1993), although the way in which it is used may differ between packages. The battery is viewed as a voltage source in series with a resistance: the internal voltage varies with the current and the state of charge, while the internal resistance is assumed constant. The model accounts for voltage level as a function of state of charge and charge/discharge rate, as well as for the apparent change of capacity as affected by the charge/discharge rate. The physical properties of the battery are its nominal voltage, capacity curve, lifetime curve, minimum state of charge, and round-trip efficiency. The capacity curve shows the discharge capacity of the battery in ampere-hours versus the discharge current in amperes; it typically decreases with increasing discharge current. The lifetime curve shows the number of discharge-charge cycles the battery can withstand versus the cycle depth; the number of cycles to failure decreases with increasing cycle depth. The minimum state of charge is the state of charge below which the battery must not be discharged to avoid damage. The round-trip efficiency is the percentage of energy going into the battery that can be drawn back out, and is a measure of energy losses. Other parameters that may be used to characterize a battery bank are the number of batteries (in series and in parallel), the bank scale factor, and the initial capacity.

Variables that are entered each time step are the required power from the battery, the available power to be used for charging and the state of charge from the previous time step. Simulation results may be the bank voltage during each time step, the storage capacity, the initial energy stored in the bank, the losses associated with charging and discharging, and the cumulative damage done to the batteries due to charge/discharge cycles over the complete simulation.

The kinetic model may be explained by means of an analogy to a system with two tanks linked by a pipe. According to this model, part of the storage capacity is available for charging or discharging in the first tank and the rest is chemically bound in the second tank. The rate of conversion between available energy and bound energy depends on the difference in height between the two tanks: at high discharge rates the available tank empties quickly, and very little of the bound energy can be converted to available energy before the available tank is empty, at which time the battery can no longer withstand the high discharge rate and appears fully discharged; at slower discharge rates, more bound energy can be converted to available energy before the available tank empties, and the apparent capacity increases. The lifetime throughput (i.e., the amount of energy cycled through the battery before failure) can be calculated by finding the product of the number of cycles, the depth of discharge, the nominal voltage of the battery, and the maximum capacity of the battery. The lifetime throughput curve exhibits a weak dependence on the cycle depth, and the battery bank life may be estimated by monitoring the amount of energy cycling through it, without having to consider the depth of the various charge-discharge cycles.

Converter — It is a device that converts electric power from DC to AC in a process called inversion, and/or from AC to DC in a process called rectification. It may be either unidirectional (AC to DC, DC to AC) or bi-directional, capable of converting power in both directions. It may also be either electronic or electromechanical, and may or may not be capable of operating in parallel with diesel generators. There are losses associated with all converters. Other physical properties of a converter are its inversion and rectification efficiencies, which are assumed constant. The specification of converter models depend on the power converters included in the system under consideration. Converter parameters are the full load power for each direction of possible power flow, the no-load loss, the full-load efficiency, and an indication if the device is capable of parallel operation.

Electrolyzer — It consumes electricity to generate hydrogen via the electrolysis of water. An electrolyzer is characterized by its size (in terms of its maximum electrical input), the type of bus (AC or DC) to which it must be connected, and the efficiency with which it converts electric energy to hydrogen. The electrolyzer efficiency is the energy content of the hydrogen produced divided by the amount of electricity consumed. Another property is the minimum load ratio, which is the minimum power input at which an electrolyzer can operate, expressed as a percentage of its maximum power input.

Hydrogen tank — It stores hydrogen produced by the electrolyzer for later use in a hydrogen-fuelled generator; e.g., a fuel cell. The hydrogen tank is characterized by the mass of hydrogen it can contain and the initial amount of hydrogen. If the year-end tank level has to be specified, the code will consider infeasible any system configuration whose hydrogen tank contains less hydrogen than specified at the beginning of the simulation. An idealized model assumes that the process of adding hydrogen to the tank requires no electricity and the tank experiences no leakage.

Table 21.2 summarizes the approaches detailed in this section to represent components of a microgeneration system in feasibility studies. The table shows also the parameters required for the economic analysis, see Sec. 21.3.

21.2.4 *System dispatch*

The way in which components work together is a very important aspect since decisions are required every time step as to how dispatchable components must operate. The fundamental principle is cost minimization. The code will always search for the

Table 21.2.   Characteristic parameters of component models.

Component	Technical parameters	Economic parameters	Outputs
Wind turbine	Type (AC/DC), Power curve, Hub height	Expected lifetime, Capital cost, Replacement cost, Annual O&M cost	Electricity production, Average output, Wind penetration, Capacity factor
PV array	Rated capacity, Derating factor, Tracking system , Array slope and azimuth, Ground reflectance	Expected lifetime, Capital cost, Replacement cost, Annual O&M cost,	Electricity production, Average production, Min/max output, Capacity factor, Hours of operation
Hydro turbine	Type (AC/DC), Design flow rate, Max/min flow ratio, Turbine efficiency, Available stream flow, Available head, Head loss	Expected lifetime, Capital cost, Replacement cost, Annual O&M cost,	Electricity production, Average output, Min/max output, Hydro penetration, Capacity factor, Hours of operation
Generator	Type Max/min power output, Type of fuel — Fuel curve, Heat recovery ratio, Schedule Emission factors	Expected lifetime, Capital cost, Replacement cost, Annual O&M cost,	Electricity production, Min/max output, Emission of pollutants, Hours of operation, Number of starts, Operational life, Fuel usage
Battery bank	Nominal voltage, Nominal capacity, Round-trip energy efficiency, Minimum state of charge, Lifetime throughput, Maximum charge rate, Maximum charge current, Capacity curve, Lifetime curve,	Minimum battery life, Capital cost, Replacement cost, Annual O&M cost,	Electricity production, Battery life, Battery throughput, Battery power, Battery state of charge, Battery energy cost
Grid	Net metering, Maximum power sale, Maximum grid demand, Grid emission factors	Grid power price, Demand rate, Sellback rate, Standby charge	Electricity sold, Energy purchased, Emission of pollutants

(Continued)

Table 21.2. (*Continued*)

Component	Technical parameters	Economic parameters	Outputs
Converter	Type Size, Rectifier capacity, Efficiencies	Expected lifetime, Capital cost, Replacement cost, Annual O&M cost,	Inverter Power, Rectifier Power
Electrolyzer	Type (AC/DC power), Size Efficiency, Minimum load ratio	Expected lifetime, Capital cost, Replacement cost, Annual O&M cost,	Energy consumption
Hydrogen tank	Size, Initial amount of hydrogen, Year-end tank level	Expected lifetime, Capital cost, Replacement cost, Annual O&M cost	Hydrogen generation, Hydrogen consumption, Tank autonomy, Stored hydrogen

combination of dispatchable sources that can serve the load at the lowest cost. Since capital costs of renewable energy sources are generally higher and their marginal costs are much lower than the corresponding costs of conventional generators, the code assumes that renewable energy sources must operate to produce as much power as the resources allow. Energy exceeding the primary load may be sent to deferrable loads, storage devices, or dumped.

Operating reserve — It is the operating capacity minus the electrical load (see "Micropower System Modeling with HOMER", Chap. 15 of *Integration of Alternative Sources of Energy*). This concept is used to obtain a safety margin. It is not honored by all packages. Both dispatchable and non-dispatchable power sources provide operating capacity. A dispatchable power source provides operating capacity in an amount equal to the maximum amount of power it could produce at a given time step. The operating capacity of a non-dispatchable power source is equal to the amount of power the source is currently producing, not the maximum amount of power it could produce. The required amount of operating reserve is calculated in every time step as a fraction of the primary load, plus a fraction of the PV power output, plus a fraction of the wind power output, plus a fraction of the annual peak primary load.

Operation reserve is not required with grid-connected systems whose capacity is larger than the load, because the grid is always operating and its capacity is more than enough to cover the required operating reserve. Similarly, operating reserve typically has little or no effect on autonomous systems with large battery banks, since the battery capacity is also always available to the system, at no fixed cost.

There is a shortage when the system is unable to supply the required amount of load plus operating reserve. The total amount of shortages over the year divided by the total annual electric load is the capacity shortage fraction. The model discards as infeasible any system whose actual capacity shortage fraction exceeds the specified value.

Dispatch strategies — Dispatchable sources must be controlled to match supply and demand properly, and to compensate for the intermittency of the renewable power sources. Each time step, the code determines whether the (non-dispatchable) renewable power sources by themselves are capable of supplying the loads plus the required operating reserve; if not, it determines how to best operate dispatchable system components (generators, battery bank, grid) to serve the loads and the operating reserve. The logics implemented in simulation tools on how to make decisions for allocating the energy can be different. Figure 21.2 shows the priorities implemented in two packages.

(a) HOMER assumes that the electricity produced on one bus will go first to serve the primary load on the same bus, then the primary load on the opposite bus, then the deferrable load on the same bus, then the deferrable load on the opposite bus, then to charge the battery bank, then to grid sales, then to serve the electrolyzer, and then to the dump load.

(b) Whenever there is excess of energy, Hybrid2 first checks if the battery is able to accept a charge. If, after charging the battery, there is still excess energy, then the model supplies the deferrable loads, then thermal loads and, finally, the optional loads. The dump load is the last sink, and it is used when it is not possible to use the energy excess to serve other requirements.

The energy and the operating reserve provided by a battery bank to the AC load are constrained by its current discharge capacity (which depends on its state of charge and recent charge-discharge history), and the capacity and efficiency of the inverter. If a battery bank and a generator are both capable of supplying the net load and the operating reserve, the model decides how to dispatch them based on their fixed and marginal costs of energy. The charge of a battery bank cannot be based on simple economic principles, because there is no deterministic way to calculate the value of charging the battery bank; it may depend on what happens in future time steps. Dispatch strategies depend on the simulation tool. A short summary of those available in the two packages mentioned previously is given below.

(1) HOMER uses two simple strategies: *load-following* and *cycle-charging*. Under the load-following strategy, the power produced by a generator only serves the load, and does not charge the battery bank; under the cycle-charging strategy,

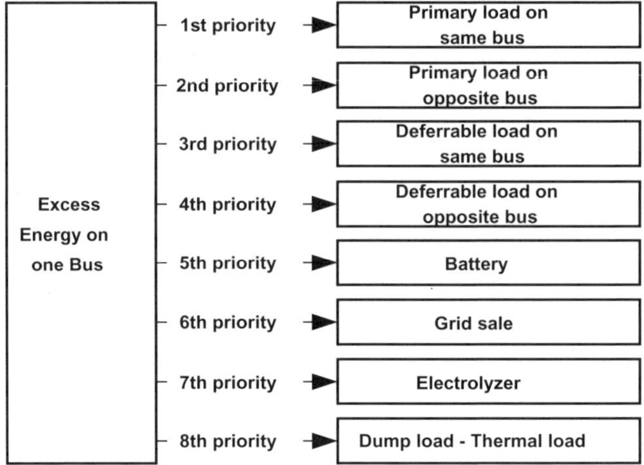

(a) Load priority in HOMER

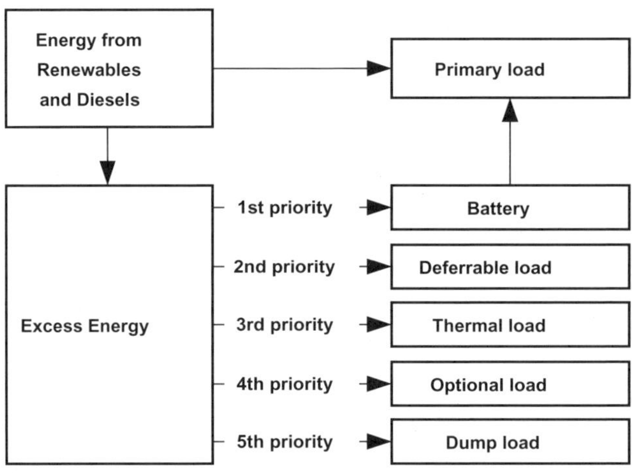

(b) Load priority in Hybrid2

Fig. 21.2.	Load management strategies.

whenever a generator operates, it runs at its maximum rated capacity and charges the battery bank with the excess. The selected strategy is applied only when the dispatchable sources are simultaneously operating during a time step. The set-point state of charge is an optional control parameter used with the cycle-charging strategy: once the generator starts charging the battery bank, it must continue to do so until the battery bank reaches the set-point state of charge; otherwise, the code may choose to discharge the battery as soon as it can supply the load. When the battery experiences shallow charge-discharge cycles near its minimum state of charge, this control avoids situations that can be harmful to battery life.

(2) Hybrid 2 uses more sophisticated control strategies with two types of dispatchable components: battery banks and diesel generators. The *battery dispatch* determines how the battery itself is charged and discharged. The *diesel dispatch* considers the minimum diesel run time, diesel shutoff criteria, forced diesel shutoff, offset in net load to force diesel to start, and dispatch order for multiple diesels. There are two possible operating conditions for a single diesel generator: above minimum load and below minimum load. The excess power produced when the load is below the minimum load will go to storage, to a deferrable load, or be dumped. A second diesel does not have to operate when the net load exceeds the rating of a diesel, if storage is used. The first diesel may operate for a significant fraction of the time at the rated power level, and the fuel consumption will include the consumption corresponding to rated load while the load is above rated load and being met, in part, by storage. For two diesels, the net load may be (1) less than the minimum of both diesels, (2) above the minimum of the two but less than the minimum of the less efficient plus the rated power of the better diesel, or (3) greater than the minimum of the less efficient plus the rated power of the better diesel. When storage is available, a fourth region would be that in which both of the diesels are running at rated power. Another strategy considers the interaction of diesels and storage, and it includes diesel operating power level (affects the power at which a diesel is run, and allows the user to select whether batteries or diesels are to be used preferentially), as well as diesel starting and stopping criteria.

21.3 Economic Modeling

The optimal design of a power plant with renewable resources may be based on the financial analysis of a wide range of system configurations comprising varying amounts of renewable and non-renewable energy sources. Economic indicators serve to rank feasible configurations; that is, configurations that are feasible from an energy point of view may be ordered using some financial figures of merit. The economic analysis must provide a reasonable estimate of system cash flow and be based on the use of life-cycle costing economics. This section summarizes some economic indicators used by current simulation packages.

The *life-cycle cost* of a system may be represented by the total *net present cost* (*NPC*), which joins all costs and revenues that occur within the project lifetime into one single quantity, with future cash flows discounted back to the present using the real interest rate. The *NPC* includes expenses (costs of initial construction, component replacements, maintenance, fuel, cost of buying power from the grid and miscellaneous costs, such as penalties resulting from pollutant emissions) and revenues (income from selling power to the grid, plus any salvage value that occurs at the end of the project lifetime). In general, it is assumed that all prices escalate at the same rate over the project lifetime, and inflation can be factored out by using the real interest rate, which is equal to the nominal interest rate minus the inflation rate. All costs are defined in terms of constant values of the currency used in calculations.

The economic properties of components are the same for all of them: capital and replacement costs, operating and maintenance costs per year, and the expected lifetime, see Table 21.2. The only exception is the grid, which is characterized by a power price, a sellback rate and a demand charge based on the peak demand within the billing period, see Sec. 21.2.3.

When calculating costs, a distinction has to be made between non-dispatchable and dispatchable components. For the first type of components, only fixed costs must be obtained; for dispatchable components, both fixed and marginal costs must be calculated. These calculations may be as follows:

(1) *Generator*: The fixed cost is the cost per hour of running the generator without producing any electricity. The marginal cost is the additional cost per kWh of producing electricity.
(2) *Grid*: The fixed cost is zero, and the marginal cost is equal to the current grid power price plus any cost resulting from emissions penalties. The marginal cost of energy may vary from hour to hour if the grid power price changes.
(3) *Battery bank*: The fixed cost is zero, while the marginal cost is the sum of the battery wear cost (the cost of cycling energy through the battery bank) and the battery energy cost (the average cost of the energy stored in the battery bank). The battery energy cost is calculated by dividing the total year-to-date cost of charging the battery bank by the total year-to-date amount of energy put into the battery bank. Under the load-following dispatch strategy, the battery bank is charged only by surplus electricity, and the cost associated with charging the battery bank is zero. Under the cycle-charging strategy, a generator will produce extra electricity to charge the battery bank, but the cost associated with charging the battery bank is not zero.

Expenses and revenues, along with the salvage value, are combined for every component to find its annualized cost. These costs, along with any miscellaneous costs, are summed to find the total annualized cost of the system, which is used to calculate the total *net present cost*:

$$NPC = \frac{C_{\text{tot}}}{\frac{i(1+i)^L}{(1+i)^L - 1}}, \tag{21.4}$$

where C_{tot} is the total annualized cost, i is the annual real interest rate (the discount rate), and L is the project lifetime. The denominator is the capital recovery factor.

Another economic indicator is the levelized *cost of energy*, which is defined as the average cost per kWh of useful electrical energy:

$$COE = \frac{C_{tot}}{E_{load} + E_{grid}}, \tag{21.5}$$

where E_{load} is the total amount of load (including primary and deferrable) that the system serves per year, and E_{grid} is the amount of energy sold to the grid per year. The denominator is, therefore, the total amount of useful energy that the system produces per year. The total *NPC* may become negative when the benefit from selling electricity to the grid exceeds the other costs of the system; in such cases, *COE* would be also negative (see Example 21.2).

A more in-depth economic analysis should include asset depreciation, taxes and loss carry forward, and consider the use of other financial indicators. Some financial feasibility indicators available in RETScreen to facilitate the project evaluation are listed below:

- *Net present value (NPV)*: It is the value of all future cash flows, discounted at the real interest rate, in present time currency.
- *Annual life cycle savings*: It is the levelized nominal yearly savings having exactly the same life and net present value as the project.
- *Benefit-cost ratio*: It is an expression of the relative profitability, and calculated as the present value of annual revenues less annual costs to the project equity.
- *Energy production cost*: It is the avoided cost of energy that brings the net present value to zero.
- *GHG emission reduction cost*: It is the levelized cost to be incurred for each tonne of GHG avoided.

This list is not complete. Readers interested in this subject can consult *Clean Energy Project Analysis. RETScreen Engineering and Case Textbook* (RETScreen International, Clean Energy Decision Support Centre, 2005, Minister of Natural Resources Canada) for more details on financial indicators and the way in which they are obtained.

21.4 Greenhouse Gas Emission Reduction

The methodology presented in this section to calculate the GHG emission reductions achieved by a plant with renewables is that proposed in *Clean Energy Project Analysis. RETScreen Engineering and Case Textbook*. The annual GHG emission reduction considering only electricity generation may be estimated as follows:

$$\Delta_{\mathrm{GHG}} = (e_{\mathrm{base}} - e_{\mathrm{prop}})E_{\mathrm{prod}}(1 - \lambda)(1 - e_r), \qquad (21.6)$$

where e_{base} is the base case GHG emission factor, e_{prop} is the proposed case GHG emission factor, E_{prod} is the annual electricity production, λ is the fraction of electricity lost in transmission and distribution, and e_r the GHG emission reduction fee. Transmission and distribution losses are negligible for on-site generation.

The above equation requires the calculation of the GHG emission factor, defined as the mass of GHG emitted per unit of energy produced. For a single fuel type or source, the following formula may be used to calculate the base case emission factor:

$$e_{\mathrm{base}} = \frac{1}{\eta}\frac{1}{1 - \lambda} \sum_{k} e_k GWP_k, \qquad (21.7)$$

where e_k are emission factors for the fuel/source considered, GWP_k are the corresponding global warming potentials, η is the fuel conversion efficiency, and λ is the fraction of electricity lost in transmission and distribution. The global warming potential of a gas describes its potency in comparison to carbon dioxide, which is assigned a GWP of 1. The GHG emission factor will vary according to the type and quality of the fuel, and the type and size of the power plant.

In cases for which there are several fuel types or sources, the GHG emission factor for the electricity mix is calculated as the weighted sum of emission factors:

$$e_{\mathrm{base}} = \sum_{i=1}^{n} f_i e_{\mathrm{base},i}, \qquad (21.8)$$

where n is the number of fuels/sources in the mix, f_i is the fraction of end-use electricity coming from fuel/source i, and $e_{\mathrm{base},i}$ is the emission factor for fuel i, calculated through a formula similar to Eq. (21.7).

The calculation of the proposed case emission factor, e_{prop}, is similar to that of the base case emission factor.

21.5 Simulation Tools

Several commercially or freely available simulation tools can be presently used for the design of renewable power plants running as either standalone or grid-connected. Their capabilities allow users to analyze the feasibility of power plants based on renewable resources in conjunction with conventional energy resources and energy storage technologies. These tools vary in terms of capabilities, structure, scale of application, and computing code/platform. Those presented in this section cover a wide range of applications; in general, they have been designed as decision support tools that can help users to select the optimal technology and size. They analyze different renewable technologies and sizes from among the available alternatives by using a multiple criteria approach to adequately address the trade-offs between economics, financial risks and environmental impacts. The capabilities of the most commonly used packages are summarized in the next paragraphs.

(1) HOMER (Hybrid Optimization Model for Electric Renewables) is an optimization model developed by the National Renewable Energy Laboratory (NREL) of the U.S. Department of Energy. This tool models systems with single or multiple sources, which can be either off-grid or grid-connected, and finds the least cost combination of components that meet electrical and thermal loads. HOMER optimizes the life-cycle cost: it is an economic model that compares different combinations of component sizes and quantities, and explores how variations in resource availability and system costs affect the cost of installing and operating different system designs. This tool allows users to perform three-level studies: simulation, optimization, and sensitivity analysis.

(2) RETScreen is a tool made available by the Government of Canada through CANMET Energy Diversification Research Laboratory. It can be used to evaluate energy production, life-cycle costs and GHG emission reduction for various renewable energy technologies. The model can be applied for each technology using the same five-step standard analysis procedure: (1) definition of the energy model; (2) cost analysis; (3) greenhouse gas analysis (optional); (4) viability (financial) analysis; (5) sensitivity and risk analysis (optional).

(3) Hybrid2 is a computer model for the simulation and analysis of hybrid power systems. It is a joint project between the University of Massachusetts and NREL. It allows users to simulate systems with several types of electrical loads, wind turbines, photovoltaics, diesel generators, battery storage, and power conversion devices. A variety of different control strategies may be implemented to incorporate detailed diesel dispatch as well as interactions between diesel generators and batteries. Hybrid2 can also analyze grid-connected systems by using the so-called pseudo-grid model. A financial model is also included to calculate the economic worth of the project.

(4) D-Gen PRO is a tool designed for economic and feasibility studies of distributed generation. This tool evaluates the cost-effective application of on-site and distributed power generation, performs economic analysis, and produces reports and graphs on the economic feasibility of on-site generation and cogeneration, including waste heat analysis.

(5) The Distributed Generation Analysis Tool performs life-cycle cost analysis and environmental impact assessment. Data input includes capital and maintenance

costs, performance data for generators (turbine, micro-turbine, fuel cell), generator usage plan, financial parameters, fuel and electricity rates, and emission factors.

The list is not complete. There are, for instance, a countless number of simulation tools for feasibility studies and design of either grid-connected or standalone photovoltaic panels. The list of packages for PV system analysis and design could include, among others, PVFORM, PVGRID, PVWATSS, PV F-CHART, PV-DesignPro, SolarPro, PV*SOL, PVSYST, GridPV, NSOL, WATSUN-PV and SAM (Solar Advisor Model). For a survey of software tools for PV applications, see D. Turcotte, M. Ross, and F. Sheriff, "Photovoltaic hybrid system sizing and simulation tools: Status and needs", *PV Horizon: Workshop on Photovoltaic Hybrid Systems*, Montreal, 2001.

The results derived from the above tools, mainly aimed at pre-feasibility and feasibility analysis, can be complemented by using other tools that have been developed to optimize costs and operating efficiencies under varying system operating conditions, or to estimate the market potential of some distributed generation technologies.

- DER-CAM (Customer Adoption Model) is an economic model implemented in GAMS (General Algebraic Modeling System) software and developed at the Lawrence Berkeley National Laboratory of the U.S. Department of Energy. Its main objective is to minimize the cost of operating on-site generation and combined heat and power (CHP) systems, for either individual customer sites or microgrids. It can be used to select which generation and/or CHP technologies should be adopted and how they should be operated based on specific site load, price information, and performance data for available equipment options. Model inputs are load profiles, electricity tariffs, capital, operating, maintenance and fuel costs of the various available technologies, together with the interest rate on customer investment, as well as the basic characteristics of generating, heat recovery and cooling technologies. Outputs are the capacities of generation and CHP technologies to be installed, as well as when and how much of the installed capacity will be running, the cost of supplying the electric and heat loads.
- Fully Integrated Dispatch and Optimization (FIDO) can optimize hourly unit commitment. It simulates the operation of utility generation, incorporating the hour-by-hour performance of intermittent renewable power, demand-side technologies, and market power transactions.
- DIStributed Power Economic Rationale SElection (DISPERSE) uses databases of industrial and commercial sites to estimate the market potential of generation technologies. This model uses electric and thermal load profiles specific to the

application and region. Combining this information with generation costs and performance data, the tool performs a life-cycle cost analysis, based on the unit life, cost and performance data, as well as fuel prices. The best generation technology is selected based on the lowest generation competing electricity price. Sensitivity analysis on important variables can be conducted.

- Clean Energy Technology Economic and Emissions Model (CETEEM) is a tool designed to assess the economics and emissions of pollutants and GHG associated with the use of clean energy technologies for distributed power generation. CETEEM can analyze PEM (proton exchange membrane) fuel cell systems powered by hydrogen produced with natural gas reformers, and can be modified to characterize other clean energy technologies and fuelling arrangements.

- Wind Deployment Systems (WinDS) is a multi-regional, multi-time-period and geographic information model, developed by NREL for analysis of wind energy penetration. WinDS uses a discrete regional structure to account for the transient variability in wind output, and consideration of ancillary services requirements and costs. An expanded version, HyDS (Hydrogen Deployment Systems model), includes the production of hydrogen from three competing technologies (wind, steam methane reforming, and distributed electrolysis powered by electricity from the grid) along with hydrogen storage and transportation.

HOMER, RETScreen and Hybrid2 are probably the most popular tools for pre-feasibility and feasibility analyses. Table 21.3 shows the list of technologies available

Table 21.3.   Available models and main capabilities.

Capabilities		HOMER	Hybrid2	RETScreen
Technology	Photovoltaics	Yes	Yes	Yes
	Wind	Yes	Yes	Yes
	Biomass	Yes	No	Yes
	Hydraulic	Yes	No	Yes
	Diesel	Yes	Yes	No
	Cogeneration	Yes	No	No
	Microturbine	Yes	No	No
	Fuel cell	Yes	Yes	No
	Battery bank	Yes	Yes	No
	Electrolyzer	Yes	No	No
Main features	Time-step	Variable	Variable	Annual balance
	Dispatch strategies	Yes	Yes	No
	Economic analysis	Yes	Yes	Yes
	Sensitivity analysis	Yes	No	Yes

in the three tools and provides a comparison of the main features. The list of components shown in the table is incomplete since only technologies for electric energy generation could be considered.

From the comparison of capabilities and solution methods implemented in each package, the following conclusions are derived:

(a) HOMER and Hybrid2 may be used to analyze hybrid systems, with more than one resource. RETScreen can analyze single-technology power plants (e.g., only wind or only solar power plants, but not both simultaneously).
(b) HOMER can simulate both on- and off-grid systems; Hybrid2 can simulate both alternatives, but its grid model is limited. RETScreen can simulate autonomous off-grid systems only, except for PV applications, for which a grid connection is possible.
(c) The three tools can perform financial analysis with a different degree of sophistication. As for technical analysis, HOMER and RETScreen calculations are based on a power/energy balance, while capabilities implemented in Hybrid2 can evaluate the state of the electrical variables of the system components.
(d) The three tools can assess GHG emissions or emission reductions.

21.6 Application Examples

This section includes two examples of feasibility studies. Both cases were implemented in HOMER. Readers are encouraged to compare these results with those obtained from other packages. The systems analyzed in these examples are not real, and for some studies unrealistic costs were used just to force the feasibility of the analyzed configurations.

21.6.1 *Example 1: Off-grid PV-wind system*

Figure 21.3 shows the schematic diagram of the first example. Only renewables are used to serve the load, which has a small fraction supplied from the DC bus. Note that in this case only intermittent non-dispatchable resources are used, and the role of the battery bank can be crucial since it will be the energy stored in this bank, the only source of available energy to attend the demand when neither the solar nor the wind resource are supplying energy. The main objective of this example is to analyze the physical feasibility of this plant from the selected components.

The study is performed considering the following aspects:

(1) The load is of primary type and supplied from both the AC and the DC buses. The average annual demand of the AC load varies between 200 and

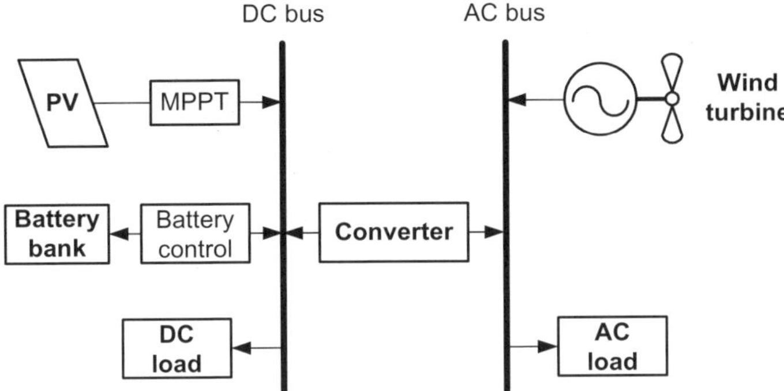

Fig. 21.3.   Example 1: Schematic diagram of an off-grid photovoltaic-wind system.

400 kWh/day, while the average annual demand of the DC load varies between 25 and 50 kWh/day. The load profiles are those depicted in Figs. 21.4 and 21.5. In fact, these figures show the load curves for January only. The profiles for other months are slightly different from those shown in these figures.

(2) Figure 21.6 shows the characteristics of the solar resource (average monthly radiation and clearness index) at the plant location. Figure 21.7 shows the characteristics of the wind resource (average monthly wind speed) at the plant location. The annual average wind speed is 5.376 m/s. The anemometer height is 10 m. The wind speed as a function of height varies according to a power law with an exponent of 0.14. The plant is located at sea level.

(3) The main characteristics of the components considered for this example are presented in Table 21.4. Figures 21.8 and 21.9 show respectively the power curve of the wind turbines and the properties of the selected batteries. The maximum number of wind turbines can be 3.

(4) The economic parameters of the components are presented in Table 21.5.

(5) The maximum annual capacity shortage should be 0%, while the operating reserve is as follows:

a.	As percent of load	Hourly load: 10%; Annual peak load: 10%
b.	As percent of renewable input	Solar power output: 20% Wind power output: 20%

(6) The project lifetime is 20 years and the real interest rate is 4%.

Table 21.6 presents a summary of the results derived from the optimization analysis. The battery life is 10 years in all cases. Actually, the table shows the optimum combination of components from an economical point of view for some

                    J. A. Martinez-Velasco and J. Martin-Arnedo

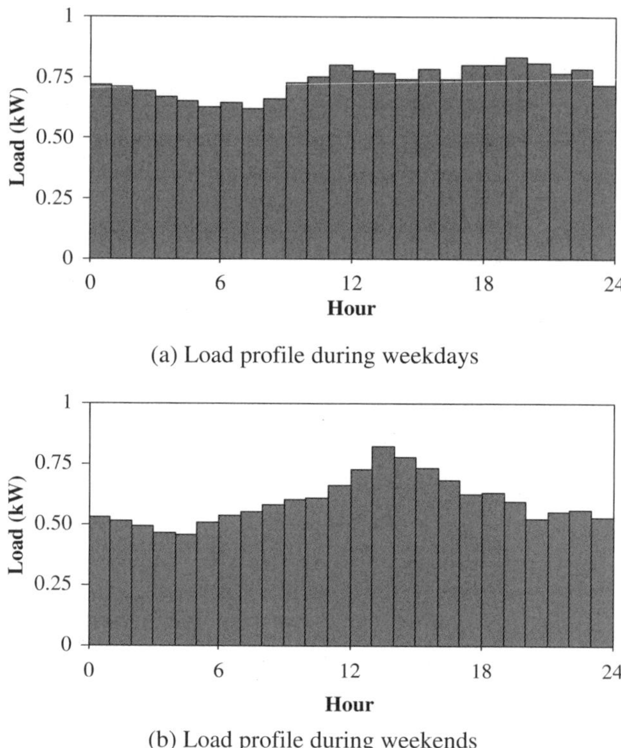

(a) Load profile during weekdays

(b) Load profile during weekends

Fig. 21.4.   Example 1: AC Load profiles.

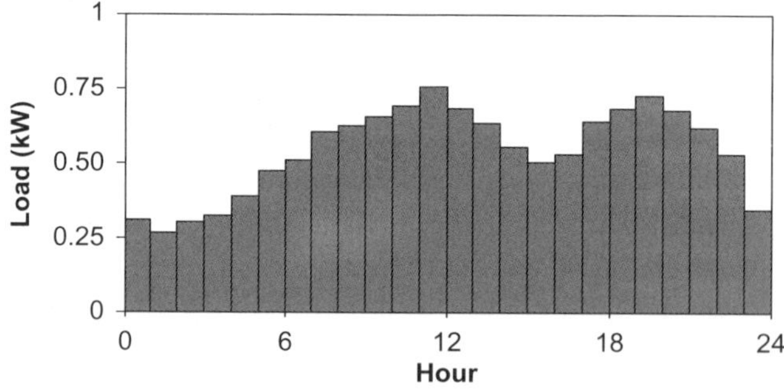

Fig. 21.5.   Example 1: DC load — Load profile during weekdays and weekends.

combinations of AC and DC demands; that is, there are many other combinations of components for which the power plant is technically feasible.

Taking into account the profiles of loads and renewable resources, one of the main conclusions is that there will be an excess of electricity production, which for

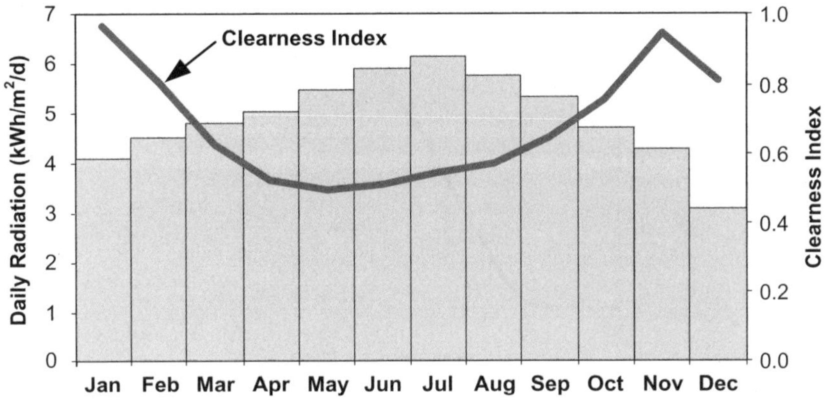

Fig. 21.6. Example 1: Characteristics of the solar resource.

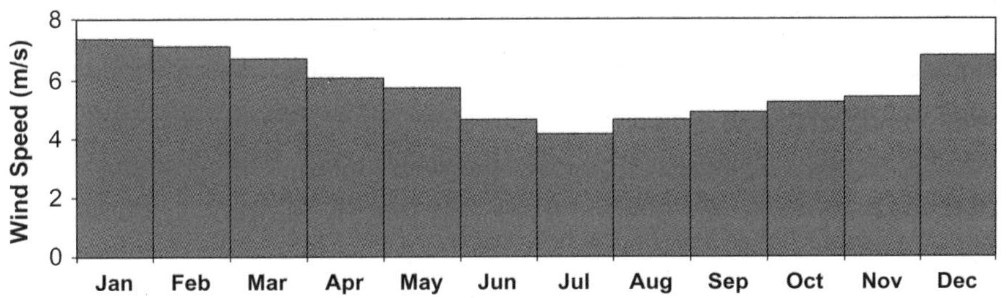

Fig. 21.7. Example 1: Characteristics of the wind resource.

Table 21.4. Example 1: Main characteristics of the plant components.

Component	Main characteristics
Wind turbines	15 m diameter, three-bladed, passive yaw downwind rotor, Tower height: 25 m, Lifetime: 20 years
PV panel	Nominal sizes: 10, 20, 30, 40, 50, 60 kW, Derating factor: 80%, Tracking system: Horizontal axis, daily adjustment, Ground reflectance: 20%, Lifetime: 20 years
Batteries	Nominal voltage: 6 V, Nominal capacity: 360 Ah (2.16 kWh), Bank size: Made in steps of 400 batteries (400, 800, 1200), Lifetime throughput: 1075 kWh, Minimum life: 5 years
Converter	Nominal size: 20, 40, 60, 80 kW, Inverter efficiency: 90%, Rectifier efficiency: 85%, Rectifier capacity (relative to inverter): 75%, Lifetime: 10 years

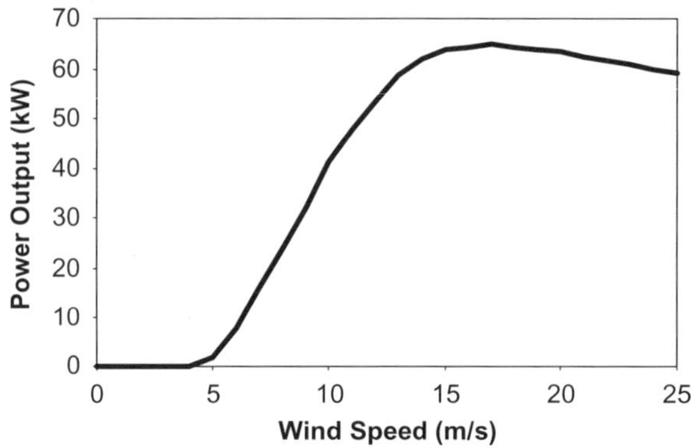

Fig. 21.8.   Example 1: Power curve of the wind turbine.

minimum average loads can be more than twice the served load. This is obviously due to the random nature of wind and solar resources, which require the support of the battery bank during a significant number of hours over a year and to the fact that no capacity shortage is accepted. Figure 21.10 shows some simulation results from the first scenario presented in Table 21.6 (average AC demand = 200 kWh/d, average DC demand = 25 kWh/d). One can observe that there are intervals during which there is no input from either wind or sun, so the energy must be supplied from the battery bank, whose state of charge reaches a value of about 80%.

With the optimum design (1 turbine, 20 kW PV capacity, 800 batteries, 20 kW converter capacity) the excess of electricity is more than 94 MWh/year while the served load is only 82 MWh/year. Another feasible solution could be based on only 1 turbine and 10 kW PV capacity, with an excess of electricity of 71 MWh/year. However, this second solution will require 1200 batteries and a 40 kW converter size, which according to the assumed costs (see Table 21.5) would increase both the initial capital and the *NPC* at about 15%.

Obviously, a different breakdown of costs would provide different solutions. For instance, with a significant reduction of PV panel and battery costs, the optimum design could be based on only one turbine. HOMER can be used to find out what cost breakdown would be required to select this alternative. Another option is to consider the possibility of selling electricity by adding a grid connection: the excess of electricity could be send to the grid and the battery bank would not be required.

21.6.2   *Example 2: Grid-connected wind-fuel cell-hydrogen system*

Figure 21.11 shows the schematic diagram of a grid-connected power plant. The system includes the wind power generation, which can be used for supplying the electric energy demand and for producing hydrogen. The hydrogen, obtained by

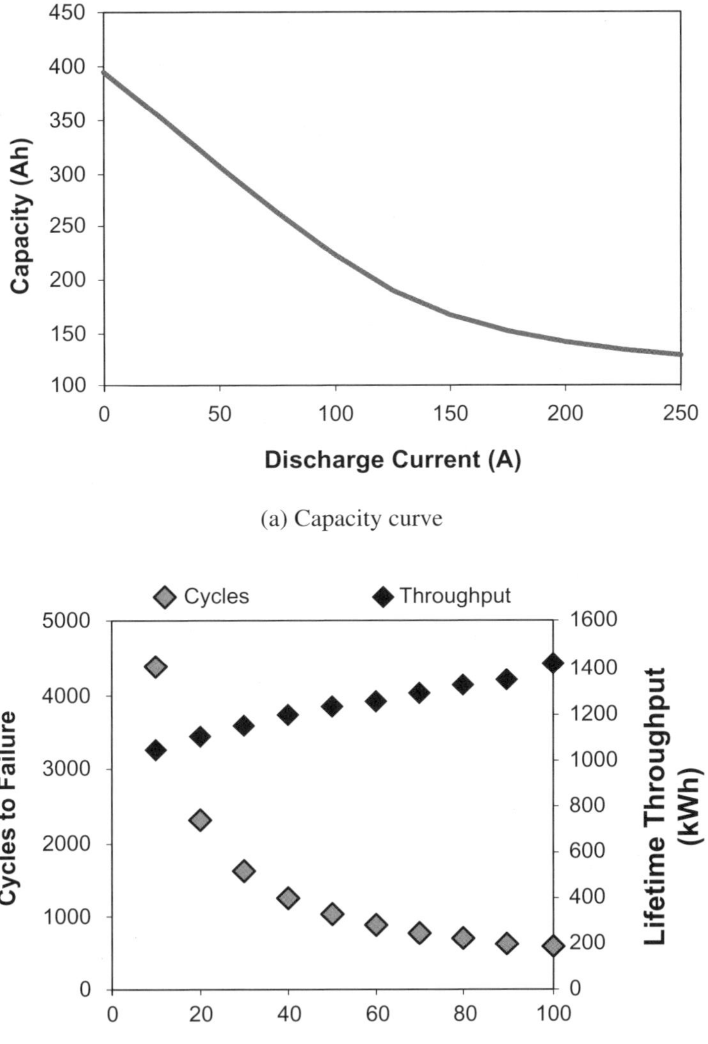

(a) Capacity curve

(b) Lifetime curve

Fig. 21.9.   Example 1: Characteristics of the batteries.

means of an electrolyzer and stored in a tank, is used by a fuel cell for electricity production. All the energy demand is supplied from an AC bus and assumed of primary type. The study will be performed considering that there will be additional energy storage in a battery bank, it will be possible to sell energy to the grid and there will not be emission penalties. The goal is to explore the influence of the sellback rate on the optimum design.

Table 21.5.   Example 1: Economic parameters of components.

Component	Capital costs	Replacement costs	O&M costs
Wind turbine	1200 €	1100 €	50 €/year
PV panel	8000 €/kW	8000 €/kW	0
Batteries	130 €/battery	117 €/battery	4 €/battery/year
Converter	1000 €/kW	1000 €/kW	10 €/year

Table 21.6.   Example 1: Summary of sensitivity analysis.

| Average demand | | Wind | | | Converter |
AC	DC	turbines	PV size (kW)	Battery bank	size (kW)
200 kWh/d	25 kWh/d	1	20	800	20
	50 kWh/d	1	40	400	20
300 kWh/d	25 kWh/d	2	20	1200	40
	50 kWh/d	1	40	1200	40
400 kWh/d	25 kWh/d	2	60	800	40
	50 kWh/d	2	50	1200	40

The study is performed considering the following aspects:

1. The average annual demand of the AC load is 2000 kWh/day, and the peak demand is 400 kW. The load profile during weekdays and weekends is shown in Fig. 21.12. In this case, it is assumed that the monthly load curves remain constant throughout the year.

2. The characteristics of the wind resource at the plant location are shown in Fig. 21.13, with the anemometer at a height of 10 m. It is assumed that the wind speed increases with height above ground according to the power low profile with an exponent of 0.14. The plant is located at sea level.

3. The main characteristics of the plant components are presented in Table 21.7. Figures 21.14 and 21.15 show, respectively, the power curve of the wind turbines and the properties of the selected batteries. The charging of the battery bank is made using cycle-charging with a set point state of charge of 80%.

4. The economic parameters of the components are presented in Table 21.8.

5. The grid rates are as follows: (1) power price: 0.1 €/kWh; (2) demand rate: 5 €/kW/month. The sellback rate is a value whose influence is explored, and it can vary from 25% (i.e., 0.025 €/kWh) to 100% (i.e., 0.1 €/kWh) of the power price. Grid emissions penalties are neglected.

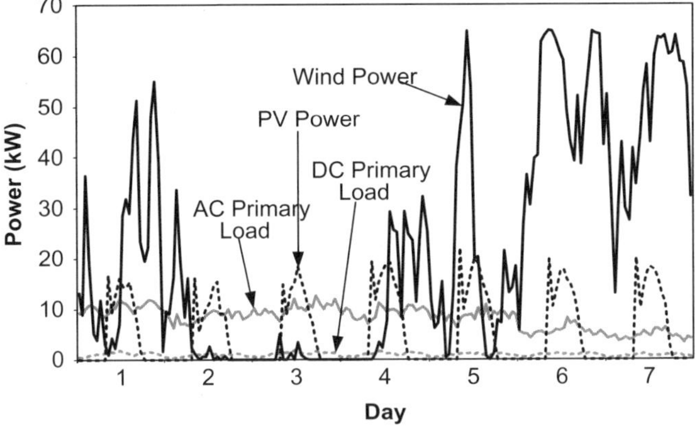

(a) Renewable resources vs. AC and DC loads

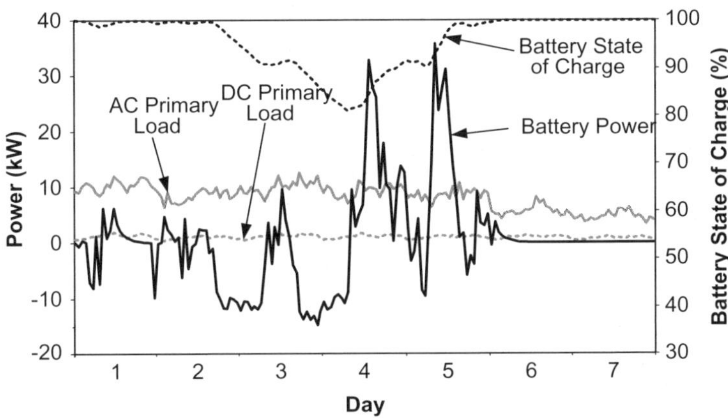

(b) Battery bank power and state of charge vs. AC and DC loads

Fig. 21.10.   Example 1: Simulation results (Average AC load: 200 kWh/d, Average DC load: 25 kWh/d — Optimum design).

6. The operating reserve as percent of load is as follows: (a) Hourly load: 10%; (b) Annual peak load: 5%. The acceptable capacity shortage is always 0%.

7. The project lifetime is 20 years and the real interest rate is 4%.

Table 21.9 summarizes the main results of the study. In all cases, the optimum size of the converter was 200 kW and the optimum number of batteries was 200. It is obvious that the *NPC* decrease as both the average wind speed and the sellback rate increase. When the cost of the electricity sold to the grid equals the cost of the electricity purchased from the grid the *NPC* decreases significantly, which means that the power plant could reach a net profit by selling electricity to the grid if the wind speed and the sellback rate were greater. In any case, the energy provided from

     J. A. Martinez-Velasco and J. Martin-Arnedo

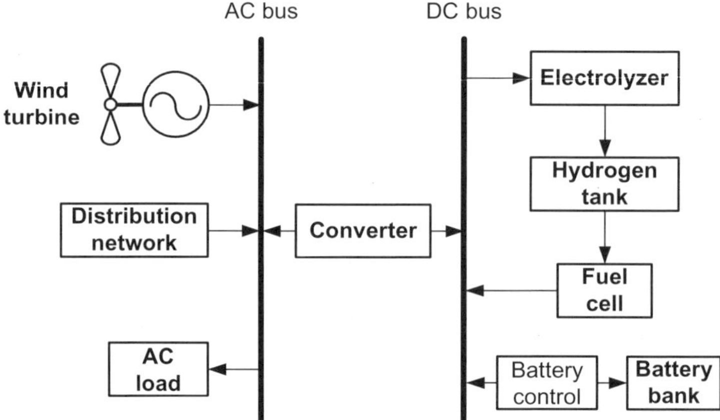

Fig. 21.11.   Example 2: Schematic diagram of a grid-connected wind-fuel cell-hydrogen system.

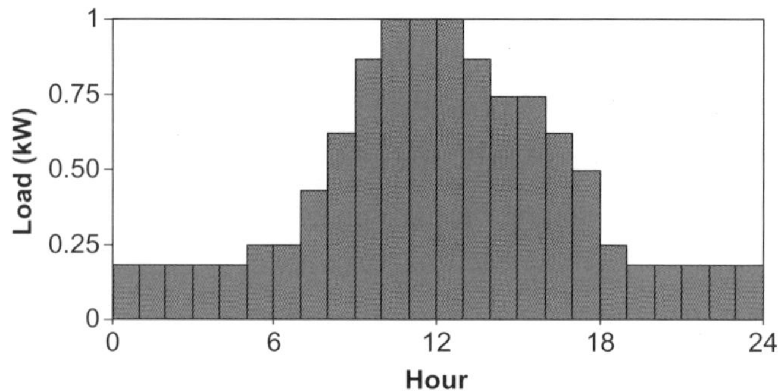

(a) Load profile during weekdays

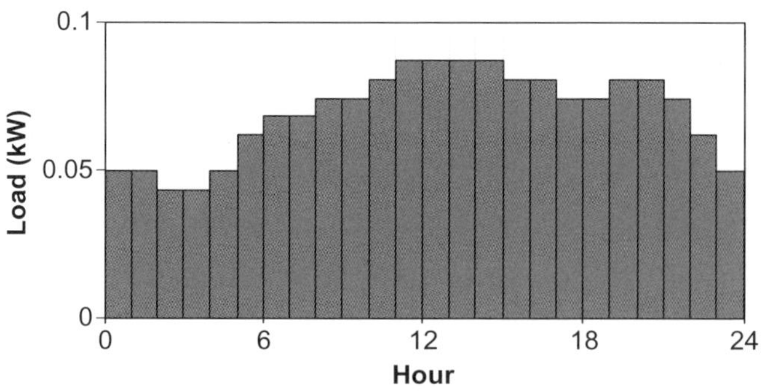

(b) Load profile during weekends

Fig. 21.12.   Example 2: Load profiles.

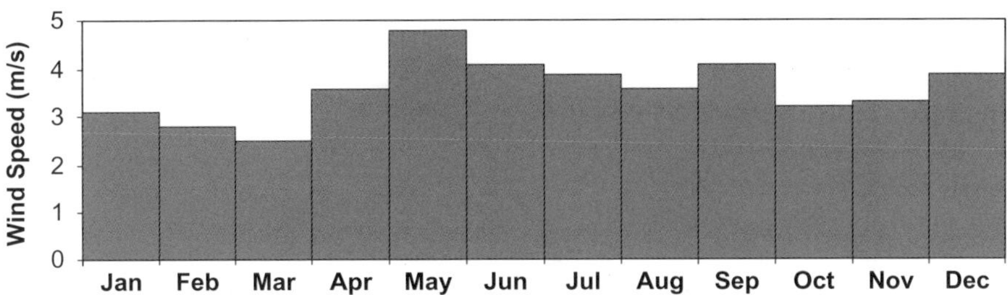

Fig. 21.13.   Example 2: Characteristics of the wind resource.

Table 21.7.   Example 2: Main characteristics of the plant components.

Component	Main characteristics
Wind turbines	96 m diameter, Tower height: 40 m, Number of turbines: 1 to 6, Lifetime: 15 years
Grid	Maximum grid demand/sale: 200 kW
Fuel cell	Nominal size: 0, 50, 100 kW, Minimum load ratio: 5%, Efficiency: 50%, Fuel: Stored hydrogen, Lifetime: 40000 hours
Electrolyzer	Nominal size: 0, 50, 100 kW, Minimum load ratio: 0%, Efficiency: 75%, Lifetime: 20 years
Hydrogen tank	Nominal size: 0, 100, 200 kg, Minimum load ratio: 0%, Initial tank level relative to tank size: 10%, Year-end tank level must be equal or exceed initial tank level, Lifetime: 20 years
Batteries	Nominal voltage: 6 V, Nominal capacity: 1156 Ah (6.94 kWh), Lifetime throughput: 9645 kWh, Bank size: 100, 200 batteries, Minimum life: 5 years
Converter	Nominal size: 100, 200, 300 kW, Inverter efficiency: 90%, Rectifier efficiency: 85%, Rectifier capacity (relative to inverter): 75%, Lifetime: 10 years

the fuel cell was not required. This is due to the cost of producing, storing and using hydrogen, which is altogether much higher than the cost of storing the energy in the battery bank.

A new study with wind turbine costs (initial cost, replacement cost, O&M cost) doubled provides similar results: fuel cell is again unnecessary, the optimum size of the converter is always 200 kW, and the optimum number of batteries is 200, although the number of wind turbines is now smaller with most scenarios. The *NPC* decreases again as both the average wind speed and the sellback rate increase. The percentage of the renewable fraction of the electricity produced in one year is lower.

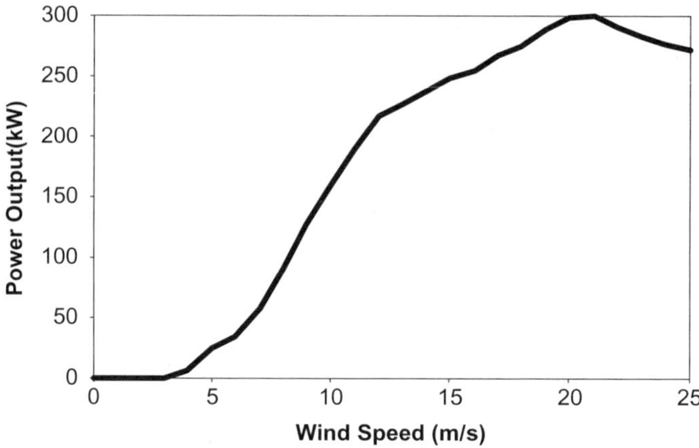

Fig. 21.14.   Example 2: Power curve of the wind turbine.

Therefore, only when the costs of producing and storing hydrogen energy, as well as the cost of producing electricity from the fuel cell, were significantly reduced, the energy from the fuel cell would be needed. A new study has been performed considering that the excess of electricity produced by the wind turbines is stored and used to produce electricity by a fuel cell, but without including a battery bank. In addition, the costs of the components needed to produce and use the hydrogen (electrolyzer, hydrogen tank, fuel cell) have been reduced. Tables 21.10 and 21.11 show, respectively, the new characteristics and the new economic parameters of the plant components.

The new results, shown in Table 21.12, prove that the new configuration of the power plant can be economically feasible; that is, if the cost of producing electricity from a fuel cell is significantly reduced. However, when comparing financial indicators (*NPC* and *COE*), the new configuration is always behind the previous one; that is both *NPC* and *COE* are now higher than without the battery bank. On the contrary, the percentage of energy from renewable resources increases. In addition, the configuration, considering the capacity of the components, is not always feasible if no shortage capacity is allowed. In fact, it could be feasible if the number of wind turbines was increased.

An important aspect to consider from the results derived with the optimum configurations obtained from this new study is that in all cases there is a significant excess of electricity produced by the wind turbines. This can be seen as an indication of overdimensioned power plants. That is, due to the intermittent nature of the wind resource and the assumed costs for components, the design of the power plant has to be overdimensioned.

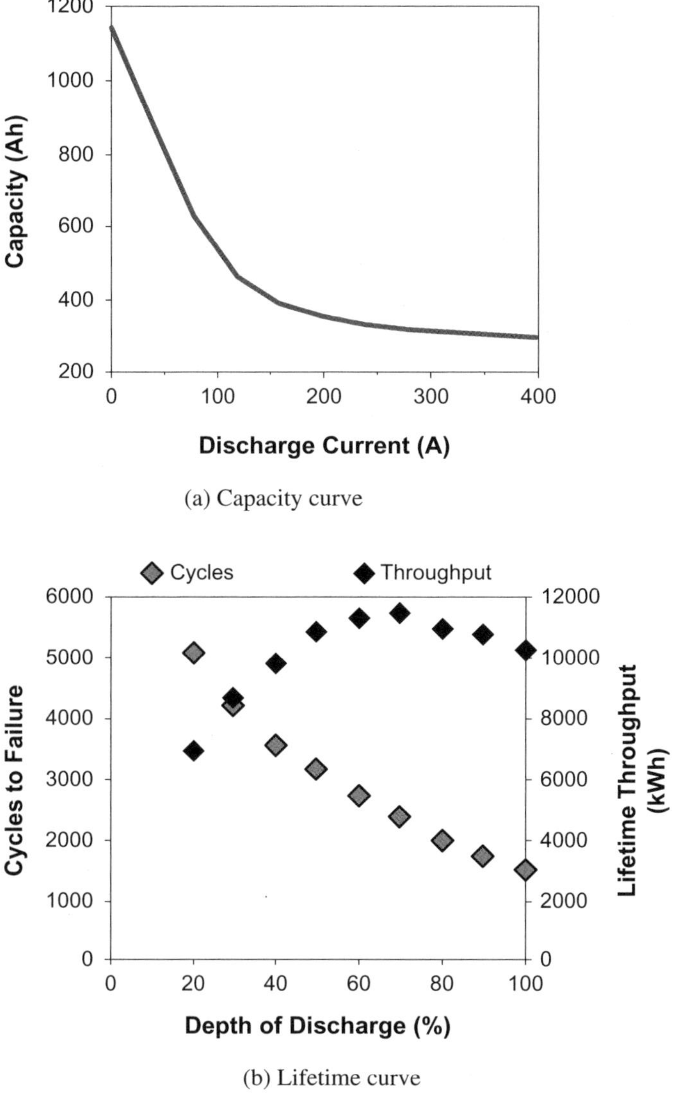

(a) Capacity curve

(b) Lifetime curve

Fig. 21.15.   Example 2: Characteristics of the batteries.

Since the calculations were made without accepting capacity shortage, they were repeated by allowing now a certain capacity shortage. The results shown also in Table 21.12 were obtained by accepting a 2% of capacity shortage and prove that the power plant is always feasible and the optimum configurations are now very different from those obtained without capacity shortage, which is an indication that the design is very sensitive to this parameter.

Table 21.8.   Example 2: Economic parameters of components.

Component	Capital costs	Replacement costs	O&M costs
Wind turbine	150000 €	120000 €	3000 €/year
Fuel cell	3000 €/kW	2700 €/kW	0.2 €/kW/h
Electrolyzer	2000 €/kW	1800 €/kW	30 €/kW/year
Hydrogen tank	1300 €/kW	1200 €/kW	10 €/kW/year
Batteries	1100 €/battery	1000 €/battery	100 €per battery/year
Converter	950 €/kW	850 €/kW	10 €/kW/year

Table 21.9.   Example 2: Summary of sensitivity analysis.

Sellback rate (€/kWh)	Wind speed (ms)	Wind turbines	Initial capital (€)	Total NPC (€)	COE (€/kWh)	Renewable fraction
0.025	6	2	710000	1687922	0.088	84%
	8	2	710000	1453073	0.062	92%
	10	2	710000	1308549	0.050	95%
0.050	6	2	710000	1455518	0.076	84%
	8	2	710000	1118017	0.048	92%
	10	2	710000	905700	0.035	95%
0.100	6	3	860000	944794	0.044	90%
	8	3	860000	440032	0.017	95%
	10	2	710000	100001	0.004	95%

Table 21.10.   Example 2: Main characteristics of the plant components (without battery bank).

Component	Main characteristics
Wind turbines	See Table 21.7
Grid	Maximum grid demand/sale: 200 kW
Fuel cell	Nominal size: 200, 250, 300, 350 kW, Minimum load ratio: 5%, Efficiency: 50%, Fuel: Stored hydrogen, Lifetime: 40000 hours
Electrolyzer	Nominal size: 200, 400, 600, 800 kW, Minimum load ratio: 0%, Efficiency: 75%, Lifetime: 20 years
Hydrogen tank	Nominal size: 800, 1000, 1200 kg, Minimum load ratio: 0%, Initial tank level relative to tank size: 10%, Year-end tank level must be equal or exceed initial tank level, Lifetime: 20 years
Converter	Nominal size: 200, 400, 600, 800 kW, Inverter efficiency: 90%, Rectifier efficiency: 85%, Rectifier capacity (relative to inverter): 75%, Lifetime: 10 years

Table 21.11. Example 2: Economic parameters of components (without battery bank).

Component	Capital costs	Replacement costs	O&M costs
Wind turbine	300000 €	240000 €	6000 €/year
Fuel cell	750 €/kW	675 €/kW	0.05 €/kW/h
Electrolyzer	500 €/kW	450 €/kW	7.5 €/kW/year
Hydrogen tank	325 €/kW	300 €/kW	2.5 €/kW/year
Converter	950 €/kW	850 €/kW	10 €/kW/year

Figure 21.16 shows some simulation results derived from the last scenario. These results exhibit a lower total electrical production and a lower energy production from the wind energy system when the wind speed is higher. This is due to the fact that the optimum number of wind turbines decreases when the average wind speed increases, as shown at the bottom side of Table 21.12.

Another interesting conclusion is that the wind energy production does not depend on the sellback rate above an average wind speed of 8 m/s; that is, the optimum number of wind turbines when the average wind speed is greater or equal that 8 m/s does not depend on the sellback rate.

As for the electricity production from fuel cell, it increases as the average wind speed decreases. When the number of required wind turbines increases as the wind speed increases, an important percentage of wind energy cannot be delivered to the grid, whose maximum sale is limited to 200 kW, and a greater percentage of wind energy is dedicated to produce hydrogen, which is later converted to electricity.

The application of HOMER to the study of some actual cases in which hydrogen technology is involved has been presented in E.I Zoulias, "Techno-economic Analysis of Hydrogen Technologies Integration in Existing Conventional Autonomous Power Systems — Case Studies" (Chap. 5 of *Hydrogen-based Autonomous Power Systems*, E.I. Zoulias and N. Lymberopoulos, Eds., Springer, 2008).

21.7 Discussion

This chapter has summarized the main capabilities of simulation tools for feasibility analysis of microgeneration systems. Although the chapter is focused on the application of renewable energy resources for electricity production, some of these tools include capabilities to model non-renewable generators and to analyze thermal systems.

Table 21.12.   Example 2: Summary of sensitivity analysis (without battery bank).

Sellback rate (€/kWh)	Wind speed (ms)	Wind turbines	Fuel cell (kW)	Converter (kW)	Electrolyzer (kW)	Hydrogen tank (kg)	Initial capital (€)	Total NPC (€)	COE (€/kWh)	Renewable fraction	Fuel cell (hours)
					Capacity shortage $= 0\%$						
0.025	6	6	250	600	600	1000	3182500	4671574	0.183	0.98	1738
	8	4	200	400	400	800	2190000	3104035	0.115	0.99	1789
	10	3	200	200	200	1200	1730000	2295516	0.081	0.99	1773
0.050	6	6	250	800	600	1200	3437500	4657055	0.181	0.99	1873
	8	4	250	400	400	1000	2292500	2801087	0.102	0.99	1817
	10	3	250	200	200	1000	1702500	1858912	0.066	0.99	1742
0.100				Not feasible unless the number of wind turbines is increased							
	8	5	300	200	200	800	2275000	1933671	0.070	0.99	1694
	10	3	300	200	200	800	1675000	984204	0.035	0.99	1695
					Capacity shortage $= 2\%$						
0.025	6	4	200	400	400	800	2190000	3256304	0.139	0.97	1655
	8	3	200	200	200	800	1600000	2258162	0.087	0.98	1653
	10	2	200	200	200	800	1300000	1773127	0.066	0.98	1639
0.050	6	4	200	400	400	800	2190000	2911568	0.123	0.97	1722
	8	3	200	200	200	800	1600000	1852548	0.071	0.98	1708
	10	2	200	200	200	800	1300000	1342954	0.050	0.98	1693
0.100	6	5	300	200	200	1200	2405000	2448884	0.100	0.97	1477
	8	3	300	200	200	800	1675000	1238888	0.048	0.98	1577
	10	2	300	200	200	800	1375000	683733	0.025	0.98	1577

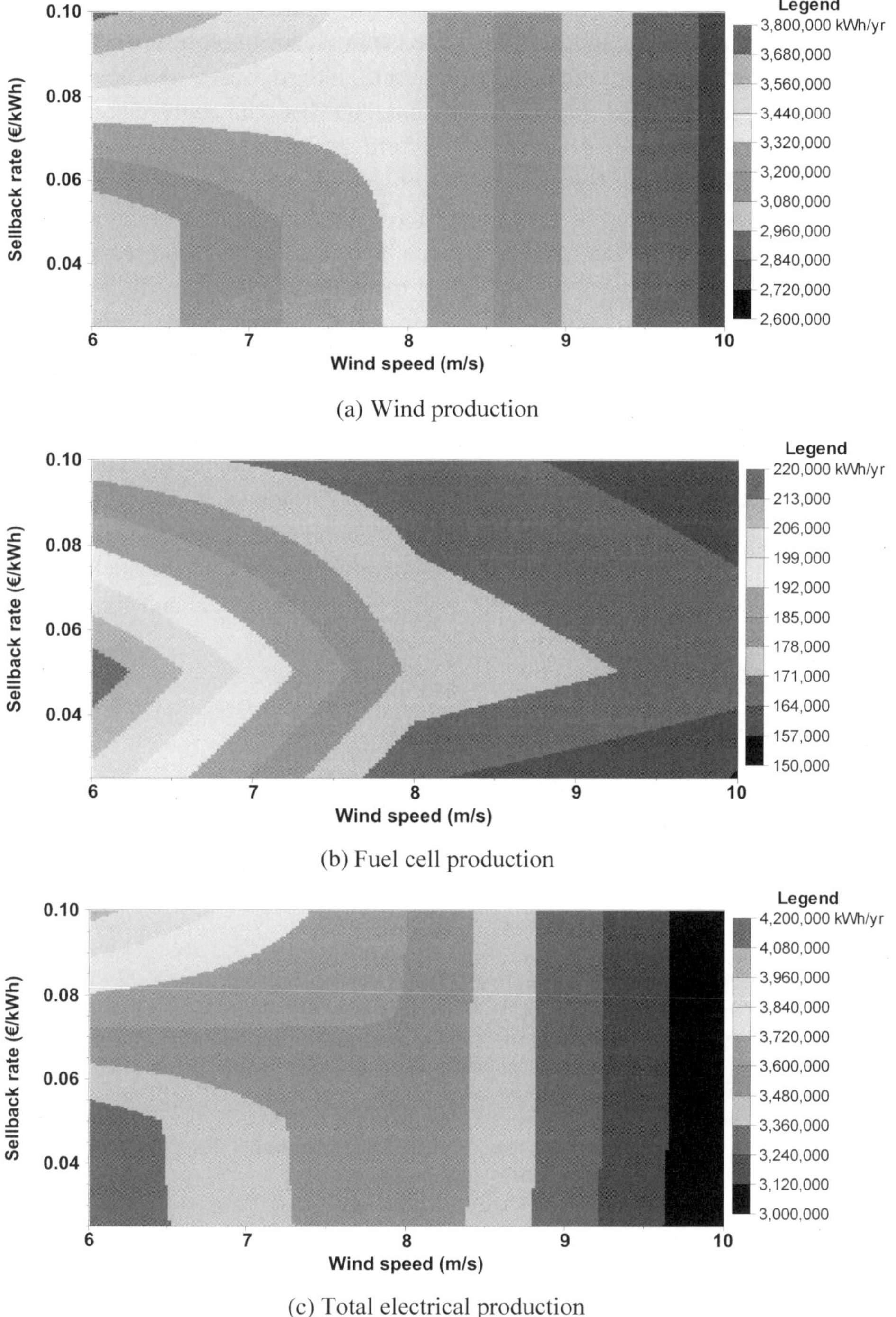

(a) Wind production

(b) Fuel cell production

(c) Total electrical production

Fig. 21.16.   Example 2: Simulation results — Maximum capacity shortage = 2%.

Technical and economic models were presented taking those implemented in HOMER as a basis. Some more sophisticated approaches to represent components and dispatch strategies are available in Hybrid2, although this tool covers fewer technologies than HOMER. On the other hand, HOMER can analyze both on- and off-grid systems, while RETScreen and Hybrid 2 can only analyze autonomous off-grid systems, although Hybrid 2 has a limited representation of the grid.

The test cases included in the chapter have illustrated just a small sample of the potential applications and studies that can be carried out with these tools. GHG emission reduction and penalties, the effect of deferrable loads or the dispatch of off-grid hybrid generation systems are scenarios not covered by these examples.

Finally, it is worth mentioning that, although these tools are powerful enough and easy to use, they have some limitations. For instance, the load growth rate and the failure rate of the various components of a power plant are aspect that can significantly affect its design, construction and performance. None of the tools analyzed in this chapter, except RETScreen for some modules, include such capabilities. Since life-cycles as long as 15–20 years are frequently used in feasibility studies, the probability that most components could fail one or several times during such a period is not negligible, and this can affect the availability of the system and consequently the results of a feasibility analysis.

Acknowledgment

This chapter has been supported by the Spanish Ministerio de Educación y Ciencia, Reference ENE2005-08568/CON.

References

1. F. Blaabjerg, Z. Chen and S. Baekhoej Kjaer, "Power electronics as efficient interface in dispersed power generation systems," *IEEE Trans. Power Electronics* **19** (2004) 1184–1194.
2. J.M. Carrasco *et al.* "Power-electronic systems for the grid integration of renewable energy sources: A survey," *IEEE Trans. Industrial Electronics* **53** (2006) 1002–1016.
3. D-Gen PRO, www.interenergysoftware.com.
4. DISPERSE, www.distributed-generation.com/disperse.htm.
5. Distributed generation analysis tool user manual, Version 1.0 (2002), http://www.naseo.org/energy_sectors/power/distributed/default.htm.
6. J.A. Duffie and W.A. Beckman, *Solar Engineering of Thermal Processes*, 2nd Edition (John Wiley, New York, 1991).
7. J.L. Edwards, C. Marnay, E. Bartholomew, B. Ouaglal, A.S. Siddiqui and K.S.H. LaCommare, "Assessment of μ grid distributed energy resource potential using DER-CAM and GIS," Lawrence Berkeley National Laboratory, LBNL-50132.
8. W. El-Khattam and M.M.A. Salama, "Distributed generation technologies, definitions and benefits," *Electric Power Systems Research* **71** (2004) 119–128.

9. D.L. Evans, "Simplified method for predicting photovoltaic array output," *Solar Energy* **27** (1981) 555–560.

10. P.S. Georgilakis and N. Hatziargyriou, "Survey of the state-of-the-art of decision support systems for renewable energy sources in isolated regions," Report D3-1, Renewables for Isolated Systems — Energy Supply and Waste Water Treatment (Project RISE), European Union (2005).

11. M. Godoy Simões and F.A. Farret, *Renewable Energy Systems* (CRC Press, 2004).

12. V.A. Graham and K.G.T. Hollands, "A method to generate synthetic hourly solar radiation globally," *Solar Energy* **44** (1990) 333–341.

13. HOMER Software, National Renewable Energy Laboratory (NREL), http://www.nrel.gov/homer (2003).

14. T. Lambert, P. Gilman and P. Lilienthal, "Micropower system modeling with HOMER," in *Integration of Alternative Sources of Energy* (John Wiley, 2006).

15. T.E. Lipman, J.L. Edwards and D.M. Kammen, "Fuel cell system economics: Comparing the costs of generating power with stationary and motor vehicle PEM fuel cell systems," *Energy Policy* **32** (2004) 101–125.

16. J.F. Manwell and J.G. McGowan, "Lead acid battery storage model for hybrid energy systems," *Solar Energy* **50** (1993) 399–405.

17. J.F. Manwell, A. Rogers, G. Hayman, C.T. Avelar and J.G. McGowan, "Hybrid2 — A hybrid system simulation model, theory manual," Renewable Energy Research Laboratory, Dept. of Mechanical Engineering, University of Massachusetts (1998).

18. G.M. Masters, *Renewable and Efficient Electric Power Systems* (John Wiley, 2004).

19. P.J. Meier, "Fully Integrated Dispatch and Optimization (FIDO) methods and applications. Net benefits analysis for energy efficiency and renewable resources," http://merllc.com.

20. RETScreen International, Clean Energy Decision Support Centre, *Clean Energy Project Analysis. RETScreen Engineering and Case Textbook*, Minister of Natural Resources Canada, http://www.retscreen.net (2005).

21. F.J. Rubio, A.S. Siddiqui, C. Marnay and K.S. Hamachi, "CERTS customer adoption model," Lawrence Berkeley National Laboratory, LBNL-47772 2001.

22. D. Turcotte, M. Ross and F. Sheriff, "Photovoltaic hybrid system sizing and simulation tools: Status and needs," *PV Horizon: Workshop on Photovoltaic Hybrid Systems*, Montreal (2001).

23. W. Short, N. Blair, D. Heimiller and V. Singh, "Modeling the long-term market penetration of wind in the United States," *Proc. WindPower Conf.*, Austin (2003).

24. W. Short, N. Blair and D. Heimiller, "Modeling the market potential of hydrogen from wind and competing sources," *Proc. WindPower Conf.*, Austin (2005).

25. A.S. Siddiqui, R.M. Firestone, S. Ghosh, M. Stadler, J. Edwards and C. Marnay, "Distributed energy resources customer adoption modeling with combined heat and power applications," Lawrence Berkeley National Laboratory, LBNL-52718 (2003).

26. E.I. Zoulias, "Techno-economic analysis of hydrogen technologies integration in existing conventional autonomous power systems — case studies," in *Hydrogen-based Autonomous Power Systems*, (eds.) Zoulias E. I. and Lymberopoulos, N. (Springer, 2008).

Chapter 22

Distributed Generation: A Power System Perspective

Hitesh D. Mathur

Electrical and Electronics Engineering Group,
Birla Institute of Technology and Science,
Pilani, Rajasthan, India
mathurhd@gmail.com

Nguyen Cong Hien

Electric Power System Development Department,
Institute of Energy, Vietnam
hienhtd3@gmail.com

Nadarajah Mithulananthan

School of Information Technology and Electrical Engineering,
The University of Queensland, Australia
mithulan@itee.uq.edu.au

Dheeraj Joshi

Electrical Engineering Department,
National Institute of Technology,
Kurukshetra, Haryana, India
dheeraj_joshi@rediffmail.com

Ramesh C. Bansal

School of Information Technology and Electrical Engineering,
The University of Queensland, Australia
rcbansal@ieee.org

The current power grid is going through tremendous changes in the way the energy is produced, transmitted and consumed. The increasing number of factors and the demand for more and more complex services to be provided by the grid exceed the capabilities of today's control systems. This chapter gives an overview of one of the major changes in power generation, i.e., distributed generation. Different aspects of distributed generation on electrical power systems such as ancillary services, voltage regulation, harmonics and loadability are also presented in the chapter.

22.1 Introduction

Owing to the growing population, increasing mechanization and automation, the global demand for energy is increasing at a breathtaking pace. This sharp increase in world energy demand will require significant investment in new power generating capacity and grid infrastructure. Just as energy demand continues to increase, supplies of the main fossil fuels used in power generation, are becoming more expensive and more difficult to extract. Considering the present energy scenario and the degrading environmental conditions, distributed generation (DG) seems to be a promising option. Distributed Generation generally refers to small-scale (typically 1 kW–50 MW) electric power generators that produce electricity at a site close to customers or that are tied to an electric distribution system. The interest in DG is the result of the opening of the energy markets under deregulation and of recent technological advances in electrical and mechanical power conversion systems. These include cheaper and more efficient static power converters, gas and wind turbines and photovoltaic and fuel cells. In order to be more effective, distributed generation which provides variable power, such as wind energy and photovoltaic energy, can be associated with energy storage, such as batteries.[1,2]

The introduction of generation sources on the distribution system can significantly impact the flow of power and voltage conditions at customers and utility equipment. These impacts may manifest themselves either positively or negatively depending on the distribution system operating characteristics and the DG characteristics. Positive impacts are generally called system support benefits like voltage support and improved power quality, loss reduction, transmission and distribution capacity release.

Achieving the above benefits is in practice much more difficult than is often realized. The DG sources must be reliable, dispatchable, of the proper size and at the proper locations. For DG to have a positive benefits, it must at least be suitably "coordinated" with the system operating philosophy and feeder design. This means addressing some of the issues related to voltage regulation, voltage flicker, harmonic distortion, islanding, grounding compatibility, overcurrent protection, capacity limits, reliability and other factors. The larger the aggregate DG capacity on a circuit relative to the feeder capacity and demand, the more critical is this "coordination" with these factors.[3–5] They must also meet various other operating criteria. Since many DG units will not be utility owned or will be variable energy sources such as solar and wind, there is no guarantee that these conditions will be satisfied and that the full system support benefits will be realized. In fact, power system operations may be adversely impacted by the introduction of DG if certain minimum standards for control, installation and placement are not maintained. The focus of this chapter is on impact of distributed generation on the issues related to power flow, power quality and loadability.

22.2 Distributed Generation Systems

Some of the important DG systems that hold the greatest technical potential are described below.

22.2.1 *Wind turbine*

Wind turbines convert the kinetic energy from moving air and use an electric generator to produce electricity. Distributed wind energy systems provide clean, renewable power for on-site use and also support the existing conventional resources to cater to the ever increasing energy demands. In the process, energy security is guaranteed and also large employment opportunities are created.

The major components of a typical wind energy conversion system (WECS) include a wind turbine, an electrical generator, a speed control system and a tower. Wind turbines can be classified into the vertical and the horizontal axis type. The drive train in the turbine consists of a low-speed shaft connecting the rotor to the gearbox, a 2- or 3-stage speed-increasing gearbox, and a high-speed shaft connecting the gearbox to the generator. The drive train converts the wind speed to a predetermined speed required for generation shaft. Some turbines are equipped with an additional small generator to generate power in low wind speeds. A wind turbine can be designed for a constant speed or variable speed operation. Variable speed wind turbines can produce 8–15% more energy output as compared to their constant speed counterparts, however, they necessitate power electronic converters to provide a fixed frequency and fixed voltage power to their loads. Most turbine manufacturers have opted for reduction gears between the low speed turbine rotor and the high speed three-phase generators. Direct drive configuration, where a generator is coupled to the rotor of a wind turbine directly, offers high reliability, low maintenance, and possibly low cost for certain turbines.[6,7]

Apart from being a renewable source of energy, wind power has following advantages over conventional methods of power generation.

- No emissions of mercury or other heavy metals into the air.
- Emissions associated with extracting and transporting fuels.
- Huge amount of water associated with mining and traditional power generation is saved.

- No waste products like toxic solid wastes, ash, or slurry are generated in the process.
- As it is a clean source of energy, no greenhouse gas (GHG) emissions are present.

22.2.2 *Solar energy*

Solar energy is radiant energy that is produced by the sun. Every day the sun radiates an enormous amount of energy. Though only a small portion of the energy radiated by the sun into space strikes the earth, one part in two billion, yet this small fraction of received per day is more than the total energy that the 6 billion inhabitants of the planet would consume in 27 years. Clearly, solar energy is a very promising and abundant form of energy but the major problem with it is to convert this energy into useful form or to trap it.

Solar power is produced through two main technologies: photovoltaic (PV) cells and concentrating solar thermal (CST) power. Optics and photonics define the fundamental mechanisms involved in the conversion of light into electricity. Here both of these technologies are discussed briefly.

A. Photovoltaic cells

Photovoltaic cells convert sunlight directly into electricity. Photovoltaic cells are made from appropriately doped silicon. Every PV cell has at least one electric field which forms when the N-type and P-type silicon are in contact. The free electrons in the N side, which are in search of holes to fall into, see all the free holes on the P side, and there is a rush to fill them in. When photons hit the solar cell, freed electrons attempt to unite with holes on the p-type layer. The pn-junction, a one-way road, only allows the electrons to move in one direction. If we provide an external conductive path, electrons will flow through this path to their original (p-type) side to unite with holes. The electron flow provides the current (I), and the cell's electric field causes a voltage (V). With both current and voltage, we have power, $P = I \times V$.[8,9]

Therefore, when an external load is connected between the front and back contacts, electricity flows in the cell. The photovoltaic cell is shown in Fig. 22.1 with its various layers labeled.

B. Concentrated Solar Thermal (CST)

Concentrated solar thermal power is a utility-scale technology that uses mirrors and lenses to focus sunlight into a concentrated beam, which is converted into steam that generates electricity. A myth about CST is that constant sunlight is required for solar power to be produced. However this is not true.

On partly cloudy days, PV systems can produce up to 80% of their potential electrical capacity and up to 25% on very overcast days. CST systems have the

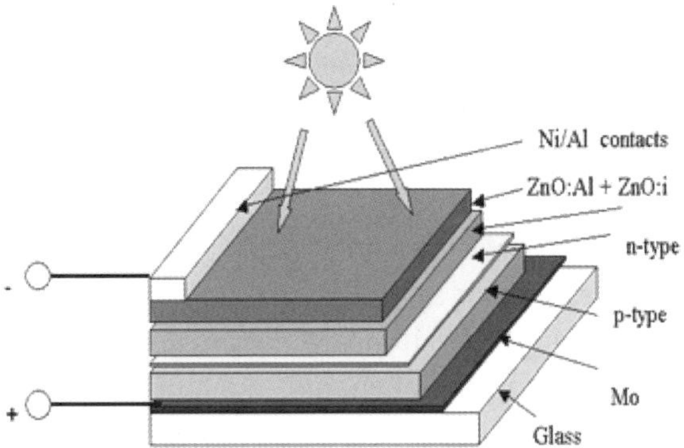

a. Glass b. Contact Grid c. The Antireflective Coating (AR Coating) d. N-Type Silicon e. P-Type Silicon f. Back Contact (not visible)

Fig. 22.1.   Photovoltaic cell with various layers, http://www.ssd.phys.strath.ac.uk.

ability to run overnight or in bad weather by storing heated transfer fluid in a hyper-efficient thermos bottle. Figure 22.2 shows the concentrated solar thermal power system arrangement. As any other technology, even power generation from solar energy has its own pros and cons. They are discussed here briefly.

Advantages:

(1) Clean source: Production of electricity from solar energy does not emit any pollutants in the atmosphere. A 100 MW solar thermal electric power plant, over its 20-year life, will avoid more than 3 million tons of carbon dioxide (CO_2) emissions when compared with the cleanest conventional fossil fuel-powered electric plants available today.

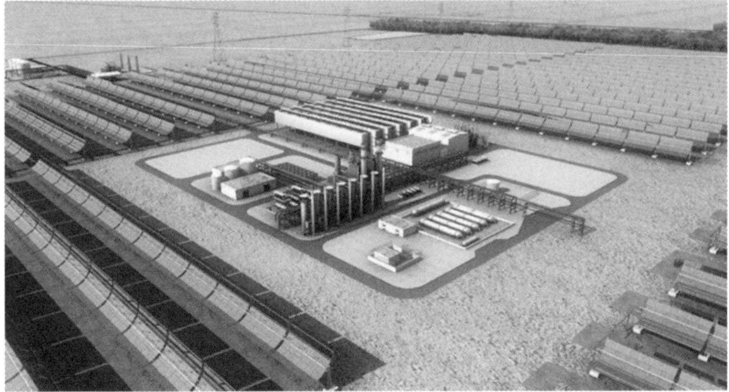

Fig. 22.2.   Concentrated solar thermal power arrangement, http://www.greentechnology. com.

(2) Simple to access: Most of the people who live far from electricity grids are in rural areas where it is difficult to build conventional power systems due to limited access or funds. Solar power offers a viable solution to these problems. Quiet, reliable and cost-effective, it guarantees simple-to-implement access to electricity.

(3) Grid free energy: The solar energy generated at the place of requirement is very valuable as it provides a back-up incase of grid failure.

(4) Use in livestock and dairy operations: Many pig and poultry farms need to heat air, which require large amounts of energy, to maintain hygiene. Solar air heaters have been incorporated into farm buildings to preheat incoming fresh air.

Limitations

- Limited use of power owing to its high cost of generation which is 3–4 times the cost from conventional sources
- Poor efficiency, i.e., 10–20% limits its wide application

22.2.3 *Micro-turbine*

A micro-turbine is a relatively small machine which utilizes a fuel such as propane for driving a small turbine that powers an electric generator. Microturbines can run on a variety of fuels, including natural gas, propane, and fuel oil. Microturbines gathered a lot of attention in the late 1990s, but initial sales goals were not met, and then some manufacturers had reliability issues. This simple design results in a system with a high power output, minimal noise generation, and efficient operation. Diesel, gasoline or kerosene can be used as alternate fuels to insure continuous electricity production in the event that the methane supply is disrupted.

Micro turbines are a compact, quiet, clean, and reliable power source. Because the generating capacity can be sized from 30 kW to 2000 kW, by integrating multiple-unit systems a mine can easily scale the project according to its needs. Existing micro-turbines have energy ranging from 20% to 35%. Microturbines are installed commercially in many applications, especially in landfills where the quality of natural gas is low.

Micro-turbines are small combustion turbines approximately the size of a refrigerator. They are evolved from automotive and truck turbochargers, auxiliary power units (APUs) for airplanes, and small jet engines. Most microturbines are comprised of a compressor, combustor, turbine, alternator, recuperate (a device that captures waste heat to improve the efficiency of the compressor stage), and generator. Figure 22.3 illustrates how a micro-turbine works. Fuel enters the combustion chamber. The turbine can run on natural gas, gasoline, kerosene-virtually anything that burns. The hot combustion gases spin a turbine, which is connected to the shaft of an electric generator. The exhaust transfers heat to the incoming air. Air

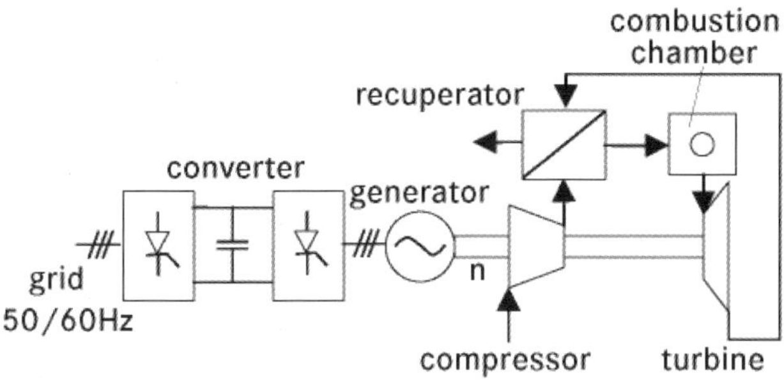

Fig. 22.3.   Schematic diagram of the micro-turbine, http://www.td.mw.tum.de.

passes through a compressor and is warmed by the exhaust gases before entering the chamber.

Microturbines are classified by the physical arrangement of the component parts: single shaft or two-shaft, simple cycle, or recuperated, inter-cooled, and reheat. The machines generally rotate over 40,000 rpm. The bearing selection — oil or air — is dependent on usage. Conversely, the split shaft is necessary for machine drive applications, which does not require an inverter to change the frequency of the AC power.

Micro-turbine generators can also be divided into two general classes:

- *Unrecuperated (or simple cycle) micro-turbines* — In a simple cycle, or unre-cuperated, turbine, compressed air is mixed with fuel and burned under constant pressure conditions. The resulting hot gas is allowed to expand through a turbine to perform work. Simple cycle microturbines have lower efficiencies at around 15%, but also lower capital costs, higher reliability, and more heat available for cogeneration applications than recuperated units.

- *Recuperated micro-turbines* — Recuperated units use a sheet-metal heat exchanger that recovers some of the heat from an exhaust stream and transfers it to the incoming air stream, boosting the temperature of the air stream supplied to the combustor. Further exhaust heat recovery can be used in a cogeneration con-figuration. The fuel-energy-to-electrical-conversion efficiencies are in the range of 20 to 30%. In addition, recuperated units can produce 30 to 40% fuel savings from preheating.

Advantages

Microturbines offer many potential advantages for distributed power generation. According to current estimates, microturbine unit costs around $1500 to $2500 per installed kW. That could lead to niche applications in areas with high energy costs for power quality, peak shaving or replacement energy. They can run on a variety of fuels, including natural gas, propane, and fuel oil. The other advantages over other types of DG technologies are small number of moving parts, compact size, lightweight, good efficiencies in cogeneration, low emissions and can utilize waste fuels.

Maintenance costs for micro-turbine units are still based on forecasts with minimal real-life situations. Estimates range from \$0.005–\$0.016 per kWh, which would be comparable to that for small reciprocating engine systems. The major issues involved in microturbines are the low fuel to electricity efficiencies and loss of power output and efficiency with higher ambient temperatures and elevation. Microturbines are less efficient than the grid to which they may be connected for grid support. Microturbines are going to play a major role in future electrical generation technologies.[10,11]

22.2.4 *Fuel cells*

A fuel cell is an electrochemical device that combines hydrogen and oxygen to produce electricity, with water and heat as its by-product. The fuel cell can operate continuously, producing electricity as long as a fuel and air are supplied. Fuel cells can operate indefinitely provided the availability of a continuous fuel source. The process is an electrochemical process and unlike combustion it is clean, quiet and highly efficient. Two electrodes pass charged ions in an electrolyte to generate electricity and heat. A catalyst enhances the process.

A typical single-family home would require a fuel cell around 3 to 8 kW. Initial installation costs will run high, around \$3000–\$10000 per kW. To make the economics of fuel cells favorable, applications that require combined heat and power as in the case of hotels, dairies and process industries are most suitable for plant size above 100 kW. It is felt that in size range 1–10 kW fuel cells will be mostly used for standby power or specialty power where as in case of 10 kW and above the main applications could be for base load, standby power and other related options.

There are five types of fuel cells under development. These are: (1) phosphoric acid (PAFC), (2) proton exchange membrane (PEMFC), (3) molten carbonate (MCFC), (4) solid oxide (SOFC), and (5) alkaline (AFC). The electrolyte and operating temperatures distinguish each type. Operating temperatures range from near ambient to 1800°F and electrical generating efficiencies range from 30 to over 50% higher heating value (HHV). As a result, they can have different performance characteristics, advantages and limitations, and therefore will be suited to distributed generation applications in a variety of approaches.

The different fuel cell types share certain important characteristics. First, fuel cells are not Carnot cycle (thermal energy based) engines. Instead, they use an electrochemical or battery like process to convert the chemical energy of hydrogen into water and electricity and can achieve high electrical efficiencies. The second shared feature is that they use hydrogen as their fuel, which is typically derived from a hydrocarbon fuel such as natural gas. Third, each fuel cell system is composed of three primary subsystems: (1) the fuel cell stack that generates direct current electricity; (2) the fuel processor that converts the natural gas into a hydrogen-rich feed stream; and (3) the power conditioner that processes the electric energy into alternating current or regulated direct current. All types of fuel cells have low emissions profiles. This is because the only combustion processes are the reforming of natural gas or other fuels to produce hydrogen and the burning of a low energy hydrogen exhaust stream that is used to provide heat to the fuel processor.

The leading fuel cell technology at the moment is generally considered to be the solid polymer, also known as Proton Exchange Membrane (PEM) cell. The PEM cell is the focus of the car industry.

The most important advantage of fuel cells is that fuel cells provide a way of generating electricity without combustion and without air and water pollution in spite of being non-renewable source of energy. Future fuel cell systems are projected with electric generation effectiveness of 50 to 60%. Fuel cell systems have a great potential in DG applications which include combined heat and power (CHP), premium power, remote power, grid support, and a variety of specialty applications.

The cost of electricity production from fuel cells depends on key input variables such as the price of natural gas, electricity prices, fuel cell and reformer system costs, and fuel cell system durability levels.

The high capital cost for fuel cells is by far the largest factor contributing to the limited market penetration of fuel cell technology. In order for fuel cells to compete realistically with contemporary power generation technology, they must become more competitive from the standpoint of both capital and installed cost. Fuel cells must be developed to use widely available fossil fuels, handle variations in fuel composition, and operate without detrimental impact to the environment or the fuel cell. The capability of running on renewable and waste fuels is essential to capturing market opportunities for fuel cells. Fuel cells could be great sources of premium power if demonstrated to have superior reliability, power quality, and if they could be shown to provide power for long continuous periods of time. The high-quality power of fuel cells alone could provide the most important marketing factor in some applications. Coupled with longevity and reliability this could greatly advance fuel cell technology.[12–14]

22.3 Impact of Distributed Generation on Electrical Power System

In the past, power systems were owned and operated by monopolists, often under the control of governments. The segments of electricity generation, transmission, distribution and supply were integrated within individual electric utilities. This made the operation of the grid less complicated because the system operator had full knowledge of the grid status and total control over it. Liberalization and deregulation of the industry led to the introduction of competition in the segments of generation and supply. In transmission and distribution, the natural monopoly element has been maintained subject to network regulation.

Electricity exhibits a combination of attributes that make it distinct from other products: non-storability (in economic terms), real time variations in demand, low demand elasticity, random real time failures of generation and transmission, and the need to meet the physical constraints on reliable network operations. One of the consequences of liberalization is the new way in which the now separated entities interact with each other.

In order to ensure instantaneous balancing of supply and demand, real-time markets are run as centralized markets, even in fully deregulated systems. The system operator acts as a single buyer and is responsible for upward and/or downward regulation, which may be done via regulating bids under an exchange or pool approach. Economic decisions are made individually by market participants and system-wide reliability is achieved through coordination among parties belonging to different companies.

In other words, in the past all grid participants pursued the same goal: the objectives of the individual entities were congruent with the objectives of the system. This has changed: today, the multitude of independent agendas does not necessarily guarantee decisions that are effective and sustainable for the power system as a whole. Coordination is therefore necessary. In addition to the provision of active power, ancillary services are required to maintain a sufficient level of system reliability and power quality. At present no uniform definition exists of the individual ancillary sub-services to attain these system objectives.[15–17]

Commonly, frequency control, voltage control, spinning and non-spinning reserve, black-start capability, islanding support and remote automatic generation control are comprised in the definition of ancillary sub-services; however, the sub-services included in the definition even vary between countries. Four major methods through which system operators procure ancillary services can be distinguished: compulsory provision, bilateral contracts, tendering, and via a spot market. With the increasing pressure of the newly created market to increase productive efficiency and minimize cost, electric utilities are looking for ways to increase profit for their stake holders. Asset management at the core of a new management strategy, combined with deregulation, has the consequence of increasing the stress on existing grid components and to reduce investments in new infrastructures. This new way of operating the power grid closer to its physical limits certainly generates more profit, but it also reduces the stability of the grid, making it more prone to blackouts. This poses a challenge to the current design and regulation of electricity networks.

When the electrical power system was conceived in the way it is today, the grid was based on large-scale generation facilities. In most countries, the topology of the transmission grid reflects the locations of these large power plants, and the large load centers. Liberalization coincided with an increasing awareness for environmental concerns, technological progress, and security of supply considerations as well as an increased need for reliable and high-quality power.

All these factors have been the drivers for an increase in DG in Europe and North America, With the help of political incentives and due to the rise in energy costs, small energy producers have begun to emerge: wind farms, solar and geothermal plants, fuel cells and micro turbines, often operated in the countryside and far away from the main transmission corridors. These small-scale producers feed the energy directly into the distribution grid.[18–20]

To provide insight into some of the major changes the power system is undergoing, in the following the impact of these developments on ancillary services, voltage and harmonics, change in power flow, protection, reactive power and loadability of radial distribution system is presented. Both the technical and economic factors inducing the changes are discussed.

22.3.1 *Impact of ancillary services*

Compared to large, fossil-fueled power plants, small generation units connected to the distribution grid will typically have a lower capacity factor, i.e., a higher ratio of peak to average generation. The reasons for this are either properties of the primary energy source — intermittent production from wind turbines and photovoltaic arrays — or operational and economical constraints, such as the heat-bound limitations of combined heat-and-power (CHP) plants. With their share of peak capacity growing even faster than the share of energy production, DG will have to participate in the provision of ancillary services to the grid to ensure reliable system operation. Functions traditionally done at transmission level will have to be provided where these DG resources are connected, i.e., within the distribution grid itself: primary, secondary and tertiary reserve, voltage/var control, black start and islanding capability.[21,22]

The underlying rationale for the creation of markets for ancillary services is to achieve the procurement of these services at least cost through the extension of competition between providers of active power and loads to this segment. For loads and generators of active power this implies the opening up of a second revenue stream. Ancillary services encompass a wide range of services with different characteristics; e.g., voltage control has to be supplied locally whereas frequency control is a system-wide service. Also, due to their diversity, different market arrangements may be chosen for the individual services. In their comparative analysis, there are many variations in ancillary services market design across countries with regard to the procurement methods applied.

The capability for the delivery of ancillary services is strongly dependent on the type of generation technology. An analysis conducted suggests that the value of most feasible ancillary services provided by DG will be low and thus only provide incremental revenue opportunities. The incentives to invest in DG to exploit this second revenue stream are thus rather small.[17,23,24]

22.3.2 *Impact on voltage and harmonics*

Noticeable voltage fluctuation may be caused by DG. Fluctuation can be either a simple issue or a complex issue as far as its analysis and mitigation are concerned. From the simple perspective, it can be the result of starting a machine (e.g., induction generator) or step changes in DG output which result in a significant voltage change on the feeder. If a generator starts, or its output fluctuates frequently enough, fluctuation of lighting loads may be noticeable to customers.

An approach to reduce fluctuation involves placing constraints on when and how often DG operators may start and change the output of DG systems. In the case of wind and solar energy systems, the outputs will fluctuate significantly as the sun and wind intensity change. The dynamic behavior of machines and their interactions with upstream voltage regulators and generators can complicate matters considerably. For example, it is possible for output fluctuations of a DG to cause hunting of an upstream regulator and, while the DG fluctuations alone may not create visible flicker, the hunting regulator may create visible fluctuation.

Thus, fluctuation can involve factors beyond simply starting and stopping of generation machines or their basic fluctuations. Dealing with these interactions requires an analysis that is far beyond the ordinary voltage drop calculation performed for generator starting. Identifying and solving these types of fluctuation problems when they arise can be difficult and the engineer must have a keen understanding of the interactions between the DG unit and the system. To model this on a computer requires good models of the distributed generators (which are often not available) and their interactions with utility system equipment. A software analysis package with the ability to analyze the dynamic behavior of systems is helpful for this type of study. It may also be necessary to perform system measurements to assess voltage and power flow oscillations and to identify how equipment controls can be "tuned" or modified to reduce flicker. In some cases, these dynamic flicker problems can be solved without a detailed study by simply performing an adjustment of a control element until the measured flicker disappears. In other cases, the fix is allusive and requires considerable investigation to solve.

Distributed generators may also introduce harmonics. The type and severity will depend on the power converter technology and interconnection configuration. In the case of inverters, there has been particular concern over the possible harmonic current contributions they may make to the utility system. Fortunately, these concerns are in part due to older SCR type power inverters that are line commutated and produce high levels of harmonic current. Most new inverter designs are based on IGBTs that use pulse width modulation to generate the injected "sine" wave.[25,26]

Rotating generators such as synchronous generators can be another source of harmonics. Depending on the design of the generator windings (pitch of the coils), core non-linearity's, grounding and other factors, there can be significant harmonics. In extreme cases, equipment at the DG site may need to be derated due to added heating caused by harmonics. Any DG installation design should be reviewed to determine whether harmonics will be confined to the DG site or also injected into the utility system. For larger DG units or cases involving complex harmonic problems, measurements and modeling of the system harmonics may be required to assess conditions. Any analysis should consider the impact of DG currents on the background utility voltage distortion levels. The limits for utility system voltage distortion are 5% for total harmonic distortion (THD) and 3% for any individual harmonic.[27–29]

22.3.3 *Impact on change in power flow*

As the voltage rise effect is the primary factor that limits the amount of additional DG capacity that can be connected, an in-depth theory review regarding this issue will be evaluated in the following part of this chapter.

Consider a two bus power system as shown in Fig. 22.4 where, Z is impedance of transmission line and R and X are resistance and reactance, respectively. S is complex power while P and Q are real power and reactive power.

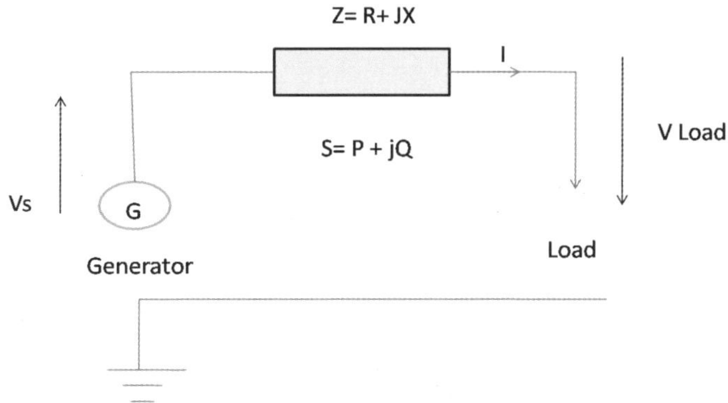

Fig. 22.4.   Two bus system example.

The apparent power transferred is given by:

$$S = P + jQ. \tag{22.1}$$

Dividing both sides by voltage current can be obtained:

$$I = \frac{P - jQ}{V_s}. \tag{22.2}$$

Load end voltage is lower than the generator side voltage due to the voltage drop across the line.

$$V_{\text{Load}} = V_s - (R + jX)I. \tag{22.3}$$

Combining the above equations for V_{Load} and I

$$V_{\text{Load}} = V_s - \frac{(R + jX)(P - jQ)}{V_s}, \tag{22.4}$$

$$V_{\text{Load}} = V_s - \frac{(R + jX)}{V_s} - \frac{P - jQ)}{V_s}. \tag{22.5}$$

It can be shown that vice-versa for the generator the voltage in terms of load and load voltage is:

$$V_S = V_{\text{Load}} - \frac{RP_{\text{Load}} + XQ_{\text{Load}}}{V_{\text{Load}}} + j\frac{XP_{\text{Load}} - RQ_{\text{Load}}}{V_{\text{Load}}}. \tag{22.6}$$

It must be noted that some of the real and reactive power entering the line will be consumed by the line itself. Using real and reactive power at the load we can derive another equation using the same process as above.

$$S = P_{\text{Load}} + jQ_{\text{Load}}. \tag{22.7}$$

Dividing both sides by voltage as in Eq. (22.2) and inserting into Eq. (22.3) we get

$$V_{\text{Load}} = \frac{V_S - (R + jX)(P_{\text{Load}} - jQ_{\text{Load}})}{V_{\text{Load}}}. \tag{22.8}$$

The above equation is used in power simulation software programs for solving power flows.

Significance of power flows.

Another way of writing Eq. (22.6) is

$$V_{\text{Load}} = V_S - \Delta V - j\delta V, \tag{22.9}$$

where the in phase component is

$$\Delta V = \frac{RP + XQ}{V_S} \tag{22.10}$$

and the quadrature component is

$$\delta V = \frac{XP + RQ}{V_S}. \tag{22.11}$$

Power Systems typically operate at load angles below 30 degrees. It can be seen that the voltage drop between the different bus bars ΔV is caused by a combination of the product of real power and resistance and the product of reactive power and reactance. Conversely, the phase angle difference is caused by a combination of the product of real power and reactance and the product of reactive power and resistance. These comparisons are more useful when considering transmission lines which have considerably higher reactance than resistance. Hence they can be simplified to the below approximations:

$$\Delta V = \frac{XQ}{V_S}, \tag{22.12}$$

$$\delta V = \frac{XP + RQ}{V_S}. \tag{22.13}$$

Thus for transmission systems voltage drop is largely determined by reactive power transfer and load angle is determined largely by real power transfer.

Since DG is likely to have an equal contribution to voltage drop from real and reactive power flows, both real power and reactive power control can be effective. The effect of controlling the net flows of real and reactive power on a bus can be approximated in p.u. by dividing Eq. (22.10) by V_{base}.

$$\Delta V_{p.u} = \frac{RP + XQ}{V_S / V_{\text{base}}}, \tag{22.14}$$

$$\Delta V_{p.u} = RP + XQ. \tag{22.15}$$

Power can flow bidirectional within a certain voltage level, but it usually flows unidirectional from higher to lower voltage levels, i.e., from the transmission to the distribution grid. An increased share of distributed generation units may induce power flows from the low voltage into the medium-voltage grid. Thus, different protection schemes at both voltage levels may be required.

Distributed generation flows can reduce the effectiveness of protection equipment. Customers wanting to operate in "islanding" mode during an outage must take into account important technical (for instance the capability to provide their own ancillary services) and safety considerations, such that no power is supplied to the grid during the time of the outage. Once the distribution grid is back into operation, the distributed generation unit must be resynchronized with the grid voltage.[30,31]

22.3.4 *Impact on loadability of distribution systems*

Distribution systems are well known for higher R/X ratio compared to transmission systems and significant voltage drops that could cause unacceptable voltage profile at the end of feeders. As the penetration of DG into power networks are increasing, sizing and location of DG could be selected to be optimal in order to maximize benefit of DG installed in the system. Poor selection of location combined with inappropriate size can reduce benefits and even jeopardize the system operation.[32]

By injecting real or reactive or real and reactive power, DG units help to reduce loss and improve voltage profile of the system. Moreover, loadability enhancement is also obtained in conjunction with voltage profile improvement. In this section, the impact of DG on loadability of distribution systems has been presented. By enhancing loading margin of the system, distribution companies or power utilities can optimize their resources and maximize their profit. In practice, loadability of distribution system is limited by voltage drop as most of distribution feeders are long and operating at low voltage level. In other words, at maximum loading point, buses or loading nodes would experience a severe voltage drop when the loading is increased. The node which would experience a maximum rate of change of voltage with respect to load increase is called the "Weakest Node" of the system.[33] Therefore, the weakest node will have the highest voltage gradient.

The maximum loading point in transmission and distribution system can be associated with saddle-node bifurcations.[33,34] At this bifurcation point, a real eigenvalue of the load flow Jacobian becomes zero, i.e., the Jacobian becomes singular. Direct and continuation techniques are typically used to study saddle-node bifurcation theory and voltage collapse in power system models. Direct methods to find the bifurcation point directly from a known operating condition by modified Newton–Raphson iterations. Continuation methods,[35] on the other hand, find the bifurcation point of the nonlinear system as the system parameter changes, yielding an adequate approximation of the location of the bifurcation point by using predictor and corrector steps. Another benefit of the continuation method is that it gives the information about the weakest bus from the predictor steps. However, the right eigenvector corresponding with the minimum eigenvalue of power flow Jacobian at the maximum loading point can be also be used to identify the weakest bus.

The authors in Ref. 36 identified the weakest bus of distribution system using a tangent vector index. Then, the DG unit, which can generate both real and reactive power, is used to control voltage at the weakest bus to maximize loadability of the system. The authors in Ref. 37 identify the weakest bus in the system by using L-index[38] and controlled the voltage at this node by DG, with 2 scenarios: DG can generate only real power and DG can generate both real and reactive power, to increase loadability of the system. The papers[36,37] showed that the weakest bus will be the best location to maximize loadability of the system with DG, which can generate both real and reactive power, into the system.

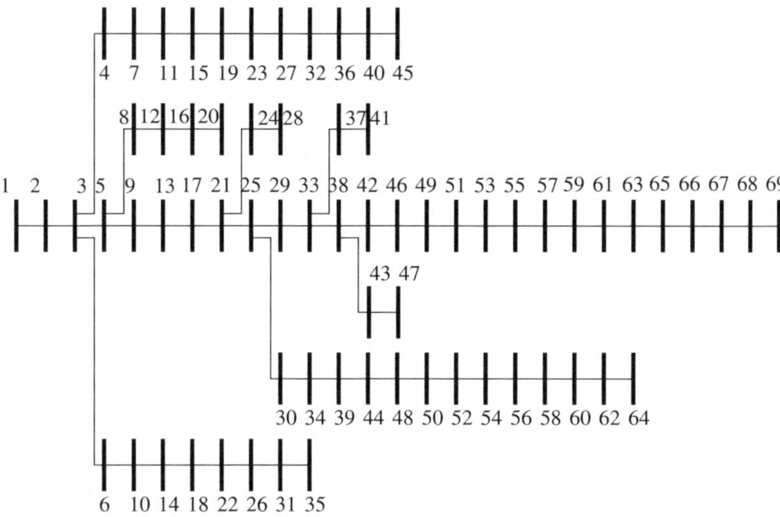

Fig. 22.5.   The 69-bus radial distribution system.

In this section, four types of DG are considered for loadability enhancement in a 69-bus test distribution system[39]:

Type 1: DG unit that injects active power (P) only, e.g., photovoltaic.

Type 2: DG unit that injects reactive power (Q) only, e.g., synchronous compensators.

Type 3: DG unit that injects active power but absorb reactive power, e.g., induction generator. The reactive power consumed in induction generator in simple form is given in Eq. (16).[40]

$$Q_{\text{DG}} = -(0.5 + 0.04P^2). \tag{22.16}$$

Type 4: DG unit that injects both active and reactive power, e.g., synchronous generators.

The PSO algorithm has been used to identify proper location and appropriate size of different types of DG units by minimizing reactive power losses of the system. Loadability factors of various scenarios are compared with each other and compared with the case of minimizing real power losses.[40]

A. Problem formulation

The objective function of the placement problem is to minimize the total reactive power loss as given in Eq. (17), where the exact loss formula of reactive power loss is used.[41]

Minimize objective function:

$$Q_L = \sum_{j=1}^{N} \sum_{k=1}^{N} [\gamma_{jk}(P_j P_k + Q_j Q_k) + \xi_{jk}(Q_j P_k - P_j Q_k)], \qquad (22.17)$$

where

$$\gamma_{jk} = \frac{x_{jk}}{V_j V_k} \cos(\delta_j - \delta_k), \quad \xi_{jk} = \frac{x_{jk}}{V_j V_k} \sin(\delta_j - \delta_k), \qquad (22.18)$$

$V_j \angle \delta_j$ is voltage at bus j, $r_{jk} + j x_{jk} = Z_{jk}$ is jkth element of Z_{bus}.

Subject to

equality constraints:

$$P_{gi} - P_{mi} = |V_i|^2 G_{ii} + \sum_{\substack{n=1 \\ n \neq i}}^{N} |Y_{in}||V_i||V_n| \cos(\theta_{in} + \delta_n - \delta_i), \qquad (22.19)$$

$$Q_{gi} - Q_{mi} = -|V_i|^2 B_{ii} - \sum_{\substack{n=1 \\ n \neq i}}^{N} |Y_{in}||V_i||V_n| \sin(\theta_{in} + \delta_n - \delta_i), \qquad (22.20)$$

where voltage at typical bus i_{th} bus, $V_i = |V_i| \angle \delta_i$ and $Y_{ij} = |Y_{ij}| \angle \theta_{ij} = |Y_{ij}|(\cos \theta_{ij} + j \sin \theta_{ij}) = G_{ij} + j B_{ij}$ bus admittance matrix element.

inequality constraints:

$$V_{\min} \leq V_i \leq V_{\max}; \qquad i = 1, \ldots, N, \qquad (22.21)$$

$$P_{g \min i} \leq P_{gi} \leq P_{g \max i}; \qquad i = 1, \ldots, N, \qquad (22.22)$$

$$Q_{g \min i} \leq Q_{gi} \leq Q_{g \max i}; \qquad i = 1, \ldots, N. \qquad (22.23)$$

The outcome of some results is presented below.

Table 22.1 shows the most appropriate location, suitable size of DG unit, reactive and real power losses and load factor for the 69-bus radial system for different types of DG units.

Table 22.2 presents a comparison of DG location and size, real and reactive power loss, and load factor between "minimizing reactive power loss" and "minimizing real power loss" objectives. The reactive power loss in the former case is lower than that in the later case as expected. The DG location is the same for the former and later case; however, the DG sizes are definitely different for each case. The table also shows the better load factor in the former case for all types of DG, except for DG type 3. However, the load factor in these two scenarios for DG type 3 installed is nearly the same.

Table 22.1. Optimal DG placement and LF of 69-bus radial distribution system.

1 DG Installed	Base case	DG Type 1	DG Type 2	DG Type 3	DG Type 4
Location	1	56	56	56	56
P_{DG} size, MW	4.02	1.8561	—	1.8291	1.8561
Q_{DG} size, MVAr	2.79	—	1.2366	-0.634	1.578
Q_{loss}, kVAr	99.5401	39.617	69.000	70.957	14.083
P_{loss}, kW	219.279	81.308	148.660	155.344	22.699
Q_{loss} reduction, %	—	60.20	30.68	28.72	85.85
P_{loss} reduction, %	—	62.92	32.20	29.16	89.65
LF	0.24072	0.65368	0.67521	0.62089	0.7323

Table 22.2. Comparison with minimize real power loss.[42]

Objective	1 DG installed	DG Type 1	DG Type 2	DG Type 3	DG Type 4
Minimize reactive power loss	Location	56	56	56	56
	DG (MW) Size	1.8561	1.2366	1.8291	1.8561
	Q_{loss}, kVAr	39.617	69.000	70.957	14.083
	P_{loss}, kW	81.308	148.660	155.344	22.699
	LF	0.65368	0.67521	0.62089	0.7323
Minimize real power loss	Location	56	56	56	56
	DG capacity	1.8074	1.3266	1.8888	2.2215
	P_{loss}, kW	84.980	155.293	161.708	23.594
	Q_{loss}, kVAr	41.464	71.970	73.945	14.685
	LF	0.65276	0.67330	0.62163	0.72656

Figure 22.6 shows the P-V curves of the system for the base case and cases with different DGs. The improvement of the voltage profile of the main feeder at zero LF for different types of DG and the base case is shown in Fig. 22.7.

It is clearly to shown that with DG units, the loadability of the system can be significant improved. DG unit (Type 4) capable of delivering reactive and real power gives the maximum reduction of reactive power losses and maximum loading margin. The DG unit (Type 3) capable of delivering real power and absorbing reactive power gives the lowest reduction of reactive power losses and lowest loading margin. Comparing DG types 2 and 1, in all the cases, DG type 2 gives a better loading factor and lower reactive power losses. Real power loss minimization objective gives lower reactive power losses and lower loading margins compared to reactive power loss minimization objective as expected.

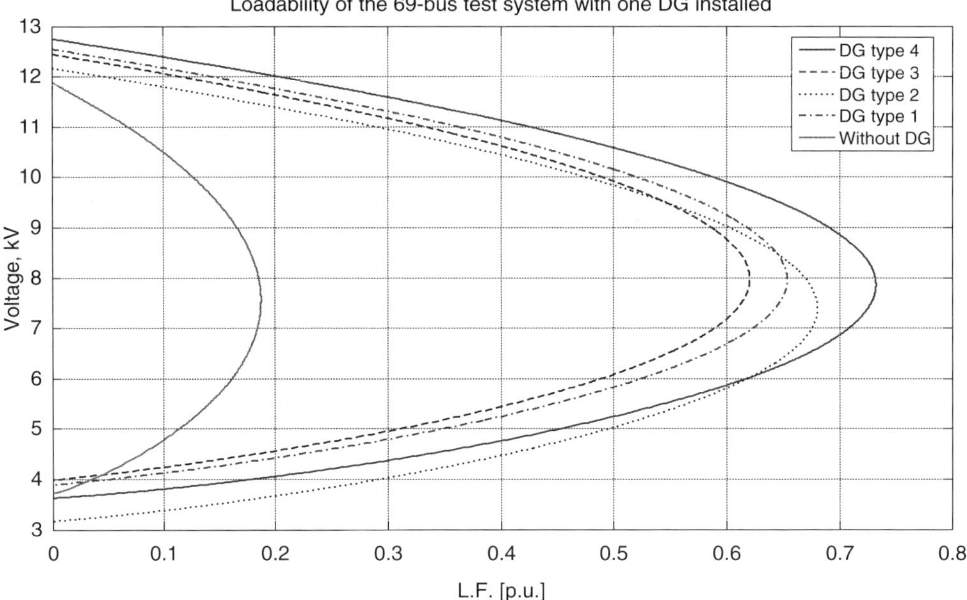

Fig. 22.6. Loadability curves of the 69-bus radial distribution system for different types of DG installed.

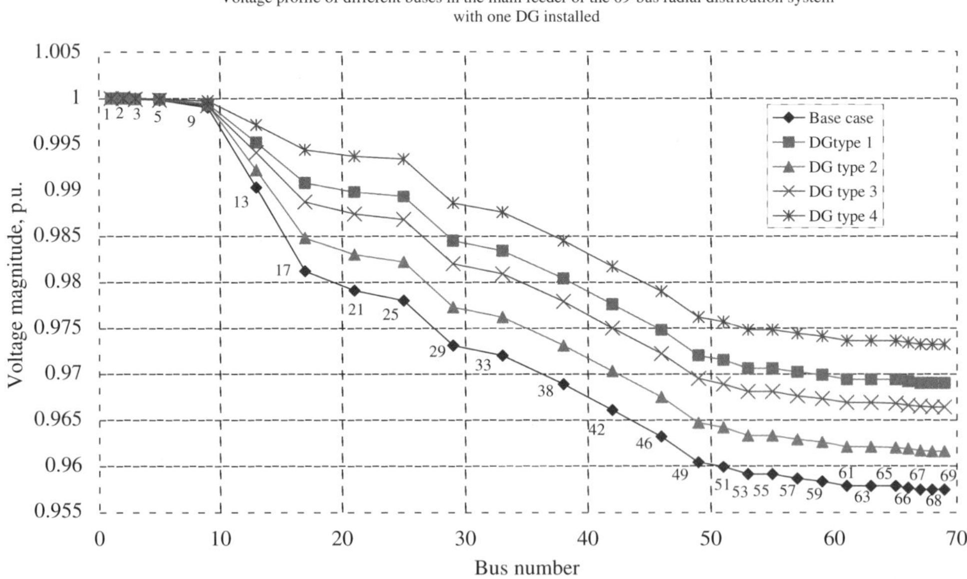

Fig. 22.7. Voltage profile in the main feeder of the 69-bus radial distribution system (LF = 0).

22.4 Conclusions

This chapter started from the observed renewed interest in small-scale electricity generation. Existing small-scale generation technologies are described and the major benefits and issues of using small-scale distributed generation are discussed. The different technologies are evaluated in terms of their contribution to the listed benefits and issues. It can be concluded that Distributed generation has the potential to offer improvements in efficiency of power system, reliability, loadability and diversification of energy, as well as to give us an opportunity to make use of renewable energy in our current generation. However, the overall shift to DG will greatly depend upon the final efficiency of the technology. In spite of these uncertainties and limitations, the future growth trend of DG is optimistic since it offers the right mix of technical and economic solutions to meet energy needs at the grass-root levels while addressing global climate change.

References

1. R. Ramakumar and P. Chiradeja, "Distributed generation and renewable energy systems," *37th Intersociety Energy Conversion Engineering Conf.*, 29–31 July 2004, pp. 716–724.
2. T. Ackermann, G. Andersson and L. Söder, "Distributed generation: A definition," *Electric Power Systems Research* **57** (2001) 195–204.
3. P. Dondi, D. Bayoumi, C. Haederli, D. Julian and M. Suter, "Network integration of distributed power generation," *J. Power Sources* **106** (2002) 1–9.
4. J. Eto, J. Koomey, B. Lehman and N. Martin, "Scoping study on trends in the economic value of electricity reliability to the US economy," LBLN-47911, Berkeley (2001), p. 134.
5. C.D. Feinstein, P.A. Morris and S.W. Chapel, "Capacity planning under uncertainty: Developing local area strategies for integrating distributed resources," *Energy J.* **18** (1997) 85–110.
6. N. Hatziargyriou *et al.*, "The CARE system overview: Advanced control advice for power systems with large scale integration of renewable energy sources," *Wind Engineering* **23** (1999) 57–68.
7. A.-M. Borbely and J.F. Kreider (eds.) *Distributed Generation: The Power Paradigm for the New Millennium* (CRC Press, USA).
8. H. Gruenspecht, "Climate and electricity policies affecting the future of distributed and renewable generation," *Environmental Electric Energy Opportunities for the Next Century, IEEE/EPRI Vision-21*, Washington, DC, April 1998.
9. J. Iannucci, "Opportunities and barriers to distributed resources in utility systems," *Environmental Electric Energy Opportunities for the Next Century, IEEE/EPRI Vision-21*, Washington, DC, April 1998.
10. R. Friedman, "Microturbine power generation: Technology development needs and challenges," *Environmental Electric Energy Opportunities for the Next Century, IEEE/EPRI Vision-21*, Washington, DC, April, 1998.
11. W.G. Scott, Micro-turbine generators for distributed systems, *IEEE Industrial Application Magazine*, May/June 1998, pp. 57–62.
12. M. Farooque and H.C. Maru, "Fuel Cells — The clean and efficient power generators," *Proc. IEEE* **89** (2001) 1819–1829.

13. M.W. Ellis, R.M. Spakovsky and J.D. Nelson, "Fuel cell systems: Efficient, flexible energy conversion for 21st century," *Proc. IEEE*, vol. 89, December, 2001, pp. 1819–1829.

14. S. Rahman, "Fuel Cell as a distributed generation technology," *Proc. IEEE Power Engineering Society Summer Meeting* **1** (2001) 551–552.

15. J. Frayer and N. Uludere, "What is it worth? Applications of real options theory to the valuation of generation assets," *Electricity J* **13** (2001) 40–51.

16. O. Gehrke, S. Ropenus and P. Venne, "Distributed energy resources and control: A power system point of view," www.risoe.dk/rispubl/reports/ris-r-1608.

17. G. Joos, B.T. Ooi, D. McGillis, F.D. Galiana and R. Marceau, "The potential of distributed generation to provide ancillary services," *IEEE Power Engineering Society Summer Meeting*, vol. 3, 16–20 July 2000, pp. 1762–1767.

18. G. Pepermans, "Distributed generation: Definition, benefits and issues," *Energy Policy* **33** (2005) 787–798.

19. P.P. Barker and R.W. De Mello, "Determining the impact of distributed generation on power systems. I. Radial distribution systems," *IEEE Power Engineering Society Summer Meeting*, **3** (2000) 1645–1656.

20. K. Voorspools and W. D'haeseleer, "The impact of the implementation of cogeneration in a given energetic context," *IEEE Trans. Energy Conversion* **18** (2003) 135–141.

21. Y. Rebours, D.S. Kirschen, M. Trotignon and S. Rossignol, "A survey of frequency and voltage control ancillary services — Part II: Economic features," *IEEE Trans. Power Systems* **22** (2007) 358–366.

22. P.J.A. Lopes *et al.*, "Integrating distributed generation into electric power systems: A review of drivers, challenges and opportunities," *Electric Power Systems Research* (2006).

23. B. Tyagi and S. Srivastava, "A decentralized automatic generation control scheme for competitive electricity markets," *IEEE Trans. Power Systems* **21** (2006) 312–320.

24. M.R. Patel, *Wind and Solar Power Systems* (CRC Press, 1999).

25. F.F. Wu, K. Moslehi and A. Bose, "Power system control centers: Past, present, and future," *Proc. IEEE* **93** (2005) 1890–1907.

26. M. Shahidehpour and Y. Wang, *Communication and Control in Electric Power Systems: Applications of Parallel and Distributed Processing* (Wiley-IEEE Press, 2003).

27. L. Philipson and H.L. Willis, *Understanding Electric Utilities and Deregulation* (Marcel Dekker, Inc. 1999).

28. "IEEE recommended practices and requirements for harmonic control in electric power systems," IEEE Standard 519-1992, Institute of Electrical and Electronics Engineers (1992).

29. J.B. Cardell and M. Llic, *Power System Restructuring Engineering and Economics* (Kluwer Academic Publisher, Boston, 1998).

30. N. Hadjsaid, J.F. Canard and F. Dumas, "Dispersed generation impact on distribution networks," *IEEE Computational Application Power* **12** (1999) 22–28.

31. R. Caire, N. Retiere, N. Martino, N. Andrieu and N. Hadjsaid, "Impact assessment of LV distributed generatin on MV distribution network," *IEEE Power Engineering Society Summer Meeting*, Paper no. 02SM152, July 2002.

32. D. Gautam and N. Mithulananthan, "Optimal DG placement in deregulated electricity market," *Electrical Power Systems Research* **77** (2007) 1627–1636.

33. C.A. Canizares, A.C.Z.D. Souza and V.H. Quintana, "Comparison of performance indices for dectection of proximity to voltage collapse," *IEEE Trans. Power Systems* **11** (1996) 1441–1450.

34. C.A. Canizares *et al.*, "UWPFLOW: Continuation and direct methods to locate fold bifurcations in AC/DC/FACTS power systems," University of Waterloo, http://www.power.uwaterloo.ca.

35. V. Ajjarapu and C. Christy, "The continuation power flow: A tool for steady state voltage stability analysis," *IEEE Trans. Power Systems* **7** (1992) 416–423.

36. N. Mithulananthan and T. Oo, "Distributed generator placement to maximize the loadability of a distribution sytem," *Int. J. Electrical Engineering Education* **43** (2006) 107–118.
37. L. Abraham, N.C. Hien and N. Mithulananthan, "Application of distributed generation to enhance loadability of distribution system, a case study," *IEEE PES/IAS Conf. Sustainable Alternative Energy*, Valencia (2009).
38. P. Kessel and H. Glavitsch, "Estimating the voltage and loadability of a power system," *IEEE Trans. Power Delivery* **PWRD-1** (1986) 346–354.
39. M.E. Baran and F.F. Wu, "Optimal sizing of capacitor placed on radial distribution systems," *IEEE Trans. Power Delivery* **4** (1989) 735–743.
40. DTI, "Network performance benefits of energy storage for a large wind farm," www.berr.gov.uk/files/file20402.pdf (2004).
41. I.O. Elgerd, *Electric Energy System Theory: An Introduction* (McGraw Hill, New York, 1971).
42. W. Krueasuk and W. Ongsakul, "Optimal placement of distributed generation using partical swarm optimization," www.itee.uq.edu.au/~aupec/aupec06/htdocs/content/pdf/163.pdf (2006).

Chapter 23

DG Allocation in Primary Distribution Systems Considering Loss Reduction

Duong Quoc Hung and Nadarajah Mithulananthan[*]

School of Information Technology and Electrical Engineering,
The University of Queensland, Brisbane, Qld 4072, Australia
[*]mithulan@itee.uq.edu.au

Loss reduction in distribution systems has been a subject of great concern since the evolution of the interconnected power system. In the recent past, with increasing interest in climate change and energy security, renewable energy integration and energy efficiency, including loss reduction, have been considered as twin-pillars of sustainable energy solutions. When renewable energy is integrated by considering loss reduction as an additional goal, it would lead to multi-fold benefits. This chapter presents the application of distributed generation for loss reduction. The two key issues of the most suitable location and appropriate size of distributed generation for loss reduction have been discussed. Analytical expressions have been developed for finding the appropriate size of different types of distributed generations. Methodologies are presented for locating the DG in primary distribution feeders, assuming primary energy resources are evenly distributed along the feeder. The analytical expressions and placement methodologies have been tested in three test distribution systems of varying sizes and complexity.

23.1 Introduction

The key aim of a power system is to ensure continuity of power supply to all consumers with a lowest possible power loss and reasonable quality. Power loss minimization can lead to energy savings, increase in power capacity, alleviation of electricity shortages, reduction in investment costs and fossil fuel usage. That would also reduce green house gas emissions (GHG) and consequently global warming and climate change.

Power systems typically consist of four major parts, namely generation, transmission, distribution and loads. The transmission and distribution systems share similar functionality; i.e., both transfer electric energy at different voltage levels from

587

one point to another. While transmission systems transmit electricity in bulk, from a generating station to load centers at distribution substations, distribution systems link distribution substations and loads where the energy is ultimately consumed. However, distribution systems are rather complex, given the number of voltage levels, unbalanced phases and different types to customers connected with. Electrical energy is continuously wasted in power systems due to electrical resistance in transmission and distribution lines. It is estimated that the distribution loss accounts for about 70 per cent, leaving the transmission loss at only 30 per cent.[1] Moreover, distribution systems are well known for higher R/X ratio compared to transmission systems and significant voltage drops that could cause substantial power loss along feeders. A study in Ref. 2 indicates that as much as 13 per cent of the total power generation is dissipated in distribution systems. As a result, loss reduction in distribution systems is one of the useful ways to increase energy efficiency to most utilities around the globe, especially in developing countries.

Utilities lose in two ways due to electrical losses in power systems. Firstly, losses cause an increase in demand of power and energy. That can lead to a rise in the cost of purchase and/or generation of electricity. Secondly, they also pose an increase in load currents across individual components of the system, thereby requiring extra costs used to increase the rating of components. Consequently, research on loss reduction is considered as part of the energy savings strategy in most utilities. Besides, it is also considered as an increase in the power capacity to reduce or defer significantly the need for upgrading existing systems or building new facilities.

Traditionally, reconfiguration has been used for load balancing, loss reduction and enhancement of voltages and reliability in distribution systems. Reconfiguration is the process of changing topology of a network by alternating open/closed status of switches. Capacitors have also been used to inject reactive power in distribution systems to reduce losses, improve voltage profiles, correct power factors and effectively use the existing facilities. To utilize the smallest amount of capacitors to achieve the best benefits, determining their appropriate locations and sizes should be considered.

In recent years, penetration of distribution generation (DG) into distribution systems has been increasing rapidly in many parts of the world. The main reasons for increasing penetration are the liberalization of electricity markets, development in DG technology, constraints on building new transmission and distribution lines, an increase in customer demands for highly reliable electricity, and environmental concerns.[3] For instance, a research by EPRI estimates that DG will be about 25 per cent of the new generation by 2010; while a study by National Gas Foundation shows that this figure could account for nearly 30 per cent by that time.[4] The numbers may vary as different agencies define DG in different ways, however, with the Kyoto protocol put in place where there will be a favorable market for DG that are coming from "Green Technologies", the share of DG would increase and there is no sign that it would decrease in near future. Moreover, the policy initiatives to promote DG throughout the world also indicate that the number will grow rapidly. As the penetration of DG in distribution system increases, it is in the best interest of all players involved to allocate DG in an optimal way such that it will reduce system losses and hence improve voltage profiles while saving the primary goal of energy injection.

At present, there are several technologies used for DG application that range from traditional to non-traditional ones. The former are non-renewable technologies such as internal combustion engines, combined cycles, combustion turbines and micro-turbines. The latter is based on renewable energy such as solar, photovoltaic, wind, geothermal, ocean, etc., and fuel cells. When renewable energy based DG units are placed for loss reduction, benefits are doubled and both aspects of sustainable energy are addressed. The challenges in DG applications for loss reduction are proper locations, appropriate sizes and operating strategies. Even if the location is fixed due to some other reasons, improper sizes would increase losses beyond the losses for case without DG units. Optimal sizing and locations depend on the type of DG units and the technology used for energy conversion as well. Utilities already facing the problem of high power losses and poor voltage profiles, especially, in the developing countries cannot tolerate any increase in losses. By optimum allocation of DG units, utilities take advantage of reduction in system losses; improve voltage regulation and reliability of supply. It will also relieve capacity from transmission and distribution system and hence, defer new investments, which have a long lead-time.

DG could also be considered as one of the viable options to ease issues related to poor power quality, congestion in transmission system, apart from meeting the energy demand of ever growing loads. In addition, the electrical power and Energy modularity and small size of DG units will facilitate planners to install it in a shorter time frame compared to the conventional solution. It would be more beneficial to install DG units in the present utility setup, which is moving towards a more decentralized environment, where there is a larger uncertainty in demand and supply. However, given the choices they should be placed in appropriate locations with suitable sizes to enjoy system-wide benefits. Hence, in this chapter, an attempt is made to address the issues associated with DG allocation in primary distribution systems by considering loss reduction.

The rest of the chapter is set out as follows: Sec. 23.2 gives a brief introduction to DG, including its definitions, technologies used and types. A brief summary of literatures on loss reduction techniques is summarized in Sec. 23.3. Section 23.4 describes the placement and sizing issues of DG units. Analytical expressions for finding optimal sizes for four DG unit types are also presented in the section. Besides, an approach to select a power factor of DG unit close to the optimal power factor is also discussed. For DG unit placement, loss sensitivity factor (LSF), improved analytical (IA) and exhaustive load flow (ELF) methods are introduced. Section 23.5 portrays three test distribution systems used in the chapter. Numerical results along with some interesting observations and discussions are presented in this section. Finally, the major contributions and conclusions are summarized in Sec. 23.6.

23.2 Distributed Generation

In Ref. 5, the authors have reported that supplying peaking power to reduce the cost of electricity, reduce environmental emissions through clean and renewable technologies (Green Power), combined heat and power (CHP) are major aims of DG. It was also pointed out that a high level of reliability and quality of supplied power and deferral of the transmission and distribution line investment through improved loadability are secondary applications. Other than these applications, the major application of DG in a deregulated electricity market environment lies in the form of ancillary services. These ancillary services include spinning and non-spinning reserves, reactive power supply and voltage control, etc. DG also has several benefits like reducing energy costs through combined heat and power generation, avoiding electricity transmission costs and less exposure to price volatility. Though the DG is considered as a viable solution to most of the problems that today's utilities are facing, there are many problems that need to be addressed. Furthermore, the type of DG technology adopted will have a significant bearing on the solution approach.

23.2.1 *Definitions of DG*

There is a wide variety of terminologies used in literatures for DG, such as "distributed generation", "embedded generation", "dispersed generation" or "decentralized generation".[4]

In general, DG means a small-scale power station different from a traditional or large central power plant. The authors in Ref. 3 have reported that there is no generally accepted definition of DG in literature as confirmed by the International Conference of Electricity Distributors (CIRED) in 1999, on the basis of a questionnaire submitted to the member countries. Some countries define distributed generation on the basis of the voltage level at which it is interconnected, whereas others start from the principle that DG is directly supplying consumer loads. Other countries define DG through some of its basic characteristic (e.g., using renewable sources, cogeneration, being non-dispatched). Following is a collection of definitions commonly appeared in some literatures for DG:

> "Distributed generation or DG, includes the application of small generations in range of 15 kW to 10 MW, scattered throughout a power system, to provide electric power needed by electrical consumers. The term DG includes all use of small electric power generators whether located in utility system at the site of a utility customer, or at an isolated site not connected to the power grid. By contrast, dispersed generation (capacity

ranges from 10 to 250 kW), a subset of distributed generation, refers to generation that is located at customer facilities or off the utility system."

ABB Electric Systems Technology Institute[6]

"Distributed generation as all generation units with a maximum capacity of 50 MW to 100 MW usually connected to the distribution network and neither centrally planned nor dispatched."

CIGRE[3]

"The generation of electricity by facilities that are sufficiently smaller than central generating plants so as to allow interconnection at nearly any point in a power system."

IEEE[3]

"Small dispersed generators are less than 5 MW and normally connected to the utility distribution system."

ANSI/IEEE std. 1001–1988[7]

"A generating plant serving a customer on-site or providing support to a distribution network, connected to grid at distribution-level voltages."

IEA[8]

In Ref. 4, the authors define DG as "an electric power source connected directly to the distribution network or on the customer site of the meter."

23.2.2 *DG technologies*

There are a number of DG technologies available in the market today and few are still in the research and development stage. Some currently available technologies are reciprocating engines, micro turbines, combustion gas turbines, fuel cells, photovoltaic, and wind turbines. Each one of these technologies has its own characteristics, i.e., benefits and limitations. Among all the technologies, diesel or gas reciprocating engines and gas turbines make up most of capacity installed so far. Simultaneously, new DG technology like micro turbine is being introduced and an older technology like reciprocating engine is being improved.[8] Fuel cell could be another technology of the future. However, as of now, there are only some prototype demonstration fuel cell projects. The costs of photovoltaic systems are expected to fall continuously over the next decade and wind turbines cost has been falling very rapidly as a result of exponential grow due to technological advancement. All these underline the statement that the future of power generation could be dominated by DG which can come from a variety of technologies. DG technologies can be classified into two broad categories, namely Combined Heat and Power (CHP) and Renewable Energy Generation.[3]

23.2.2.1 *Combined heat and power*

CHP plants or cogeneration are power plants where either electricity is the primary product and heat is a byproduct, or heat is the primary product and electricity is generated as a byproduct. The overall energy efficiency is then increased by effectively using the heat contained in primary fuel. Many DG technologies, such as reciprocating engines, micro-turbines and fuel cells can be used as CHP plants.

23.2.2.2 *Renewable energy generation*

This refers to as DG units that use renewable energy resources such as heat and light from sun, wind, falling water, ocean energy and geothermal heat biomass, etc. The main DG technologies falling under this category are wind turbines, small and micro hydro power, photovoltaic arrays, solar thermal power, and geothermal power.

Table 23.1 shows a brief description of DG technologies used nowadays, including reciprocating engines, gas and micro turbines, fuel cells, photovoltaic arrays, wind, small and hydro power. The comparison involves typical size, efficiency, fuel used, CO_2 and NO_x emissions, generation and operating costs, applications, advantages and drawbacks of each technology.[3,4,9]

23.2.3 *Interfaces to the utility system*

DG units are interconnected with the utility to operate in parallel with its distribution system. The main type of electrical system machines/interfaces are synchronous generators, asynchronous generators (induction machines) and power electronic inverters.[3]

23.2.3.1 *Synchronous machines*

The majority of DG units interconnected for parallel operation with utility distribution system are three-phase synchronous machines. Synchronous generators use DC field for excitation, and hence they can produce both active and reactive power. Emergency back-up generators using fossil-fuels combustion engines are normally synchronous machines. The machine can follow any load within its design capability with suitable field control. Besides, the inherent inertia, relatively small in the case of DG units compared to traditional generators, allows it to tolerate any step changes in the load.

Being capable of producing reactive power, large synchronous generators, relative to utility system capacity, might act as voltage regulators to improve voltage profile across the distribution feeder to which they are connected. This is considered as one major advantage of this type of DG units in weak systems. However, these generators should be coordinated with utility voltage regulators and protection

Table 23.1.   DG technologies.

DG technology	Reciprocating engines	Gas turbines	Micro-turbines	Fuel cells	Photovoltaic	Wind	Small/Micro hydro
Size (kW)	Diesel: 20–10,000+ Gas: 50–5,000+	1–20,000	30–200	50–1000+	1–20	200–3000	Small: 1000–100000 Micro: 25–1000
Efficiency (%)	36–43 28–42	21–40	25–30	35–54	6–15	N/A	
Fuel	Heavy fuel oil and bio-diesel, natural gas, biogas and landfill gas	Gas, Kerosene	Mainly natural gas but also landfill and biogas	Methanol, hydrogen or natural gas	Sun light	Wind	Water
CO_2 emissions (kg/MWh)	650 500–620	580–680	720	430–490	No direct emissions	No direct emissions	Small: 10–12 Micro: 16–20
NO_x emissions (kg/MWh)	10 0.2–1	0.3–0.5	0.1	0.005–0.01	No direct emissions	No direct emissions	Small: 0.046–0.056 Micro: 0.071–0.086
Gen. cost (USD/kW)	Diesel: 125–300 Gas: 250–600	300–600	500–750	1500–3000	5000–7000	900–1400	
O&M cost (USD/MWh)	5–10 7–15	3–8	5–10	5–10	1–4	10	

Table 23.1. (*Continued*)

DG technology	Reciprocating engines	Gas turbines	Micro-turbines	Fuel cells	Photovoltaic	Wind	Small/Micro hydro
Applications	Emergency, standby services and CHP	CHP and peak power supply	Transportation sector, power generation and CHP	Transportation sector, power generation and CHP	Household, small scale and off-grid applications	Central generation more than DG	Household, off-grid, grid connection and reliable back-up power
Advantages	Low capital cost, large size, fast start up, good efficiency and high reliability	Low maintenance cost and lower NOx emissions	High speed, less noise, low NO_x emissions	Compact, high efficiency and reliability low emissions	Low operating cost and no emissions	No emissions	Limited maintenance. Almost no environmental impact. Reliable and flexible operation. Likely to serve as reliable back-up power
Drawbacks	Noise, costly maintenance and high NO_x emissions	Noise and lower efficiency	High cost and recently commer-cialized	High cost and recently commer-cialized	High cost and unpre-dictable output	High cost and unpre-dictable output	The impact of variable water flow reducing the average power output as compared to its peak generating capacity

equipment to avoid any operating conflicts such as over-current protection, voltage regulation and others.[3]

23.2.3.2 *Asynchronous machines*

Asynchronous machines, also known as induction machines can be used as induction generators by driving the rotor slightly faster than synchronous speed. They are often started as a motor using the utility power line. For weak systems, the prime mover is started and brought to near synchronous speed before the machine is interconnected. Induction generators are typically smaller than 500 kW and they are suitable for wind DG units. They are easily interfaced to the utility as no special synchronizing equipment is required.

Unlike synchronous generators, induction generators are capable of producing active power only, and not reactive power. They require reactive power, from the power system to which it is connected, to provide excitation. This might affect the utility voltage and result in a low-voltage problem. Capacitors are then installed on the induction generator side to supply required reactive power to avoid any problems.

23.2.3.3 *Power electronic inverters*

An inverter is a solid state device that converts DC to AC at a desired voltage and frequency. DG technologies that generate either DC (wind, fuel cells and photovoltaic) or non-power frequency AC (micro turbines) must use an inverter to interface with power systems. The inverter technology has changed from the early thyristor-based, line commutated inverters to switched PWM inverters using insulated gate bipolar transistor (IGBT) switches. The shift in technology has greatly reduced the amount of harmonics injected by these inverters to the utility system. Power electronic inverters produce power at unity power factor to allow the full current-carrying capability of the switch to be used for delivering active power. When trouble is detected, the inverter can be switched off very quickly (in milliseconds) unlike the rotating machines which may require several cycles to respond. Table 23.2 shows a brief description of electrical interfaces required by different DG technology. Synchronous machines still dominate in most of the DG technology followed by asynchronous machine and power electronic converters, respectively.[7,10,11]

23.3 Loss Reduction in Distribution Systems

So far, there have been various techniques of loss reduction applied in distribution systems. In this section, a literature review for commonly used techniques, namely load balancing, network reconfiguration, capacitor placement and DG unit allocation is presented.

Table 23.2.   Summary of DG classification for power system interfaces.

DG technology	Synchronous machine To deliver P and absorb or deliver Q	Asynchronous machine To deliver P but likely to absorb Q as a source to operate	Power electronic inverter To deliver P at unity power factor, but likely to introduce harmonic currents
Small Hydro	×	×	
Solar, Photovoltaic			×
Wind	×	×	× (variable frequency generation + static power converter, or dc generation + static power converter)
Ocean	× (Four-quadr. synchronous machine)		
Geothermal	×		
Fuel cells			× (dc to ac converter)
CHP	×	×	
Micro-turbines			× (ac to ac converter (non-power frequency ac generation))
Combustion turbine	×		
Combined cycle	×		

23.3.1 *Load balancing*

Distribution systems are normally configured radially. Most distribution networks use sectionalizing switches that are normally closed, and tie-switches that are normally opened. These switches are used for both protection and network reconfiguration. Due to changing operating conditions, networks are reconfigured to reduce the system power loss (i.e., network reconfiguration for loss reduction), and to relieve overloads in the networks (i.e., network reconfiguration for load balancing). By modifying the radial structure of the feeders periodically may significantly improve the operating conditions of the overall system. Feeders in a distribution system normally have a mixture of industrial, commercial, residential, lighting, type of loads, etc. The peak load on the substation transformers and feeders occur at different times of the day. Therefore, the distribution system becomes heavily loaded at certain times of the day, and lightly loaded at other times. Load balancing is obtained by transferring loads from the heavily loaded feeders to lightly loaded feeders by reconfiguring the network. This is done to reschedule the loads for efficient operation of power systems.

The authors in Ref. 12 devised the problem of loss minimization and load balancing as an integer-programming problem. A correlation existed between load balancing and loss reduction was described. As the objective functions for load balancing and loss reduction are very similar, the calculation for load balancing are similar to that of loss reduction; therefore, the search for loss reduction can also be applied to improve load balancing in distribution networks. In Ref. 13, a constrained multi-objective and non-differentiable optimization problem, with equality and inequality constraints for both loss reduction and load balancing was introduced. In Ref. 14, the authors presented a three-phase load balancing in distribution system using index measurement techniques. In the initial stage of the technique, a loop giving the maximum improvement in load balancing is determined. In the next stage, the switching operation to be executed in that loop to get the maximum improvement in load balancing in the network is identified. In this technique, various indices to determine the optimal or near optimal solution for load balancing were developed. In Ref. 15, the authors proposed a fuzzy logic-based load balancing system along with a combinatorial optimization-based implementation system for changing loads.

23.3.2 *Reconfiguration*

There have been many algorithms proposed to solve optimal network reconfiguration problems and they can be classified into two categories. The first category is traditional optimization algorithms,[12,16,17] such as linear programming algorithms, branch exchange algorithms, optimal flow pattern algorithms, etc. The second category is artificial intelligence-based algorithms,[13,18,25] such as genetic algorithms (GA), simulated annealing (SA) algorithms, Tabu search (TS) algorithms, etc.

(1) Traditional optimization algorithms: In Ref. 16, a solution method based on a switch exchange algorithm was proposed, in which a simple formula for estimation of reducing losses by means of a switch exchange was used. In Ref. 12, an approximate power-flow method for loss minimization due to a switch operation was introduced. In Ref. 17, the authors formulated the network reconfiguration problem as a

linear programming problem and applied a single-loop optimization method to deal with optimal network reconfiguration for loss minimization. In general, traditional optimization algorithms are easy to implement and require rather less computational burden; however, they generally cannot converge to a global optimum solution.

(2) Artificial intelligence-based algorithms: In Ref. 18, the authors used GA to successfully solve the combinatorial optimization network reconfiguration problem to minimize losses. In Refs. 19 and 20, the authors presented a solution approach based on SA to solve reconfiguration problem in large-scale distribution systems, but the SA requires excessive computation time. In Ref. 21, the authors proposed a solution algorithm based on a parallel Tabu search (PTS) technique, and results showed that PTS is better than SA, parallel SA, GA, parallel GA and TS algorithms in terms of the total system cost and computational time. Furthermore, an improved Tabu search (ITS) technique for network reconfiguration was also presented to effectively minimize power losses in large-scale distribution systems.[22] In Ref. 23, the authors also improved the computation time and convergence property by using a hybrid algorithm of SA and TS to deal with reconfiguration for loss reduction in large-scale distribution systems. In Ref. 24, the authors introduced the ant colony search algorithm (ACSA) which is a relatively new and powerful intelligence evolution method to solve the optimal network reconfiguration problem for power loss reduction. The test results demonstrated that ACSA is better than both GA and SA in terms of loss reduction and computation time. ACSA is a population-based technique that uses exploration of positive feedback as well as greedy search. ACSA was inspired from natural behavior of ant colonies on how they find food sources and bring them back to their nest by building the unique trail formation. Based on ACSA, the near-optimal solution to network reconfiguration can be effectively attained. ACSA uses the state transition, local pheromone-updating, and global pheromone-updating rules to facilitate the computation. In addition, in Ref. 25, a modified particle swarm optimization (MPSO), which can effectively solve the complex network reconfiguration problem, was presented. The simulated results showed that MPSO is better than the optimal power flow pattern algorithm and GA in terms of the total power loss.

23.3.3 *Capacitor placement*

In the past decades, many optimization techniques have been proposed to solve the capacitor placement problem. Solution techniques for this problem can be divided into four categories: (1) analytical, (2) numerical programming, (3) heuristic, and (4) artificial intelligence-based algorithms.

All the early works of optimal capacitor allocation used ***analytical methods*** which involves the use of calculus to determine the maximum of a capacitor's savings function. From the early study in 1950s, the famous "2/3" rule was developed. It states that for an optimal loss reduction, a capacitor rated 2/3 of the total peak reactive demands needs to be installed at a distance of 2/3 along the total feeder length away from substation feeding the feeder. Although based on unrealistic assumptions of a feeder with a fixed conductor size and uniform load, today this rule is still being used by many utilities because it is easy to understand and implement. A shortcoming of the analytical techniques is the modeling of capacitor distributed locations and capacity as continuous variables. Consequently, the calculated capacitor sizes may not match the standard sizes available and calculated sites may also not match the locations in distribution systems.[26]

With the advent of computing systems, ***numerical programming*** techniques were proposed to deal with optimization problems. They are iterative techniques used to minimize (or maximize) an objective function of decision variables. For optimal capacitor allocation, the savings function would be the objective function and the locations, sizes, number of capacitors, bus voltages, and currents would be the decision variables which must satisfy operational constraints.[26] In Ref. 27, Duran (1968) was the first to utilize a dynamic programming technique to capacitor allocation. This technique is simple and only considers energy loss reduction and accounts for discrete capacitor sizes. In Ref. 28, the authors presented the capacitor allocation problem using a mixed integer programming. In general, merits of numerical programming over the analytical techniques is that it only considers feeder bus locations and capacitor sizes as discrete variables. However, data preparation, and interface development for numerical programming techniques may need more time than for analytical ones.

Heuristic methods are based on rules developed through intuition, experience, and judgment. Heuristic rules produce fast and practical strategies which reduce an exhaustive search space and can lead to a solution that is nearly optimal with confidence.[26] In Ref. 28, the authors proposed a heuristic technique to identify a section in the distribution system with the highest loss owing to reactive load currents and then find the sensitive buses which have the greatest influence on system loss reduction. The sizes of capacitors located on the sensitive buses are then determined by maximizing the power loss reduction from capacitor compensation. In Ref. 29, the authors improved the work of Ref. 28 by determining the sensitive buses that have the greatest impact on loss reduction for the whole distribution system directly and by optimizing the size of capacitors based on maximizing the net economic savings from both energy and peak power loss reductions. Both of the above heuristic methods are intuitive, easy to understand, and simple to implement as compared with analytical and DG numerical programming techniques. However, the results based on heuristic algorithms are not guaranteed to be optimal.

Recently, there have been many researches based on ***artificial intelligence-based*** techniques (AI), namely GA, SA, expert systems (ES) algorithms, artificial neural networks (ANN) algorithms, fuzzy set theory (FST) and discrete particle swarm optimization (DPSO) algorithms to solve optimal capacitor allocation problems.

In Ref. 30, the authors proposed a solution algorithm based on GA to optimize capacitor sizes and sites. The parameter sets including the capacitor sizes and locations are coded. GA operates by choosing a population of the coded parameters with the highest fitness levels (i.e., minimum costs of capacitors, maximum reduction of peak power losses), and carrying out a combination of mating, crossover, and mutation operations on them to generate a better set of the coded parameters.

ES consists of a collection of rules, facts (knowledge), and an inference engine to perform logical reasoning. In Refs. 31 and 32, the authors developed ES to solve capacitor placement problems for maximum savings from peak power and energy loss reduction.

SA is an iterative optimization algorithm based on the annealing of solids. When a material is annealed, it is heated to a high temperature and slowly cooled according to a cooling schedule to reach a desired state. At the highest temperature, particles of the material are arranged in a random formation. As the material is cooled, the particles become organized into a lattice-like structure which is a minimum energy state.[26]

ANN is the connection of artificial neurons which simulates the nervous system of a human brain. In Ref. 33, the authors conducted optimal switched capacitor control with two neural networks utilized. A network is utilized to predict a load profile from a set of previous load values obtained from direct measurement at various buses. And another is utilized to choose optimal capacitor tap positions based on a load profile as predicted by the first network to maximize the energy loss reduction for a given load condition. Once both networks are used, iterative calculations are no longer required and a fast solution for a given set of inputs can be provided. Although this technique was suitable for on-line implementation of small systems, it may not be appropriate for much larger realistic distribution systems.

In Ref. 34, a solution algorithm based on FST was also used to solve the capacitor allocation problem. In this technique, voltage and power loss indices are modeled by membership functions and a fuzzy expert system containing a set of heuristic rules which performs the inference to find a capacitor allocation suitability index of each bus. Capacitors are located at the bus with the highest suitability.

In addition, in Ref. 35, the authors presented DPSO that solves the problems of finding the optimal fixed capacitor placement and sizing of a fixed capacitor simultaneously. The discrete nature of placement and sizing are incorporated in the proposed algorithm to provide a more realistic model.

23.3.4 *DG unit placement*

In a radial feeder, DG units can deliver a portion of real and reactive power to the loads so that the feeder current reduces from the sources to the location of DG units. However, the studies in Refs. 5 and 36 presented that poor location and size of DG unit can result in larger system losses. Hence, to minimize losses, it is important to find the optimal location and sizing of DG units assuming primary fuel resources for DG units are not constraints by location in distribution systems.

Unlike capacitor placement, optimal DG unit allocation studies of loss reduction and voltage profile enhancement in distribution systems are relatively new. However, in recent years, there have been many researches on this problem. Normally DG units are located close to consumption or at the end of the most heavily loaded feeder[36] or at the most heavily loaded nodes to reduce losses. However, such techniques may not yield the lowest loss.

Another technique for DG unit placement using "2/3 rule" has been presented.[36] Although the 2/3 rule is simple and easy to implement, this technique may not be effective in distribution systems with no uniformly distributed loads. Besides, if a DG unit is capable of delivering real and reactive power, applying the method that was developed for capacitor placement may not work. This method cannot be applied in meshed distribution systems as well.

In Ref. 37, an analytical approach has been presented to identify the optimal location to place a DG unit in radial as well as meshed networks to minimize losses. But, in this approach, the optimal sizing is not considered. GA was applied to determine the size and location of DG unit in Refs. 38 and 39. Though GA is suitable for multi-objective optimization problems, it can lead to a near optimal solution with higher computational time.

Recently, an analytical approach based on an exact loss formula was presented to find the optimal size and location of a DG unit.[5] In this method, a new methodology was proposed to quickly calculate approximate losses for identifying the best location; the load flow is required to be conducted only twice. The first load flow calculation is needed to calculate the loss of base case. The second load flow solution is required to find the minimum total loss after DG unit placement. The technique requires less computation. However, the analytical approach was limited to DG unit capable of delivering real power only.

Most of the approaches presented so far model DG unit as a machine that is capable of delivering real power only. However, there are other types of DG units (e.g., DG unit capable of injecting both P and Q or injecting P but consuming Q) being integrated into distribution systems. In Refs. 40 and 41, the authors developed a comprehensive formula by improving the analytical method proposed in Ref. 5 to find the optimal size, location and power factor of various types of single or multiple-DG units for loss reduction in large-scale distribution systems. This method discussed in Sec. 23.4 is referred to as an improved analytical method (IA).

23.4   Loss Reduction Using DG

In this section, the IA method uses a comprehensive analytical formula to find the optimal sizing, location and power factor of different types of DG units as defined in Sec. 23.4.1 to achieve a highest loss reduction in large-scale distribution systems that were presented in Refs. 40 and 41. In this method, a "fast method" to identify an optimal or near optimal power factor of DG units is also presented.

To validate the effectiveness and applicability of method, other methods such as loss sensitivity factor (LSF) and exhaustive load flow (ELF) methods are also presented. LSF cannot lead to the best placement for single DG unit to reduce losses in Ref. 5; but it has been used to select the location of DG units to reduce search space. Hence, it is a good tool to compare with IA in terms of computational time. Here, only the most suitable bus for each method is considered to place a DG unit. The optimal size of DG units is identified by repeating the load flow at only that bus to save computational time. These results were verified by ELF solutions. ELF demands a high computational effort since all buses are considered and several simulations are made in the process. But, it has been used to find the optimal location and sizing of a single DG unit in Ref. 5. Hence, it is a good tool to compare with other approaches in terms of, optimal location, size and amount of loss. Effect of size and location of DG unit with respect to losses in the network is also examined in detail.

23.4.1 *Types of DG units*

DG units are modeled as synchronous generators for small hydro power, geothermal power, combined cycles and combustion turbines. Induction generators are used in wind and micro hydro power generation and DG units are considered as power electronic inverter generators or static generators for technologies such as micro gas turbines, solar power photovoltaic power and fuel cells.[11] In general, DG units can be classified into four types as follows:

— Type 1: DG unit capable of injecting P only.
— Type 2: DG unit capable of injecting both P and Q.
— Type 3: DG unit capable of injecting P but consuming Q.
— Type 4: DG unit capable of injecting Q only.

For instance, photovoltaic, micro turbines, fuel cells which are integrated to the main grid with the help of converters/inverters are good examples of Type 1, if they are running at unity power factor. DG units that are based on synchronous machine (cogeneration, gas turbine, etc.) fall in Type 2. Type 3 is mainly induction generators that are used in wind farms. Type 4 could be synchronous compensators such as gas turbines.

23.4.2 *Power losses*

The real power loss in a system is represented by Eq. (23.1). This is popularly known as the "exact loss" formula.[42]

$$P_L = \sum_{i=1}^{N} \sum_{j=1}^{N} [\alpha_{ij}(P_i P_j + Q_i Q_j) + \beta_{ij}(Q_i P_j - P_i Q_j)], \qquad (23.1)$$

where

$$\alpha_{ij} = \frac{r_{ij}}{V_i V_j} \cos(\delta_i - \delta_j),$$

$$\beta_{ij} = \frac{r_{ij}}{V_i V_j} \sin(\delta_i - \delta_j),$$

$V_i \angle \delta_i =$ the complex voltage at the bus ith,

$r_{ij} + jx_{ij} = Z_{ij}, \quad$ the ijth element of $[Zbus]$ impedance matrix,

P_i and $P_j =$ the active power injections at the ith and jth buses, respectively,

Q_i and $Q_j =$ the reactive power injections at the ith and jth buses, respectively,

$N =$ the number of buses.

23.4.3 *Location and sizing issues*

Figure 23.1 shows a 3D plot of typical power loss versus size of DG unit at each bus in a distribution system. From the figure, it is obvious that for a particular bus, as the size of DG unit is increased, the losses are reduced to a minimum value and increased beyond a size of DG unit (i.e., the optimal DG unit size) at that location. If the size of DG is further increased, the losses start to increase and it is likely that it may overshoot the losses of the base case. Also notice that the location of DG unit plays an important role in minimizing the losses.

The important conclusion that can be drawn from the figure is that, given the characteristics of a distribution system, it is not advisable to construct a sufficiently high DG unit in the network. At most the size should be such that it is consumable within the distribution substation boundary. Any attempt to install the high capacity DG unit with the purpose of exporting power beyond the substation (reverse flow of

     D. Q. Hung and N. Mithulananthan

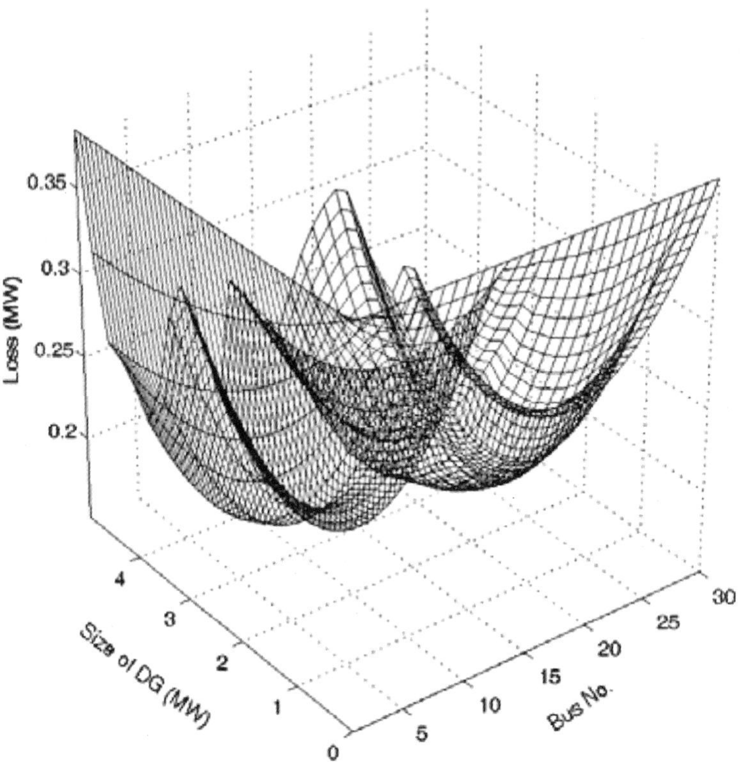

Fig. 23.1.   Effect of size and location of DG unit on system loss.[5]

power though distribution substation), will lead to very high losses. So, the size of
distribution system in term of load (MW) will play an important role in selecting the
size of the DG unit. The reason for higher losses and high capacity of the DG unit can
be explained by the fact that the distribution system was initially designed such that
power flows from the sending end (source substation) to the load and conductor sizes
are gradually decreased from the substation to consumer point. Thus without rein-
forcement of the system, the use of the high capacity DG unit will lead to excessive
power flow through small-sized conductors and hence results in higher losses.

To avoid exhaustive computation and to save time, existing techniques, such as
the LSF method, find the location issue before the sizing issue. This may not result
in the best choice. A brief description of this technique and associated problems are
presented below.

The DG unit allocation can be handled by resolving the sizing issue first followed
by the location issue. For such DG unit placement, a recent IA method for large-scale
systems in Refs. 35, 40 and 41 as described below, can lead to an optimal or a near
optimal solution with less computational effort and time.

A simple technique known as an exhaustive load flow (ELF) or a repeated load flow can find the sizing and location by repeating the load flow at every bus with a change of the DG unit size in "small" steps (normally 10%) and calculating the loss for each. The minimum loss can be identified. Such a technique can lead to a completely optimal solution but can demand excessive computational time and is unsuitable for large-scale systems.

23.4.4 *Placement*

23.4.4.1 *Loss sensitivity factor method*

Loss sensitivity factor (LSF) method is based on the principle of linearization of original nonlinear equation around the initial operating point, which helps to reduce solution space. The loss sensitivity factor method has been widely used to solve the capacitor allocation problem.[43]

The sensitivity factor of real power loss with respect to real power injection from a DG unit is given by

$$\alpha_i = \frac{\partial P_L}{\partial P_i} = 2 \sum_{j=1}^{N} (\alpha_{ij} P_j - \beta_{ij} Q_j). \qquad (23.2)$$

Sensitivity factors are evaluated at each bus, firstly using the values obtained from the base case power flow. Then the buses are ranked in descending order of the values of their sensitivity factors to form a priority list. The top-ranked buses in the priority list will be studied to identify the best location. This is generally done to take into account the effect of nonlinearities in the system. The first order sensitivity factor is based on linearization of the original nonlinear equation around the initial operating condition and is biased towards a function which has higher slope at the initial condition, that might not identify the global optimum solution. This condition is depicted in Fig. 23.2. Therefore, the priority list of candidate location is a prerequisite to get a near optimal solution.

Figure 23.2 shows a probable case, captured from the trend of losses in Sec. 23.4.3. The curve with solid line has the highest sensitivity factor at the initial operating condition than dotted curve, but does not give the lowest loss, as PL1 > PL2. It shows why the sensitivity factor may not give the optimum result if a number of alternative locations are not taken into account.

23.4.4.2 *Priority list*

The sensitivity factor will reduce the solution space to few buses, which constitute the top ranked buses in the priority list. The effect of number of buses taken in priority will have the effect of the optimum solution obtained for some systems.

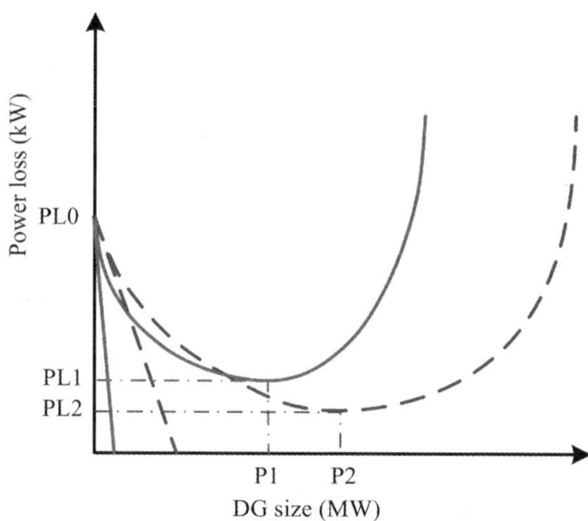

Fig. 23.2. Nonlinearity in loss curve.[5]

For each bus in the priority list, the DG unit is placed and the size is varied from minimum (0 MW) to a higher value until the minimum system loss is found with the DG unit size. The process is computationally demanding as one needs a large number of load flow solution.

23.4.4.3 *Optimization technique for multiple DG units*

In this technique,[41] DG units are modeled as type 1 generators, which are capable of injecting active power only. The computational procedure to find the optimal locations, and sizes of multiple DG units is described below. In fact, this technique has been presented to solve single DG unit placement with a unity power factor in Ref. 5. In this work, based on an idea of updating the load data after each time of DG unit placement, the technique is presented to solve optimal multiple DG unit placement. After updating the load data, the algorithm will continue to optimize other DG unit placement until it does not satisfy at least one of the constraints in step 9 described below.

Step 1: Enter the number of DG units.
Step 2: Run load flow for the base case.
Step 3: Find the base case loss using Eq. (23.1).
Step 4: Find the sensitivity factor using Eq. (23.2). Rank buses in descending order of the values of their sensitivity factors to form a priority list.
Step 5: Select the bus with the highest priority. Place a DG unit (active power) at that bus.

Step 6: Change the DG size obtained in step 5 in "small" steps, update the values α and β, and calculate the loss for each case using Eq. (23.1) by running the load flow.

Step 7: Select and store the optimal size of DG unit that gives the minimum loss.

Step 8: Update load data after placing DG unit with the optimal size obtained in step 7.

Step 9: Stop if either:

— the voltage at a particular bus is over the upper limit,
— the total size of DG units is over the total load plus loss,
— the maximum number of DG units is unavailable, or
— new iteration loss is greater than the previous iteration loss. The previous iteration loss is retained; otherwise, repeat steps 2 to 9.

23.4.5 *Sizing*

23.4.5.1 *Analytical expressions for loss reduction*

In this section, analytical expressions to find the optimal sizes of four DG unit types are shown for maximum loss reduction.[40,41] Besides, the importance of DG operation (i.e., real and reactive power dispatch) for loss minimization along with a simple way to select a power factor of DG unit close to the optimal power factor is presented.

23.4.5.1.1 Sizing at various locations

Assuming $a = (sign)\,\tan(\cos^{-1}(PF_{DG}))$, the reactive power output of a DG unit is expressed by Eq. (23.3):

$$Q_{DGi} = aP_{DGi} \tag{23.3}$$

in which,

$sign = +1$: DG unit injecting reactive power,

$sign = -1$: DG unit consuming reactive power,

PF_{DG} is the power factor of DG unit.

The active and reactive power injected at bus i, where the DG unit located, is given by Eqs. (23.4) and (23.5), respectively:

$$P_i = P_{DGi} - P_{Di}, \tag{23.4}$$

$$Q_i = Q_{DGi} - Q_{Di} = aP_{DGi} - Q_{Di}. \tag{23.5}$$

From Eqs. (23.1), (23.4) and (23.5), the active power loss can be rewritten as

$$P_L = \sum_{i=1}^{N} \sum_{j=1}^{N} \begin{bmatrix} \alpha_{ij}[(P_{DGi} - P_{Di})P_j + (aP_{DGi} - Q_{Di})Q_j] \\ + \beta_{ij}[(aP_{DGi} - Q_{Di})P_j - (P_{DGi} - P_{Di})Q_j] \end{bmatrix}. \tag{23.6}$$

The total active power loss of the system is minimum if the partial derivative of Eq. (23.6) with respect to the active power injection from a DG unit at bus i becomes zero. Following simplification and re arrangement the equation can be written as

$$\frac{\partial P_L}{\partial P_{DGi}} = 2 \sum_{j=1}^{N} [\alpha_{ij}(P_j + aQ_j) + \beta_{ij}(aPj - Qj)] = 0. \tag{23.7}$$

Equation (23.7) can be rewritten as

$$\alpha_{ii}(P_i + aQ_i) + \beta_{ii}(aP_i - Q_i) + \sum_{\substack{j=1 \\ j \neq i}}^{N} (\alpha_{ij}P_j - \beta_{ij}Q_j)$$

$$+ a \sum_{\substack{j=1 \\ j \neq i}}^{N} (\alpha_{ij}Q_j + \beta_{ij}P_j) = 0, \tag{23.8}$$

$$\text{Let:} \begin{cases} X_i = \sum_{\substack{j=1 \\ j \neq i}}^{n} (\alpha_{ij}P_j - \beta_{ij}Q_j) \\ \\ Y_i = \sum_{\substack{j=1 \\ j \neq 1}}^{n} (\alpha_{ij}Q_j + \beta_{ij}P_j) \end{cases}. \tag{23.9}$$

From Eqs. (23.4), (23.5), (23.8) and (23.9), Eq. (23.10) can be developed:

$$\alpha_{ii}(P_{DGi} - P_{Di} + a^2 P_{DGi} - aQ_{Di})$$

$$+ \beta_{ii}(Q_{Di} - aP_{Di}) + X_i + aY_i = 0. \tag{23.10}$$

From Eq. (23.10), the optimal size of a DG unit at each bus i for minimizing loss can be written as

$$P_{DGi} = \frac{\alpha_{ii}(P_{Di} + aQ_{Di}) + \beta_{ii}(aP_{Di} - Q_{Di}) - X_i - aY_i}{a^2\alpha_{ii} + \alpha_{ii}}. \tag{23.11}$$

The above equation gives the optimum size of a DG unit for each bus i, for the loss to be minimum. Any size of a DG unit other than P_{DGi} placed at bus i, will lead to a higher loss. This loss, however, is a function of loss coefficient α and β. When a DG unit is installed in the system, the values of loss coefficients will change, as it depends on the state variable voltage and angle. Updating values of α and β again requires another load flow calculation. But numerical result shows that the accuracy gained in the size of DG units by updating α and β is small and is negligible. With this assumption, the optimum size of a DG unit for each bus, given by relation (23.11) can be calculated from the base case load flow (i.e., without any DG case).

The power factor of DG units depends on operating conditions and the type of DG units. When the power factor of a DG unit is given, the optimal size of the DG unit at each bus i for minimizing losses can be found in the following ways.

(i) *Type 1 DG unit*:
For type 1 DG unit, power factor is at unity, i.e., $PF_{DG} = 1$, a $=$ zero. From Eqs. (23.9) and (23.11), the optimal size of a DG unit at each bus-i for minimizing losses can be given by reduced Eq. (23.12):

$$P_{DGi} = P_{Di} - \frac{1}{\alpha_{ii}} \left[\beta_{ii} Q_{Di} + \sum_{\substack{j=1 \\ j \neq i}}^{N} (\alpha_{ij} P_j - \beta_{ij} Q_j) \right]. \tag{23.12}$$

This equation is similar to what is presented in Ref. 5.

(ii) *Type 2 DG unit*:
Assuming $0 < PF_{DG} < 1$, sign $= +1$ and "a" is a constant, the optimal size of a DG unit at each bus i for the minimum loss is given by Eqs. (23.11) and (23.3).

(iii) *Type 3 DG unit*:
Assuming $0 < PF_{DG} < 1$, sign $= -1$ and "a" is a constant, the optimal size of a DG unit at each bus i for the minimum loss is given by Eqs. (23.11) and (23.3).

(iv) *Type 4 DG unit*:
Assuming $PF_{DG} = 0$ and $a = \infty$, from Eqs. (23.3), (23.9) and (23.11) the optimal size of a DG unit at each bus-i for minimizing losses is given by reduced Eq. (23.13):

$$Q_{DGi} = Q_{Di} + \frac{1}{\alpha_{ii}} \left[\beta_{ii} P_{Di} - \sum_{\substack{j=1 \\ j \neq i}}^{N} (\alpha_{ij} Q_j + \beta_{ij} P_j) \right]. \tag{23.13}$$

23.4.5.1.2   Optimal location

The next step is to find the optimum DG unit location, which will give the lowest possible total losses. Calculation of loss with the DG unit one at a time at each bus again requires several load flow solutions, as many as the number of buses in the system. Therefore a new methodology in Ref. 5 is proposed to quickly calculate approximate loss, which would be used for the purpose of identifying the best location. Numerical result shows that approximate loss follows the same pattern as that calculated by accurate load flow. It means that, if accurate loss calculation from the load flow gives a minimum for a particular bus, then loss calculated by the approximate loss method will also be minimum at that bus. This is verified by the simulation results and presented later in the chapter. With this methodology one can avoid exhaustive computation and save time, especially for large-scale distribution systems as trend of loss reduction can be captured with α and β coefficients from the base case, i.e., without calculating α and β every time a DG unit is placed in a different location.

23.4.5.1.3   Optimal power factor

Consider a simple distribution system with two buses, a source, a load and a DG unit connected through a transmission line shown in Fig. 23.3.

The power factor of the single load (PF_D) is given by Eq. (23.14):

$$PF_D = \frac{P_D}{\sqrt{P_D^2 + Q_D^2}}. \tag{23.14}$$

It is obvious that minimum loss occurs when the power factor of a DG unit is equal to the power factor of load as given by Eq. (23.15).

$$PF_D = PF_{DG} = \frac{P_{DG}}{\sqrt{P_{DG}^2 + Q_{DG}^2}}. \tag{23.15}$$

In practice, a complex distribution system includes a few sources, many buses, many lines and loads. The power factors of loads are different. If each load is supplied by each local DG unit, at which the power factor of each DG unit is equal to that of each load, there is no current in the lines. The total line power loss is zero. The transmission lines are also unnecessary. However, that is unrealistic since the capital investment cost for DG units is too high. Therefore, the number of installed DG units should be limited.

To find the optimal power factor of DG units for a radial complex distribution system, a fast approach is proposed. A repeated approach is also introduced and

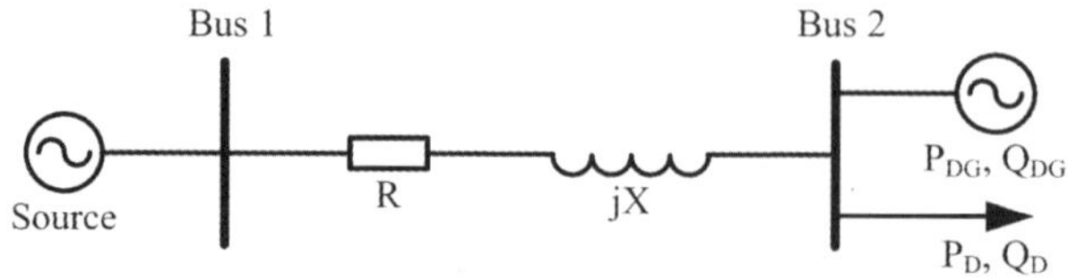

Fig. 23.3.   A simple distribution system with a DG unit.

compared to the proposed one for checking the effectiveness and applicability. It is interesting to note that in all the three test systems the optimal power factor of type 3 DG units placed for loss reduction was found to be closer to the power factor of the combined load of the respective system.

(i) *Fast approach*:

The power factor of the combined total load of the system (PF_D) can be expressed by Eq. (23.14); in which the total active and reactive power of the load demand are expressed as

$$P_D = \sum_{i=1}^{N} P_{Di}$$

$$Q_D = \sum_{i=1}^{N} Q_{Di}$$

The "possible minimum" total loss can be achieved if the power factor of a DG unit (PF_{DG}) is quickly selected to be equal to that of the total load (PF_D). That can be expressed by Eq. (23.16)

$$PF_{DG} = PF_D. \tag{23.16}$$

(ii) *Repeated approach*:

In this method, the optimal power factor is found by calculating power factors of a DG unit (change in a small step of 0.01) near to the power factor of combined load. The sizes and locations of DG units at various power factors with respect to losses are identified. The losses are compared together. The optimal power factor of a DG unit for which the total loss is at minimum is determined.

23.4.5.2 *Optimization technique for single-DG unit*

This technique is developed using the above analytical expressions to find the optimum size, location and power factor of a DG unit in the distribution system. It requires load flow to be carried out only two times, the first time for the base case and second load flow at the end with a DG unit included to obtain the final solution.[40]

When the power factor of DG units is set to be equal to that of combined total loads, a computational procedure to find the optimal size and location of four different DG unit types is described below.

Step 1: Run load flow for the base case.
Step 2: Find the base case loss using Eq. (23.1).
Step 3: Calculate the power factor of a DG unit using Eq. (23.16).

Step 4: Find the optimal size of a DG unit for each bus using Eqs. (23.3) and (23.11).

Step 5: Place the DG unit with the optimal size obtained in step 4 at each bus, one at a time. Calculate the approximate loss for each case using Eq. (23.1) with the values α and β of base case.

Step 6: Locate the optimal bus at which the total loss is at minimum with corresponding optimal size at that bus.

Step 7: Run load flow with the optimal size at the optimal location obtained in step 6. Calculate the exact loss using Eq. (23.1) and the values α and β after DG placement.

It is noted that when the type and power factor of DG units is given, the computational procedure to find the optimal size and location of DG units is as above — apart from step 3. At this step, the power factor of DG units is entered rather than calculating it using Eq. (23.16). In the ELF method, optimal sizes, optimal location and power factor are obtained with a number of load flow solutions.

23.4.5.3 *Optimization technique for multiple-DG units*

As presented in Sec. 23.4.4.3, based on an idea of updating the load data after each time of DG unit placement, this technique is used to solve different types of multiple-DG unit placement. In this work, the technique for the single-DG unit mentioned in Sec. 23.4.5.2 is applied to quickly calculate approximate loss for identifying the best location after one load flow run to reduce the size of the search space. When the power factor of DG units is selected to be equal to that of the combined load, the computational procedure to find the optimal size, location and power factor of multiple DG units is as follows:

Step 1: Enter the number of DG units.

Step 2: Run load flow for the base case.

Step 3: Find the base case loss using Eq. (23.1).

Step 4: Calculate the power factor of DG units using Eq. (23.16).

Step 5: Find the optimal DG unit size for each bus using Eqs. (23.3) and (23.11).

Step 6: Place the DG unit with the optimal size obtained in step 4 at each bus, one at a time. Calculate the approximate loss for each case using Eq. (23.1) with the values α and β of base case.

Step 7: Locate the optimal bus at which the loss is at minimum. Place a DG unit at this bus.

Step 8: Change the DG unit size obtained in step 7 in "small" steps, update the values α and β, and calculate the loss for each case using Eq. (23.1) by running the load flow.

Step 9: Select and store the optimal size of the DG unit that gives the minimum loss.

Step 10: Update load data after placing the DG unit with the optimal size obtained in step 9.

Step 11: Stop if either:

— the voltage at a particular bus is over the upper limit,
— the total size of DG units is over the total load plus loss,
— the maximum number of DG units is unavailable, or
— the new iteration loss is greater than the previous iteration loss.

The previous iteration loss is retained; otherwise, repeat steps 2 to 11.

When the type and power factor of DG units is given; the computational procedure to find the optimal size, location of DG units is as above, except step 4. At this step, the power factor of DG units is entered instead of calculating it using Eq. (23.16).

23.4.6 *ELF for multiple-DG units*

This technique demands so much computational time with a number of load flow runs because all the buses are considered in the calculation; but it can lead to an optimal solution. When the DG power factor is selected to be equal to that of the total load, the computational procedure to find the optimal size, location and power factor of different types of multiple DG units is as follows[40,41]:

Step 1: Enter the number of DG units.
Step 2: Run load flow for the base case.
Step 3: Find the base case loss using Eq. (23.1).
Step 4: Calculate the power factor of DG units using Eq. (23.16).
Step 5: Place a DG unit at the first bus (normally bus 2).
Step 6: Change the DG unit size obtained in step 5 in "small" steps, update the values α and β, and calculate the loss for each case using Eq. (23.1) by running the load flow. Repeat steps 5 and 6 for all the rest buses.
Step 7: Locate the optimal bus at which the total loss is at minimum with the corresponding optimal size. Store the result.
Step 8: Update load data after placing the DG unit with the optimal size obtained in step 7.

614   *D. Q. Hung and N. Mithulananthan*

Step 9:   Stop if either:

     — the voltage at a particular bus is over the upper limit,
     — the total size of DG units is over the total load plus loss,
     — the maximum number of DG units is unavailable, or
     — the new iteration loss is greater than the previous iteration loss.

The previous iteration loss is retained; otherwise, repeat steps 2 to 9.

When the type and power factor of DG units is given; computational procedure to find the optimal size, and location of DG units is as above except step 4. At this step, the power factor of DG units is entered instead of calculating it using Eq. (23.16).

23.5 Numerical Results

This section provides numerical results on 16-, 33- and 69-bus test distribution systems to validate the methodologies that were introduced in Sec. 23.4. Section 23.5.1 portrays the test systems. Section 23.5.2 presents simulation results for four types of single-DG unit placement for loss reduction. Results of multiple-DG unit placement are also shown in Sec. 23.5.3.

23.5.1 *Test systems*

Three test systems with varying sizes and complexities have been used to test and validate the proposed analytical expressions and algorithms. The first test system illustrated in Fig. 23.4 is a 16-bus radial test distribution system with a total load of 28.7 MW and 5.9 MV Ar real and reactive power, respectively.[12] In this test system, it is assumed that the tie lines are always switched out. The second system depicted

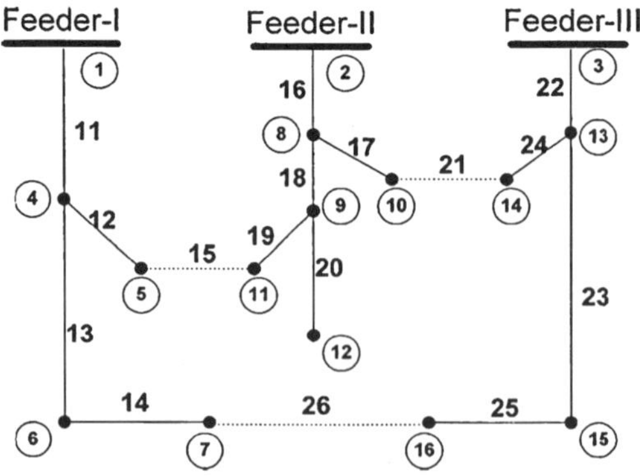

Fig. 23.4.   Single line diagram of a 16-bus test distribution system.

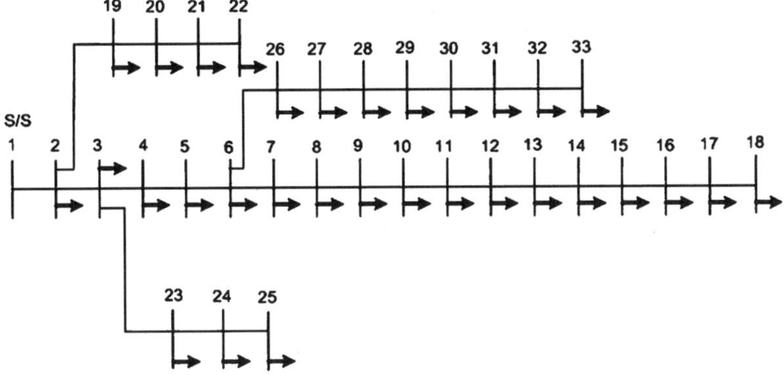

Fig. 23.5. Single line diagram of a 33 bus-test distribution system.

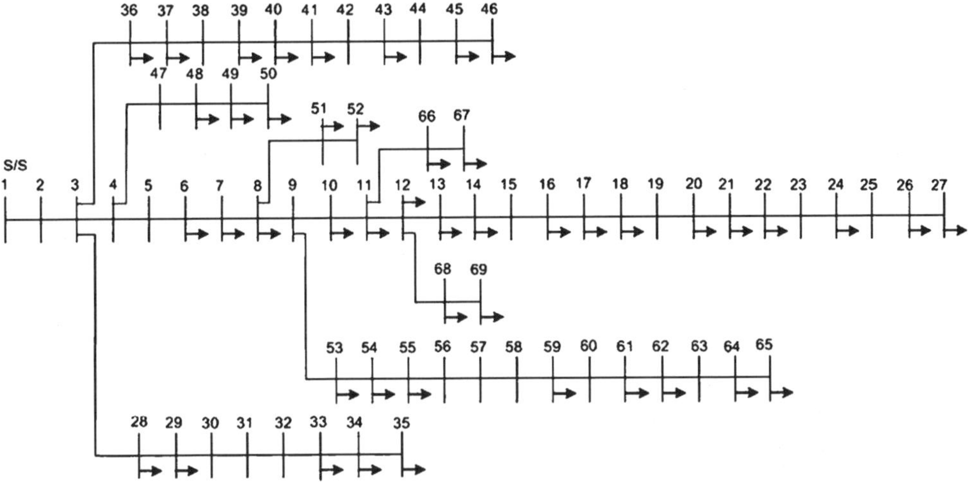

Fig. 23.6. Single line diagram of a 69-bus test distribution system.

in Fig. 23.5 is a 33-bus test radial distribution system with a total load of 3.7 MW and 2.3 MV Ar, real and reactive power, respectively.[44] The last system shown in Fig. 23.6 is a 69-bus test radial distribution system with a total load of 3.8 MW and 2.69 MV Ar, real and reactive power, respectively.[45]

Tables 23.3 to 23.5 present the data of the above systems.

An analytical software tool has been developed in MATLAB environment to run load flow, calculate losses and find optimal sizes, locations and power factor of DG units. Though the proposed methods can handle four different types of DG units, results of type 1-DG units (i.e., DG unit capable of delivering real power) and type 2-DG (i.e., DG unit capable of delivering real and reactive power) are presented.

Table 23.3. Data for a 16-bus test distribution system.

Line number	Sending bus no.	Receiving bus no.	Section resistance (p.u.)	Section reactance (p.u.)	Load at receiving end bus		End bus fixed capacitor (MVAr)
					P (MW)	Q (MVAr)	
11	1	4	0.075	0.1	2.0	1.6	
12	4	5	0.08	0.11	3.0	1.5	1.1
13	4	6	0.09	0.18	2.0	0.8	1.2
14	6	7	0.04	0.04	1.5	1.2	
16	2	8	0.11	0.11	4.0	2.7	
18	8	9	0.08	0.11	5.0	3.0	1.2
17	8	10	0.11	0.11	1.0	0.9	
19	9	11	0.11	0.11	0.6	0.1	0.6
20	9	12	0.08	0.11	4.5	2.0	3.7
22	3	13	0.11	0.11	1.0	0.9	
24	13	14	0.09	0.12	1.0	0.7	1.8
23	13	15	0.08	0.11	1.0	0.9	
25	15	16	0.04	0.04	2.1	1.0	1.8
*Tie lines							
15*	5	11	0.04	0.04			
21*	10	14	0.04	0.04			
26*	7	16	0.09	0.12			

Substation voltage = 10 kV, MVA base = 10 MVA

Followings are the assumptions and constraints for this simulation:

— The systems are considered at the peak load only.

— The lower and upper voltage thresholds are set at 0.90 pu and 1.05 pu, respectively.

— The maximum number of DG units considered in the study is three, with the maximum size equal to the total load plus loss.

23.5.2 *Simulation results of single-DG unit*

23.5.2.1 *Sizing allocation*

Figures 23.7 to 23.9 show optimal sizes and total losses with a DG unit one at a time at the respective location for 16-, 33- and 69-bus test systems, respectively. In all the figures, the lower half shows the optimal sizes obtained using the proposed method (IA) and exhaustive load flow (ELF) method. Notice that in all the cases, optimal sizes of the DG unit obtained from the IA method is closer to actual optimal size obtained from ELF. Similarly, total losses obtained from the method without

Table 23.4.   Data for a 33-bus test distribution system.

Line number	Sending bus no.	Receiving bus no.	Resistance (Ω)	Reactance (Ω)	Load at receiving end bus	
					P (kW)	Q (kVAr)
1	1	2	0.0922	0.0477	100.0	60.0
2	2	3	0.4930	0.2511	90.0	40.0
3	3	4	0.3660	0.1864	120.0	80.0
4	4	5	0.3811	0.1941	60.0	30.0
5	5	6	0.8190	0.7070	60.0	20.0
6	6	7	0.1872	0.6188	200.0	100.0
7	7	8	1.7114	1.2351	200.0	100.0
8	8	9	1.0300	0.7400	60.0	20.0
9	9	10	1.0400	0.7400	60.0	20.0
10	10	11	0.1966	0.0650	45.0	30.0
11	11	12	0.3744	0.1238	60.0	35.0
12	12	13	1.4680	1.1550	60.0	35.0
13	13	14	0.5416	0.7129	120.0	80.0
14	14	15	0.5910	0.5260	60.0	10.0
15	15	16	0.7463	0.5450	60.0	20.0
16	16	17	1.2890	1.7210	60.0	20.0
17	17	18	0.7320	0.5740	90.0	40.0
18	2	19	0.1640	0.1565	90.0	40.0
19	19	20	1.5042	1.3554	90.0	40.0
20	20	21	0.4095	0.4784	90.0	40.0
21	21	22	0.7089	0.9373	90.0	40.0
22	3	23	0.4512	0.3083	90.0	50.0
23	23	24	0.8980	0.7091	420.0	200.0
24	24	25	0.8960	0.7011	420.0	200.0
25	6	26	0.2030	0.1034	60.0	25.0
26	26	27	0.2842	0.1447	60.0	25.0
27	27	28	1.0590	0.9337	60.0	20.0
28	28	29	0.8042	0.7006	120.0	70.0
29	29	30	0.5075	0.2585	200.0	600.0
30	30	31	0.9744	0.9630	150.0	70.0
31	31	32	0.3105	0.3619	210.0	100.0
32	32	33	0.3410	0.5302	60.0	40.0

Substation voltage = 12.66 kV, MVA base = 10 MVA

updating α and β for the purpose of identifying best location and exhaustive methods give a comparable result. In addition, observe that when a DG unit with a large size is placed at any locations near the substation, the loss reduction of each system is quite low. In contrast, only a small size of DG unit is added at the other locations, a higher loss reduction can be achieved. As a result, finding a location at which the

Table 23.5. Data for a 69-bus test distribution system.

Line number	Sending bus no.	Receiving bus no.	Resistance (Ω)	Reactance (Ω)	Load at receiving end bus	
					P (kW)	Q (kVAr)
1	1	2	0.0005	0.0012	0.00	0.00
2	2	3	0.0005	0.0012	0.00	0.00
3	3	4	0.0015	0.0036	0.00	0.00
4	4	5	0.0251	0.0294	0.00	0.00
5	5	6	0.3660	0.1864	2.60	2.20
6	6	7	0.3811	0.1941	40.40	30.00
7	7	8	0.0922	0.0470	75.00	54.00
8	8	9	0.0493	0.0251	30.00	22.00
9	9	10	0.8190	0.2707	28.00	19.00
10	10	11	0.1872	0.0691	145.00	104.00
11	11	12	0.7114	0.2351	145.00	104.00
12	12	13	1.0300	0.3400	8.00	5.50
13	13	14	1.0440	0.3450	8.00	5.50
14	14	15	1.0580	0.3496	0.00	0.00
15	15	16	0.1966	0.0650	45.5	30.00
16	16	17	0.3744	0.1238	60.00	35.00
17	17	18	0.0047	0.0016	60.00	35.00
18	18	19	0.3276	0.1083	0.00	0.00
19	19	20	0.2106	0.0690	1.00	0.60
20	20	21	0.3416	0.1129	114.00	81.00
21	21	22	0.0140	0.0046	5.30	3.50
22	22	23	0.1591	0.0526	0.00	0.00
23	23	24	0.3463	0.1145	28.0	20.00
24	24	25	0.7488	0.2745	0.00	0.00
25	25	26	0.3089	0.1021	14.00	10.00
26	26	27	0.1732	0.0572	14.00	10.00
27	3	28	0.0044	0.0108	26.00	18.60
28	28	29	0.0640	0.1565	26.00	18.60
29	29	30	0.3978	0.1315	0.00	0.00
30	30	31	0.0702	0.0232	0.00	0.00
31	31	32	0.3510	0.1160	0.00	0.00
32	32	33	0.8390	0.2816	14.00	10.00
33	33	34	1.7080	0.5646	19.50	14.00
34	34	35	1.4740	0.4673	6.00	4.00
35	3	36	0.0044	0.0108	26.00	18.55
36	36	37	0.0640	0.1565	26.00	18.55
37	37	38	0.1053	0.1230	0.00	0.00
38	38	39	0.0304	0.0355	24.00	17.00
39	39	40	0.0018	0.0021	24.00	17.00

Table 23.5.   (*Continued*)

Line number	Sending bus no.	Receiving bus no.	Resistance (Ω)	Reactance (Ω)	Load at receiving end bus	
					P (kW)	Q (kVAr)
40	40	41	0.7283	0.8509	1.20	1.00
41	41	42	0.3100	0.3623	0.00	0.00
42	42	43	0.0410	0.0478	6.00	4.30
43	43	44	0.0092	0.0116	0.00	0.00
44	44	45	0.1089	0.1373	39.22	26.30
45	45	46	0.0009	0.0012	39.22	26.30
46	4	47	0.0034	0.0084	0.00	0.00
47	47	48	0.0851	0.2083	79.00	56.40
48	48	49	0.2898	0.7091	384.70	274.50
49	49	50	0.0822	0.2011	384.00	274.50
50	8	51	0.0928	0.0473	40.50	28.30
51	51	52	0.3319	0.1114	3.60	2.70
52	9	53	0.1740	0.0886	4.35	3.50
53	53	54	0.2030	0.1034	26.40	19.00
54	54	55	0.2842	0.1447	24.00	17.20
55	55	56	0.2813	0.1433	0.00	0.00
56	56	57	1.5900	0.5337	0.00	0.00
57	57	58	0.7837	0.2630	0.00	0.00
58	58	59	0.3042	0.1006	100.00	72.00
59	59	60	0.3861	0.1172	0.00	0.00
60	60	61	0.5075	0.2585	1244.00	888.00
61	61	62	0.0974	0.0496	32.00	23.00
62	62	63	0.1450	0.0738	0.00	0.00
63	63	64	0.7105	0.3619	227.00	162.00
64	64	65	1.0410	0.5302	59.00	42.00
65	11	66	0.2012	0.0611	18.00	13.00
66	66	67	0.0047	0.0014	18.00	13.00
67	12	68	0.7394	0.2444	28.00	20.00
68	68	69	0.0047	0.0016	28.00	20.00

Substation voltage = 12.80 kV, MVA base = 10 MVA

total loss at minimum is important. That can be implemented by the support of IA as presented below.

23.5.2.2   *Location selection*

Figures 23.7 to 23.9 also show (upper half) the total losses with the optimal DG unit at various buses, one at a time using both methodologies. The total losses by IA are

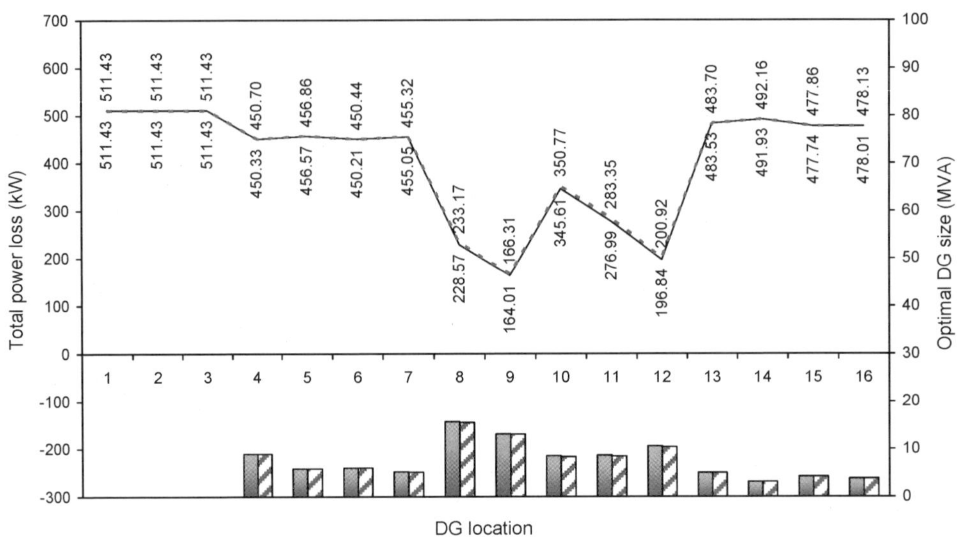

Fig. 23.7.   Optimal sizes of DG units at 0.98 lagging load power factor at various locations with respect to losses for a 16-bus system.

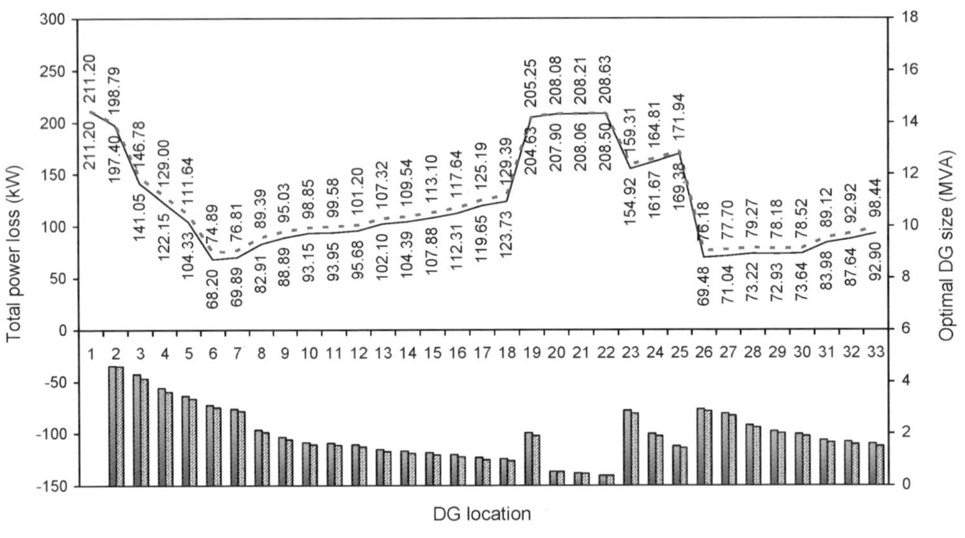

Fig. 23.8.   Optimal sizes of DG units at 0.85 lagging load power factor at various locations with respect to losses for a 33-bus system.

slightly higher than the exact losses by ELF. However, the trend of losses with IA and ELF are completely coinciding and for the purpose of identifying location, a fast method would be sufficient. In the other words, IA can find the optimal locations with less computational efforts.

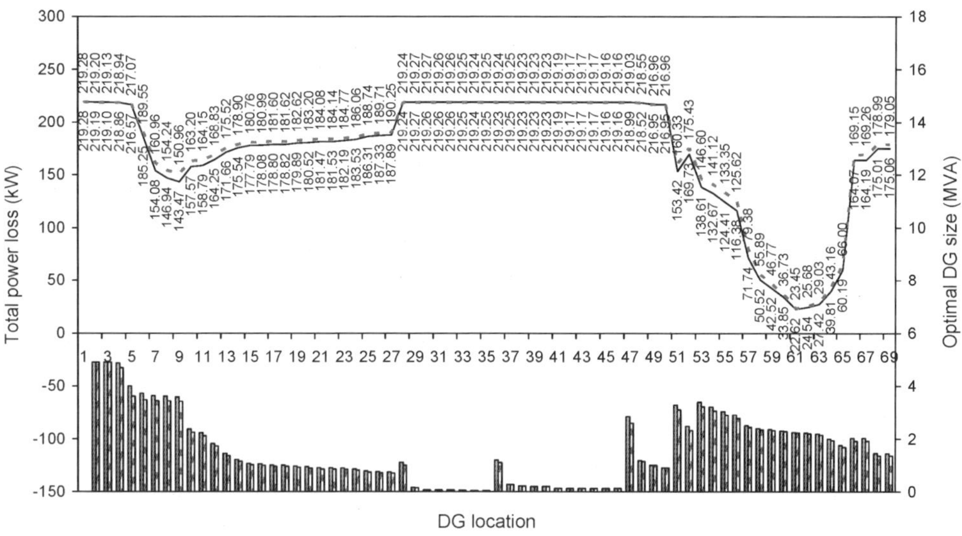

Fig. 23.9. Optimal sizes of DG units at 0.82 lagging load power factor at various locations with respect to losses for a 69-bus system.

In 16-, 33- and 69-bus test systems, the best locations for DG units for minimizing losses were found to be buses 9, 6 and 61, respectively.

23.5.2.3 *Power factor selection*

Figures 23.10 to 23.12 show the optimal sizes and locations of DG units at various power factors and total losses by IA for 16-, 33- and 69-bus test systems, respectively. The power factors of DG units varied from zero to 1.00, both in leading and lagging operation in small steps of 0.01. This study was carried out to see the optimal power factor of DG units that would give minimum losses.

In a 16-bus system, buses 8, 9 and 12 were found to be good candidate locations for DG unit placement for minimizing losses. These results are in agreement with the results obtained from a previous section "site selection". Notice at bus 9, the best location, power factor plays an important role in loss reduction. The range of power factors good for minimizing losses at bus 9 is 0.87 leading to 0.46 lagging with optimal sizes ranging from 7.46 to 13.12 MVA.

Though in reality the sizes will be fixed and the power factor can be allowed to vary to observe the impact of the DG unit power factor on loss reduction. If the power factor is fixed due to the limitation of the technology, it is important to select the appropriate size and location to achieve minimum possible reduction of losses. If the technology available for DG unit is to be type 1, i.e., around unity power factor the best location is bus 9. On the other hand, if DG unit is type 2, i.e., zero

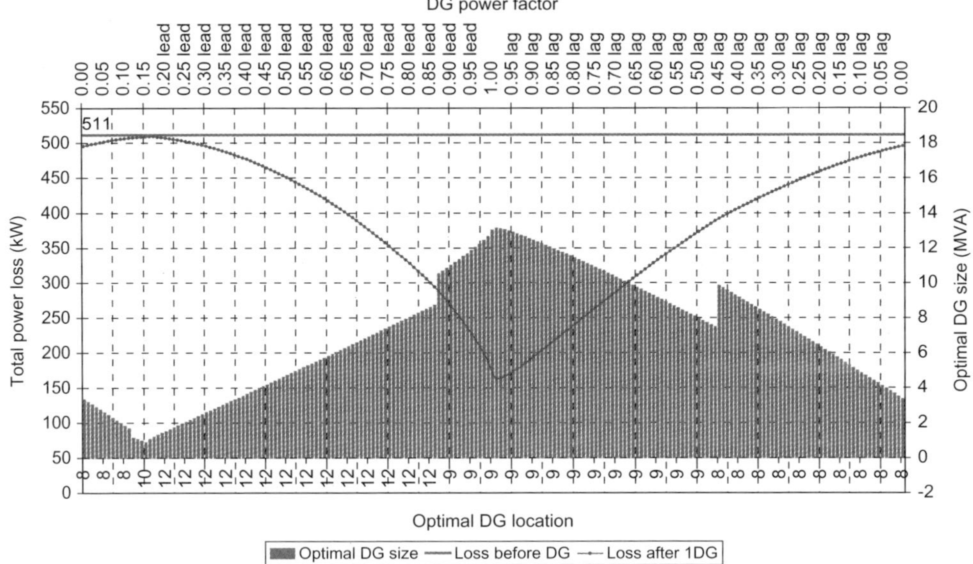

Fig. 23.10.   Plot of optimum sizes and locations of DG units at various power factors versus losses for a 16-bus test system.

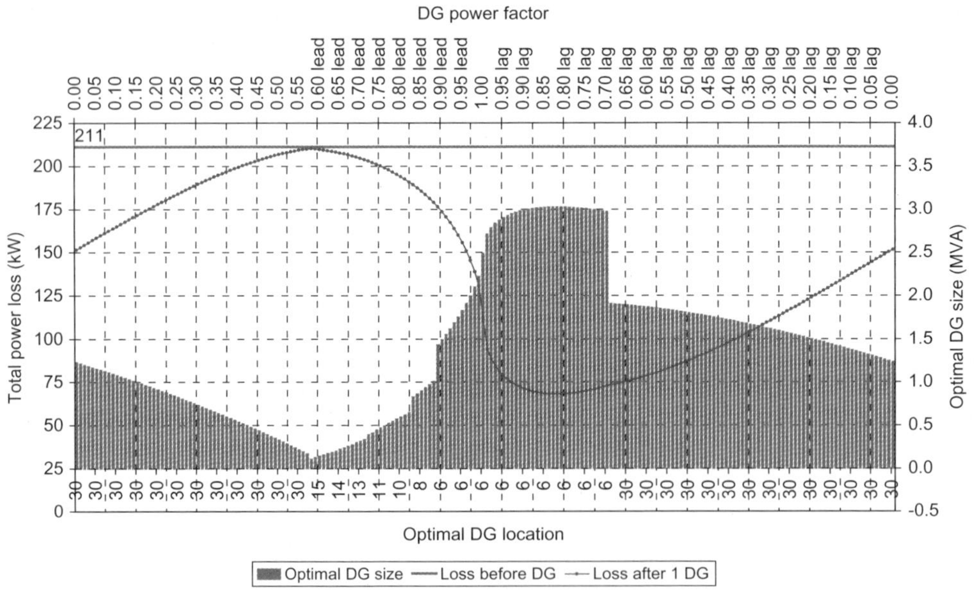

Fig. 23.11.   Plot of optimum sizes and locations of DG units at various power factors versus losses for a 33-bus test system.

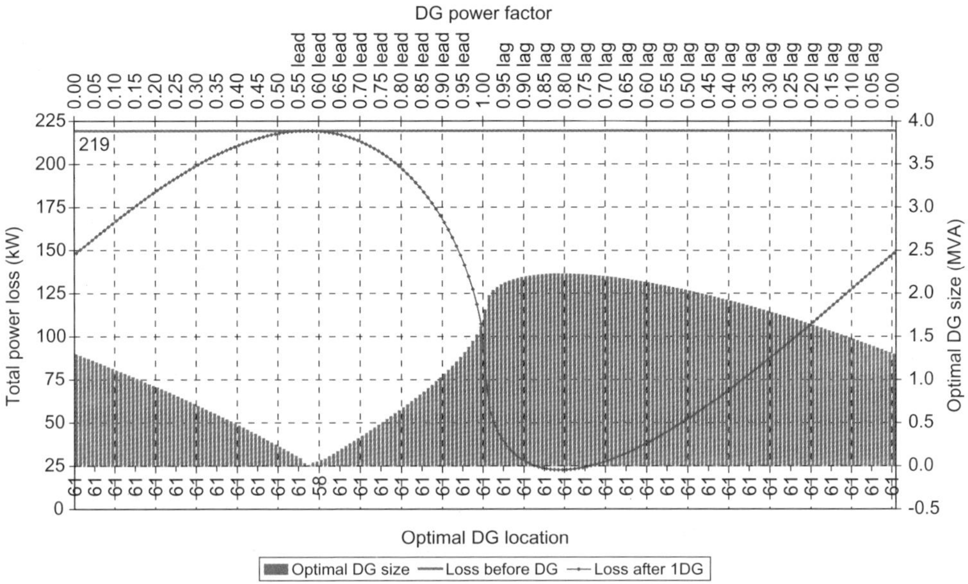

Fig. 23.12. Plot of optimum sizes and locations of DG units at various power factors versus losses for a 69-bus test system.

power factor then the best location is at bus 8, for reducing losses and so on. In this test system, the best power factor found to be closer to unity (i.e., 0.99) and corresponding size and losses are 13.12 MVA and 1642.58 kW, respectively. Unity power factor means this system needs only real power injections to reduce losses. However, this is not the case with other systems as presented below.

In a 33-bus test system, bus 6 was found to be the best location. This result too coincides with the result obtained from the previous section. The range of power factors for that location is 0.89 leading to 0.7 lagging, and corresponding optimal sizes are 1.43 to 3.03 MVA. The best power factor of the DG unit located at 6 is 0.82 lagging, with an optimal size of 3.02 MVA. This means that the system needs both real and reactive power injection to reduce losses. However, more real power injection is needed compared to reactive power. The total losses corresponding to the DG unit at the optimal location, optimal size and best power factor is 67.98 kW.

In a 69-bus test system, bus 61 is the obvious location for a range of power factors compared to other locations. The best power factor of the DG unit located at bus 61 is 0.82 lagging, with an optimal size of 2.2 MVA. In this case too, both real and reactive power injections are needed to minimize the losses. The total loss corresponding to these optimal settings is 22.64 kW.

Table 23.6 shows the optimal power factor of DG unit for different test systems using IA with repeated approach and combined load power factor of each system. It

Table 23.6. Power factors of DG units and combined loads.

Systems	Power factor	
	DG unit	Combined loads
16 bus	0.99 (lagging)	0.98
33 bus	0.82 (lagging)	0.85
69 bus	0.82 (lagging)	0.82

is interesting to see that the optimal power factor of DG unit for minimizing losses is in close agreement with the load power factor. This could be a guidance to select the appropriate technological option for minimizing losses in the system, given a choice. However, it is important to notice that optimal power factor depends on the location as well.

It should be noted that, in the DG unit case, leading power factor means that the DG unit is absorbing reactive power. On the other hand, lagging means the DG unit is supplying reactive power and this is just opposite to load power factor conventions.

23.5.2.4 *Summary*

Table 23.7 shows a summary and comparison of the simulation results for 16-, 33- and 69-bus test systems for two approaches for calculating the power factor of the DG unit, i.e., fast and repeated approaches. In the fast approach, the DG unit power factor is set at the combined load power factor. In the repeated approach the best power factor of the DG unit is calculated by checking all the power factors step by step (change in a small step of 0.01) near to the power factor of the combined load. The base case losses of all the test systems are too presented to show the loss reduction by DG unit.

The loss reduction by IA and ELF is almost the same. But ELF can demand excessively computational time since it runs so many load flow runs in calculation around the solution obtained for the IA method which needs only a couple of load flow runs. Hence, the IA method requires shorter computational time. This technique is appropriate and useful for large-scale distribution systems.

Selection of the power factor of the DG unit equal to that of the load is feasible, and it can lead to an optimal or near-optimal solution as shown in Table 23.8.

With the DG unit the total losses can be reduced significantly while satisfying all the power and voltage constraints. When only one DG unit is considered in terms of the optimal size, location and power factor simultaneously; the loss reduction for the three systems is achieved at the highest levels of 68.21%, 67.81% and 89.68%, respectively.

Table 23.7. Comparison of results of different techniques.

		Base	Fast approach				Repeated approach			
Systems	Techniques	Loss (kW)	Power factor	Bus	DG size (MVA)	Loss (kW)	Power factor	Bus	DG size (MVA)	Loss (kW)
16 bus	ELF	511.43	0.98	9	13.1551	164.01	0.99	9	13.1759	162.58
	IA	511.43	0.98	9	13.0877	164.02	0.99	9	13.1216	162.58
33 bus	ELF	211.20	0.85	6	3.1034	68.20	0.82	6	3.1070	67.90
	IA	211.20	0.85	6	3.0247	68.28	0.82	6	3.0284	67.98
69 bus	ELF	219.28	0.82	61	2.2429	22.62	0.82	61	2.2429	22.62
	IA	219.28	0.82	61	2.2219	22.64	0.82	61	2.2219	22.64

Table 23.8. Voltages before and after DG unit.

	Voltage @ bus before DG		Voltage @ bus after DG	
Systems	Min	Max	Min	Max
16 bus	0.9693 @12	1.0000 @1	0.9849 @7	1.0008 @9
33 bus	0.9037 @18	1.0000 @1	0.9570 @18	1.0002 @6
69 bus	0.9113 @65	1.0000 @1	0.9730 @27	1.0000 @1

Table 23.8 shows the minimum and maximum voltages before and after the DG unit by IA for the three test systems. After the DG unit, voltage profiles improve significantly. The voltages at various buses are maintained within the acceptable constraints.

23.5.3 *Simulation results of multiple-DG units*

23.5.3.1 *Type 1 DG placement*

23.5.3.1.1 16-bus test system

Table 23.9 presents the simulation results of placing DG units by various techniques. The results of the base case and three cases with the DG unit numbers ranging from one to three are compared. They include the optimal size and location of DG units with respect to the total losses by each technique. The loss reduction, computational time and schedule of the installed DG units of each technique are also presented.

For all the cases, IA leads to a completely optimal solution as corroborated with ELF; particularly the loss reduction, optimal location and size of the DG units by IA is the same as those by ELF. Among all the cases, LSF yields the lowest loss

Table 23.9. DG placement by various techniques for a 16-bus test system.

Cases	Techniques	Installed DG schedule				DG (MW)	Ploss (kW)	Loss reduction (%)	Time (s)
No DG						—	511.43	0.00	0.01
1 DG	LSF	Bus	12						
		Size	10.40			10.40	193.60	62.15	0.12
	IA	Bus	9						
		Size	13.10			13.10	168.49	67.06	0.16
	ELF	Bus	9						
		Size	13.10			13.10	168.49	67.06	0.24
2 DGs	LSF	Bus	12	7					
		Size	10.40	5.20		15.60	142.12	72.21	0.20
	IA	Bus	9	6					
		Size	13.10	5.24		18.34	112.29	78.04	0.28
	ELF	Bus	9	6					
		Size	13.10	5.24		18.34	112.29	78.04	0.38
3 DGs	LSF	Bus	12	7	16				
		Size	10.40	5.20	3.90	19.50	106.82	79.11	0.27
	IA	Bus	9	6	16				
		Size	13.10	5.24	3.93	22.27	76.99	84.95	0.38
	ELF	Bus	9	6	16				
		Size	13.10	5.24	3.93	22.27	76.99	84.95	0.54

reduction. For instance, placing a single DG unit by IA, ELF and LSF is 67.06%, 67.06% and 62.15%, respectively.

IA demands shorter computational time, as compared to ELF method as expected. However, the LSF method is the quickest among all the methods.

23.5.3.1.2 33-bus test system

Similar to the 16-bus system, Table 23.10 presents the results of the optimal size and location of DG units for a 33-bus test system by various techniques.

For a single DG unit, the loss reduction by IA, at 47.39% is the same as that by ELF. Among all the cases, LSF produces a loss reduction of only 30.48%. For two DG units, loss reduction by IA, at 56.61% is lower than that by ELF, at 58.51%. By contrast, it is higher than the loss reduction by LSF at only 52.32%. For three DG units, IA achieves a loss reduction of 61.62%, compared with ELF at 64.83%. However, it is better than LSF that yields a loss reduction of 59.72%. IA needs a

Table 23.10.   DG placement by various techniques for a 33-bus test system.

Cases	Techniques	Installed DG schedule				DG (MW)	Ploss (kW)	Loss reduction (%)	Time (s)
No DG						—	211.20	0.00	0.02
1 DG	LSF	Bus	18						
		Size	743			743	146.82	30.48	0.11
	IA	Bus	6						
		Size	2601			2601	111.10	47.39	0.16
	ELF	Bus	6						
		Size	2601			2601	111.10	47.39	1.06
2 DGs	LSF	Bus	18	33					
		Size	720	900		1620	100.69	52.32	0.18
	IA	Bus	6	14					
		Size	1800	720		2520	91.63	56.61	0.27
	ELF	Bus	12	30					
		Size	1020	1020		2040	87.63	58.51	2.03
3 DGs	LSF	Bus	18	33	25				
		Size	720	810	900	2430	85.07	59.72	0.23
	IA	Bus	6	12	31				
		Size	900	900	720	2520	81.05	61.62	0.40
	ELF	Bus	13	30	24				
		Size	900	900	900	2700	74.27	64.83	3.06

short computational time. Particularly, for three DG units, the time by IA is 0.40 seconds, nearly twice longer than that by LSF, at 0.23 seconds. By contrast, it is approximately eight times shorter than the time by ELF, at 3.06 seconds.

23.5.3.1.3   69-bus test system

Table 23.11 presents the results of the optimal size and location of DG units by various techniques for a 69-bus test system.

For all the cases, IA leads to an optimal solution as compared with ELF; particularly the results by IA are the same as those by ELF in terms of the loss reduction, optimal location and size of DG units. However, it is better than ELF in terms of computational time. Particularly, for the three DG units case, the time by IA is 0.71 seconds, nearly 33 times shorter than that by ELF at 23.16 seconds. It is longer than the time by LSF at 0.52 seconds. In addition, IA achieves better loss reduction than LSF. For instance, for two DG units, IA reaches a loss reduction of 67.94%;

Table 23.11. DG placement by various techniques for a 69-bus test system.

Cases	Techniques	Installed DG schedule				DG (MW)	Ploss (kW)	Loss reduction (%)	Time (s)
No DG						—	219.28	0.00	0.03
1 DG	LSF	Bus	65						
		Size	1520			1520	109.77	49.94	0.15
	IA	Bus	61						
		Size	1900			1900	81.33	62.91	0.28
	ELF	Bus	61						
		Size	1900			1900	81.33	62.91	7.75
2 DGs	LSF	Bus	65	27					
		Size	1440	540		1980	98.74	54.97	0.30
	IA	Bus	61	17					
		Size	1700	510		2210	70.30	67.94	0.52
	ELF	Bus	61	17					
		Size	1700	510		2210	70.30	67.94	15.53
3 DGs	LSF	Bus	65	27	61				
		Size	1360	510	510	2380	90.84	58.57	0.52
	IA	Bus	61	17	11				
		Size	1700	510	340	2550	68.38	68.82	0.71
	ELF	Bus	61	17	11				
		Size	1700	510	340	2550	68.38	68.82	23.16

while LSF obtains that of 54.97%. The differences in loss between the LSF and IA methods are due to different locations and optimal sizes at those locations.

23.5.3.2 *Type 2 DG placement*

23.5.3.2.1 16-bus test system

Table 23.12 shows the results of the optimal size, location and power factor of DG units by IA for this system. The results of the base case and three cases with DG units at the optimal and combined load power factors are compared. The power factor of the combined load is 0.98 lagging. The optimal power factor of DG units is identified at 0.99 lagging. All the cases, the results of loss reduction at the optimal power factor are slightly higher compared to those at the combined load factor power. As a result, selection of the power factor of DG units can be based on combined load power factor.

Table 23.12. DG placement at optimal and combined load power factors for a 16-bus system.

Cases	DG power factors @	Installed DG schedule				DG (MVA)	Ploss (kW)	Loss reduction (%)
No DG						—	511.43	0.00
1 DG	Combined Load P.F. = 0.98 lagging	Bus Size	9 13.09			13.09	164.02	67.93
	Optimal P.F. = 0.99 lagging	Bus Size	9 13.12			13.12	162.58	68.21
2 DGs	Combined Load P.F. = 0.98 lagging	Bus Size	9 13.27	6 5.97		19.23	102.82	79.90
	Optimal P.F = 0.99 lagging	Bus Size	9 13.13	6 5.91		19.04	102.22	80.01
3 DGs	Combined Load P.F. = 0.98 lagging	Bus Size	9 13.27	6 5.97	15 3.98	23.21	69.20	86.47
	Optimal P.F. = 0.99 lagging	Bus Size	9 13.13	6 5.91	15 3.94	22.98	68.00	86.70

Among the cases, three DG units' case at the optimal power factor yields a maximum loss reduction of 86.70%, while one DG unit at this factor obtains a minimum loss reduction of only 68.21%. As the number of DG units increase, the loss reduction becomes more effective.

These results are obtained with the help of the proposed method and verified by exhaustive load flow solutions.

23.5.3.2.2 33-bus test system

Table 23.13 shows the simulation results of the optimal size, location and power factor of DG units by IA for a 33-bus test system. The power factor of the combined load is 0.85 lagging. The optimal power factor of DG units is identified at 0.82 lagging. All the cases, the results of loss reduction at the optimal power factor are slightly higher as compared with those at the combined load factor power. Therefore,

Table 23.13. DG placement at optimal and combined load power factors for a 33-bus system.

Cases	DG power factors @	Installed DG schedule				DG (MVA)	Ploss (kW)	Loss reduction (%)
No DG						—	211.20	0.00
1 DG	Combined Load P.F. = 0.85 lagging	Bus Size	6 3103			3103	68.20	67.71
	Optimal P.F. = 0.82 lagging	Bus Size	6 3107			3107	67.90	67.85
2 DGs	Combined Load P.F. = 0.85 lagging	Bus Size	6 2118	30 1059		3176	44.84	78.77
	Optimal P.F. = 0.82 lagging	Bus Size	6 2195	30 1098		3293	44.39	78.98
3 DGs	Combined Load P.F. = 0.85 lagging	Bus Size	6 1059	30 1059	14 741	2859	23.05	89.09
	Optimal P.F. = 0.82 lagging	Bus Size	6 1098	30 1098	14 768	2963	22.29	89.45

selection of the power factor of DG units equal to that of the combined load is feasible for this case.

Similar to a 16-bus test system, three DG units at the optimal power factor produces a maximum loss reduction of 89.45%; while one DG unit at this factor obtains a minimum loss reduction of only 67.83%. The more the number of DG units is placed, the better the loss reduction can be achieved.

23.5.3.2.3   69-bus test system

Table 23.14 shows the results of the optimal size, location and power factor of DG units by IA for a 69-bus test system. The results of the base case and three cases with DG units at the optimal power factor are compared. The optimal power factor of DG units is determined to be equal to the power factor of the combined load at 0.82 lagging. As a result, selection of the power factor of DG units equal to that of the combined load can lead to an optimal solution for this system.

Table 23.14.   DG placement at optimal power factor for a 69-bus system.

Cases	DG power factors @	Installed DG schedule				DG (MVA)	Ploss (kW)	Loss reduction (%)
No DG						—	219.28	0.00
1 DG	Optimal P.F. = 0.82 lagging	Bus Size	61 2243			2243	22.62	89.68
2 DGs	Optimal P.F. = 0.82 lagging	Bus Size	61 2195	17 659		2854	7.25	96.69
3 DGs	Optimal P.F. = 0.82 lagging	Bus Size	61 2073	17 622	50 829	3524	4.95	97.74

Similar to a 33-bus test system, three DG units at the optimal power factor results in a maximum loss reduction of 97.74%. By contrast, one DG unit at that factor yields a minimum loss reduction of only 89.68%. The number of DG units becomes larger, the loss reduction increases.

23.5.3.3   *Results of voltages*

Tables 23.15 to 23.17 indicate the minimum and maximum voltages for all the cases of 16-, 33- and 69-bus test systems, respectively. In all the cases, after DG units are

Table 23.15.   Voltages of cases for a 16-bus test system.

DG power factors @	Cases	Min voltage @ bus	Max voltage @ bus
	No DG	0.9693 @ 12	1.0000 @ 1
Unity	1 DG	0.9849 @ 7	1.0000 @ 1
	2 DGs	0.9913 @ 16	1.0000 @ 1
	3 DGs	0.9918 @ 5	1.0004 @ 16
Combined Load P.F. = 0.98 lagging	1 DG	0.9849 @ 7	1.0011 @ 9
	2 DGs	0.9913 @ 16	1.0026 @ 9
	3 DGs	0.9941 @ 5	1.0026 @ 9
Optimal P.F. = 0.99 lagging	1 DG	0.9849 @ 7	1.0025 @ 9
	2 DGs	0.9913 @ 7	1.0009 @ 9
	3 DGs	0.9937 @ 5	1.0009 @ 9

Table 23.16.   Voltages of cases for a 33-bus test system.

DG power factors @	Cases	Min voltage @ bus	Max voltage @ bus
	No DG	0.9037 @ 18	1.0000 @ 1
Unity	1 DG	0.9425 @ 18	1.0000 @ 1
	2 DGs	0.9539 @ 33	1.0000 @ 1
	3 DGs	0.9690 @ 18	1.0000 @ 1
Combined Load P.F. = 0.98 lag.	1 DG	0.9575 @ 18	1.0007 @ 6
	2 DGs	0.9619 @ 18	1.0049 @ 6
	3 DGs	0.9824 @ 25	1.0038 @ 14
Optimal P.F. = 0.99 lagging	1 DG	0.9575 @ 18	1.0007 @ 6
	2 DGs	0.9600 @ 18	1.0031 @ 6
	3 DGs	0.9821 @ 25	1.0006 @ 14

Table 23.17.   Voltages of cases for a 69-bus test system.

DG power factors @	Cases	Min voltage @ bus	Max voltage @ bus
	No DG	0.9113 @ 65	1.0000 @ 1
Unity	1 DG	0.9692 @ 27	1.0000 @ 1
	2 DGs	0.9765 @ 65	1.0000 @ 1
	3 DGs	0.9785 @ 65	1.0000 @ 1
Optimal P.F. = 0.99 lagging	1 DG	0.9732 @ 27	1.0000 @ 1
	2 DGs	0.9944 @ 50	1.0024 @ 61
	3 DGs	0.9939 @ 69	1.0000 @ 1

added, the total losses can reduce significantly while satisfying all the power and voltage constraints. This was checked with exhaustive power flow. It is interesting to note that the voltage profile improves with the number of DG units installed in the system.

23.6   Conclusions

This chapter has discussed the application of DG units for loss reduction while fulfilling the main objective of energy injection. Analytical expressions for finding the optimal size of four different types of DG units for minimizing losses in large scale primary distribution systems have been presented. A technique to obtain the optimal power factor is also presented for a DG unit capable of delivering real and reactive power. For DG unit placement, three methods, namely loss sensitivity factor (LSF), improved analytical (IA) and exhaustive load flow (ELF) methods are introduced.

Validity of the proposed analytical expressions for finding the optimal size are tested and verified on three test distribution systems with varying sizes and complexity using exhaustive load flow solutions. The results show that locations, sizes and operating power factor of DG units are crucial factors in loss reduction. If properly placed, appropriately sized and operated DG units can reduce losses significantly.

The number of DG units, their location and sizes play an important role in loss reduction. The number of DG units with optimal sizes and appropriate location reduce the losses to a considerable amount. Given the choice, DG units should be allocated to enjoy other benefits as well, such as loss reduction. Among different types of DG units, a DG unit capable of delivering both real and reactive power reduce losses more than that of a DG unit capable of delivering real power only in single DG or multiple DG units cases. For a DG unit capable of delivering real and reactive power, its power factor too plays a crucial role in loss reduction and optimal loss.

In all the test systems used in this work, the operating power factor of DG units for minimizing losses was found to be closer to the power factor of the combined load of the respective system. This could be a good guidance for operating DG units that have the capability to deliver both real and reactive power for minimizing losses.

References

1. J. Federico, V. Gonzalez and C. Lyra, "Learning classifiers shape reactive power to decrease losses in power distribution networks," *IEEE Power Engineering Society General Meeting* **1** (2005) 557–562.
2. B. Radha, R.T.F. Ah King and H.C.S. Rughooputh, "Optimal network reconfiguration of electrical distribution systems," *IEEE Int. Conf.* **1** (2003) 66–71.
3. I. El-Samahy and E. El-Saadany, "The effect of DG on power quality in a deregulated environment," *IEEE Power Engineering Society General Meeting* **3** (2005) 2969–2976.
4. T. Ackermann, G. Andersson and L. Soder, "Distributed generation: A definition," *Electric Power Systems Research* **57** (2001) 195–204.
5. N. Acharya, P. Mahat and N. Mithulananthan, "An analytical approach for DG allocation in primary distribution network," *Int. J. Electrical Power & Energy Systems* **28** (2006) 669–678.
6. H.L. Willis and W.G. Scott, *Distributed Power Generation, Planning and Evaluation* (Marcel Dekker, Inc., New York, 2000).
7. J.L. Koepfinger, B.L. Beckwith and M.C. Lindsey, "IEEE guide for interfacing dispersed storage and generation facilities with electric utility systems (IEEE Std 1001–1988)," The Institute of Electrical and Electronics Engineers, Inc. (1989).
8. P. Fraser and S. Morita, *Distributed Generation in Liberalized Electricity Markets* (OECD/IEA, Paris, 2002), http://www.iea.org.

9. O. Paish, "Small hydro power: Technology and current status," *Renewable and Sustainable Energy Reviews* **6** (2002) 537–556.

10. N. Jenkins, R. Allan, P. Crossley, D. Kirschen and G. Strbac, *Embedded Generation* (London: The Institution of Electrical Engineers, 2000).

11. H.B. Puttgen, P.R. MacGregor and F.C. Lambert, "Distributed generation: Semantic hype or the dawn of a new era?" *IEEE Power and Energy Magazine* **1** (2003) 22–29.

12. M.E. Baran and F.F. Wu, "Network reconfiguration in distribution systems for loss reduction and load balancing," *IEEE Trans. Power Delivery* **4** (1989) 1401–1407.

13. H.D. Chiang and R.J. Jumeau, "Optimal network reconfigurations in distribution systems: Part 1: A new formulation and a solution methodology," *IEEE Trans. Power Delivery* **5** (1990) 1902–1909.

14. M.A. Kashem and V. Ganapathy, "Three-phase load balancing in distribution system using index measurement technique," *Electrical Power and Energy Systems* **24** (2002) 31–40.

15. A. Ukil and W. Siti, "Feeder load balancing using fuzzy logic and combinational optimization-based implementation," *Electric Power Systems Research* **78** (2008) 1922–1932.

16. S. Civanlar, J.J. Grainger, H. Yin and S.S.H. Lee, "Distribution feeder reconfiguration for loss reduction," *IEEE Trans. Power Delivery* **3** (1988) 1217–1223.

17. J.Y. Fan, L. Zhang and J.D. McDonald, "Distribution network reconfiguration: Single loop optimization," *IEEE Trans. Power Systems* **11** (1996) 1643–1647.

18. K. Nara, A. Shiose, M. Kitagawa and T. Ishihara, "Implementation of genetic algorithm for distribution system loss minimum re-configuration," *IEEE Trans. Power Systems* **8** (1992) 1044–1051.

19. D. Jiang and R. Baldick, "Optimal electric distribution system switch reconfiguration and capacitor control," *IEEE Trans. Power Systems* **11** (1996) 890–897.

20. Y.J. Jean and J.C. Kim, "An efficient simulated annealing algorithm for network reconfiguration in large-scale distribution systems," *IEEE Trans. Power Delivery* **17** (2002) 1070–1078.

21. H. Mori and Y. Ogita, "A parallel tabu search based method for reconfigurations of distribution systems," *IEEE Power Engineering Society Summer Meeting* **1** (2000) 73–78.

22. D. Zhang, Z. Fu and L. Zhang, "An improved TS algorithm for loss-minimum reconfiguration in large-scale distribution systems," *Electric Power Systems Research* **77** (2007) 685–694.

23. Y.-J. Jeon and J.-C. Kim, "Application of simulated annealing and tabu search for loss minimization in distribution systems," *Electrical Power Energy Systems* **26** (2004) 9–18.

24. C.-T. Su, C.-F. Chang and J.-P. Chiou, "Distribution network reconfiguration for loss reduction by ant colony search algorithm," *Electric Power Systems Research* **75** (2005) 190–199.

25. C.-R. Wang and Y.-E. Zhang, "Distribution network reconfiguration based on modified particle swarm optimization algorithm," *IEEE Int. Conf., Machine Learning and Cybernetics* (2006), pp. 2076–2080.

26. H.N. Ng, M.M.A. Salama and A.Y. Chikhani, "Classification of capacitor allocation techniques," *IEEE Trans. Power Delivery* **15** (2000) 387–392.

27. H. Duran, "Optimum number, location, and size of shunt capacitors in radial distribution feeders, A dynamic programming approach," *IEEE Trans. Power Apparatus Systems* **87** (1962) 1769–1774.

28. T.S. Abdel-Salam, A.Y. Chikhani and R. Hackam, "A new technique for loss reduction using compensating capacitors applied to distribution systems with varying load condition," *IEEE Trans. Power Delivery* **9** (1994) 819–827.

29. M. Chis, M.M.A. Salama and S. Jayaram, "Capacitor placement in distribution systems using heuristic search strategies," *IEE Proc. Generation, Transmission and Distribution* **144** (1997) 225–230.

30. S. Sundhararajan and A. Pahwa, "Optimal selection of capacitors for radial distribution systems using a genetic algorithm," *IEEE Trans. Power Systems* **9** (1994) 1499–1507.

31. M.M.A. Salama and A.Y. Chikhani, "An expert system for reactive power control of a distribution systems, Part 1: System configuration," *IEEE Trans. Power Delivery* **7** (1992) 940–945.

32. J.R.P.R. Laframboise, G. Ferland, A.Y. Chikhani and M.M.A. Salama, "An expert system for reactive power control of a distribution system, Part 2: System implementation," *IEEE Trans. Power Systems* **10** (1995) 1433–1441.

33. N.I. Santoso and O.T. Tan, "Neural-net based real-time control of capacitors installed on distribution systems," *IEEE Trans. Power Delivery* **5** (1990) 266–272.

34. H.N. Ng, M.M.A. Salama and A.Y. Chikhani, "Capacitor allocation by approximate reasoning: Fuzzy capacitor placement," *IEEE Trans. Power Delivery* **15** (2000) 393–398.

35. M.M.F. AlHajri, M.R. AlRashidi and M.E. El-Hawary, "A novel discrete particle swarm optimization algorithm for optimal capacitor placement and sizing," *IEEE* (2007), pp. 1286–1289.

36. H.L. Willis, "Analytical methods and rules of thumb for modeling DG-distribution interaction," *IEEE Power Engineering Society Summer Meeting* **3** (2000) 1643–1644.

37. K. Nara, Y. Hayashi, K. Ikeda and T. Ashizawa, "Application of tabu search to optimal placement of distributed generators," *Proc. IEEE Power Engineering Society Winter Meeting* (2001), pp. 918–923.

38. A. Silvestri, A. Berizzi and S. Buonanno, "Distributed generation planning using genetic algorithms," *Proc. Int. Conf. Electric Power Engineering*, Budapest (1999), p. 257.

39. K.-H. Kim, Y.-J. Lee, S.-B. Rhee, S.-K. Lee and S.-K. You, "Dispersed generator placement using fuzzy-GA in distribution systems," *Proc. 2002 IEEE Power Engineering Society Summer Meeting* **3** (2002) 1148–1153.

40. D.Q. Hung, N. Mithulananthan and R.C. Bansal, "Analytical expressions for DG allocation in primary distribution networks," *IEEE Trans. Energy Conversion*, February 2010.

41. D.Q. Hung, N. Mithulananthan and R.C. Bansal, "Multiple-DG placement in primary distribution networks for loss reduction using analytical approach," *IEEE Trans. Power Delivery*, March 2010.

42. D.P. Kothari and J.S. Dhillon, *Power System Optimization* (Prentice-Hall, New Delhi, 2006).

43. J.L. Bala, P.A. Kuntz and M.N. Pebles, "Optimum capacitor allocation using a distribution-analyzer-recorder," *IEEE Trans. PWRD* **12** (1997) 464–469.

44. M.A. Kashem, V. Ganapathy, G.B. Jasmon and M.I. Buhari, "A novel method for loss minimization in distribution networks," *Int. Conf. Electric Utility Deregulation and Restructuring and Power Technologies, Proc.*, April 2000, pp. 251–256.

45. M.E. Baran and F.F. Wu, "Optimum sizing of capacitor placed on radial distribution systems," *IEEE Trans. Power Delivery* **4** (1989) 735–743.

Chapter 24

Renewable-Based Generation Integration in Electricity Markets with Virtual Power Producers

Zita A. Vale[*], Hugo Morais[†] and Hussein Khodr[‡]

*GECAD — Knowledge Engineering and Decision-Support,
Research Center Engineering,*

*Institute of Porto — Polytechnic Institute of Porto (ISEP/IPP),
Rua Dr. António Bernardino de Almeida 431,
4200-4072 Porto, Portugal*
[*]*zav@isep.ipp.pt*
[†]*hgrm@isep.ipp.pt*
[‡]*hmk@isep.ipp.pt*

The development of renewable energy sources and Distributed Generation (DG) of electricity is of main importance in the way towards a sustainable development. However, the management, in large scale, of these technologies is complicated because of the intermittency of primary resources (wind, sunshine, etc.) and small scale of some plants.

The aggregation of DG plants gives place to a new concept: the Virtual Power Producer (VPP). VPPs can reinforce the importance of these generation technologies making them valuable in electricity markets.

VPPs can ensure a secure, environmentally friendly generation and optimal management of heat, electricity and cold as well as optimal operation and maintenance of electrical equipment, including the sale of electricity in the energy market. For attaining these goals, there are important issues to deal with, such as reserve management strategies, strategies for bids formulation, the producers' remuneration, and the producers' characterization for coalition formation.

This chapter presents the most important concepts related with renewable-based generation integration in electricity markets, using VPP paradigm. The presented case studies make use of two main computer applications: ViProd and MASCEM. ViProd simulates VPP operation, including the management of plants in operation. MASCEM is a multi-agent based electricity market simulator that supports the inclusion of VPPs in the players set.

24.1   Introduction

The physical impact of wind power in the Electric Power System has been widely researched and many studies have been published on this subject. In this introduction only a small selection is considered, among which Ref. 1 presents a comparative analysis of the impact of several wind generator technologies and electrical grid configurations. The advantage of variable speed turbines is demonstrated, as they permit voltage control.

In Ref. 2, the positive effect of a wind farm on the voltage level of a weak grid is shown. The importance of the capability of reactive power generation, which enables the wind farm helping to maintain the voltage level, is highlighted. On the other side, in Ref. 3, negative effects of wind power are analyzed, such as:

- Steady state voltage level;
- Voltage fluctuations;
- Wave form quality.

To compensate these impacts, the following solutions are discussed:

- Installation of new power lines;
- Reactive power control;
- Load management;
- Energy dissipation (e.g., water heating);
- Energy storage (water pumping, batteries, fly wheels).

While the installation of new power lines is too costly, the other solutions are considered viable, but always depending on the local conditions. It is widely accepted that high penetration levels (far beyond 10% of the short circuit power at the point of common coupling) only can be achieved with additional technical measures.[1] These measures can refer to the wind turbine itself or to the surrounding power system. It is commonly stated that power systems must have an elevated flexibility. This means more regulation reserve which can be achieved by relatively small gas fired power stations and new energy storage systems such as advanced batteries, hydrogen (storage and fuel cell), double layer capacitors ("Ultra-Caps") and others. A method to quantify the need of reserve in case of an elevated wind power penetration is presented in Ref. 4. Anyway, if the power system cannot absorb the generated energy, wind curtailment is required giving place to large amounts of wind energy loss, which is especially common in island grids, for example in Greece.[5]

Renewable electricity generation demands additional flexibility of the power system, which creates a great challenge for the established centralized system control and points to distributed generation and control. In Refs. 6 and 7 the application of wide area measurement systems (WAMS) to distributed generation control is shown. The basic idea of this approach is the creation of a control hierarchy having several layers. Autonomous agents take local decisions and only limited information is exchanged which is relevant for higher control layers. As a result, the complexity in each layer is limited and a stable and very flexible system control is achieved. Synchronized measurements are very important for this method, which are available from so-called phasor measurement units (PMU).

To get a rapid overview of the situation of wind power balancing in Europe, the "European Balancing Act" published in 2007 by Thomas Ackermann *et al.*[8] can be recommended. In this work, issues like increased reserve requirements, balancing cost and fault-ride-through (FRT) capability are treated. The cases of Spain, Ireland and Germany are discussed in more detail. In addition, results of studies from Finland, Sweden and the UK are presented. The authors concluded that wind power integration has an economical and a technical dimension. The economical dimension is mainly related to the allocation of additional reserves which may increase costs for power system operation.

To reduce these costs, three aspects are identified: (a) aggregating wind generation over large geographical areas, (b) larger balancing areas and (c) more flexible power system operation, with a reduced closure delay. Referring to the technical dimension, FRT and frequency control are mentioned as essential for wind penetration levels beyond 15% (of total energy consumption). First, results from Horns Rev offshore wind farm were also presented in Ref. 8, highlighting the integration of the main controller of the wind farm into the central control of the Danish power system as a model for the future.

One of the first places worldwide, where wind power reached very high penetration is Denmark. As described in Ref. 9, already in 2003, wind power covered occasionally almost 100% of total power consumption in western Denmark. Nowadays it is common that wind power exceeds sometimes the regional consumption. As stated in Ref. 10, in 2001 installed wind power exceeded off peak load level and wind power production accounted for 16% of the total demand. In 2007 this figure grew to almost 20% and with the planned offshore wind farms it is estimated to reach 50% by the year 2025.[8]

For this reason, many publications concerning wind power are based on data from Denmark. This country belongs to the NordPool power exchange which is geographically bound to Norway, Sweden, Finland and Denmark. A good introduction to the NordPool market is given in Ref. 9 and it is described in more detail in Ref. 11.

Two energy markets are of special importance for wind power: the spot market and the regulating (or balancing) market. The spot market is the central energy market, and the regulating market only comes into force if the bids on the spot market are not fulfilled.

The spot price or marginal price is determined at the daily spot market by supply and demand. On the NordPool market, for example, producers and customers give their bids to the market 12 to 36 hours in advance, providing price and quantity of electricity to buy or sell. At the power exchange (e.g., Elspot in Scandinavia, MIBEL in Spain and Portugal, EEX in Germany) marginal prices (i.e., prices that result from market clearance) are fixed for each hour.

In this context, VPPs may aggregate several technologies in distinct places. In the scope of a VPP, producers can make sure their generators are optimally operated and that the power that is not consumed in their installations has good chances to be sold on the market. At the same time, VPPs will be able to commit to a more robust generation profile, raising the value of non-dispatchable generation technologies.

In the scheme of Fig. 24.1 (adapted from Ref. 12), the electricity physical flow is represented by the full thick lines and is assured by the GENCOs (Generating Companies), the TRANSCOs (Transmission Companies) and the DISTCOs (Distribution Companies). Electric energy property is represented by the dashed lines. In this case GENCOs sell the electricity to the Trading floors that, later, sell it either to brokers, directly (bilateral contracts), or in the energy markets. Brokers sell this energy to the consumers directly or through the aggregators' mediation. The financial flows are represented by the thin full lines and are much more complex. Beyond the inherent sale costs of the electricity, there are also costs associated with the payment of services: These can include the use of transmission (TRANSCO)

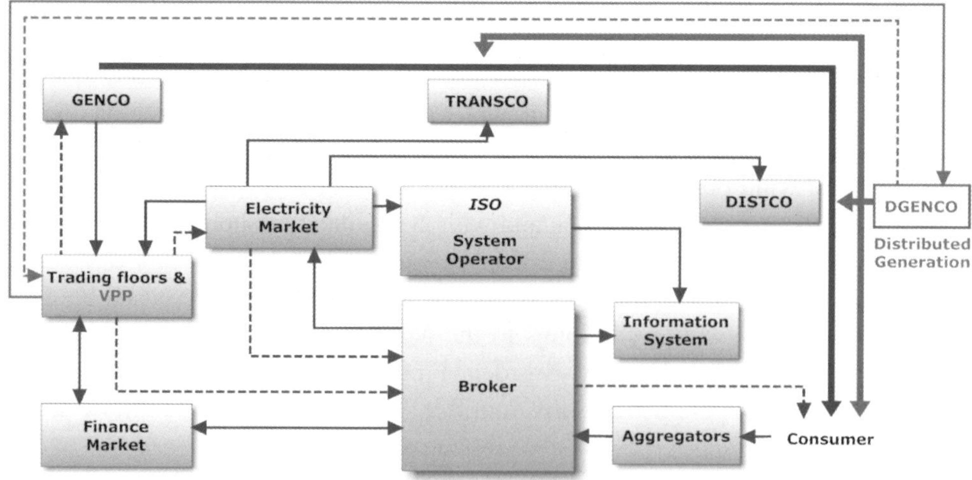

Fig. 24.1.　Market structure (adapted from Flechener[12]).

and distribution (DISTCO) grids, commissions to the Independent System Operator (ISO), the information systems, the aggregators, the energy markets commission and the costs associated to the financial relations between the financial markets and the trading floors and brokers. These flows can present variations, in accordance with the legislation and the regulation of each market.

The energy produced by distributed producers (DGENCO) in different locations can be delivered to the grid in any voltage level. This energy can be produced by a micro generation unit, with several ten of kW, where the customer uses the LV grid or, alternatively, can be directly injected in transmission or distribution networks by a larger generation plant, with many MW.

24.2 Electricity Markets and DG

The aim to provide power systems that can fulfill their goals and guarantee the required conditions for sustainable development requires significantly increasing electricity generation based on renewable resources. The intensification of Distributed Generation (DG), strongly associated with the use of generation technologies based on renewable resources, is a key issue for sustainable development.

Ideally, this energy is produced locally, close to consumption centers, and is not intended to be transmitted over long distances. The size of the generation in each site varies from kW to dozens of MW and can be coupled with heat generation and even cooling systems. Such generation is generally connected to distribution systems. Furthermore, it is encouraged by a regulatory framework favorable to co-generation and renewable energy as well as by the attractive cost of some equipment and the low cost and flexibility of some primary energy. At this stage the main technologies are gas, steam, micro-turbines and hydro turbines, fuel cells, wind generation, and solar cells.[13]

Currently, DG is introduced through government-level policies and subsidies. Ultimately, however, DG has to stand on its own feet within the market without such measures and become a normal profitable business thriving in a competitive market environment.

This requires not only adequate technological means but also intelligent interaction between the decentralized grid of the future, new electrical energy technologies including renewable generation and storage devices, the system operator, the market and, last but not the least, the customer.

A deregulated market where low power rating generators, sell their power can present economic and technical benefits.[14] However, there are serious barriers to the successful participation of these generators in the market, both of economic and technical nature. An aggregating strategy can enable owners of DG to gain technical and commercial advantages, making profit of the specific advantages of a mix of several generation technologies.[15] In this context, serious disadvantages of some technologies can be overcome.

In competitive electrical energy markets, market players have to operate with well supported strategic behavior in order to attain their goals. These markets are relatively recent and exhibit significantly different characteristics from other commodities markets. Due to this, electricity market players lack experience in market participation.

The success of electricity markets relies not only on adequate regulation models and on efficient and fair market and system management, but also on market participants' success. In the context of competitive markets, market players experience a high risky business which consequences can be suffered by the power sector. Derivatives markets allow mitigating the risks but they can only mean a significant improvement if adequately used by market players.

One of the main objectives of electricity markets is to decrease electricity costs through competition. There are several market structure models that could help achieving this goal. The market environment typically consists of a Pool, as well as a floor for Bilateral Contracts.[16]

A Pool is a marketplace where electricity-generating companies submit production bids and their corresponding market prices, and consumer companies submit consumption bids. A Market Operator regulates the Pool. The Market Operator uses a market-clearing tool to set market price and a set of accepted production and consumption bids for every period (usually one hour) in a daily basis. In Pools, an appropriate market-clearing tool is an auction mechanism. Bilateral Contracts are negotiable agreements between two traders about power delivery and consumption. The Bilateral-Contract model is flexible; negotiating parties can specify their own contract terms. The Hybrid model combines features of Pools and Bilateral Contracts. In this model, a Pool is not mandatory, and customers can either negotiate a power supply agreement directly with suppliers or accept power at the established market price. This model therefore offers customer choice.

The success of each player's participation in the market can be determined by the use of adequate decision-support tools. These tools can be seen as part of a complete electricity market simulation tool. Competitive electricity markets are dynamic and complex environments in which a lot of different kinds of players participate. These players have to take decisions with impacts that can be enormous both for individual players, but also for the whole market. To adequately simulating such an environment, multi-agent systems show an important set of advantages.

Agents are autonomous or semi-autonomous entities that can perform tasks in complex and dynamically changing environments. Agent technology seems to be an appropriate paradigm for use in modeling individual participants of electricity markets since they exhibit some relevant capabilities like autonomy, adaptability and ability to interact with others.

A multi-agent system consists of a group of agents that combine their specific competencies and cooperate in order to achieve a common goal. Efficient cooperation as well as coordination procedures between agents endows a multi-agent system with a capability higher than the sum of the individual agent capabilities.

Agents and multi-agent systems that adequately simulate electricity markets behavior are essential tools to gather knowledge and provide market agents with decision-support to strategic behavior. A wide variety of existing tools and services are available concerning agent-base research on restructured electricity markets.[17]

24.3 Virtual Power Producers (VPP)

The aggregation of DG plants gives place to a new concept: the Virtual Power Producer (VPP). VPPs are multi-technology and multi-site heterogeneous entities. In the scope of a VPP, producers can make sure their generators are optimally operated and that the power that is not consumed in their installation has good chances to be sold in the market. At the same time, VPPs will be able to commit to a more robust generation profile, increasing the value of non-dispatchable generation technologies.

In this context, VPPs can ensure a secure, environmentally friendly generation and optimal management of heat, electricity and cold as well as optimal operation and maintenance of electrical equipment, including the sale of electricity in the energy market.

In a deregulated market, generation is scheduled through an open wholesale market where large amounts of electrical energy are traded daily. Nowadays market places are organized over one or several countries, and each market place has its own set of rules.

Current electricity markets have effectively implemented real-time and day-ahead markets. VPPs should adopt organization and management methodologies so that they can make DG a really profitable activity able to participate in these markets.

In order to operate in an efficient way, VPPs should have detailed knowledge about the aggregated producers in order to coordinate the several generation technologies towards a common goal. This knowledge includes characteristics such as technology ripeness, profitability, availability and reliability, dispatchability, greenhouse effect gases, relation with external factors and lifetime.

24.3.1 *Producers' aggregation*

VPP formation can be seen as a coalition formation. For this purpose, the VPP needs to have adequate knowledge of each potential associated producer's characteristics. The most important characteristics to be considered are:

- Nominal power — The total nominal power installed in each producer;

- Available power — The power that the VPP can buy to the producer. The available power can be different from the installed power when there is overcapacity, or when the producer has already tied part of its power capacity with another market agent;

- Overload power — In same units it is possible to produce overload power for limited periods. The VPP may use this power in critical situations;

- Optimal operation point — This is an important characteristic for producers that use generation technologies that allow storage. The VPP must manage the generation and the reserves in such a way that the generators work close to their optimal operation point and in agreement with the VPP's needs, increasing overall efficiency;

- Equipment characteristics — Information concerning producers' equipment allows the VPP to know the power characteristic, reliability, maintenance periods, lifetime, relation with external factors, possible variations on the energy price taking into account the cost of the primary resources (oil or natural gas), etc;

- Operating limits — For the units which depend on natural resources, it is possible that the primary resource is below or above of the equipment operating limits. This must be considered in risk analysis of the generation forecast. Usually when the resources forecast is near to the minimum machines operating limit, the risk is small, but when it approaches the maximum limit the risk can be enormous. Let us consider the example of a wind farm with nominal power $P = 50\,\text{MW}$ to operate with wind speed of 25 m/s delivering 50 MW to the grid. If the speed of the wind increases to 26 m/s the wind turbines must be turned off and the delivered power drops instantaneously to zero;

- Grid connection characteristics — This is an important aspect if it is necessary to pay the losses due to the power circulating in the lines. As an example, in Portugal, if the unit is connected to a distribution utility substation, the producer does not pay the losses. However, if the producer is connected to a transmission utility substation it will have to pay for the losses. Another aspect to consider will be the existence of two or more producers connected to the same electric substation. In this case, the VPP will be able to coordinate the generation of his own associated producers to optimize the power generation strategies, for example the use of the overcapacity of a wind farm;

- Historic generation data — The availability of historic generation data can enable the VPP to use helpful forecasting tools.

The knowledge of these characteristics will allow the VPP to evaluate the possibility of integrating a new producer into the group and to define the financial conditions in the adequate way. It may also be useful to establish the contract period.

The producer can choose between working with a VPP or negotiating directly in the market; on the other hand, the producer can choose the VPP that intends to be associated with. Therefore, the producer also has interest in having knowledge concerning VPP characteristics.

The most important aspects of a VPP are:

- Generation and reserves remuneration — The most important aspect for the producer concerns the generation mechanisms and reserves remuneration. A possible and simple strategy is to fix, for a large period of time (6 to 12 months), values for each kWh in different periods of time during the day; for example, consider peak, partial peak and off peak periods. However, the VPP will have to assure a minimum level of generation. Another strategy is to do an internal pool with the VPP's associated producers. However, this pool would have to occur two days ahead, which would reduce the reliability of the generation forecasts and the VPP risk associated with energy purchases above the real market price. A VPP can choose an auction strategy with which, after the market closing and considering the energy sold, it will be possible to carry out an auction between the associated producers. Finally, the VPP could also select a solution that combines the paid price of the producers with the selling value of the energy market. This strategy, probably, is the most correct as it implies least risks for the producer as well as for the VPP. The producer will have to limit the price below which it would not be willing to produce;
- Penalties — If a VPP does not accomplish the established contracts, the market operator may impose several penalties on him. Therefore, the VPP has to assure an adequate reserve level to prevent these penalties. This aspect is of a great value for the producers dependent on natural resources; so the uncertainty of the generation is raised. However, the VPP must fix maximum levels of penalties, thus, limiting the reserve levels and obtaining lower sell electricity price;
- Services — The VPP assures some services, such as generation management, forecasts' support, maintenance, marketing and energy selling in the electricity markets. These services will be important for the micro generation producers.

     Z. A. Vale, H. Morais and H. Khodr

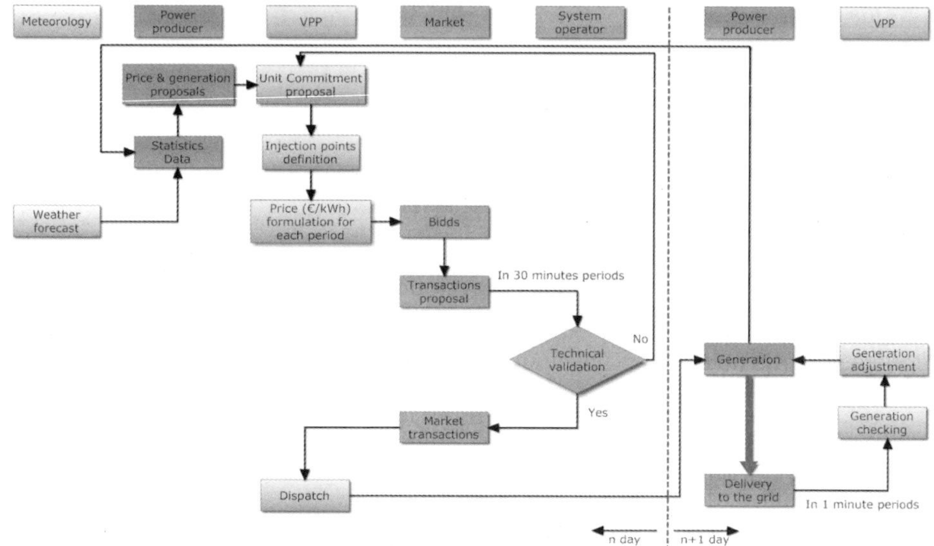

Fig. 24.2.   Schematic representation of the operation of a VPP.

24.3.2 *VPP operation and management*

The scheme presented in Fig. 24.2 represents the operation of a VPP.

In this operation scheme, the VPP performance can easily be analyzed based on the following:

- The producers present energy sale proposals based on the generation of the previous days, on the meteorological forecasts and on the existing reserves. The considered price for each producer must be agreed with the VPP so that competitive prices can be obtained, to allow the producers to have revenues from their investments in reasonable periods of time. If subsidies exist, these will have to be included in the prices calculation. The reserve price will also have to be previously agreed between the VPP and the producers;
- VPP can be stated as a unit commitment with the prices presented by the producers to the next day. This unit commitment based on the economic merit, having as orientation several criteria, such as price, quantity of the available power, generation guarantee and existence of generation reserve;
- Besides the ideal scheduling, an alternative scheduling of generation must be fixed that considers the shortage and surplus generation. The values of the considered variations must be based on the probabilities of the generation given by the producers who have technologies that do not guarantee its generation;
- When the generation scheduling is carried out, it is necessary to identify the amount of energy injected and the network injection points;

- With generation scheduling, prices and amounts of power presented by the producers, the VPP formulates the energy price;
- After the negotiation period, the commercial transactions are presented by the market operator and subjected to technical validation by the system operator;
- The system operator has to make technical validation for all the transactions made in the market. This can be made using load flow and contingency analysis algorithms;
- In situations when technical constraints are violated, it is interesting to show the critical points to the producers and costumers to allow them to adjudicate the supply to another producer;
- If there are no problems in the technical analysis, the system operator will communicate this to the market operator to allow him to make the transactions effective. This is in case of the Market Operator being separate from the System Operator;
- Knowing the amount of sold energy, VPP will do the announcement to the next day, according to the scheduling previously determined, which will be communicated to the producers;
- In the next day, VPP must undertake a permanent control (each minute) of the energy delivered to the network by each partner producer, to compensate the variation of the generation of the non-dispatchable technologies;
- The produced energy and the climatic conditions must be used to update the database of the producers to forecast more and more efficiently the energy that they will be able to provide to the VPP.

With the developed simulation tool, several studies were done, using different levels of reserve. The goal was to test if the amount of power reserve was enough to allow the VPP supplying the sold energy, preventing penalty costs.

In this section we present a case study, simulating a VPP that aggregates three wind power producers. Each one of these producers has a co-generation unit with

Table 24.1. Producers data.

Producer	Technology	P (kW)	P_T (kW)
1	Wind farm	30000	33.030
	Co-generation	3030	
2	Wind farm	20000	22.100
	Co-generation	2100	
3	Wind farm	13000	14.250
	Co-generation	1250	

a nominal power of about 10% of the wind farm capacity. Table 24.1 presents the most important characteristics of these producers.

The co-generation will serve as storage system to support the wind farms forecast errors. In this case study, all the required reserve is assured by VPP own means.

In this case, the simulated scenarios are the following:

Scenario 1 — Reserve assured by the producers, with all the co-generation production as reserve;

Scenario 2 — Reserve assured by the producers, with 75% of the co-generation production as reserve;

Scenario 3 — Reserve assured by the producers, with 50% of the co-generation production as reserve;

Scenario 4 — Reserve assured by the producers, with 25% of the co-generation production as reserve;

Scenario 5 — Without reserve.

Table 24.2 summarizes the characteristics of these scenarios.

The simulation considered two different days, one of September and another of November of 2005, in two different markets (UK and France). All prices are in Euros, considering 1£ = 1.45€

Figure 24.3 presents the VPP profits for each scenario.

The simulation has considered the values provided by the entities responsible for the market management and/or for the balancing mechanisms that, in the case of UK and France, are respectively UKPX/Elexon and PowerNext/RTE.[19–22]

The cost evaluation associated with the operation of the producers has been done with software available in the RETScreen web site.[23] With this software, it is possible to determine the costs of generation for the different technologies.

VPP costs are calculated using the values proposed by the producers, the cost of the reserve, and costs associated with VPP management and operation services.

Analyzing the values presented in Fig. 24.3, it can be concluded that during a two months period, between September and November of 2005, the VPPs results of the VPP can vary a lot.

Table 24.2.   Energy balancing.

Scenario	E_S (MWh)	E_D (MWh)	E_U (MWh)	E_E (MWh)
1	529.41	570.29	3.17	44.06
2	567.69	574.47	0.92	7.70
3	605.97	614.64	14.05	22.72
4	644.25	633.77	11.85	1.38
5	682.53	659.80	32.00	9.27

Profit

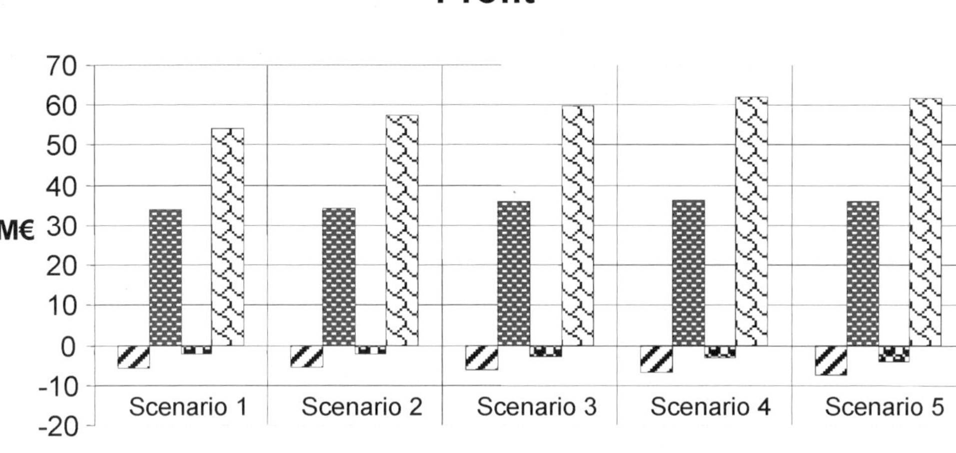

Fig. 24.3.   VPP profits in considered scenarios.

In both markets, we got results from about ten thousand Euros of losses in September to hundreds thousand Euros of profits in November. These variations reflect the instability of the oil markets, in this time period.

It is important to point out that the undertaken calculations did not take into account any type of subsidies or benefits that usually are attributed to DG by government policies.

The results of the considered September day have been negative. For the considered November day, the results have been positive, and the scenario that presents greater profit is scenario 4. In this scenario the reserve is only 2.5% of the wind generation; however, this reserve is enough to reduce the penalties by half and to have some profits.

Comparing scenarios 3 and 5, we can see that, in France, scenario 5 is more advantageous; however, in UK, it is the inverse. On the other hand, in scenario 3, in France, the penalties are positive and in England the penalties are negative. While in the first one we have energy surplus, in the second we have energy shortage, modifying the penalty regime.

The results of this case study are according to the results of all the simulations we have undertaken, leading to the most important conclusion being that VPPs' reserve strategies should take into account the context, namely in what concerns market model and characteristics and the considered situation. This is why the inclusion of VPP models in MASCEM, an electricity market simulator with dynamic strategies is producing interesting and valuable results.[24]

24.3.3 *VPP's reserve strategies*

VPPs must identify the characteristics of each one of their associates and try to optimize the selling activity so that each associate delivers the biggest possible amount of energy. The ideal situation would be to sell all the energy that its associates are able to produce at each instant. The problem is that this is not possible due to the uncertainty of generation of the technologies that depend from natural resources as the wind, sunshine, waves or water flows.

The method used by the markets to penalize the violation of the established contracts is not uniform, which implies different forms of the VPPs action.

In markets where penalty mechanisms exist, such as the French, the Italian, the Finnish, the Swedish or the British, the VPPs have to conveniently manage the generation capacity of the associated producers to assure enough reserves, to compensate generation oscillations.

The energy reserves will have to be assured, in first place, by producers using technologies that allow them to control the net injected power, such as co-generation, fuel cells, or gas turbines. These producers may establish contracts with the VPP, for supplying the imbalance settlement energy, below their nominal capacity.

Another possibility for the VPP to assure some level of reserve will be to have specific plants for that. These plants can be managed by the VPP or by another entity.

However, it will have higher costs, essentially for the initial investment. Start-up costs must be considered to decide what units should be in spinning reserve.

It is also possible to establish contracts with large power plants. These contracts may be bilateral for large periods of time, for example, 6 or 12 months, with a fixed power contract, or daily in the spot market, in function of the necessities forecast. If the energy is bought in the market, the purchase price will be equal to the selling price, resulting in profits due to the penalties inexistence. If the energy is acquired by bilateral contracts, it will have a fixed value every day and it can be used for reserve or as a part of the production after checking that it is not necessary to use all power as reserve.

The VPP can also adopt a hybrid solution, combing more than one option of reserve, in order to minimize the costs and to diversify options.

Whichever the strategy to adopt, the VPP must define, previously, who are the producers that will compensate these generation variations, taking the merit order account.

The reserve value that the VPP will have to guarantee depends basically in the Market, and on the generation technologies that it can use.

A possible strategy is to have as reserve the amount of power that guarantees $n-1$ security criterion. In this case, if the biggest generation unit gets out of service and if all the other producers are operating in accordance with the foreseen, the VPP will be able to continue to guarantee all the established contracts.

Let us consider a VPP with n associated producers. If this VPP contracts a fixed power value with a thermal producer on an annual basis to use for reserve proposes, we have:

$$P_{\text{therm}} = P_i^{\text{max}} - \sum_{j=1}^{n} R_j, \tag{24.1}$$

where

P_{therm} — Contracted thermal power
P_i^{max} — Nominal power ob biggest generation unit
R_j — Contract reserve with aggregated producer.

The VPP can keep this constant reserve value, resulting in a higher security when the sold energy decreases.

Because of that, the VPP can reformulate its strategy, in function of the sold energy, changing the scheduled generation.

The VPP can negotiate the excess reserve with other market agents or the system operator.

When the forecasted generation for the bigger generation unit is lower than its nominal power, the required reserve is also lower. In this case, the VPP can sell the exceeding energy in the spot market.

In this situation, we have:

$$P_{\text{VPP}} = \sum_{i=1}^{n} P_i^f + P_{\text{therm}} - R_{\text{VPP}}, \tag{24.2}$$

$$R_{\text{VPP}} = (P_i^f)^{\text{max}}, \tag{24.3}$$

where

P_{VPP} — Power that VPP can negotiate in the market
R_{VPP} — VPP reserve
P_i^f — Forecasted power for generation unit i.

Generation and reserve values can change dynamically.

In some situations, the $n - 1$ security level cannot be assured. The bigger penetration of wind energy (and other technologies with similar characteristics) the less the generation stability causing oscillations around the values foreseen for the reserve and generation.

Even if the required security level is assured, this can bring higher costs to VPP. However, these solutions may be useful to avoid penalties by the market operator such as: payments, warning, suspension or exclusion.

Another solution is to guarantee as reserve a percentage of the foreseen generation in each moment. In this situation we have:

$$P_{\text{VPP}} = \frac{\sum_{i=1}^{n} P_i^f + P_{\text{therm}}}{1 + P_{\text{erc}}}, \tag{24.4}$$

$$R_{\text{VPP}} = \frac{\left(\sum_{i=1}^{n} P_i^f + P_{\text{therm}}\right) \cdot P_{\text{erc}}}{1 + P_{\text{erc}}}, \tag{24.5}$$

where

P_{VPP} — Power that VPP can negotiate in the market.

The VPP can adopt a hybrid strategy to determine the reserve value; it must consider the following aspects:

- Resource forecast — if the forecast is made too early, bigger is the probability of error, what implies more reserve. There are two different short-term prediction models: the physical and the statistical approach. In some models, a combination of both is used, as both approaches can be needed for successful forecasts. In short, the physical models try to use physical considerations as long as possible to reach the best possible estimate of the local resources before using Model Output Statistics (MOS) to reduce the remaining error. Statistical models in their pure form try to find the relationships between a wealth of explanatory variables including NWP (HIRLAM/WAsP/Park, nowadays called Prediktor) results, and online measured power data, usually employing recursive techniques. Often, black-box models like advanced Recursive Least Squares or Artificial Neural Networks, the statistical β fit distribution, the Weibull distribution and other techniques. The most successful statistical models employ grey-box models, where some knowledge of the wind power properties is used to tune the models to the specific domain[25];
- Sold energy forecast — The reserve has to be adjusted to the prediction of the energy to be sold in the market. This prediction is based on the VPP selling strategies and on historic data;
- Forecasted price — The importance of reserve depends on the power selling and buying price, the reserve cost, and the penalties cost. Several techniques are used for market price forecast: soft computing methods are applied to forecast electricity market prices. Our VPP model uses a price forecasting methodology based on Neural Networks, Data Mining and Particle Swarm Optimization[26];
- Generation units operating point — when considering cogeneration, fuel cells or gas turbines, if they allow some overcharge or if they are working below their nominal charge, they can compensate some generation oscillation that could appear. In what concerns the wind power generation, it is known that the

delivered power of a wind turbine is almost constant when the velocity of the wind is between 15 and 25 m/s. If the wind forecast is at 20 m/s the wind turbine can stand a wind variation of $\pm 25\%$. On the other hand, between 5 and 15 m/s, the smallest velocity variation can severely affect the production;

- Risk Management — The VPP can take more or less risk in its market bids. If a high generation quantity bid is submitted, the penalty risk is higher than if the VPP assures some level of reserve. The VPP manages the risk considering the results of past market sessions results and the primary resources. In this strategy the VPP uses the following method:

(1) Determination of the weather forecast and its range;
(2) Determination of the generation forecast for each generation technology (using the forecasted resource values) and the forecasted generation range (considering the forecasted resource range);
(3) Determination of the maximum market penalties for each time slice;
(4) Computation of the minimum reserve amount to limit the penalties, according to Eqs. (24.6) and (24.7):

$$F_p = M_p - M_{\text{pen}} \cdot VPP_{\text{pen}}, \tag{24.6}$$

$$R_{\text{VPP}} = \sum (P_f) - \sum \left(P_f \cdot \frac{(1 - \varepsilon)}{F_p} \right), \tag{24.7}$$

where

$$
\begin{aligned}
F_p &\;—\; \text{Penalties factor} \\
M_p &\;—\; \text{Forecasted market price} \\
M_{pen} &\;—\; \text{Forecasted market penalties price} \\
VPP_{\text{pen}} &\;—\; \text{VPP maximum penalties} \\
P_f &\;—\; \text{Power forecast} \\
\varepsilon &\;—\; \text{Forecast error.}
\end{aligned}
$$

(5) Determination of the producer who guarantees the reserve.

$$F_r = P_p - R_p + M_{\text{pen}} - M_p, \tag{24.8}$$

where

$$
\begin{aligned}
F_r &\;—\; \text{Reserve factor} \\
P_p &\;—\; \text{Energy price by producer} \\
R_p &\;—\; \text{Reserve price producer.}
\end{aligned}
$$

The producer with the higher reserve factor is chosen to assure the reserve.

There are some markets where contract violations do not imply direct penalties, fixed taxes being paid by each seller agent to allow the system operator to handle the system reserve. This methodology does not prevent the market from inadequate behavior of seller agents who can impose taxes rise. On the other hand, this situation can promote speculations with the sellers negotiating, deliberately, energy without any type of guarantee of supply.

To illustrate this methodology let us consider a case study with three wind power producers and three co-generation unit associations. The co-generation unit has a nominal power of about 10% of the wind farm capacity (see Table 24.3).

Market price and the producer's generation costs are based on data of OMEL, Spanish market. We considered the values of January of 2008.

With these values the VPP can assure the best strategy to operate in the competitive market with few penalties and the adequate reserve. The results can be seen in Fig. 24.4 In this figure, P_Wind1, P_Wind2 and P_Wind3 represent the

Table 24.3. Energy and reserve prices.

Hour	Market price	Market penalties	Producer 1 price	Producer 2 price	Producer 3 price	Producer 4 price/ reserve	Producer 5 price/ reserve	Producer 6 price/ reserve
1	6.84	9.57	3.5	3.0	2.8	2.0/1.5	3.2/1.6	2.5/1.0
2	6.26	8.76	3.5	3.0	2.8	2.0/1.5	3.2/1.6	2.5/1.0
3	5.65	7.91	3.5	2.5	2.8	2.0/1.5	3.2/1.6	2.5/1.0
4	5.20	7.27	3.5	2.5	2.8	2.0/1.5	3.2/1.6	2.5/1.0
5	4.83	6.76	3.5	2.5	2.8	2.0/1.5	3.2/1.6	2.5/1.0
6	4.89	6.85	3.5	3.0	2.8	2.0/1.5	3.2/1.6	2.5/1.0
7	5.45	7.63	3.5	4.0	3.8	3.0/2.5	3.2/1.6	2.5/1.0
8	6.33	6.33	4.5	4.0	3.8	3.0/2.5	3.2/1.6	2.5/1.0
9	7.20	7.20	4.5	4.0	3.8	3.0/2.5	3.2/1.6	2.5/1.0
10	7.49	7.49	4.5	4.0	3.8	3.0/2.5	3.2/1.6	2.5/1.0
11	7.98	7.98	4.5	4.0	3.8	3.0/2.5	3.2/1.6	2.5/1.0
12	7.97	11.2	4.5	4.0	3.8	3.5/2.5	3.2/1.6	2.5/1.0
13	7.91	11.1	4.5	4.0	3.8	3.5/2.5	3.2/1.6	2.5/1.0
14	7.62	10.7	4.5	4.0	3.8	3.0/2.5	3.2/1.6	2.5/1.0
15	6.97	9.76	4.5	4.0	3.8	3.0/2.5	3.2/1.6	2.5/1.0
16	6.77	9.47	4.5	4.0	3.8	3.0/2.5	3.2/1.6	2.5/1.0
17	6.69	9.36	4.5	4.0	3.8	3.0/2.5	3.2/1.6	2.5/1.0
18	7.19	10.1	4.5	4.0	3.8	3.0/2.5	3.2/1.6	2.5/1.0
19	8.58	12.0	4.5	4.0	3.8	3.5/2.5	3.2/1.6	2.5/1.0
20	8.82	12.4	4.5	4.0	3.8	3.5/2.5	3.2/1.6	2.5/1.0
21	8.64	12.1	4.5	4.0	3.8	3.5/2.5	3.2/1.6	2.5/1.0
22	8.36	11.7	4.5	4.0	3.8	3.5/2.5	3.2/1.6	2.5/1.0
23	7.52	10.5	3.5	3.0	3.8	3.0/1.5	3.2/1.6	2.5/1.0
24	7.38	10.3	3.5	3.0	2.8	3.0/1.5	3.2/1.6	2.5/1.0

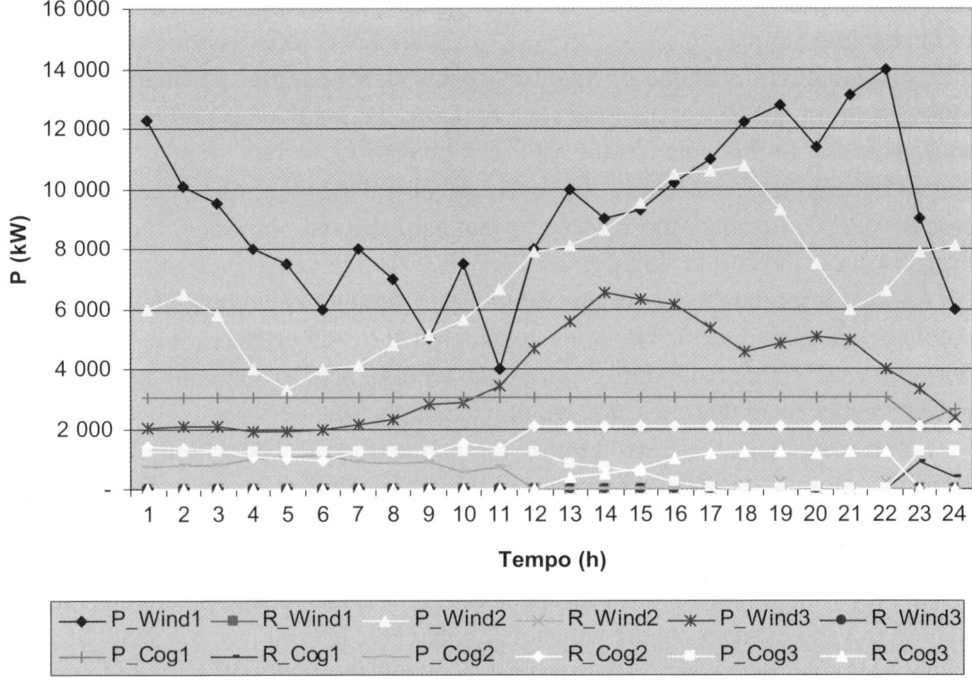

Fig. 24.4.   Dispatch of generation units by VPP.

energy delivered by the wind farm producers. P_Cog1, P_Cog2 e P_Cog3 represents the energy delivered by co-generation units. R_Wind1, R_Wind2 and R_Wind3 represent the energy of wind farm producers not dispatched. R_Cog1, R_Cog2 e R_Cog3 represents the energy reserve by co-generation units.

24.3.4 *Bids formulation*

Virtual Power Producers (VPPs) are multi-technology and multi-site heterogeneous entities, being relationships among aggregated producers and among VPPs and the remaining Electricity Market agents a key factor for their success.

To sell energy in the market VPP must forecast the generation of aggregated producers and "save" some power capacity to assure a reserve to compensate a generation oscillation of producers with natural resources technologies dependent.

The VPP can use different market strategies in the S-period, considering specific aspects such as producers established contracts and range of generation forecast. The prediction errors increase with the distance between the forecasting and the forecast times. The standard error figures are given as a percentage of the installed capacity, since this is what the utilities are most interested in (installed capacity is easy to measure); sometimes they are given as a percentage of the mean production or in absolute numbers.

Considering an example of Spanish market (OMEL), the spot market session closes at 11am, therefore, the time slice between the predictions and real day is 13 to 37 hours.[27] In this context, the VPP can change its market strategy during the day to manage the risk. These strategies are also depending of reserves, in other words, VPP can change the reserve to maintain the risk, however, if VPP has a bigger reserve the costs is higher.

Another important factor to the VPP market strategy is the buy energy price to the aggregated producers. The price considered for each producer must be agreed with the VPP so that competitive prices can be obtained, to allow the producers to have revenues from their investments in reasonable time periods.

If subsidies exist, these will have to be included in the calculation of the prices considered for the producers. The price of the reserve will also have to be previously agreed between the VPP and the producers.

This way of working implies a great complicity between the VPP and the associated producers in order to prevent speculations. The profit edge is variable, but the minimum value is given by the value of sell percentage.

In markets where price variations are frequent, the VPP will be able to define different strategies, for example, the prices elaboration depending on the generating technologies, to obtain prices that can easily be adapted to the market.

In this scenario the technologies can be divided in 4 groups. The first group will include the technologies for which the primary energy cannot be stored, for example, wind, solar, and co-generation.

The second group includes the technologies for which it is possible to store the primary resource, and for strategically interests it could be convenient to use the resources in the periods when the price of energy is higher. The technologies that belong to this group are the hydroelectric with dam, the biomass and biogas energy.

The third group includes the VPP reserve, either the production part that the VPP has contracted as reserve with the producers or with thermal central offices.

The fourth group includes the technologies for which the primary resources are storable, but expensive, for example, gas turbines and fuel cells.

For the first group, whatever is the final price, it is always better than to waste the existing resources, therefore, the VPP will have to adjust its strategy to obtain the best final sell price, but having in mind that even when the price is low, it may be interesting for the producers.

For the second group the way of using the resources must be managed with care to valuing them as much as possible. However, if the resources are in excess and it is not possible to store them, these could be used in periods in which the prices are lower. In these situations, the VPP will have to consider these producers in the same way it considers those in group 1.

For the third group the VPP will have to consider the difference between the price to be paid for the produced energy and the price to be paid for the reserve.

In the fourth group, the considered price will be constant during the day. The factor that could be changeable is the amortization of the investment; it should consider at least the marginal cost.

In countries where subsidies exist, the VPP will be able to obtain very competitive prices, which makes its participation in the market easier.

A model has been developed to simulate the operation of a VPP, taking into account the characteristics of the technologies used by the power producers and can be used to provide decision support to VPPs.

With the MASCEM simulation tool, several studies were done, using different levels of reserve. The goal was to verify which was the most advantageous strategy to pay the energy to producers. On the other hand, this tool will be very useful for the producers because it can verify, in different conditions, how much it receives when aggregated to the VPP.[28]

Figures 24.5 and 24.6 present some results of a study of a producer's remuneration by VPP, considering wind farm and co-generation technologies.

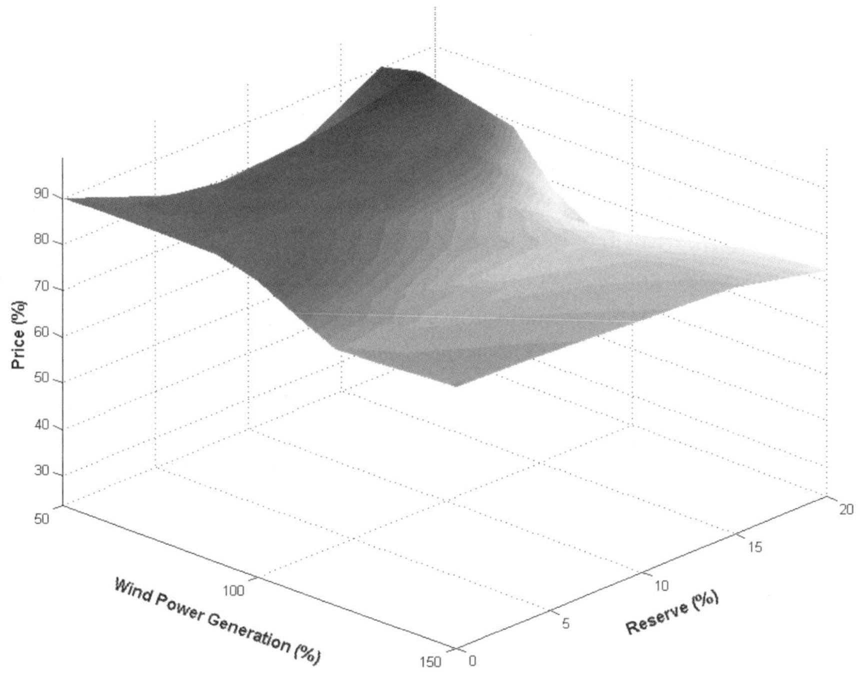

Fig. 24.5.   Price variation for wind farm.

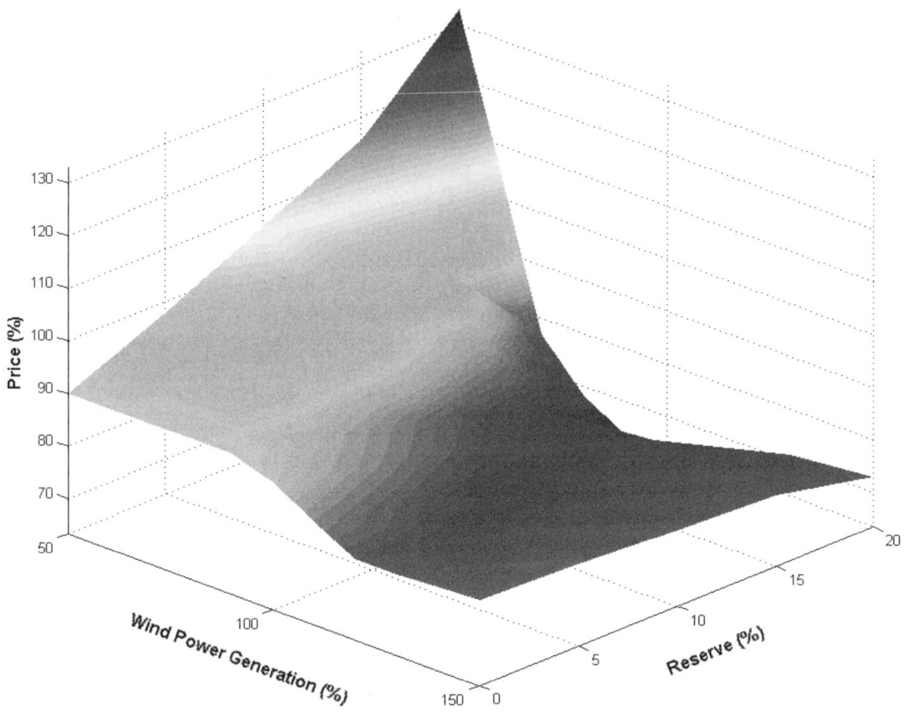

Fig. 24.6.   Price variation for co-generation.

These results show the advantages that each producers has when aggregated to the VPP in what concerns its remuneration. In general, when the produced energy is lower than forecasted, this advantage can be traduced in higher profits. On the other hand, when the produced energy is higher that what is forecast, the producer's remuneration can be lower when associated with a VPP. In what concerns remuneration values, wind generation is the one that most benefits from the aggregation. However, there are other significant advantages besides remuneration, namely the providing by the VPP of relevant services (market bidding, maintenance...). These are especially important for other production technologies with smaller size.

24.3.5   *Producers' remuneration*

After the closing of the market, the VPP will have to forward the production scheduling to the associated producers, as well as determining the value to be paid to them for the seller energy.

Despite the prices presented in the market being calculated on the basis of the prices presented by the producers, the producers' remuneration strategy must take into account the market price for which the energy has been sold.

Proportionality method — As a first approach, we can consider that each producer's remuneration should be proportional to the value of the produced energy relative to the total produced energy by associated producers.

$$p_t = (p_m \cdot E_p \pm I_{\text{VPP}}) \cdot (1 - L) - p_r, \tag{24.9}$$

$$p_i = \frac{p_t}{\sum\limits_{i=1}^{n} E_i} \cdot E_i + p_r \cdot \frac{E_{ri}}{E_r}, \tag{24.10}$$

where

p_t — Total payment to the producers

p_m — Energy market price

E_p — Delivered energy

I_{VPP} — VPP penalties

L — VPP profit

p_r — Reserve cost

p_i — Payment to the producer i

E_i — Delivered energy by the producer i

E_{ri} — Reserve of the producer i

E_r — VPP reserve.

This method consists in paying to the producers a value which is directly proportional to the produced energy, considering a previously defined value to pay for the reserve.

This method is somehow unfair because it is mainly advantageous for the producers that cannot guarantee the production, because they usually sell all the produced energy. On the other hand, the producers that accept to adapt their production to the general necessities of the VPP are impaired because their contribution to the producer aggregation is not recognized.

Equal percentage method — Another form of remuneration of the produced energy consists in providing equal percentages of the profits to the producers, the same that they would have in case of no association with VPP.

$$p_t = (p_m \times E_p \pm I_{\text{VPP}}) \times (1 - L) - p_{re}, \tag{24.11}$$

$$p_i = p_t \times \frac{(p_m \times (E_i + E_{di}) \pm I_i)}{\sum\limits_{i=1}^{n} (p_m \times (E_i + E_{di}) \pm I_i)}, \tag{24.12}$$

where

p_{re} — External reserve cost

E_{di} — Energy that could be delivered by producer i

l_i — Penalties that the producer i could have.

Factor "G" — Factor "G" determines the distribution of remuneration between the dispatchable and non-dispatchable producers. We can optimize this value for the different objective functions. In this method the price to pay to the producers will be divided in 3 parts: the relative value of energy, the reserve capacity, and the use of the reserve.

The value to pay for the produced energy (foreseen) is given by he following expression:

$$p_{i1} = p_m \times E_{pi} \times (1 - L), \qquad (24.13)$$

where

p_{i1} — Payment to the producer I (part 1)
E_{pi} — Foreseen delivered energy by the producer i.

The second part is related with the existing reserves:

$$p_{i2} = p_m \times G \times \left(E_{ri} - E_r \times \frac{E_{pi}}{E_p} \right), \qquad (24.14)$$

where

p_{i2} — Payment to the producer i (part 2)
G — Factor G.

In this expression all the producers pay for the reserve, but the producers that make use of technologies with controllable production can be relieved from reserve payment, since they have a strong probability of not using it. If the VPP opts for this type of management, we will have as:

$$\begin{cases} p_{i2d} = p_m \times G \times E_{ri}, \\ p_{i2u} = -p_m \times G \times E_r \times \dfrac{E_{pi}}{E_{pu}}, \end{cases} \qquad (24.15)$$

where

p_{i2d} — Payment to the producer i (part 2) with despatchable technology
p_{i2u} — Payment to the producer i (part 2) with non-dispatchable technology
E_{pu} — Energy sold by non-dispatchable technology.

The third part deals with the use of the reserve in function of the production technologies. For production technologies for which, it is not possible to control the production, the use of reserves can be significantly different from the foreseen. If the used reserve is larger than foreseen the use is negative, if the used reserve is smaller than foreseen the use is positive.

The expressions of calculation of this part are different for the two situations. If the production is larger than foreseen, the expressions are the following ones:

$$\begin{cases} p_{i3e} = p_m(\Delta E) \times (1 - G) \\ \qquad + (\Delta P) \times Ep_i/Ep_1 \times (1 - L) \times G, \\ p_{i3f} = p_m[(-\Delta E) \times (G + L - 1)] \\ \qquad + (\Delta P) * Ep_i/Ep_2 \times (1 - L) \times (1 - G), \end{cases} \tag{24.16}$$

where

p_{i3e} — Payment to the producer i (part 3) with surplus generation
p_{i3f} — Payment to the producer i (part 3) with shorted generation
ΔE — Difference between the forecasted and produced energy
ΔP — Difference between expected remuneration and real remuneration
Ep_1 — VPP Surplus Energy
Ep_2 — VPP Energy Shortage.

If the production is smaller than the foreseen, the expressions are as follows:

$$\begin{cases} p_{i3e} = p_m(\Delta E) \times (1 - G) \\ \qquad + (\Delta P) * Ep_i/Ep_1 \times (1 - L) \times (1 - G), \\ p_{i3f} = p_m[(-\Delta E) \times (G + L - 1)] \\ \qquad + (\Delta P) * Ep_i/Ep_2 \times (1 - L) \times G. \end{cases} \tag{24.17}$$

24.4 VPP and Electricity Market Simulation

MASCEM — Multi-Agent Simulator of Competitive Electricity Markets — has been developed to study several negotiation mechanisms usually found in electricity markets. In the MASCEM market participants have strategic behavior and a scenario decision algorithm to support their decisions.[24]

MASCEM provides users with dynamic strategies which can be specifically tailored to adapt themselves to each agent characteristics and to each situation. Moreover, MASCEM provides decision-support tools that can be accessed by each agent for simulation purposes.

Unlike traditional tools, MASCEM does not postulate a single decision-maker with a single objective for the entire system. Rather, agents, representing the different independent entities in Electricity Markets, are allowed to establish their own objectives and decision rules. Moreover, as the simulation progresses, agents can adapt their strategies, based on the success or failure of previous efforts. Learning capabilities enable market agents to update their knowledge, according to their own past behavior and with other agents' behavior. In each situation, agents dynamically adapt their strategies, according to the present context and using the dynamically updated detained knowledge.

Power exchanges established the trade of forward and futures contracts early on and, by now, large volumes are being traded. A power forward contract is characterized by a fixed delivery price per MW, a delivery period and the total amount of energy to be delivered. One can observe that contracts with a long delivery period show less volatile prices than those with short delivery periods.

VPP Forward Market operation will be limited by the aggregated producers. If the VPP has many producers whose generation depends on natural resources, it is complicated to establish forward contracts because the guarantee of the supply energy is low.

In electricity spot markets electricity is traded for each hour, or mid-hour of the next day.

24.4.1   *MASCEM multi-agent model*

Competitive electricity markets involve several entities interacting in complex negotiations (Fig. 24.7). These entities are of very different types and can be complex organizations involved in difficult decision making processes. Attending to these, we propose a multi-agent model to represent all the involved entities and their relationships.[29]

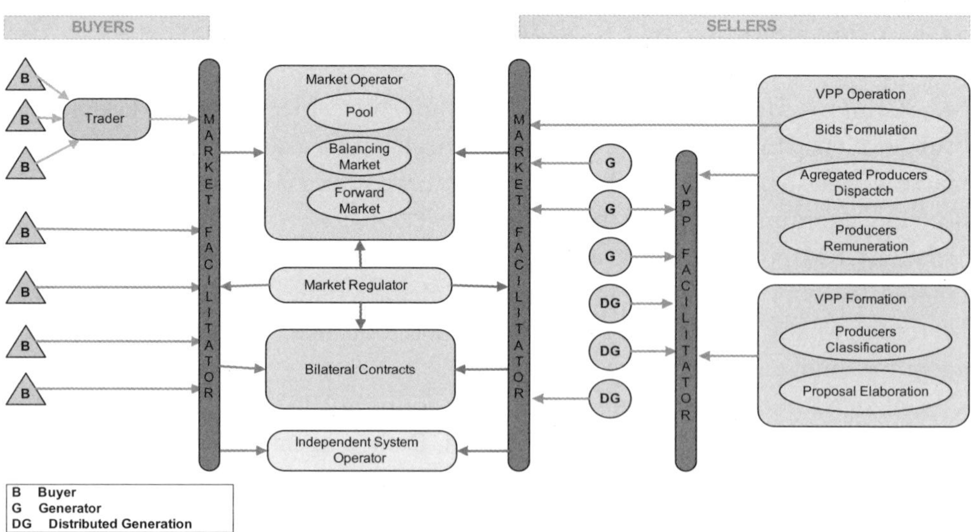

Fig. 24.7.   MASCEM architecture.

MASCEM includes the following types of agents:

- Market Facilitator Agent;
- VPP Facilitator Agent;
- Seller Agents;
- Buyer Agents;
- Trader Agents;
- VPP Agents;
- Market Operator Agent;
- System Operator Agent;
- Market Regulator Agent.

This architecture has two facilitators, the Market Facilitator and the VPP Facilitator, and allows the evaluation of several scenarios, according to market agents and the VPP's strategies. Moreover, as the simulation progresses, agents can adapt their strategies, based on the success or failure of their previous actions.

Facilitators are used in the scope of MAS in order to ease relationship management; they do not correspond to physical entities in the market.[30]

The Market Facilitator coordinates the market. It knows the identities of all agents present in the market, regulates the negotiation process and ensures the market is functioning according to the established rules. The first step agents have to make in order to participate in the market is to register them at the Market Facilitator, specifying their market role and services.

The VPP Facilitator supports VPPs business. It gathers information about generation agents, both those who are playing in the market and those who are not. This information is relevant for VPP business because some producers of reduced dimension cannot participate separately in the market, but can be associated to a VPP.

The VPP Facilitator manages the information between the producers, and the VPP in the VPP aggregation process and in its operation process. The VPP Facilitator informs the Market Facilitator about new agreements between producers and VPPs.

Seller and Buyer Agents are the two key players in the market. Sellers represent entities able to sell electricity in the market (e.g., companies holding electricity generation units and VPPs). Buyers may represent electricity consumers, consumer aggregators, retailers, etc.

MASCEM can be used by any type of market agent to simulate the envisaged scenarios and to obtain decision support for operations in the market. The user must specify his intrinsic and strategic characteristics and defines the number of Sellers, Buyers and VPPs in each scenario. By intrinsic characteristics we mean the individual knowledge related to reserve and preferred prices, and also to the available capacity. By strategic characteristics we mean the strategies the agent envisages to employ to reach his objectives.

In the case of use by a regulator, the regulator agent assumes itself as a entity able to introduce regulation changes in the market and MASCEM simulates the behavior of the other agents according to these changes. Regulators can also use MASCEM as any other type of agent in order to analyze market operation, in face of established rules and according to each agent strategy.

Sellers will compete with each other, since they aim to obtain the highest possible profits. On the other hand, Sellers will cooperate with Buyers while trying to establish some agreement that is profitable for both. This is a very rich domain where it is possible to develop and test several algorithms and negotiation mechanisms for both cooperation and competition.

The Independent System Operator Agent is responsible for the transmission grid and all the involved technical constraints. Every contract established, either through Bilateral Contracts or through the Market, must be communicated to the System Operator, who analyzes its technical viability from the power system point of view (e.g., feasibility of power flow without technical constraint violation, congestion management).

The Market Operator Agent represents the entity that is responsible for the Forward, Pool and Balancing mechanism. The Market Operator will receive the bids of the Sellers, Buyers, Traders and VPPs, analyze them and establish the market price and accepted bids (using the offer and demand aggregated curves).

The increase in competitiveness creates opportunities for many new players to enter the market; one of these players being the Trader. The introduction of this entity allows liberalization and competition in the electricity industry to be developed. It can also simplify the whole process for some entities, freeing some producers and customers from participating directly in the market, allowing them to focus on their core business. Traders participate in the market on behalf of customers acting as an intermediary between them and the market. Customers delegate, to the trader, the selling of their production or the purchasing of their needs. The increasing role of this type of entity in Electricity Spot Markets, turns it into an important feature of our simulator.

VPP agents

From the point of view of the multi-agent system, VPP are seen as coalitions of agents, requiring specific procedures for coalition formation. Once a coalition is established, it can aggregate more agents or even discard some agents. This model allows the modeling of all the decision making concerning VPP formation and also subsequence aggregation of more producers.

To take decisions about these issues, VPPs have to detain knowledge related with the existing producers, which can eventually be aggregated. Decision concerning VPP formation and aggregation of new producers results mainly from two distinct matters. On one hand, each VPP classifies the producers according to criteria defined by itself. On the other hand, it establishes the goals of VPP formation or of VPP aggregation of more producers, according to its operating strategies and to its necessities at the moment. Aggregation proposals are then elaborated on in function of the resulting knowledge.

MASCEM considers that VPPs can also aggregate consumers, what this means is that they can act in the market as both sellers and buyers, according to their needs in each period.

Considering the VPP formation process as finished, the VPP needs to co-ordinate its operation. The VPP must place bids in the market, considering the contracts with producers, the generation forecast, the reserves and its market strategy.

VPP agents have the same market interface as Seller or Buyer agents. According to its members, generation capabilities and consumption needs for a given period the agent will need to sell or buy electricity.

However, as VPPs are themselves a set of other agents, there are some preliminary steps to define its bids.

Firstly, all the capacity available from the different aggregated distributed energy resources must be gathered to establish the electricity amount to trade on the market. The different generation costs must be analyzed to define the interval for envisaged proposals. This means VPP agents will have an utility function that aggregates all the involved units' characteristics. The analysis of the aggregated producers' proposals will be done according to each unit capabilities and costs.

After the market session, the VPP agent undertakes an internal dispatch, analyzing and adjusting its generation and reserve to maximize profits. VPP informs the aggregated producers about their dispatch.

Finally, in function of the generation, the used and unused reserve of each producer and the established contracts of the VPP fulfillment, the VPP determines the producers' remuneration.

MASCEM negotiation

On the basis of the results obtained in a period, Sellers and Buyers revise their strategies for the next S-period. Seller and Buyer Agents have strategic behavior to define their desired price. These agents have time-dependent strategies, and behavior-dependent strategies, to define the price in the following periods according to the results obtained in the previous ones.

To adjust price between S-periods, also referred to as behavior-dependent strategies, MASCEM provides two basic strategies: one called Composed Goal Directed and another called Adapted Derivative Following. These are important strategies that use the knowledge obtained with past experiences to define bid prices for next periods.

The Composed Goal Directed strategy is based on two consecutive objectives, the first one is selling (or buying) all the available capacity (power needed) and then increase the profit (reduce the payoff).

The Adapted Derivative Following strategy is based on a Derivative Following strategy proposed by Greenwald.[31] The Adapted Derivative-Following strategy adjusts its price by looking to the amount of revenue earned in the previous S-period as a result of the previous period's price change. If the last period's price change produced more revenue per good than the previous period, then the strategy makes a similar change in price. If the previous change produced less revenue per good, then the strategy makes a different price change.

The price adjustment is based on the same calculation for both strategies and takes into account the difference between the desired results and the obtained results in the previous period (details and case-study in Ref. 32).

In general, using a very simplified approach, one can say that the only goal of the player that participates in a market is to maximize its profits or minimize its costs. However, this does not correspond to the reality when there are more factors to consider than the economic ones or when these must be considered over relatively long periods of time.

In the case of electricity markets, buyers, when representing electricity consumers or consumer aggregations, although aiming at minimizing their costs, can have as a primary goal to acquire all the required amount of electrical energy. In this case, the Composed Goal Directed strategy is more adapted to their objectives.

On the contrast, sellers, representing directly a producer or an aggregation of producers, have profit maximization as the primary goal. Although this can seem very clear, there are some intrinsic factors that can make this problem more complex. For instance, profits must take into account not only the amount of produced energy but also the conditions of this production. According to the used production technology, it may be better to produce less in certain periods to attain better results in alternative periods. Moreover, penalties, which are applied in some market models, must also be considered.

According to each player's model and knowledge, these strategies are composed with more specific strategies, giving place to specially tailored strategies for each agent. For example, in the case of producers, the specific strategies take into account the production technology. In the case of production technologies based on renewable sources, highly dependent from weather factors, these are considered. For each player, all relevant strategies are composed, according to the player defined goals and to the identified situation. In this way, player strategic behavior depends on several aspects, namely the following:

- Player defined goals;
- Player model (including technical characteristics);
- Player knowledge (namely concerning other players' models);
- Context (taking into account factors of different nature, including market regulation, external factors such as oil prices, weather, which is considered in the player model but also in a more general context, namely for load forecasting, ...).

This approach makes players' strategies adaptive both to each player and to each situation.

24.4.2 *Electricity markets strategies*

MASCEM implements four types of strategies to change the price during a negotiation period: Determined, Anxious, Moderate and Gluttonous. The difference between these strategies is the time instant at which the agent starts to modify the price and the amount it changes. Although time-dependent strategies are simple to understand and implement,[33] they are very important since they allow the simulation of important issues such as: emotional aspects and different risk behaviors. For example: an agent using a determined strategy is a risk indifferent one; while Gluttonous agents exhibit the behavior of being more risk disposable, since they maintain the same price until very close to the end of the negotiation period, taking the risk of not selling.

To adjust price between negotiation periods, also referred to as behavior-dependent strategies, two different strategies were implemented: one called Composed Goal Directed and another called Adapted Derivative Following. These are important strategies that use the knowledge obtained with past experiences to define bid prices for next periods.

To obtain an efficient decision support, Seller and Buyer agents also have the capability of using the Scenario Analysis Algorithm.

This algorithm provides more complex support to develop and implement dynamic pricing strategies since each agent analyzes and develops a strategic bid, for the next period, taking into account not only their previous results but also other players results and expected future reactions. The algorithm is based on analyzing several bids under different scenarios and applying a decision method, based on the Game-Theory, to select the bid to propose.

24.5 Conclusions and Future Perspectives

The usage of renewable energies integration in power systems have been extended widely in the last decade, given that diversity of this renewable energy has increasingly been the use of the optimization techniques. The problem exposed in this work can be formulated as a mixed-integer linear problem. This problem is very hard to solve due to the binary decision variables, and the several features that have to be taken into account in the problem formulation.

In future works we propose an integral methodology that might optimally schedule reliable integration of renewable energy into power systems. The main idea is the correction of the wind or solar forecast error and therefore, the intermittency of wind power generation in open electricity market. This methodology will consider the weather condition data base from the satellite, which can be used into a wind power forecast process. It should be noted that the wind power forecast is not exact at all. This error can be corrected using storage energy systems. This system can be used as well as several other technologies, such as hydrogen, batteries, cogeneration, flywheels, capacitors, pumping, etc. Some of these storage technologies can be used not only for the correction of the wind power forecast error, but also for industrial processes or for selling hydrogen.

The VPPs stated in this work function as Generation Control Center (GCC). By this way, the results of the optimization problem can be adopted by the GCC to control efficiently all the parameters of the grid and generation elements, increasing the profitability of the whole power system.

Finally, the problem can be formulated as an optimization model, taking into account the total cost of each technology investment, maintenance and operation cost. The results will be the type of technology, the capacity and the location which can be obtained by the aforementioned optimization technique.

Acknowledgments

The authors would like to acknowledge the Portuguese Science and Technology Foundation (FCT), the European Fund for Regional Development (FEDER), the Operational Program for Science Technology and Innovation (POCTI), the Information Society Operational Program (POSI), the Operational Program Science and Innovation (POCI), the Operational Program for the Knowledge Society (POSC) and the Program of Projects in all Scientific Areas (PTDC) for their support to R&D Projects and GECAD Unit.

References

1. J.G. Slootweg, "Wind power — modelling and impact on power system dynamics," Ph.D. thesis, Technical University of Delft, Netherlands (2003).
2. T. Ackermann, K. Garner and A. Gardiner, "Embedded wind generation in weak grids–economic optimisation and power quality simulation," *Renewable Energy* **18** (1999) 205–221.
3. J.O.G. Tande, "Exploitation of wind-energy resources in proximity to weak electric grids," *Applied Energy* **65** (2000) 395–401.
4. R. Doherty and M. O'Malley, "A new approach to quantify reserve demand in systems with significant installed wind capacity," *IEEE Trans. Power Systems* **20** (2005) 587–595.
5. J.K. Kaldellis, K.A. Kavadias, A.E. Filio and S. Garofallakis, "Income loss due to wind energy rejected by the Crete island electrical network–the present situation," *Applied Energy* **79** (2004) 127–144.
6. C. Rehtanz, *Autonomous Systems and Intelligent Agents in Power System Control and Operation* (Springer, 2003).
7. J. Bertsch *et al.*, "Wide-area protection and power system utilization," *Proc. IEEE* **93** (2005) 997–1003.
8. T. Ackermann *et al.*, "European balancing act," *IEEE Power and Energy Magazine* **5** (2007) 90–103.
9. P.E. Morthorst, "Wind power and the conditions at a liberalized power market," *Wind Energy* **6** (2003) 297–308.
10. H. Holttinen, "Optimal electricity market for wind power," *Energy Policy* 33 (2005) 2052–2063.
11. T. J'onsson, "Forecasting of electricity prices accounting for wind power predictions," Master's Thesis, Technical University of Denmark, Kongens Lyngby, Denmark (2008), http://www.imm.dtu.dk.
12. B. Flechener, "Business management systems for energy trading in an open electricity market," (1998).
13. IEA, International Energy Agency, "Distributed generation in liberalised electricity markets," (2002).
14. K.W. El-Khattam, K. Bhattacharya, Y. Hegazy and M.M.A. Salama, "Optimal investment planning for distributed generation in a competitive electricity market," *IEEE Trans. Power Systems*, vol. 19, August 2004.
15. A. Bertani, C. Bossi, F. Fornari, S. Massucco, S. Spelta and F. Tivegna, "A microturbine generation system for grid connected and islanding operation," *Power Systems Conf. and Exposition, 2004. IEEE PES*, 10–13 October 2004, pp. 360–365.
16. M. Shahidehpour, H. Yamin and L. Zuyi, *Market Operations in Electric Power Systems: Forecasting, Scheduling and Risk Management* (John Wiley & Sons, 2002).
17. L. Tesfatsion, "General software and toolkits: Agent-based computational economics," Economics Department, Iowa State University, http://www.econ.iastate.edu/tesfatsi/acecode.htm.
18. H. Morais, M. Cardoso, L. Castanheira, Z. Vale and I. Praça, "A decision-support simulation tool for virtual power producers," *Future Power Systems* (2005).
19. PowerNext, www.powernext.fr.
20. Elexon, www.elexon.co.uk.
21. UKPX, www.apxgroup.com/index.php?id=39.

22. RTE, www.rte-france.com.
23. RETScreen, www.retscreen.net.
24. I. Praça, C. Ramos, Z.A. Vale and M. Cordeiro, "MASCEM: A multi-agent system that simulates competitive electricity markets," *IEEE Intelligent Systems* **18** (2003) 54–60.
25. G. Giebel, "The State-of-art in short-term prediction of wind power," Anemos Deliverable Report (2003).
26. F. Azevedo and Z. Vale, "Forecasting electricity prices with historical statistical information using neural networks and clustering techniques," *IEEE Power Systems Conf. and Exposition (PSCE'06)*, Atlanta, November 2006.
27. OXERA Consulting Ltd, "Electricity liberalisation — indicators in Europe," (2001).
28. H. Morais, M. Cardoso, L. Castanheira, Z. Vale and I. Praça, "Producers remuneration by virtual power producers," *WSEAS Trans. Power Systems* **1** (2006) 1358–1365.
29. I. Praça, H. Morais, M. Cardoso, C. Ramos and Z. Vale, "Virtual power producers integration into mascem," in *Establishing The Foundation Of Collaborative Networks* (Springer, 2007).
30. A. Greenwald, J. Kephart and G. Tesauro, "Strategic pricebots dynamics," *Proc. 1st ACM Conf. on Electronic Commerce* (1999), pp. 58–67.
31. M. Wellman, A. Greenwald and P. Stone, *Autonomous Bidding Agents: Strategies and Lessons from the Trading Agent Competition* (MIT Press, 2007).
32. I. Praça, C. Ramos, Z. Vale and M. Cordeiro, "Intelligent agents for negotiation and game-based decision support in electricity markets," *Int. J. Engineering Intelligent Systems* **13** (2005) 147–154.
33. J. Morris, A. Greenwald and P. Maes, "Learning curve: A simulationbased approach to dynamic pricing," *Electronic Commerce Research* **3** (2003) 245–276.

Induction Generators, Power Quality, Power Electronics and Energy Planning for Renewable Energy Systems

Chapter 25

Modern Power Electronic Technology for the Integration of Renewable Energy Sources

Vinod Kumar[*], Ramesh C. Bansal[†], Raghuveer R. Joshi[*], Rajendrasinh B. Jadeja[‡] and Uday P. Mhaskar[†]

[*]*Department of Electrical Engineering,*
Maharana Pratap University of Agriculture & Technology,
Udaipur, India
vinodcte@yahoo.co.in

[†]*School of Information Technology and Electrical Engineering,*
The University of Queensland, St. Lucia,
Queensland 4072, Australia

[‡]*Department of Electrical and Electronics Engineering,*
CU Shah College of Engineering & Technology,
Wadhwan, India

In this chapter, new trends in power-electronic technology for the integration of renewable energy sources like wind/photovoltaic and energy-storage systems are presented along with the current technology and future trends in variable-speed wind turbines. Also, the research trends in energy-storage systems used for the grid integration of intermittent renewable energy sources are discussed in detail.

25.1 Introduction

During the last few years, power electronics have undergone a fast evolution, which is mainly due to two factors. The first one is the development of fast semiconductor switches that are capable of switching quickly and handling high current/voltage. The second factor is the introduction of real-time computer controllers that can implement advanced and complex control algorithms. These factors together have led to the development of cost-effective and grid-friendly converters. It is expected that current developments in gearless energy transmission with power-electronic grid interface will lead to a new generation of quiet, efficient, and economical wind turbines of size more than 7 MW.

Energy storage in an electricity generation and supply system enables the decoupling of electricity generation from demand. In other words, the electricity that can be produced at times of either low-demand/low-generation cost or from intermittent renewable energy sources is shifted for release at the time of high-demand/high-generation cost or when no other generation is available. Appropriate integration of renewable energy sources with storage systems allows for a greater market penetration and results in primary energy and emission savings.

In this chapter, most common topologies are selected and discussed in respect to advantages and drawbacks along with current and new trends in power-electronic technology for the integration of renewable energy sources and energy-storage systems.

25.2   Various Topologies of Power Electronic Converters

Different power converters can be used in wind turbine applications. In the case of an induction generator, the power converter converts from a variable voltage and frequency to a fixed voltage and frequency. This may be implemented in different ways, as it will be seen in the next section. The most commonly used topology so far is a soft-starter, which is used during start up in order to limit the in-rush current and thereby reduce the disturbances to the grid.

25.2.1   *Soft starter*

The soft starter is a power converter, which has been introduced to fixed speed wind turbines to reduce the transient current during connection or disconnection of the generator to/from the grid. When the generator speed exceeds the synchronous speed, the soft-starter is connected. Using firing angle control of the thyristors in the soft starter the generator is smoothly connected to the grid over a predefined number of grid periods (see Fig. 25.1).

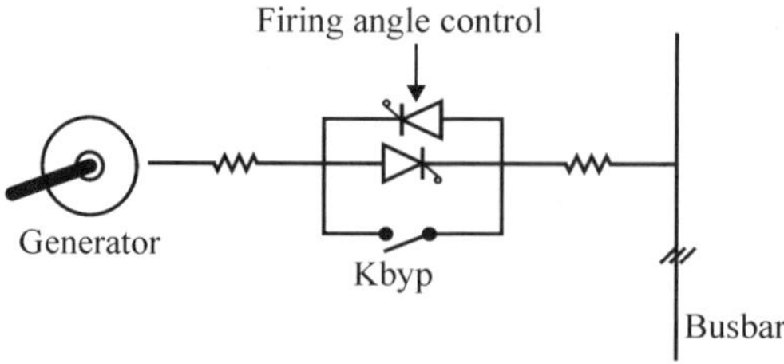

Fig. 25.1.   Connection diagram of soft starter with generators.

The commutating devices are two thyristors for each phase. These are connected in anti-parallel. The relationship between the firing angle (a) and the resulting amplification of the soft starter is voltage and depends on the power factor of the connected element.

When the generator is connected to the grid a contactor bypasses the soft-starter in order to reduce the losses during normal operation. The soft-starter is very cheap and it is a standard converter in many fixed wind turbines.

25.2.2 *Diode rectifier*

The diode rectifier is the most commonly used topology in power electronic applications. For a three-phase system it consists of six diodes. It is shown in Fig. 25.2. The diode rectifier can only be used in one quadrant, it is simple and it is not possible to control it. It can be used in some applications such as pre-charging.

25.2.3 *The back-to-back Pulse Width Modulated-Voltage Source Inverter (PWM-VSI)*

The back-to-back PWM-VSI is a bi-directional power converter consisting of two conventional PWM-VSI as shown in Fig. 25.3.

To achieve full control of the grid current, the DC-link voltage must be boosted to a level higher than the amplitude of the grid line-line voltage. The power flow of the grid side converter is controlled in order to keep the DC-link voltage constant, while the control of the generator side is set to suit the magnetization demand and

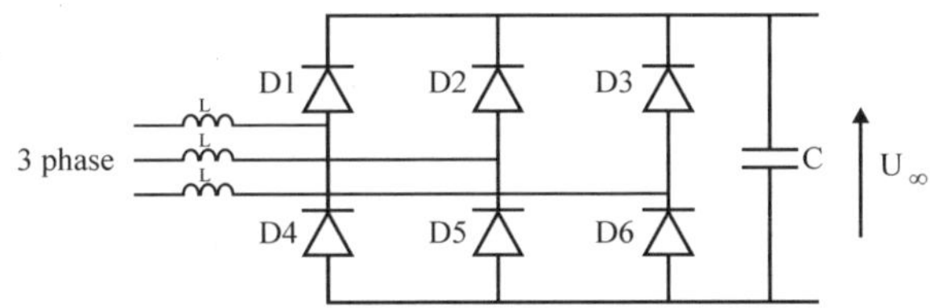

Fig. 25.2.   Diode rectifier for three-phase ac/dc conversion.

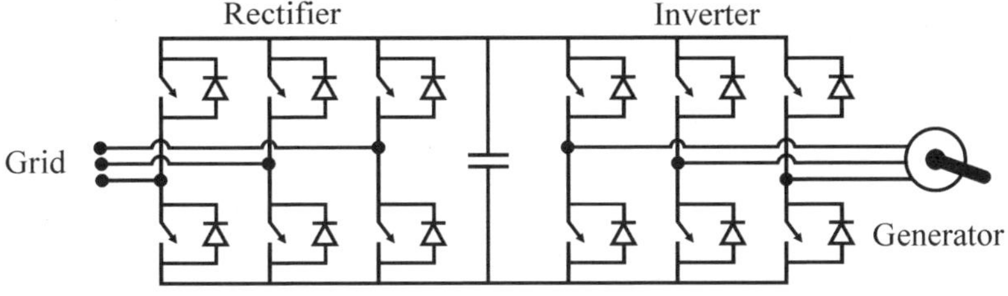

Fig. 25.3.   The back-to-back PWM-VSI converter topology.

the reference speed. The control of the back-to-back PWM-VSI in the wind turbine application is described in several papers.[1-4]

Advantages related to the use of the back-to-back PWM-VSI: The PWM-VSI is the most frequently used three-phase frequency converter. As a consequence of this, the knowledge available in the field is extensive and well established. The literature and the available documentation exceed that for any of the other converters considered in this survey. Furthermore, many manufacturers produce components especially designed for use in this type of converter (e.g., a transistor-pack comprising six bridge coupled transistors and anti paralleled diodes). Due to this, the component costs is low compared to converters requiring components designed for a niche production.

A technical advantage of the PWM-VSI is the capacitor decoupling between the grid inverter and the generator inverter. Besides some protection, this decoupling offers separate control of the two inverters, allowing compensation of asymmetry both on the generator side and on the grid side, independently. The inclusion of a boost inductance in the DC-link circuit increases the component count, but a positive effect is that the boost inductance reduces the demands on the performance of the grid side harmonic filter, and offers some protection of the converter against abnormal conditions on the grid.

Disadvantages of applying the back-to-back PWM-VSI: Some of the reported disadvantages of the back-to-back PWM-VSI which justify the search for a more suitable alternative converter. In several papers concerning adjustable speed drives, the presence of the DC-link capacitor is mentioned as a drawback, since it is heavy and bulky, it increases the costs and may be of most importance — it reduces the overall life of the system.[5,6]

Another important drawback of a back-to-back PWM-VSI is the switching losses. Every commutation in both the grid inverter and the generator inverter between the upper and lower DC-link branch is associated with a hard switching and a natural commutation. Since the back-to-back PWM-VSI consists of two inverters, switching losses might be even more pronounced. The high switching speed to the grid may also require extra EMI-filters. To prevent high stresses on the generator insulation and to avoid bearing current problems,[13] the voltage gradient may have to be limited by applying an output filter.

25.2.4 *Tandem converter*

The tandem converter is quite a new topology and a few papers only have treated it up till now.[7-9] However, the idea behind the converter is similar to those presented in Ref. 10, where the PWM-VSI is used as an active harmonic filter to compensate harmonic distortion. The topology of the tandem converter is shown in Fig. 25.4.

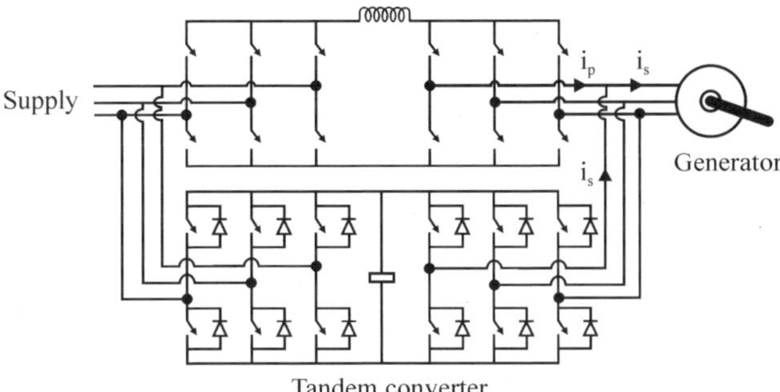

Fig. 25.4.   The tandem converter topology used in an induction generator wind turbine system.

The tandem converter consists of a current source converter (CSC), designated as the primary converter, and a back-to-back PWM-VSI, designated as the secondary converter. Since the tandem converter consists of four controllable inverters, several degrees of freedom exist which enable sinusoidal input and sinusoidal output currents. However, in this context it is believed that the most advantageous control of the inverters is to control the primary converter to operate in square-wave current mode. Here, the switches in the CSC are turned on and off only once per fundamental period of the input and output current, respectively. In square wave current mode, the switches in the primary converter may either be GTO's, or a series connection of an IGBT and a diode.

In order to achieve full control of the current to/from the back-to-back PWM-VSI, the DC-link voltage is boosted to a level above the grid voltage. As mentioned, the control of the tandem converter is treated only in few papers. However, the independent control of the CSC and the back-to-back PWM-VSI are both well established.[11−14]

Advantages in the use of the tandem converter: The investigation of new converter topologies is commonly justified by the search for higher converter efficiency. Advantages of the tandem converter are the low switching frequency of the primary converter, and the low level of the switched current in the secondary converter. It is stated that the switching losses of a tandem inverter may be reduced by 70%[9] in comparison with those of an equivalent voltage source inverter (VSI), and even though the conduction losses are higher for the tandem converter, the overall converter efficiency may be increased.

Compared to the current source inverter (CSI), the voltage across the terminals of the tandem converter contains no voltage spikes since the DC-link capacitor of the secondary converter is always connected between each pair of input and output lines.[8] Concerning the dynamic properties[9], it states that the overall performance of the tandem converter is superior to both the CSC and the VSI. This is because current magnitude commands are handled by the voltage source converter, while phase-shift current commands are handled by the current source converter.[10]

Besides the main function, which is to compensate the current distortion introduced by the primary converter, the secondary converter may also act like an active resistor, providing damping of the primary inverter in light load conditions.[10]

Disadvantages of using the tandem converter: An inherent obstacle to applying the tandem converter is the high number of components and sensors required. This increases the costs and complexity of both hardware and software. The complexity is justified by the redundancy of the system,[8] however the system is only truly redundant if a reduction in power capability and performance is acceptable. Since the voltage across the generator terminals is set by the secondary inverter, the voltage stresses at the converter are high. Therefore the demands on the output filter are comparable to those when applying the back-to-back PWM-VSI.

By applying the CSI as the primary converter, only 0.866% of the grid voltage can be utilized. This means that the generator currents (and also the current through the switches) for the tandem converter must be higher in order to achieve the same power.

25.2.5   *Matrix converter*

Ideally, the matrix converter should be an all silicon solution with no passive components in the power circuit. The conventional matrix converter topology is shown in Fig. 25.5.

The basic idea of the matrix converter is to obtain a desired input current (to/from the supply), a desired output voltage and a desired output frequency may be obtained by properly connecting the output terminals of the converter to the input terminals of the converter. In order to protect the converter, the following two control rules must be complied with: Two (or three) switches in an output leg are never allowed

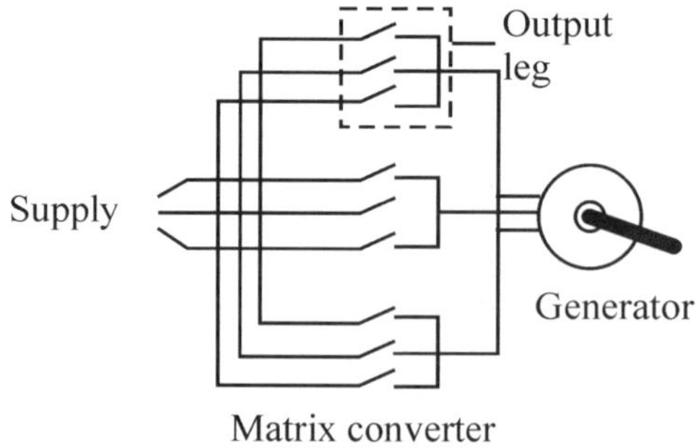

Fig. 25.5.   The conventional matrix converter topology.

to be on at the same time. All of the three output phases must be connected to an input phase at any instant of time. The actual combination of the switches depends on the modulation strategy.

Advantages of using the matrix converter: This section summarizes some of the advantages of using the matrix converter in the control of an induction wind turbine generator. For a low output frequency of the converter the thermal stresses of the semiconductors in a conventional inverter are higher than those in a matrix converter. This arises from the fact that the semiconductors in a matrix converter are equally stressed, at least during every period of the grid voltage, while the period for the conventional inverter equals the output frequency. This reduces the thermal design problems for the matrix converter.

Although the matrix converter includes six additional power switches compared to the back-to-back PWM-VSI, the absence of the DC-link capacitor may increase the efficiency and the life time for the converter.[15,16] Depending on the realization of the bi-directional switches, the switching losses of the matrix inverter may be less than those of the PWM-VSI, because the half of the switching become natural commutations.[16]

Disadvantages and problems of the matrix converter: A disadvantage of the matrix converter is the intrinsic limitation of the output voltage. Without entering the over-modulation range, the maximum output voltage of the matrix converter is 0.866 times the input voltage. To achieve the same output power as the back-to-back PWM-VSI, the output current of the matrix converter has to be 1.15 times higher, giving rise to higher conducting.[16]

In many papers concerning the matrix converter, the unavailability of a true bi-directional switch is mentioned as one of the major obstacles for the propagation of the matrix converter. In the literature, three proposals for realizing a bi-directional switch exists. The diode embedded switch[17] which acts like a true bi-directional switch, the common emitter switch and the common collector switch.[18] The latter two are able to control the current direction, which is preferable in the phase commutations.

Since real switches do not have infinite signal switching times (which is not desirable either) the commutation between two input phases constitutes a contradiction between the two basic control rules of the matrix converter. In the literature at least six different commutation strategies are reported.[19–21] The most simple of the commutation strategies are those reported in Ref. 19 but neither of these strategies comply with the basic control rules. The most commonly used commutation strategies are those reported in Refs. 20 and 22. In either reference half of the switching become soft switching, thereby reducing the switching losses. The solutions in Ref. 21 require a more complex hardware structure of the converter.

Due to the absence of the DC-link, there is no decoupling between the input and output of the converter. In ideal terms, this is not a problem but in the case of unbalanced or distorted input voltages, or unbalanced load, the input current and the output voltage also become distorted. Several papers have dealt the problems of unbalanced input voltages and various solutions have been proposed.[23–28] Each of the solutions has superiorities and drawbacks, and the choice of solution depends on which performance indicator has the highest priority (input disturbance, line losses, controllability of the input power factor, etc.).

Finally, the protection of the matrix converter in a fault situation presents a problem. The protection of the matrix converter is treated in Ref. 27. This protection circuit adds an extra 12 diodes and a DC-link capacitor to the component list, although these components are rated much smaller than the components in the power part of the matrix converter.

25.2.6 *Multilevel converter*

Since the development of the neutral-point clamped three-level converter,[29] several alternative multilevel converter topologies have been reported in the literature. The general idea behind the multilevel converter technology is to create a sinusoidal voltage from several levels of voltages, typically obtained from capacitor voltage sources. The multilevel converter topologies can be classified with the following five categories[30–32]:

- Multilevel configuration with diode clamps.[29,33–35]
- Multilevel configuration with bi-directional switch interconnection.[36,29]
- Multilevel configurations with flying capacitors.[37]
- Multilevel configurations with multiple three-phase inverters.[38]
- Multilevel configurations with cascaded single phase H-bridge inverters.[39]

A common feature of the five different multilevel converter concepts is, that in theory, all the topologies may be constructed to have an arbitrary number of voltage levels, although in practice some topologies are easier to realize than others. The principle of the five topologies is illustrated in Fig. 25.6.

Below is a brief comment on the different multi-level converter topologies:

(a) Despite the more complex structure, the diode clamped multilevel converter is very similar to the well known back-to-back PWM-VSI. Unlike the multilevel topologies shown in Figs. 25.6(d) and 25.6(e), the diode clamped multilevel

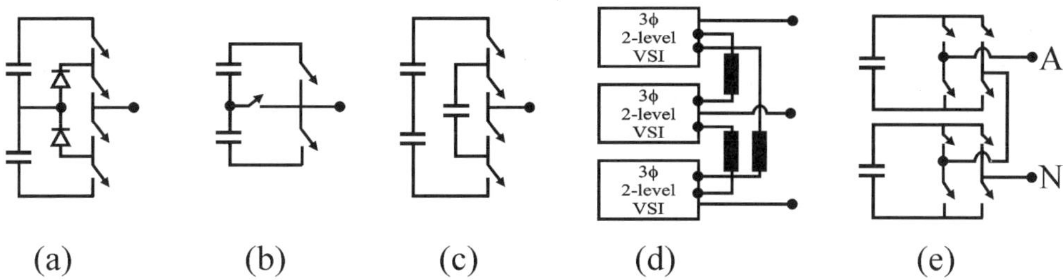

Fig. 25.6.   Multilevel topologies, (a) one inverter leg of a three-level diode clamped multilevel converter, (b) one inverter leg of a three-level multilevel converter with bidirectional switch interconnection, (c) one inverter leg of a three level flying capacitor multilevel converter, (d) schematic presentation of a three-level converter consisting of three three-phase inverters, (e) one inverter leg of a three-level converter consisting of H-bridge inverters.

converter may be coupled directly to the grid without transformers. For a converter based on the diode clamped multilevel converter a voltage-balancing problem occurs for levels higher than three, but for a three level converter this is only a minor problem.

(b) For a three level structure, the topology of the multilevel converter with bidirectional switch interconnection requires the same number of switches as the diode clamped three-level converter and the three-level flying capacitor converter. However, half of the switches have to block the full DC-link voltage.

(c) The topology of the flying capacitor multilevel converter is very similar to that of the diode clamped multilevel converter shown in Fig. 25.6(a). In the literature it is stated, that the voltage-balancing problem is relatively easy to solve, compared to the diode clamped converter. The difference in component count between the diode clamped multilevel converter and the flying capacitor multilevel converter is, that two diodes per phase may be substituted by one capacitor.

(d) The number of switches (and other components) required realizing a three level converter is very high compared to the concepts a-c, but the converter can be realized from the well-proven VSI technology.

(e) The multilevel converter based on multiple H-bridge inverters is heavy, bulky and complex,[33] and may be of most importance, connecting separated DC-sources between two converters in a back-to-back fashion is very difficult because a short circuit will occur when two back-to-back converters are not switching synchronously.

Advantages of using multilevel converter: Main purpose of the multilevel converter is to achieve a higher voltage capability of the converters. As the ratings of the components increases and the switching- and conducting properties improve, the secondary effects of applying multilevel converters become more and more advantageous. In recent papers, the reduced content of harmonics in the input and output voltage is highlighted, together with the reduced EMI. The multilevel converter distinguishes itself with the lowest demands to the input filters. (or alternatively reduced number of switching).

The switching losses of the multilevel converter are another feature, which is often accentuated,[40] it is stated, that for the same harmonic performance the switching frequency can be reduced to 25% of the switching frequency of a two-level converter. Even though the conducting losses are higher for the multilevel converter, it is stated in Ref. 41 that the overall efficiency for the diode clamped multilevel converter is higher than the efficiency for a comparable two-level converter. Of course, the truth in this assertion depends on the ratio between the switching losses and the conducting losses.

Disadvantages concerning the multilevel converter: The most commonly reported disadvantage of the three level converters with split DC-link is the voltage imbalance between the upper and the lower DC-link capacitor. However, for a three-level converter this problem is not very serious, and the problem in the three-level converter is mainly caused by differences in the real capacitance of each capacitor, inaccuracies in the dead-time implementation or an unbalanced load. By a proper modulation control of the switches, the imbalance problem can be solved. However, whether the voltage-balancing problem is solved by hardware or software, it is necessary to measure the voltage across the capacitors in the DC-link.

For converters based on the topology in Figs. 25.6(a) to 25.6(c), another problem is the unequal current stress on the semiconductors. It appears that the upper and lower switches in an inverter branch might be de-rated compared to the switches in the middle. For an appropriate design of the converter, different devices are required. For the topology in Fig. 25.6(b) it appears that both the unequal current stress and the unequal voltage stress might constitute a design problem.

From the topologies in Fig. 25.6, it is evident that the number of semiconductors in the conducting path is higher than for the other converters treated in this survey; this might increase the conduction losses of the converter. On the other hand, each of the semiconductors need only block half the total DC-link voltage and for lower voltage ratings, the on-state loss per switch decreases, which to a certain extent to justify the higher number of semiconductors in the conducting path.

25.2.7 *Resonant converter*

The efforts towards reducing the switching losses in power converters, several resonant converter topologies have been proposed. A common drawback of these converter topologies is that they suffer from one or more of the following properties:

- Complex hardware structure and complex control
- High peak voltage in the DC-link and across the load
- High power flow through the resonant circuit

The only resonant converter treated in this section is a topology, which does not suffer from the disadvantages mentioned above. The converter topology is termed *Natural Clamped Converter* (NCC). The NCC topology is shown in Fig. 25.7. The NCC consists of the conventional back-to-back PWM-VSI and in addition a circuit to obtain the resonance. To start and to maintain the resonance an inductive energy transfer is used. In order to control the current to/from the supply grid, the DC-link voltage is boosted. Like for the back-to-back PWM-VSI, it is assumed that the voltage is boosted to 800 V. In Fig. 25.7, the boost inductances are shown as a part of the line filter. The used switches in the resonance circuit are bi-directional switches and as in the case of the matrix converter, these have to be built from several discrete devices.

The bi-directional switch shown in Fig. 25.7 is made from a standard two-pack transistor module and additional two diodes, but may as well be constructed as a common-emitter, a common-collector or a diode-embedded switch.

Advantages in the use of the NCC: Compared to other resonant converters, the NCC possesses the advantageous properties of the different resonant converters while the unsuitable properties are avoided. The advantageous properties of the NCC are: the resonant voltage is clamped and never exceeds the DC-link voltage, the resonant circuit is not in the power part of the converter, and only one resonant circuit is needed for the entire converter. By inspection of the topology of Fig. 25.7, it appears that the NCC also supports three level operations. This implies however, that the

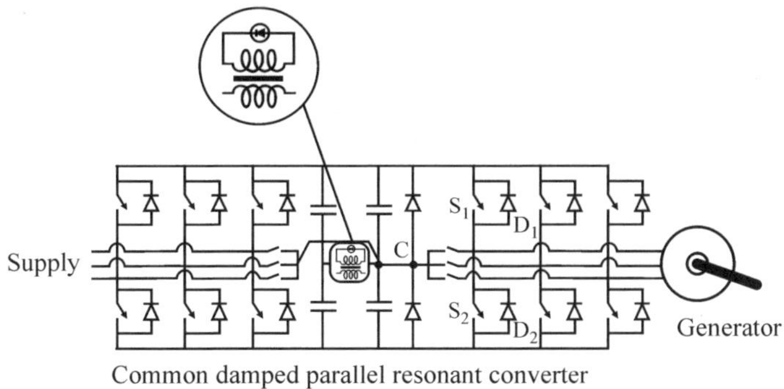

Fig. 25.7.   The natural clamped converter topology used in the induction generator wind turbine system.

bi-directional switches must be rated for a higher power level. In comparison with the back-to-back PWM-VSI, the main advantage of the NCC is the reduced switching losses, which in applications where the switching losses dominate might increase the overall converter efficiency. Furthermore, due to the lower rate of change of applied voltage (du/dt), the radiated EMI is reduced, and the demands to the output filter inductance reduced as well. The conduction losses of the NCC are similar to those of the back-to-back PWM-VSI.

The modulation of the NCC may be either PWM or Direct Pulse Modulation (DPM). It may also be possible to use a different modulation strategy in the machine inverter and in the grid inverter. The use of DPM makes the response of the converter very fast, while the use of PWM more or less gives a harmonic spectrum comparable to that of a back-to-back PWM-VSI.

Disadvantages of the NCC: Although the complexity is less than the pole commutated resonant converter, the converter is still more complex (hardware and control) than the back-to-back PWM-VSI and several additional components are needed. Furthermore, sensors are needed to detect the zero voltage events and to determine the energy needed to maintain the resonance.

As mentioned above, the use of the DPM technique yields fast response time, but a consequence of the DPM is that the switching harmonics are distributed over a broader spectrum. The broad switching harmonic spectrum makes the grid filter design procedure a quite difficult task since the harmonics may coincide with the resonant frequency of the filters. However, this increases the complexity of the converter. The voltage imbalance of the two main capacitors may constitute a problem for the NCC. For every commutation, only one of the main capacitors is discharged. In the case of asymmetrical commutations due to small differences in the switch characteristics, the voltage unbalance has to be compensated actively.

25.2.8 *Comparison of five types of frequency converters*

The five converters presented above: the back-to-back PWM VSI, the tandem converter, the matrix converter, the multilevel converter and the NCC, may be evaluated in terms of their applicability to wind turbine systems. For each converter, a presentation of the topology and the working principles has been presented, combined with a discussion of advantages and disadvantages. It is evident that the back-to-back PWM-VSI converter is highly relevant, as this converter is the one used in wind turbines today. Therefore, this converter type may be used as a reference in a benchmark of the other converter topologies:

- *Components and their ratings*. Considering the number of used components, the matrix converter differs from the other converters, because it consists only of active components in the power part. The back-to-back PWM-VSI converter includes a moderate number of passive components, while it has the least number of active components. The latter three converters include a high number of both active and passive components.

- *Auxiliary components.* Two converters distinguish themselves, namely the matrix converter and the multilevel converter. The main reason for this is that in both converters the transformer may be omitted, without requiring excessively high voltage ratings for the components. On the other hand, the tandem converter and the NCC both require a greater number of transducers than the other three converters.

- *Efficiency.* Assessment of the efficiency of the five converters is based on the literature and includes the reported conducting losses and switching losses. The NCC is assessed to have potentially the highest efficiency of the five converters followed by the multilevel converter and the tandem converter.

- *Harmonic performance.* Considering the harmonic performance, and requirements for filters, the multilevel converter shows the best spectra on both the grid side and the generator side. The only converter that really seems to constitute a problem for the filter design is the DPM modulated NCC. Due to a broad switching spectrum, the filter design may be complicated.

- *Implementation.* Because the back-to-back converter and the multilevel converter are the most used commercial converters of the five, it is believed that these two converters are the least troublesome converters to implement, while the latter three are all quite undiscovered in a commercial sense.

Summarizing the findings on the converters presented, it is concluded that the back-to-back converter, the matrix converter and the multilevel converter are to be recommended for further studies in different generator topologies.

25.3 Current Wind Power Technology

A new technology has been developed in the wind power market introducing variable-speed working conditions depending on the wind speed in order to optimize the energy captured from the wind. The advantages of variable-speed turbines are that their annual energy capture is about 5% greater than the fixed-speed technology, and that the active and reactive powers generated can be easily controlled. There is also less mechanical stress, and rapid power fluctuations are scarce because the rotor acts as a flywheel (storing energy in kinetic form). In general, no flicker problems occur with variable-speed turbines. Variable-speed turbines also allow the grid voltage to be controlled, as the reactive-power generation can be varied. As disadvantages, variable-speed wind turbines need a power converter that increases the component count and make the control more complex. The overall cost of the power electronics is about 7% of the whole wind turbine.

Variable-speed operation can only be achieved by decoupling the electrical grid frequency and mechanical rotor frequency. To this end, power-electronic converters are used, such as an ac–dc–ac converter, matrix converter combined with advanced control systems.

25.3.1 *Variable-speed concept utilizing Doubly Fed Induction Generator (DFIG)*

In a variable-speed wind turbine with DFIG[43,44] the converter feeds the rotor winding, while the stator winding is connected directly to the grid. This converter, thus decoupling mechanical and electrical frequencies and making variable-speed operation possible, can vary the electrical rotor frequency. This turbine cannot operate in the full range from zero to the rated speed, but the speed range is quite sufficient.

This limited speed range is caused by the fact that a converter is considerably smaller than the rated power of the machine. In principle, one can say that the ratio between the size of the converter and the wind-turbine rating is half of the rotor-speed span. In addition to the fact that the converter is smaller, the losses are also lower. The control possibilities of the reactive power are similar to the full power-converter system. For instance, the Spanish company Gamesa supplies this kind of variable-speed wind turbines to the market.

The forced switched power-converter scheme is shown in Fig. 25.8. The converter includes two three-phase ac–dc converters linked by a dc capacitor battery. This

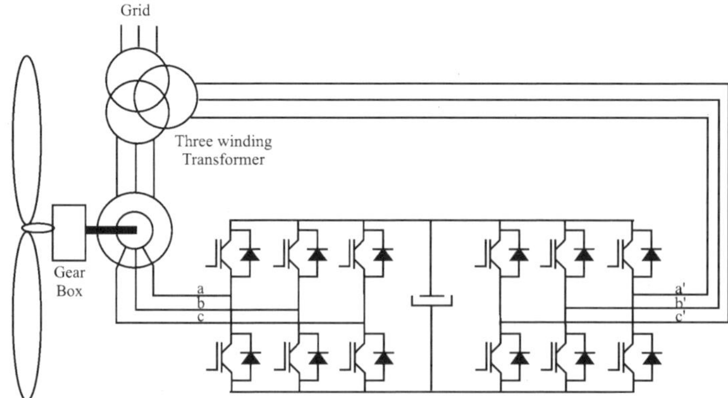

Fig. 25.8. Single doubly fed induction machine with two fully controlled ac–dc power converters.

scheme allows, on one hand, a vector control of the active and reactive powers of the machine, and on the other hand, a decrease by a high percentage of the harmonic content injected into the grid by the power converter.

Vestas and Nordic Wind power supply a variation of this design, which is the semi variable-speed turbine, in which the rotor resistance of the squirrel cage generator can be varied instantly using fast power electronics. So far, Vestas alone has succeeded in commercializing this system under the trade name Opti Slip.

A number of turbines, ranging from 600 kW to 2.75 MW, have now been equipped with this system, which allows transient rotor speed increase of up to 10% of the nominal value. In that case, the variable-speed conditions are achieved dissipating the energy within a resistor placed in the rotor, as shown in Fig. 25.9. Using that technology, the efficiency of the system decreases when the slip increases, and the speed control is limited to a narrow margin. This scheme includes the power converter and the resistors in the rotor. Trigger signals to the power switches are accomplished by optical coupling.

25.3.2 *Variable-speed concept utilizing full-power converter*

In this concept, the generator is completely decoupled from the grid.[45] The energy from the generator is rectified to a dc link and there after it is converted to suitable ac energy for the grid. The majority of these wind turbines are equipped with a multipole synchronous generator, although it is quite possible (but rather rare) to use an induction generator and a gearbox. There are several benefits of removing the gearbox: reduced losses, lower costs due to the elimination of this expensive component, and increased reliability due to the elimination of rotating mechanical components. Enercon supplies such technology.

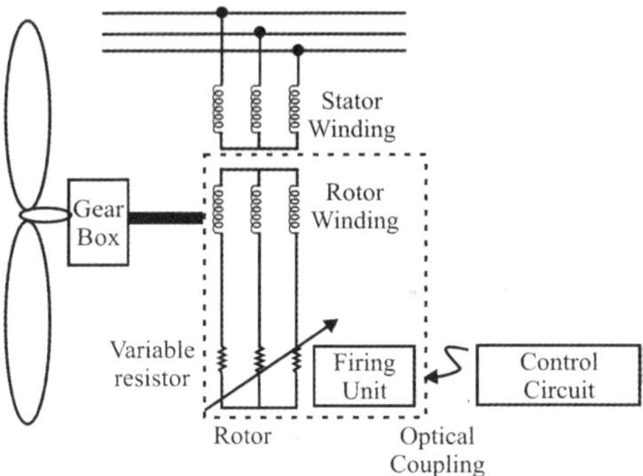

Fig. 25.9.   Single doubly fed induction machine controlled with slip power dissipation in an internal resistor.

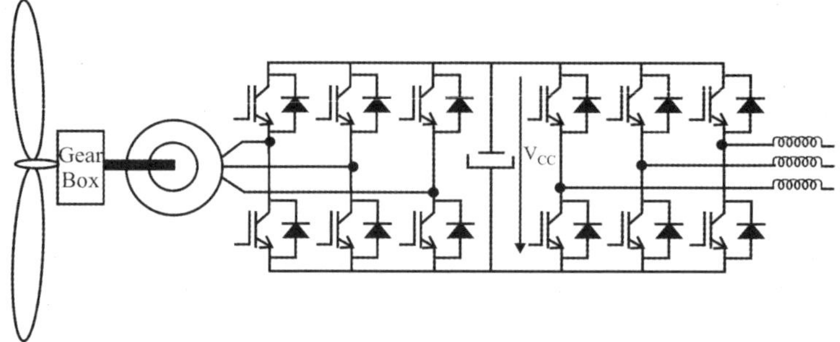

Fig. 25.10.   Double three-phase VSI.

Figure 25.10 shows the scheme of a full power converter for a wind turbine. The machine-side three-phase converter works as a driver controlling the torque generator, using a vector control strategy. The grid-side three-phase converter permits wind energy transfer into the grid and enables to control the amount of the active and reactive powers delivered to the grid. It also keeps the total-harmonic-distortion (THD) coefficient as low as possible, improving the quality of the energy injected into the public grid. The objective of the dc link is to act as energy storage, so that the captured energy from the wind is stored as a charge in the capacitors and may be instantaneously injected into the grid. The control signal is set to maintain a constant reference to the voltage of the dc link V_{dc}. An alternative to the power-conditioning system of a wind turbine is to use a synchronous generator instead of an induction one and to replace a three-phase converter (connected to the generator) by a three-phase diode rectifier and a chopper, as shown in Fig. 25.11. Such choice is based on the low cost as compared to an induction generator connected to a VSI used as a rectifier. When the speed of the synchronous generator alters, the voltage on the dc side of the diode rectifier will change. A step-up chopper is used to adapt the rectifier voltage to the dc-link voltage of the inverter. When the inverter system is analyzed, the generator/rectifier system can be modeled as an ideal current source. The step-up

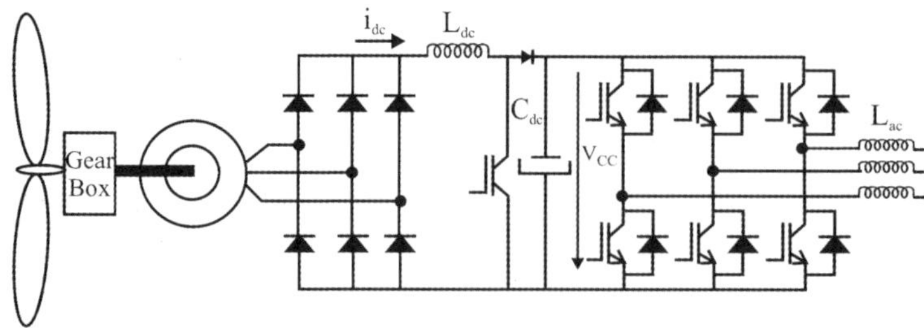

Fig. 25.11.   Step-up converter in the rectifier circuit and full power inverter topology used in wind-turbine applications.

chopper used as a rectifier utilizes a high switching frequency, so the bandwidth of these components is much higher than the bandwidth of the generator. Controlling the inductance current in the step-up converter can control the machine torque and, therefore, its speed. The Spanish Company MADE has marketed this design.

25.3.3 *Direct-drive technology for wind turbines*

Direct-drive applications are on increase because the gearbox can be eliminated. As compared to a conventional gearbox-coupled wind turbine generator, a direct-drive generator has reduced the overall size, has lower installation and maintenance cost, has a flexible control method and quick response to wind fluctuations, and load variation. For small wind turbine, permanent magnet synchronous machines are more popular because of their higher efficiency, high-power density, and robust rotor structure as compared to induction and synchronous machines.

A number of alternative concepts have been proposed for direct drive electrical generators for use in grid-connected or standalone wind turbines. In Ref. 53, the problem to adapt a standard permanent-magnet synchronous machine to a direct-drive application is presented. A complete design of a low-speed direct drive permanent-magnet generator for wind application is depicted in Refs. 54 and 55. A new trend that is very popular for propulsion systems applications is to use an axial flux machine.[56] These new machines are applied in small-scale wind and water-turbine direct-drive generators because higher torque density can be obtained in a more simple and easy way.

25.3.4 *Semiconductor-device technology*

Improvements in the performance and reliability of power-electronic variable frequency drives for wind-turbine applications have been directly related to the availability of power semiconductor devices with better electrical characteristics and lower prices because the device performance determines the size, weight, and cost of the entire power electronics used as interfaces in wind turbines. The IGBT is now the main component for power electronics and also for wind-turbine applications. They are now mature technology turn-on components adapted to a very high power (6 kV–1.2 kA), and they are in competition with GTOs for high power applications.[46]

Recently, the integrated gated control thyristor (IGCT) has been developed as a mechanical integration of a GTO plus a delicate hard drive circuit that transforms the GTO into a modern high-performance component with a large safe operation area (SOA), lower switching losses, and a short storage time.[47] The comparison between IGCT and IGBT for frequency converters that are used, especially in wind turbines, is explained below:

- IGBTs have higher switching frequency than IGCTs, so they introduce less distortion in the grid.
- IGCTs are made like disk devices. They have to be cooled with a cooling plate by electrical contact on the high-voltage side. This is a problem because high electromagnetic emission will occur. Another point of view is the number of allowed load cycles. Heating and cooling the device will always bring mechanical stress to the silicon chip, and it can be destroyed. This is a serious problem, especially in wind-turbine applications.

  On the other hand, IGBTs are built like modular devices. The silicon is isolated to the cooling plate and can be connected to the ground for low electromagnetic emission even with higher switching frequency. The base plate of this module is made of a special material that has exactly the same thermal behavior as silicon, so nearly no thermal stress occurs. This increases the lifetime of the device by ten folds approximately.
- The main advantage of IGCTs versus IGBTs is that they have a lower ON-state voltage drop, which is about 3.0 V for a 4500 V device. In this case, the power dissipation due to a voltage drop for a 1500-kW converter will be 2400 W per phase. On the other hand, in the case of IGBT, the voltage drop is higher than IGCTs. For a 1700 V device having a drop of 5 V, the power dissipation due to the voltage drop for a 1500 kW condition will be 5 kW per phase.

In conclusion, with the present semiconductor technology, IGBTs present better characteristics for frequency converters in general and especially for wind-turbine applications.

25.4   Future Trends in Wind-Power Technology

With rapid development of wind turbine technologies, future trends in the wind turbine industry will probably be focused on the gradual improvement of already known technologies, which can be summarized as follows:

- The power level of a single wind turbine will continue to increase, because this reduces the cost of placing wind turbines, especially for offshore wind farms.
- Offshore wind energy is more attractive, because of higher wind speed and more space than on shore wind energy.

- An increasing trend is to remove dispersed single wind turbine in favor of concentrated wind turbines in large wind farms.
- An increasing trend in the penetration of wind power into the power system.

25.4.1 *Transmission technology for the future — connecting wind generation to the grid*

One of the main trends in wind turbine technology is offshore installation. There are great wind resources at sea for installing wind turbines in many areas where the sea is relatively shallow. Offshore wind turbines may have slightly more favorable energy balance than onshore turbines, depending on the local wind conditions. In places where onshore wind turbines are typically placed on flat terrain, offshore wind turbines will generally yield some 50% more energy than a turbine placed on a nearby onshore site. The reason is that there is less friction on the sea surface. On the other hand, the construction and installation of a foundation requires 50% more energy than onshore turbines. It should be remembered, however, that offshore wind turbines have a longer life expectancy than onshore turbines, which is around 25–30 years. The reason is that the low turbulence at sea gives lower fatigue loads on the wind turbine.

Conventional heating-ventilation-air conditioning (HVAC) transmission systems are a simple and cost-efficient solution for the grid connection of wind farms. Unfortunately, for offshore wind parks, the distributed capacitance of undersea cables is much higher than that of overhead power lines. This implies that the maximum feasible length and power-transmission capacity of HVAC cables is limited. Grid access technology in the form of high-voltage dc (HVDC) can connect the wind-farm parks to the grid and transmit the power securely and efficiently to the load centers. Looking at the overall system economics, HVDC transmission systems are most competitive at transmission distances over 100 km or power levels of between approximately 200 and 900 MW. The HVDC transmission offers many advantages over HVAC.[48]

- Sending and receiving end frequencies are independent.
- Transmission distance using dc is not affected by cable charging current.
- Offshore installation is isolated from mainland disturbances and vice versa.
- Power flow is fully defined and controllable.
- Cable power losses are low.
- Power-transmission capability per cable is higher.

Classical HVDC transmission systems (as shown in Fig. 25.12(a)) are based on the current source converters with naturally commutated thyristors, which are the

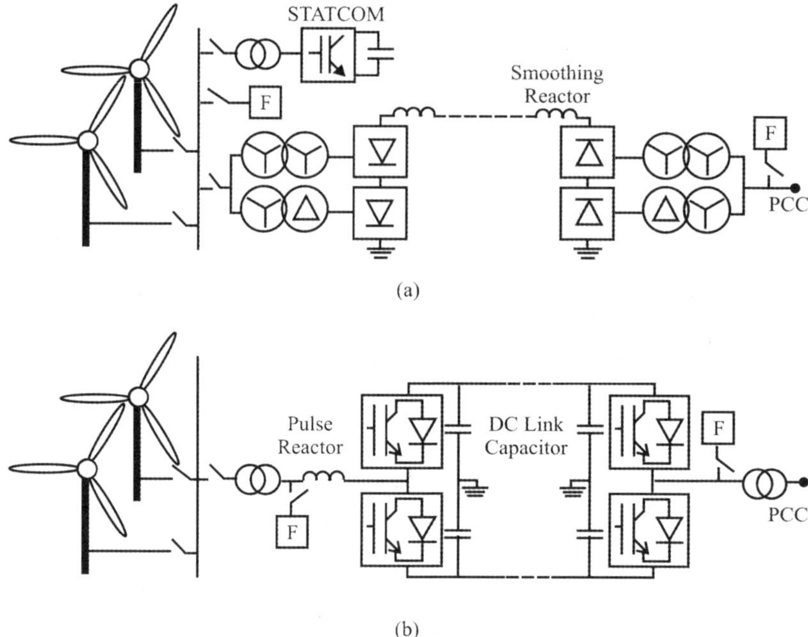

Fig. 25.12. Two HVDC transmission solutions, (a) Classical LCC-based system with STATCOM, (b) VSC-based system.

so-called line commutated converters (LCCs). This name originates from the fact that the applied thyristors need an ac voltage source in order to commutate and thus only can transfer power between two active ac networks. They are, therefore, less useful in connection with the wind farms as the offshore ac grid needs to be powered up prior to a possible startup. A further disadvantage of LCC-based HVDC transmission systems is the lack of the possibility to provide an independent control of the active and reactive powers. Furthermore, they produce large amounts of harmonics, which make the use of large filters inevitable.

Voltage-source-converter (VSC)-based HVDC transmission systems are gaining more and more attention not only for the grid connection of large offshore wind farms. Nowadays, VSC based solutions are marketed by ABB under the name "HVDC Light"[49] and by Siemens under the name "HVDC Plus". Figure 25.12(b) shows the schematic of a VSC-based HVDC transmission system. This comparatively new technology (with first commercial installation in 1999) has only become possible by the development of the IGBTs, which can switch off currents. This means that there is no need for an active commutation voltage. Therefore, VSC-based HVDC transmission does not require a strong offshore or onshore ac network and can even start up against a dead network (black-start capability). But, VSC-based systems have several other advantages. The active and reactive powers can be controlled independently, which may reduce the need for reactive-power compensation and can contribute to the stabilization of the ac network at their connection points.[50]

25.4.2 *High-power medium-voltage converter topologies*

In order to decrease the cost per megawatt and to increase the efficiency of the wind-energy conversion, nominal power of wind turbines has been continuously growing in the last years.[51] The different proposed multilevel-converter topologies can be classified into the five categories as explained in Sec. 2.6.

A common feature of the five different topologies of multilevel converters is that, in theory, all the topologies may be constructed to have an arbitrary number of levels, although in practice, some topologies are easier to realize than others. As the ratings of the components increase and the switching and conducting properties improve, the advantages of applying multilevel converters become more and more evident. In recent papers, the reduced content of harmonics in the input and output voltages is highlighted together with the reduced electromagnetic interference. Moreover, the multilevel converters have the lowest demands for the input filters or alternatively reduced number of commutations. For the same harmonic performance as a two-level converter, the switching frequency of a multilevel converter can be reduced to 25% that results in the reduction of the switching losses. Even though the conducting losses are higher in the multilevel converter, the overall efficiency depends on the ratio between the switching and the conducting losses.

The most commonly reported disadvantage of the multilevel converters with split dc link is the voltage unbalance between the capacitors that integrates it. Numerous hardware and software solutions are reported: the first one needs additional components that increase the cost of the converter and reduce its reliability; the second one needs enough computational capacity to carry out the modulation signals. Recent papers illustrate that the balance problem can be formulated in terms of the model of the converter, and this formulation permits solving the balancing problem directly by modifying the reference voltage with a relatively low computational burden. Trends on wind-turbine market are to increase the nominal power and due to the voltage and current ratings. This makes the multilevel converter suitable for modern high-power wind-turbine applications. The increase of voltage rating allows for connection of the converter of the wind turbine directly to the wind-farm distribution network, avoiding the use of a bulky transformer[52] (see Fig. 25.13).

The main drawback of some multilevel topologies is the necessity to obtain different dc-voltage independent sources needed for the multilevel modulation. The use of low-speed permanent-magnet generators that have a large number of poles allows obtaining the dc sources from the multiple wounds of this electrical machine,

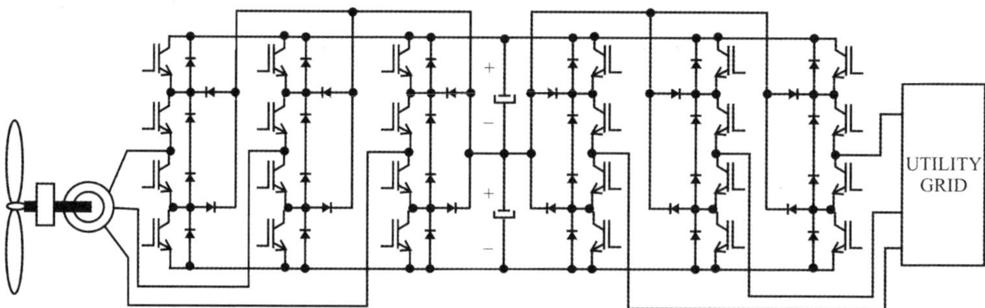

Fig. 25.13.   Multilevel back-to-back converter for a direct connection of a wind turbine to the utility grid.

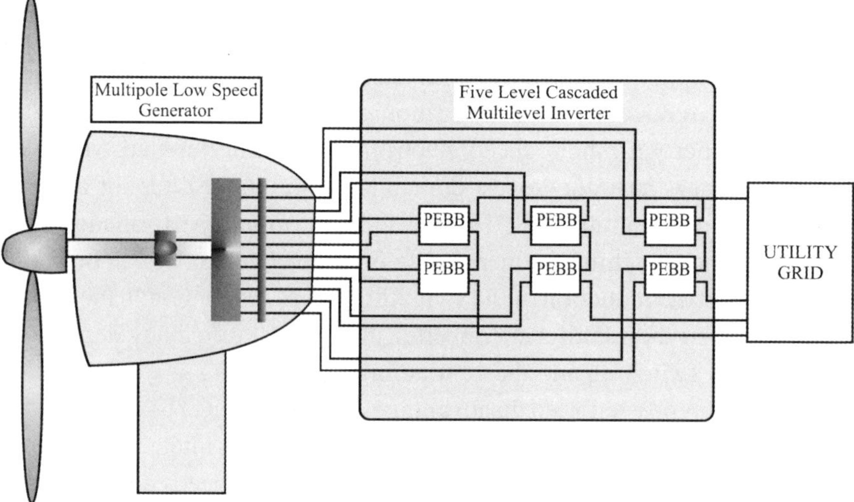

Fig. 25.14.   Five-level cascaded multilevel converter connected to a multipole low-speed wind-turbine generator.

as can be seen in Fig. 25.14. In this case, the power-electronic building block (PEBB) can be composed of a rectifier, a dc link, and an H-bridge. Another possibility is to replace the rectifier by an additional H-bridge. The continuous reduction of the cost per kilowatt of PEBBs is making the multilevel cascaded topologies to be the most commonly used by the industrial solutions.

25.5   Grid-Interconnection Requirements for Wind Farms: Overview

Large capacity wind farms have created several challenges for transmission network operators. The intermittent nature of the wind causes power quality and stability problems. Unpredictable wind power penetration affects reliability and stability of the grid. As a result the grid code/rules are being imposed on large wind farms to operate as conventional power plants. Grid code requirements vary considerably from one power system network to another and/or different connection points in a large power system network.

Accordingly to Refs. 57 to 59, and from the grid codes published for the wind based power schemes, it is found that the grid codes generally deal with the following issues:

1. Voltage and reactive power
2. Frequency and active power control
3. Fault ride through capability (LVRT)
4. Flicker, harmonics
5. Data exchange with system operator.

25.5.1 *Voltage and reactive power requirements*

The reactive power requirement defines that the wind farm should be operated within the range from 0.95 lagging to leading power factor. These limits reduce the impact of fluctuating wind power on the grid voltage. A wider range of power factor would be required for a remote area when low power is generated by the wind farm. The reactive power requirement by the national grid operator of UK is a bit complex and is as shown in Fig. 25.15. Table 25.1 summarizes the reactive power requirements for various other areas/countries/grids.

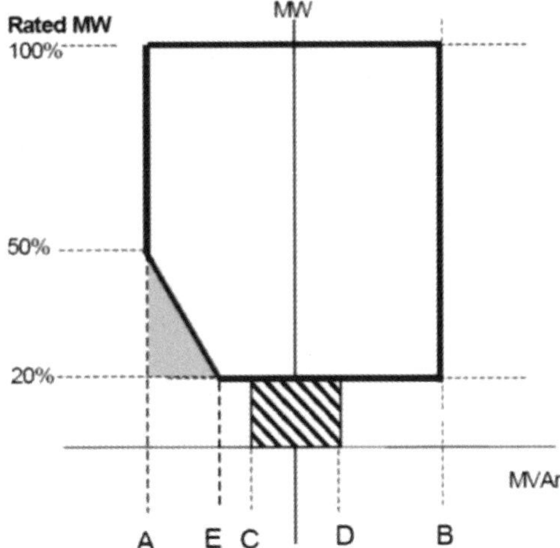

Fig. 25.15. Reactive power requirement for power plant connected to Grid (UK Grid Requirements).

Table 25.1. Reactive/voltage power requirement.

Area	Voltage/reactive power requirements
UK	0.95 lagging to 0.95 leading
Germany/	0.95 lagging to 0.95 leading for a rated active power capacity $<100\,$MW,
E.ON	For a rated active power capacity $>100\,$MW the power factor is voltage dependent
Denmark	Q/Prated $= 0$ to Q/Prated $= 0.1$ at full production and through a straight line to Q/Prated $= -0.1$ to Q/Prated $= 0$ at zero production
Quebec	The generator facilities must be designed to supply or absorb at the generating unit outlet (system side) the reactive power that corresponds to an overexcited or under excited rated power factor equal to or less than 0.95. The reactive power must be available over the entire active power generation range
Alberta	0.90 lagging to 0.95 leading
China	0.97 lagging to 0.97 leading
Australia	Constant power factor mode: 100% power at Power Factor$_{ind}$ $= 1.0$ and Power Factor$_{cap}$ $= 0.95$; 50% power at Power Factor$_{ind}$ $= 1.0$ and Power Factor$_{cap}$ $= 0.95$

The operating points of Fig. 25.15 "A, E, C, D and B" are defined as follows:

Point A is equivalent (in MVAr) to: 0.95 leading Power Factor at Rated MW output,
Point B is equivalent (in MVAr) to: 0.95 lagging Power Factor at Rated MW output,
Point C is equivalent (in MVAr) to: -5% of Rated MW output,
Point D is equivalent (in MVAr) to: $+5\%$ of Rated MW output,
Point E is equivalent (in MVAr) to: -12% of Rated MW output.

For grid operators like Alberta, an automatic voltage regulation scheme is required to satisfy the requirements. It is expected that the wind farm should be capable of maintaining its terminal voltage to a given set value, if it falls within its required reactive power range. The typical reactive power response requirement for Alberta grid code operator is shown in Fig. 25.16.

25.5.2 *Active power and frequency control requirements*

Table 25.2 summarizes typical frequency operational requirement for the wind farm. Figure 25.17 shows the power production requirement as a function of frequency for typical wind farm as per Danish grid code. In the **E.ON** grid code it is recommended to undertake detailed dynamic simulations to show that there is no dynamic interaction with the power system. This is because a large wind farm may interfere

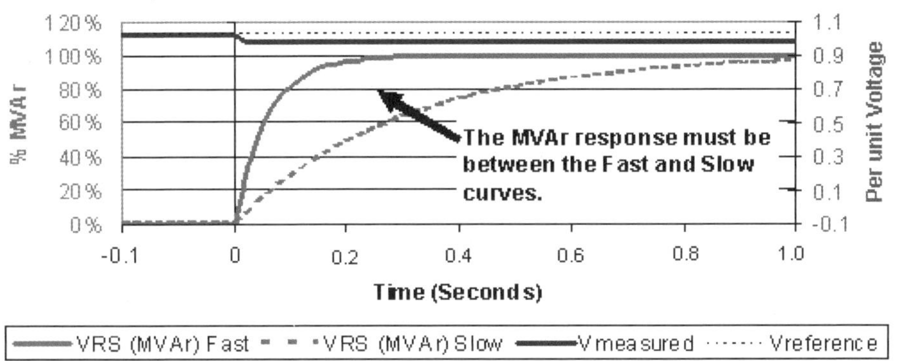

Fig. 25.16.   MVAr response requirement (Alberta grid code).

Table 25.2.   Minimum time for which wind farm remain connected to the grid without tripping due to frequency variation.

Country/region/grid	Frequency	Desired operation
UK	47.5 to 52 Hz	Continuous
	less than 47.5 Hz	20 s
Germany/EON	47.5 to 51.5 Hz	Continuous operation
	less than 47.5 Hz and greater than 51.5	Not specified
Quebec	59.4 to 60.6 Hz	Continuous
	58.5 to 59.4 Hz and 60.6 to 61.5 Hz	11 minutes (remain connected)
	57.5 to 58.5 Hz and 61.5 to 61.7 Hz	90 s
	57.0 to 57.5 Hz	10 s

with power system stabilizers in the grid. In some cases an active contribution of wind farm to damp out the oscillations is required.

25.5.3   *Voltage fault ride-through capability of wind turbines*

To enable a large-scale application of the wind energy without compromising the power-system stability, the turbines are now required to stay connected and contribute to the grid stability in case of a disturbance such as a voltage dip.

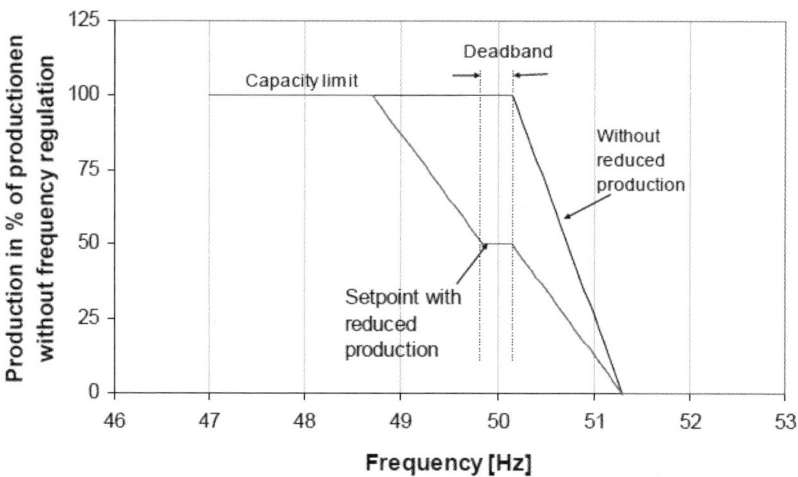

Fig. 25.17.   Power production requirements, Danish Grid Code.

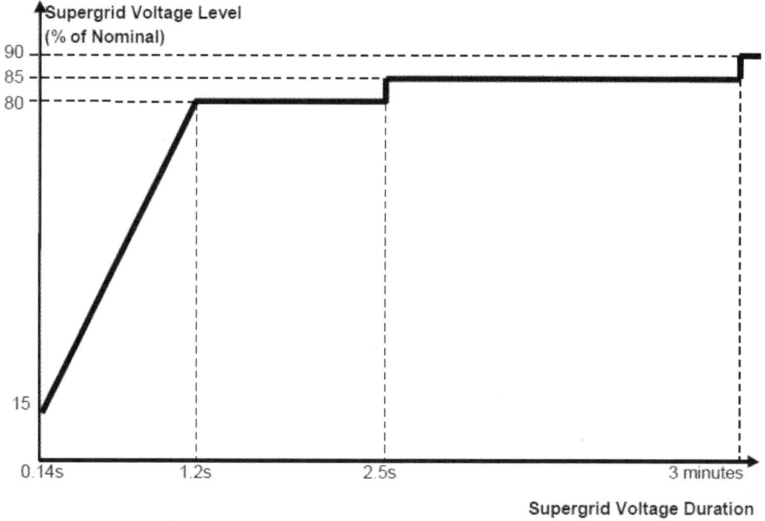

Fig. 25.18.   Fault ride through capability, UK Grid.

Wind farms should generate like conventional power plants, supplying active and reactive powers for frequency and voltage recovery, immediately after the fault occurred. Figures 25.18 and 25.19 show the typical fault ride through the requirement of national grid operators of UK and USA, respectively. The E.ON grid code requirement for large wind farms states that the wind farm should not be disconnected from the grid above 15% nominal voltage at the grid connection. During a

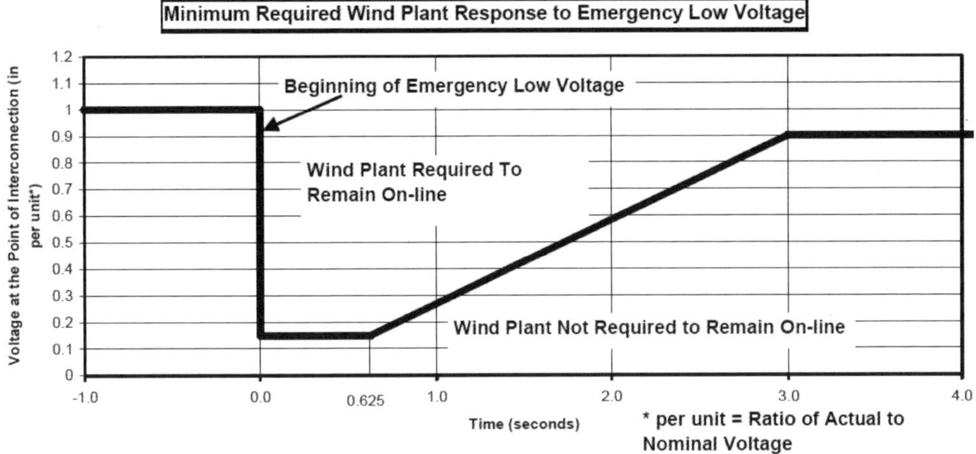

Fig. 25.19.   Fault ride through Capability USA.

fault the wind farm should supply a short circuit current to ensure the operation of protective devices in order to reduce the fault period. Following a fault, wind farms have to supply a defined reactive current depending on the grid voltage.

25.5.4 *Powers-quality requirements for grid-connected wind turbines*

Measurements show that the power quality impact of wind turbines has been improved in recent years. Especially, variable-speed wind turbines have some advantages concerning flicker. But, a new problem arose with variable-speed wind turbines. Modern forced-commutated inverters used in variable-speed wind turbines produce not only harmonics but also inter harmonics. The International Electrotechnical Commission (IEC) initiated the standardization on the power quality for wind turbines in 1995 as part of the wind-turbine standardization in TC88, and ultimately 1998 IEC issued a draft IEC-61400-21 standard for "power-quality requirements for Grid Connected Wind Turbines".[61] The methodology of IEC standard consists of three analyses. The first one is the flicker analysis.

IEC-61400-21 specifies a method that uses current and voltage time series measured at the wind-turbine terminals to simulate the voltage fluctuations on a fictitious grid with no source of voltage fluctuations other than the wind-turbine switching operation. The second one regards switching operations. Voltage and current transients are measured during the switching operations of the wind turbine (startup at cut wind speed and startup at rated wind speed). The last one is the harmonic analysis, which is carried out by the fast Fourier transform (FFT) algorithm. Rectangular windows of eight cycles of fundamental frequency width, with no gap and no overlapping between successive windows, are applied. Furthermore, the current total harmonic distortion (THD) is calculated up to the 50th harmonic order. Recently, high-frequency (HF) harmonics and inter harmonics are treated in the IEC 61000-4-7 and IEC 61000-3-6.[62,63] The methods for summing harmonics and inter harmonics in the IEC 61000-3-6 are applicable to wind turbines. Wind turbines not only produce harmonics; they also produce inter harmonics, i.e., harmonics that are not a multiple of 50 Hz. Since the switching frequency of the inverter is not constant but varies, the harmonics will also vary. Consequently, since the switching frequency is arbitrary, the harmonics are also arbitrary. Sometimes they are a multiple of 50 Hz, and sometimes they are not.

25.5.5　*Data exchange requirements with system operator*

The system operator may demand following data for power production planning and network studies. Such as power production at PCC, main transformer tap position, main transformer fault indication, circuit breaker position indicator, current measurement at PCC, frequency at PCC, status of compensation equipment, number of wind turbines in the wind farm with their status, frequency response mode signal, frequency response mode status indication, wind speed, wind direction ambient temperature atmospheric pressure on continuous basis.

25.6　Power Electronics in Photovoltaic (PV) System

This section focuses on the review of the recent developments of power-electronic converters and the state of the art of the implemented PV systems. PV systems as an alternative energy resource or an energy-resource complementary in hybrid systems are becoming feasible due to the increase of research and development work in this area.

Solar-electric-energy demand has grown consistently by 20%–25% per annum over the past 20 years, which is mainly due to the decreasing costs and prices. This decline has been driven by (1) an increasing efficiency of solar cells; (2) manufacturing-technology improvements; and (3) economies of scale. In 2001, 350 MW of solar equipment was sold to add to the solar equipment already generating clean energy. In 2003, 574 MW of PV was installed. This increased to 927 MW in 2004. The European Union is on track to fulfilling its own target of 3 GW of renewable electricity from PV sources for 2010, and in Japan, the target is 4.8 GW.

25.6.1　*Design of PV-converter families*

An overview of some existing power inverter topologies for interfacing PV modules to the grid is presented. The approaches are further discussed and evaluated in order to recognize the most suitable topologies for future PV converters, and, finally, a conclusion is given. Due to advances in transistor technology, the inverter topologies have changed from large thyristor-equipped grid connected inverters to smaller IGBT-equipped ones. These transistors permit to increase the power switching frequency in order to extract more energy and fulfill the connecting standards.

One requirement of standards is that the inverters must also be able to detect an islanding situation and take appropriate measures in order to protect persons and equipment. In this situation, the grid has been removed from the inverter, which then only supplies local loads. This can be troublesome for many high-power transformer less systems, since a single phase inverter with a neutral-to-line grid connection is a system grounded on the grid side.

In general, PV cells can be connected to the grid (grid connection application), or they can be used as isolated power supplies. These two different applications of PV systems are shown in Fig. 25.20. Several classifications of converter topologies can be done with respect to the number of power processing stages, location of power-decoupling capacitors, use of transformers, and types of grid interface. However, before discussing PV converter topologies, three designs of inverter families are defined: central inverters, module-oriented or module-integrated inverters, and string inverters.[64,65] The central converters connect in parallel and/or in series on the dc side. One converter is used for the entire PV plant (often divided into several units organized in master–slave mode). The nominal power of this topology is up to several megawatts.

The module-oriented converters with several modules usually connect in series on the dc side and in parallel on the ac side. The nominal power ratings of such PV power plants are up to several megawatts. In addition, in the module-integrated converter topology, one converter per PV module and a parallel connection on the ac side are used. In this topology, a central measure for main supervision is necessary. Although this topology optimizes the energy yield, it has a lower efficiency than the string inverter. This concept can be implemented for PV plants of about 50–100 W. In Fig. 25.21, a one-phase multistring converter (Fig. 25.21(a)) and a

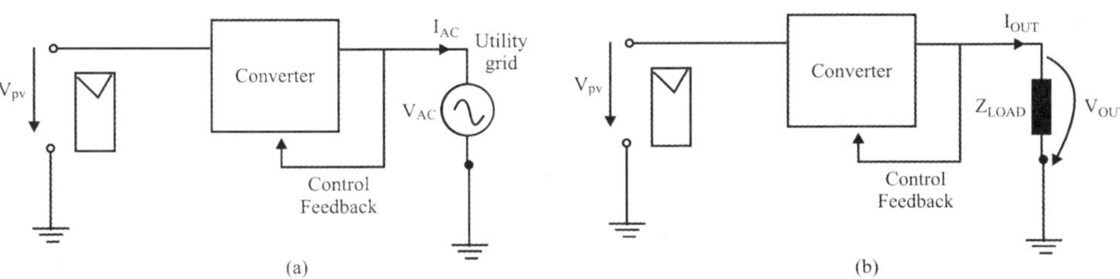

Fig. 25.20. PV energy applications, (a) grid-connection application, (b) power-supply application.

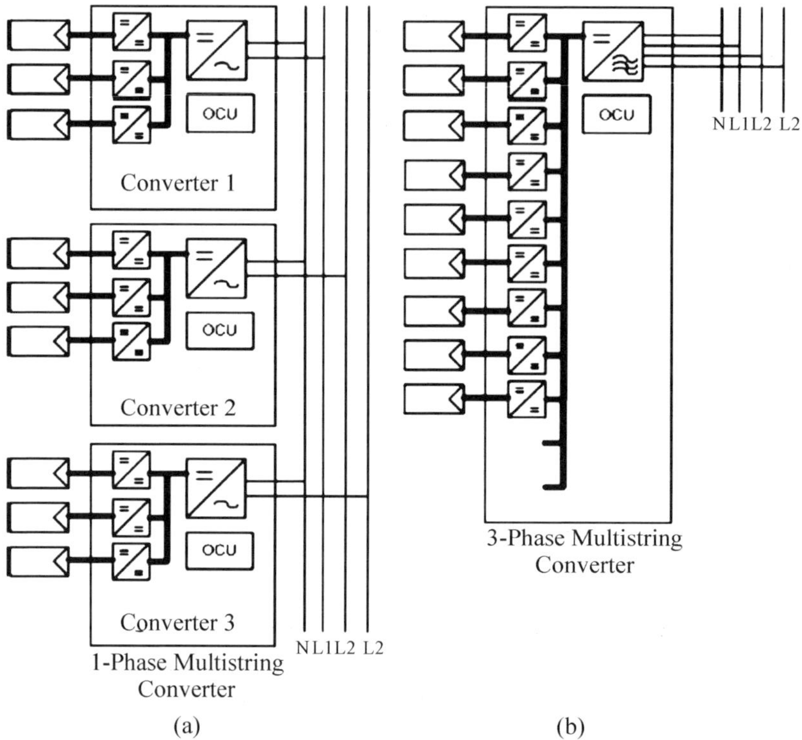

Fig. 25.21.   (a) One-phase multistring converter, (b) three-phase multistring converter.

three-phase multistring converter (Fig. 25.21(b)) are shown. A detail of a multistring converter with a single-phase inverter stage is illustrated in Fig. 25.22. The multistring topology permits the integration of PV strings of different technologies and orientations (north, south, east, and west).

25.6.2 *PV topologies*

Conventionally, a classification of PV topologies is divided into two major categories: PV inverters with dc/dc converter (with or without isolation) and PV inverters without dc/dc converter (with or without isolation).[64,66] The isolation used in both categories is acquired using a transformer that can be placed on either the grid or low frequency (LF) side or on the high frequency (HF) side. The line-frequency transformer is an important component in the system due to its size, weight, and price. The HF transformer is more compact, but special attention must be paid to reduce losses.[64] The use of a transformer leads to the necessary isolation (requirement in U.S.), and modern inverters tend to use an HF transformer. However, PV inverters

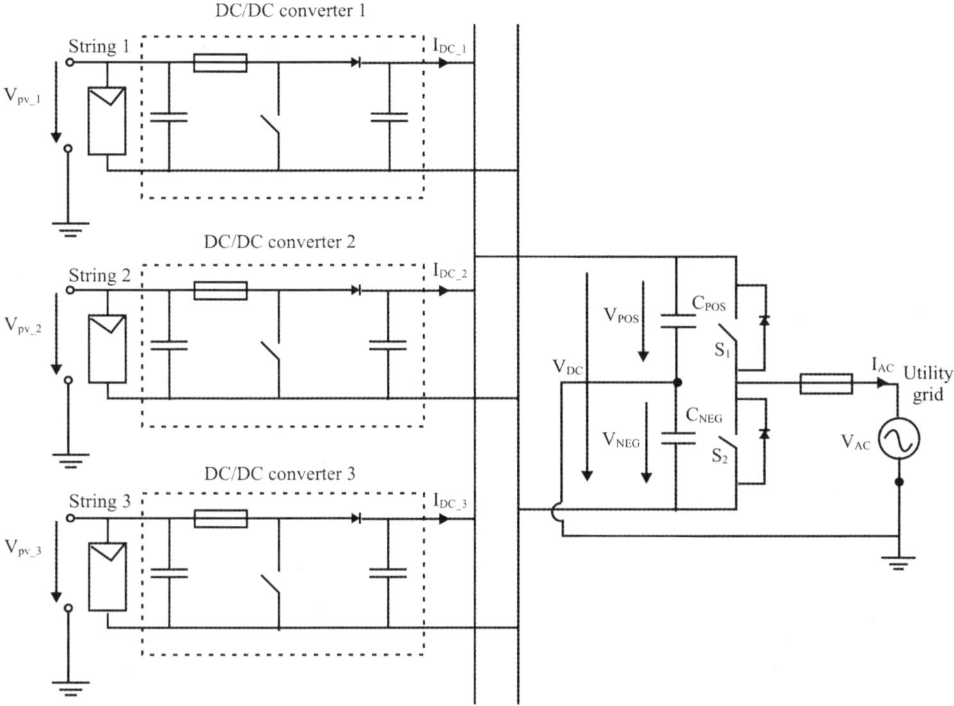

Fig. 25.22.   Detail of a multistring converter with a single-phase inverter stage.

with a dc/dc converter without isolation are usually implemented in some countries where grid-isolation is not mandatory.

Basic designs focused on solutions for HF dc/dc converter topologies with isolation such as full-bridge or single-inductor push–pull permit to reduce the transformer ratio providing a higher efficiency together with a smoother input current. However, a transformer with tap point is required. In addition, a double-inductor push–pull is implemented in other kind of applications (equivalent with two interleaved boost converters leading to a lower ripple in the input current), but extra inductor is needed. A full-bridge converter is usually used at power levels above 750 W due to its good transformer utilization.

Another possible classification of PV inverter topologies can be based on the number of cascade power processing stages. The single-stage inverter must handle all tasks such as maximum-power-point-tracking (MPPT) control, grid-current control, and voltage amplification. This configuration, which is useful for a centralized inverter, has some drawbacks because it must be designed to achieve a peak power of twice the nominal power. Another possibility is to use a dual-stage inverter. In this case, the dc/dc converter performs the MPPT (and perhaps voltage amplification), and the dc/ac inverter is dedicated to control the grid current by means of pulse width modulation (PWM), space vector modulation (SVM), or bang-bang operation. Finally, multistage inverters can be used, as mentioned above. In this case, the task for each dc/dc converter is MPPT and, normally, the increase of the dc voltage. The dc/dc converters are connected to the dc link of a common dc/ac inverter, which takes care for the grid-current control. This is beneficial since a better control of each PV module/string is achieved, and that common dc/ac inverter may be based on a standard variable speed-drive (VSD) technology.

There is not any standard PV inverter topology. Several useful proposed topologies have been presented, and some good studies regarding current PV inverters have been presented.[67,68] The current control scheme is mainly used in PV inverter applications.[69] In these converters, the current into the stage is modulated/controlled to follow a rectified sinusoidal waveform, and the task for the circuit is simply to recreate the sine wave and inject it into the grid. The circuits apply zero voltage switching (ZVS) and zero-current switching (ZCS).

Thus, only conduction losses of the semiconductors remain. If the converter has several stages, power decoupling must be achieved with a capacitor in parallel with the PV module(s). The current control scheme is employed more frequently because a high-power factor can be obtained with simple control circuits, and transient current suppression is possible when disturbances such as voltage changes occur in the utility power system. In the current control scheme, operation as an isolated power source is difficult, but there are no problems with grid interconnection operation.

PV automatic-control (AC) module inverters used to be dual stage inverters with an embedded HF transformer. Classical solutions can be applied to develop these converters: fly back converters (single or two transistors), flyback with a buck-boost converter, resonant converters, etc. For string or multistring systems, the inverters used to be single or dual-stage inverters with an embedded HF transformer. However, new solutions try to eliminate the transformer using multilevel topologies.

A very common ac/dc topology is the half-bridge two-level VSI, which can create two different voltage levels and requires double dc-link voltage and double switching frequency in order to obtain the same performance as the full bridge. In this inverter, the switching frequency must be double the previous one in order to obtain the same size of the grid inductor. A variant of this topology is the standard full-bridge three-level VSI, which can create a sinusoidal grid current by applying the positive/negative dc-link or zero voltage, to the grid plus grid inductor. This inverter can create three different voltages across the grid and inductor, the switching frequency of each transistor is reduced, and good power quality is ensured. The voltage across the grid and inductor is usually pulse width modulated but hysteresis (bang-bang) current control can also be applied.

25.6.3 *Future trends*

The increasing interest and steadily growing number of investors in solar energy stimulated research that resulted in the development of very efficient PV cells, leading to universal implementations in isolated locations. Due to the improvement of roofing PV systems, residential neighborhoods are becoming a target of solar panels, and some current projects involve installation and setup of PV modules in high building structures. PV systems without transformers would be the most suitable option in order to minimize the cost of the total system. On the other hand, the cost of the grid-connected inverter is becoming more visible in the total system price. A cost reduction per inverter watt is, therefore, important to make PV-generated power more attractive. Therefore, it seems that centralized converters would be a good option for PV systems. However, problems associated with the centralized control appear, and it can be difficult to use this type of systems. An increasing interest is being focused on ac modules that implement MPPT for PV modules improving the total system efficiency. The future of this type of topologies is to develop "plug and play systems" that are easy to install for non-expert users. This means that new ac modules may see the light in the future, and they would be the future trend in this type of technology. The inverters must guarantee that the PV module is operated at the maximum power point (MPP) owing to use MPPT control increasing the PV systems efficiency. The operation around the MPP without too much fluctuation will reduce the ripple at the terminals of the PV module.

Therefore, the control topics such as improvements of MPPT control, THD improvements, and reduction of current or voltage ripples will be the focus of researchers in the years to come. These topics have been deeply studied during the last years, but some improvements still can be done using new topologies such as multilevel converters. In particular, multilevel cascade converters seem to be a good solution to increase the voltage in the converter in order to eliminate the HF transformer. A possible drawback of this topology is control complexity and increased number of solid-state devices (transistors and diodes). It should be noticed that the increase of commutation and conduction losses has to be taken into account while selecting PWM or SVM algorithms.

Finally, it is important to remember that standards, regarding the connection of PV systems to the grid, are actually becoming more and more strict. Therefore, the future PV technology will have to fulfill them, minimizing simultaneously the cost of the system as much as possible. In addition, the incorporation of new technologies, packaging techniques, control schemes, and an extensive testing regimen must be developed. Testing is not only the part of each phase of development but also the part of validation of the final product.

25.7   Recent Trends in Energy-Storage Technologies

In order to improve the quality of the generated power, as well as to support critical loads during mains' power interruption, several energy-storage technologies have been investigated, developed, proved, and implemented in renewable energy systems which are discussed in below section.

25.7.1 *Flywheels*

Flywheels are very commonly used due to the simplicity of storing kinetic energy in a spinning mass. For approximately 20 years, it has been a primary technology used to limit power interruptions in motor/generator sets where steel wheels increase the rotating inertia providing short power interruptions protection and smoothing of delivered power. One of the first commercial uses of flywheels in conjunction with active filtering to improve frequency distortion on a high-voltage power-system line is described in Ref. 70.

There are two broad classes of flywheel-energy-storage technologies. One is a technology based on low-speed flywheels (up to 6000 r/min) with steel rotors and conventional bearings. The other one involves modern high-speed flywheel systems (up to 60000 r/min) that are just becoming commercial and make use of advanced composite wheels that have much higher energy and power density than steel wheels. This technology requires ultra low friction bearing assemblies, such as magnetic bearings, and stimulates a research trend.[71]

Most applications of flywheels in the area of renewable energy delivery are based on a typical configuration where an electrical machine (i.e., high-speed synchronous machine or induction machine) drives a flywheel, and its electrical part is connected to the grid via a back-to-back converter, as shown in Fig. 25.23. Such configuration requires an adequate control strategy to improve power smoothing. The basic operation could be summarized as follows. When there is excess in the generated power with respect to the demanded power, the difference is stored in the flywheel that is driven by the electrical machine operating as a motor. On the other hand, when a perturbation or a fluctuation in delivered power is detected in the loads, the electrical machine is driven by the flywheel and operates as a generator supplying needed extra energy. A typical control algorithm is a direct vector control with rotor-flux orientation and sensor less control using a model-reference-adaptive-system (MRAS) observer.

Experimental alternatives for wind farms include flywheel compensation systems connected to the dc link, which are the same as the systems used for power smoothing for a single or a group of wind turbines. Usually, a control strategy is applied to regulate the dc voltage against the input power surges/sags or sudden changes in

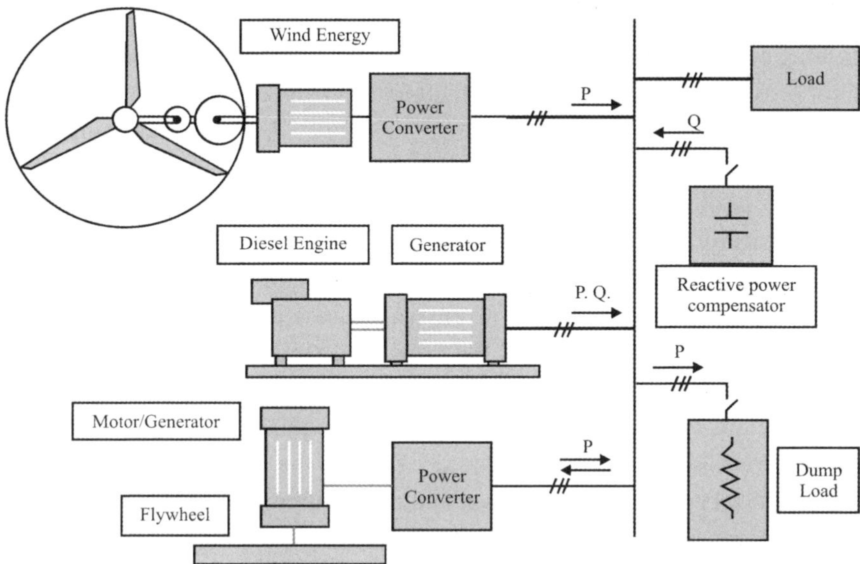

Fig. 25.23. Typical compensation system for renewable energy applications based on flywheel energy storage.

the load demand. A similar configuration can be applied to solar cells. Another renewable energy resource where power oscillations need to be smoothed is wave energy. In Ref. 72, a D-static synchronous compensator (STATCOM) is proposed, as an alternative to flywheels, to accomplish the output power smoothing on a wave energy converter where several operating conditions should be taken into account. Recent proposals on using flywheels to regulate the system frequency include the disposal of a matrix of several flywheels to compensate the difference between the network's load and the power generated. Recently, there has been research where integrated flywheel systems can be encountered. Those systems use the same steel rotor of the electrical machine as energy-storage element.

25.7.2 *Hydrogen*

This section aims to analyze new trends in hydrogen-storage systems for high-quality back-up power. The hydrogen-fuel economy has been rapidly increasing in industrial application due to the advantages of the hydrogen of being storable, transportable, highly versatile, efficient, and clean energy carrier to supplement or replace many of the current fuel options. It can be used in fuel cells to produce electricity in a versatile way, for example, in portable applications, stationary use of energy, transportation, or high-power generation. The use of fuel cells in such applications is justified since they are a very important alternative power source due to their well-known specific characteristics such as very low toxic emissions, low noise and vibrations, modular design, high efficiency (especially with partial load), easy installation, compatibility with a lot of types of fuels, and low maintenance cost. An example of the hydrogen storage application to improve the grid power quality through smoothing large and quick fluctuations of wind energy is reported in Ref. 73.

Hydrogen could be stored as compressed or liquefied gas or by using metal hydrides or carbon Nan tubes. For a particular application, the choice of a storage technology implies a trade-off between the characteristics of available technologies in terms of technical, economical, or environmental performance.[74] Applications must also include a discussion of the lifecycle efficiency and cost of the proposed storage system. This analysis should consider the total life of the proposed hydrogen storage system including raw-material requirements, manufacturing and fabrication processes, integration of the system into the vehicle or off-board configuration, useful service life, and removal and disposal processes including recycling. Recently, research and development are focused on new materials or technologies for hydrogen storage: metal hydrides (reduce the volumetric and pressure requirements for storage, but they are more complex than other solutions), chemical hydrides, carbon-based hydrogen-storage materials, compressed- and liquid-hydrogen-tank technologies, off-board hydrogen-storage systems (a typical refueling station will be delivering 200–1500 kg/day of hydrogen), and new materials and approaches for storing hydrogen on board a vehicle.

25.7.3 *Compressed-Air Energy Storage (CAES)*

Energy storage in compressed air is made using a compressor that stores it in an air reservoir (i.e., an aquifer like the ones used for natural-gas storage, natural caverns, or mechanically formed caverns, etc.). When a grid is operating off peak, the compressor stores air in the air reservoir. During discharge at peak loads, the compressed air is released to a combustor where it is mixed with oil or gas driving a gas turbine. Such systems are available for 100–300 MW and burn about one-third of the premium fuel of a conventional simple cycle combustion turbine.

An alternative to CAES is the use of compressed air in vessels (called CAS), which operates exactly in the same way as CAES except that the air is stored in pressure vessels rather than underground reservoirs. Such difference makes possible variations consisting of the use of pneumatic motor acting as compressors or driving a dc motor/generator according to the operation required by the system, i.e., storing energy when there is no extra demand of energy or delivering extra power at peak loads. Recent research is devoted to the maximum-efficiency point-tracking control or integrated technologies for power-supply applications.

25.7.4 *Supercapacitors*

Super-capacitors, which are also known as ultra-capacitors or electric double layer capacitors (EDLC), are built up with modules of single cells connected in series and packed with adjacent modules connected in parallel. Single cells are available with capacitance values from 350 to 2700 F and operate in the range of 2 V. The module voltage is usually in the range from 200 to 400 V. They have a long life cycle and are suitable for short discharge applications and are less than 100 kW. New trends focused on using ultra capacitors to cover temporary high peak power demands, integration with other energy-storage technologies, and development of high-voltage applications.

25.7.5 *Superconducting Magnetic Energy Storage (SMES)*

In an SMES, a coil of superconducting wire stores electrical energy in a magnetic field without resistive losses. Also, there is no need for conversion between chemical or mechanical forms of energy. Recent systems are based on both general configurations of the coil: solenoidal or toroidal. The second topology has a minimal external magnetic field but the cost of superconductor and coil components is higher than the first topology. Such devices require cryogenic refrigerators (to operate in liquid helium at $-269°C$) besides the solid-state power electronics. The system operates by injecting a dc current into the superconducting coil, which stores the energy in magnetic field.

When a load must be fed, the current is generated using the energy stored in the magnetic field. One of the major advantages of SMES is the ability to release large quantities of power during a fraction of a cycle. Typical applications of SMES are corrections of voltage sags and dips at industrial facilities (1 MW units) and stabilization of ring networks (2 MW units). New trends in SMES are related to the use of low temperature superconductors (liquid-nitrogen temperature), the use of secondary batteries, and the integration of STATCOM and several topologies of ac–dc–ac converters with SMES.

25.7.6 *Battery storage*

The use of batteries as a system to interchange energy with the grid is well known. There are several types of batteries used in renewable energy systems: lead acid, lithium, and nickel. Batteries provide a rapid response for either charge or discharge, although the discharge rate is limited by the chemical reactions and the type of battery. They act as a constant voltage source in the power systems. New trends in the use of batteries for renewable energy systems focused on the integration with several energy sources (wind energy, PV systems, etc.) and also on the integration with other energy-storage systems complementing them. Also, there are attempts to optimize battery cells in order to reduce maintenance and to increment its lifetime.

25.7.7 *Future energy-storage technologies applied in wind farms*

Energy-storage systems can potentially improve the technical and economic attractiveness of wind power, particularly when it exceeds about 10% of the total system energy (about 20%–25% of the system capacity). The storage system in a wind farm will be used to have a bulk power storage from the wind during the time-averaged 15-min periods of high availability and to absorb or to inject energy over shorter time periods in order to contribute to the grid-frequency stabilization. Several kinds of energy-storage technologies are being applied in wind farms. For wind-power application, the flow (zinc bromine) battery system offers the lowest cost per energy stored and delivered. The zinc–bromine battery is very different in concept and design from the more traditional batteries such as the lead–acid battery. The battery is based on the reaction between two commonly available chemicals: zinc and bromine. The zinc–bromine battery offers two to three times higher energy density (75–85 Wh per kilogram) along with the size and weight savings over the present lead/acid batteries. The power characteristics of the battery can be modified for selected applications. Moreover, zinc–bromine battery suffers no loss of performance after repeated cycling. It has a great potential for renewable energy applications.

As the wind penetration increases, the hydrogen options become most economical. Also, sales of hydrogen as a vehicle fuel are more lucrative than reconverting the hydrogen back into electricity. The industry is developing low-maintenance electrolysers to produce hydrogen fuel. Because these electrolysers require a constant minimum load, wind turbines must be integrated with grid or energy systems to provide power in the absence of wind.

25.8 Conclusions

In this chapter, a comparison of various converters is carried out with advantages and drawbacks for each one. As the back-to-back converter is state-of-the-art today in wind turbines it can be used as a reference in a benchmark of the other converter topologies regarding the number of the components and their ratings, the auxiliary components, the efficiency, the harmonic performances and implementation.

It also describes the common and future trends for renewable energy systems, specific interfacing challenges for renewable energy systems, grid-connection standards for wind farms and recent trends in energy storage technologies. As a current energy source, wind energy is the most advanced technology due to its installed power and the recent improvements of the power electronics and control.

References

1. E. Bogalecka, "Power control of a doubly fed induction generator without speed or position sensor," *EPE 5th European Conf. Power Electronics and Application*, vol. 8 (1993), pp. 224–228.
2. T. Yifan and X. Longya, "Stator field oriented control of doubly-exited induction machine in wind power generating system," *IEEE Symp. Circuits and Systems* (1992), pp. 1446–1449.
3. T. Yifan and X. Longya, "A flexible active and reactive power control for a variable speed constant frequency generating system," *IEEE Trans. Power Electronics*, vol. 10 (1995), pp. 472–478.
4. R. Pena, J.C. Clare and G.M. Asher, "Doubly fed induction generator using back-to-back PWM converters and its application to variable speed wind-energy generation," *IEE Proc. Electronic Power Application*, vol. 143 (1996), pp. 231–241.
5. J.S. Kim and S.K. Sul, "New control scheme for AC-DC-AC converter without DC link electrolytic capacitor," *IEEE Power Electronics Specialists Conf.* (1993), pp. 300–306.
6. S. Kim, S.-K. Sul and T.A. Lipo, "AC to AC power conversion based on matrix converter topology with unidirectional switches," *IEEE Applied Power Electronics Conf. and Exposition*, vol. 2 (1998).
7. G.D. Marques and P. Verdelho, "An active power for the compensation of the oscillatory torque produced by the induction machine driven by a six step current inverter," *PEMC*, vol. 7 (1998), pp. 115–119.
8. A.M. Trzynadlowski, F. Blaabjerg, J.K. Pedersen, A. Patriciu and N. Patriciu, "A tandem inverter for high-performance AC-drives," *IEEE Industry Applications Conf. Annual Meeting*, vol. 3 (1998a), pp. 500–505.
9. A.M. Trzynadlowski, J.K. Pedersen, F. Blaabjerg and N. Patriciu, "The tandem inverter: Combining the advantages of voltage-source and current-source inverters," *IEEE Applied Power Electronics Conf. and Exposition*, vol. 1 (1998b), pp. 315–320.
10. R. Zhang, F.C. Lee, D. Boroyevich and H. Mao, "New high power, high performance power converter systems," *IEEE Power Electronics Specialists Conf.*, vol. 1 (1998b), pp. 8–14.
11. P. Mutschler and M. Meinhardt, "Competitive hysteresis controllers — a control concept for inverters having oscillating DC- and AC-side state variables," *IEE Proc. Electric Power Applications*, vol. 145 (1998), pp. 569–576.
12. A.B. Nikolic and B.I. Jeftenic, "The new algorithm for vector controlled CSI fed induction motor drive," *PEMC*, vol. 5 (1998), pp. 143–148.
13. M. Salo and H. Tuusa, "Active and reactive power control of a current-source PWM-rectifier using space vectors," *Finnish Workshop on Power and Industrial Electronics* (1997), pp. 75–78.
14. M. Salo and H. Tuusa, "A high performance PWM current source inverter fed induction motor drive with a novel motor current control method," *IEEE Power Electronics Specialists Conf.*, vol. 1 (1999), pp. 506–511.
15. A. Schuster, "A matrix converter without reactive clamp elements for an induction motor drive system," *IEEE Power Electronics Specialists Conf.*, vol. 1 (1998), pp. 714–720.
16. V. Kumar, R. Joshi and R.C. Bansal, "Optimal control of matrix converter based WECS for performance enhancement and efficiency optimization," *IEEE Trans. Energy Conversion*, vol. 24 (2009), pp. 264–273.
17. C.L. Neft and C.D. Schauder, "Theory and design of a 30-Hp matrix converter," *IEEE Industrial Application Society Annual Meeting* (1988), pp. 248–253.

18. R.R. Beasant, W.C. Beatie and A. Refsum, "Current commutation problems in the Venturini converter," *Universities Power Engineering Conf.* (1989), pp. 43–46.

19. R.R. Beasant, W.C. Beatie and A. Refsum, "An approach to realisation of a high power Venturini converter," *IEEE Power Electronics Specialists Conf.*, vol. 1 (1990), pp. 291–297.

20. N. Burany, "Safe control of four-quadrant switches," *IEEE Industrial Application Society Annual Meeting*, vol. 1 (1989), pp. 1190–1194.

21. H.L. Hey, H. Pinheiro and J.R. Pinheiro, "A new soft-switching AC-AC matrix converter, with a single active commutation auxiliary circuit," *IEEE Power Electronics Specialists Conf.*, vol. 2 (1995), pp. 965–970.

22. B.H. Kwon, B.D. Min and J.H. Kim, "Novel commutation technique of AC-AC converters," *IEE Proc. Electronic Power Application*, vol. 145 (1998), pp. 295–300.

23. D. Casadei, G. Grandi, G. Serra and A. Tani, "Analysis of space vector modulated matrix converters under unbalanced supply voltages," *Symp. Power Electronics, Electrical Drives, Advanced Electrical Motors* (1994), pp. 39–44.

24. D. Casadei, G. Serra and A. Tani, "Reduction of the input current harmonic content in matrix converter under input/output unbalance," *IEEE Int. Conf. Industrial Electronics, Control and Instrumentation*, vol. 1 (1995a), pp. 457–462.

25. D. Casadei, G. Serra and A. Tani, "A general approach for the analysis of the input power quality in matrix converters," *IEEE Power Electronics Specialists Conf.*, vol. 2 (1996), pp. 1128–1134.

26. R. Vinod Kumar, C. Bansal and R.R. Joshi, "Experimental realization of matrix converter based induction motor drive under various abnormal voltage conditions," *Int. J. Control, Automation, and Systems*, **6** (2008) 670–676.

27. P. Nielsen, "The matrix converter for an induction motor drive," Ph.D. thesis, Aalborg University, Denmark (1996).

28. L. Zhang, C. Watthanasarn and W. Shepherd, "Analysis and comparison of control techniques for AC-AC matrix converters," *IEE Proc. Electronic Power Application*, vol. 145 (1998a), pp. 284–294.

29. A. Nabae, I. Takahashi and H. Akagi, "A new neutral-point-clamped PWM-inverter," *IEEE Trans. Industry Applications* **IA-17** (1981) 518–523.

30. J.S. Lai and F.Z. Peng, "Multilevel converters — a new breed of power converters," *IEEE Trans. Industry Applications* **32** (1996) 509–517.

31. J.S. Lai and F.Z. Peng," "Multilevel converters — A new breed of power converters," *IEEE Industrial Application Society Annual Meeting*, vol. 3 (1995), pp. 2348–2356.

32. B.-S. Suh, G. Sinha, M. Manjrekar and T.A. Lipo, "Multilevel power conversion — An overview of topologies and modulation strategies," *OPTIM 6th Int. Conf. Optimization of Electrical and Electronic Equipment*, vol. 2 (1998), AD1-11-AD-24.

33. S.C. Nam, G.C. Jung and H.C. Gyu, "A general circuit topology of multilevel inverter," *IEEE Power Electronics Specialists Conf.* (1991), pp. 96–103.

34. S.-K. Lim, J.-H. Kim and K. Nam, "A DC-link voltage balancing algorithm for 3-level converter using the zero sequence current," *IEEE Power Electronics Specialists Conf.*, vol. 2 (1999), pp. 1083–108.

35. W. Lixiang and L. Fahai, "A direct power feedback method of a dual PWM three-level voltage source converter system," *IEEE Power Electronics Specialists Conf.*, vol. 2 (1999), pp. 1089–1094.

36. W.E. Brumsickle, D.M. Divan and T.A. Lipo, "Reduced switching stress in high-voltage IGBT inverters via a three-level structure," *IEEE Applied Power Electronics Conf. Exposition*, vol. 2 (1998), pp. 544–550.

37. Y. Xiaoming, H. Stemmler and I. Barbi, "Investigation on the clamping voltage self-balancing of the three-level capacitor clamping inverter," *IEEE Power Electronics Specialists Conf.*, vol. 2 (1999), pp. 1059–1064.

38. E. Cengelci, S.U. Sulistiju, B.O. Woo, P. Enjeti, R. Teodorescu and F. Blaabjerg, "A new medium voltage PWM inverter topology for adjustable speed drives," *Industrial Application Society Annual Meeting*, vol. 2 (1998), pp. 1416–1423.

39. F.Z. Peng, J.S. Lai, J. McKleever and J. Van Coevering, "A multilevel voltage-source converter system with balanced DC-voltages," *IEEE Power Electronics Specialist's Conf.*, vol. 2 (1995), pp. 1144–1150.

40. M. Marchesoni and M. Mazzucchelli, "Multilevel converters for high power ac drives: A review," *IEEE Int. Symp. Industrial Electronics* (1993), pp. 38–43.

41. L.M. Tolbert and F.Z. Peng, "A multilevel converter-based universal power conditioner," *IEEE Power Electronics Specialists Conf.*, vol. 1 (1999), pp. 393–399.

42. A.D. Hansen and L.H. Hansen, "Wind turbine concept market penetration over 10 years," *Wind Energy* **10** (2007) 81–97.

43. S. Muller, M. Deicke and R.W. De Doncker, "Doubly fed induction generator systems for wind turbines," *IEEE Industrial Applications Magazine* **8** (2002) 26–33.

44. F.M. Hughes, O. Anaya-Lara, N. Jenkins and G. Strbac, "Control of DFIG-based wind generation for power network support," *IEEE Trans. Power Systems* **20** (2005) 1958–1966.

45. M. Orabi, F. El-Sousy, H. Godah and M.Z. Youssef, "High-performance induction generator-wind turbine connected to utility grid," *Proc 26th Annual INTELEC*, 19–23 September 2004, pp. 697–704.

46. J.M. Peter, "Main future trends for power semiconductors from the state of the art to future trends," *PCIM*, Nürnberg, Germany, June 1999.

47. H. Grüning *et al.*, "High power hard-driven GTO module for 4.5 kV/3 kA snubberless operations," *Proc. PCI Eur.* (1996), pp. 169–183.

48. N. Kirby, L. Xu, M. Luckett and W. Siepmann, "HVDC transmission for large offshore wind farms," *Power Engineering J.* **16** (2002) 135–141.

49. K. Eriksson, C. Liljegren and K. Söbrink, "HVDC light experiences applicable for power transmission from offshore wind power parks," *42nd AIAA Aerospace Sciences Meeting and Exhibit*, Reno, NV (2004).

50. S. Meier, "Novel voltage source converter based HVDC transmission system for offshore wind farms," Ph.D. dissertation, Department of Electrical Engineering, Royal Institute of Technology, Stockholm, Sweden (2005).

51. R. Swisher, C.R. de Azua and J. Clendenin, "Strong winds on the horizon: Wind power comes on age," *Proc. IEEE*, vol. 89 (2001), pp. 1757–1764.

52. R. Portillo, M.M. Prats, J.I. Leon, J.A. Sanchez, J.M. Carrasco, E. Galvan and L.G. Franquelo, "Modeling strategy for back-to-back three level converters applied to high power wind turbines," *IEEE Trans. Indudstrial Electronics*, to be published.

53. M.A. Khan, P. Pillay and M. Malengret, "Impact of direct-drive WEC Systems on the design of a small PM wind generator," *Proc. IEEE Power Technical Conf.*, Bologna, Italy, June 23–26 2003.

54. I. Schiemenz and M. Stiebler, "Control of a permanent magnet synchronous generator used in a variable speed wind energy system," *Proc. IEEE IEMDC* (2001), pp. 872–877.

55. L. Chang, Q. Wang and P. Song, "Application of finite element method in design of a 50 kW direct drive synchronous generator for variable speed wind turbines," *Proc. 4th IPEMC Conf.*, vol. 2 (2004), pp. 591–596.

56. J.R. Bumby and R. Martin, "Axial-flux permanent-magnet air-cored generator for small-scale wind turbines," *Proc. Inst. Electr. Eng.—Elect. Power Appl.*, vol. 152 (2005), pp. 1065–1075.

57. J.M. Carrasco, L.G. Franquelo, J.T. Bialasiewicz, E. Galvan, R.C. Portillo Guisado, M.A.M. Prats, J.I. Leon and N. Moreno-Alfonso, "Power-electronic systems for the grid integration of renewable energy sources: A survey," *IEEE Trans. Industrial Electronics*, **53** (2006) 1002–1016.

58. F. Iov, R. Teodorescu, F. Blaabjerg, B. Andresen, J. Birk and J. Miranda, "Grid code compliance of grid-side converter in wind turbine systems," *IEEE 37th Power Electronics Specialists Conf.*, June 2006, p. 7.

59. D. Andersson, A. Petersson, E. Agneholm and D. Karlsson, "Kriegers flak 640 MW off-shore wind power grid connection — A real project case study," *IEEE Trans. Energy Conversion* **22** (2007) 79–85.

60. E.ON Netz grid code, Bayreuth, Germany, 1 August 2003.

61. D. Foussekis, F. Kokkalidis, S. Tentzevakis and D. Agoris, "Power quality measurement on different type of wind turbines operating in the same wind farm," *EWEC (Session BT2.1 Grid Integration)*, Madrid, Spain, 16–19 June 2003.

62. Electromagnetic compatibility, general guide on harmonics and inter harmonics measurements and instrumentation, IEC Standard 61000-4-7 (1997).

63. Electromagnetic compatibility, assessment of emission limits for distorting loads in MV and HV power systems, IEC Standard 61000-3-6 (1996).

64. F. Blaabjerg, R. Teodorescu, Z. Chen and M. Liserre, "Power converters and control of renewable energy systems," *Proc. 6th Int. Conf. Power Electronics*, vol. 1 (2004), pp. 1–20.

65. J.M.A. Myrzik and M. Calais, "String and module integrated inverters for single-phase grid connected photovoltaic systems — A review," *Proc. IEEE Power Technical Conf.*, Bologna, Italy, 23–26 June 2003, pp. 430–437.

66. H. Haeberlin, "Evolution of inverters for grid connected PV systems from 1989 to 2000," *Proc. Photovoltaic Solar Energy Conf.* (2001), pp. 426–430.

67. S.B. Kjaer, J.K. Pedersen and F. Blaabjerg, "A review of single-phase grid-connected inverters for photovoltaic modules," *IEEE Trans. Industrial Applications* **41** (2005) 1292–1306.

68. S.B. Kjaer, J.K. Pedersen and F. Blaabjerg, "Power inverter topologies for photovoltaic modules — Areview," *Proc. 37th IEEE-IAS Annual. Meeting*, vol. 2, 13–18 October 2002, pp. 782–788.

69. "Implementing agreement on photovoltaic power systems," in "Grid connected photovoltaic power systems: Survey of inverter and related protection equipments," International Energy Agency, Central Research Institute Electrical Power Industry, Paris, France, IEA PVPS T5-05, December 2002.

70. H. Akagi, "Active filters and energy storage systems for power conditioning in Japan," *Proc. 1st Int. Conf. Power Electronic Systems and Applications* (2004), pp. 80–88.

71. R. de Andrade, Jr., A.C. Ferreira, G.G. Sotelo, J.L.S. Neto, L.G.B. Rolim, W.I. Suemitsu, M.F. Bessa, R.M. Stephan and R. Nicolsky, "Voltage sags compensation using a superconducting flywheel energy storage system," *IEEE Trans. Application Superconductors* **15** (2005) 2265–2268.

72. M. Barnes, R. El-Feres, S. Kromlides and A. Arulampalam, "Power quality improvement for wave energy converters using a D-STATCOM with real energy storage," *Proc. 1st Int. Conf. Power Electronic Systems and Applications*, November 2004, pp. 72–77.

73. M. Nitta, S. Hashimoto, N. Sekiguchi, Y. Kouchi, T. Yachi and T. Tani, "Experimental study for wind power-hydrogen energy system with energy capacitor system," *Proc. IEICE/IEEE INTELEC Conf.* (2003), pp. 451–456.

74. A. Von Jouanne, I. Husain, A. Walace and A. Yokochi, "Gone with the wind: Innovative hydrogen/fuel cell electric vehicle infrastructure based on wind energy sources," *IEEE Industrial Applications Magazine* **11** (2005) 12–19.

75. http://www.nationalgrid.com/uk/Electricity/Codes/gridcode/gridcodedocs/.

76. *E.ON Netz Grid Code*, Bayreuth, Germany: E.ON Netz GmbH, 1 August 2003.

77. http://www.energinet.dk/en/menu/System+operation/Technical+Regulations+for+electricity/Regulations+for+grid+connection/Regulations+for+grid+connection.htm.

78. http://www.hydroquebec.com/transenergie/fr/commerce/pdf/Exig_compl_eolien_transport_mai03_ang.pdf.
79. http://www.hydroquebec.com/transenergie/fr/commerce/pdf/Exig_compl_eolien_transport_mai03_ang.pdf.
80. http://www.aesoc.ca/.
81. http://www.dwed.org.cn/uploadfiles/2009102691628/2009-10-22%20e%20grid%20code.pdf.
82. http://www.aemc.gov.au/Media/docs/Rule%20To%20Be%20Made-d63e6f1b-f171-434c-a0f3-cc9877c4d730-0.pdf.
83. http://www.ferc.gov/industries/electric/indus-act/gi/wind.asp.

Chapter 26

Analysis of Induction Generators for Renewable Energy Applications

Kanwarjit S. Sandhu

National Institute of Technology Kurukshetra,
Haryana-136119, India
kjssandhu@rediffmail.com
kjssandhu@yahoo.com

This chapter includes the analysis of induction generators usually employed for wind energy conversion. A self-excited mode which is found to be useful in remote windy locations needs special treatment for its steady state analysis in contrast to grid connected mode. Efforts are made to discuss various possible approaches for the estimation of steady state performance of a self-excited induction generator. Further methodologies have been explained to control the terminal conditions of such generators. References 1 to 19 are referred to finalize this chapter in its existing form.

26.1   Introduction

Synchronous generators which are the main attraction for their utility to extract the power from conventional sources of energy are generally used in hydro and thermal power stations. These generators have undergone an impressive evolution in terms of their ratings, cooling methods and parameters. But the increasing rate of depletion of conventional sources of energy as well as tremendous increase in power demand throughout the world has led scientists to explore new energy sources. As a result, the interest in renewable energy sources such as wind energy, solar energy, tidal energy, geothermal energy, etc., have grown significantly during the past few years. Out of these, wind energy is the fastest growing area of all renewable energy resources, and is attractive and viable. It is observed that winds carry enormous amount of energy and could meet the substantial needs of the world. The regions in

717

which strong winds prevail for a sufficient time during the year may use wind energy profitably for different purposes. It has been found that the cost of wind generation is comparable to that from hydro and thermal plants. There is little doubt that while the cost of wind generation would be even low in the coming years, the prices of fossil fuels used by thermal plants would definitely go up. In addition, wind energy generation provides a clean and pollution-free environment. It does not lead to global warming and ozone depletion. No hazardous waste is created. Furthermore a wind turbine generator may be a worthwhile proposition for isolated and remote areas. To feed such an area from a power grid, long transmission lines are required. It needs huge investment. Such areas are best served by diesel plants and a wind turbine generator may be installed to work in combination with a diesel plant to meet the local demands. Such an operation will lead to savings in fuel and economy. On the whole these have given a great impetus to investment in research and technology. Bigger turbines, better blade design, advanced materials, smart electronics and micro controls have all helped to improve the wind generation technology a lot.

Presently, 80 countries are using wind energy[1] on a commercial basis. USA, Germany, Spain, China and India, account for three-quarters of the global wind installations. By the end of 2008, 120 Gigawatt of wind power capacity were installed worldwide. Today, wind comprises more than 1.5% of the global electricity consumption and the wind industry employs half a million people. North America and Asia showed the most dynamic growth rates in 2008. In the foreseeable future wind energy is expected to cover a large share of the electricity needs worldwide.

In order to harness wind energy, induction generators are found to be the most appropriate machines in contrast to other machines. Such generators do not need synchronization with existing systems as in the case of conventional generators. The other advantages are their low cost, simple and rugged construction, less maintenance and easy operation.

Based on the excitation and mode of operation, induction generators are classified into the following two categories;

- Grid Connected Induction Generator (GCIG).
- Self-Excited Induction Generator (SEIG).

26.2   Equivalent Circuit Model of Induction Machine

Conventional equivalent circuit representation for a three-phase induction motor operating at slip "s" is shown in Fig. 26.1. Here the core loss branch has been omitted with the provision that core losses are to be combined with mechanical losses and deducted out of the gross mechanical power to obtain shaft output.

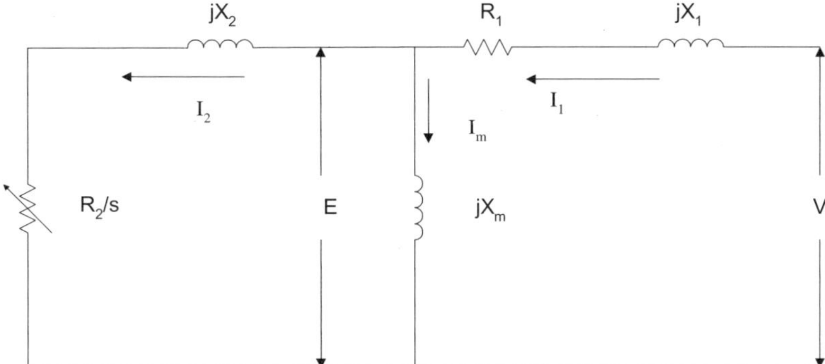

Fig. 26.1.   Conventional per phase equivalent circuit representation of three phase induction motor.

This circuit may be used for the analysis of induction generator when slip "s" becomes negative. Now the direction of rotor current will be opposite to motoring action.

Figure 26.1 gives

$$E = I_2 \left[\frac{R_2}{s} + jX_2 \right]$$

$$= I_2 \left[\left(\frac{R_2}{s} + R_2 - R_2 \right) + j(X_2 + sX_2 - sX_2) \right]$$

$$= I_2[(R_2 + jsX_2)] + I_2 \left[\left(\frac{R_2}{s} + jX_2 \right)(1 - s) \right], \qquad (26.1)$$

where

$$I_2 = \frac{sE}{R_2 + jsX_2} = \frac{E}{\left(\dfrac{R_2}{s} \right) + jX_2}. \qquad (26.2)$$

From Eqs. (26.1) to (26.2) the circuit as given in Fig. 26.1 may be modified[7] as shown in Fig. 26.2.

Here $E(1-s)$ is sink voltage and in case slip is negative, i.e., machine is operating as a generator, this sink voltage becomes source voltage as $E(1+s)$. This gives a new concept of representing the induction generator with a source in rotor circuit with change in the direction of rotor current.

26.3   Slip in Terms of Per Unit Frequency and Speed[2]

Let

$a = $ the ratio of generated frequency to rated frequency [pu frequency].

$b = $ the ratio of actual rotor speed to the synchronous speed corresponding to rated frequency [pu speed].

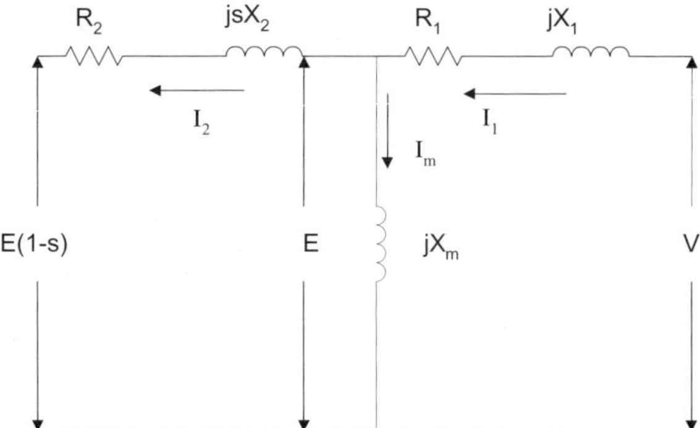

Fig. 26.2. Modified per phase equivalent circuit representation of three phase induction motor.

Thus the operating slip "s" in case of generator may be expressed as

$$= \frac{\text{Speed of rotating field} - \text{Actual rotor speed}}{\text{Speed of rotating field}}$$

$$= \frac{120\,(\text{Generated Frequency})/P) - N}{120\,(\text{Generated Frequency})/P}$$

$$= \frac{[\text{Generated Frequency}/(PN_s/120)] - (N/N_s)}{[\text{Generated Frequency}/(PN_s/120)]}$$

$$= \frac{(\text{Generated Frequency/Rated Frequency}) - (N/N_s)}{(\text{Generated Frequency/Rated Frequency}}$$

or

$$S = \frac{a - b}{a}.$$

26.4 Grid Connected Induction Generator

When the stator winding of a three-phase induction machine is excited with a three-phase supply system, a synchronous rotating field is established in the air gap. The speed of this synchronous rotating field "N_s" depends upon the supply frequency and number of poles for which machines are wound and is given by an expression as

$$N_s = \frac{120 f}{P} \text{ rpm}$$

where

P = Total number of poles

f = Supply frequency in Hz.

When the machine is operating as a motor the rotating field induces a current and voltage in the rotor conductors and the rotor rotates at a speed slightly less than that of a rotating field speed. The difference in speed of rotation between the field and rotor expressed as a percentage of synchronous speed is called a slip. It is well known that the slip of an induction machine when working as a motor lies between zero and unity, corresponding to synchronous speed and standstill. As shown in Fig. 26.3 under this operation the torque developed by the machine is positive. The positive sign of torque therefore indicates the motor action and machine supplies mechanical power.

In case a prime-mover is mechanically coupled to the induction motor, which in turn is connected to an appropriate power source, the operating slip of the machine may be made zero. This is possible only if the prime-mover is run at the synchronous speed. Under such condition, the rotor resistance R_2 divided by slip "s" results in an open circuit across the rotor side of the machine. Thus rotor current becomes zero and no mechanical power is developed. The machine will draw a current that is sufficient to meet the iron losses and excitation current. In case iron losses are

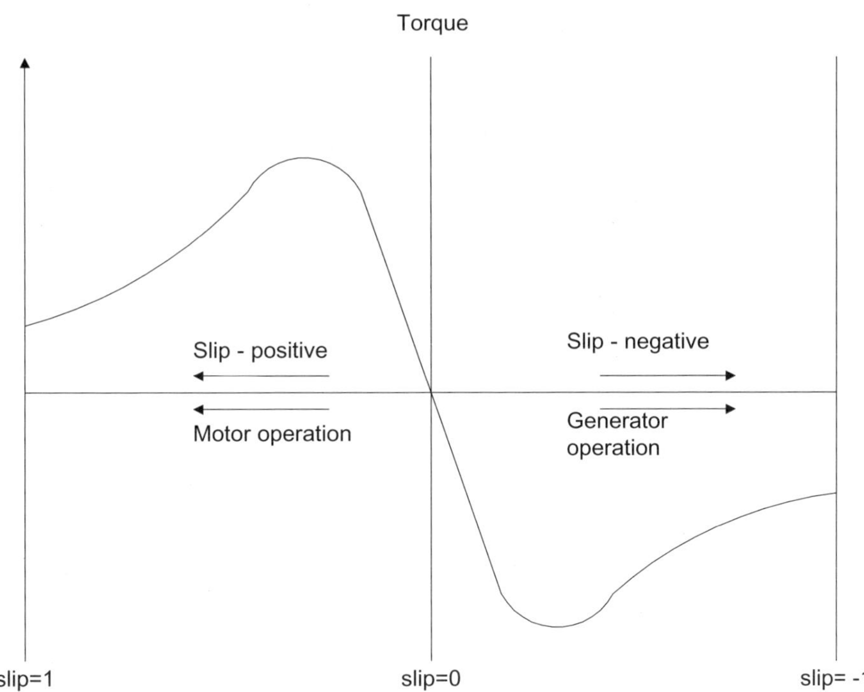

Fig. 26.3.   Variation of torque with operating slip.

                                  K. S. Sandhu

negligible the current drawn is purely a magnetizing current. Now if the prime-mover is made to run at a speed greater than the synchronous rotating field speed, it results in a negative slip. Under such an operation, the relative direction of rotation between the rotor and magnetic field has therefore been reversed, and the rotor emfs and currents are likewise reversed, indicating that the machine has changed from motor to generator action. When the slip is negative, rotor resistance divided by the slip becomes negative, which is considered as a voltage source. Now instead of producing power, the machine absorbs power from the prime-mover and delivers electrical energy[3,4] through the air gap as shown in Fig. 26.4. Under such condition the machine will draw its exciting current from the lines where the machine is connected. Such an operation is called "grid connected operation" and the induction machine operating under this mode is known as grid connected induction generator. Figure 26.5 shows the per phase equivalent circuit representation for the induction generator with the slip as negative.

Analysis of the equivalent circuit as shown in Fig. 26.5 gives;

- Output Power

$$P_{\text{out}} = 3VI_1 \cos \theta, \tag{26.3}$$

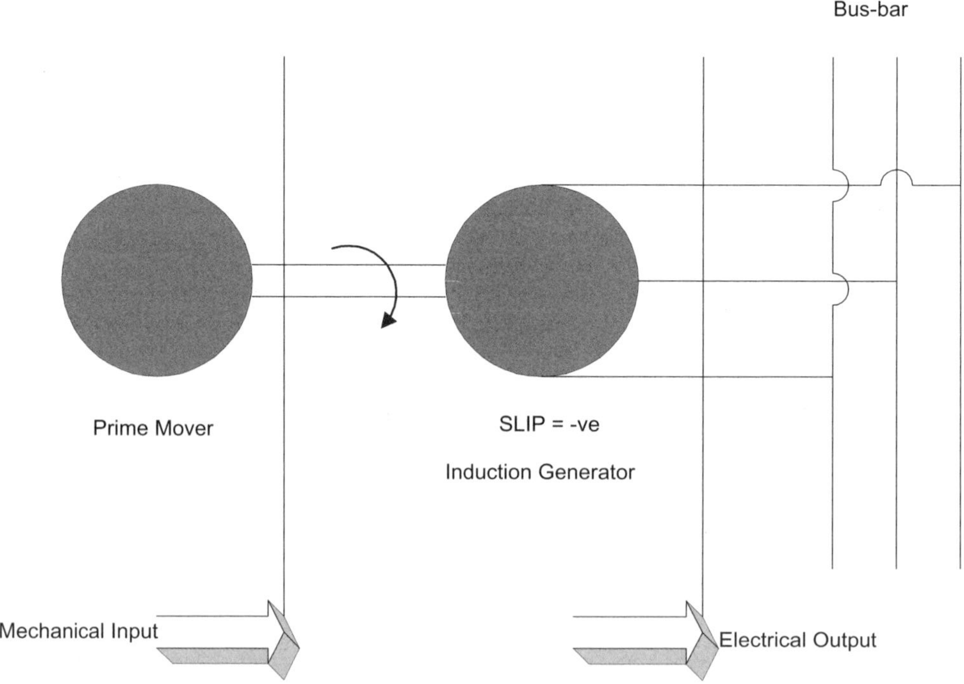

Fig. 26.4.   Grid connected induction generator.

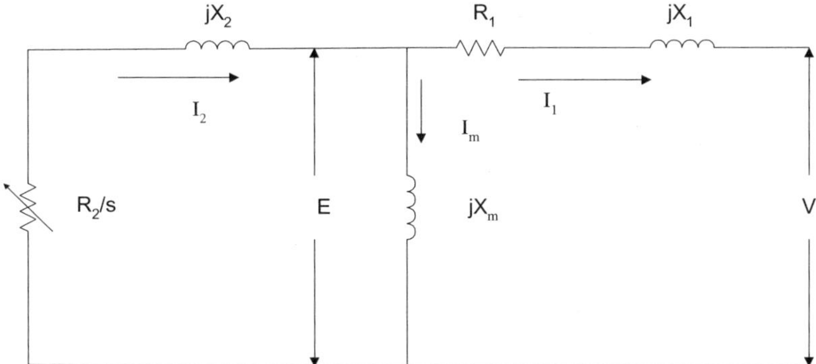

Fig. 26.5. Per phase equivalent circuit representation of grid connected induction generator.

where

V = Grid Voltage/phase

I_1 = Stator Current/phase

$\cos\theta$ = Operating power factor of the machine.

- Stator Current per phase

$$I_1 = V/Z, \tag{26.4}$$

where

Z is per phase impedance seen by source

$$Z = R_1 + jX_1 + \frac{\left(\dfrac{R_2}{s} + jX_2\right)(jX_m)}{\dfrac{R_2}{s} + j(X_2 + X_m)}$$

with

R_1 = Stator resistance/phase

X_1 = Stator reactance/phase

R_2 = Rotor resistance/phase, referred to stator

X_2 = Rotor reactance/phase, referred to stator

X_m = Magnetizing reactance/phase, at rated frequency

S = Operating slip [−ve].

If "N" is the operating speed in rpm $> N_s$, then slip "s" is negative and is as

$$S = \frac{N_s - N}{N_s}. \tag{26.5}$$

- Rotor current per phase referred to stator side

$$I_2 = \frac{E}{\dfrac{R_2}{s} + jX_2}, \tag{26.6}$$

where E is per phase rotor emf referred to stator and may be obtained by vector addition of per phase stator impedance drop to per phase terminal voltage.

- Internal torque "T_e" required to generate air gap power is as;

$$T_e = \frac{3I_2^2 R_2/s}{\omega_s},$$ (26.7)

where, $\omega_s = 2\pi N_s/60$ rad/sec.

- Input power "P_{in}" required is as;

$$P_{in} = [3I_2^2 R_2/s] + \text{Rotor Losses.}$$ (26.8)

- Efficiency = output power/input power.

Hence steady state analysis of such generators is possible by using the conventional equivalent circuit model for the induction machine as shown in Fig. 26.5, but with the slip taken as negative. In this mode of operation, the frequency and voltage remain unaffected with the change in value of slip of the machine but active power output of generator changes with the change of slip. Thus, the induction machine only contributes active power, which is a function of the slip of the machine. However, the performance of the generator is associated with the grid abnormalities, input source disturbances, protection problems of the generators, turbine and the system itself. Figures 26.6 to 26.9 show the simulated results[5] for a variation of torque, magnetizing reactance, power factor and stator current with operating slip on machine 1 [see Appendix]. From Fig. 26.6 it is observed that the load requirement can be met by increasing the operating slip of the generator. However a variation of stator current with slip as shown in Fig. 26.7 puts the rider on the operating slip of the induction generator. Observation of Fig. 26.8 indicates the improvement of p.f. on loading, which is true. Figure 26.9 shows the effect of loading on X_m. This is an indication

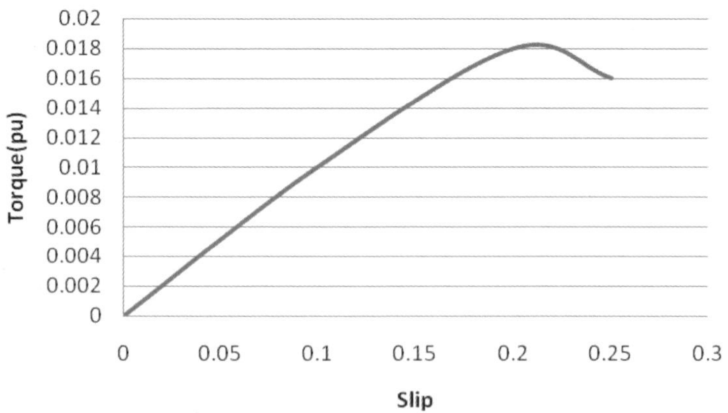

Fig. 26.6.   Variation of torque with slip.

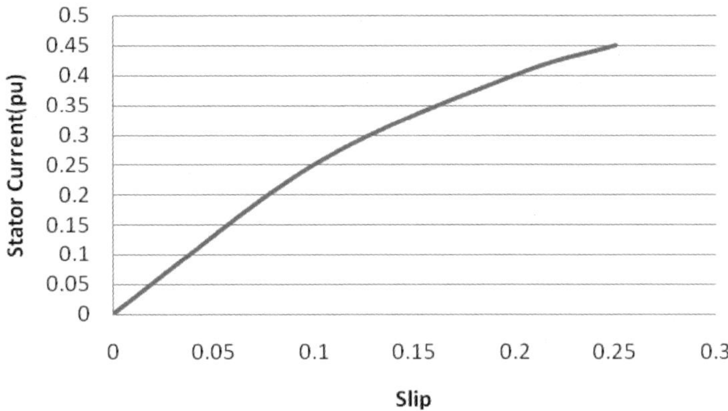

Fig. 26.7.   Variation of stator current with slip.

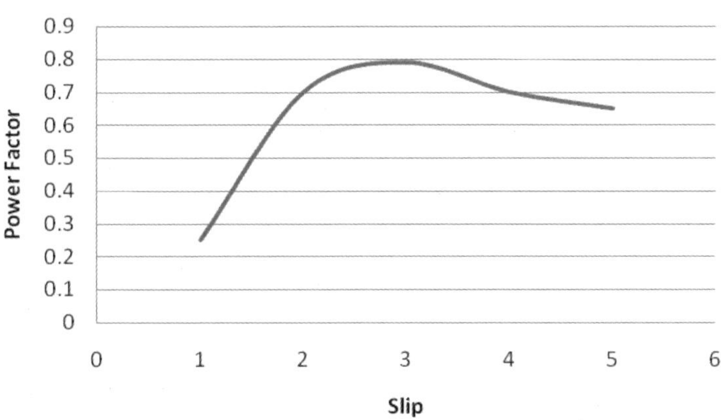

Fig. 26.8.   Variation of power factor with slip.

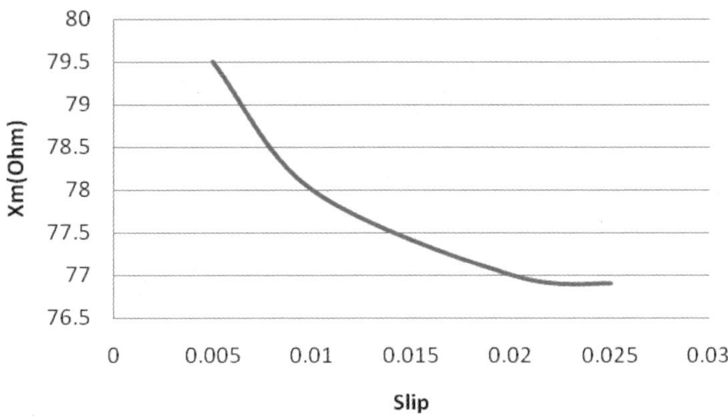

Fig. 26.9.   Variation of magnetizing reactance with slip.

that the operating point on the magnetic characteristics of induction machine moves in to a saturated region with an increase in load.

26.5   Self-Excited Induction Generators [SEIG]

The utility of the Self-Excited Induction Generator (SEIG) in power system networks started gaining importance when it was found difficult to wheel out power through transmission and distribution lines in remote areas due to geographical conditions. With the passage of time and use of non-conventional energy sources, particularly wind energy, the operation of SEIG emerged as a cost-effective alternative, where it was not economical to connect the consumer through electric lines. This has led to the intensive investigations pertaining to the performance of the self-excited induction generator by various researchers. The analysis of steady state performance of the induction machine as a generator is important for ensuring good quality power supply and assessing its suitability for specific application.

A self-excited induction generator does not require any grid system to feed reactive power. The generating operation is possible when the machine is driven at an appropriate speed and with the sufficient value of reactive power source at stator terminals of the machine. Thus operation of induction generators in self-excited mode gives the opportunity to utilize the wind energy in remote windy areas even in the absence of transmission lines. As a result, self-excited induction generators are receiving greater attention from the utilities for the cost effective power generation. These generators may be used to generate the power from a fixed as well as a variable speed prime-mover. Today most of the researchers are going for induction generators in self-excited mode due to their ability to convert mechanical power over a wide range of rotor speeds. This gives the operating flexibility to the machine in terms of operating speed. However major problems associated with the use of the self-excited induction generator are its poor "voltage" and "frequency" regulation. The use of such generators could only be made viable, if these machines are capable to generate the supply with the constant "voltage and frequency", under varying load and wind speeds. These generators could contribute to an overall cost reduction of the generation, in case free from sophisticated and complicated controls such as the governors, automatic voltage regulators and other associated auxiliary devices as in conventional generators.

An induction machine[3,4] can be made to operate as an isolated induction generator by supplying the necessary exciting or magnetizing current from capacitors connected across the stator terminals of the machine. Excitation to the induction machine when supplied by the capacitor bank in the absence of grid makes the operation of the machine as a self-excited induction generator. Grid availability is not essential for this mode of operation. The schematic arrangement of self-excited

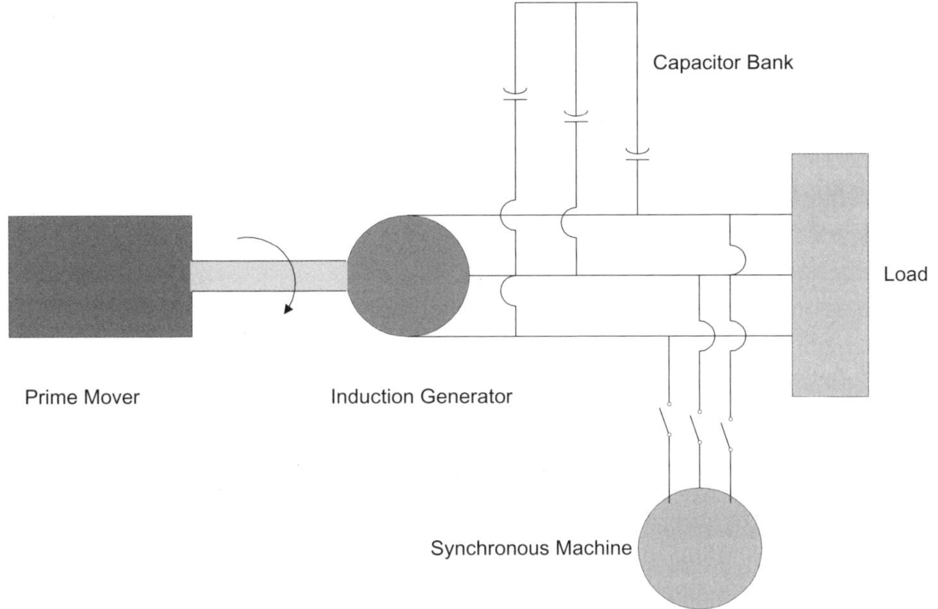

Fig. 26.10.   Self-excited induction generator.

induction generator supplying isolated power to the load is given in Fig. 26.10. The total active power of load and other losses of the machine are supplied by the induction generator. Reactive power supplied by the capacitor bank meets the reactive power requirements of the load and the machine.

The output of the self-excited induction generator is a function of speed, exciting capacitance and load on the machine. For successful voltage build up it is necessary that machine must have a residual magnetic field of correct polarity. The speed and excitation capacitance must also be sufficient to excite the machine under no load or loading conditions. This means that there are minimum speed and capacitance requirements that have to be fulfilled for self-excitation depending upon the load connected to the machine. The self-excitation process in induction generators is a complex phenomenon that has been studied extensively in the past and is still a subject of considerable attention. Acceptability of this generator as a viable unit is dictated by the fact that output frequency and voltage are highly dependent on speed, terminal capacitance and load impedance that causes certain limitations on its performance.

26.5.1 *Equivalent circuit model*

Figure 26.11 shows the equivalent circuit representation[6] of a self-excited induction generator. Here generated frequency is assumed to be rated one and $X_c = 1/(2\pi fc)$.

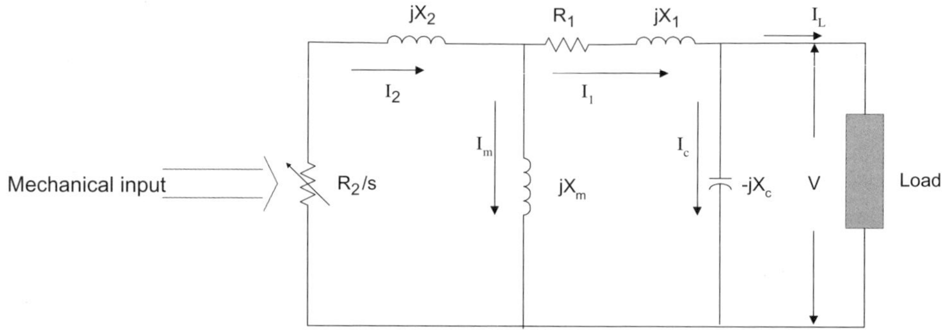

Fig. 26.11. Per phase equivalent circuit representation of SEIG at rated frequency.

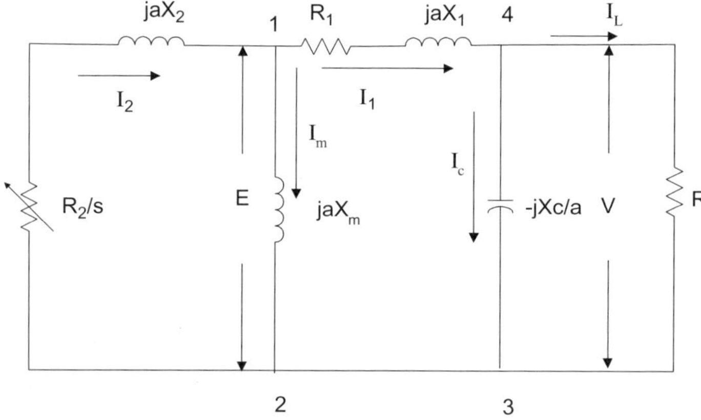

Fig. 26.12. Per phase equivalent circuit representation of SEIG at per unit frequency.

However this representation with resistive load "R" may be modified by introducing per unit frequency "*a*" and is shown in Fig. 26.12. Further using Fig. 26.2 this may be modified as shown in Fig. 26.13. This is a circuit representation which gives the feeling of generation[7,8] as usually in a generator circuit. Therefore any one of the circuit representations as shown in Figs. 26.12 and 26.13 may be used to analyze the behavior of the self-excited induction generator.

26.5.2 *Generated voltage and frequency*

Steady state analysis[6–16] of a self-excited induction generator is possible using any one of the circuit representations as discussed in the previous section. Here the problem is to determine the generated frequency and terminal voltage for a self-excited induction generator when operated with a certain value of excitation

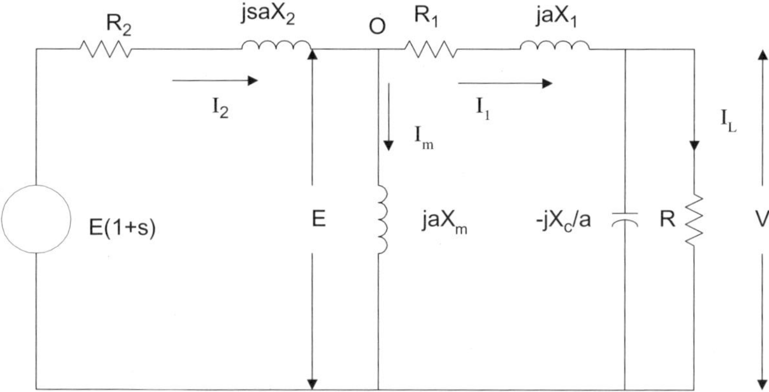

Fig. 26.13.　Per phase equivalent circuit representation of SEIG with source on rotor side.

capacitance, operating speed and load resistance. This is possible in three steps as given below.

- Determination of generated frequency and magnetizing reactance.
- Determination of air gap voltage using magnetization characteristics.
- Determination of generated voltage.

A. Determination of generated frequency and magnetizing reactance

Any one of the following methodologies may be adopted to determine the unknown values of generated frequency "a" and magnetizing reactance "X_m".

A.1.　Loop impedance approach.
A.2.　Nodal approach.
A.3.　Genetic algorithm based approach.

A.1. Loop impedance approach

An equivalent circuit representation as given by Fig. 26.12 may be adopted to proceed with this approach.[12] This figure may be modified with a single mesh as shown in Fig. 26.14,

where

$$Z_{12} = [\frac{R_2}{s} + jaX_2] \text{ in parallel with } [jaX_m]$$
$$Z_{14} = R_1 + jaX_1$$
$$Z_{34} = [R] \text{ in parallel with } [jX_c/a].$$

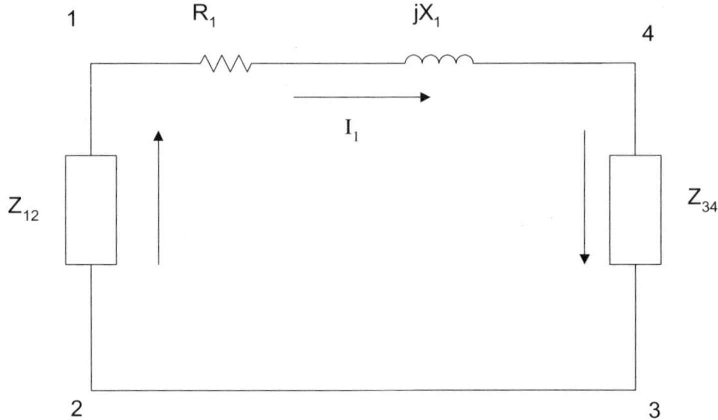

Fig. 26.14.   Per phase equivalent circuit representation of SEIG with single loop.

From Fig. 26.14, the loop equation may be written as;

$$[Z_{12} + Z_{14} + Z_{34}]I_1 = 0.$$

For successful generation I_1 cannot be zero.

Hence

$$[Z_{12} + Z_{14} + Z_{34}] = 0.$$

Let

$$[Z_{12} + Z_{14} + Z_{34}] = Z = 0. \tag{26.9}$$

Here Z comprises of real and imaginary parts. Further it is a function of machine parameters, excitation capacitance, operating speed, load resistance, frequency and magnetizing reactance. For a given machine operating as generator (i.e., with given value of excitation capacitance, operating speed and load resistance) the unknown parameters are generated frequency and magnetizing reactance.

The solution of Eq. (26.9) by equating real and imaginary parts separately to zero, and results in two unknowns, i.e., generated frequency "a" and magnetizing reactance "X_m".

A.2. Nodal approach

An Equivalent circuit representation as given by Fig. 26.13 may be modified[2,6] as shown in Fig. 26.15. In this representation the load and capacitor branch have been combined together to proceed with nodal analysis at point "O".

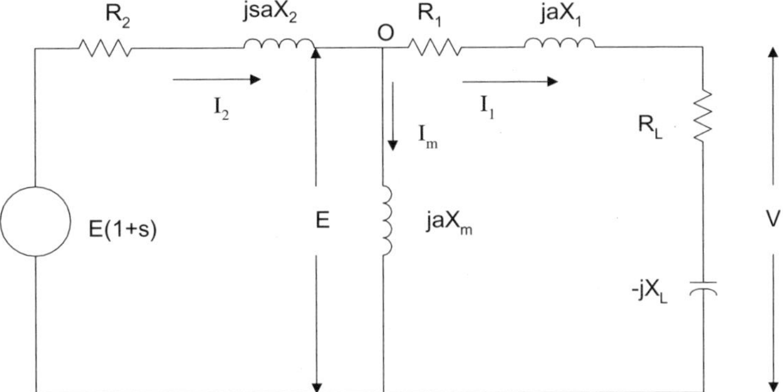

Fig. 26.15.   Per phase equivalent circuit representation of SEIG with two nodes.

From Fig. 26.15,

- In the absence of core loss branch, "I_m" the magnetizing current is purely reactive and is given as

$$\overline{I}_m = \frac{\overline{E}}{jaX_m} = -j\frac{\overline{E}}{aX_m}. \tag{26.10}$$

- Stator current per phase is

$$\overline{I}_1 = \frac{\overline{E}}{R_{1L} + jX_{1L}} \tag{26.11}$$

$$= \overline{E}\left[\frac{R_{1L}}{R_{1L}^2 + X_{1L}^2} - j\frac{X_{1L}}{R_{1L}^2 + X_{1L}^2}\right], \tag{26.12}$$

where

$$\left.\begin{aligned}
R_{1L} &= R_1 + R_L \\
X_{1L} &= aX_1 - X_L \\
R_L &= \frac{RX_c^2}{a^2R^2 + X_c^2} \\
X_L &= \frac{aR^2X_c}{a^2R^2 + X_c^2}
\end{aligned}\right\} \tag{26.13}$$

- Rotor current per phase is

$$\overline{I}_2 = \frac{s\overline{E}}{R_2 + jsaX_2}$$

$$= \overline{E}\left[\frac{sR_2}{R_2^2 + s^2a^2X_2^2} - j\frac{s^2aX_2}{R_2^2 + s^2a^2X_2^2}\right]. \tag{26.14}$$

Out of this the active part is supplied by the rotor and the reactive component is supplied by the capacitor connected across the stator.

Nodal analysis of the circuit of Fig. 26.15 at point "O" using Eqs. (26.10), (26.12) and (26.14), and by equating real and imaginary parts separately equal to zero gives the following:

$$\frac{sR_2}{R_2^2 + s^2 a^2 X_2^2} - \frac{R_{1L}}{R_{1L}^2 + X_{1L}^2} = 0, \tag{26.15}$$

$$\frac{s^2 a X_2}{R_2^2 + s^2 a^2 X_2^2} + \frac{1}{a X_m} + \frac{X_{1L}}{R_{1L}^2 + X_{1L}^2} = 0. \tag{26.16}$$

Equations (26.15) and (26.16) may be solved to determine the two unknown, i.e., generated frequency "a" and magnetizing reactance "X_m".

A.3. Genetic algorithm base approach[9]

Genetic algorithm (GA) is becoming a popular method for optimization, because it has several advantages over other optimization methods. Different from conventional optimization methods, the genetic algorithm was developed based on the Darwinian evolution theory of "survival of the fittest". It has produced good results in many practical problems and has become a powerful tool to solve nonlinear equations with a number of constraints. It is robust, able to find global minimum, and does not require accurate initial estimates. The GA manipulates strings of binary digits and measures each string's strength with a fitness value. The main idea is that stronger strings advance and mate with other strong strings to produce offspring. Finally one string emerges as the best. Another important advantage is that it offers parallel search, which can overcome local optima and then finally find the globally optimal solution. Over the past few years, many researchers have been paying attention to real-coded evolutionary algorithms, particularly for solving real-world optimization problems. Three main operators responsible for the working of the GAs are reproduction, crossover, and mutation. Selection of these three operators is very important before proceeding with the genetic algorithm approach.

In order to determine the two unknown, i.e., generated frequency "a" and magnetizing reactance "X_m", one may adopt the circuit representation of a self-excited induction generator as shown in Fig. 26.14. Loop impedance as given by Eq. (26.9) may be used as the objective function. This objective function may be minimized using a genetic algorithm with defined boundaries for generated frequency "a" and magnetizing reactance "X_m". The flowchart describing the GA optimization system implemented in the present section is shown in Fig. 26.16.

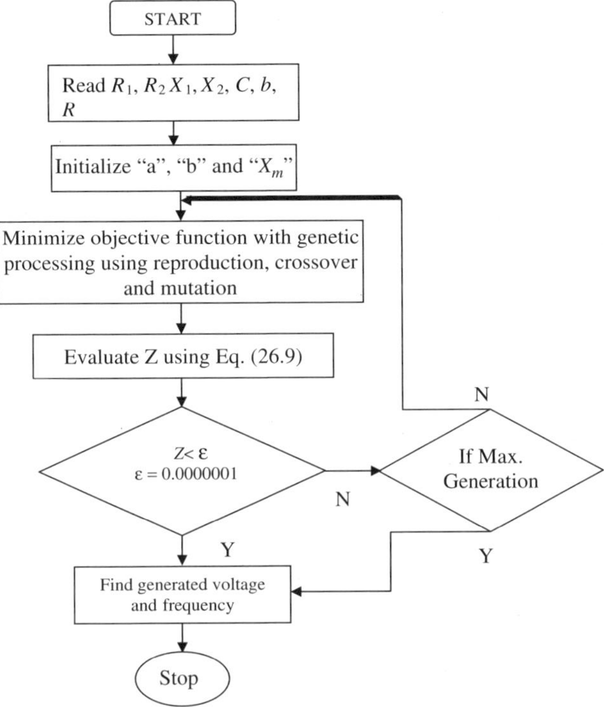

Fig. 26.16.   Flowchart for implementing GA.

B. Determination of air gap voltage "*E*" using magnetization curve

The magnetization characteristics for an induction machine, which relates the air gap emf with magnetizing current at rated frequency, can be obtained using a synchronous run test as shown in Fig. 26.17.

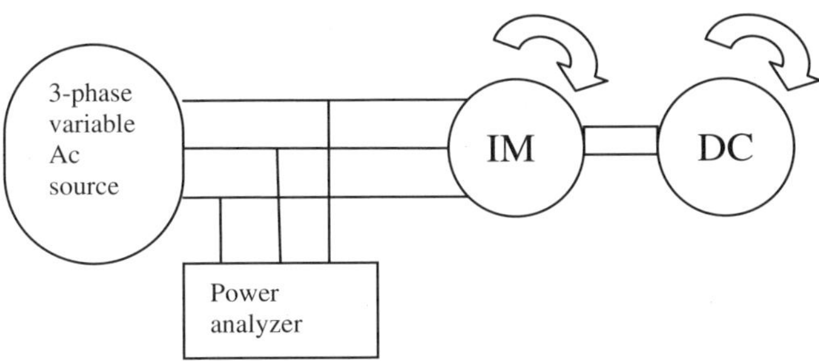

Fig. 26.17.   Synchronous run test on the induction machine.

The detailed procedure for a laboratory test to perform the synchronous run test for determination of magnetizing characteristics of induction machine is described as under:

- Three-phase induction motor which is coupled to the prime-mover is run by supplying power at rated voltage and frequency. Check the direction of rotation and switch of the supply.
- Now, supply of the DC shunt motor (prime-mover) is switched on to provide torque to the induction motor in the same direction as obtained earlier. Speed of the prime-mover is adjusted such that speed of the set is equal to synchronous speed corresponding to rated frequency.
- The input voltage of the induction motor is now varied in steps from sufficient low voltage to a voltage level slightly greater than the rated value. Measure voltage, current, power and power factor at each step with the power analyzer.

During the above test, when the induction motor is driven at synchronous speed by an external prime-mover, the mechanical losses of the set are supplied by the prime-mover. At synchronous speed, the slip of motor is zero therefore the machine will draw only magnetizing current (assuming copper loss and core loss component to be negligible). Further circuit representation under such an operation, as shown in Fig. 26.18, may be used to determine the air gap voltage air gap voltage.

After estimating the air gap voltage corresponding to different values of applied voltage the data may be used to plot the magnetization characteristics of the machine as shown in Fig. 26.19. This representation may be converted to plot a relation between air gap voltage and magnetizing reactance as shown in Fig. 26.20. Such a relationship is very useful to estimate the generated air gap voltage in case of

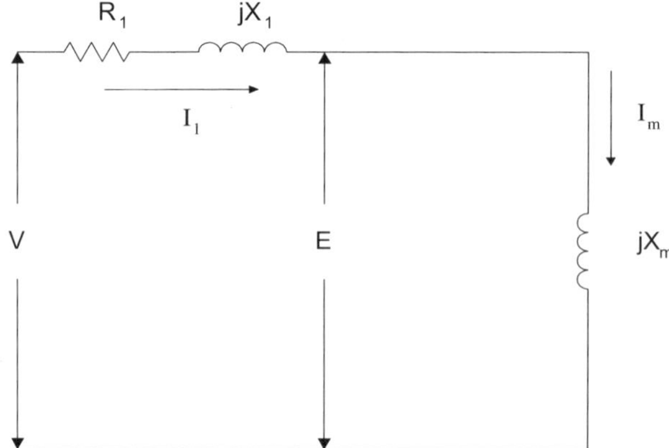

Fig. 26.18.   Per phase circuit representation of induction machine for a synchronous run test.

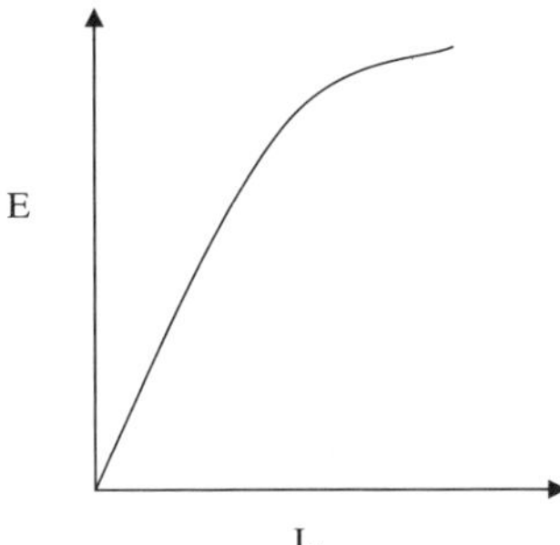

Fig. 26.19.   Magnetizing characteristics of the induction machine.

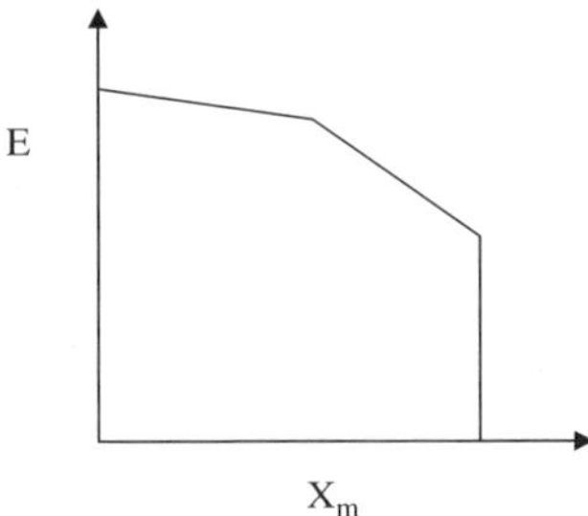

Fig. 26.20.   Variation of air gap emf with magnetizing reactance.

a induction generator corresponding to a value of "X_m" as obtained in previous sections.

C. Determination of generated voltage

With an estimated value of air gap voltage as obtained above, one can proceed with the complete solution by using the circuit representation as given in Fig. 26.15. The expression for the terminal voltage comes out to be as given below.

$$V = \frac{E}{[R_1 + R_L] + j[aX_1 - X_L]}[R_L - jX_L]. \tag{26.17}$$

26.5.3   *Effect of excitation capacitance*[2]

As X_c is inversely proportional to $C[X_c \propto (1/C)]$ an increase in the value of the capacitor will reduce the value of X_c, thus an increase in excitation component. This implies that the loading capacity of the machine increases with an increase in the value of excitation capacitance. Figures 26.21 to 26.24 give the variation of terminal voltage, magnetizing reactance, frequency and load current with excitation capacitance for machine 1. Here the operating speed of the machine is kept

                    K. S. Sandhu

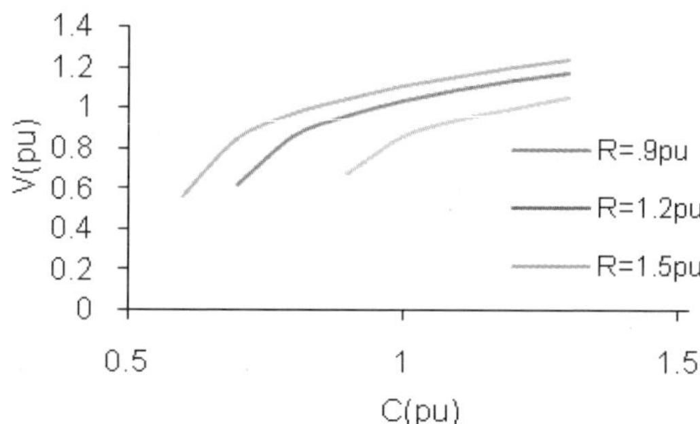

Fig. 26.21.   Variation of terminal voltage with excitation capacitance for different values of load resistance, $b = 1$ pu.

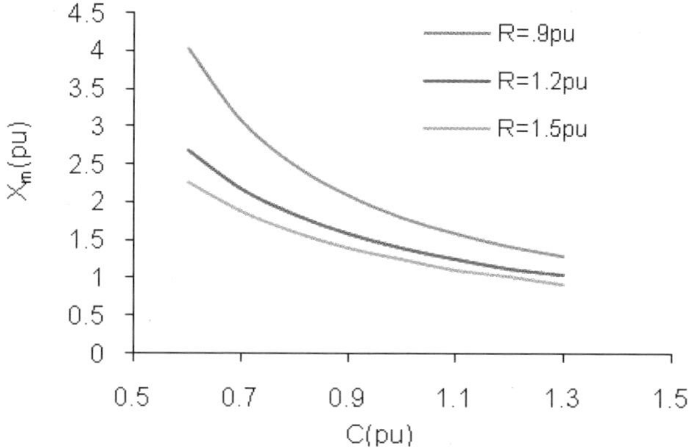

Fig. 26.22.   Variation of megnetizing reactance with excitation capacitance for different values of load resistance, $b = 1$ pu.

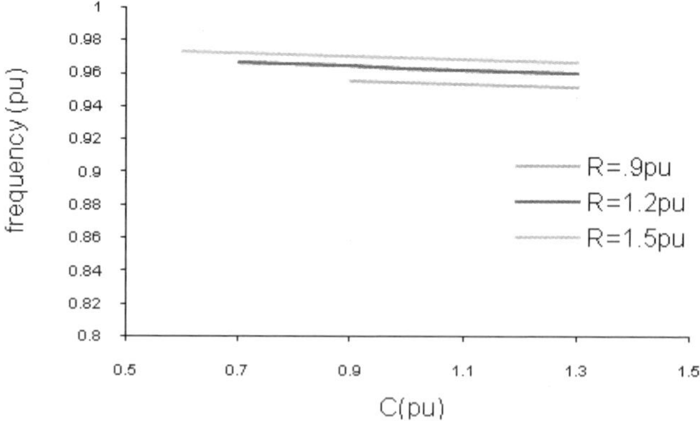

Fig. 26.23.   Variation of frequency with excitation capacitance for different values of load resistance, $b = 1$ pu.

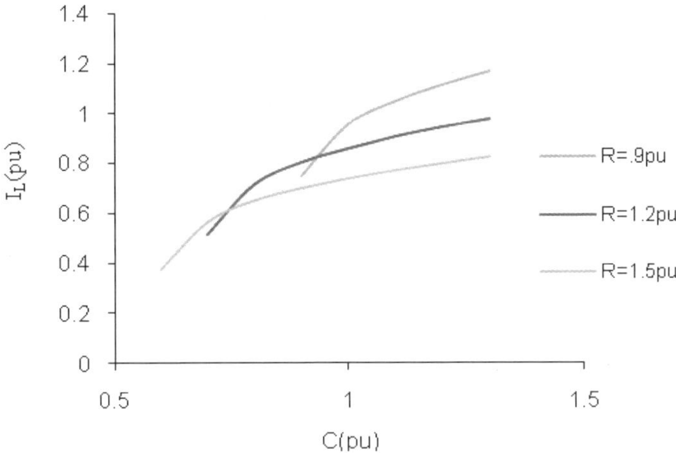

Fig. 26.24. Variation of load current with excitation capacitance for different values of load resistance, $b = 1$ pu.

constant as 1 pu. It is felt that any change in the excitation capacitance affects the terminal voltage, magnetizing reactance, generated frequency and load delivered by the machine. Thus the load carrying capacity of the machine may be controlled by a change of excitation capacitance and it may act as a control parameter in case of a self-excited induction generator.

26.5.4 *Effect of operating speed*[2]

It has been observed that the operating speed is almost linearly related to the generated frequency for a given set of operating conditions. Thus any change in the operating speed affects the generated frequency and plays an important role to control it. Further, any change in the generated frequency affects the effective value of excitation reactance. The effective value of excitation reactance decreases with an increase in the frequency, which in turn increases with an increase in the operating speed. Thus an increase in the speed will result in a reduction in the excitation reactance. This in turn is equivalent to the effect due to an increase in the capacitance. So it will affect the terminal conditions in the same manner as discussed in the previous section. Thus an increase in the operating speed with constant excitation capacitance and load resistance will result in an increase in the terminal voltage. Figures 26.25 to 26.27 give the variation in the terminal voltage, generated frequency and magnetizing reactance with the operating speed for machine 1 under a given operating condition. It is found that any change in the operating speed affects terminal voltage, generated frequency and value of magnetizing reactance. Therefore, similar to excitation capacitance, operating speed becomes another control variable.

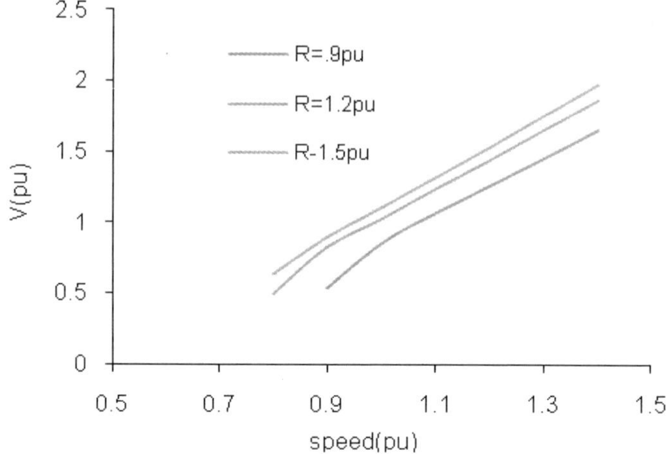

Fig. 26.25.   Variation of terminal voltage with speed for different values of load resistance, $c = 1$ pu.

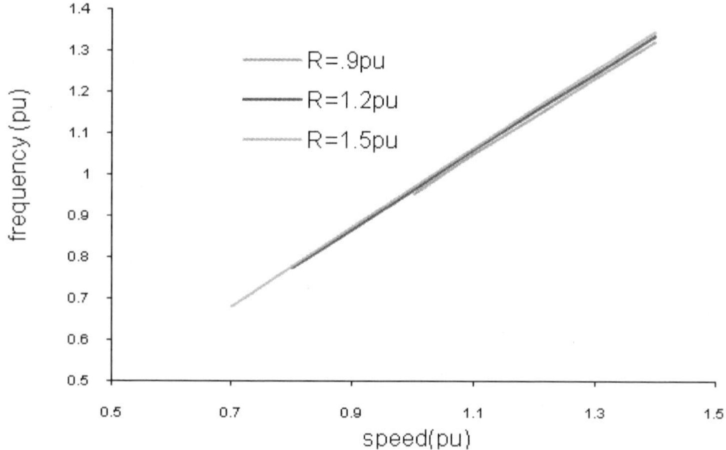

Fig. 26.26.   Variation of frequency with speed for different values of load resistance, $c = 1$ pu.

26.5.5 *Effect of stator resistance*[2]

Figure 26.28 shows the variation of the terminal voltage of machine 1 when operating at constant speed for different values of stator resistance. It is observed that there is an appreciable fall in the terminal voltage at high loads with an increase in the stator resistance. Figure 26.29 gives the variation in the generated frequency of machine 1, for different values of stator resistance with constant speed operation. It

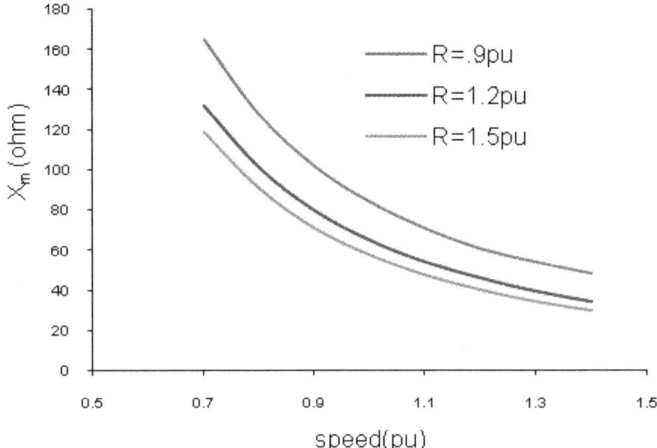

Fig. 26.27. Variation of magnetizing reactance with speed for different values of load resistance, $c = 1$ pu.

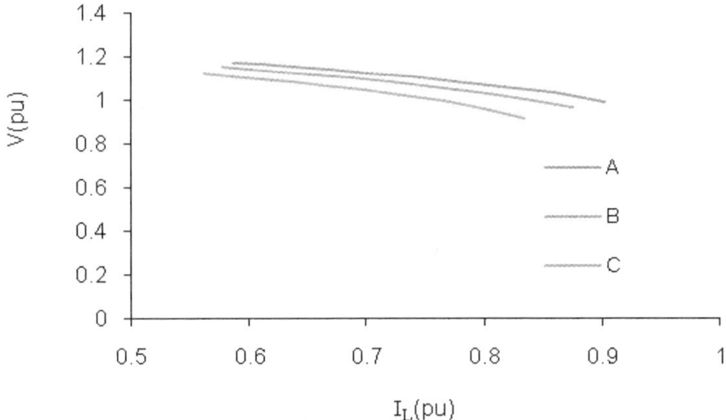

Fig. 26.28. Effect of stator resistance, $c = 1$ pu, $b = 1$ pu, curve $A-R_1 = 3.35$ ohm, curve $B-R_1 = 4.02$ ohm, curve $C-R_1 = 5.02$ ohm.

may be observed that the change in frequency is negligible for all loads due to the change in the stator resistance. To sum up, an increase in stator resistance

(a) Reduces the terminal voltage.
(b) Has no appreciable effect on frequency of the generated voltage.

To conclude, stator resistance should be as small as possible to obtain better voltage regulation of the induction generator.

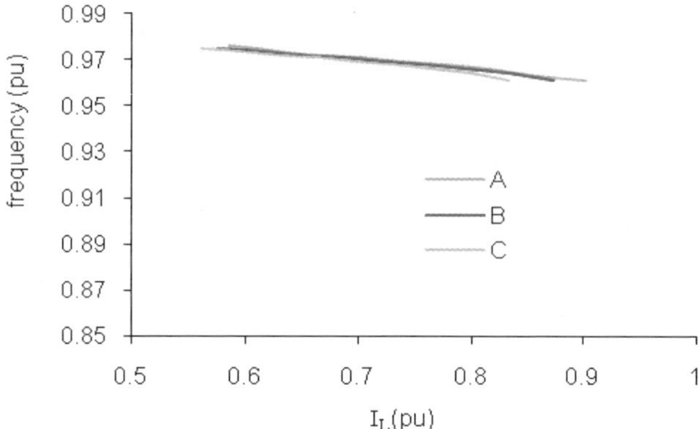

Fig. 26.29.   Effect of stator resistance, $c = 1$ pu, $b = 1$ pu, curve $A-R_1 = 3.35$ ohm, curve $B-R_1 = 4.02$ ohm, curve $C-R_1 = 5.02$ ohm.

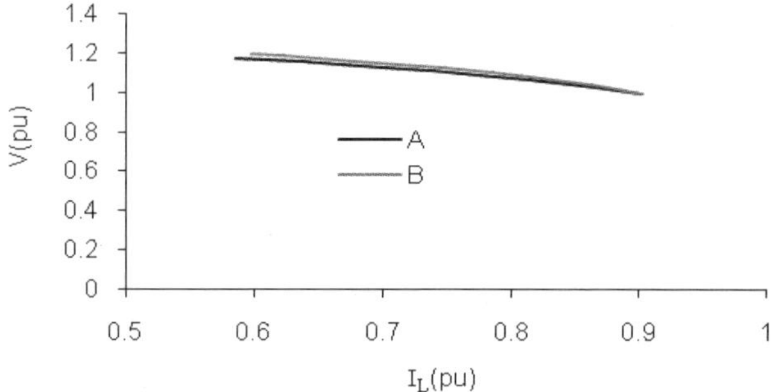

Fig. 26.30.   Effect of stator reactance, $c = 1$ pu, $b = 1$ pu, Curve $A-X_1 = 4.85$ ohm, curve $B-X_1 = 5.82$ ohm.

26.5.6   *Effect of stator leakage reactance[2]*

The effect of stator leakage reactance on the terminal voltage and frequency has been shown in Figs. 26.30 and 26.31 for machine 1. It is observed that the change in the values of the terminal voltage and frequency with any change in the stator leakage reactance is small.

26.5.7   *Effect of rotor resistance[2,8]*

Rotor resistance is a very sensitive parameter, which may act as a control parameter in case of the slip ring induction generator working in isolation. The effect of change in R_2 on the following is to be looked into:

A. On slip corresponding to Maximum torque.
B. On torque speed characteristics.
C. On terminal voltage and frequency.

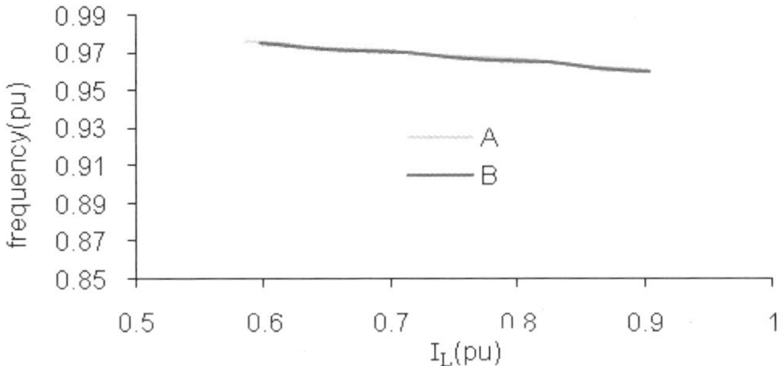

Fig. 26.31.   Effect of stator reactance, $c = 1$ pu, $b = 1$ pu, Curve $A - X_1 = 4.85$ ohm, curve $B - X_1 = 5.82$ ohm.

A. On slip corresponding to maximum torque

Equation (26.15) after certain modification gives;

$$s = \frac{R_2(R_{1L}^2 + X_{1L}^2) \pm R_2\sqrt{(R_{1L}^2 + X_{1L}^2)^2 - 4a^2 R_{1L}^2 X_2^2}}{2a^2 X_2^2 R_{1L}}. \tag{26.18}$$

The above equation gives two possible values of slip to which the stipulated operating conditions confirm to. But only the lower of the two values is relevant for generating mode.

This slip will be real only if

$$R_{1L}^2 + X_{1L}^2 \geq 2aR_{1L}X_2. \tag{26.19}$$

If the limiting value (minimum) of $(R_{1L}^2 + X_{1L}^2)$ given by Eq. (26.19) is substituted in Eq. (26.18), it gives the maximum possible value of operating slip for a given combination of exciting capacitance and rotor speed as

$$s_{\max} = \frac{R_2}{aX_2}. \tag{26.20}$$

But it is to be noted that for the limiting value given by Eq. (26.20), the load on the machine becomes so large that the operation as generator fails.

Further Eq. (26.19) gives

$$\frac{R_{1L}}{R_{1L}^2 + X_{1L}^2} < \frac{1}{2aX_2}. \tag{26.21}$$

Using Eq. (26.15) along with the above equation it may be written as

$$\frac{sR_2}{R_2^2 + s^2 a^2 X_2^2} < \frac{1}{2aX_2}. \tag{26.22}$$

With the assumption that $(saX_2)^2 \ll (R_2)^2$, Eq. (26.22) becomes

$$\frac{s}{R_2} < \frac{1}{2aX_2}$$

or

$$s < \frac{R_2}{2aX_2}. \qquad (26.23)$$

The above equation gives the limiting value of slip for the generator operation. Thus using Eqs. (26.20) and (26.23) the limiting value of the operating slip in terms of $s_{\max}$ is

$$s < \frac{s_{\max}}{2}. \qquad (26.24)$$

The above equation itself indicates that operation as generator is not possible at $s_{\max}$.

B. On torque-speed characteristics

It is well known that under induction machine operation as a generator, the slip is $-$ve. This negative slip results in a negative internal torque, which is called a generating torque. The torque speed characteristics under generating mode are shown in Fig. 26.32.

The generating torque T_G developed by the machine in synchronous-watts is given as

$$T_G = 3I_2^2 \frac{R_2}{s}$$

Substitution for I_2 gives;

$$T_G = 3 \frac{s^2 a^2 E^2}{R^2 + s^2 a^2 X_2^2} \left(\frac{R_2}{s} \right). \qquad (26.25)$$

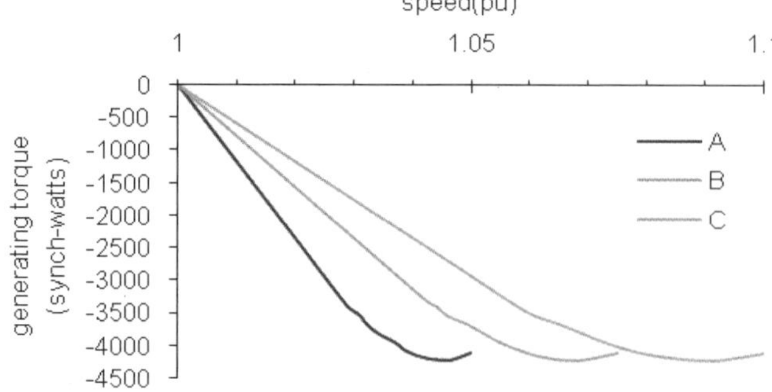

Fig. 26.32.   Torque-speed characteristics with generated frequency as rated one curve $A-R_{2\,\text{low}}$, curve $B-R_{2\,\text{medium}}$, curve $C-R_{2\,\text{high}}$.

In case $S^2 a^2 X_2^2$ is negligible as compared to R_2^2 the above expression becomes

$$T_G = \frac{3sa^2 E^2}{R_2}. \tag{26.26}$$

From the above expression it is clear that in case of the induction generator the air-gap power/torque in synchronous-watts is dependent upon the generated frequency in addition to the voltage. However, in case the machine is operating with a constant-voltage constant-frequency output, the expression becomes similar to that of a motor operation but with a power flow from the rotor to stator.

The torque given by Eq. (26.25) will be the maximum possible generating torque T_{GM}, in case the slip is corresponding to maximum torque, i.e., as given by expression (26.20).

Thus

$$T_{GM} = \frac{3s_{\max}^2 E^2}{(R_2^2 + s_{\max}^2 a^2 X_2^2)} \left(\frac{R_2}{s_{\max}} \right), \tag{26.27}$$

where I_2 at $s = s_{\max}$ is

$$= \frac{s_{\max} E}{\sqrt{(R_2^2 + s_{\max}^2 a^2 X_2^2)}}.$$

Substitution the value of slip from Eq. (26.20) into Eq. (26.27) gives the maximum generating torque T_{GM} as

$$T_{GM} = \frac{3E^2}{2aX_2}. \tag{26.28}$$

Thus it can be concluded that for a constant frequency operation, with constant air gap voltage, maximum torque is constant and is independent of rotor resistance. So from the above discussion it is clear that any change in the rotor resistance will shift the torque slip characteristics of the machine in the same manner as in the case of the induction motor, without affecting the maximum torque, provided generated frequency and terminal voltage remain constant. As shown in Fig. 26.32, simulations on machine 1 give the variation of generating torque due to any change in the rotor resistance. The torque speed characteristics give an opportunity to control the air gap power by a proper control of rotor resistance and hence it becomes an additional control parameter in case of the slip ring induction machine.

For the given operating conditions (speed, excitation capacitance and load resistance), Eq. (26.18) may be written as

$$\frac{s}{R_2} = \frac{(R_{1L}^2 + X_{1L}^2) - \sqrt{(R_{1L}^2 + X_{1L}^2)^2 - 4a^2 R_{1L}^2 X_2^2}}{2a^2 X_2^2 R_{1L}} = K_1, \tag{26.29}$$

where for a constant frequency operation K_1 is a constant with its value

$$= \frac{(R_{1L}^2 + X_{1L}^2) - \sqrt{(R_{1L}^2 + X_{1L}^2)^2 - 4a^2 R_{1L}^2 X_2^2}}{2a^2 X_2^2 R_{1L}}.$$

This implies that for a constant frequency operation the original conditions may be restored by a change in the operating slip if there is any change in the rotor resistance (for the slip ring induction machine). Thus the operating speed of the machine must rise with rotor resistance to keep the ratio s/R_2 constant for a certain value of generated frequency, as given by Eq. (26.29). Whereas any rise in the rotor resistance with constant speed operation results in a fall in the frequency. Where reduction of frequency will increase the effective capacitive reactance X_c, an increase in X_c will result in a fall of terminal voltage. Thus the machine starts operating at a point with a smaller value of generated frequency and voltage which satisfies Eq. (26.25).

Otherwise also if there is any change in the operating speed of the machine, for a given value of load resistance and capacitance, its effect may be accommodated by changing the rotor resistance accordingly. The required increase in the rotor resistance to neutralize the effect of an increase in the operating speed, must be such that it results in a constant value of $s/R_2 = K_1$ (Eq. 26.29). This phenomenon is very useful to maintain the terminal conditions for a self-excited induction generator (slip ring type) under variable speed operation.

C. On terminal voltage and frequency

As discussed in the previous section that during the constant speed operation, any increase in the rotor resistance will disturb the operating conditions. Hence the system will acquire a new value of slip and operating voltage which satisfies Eq. (26.25). Figures 26.33 and 26.34 give the variations in the terminal voltage and frequency for machine 1 due to any change in the rotor resistance. It is observed that this change affects both the voltage and frequency. However the variations in the frequency are large as compared to voltage variations.

26.5.8 *Effect of rotor leakage reactance*[2]

Figure 26.35 and Table 26.1 give the variations in the terminal voltage and generated frequency for machine 1 due to any change in the rotor reactance. It is observed that it affects the terminal voltage but the generated frequency is found to be independent of the rotor reactance.

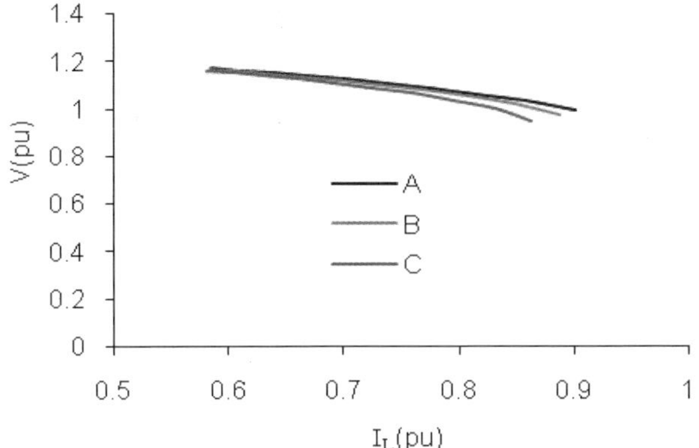

Fig. 26.33.   Effect of rotor resistance, $c = 1$ pu, $b = 1$ pu, curve $A - R_2 = 1.76$ ohm, curve $B - R_2 = 2.11$ ohm, curve $C - R_2 = 2.64$ ohm.

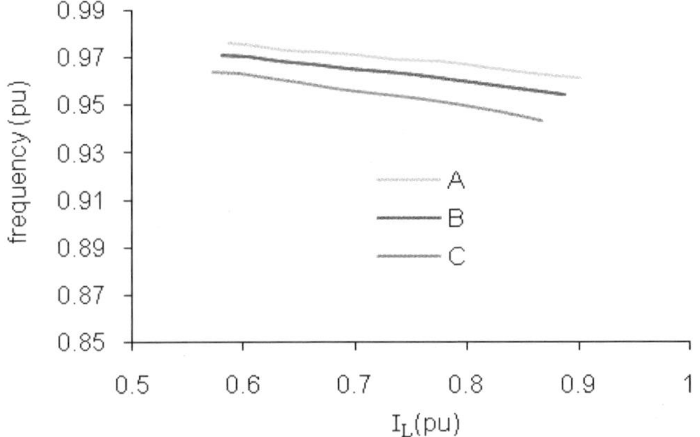

Fig. 26.34.   Effect of rotor resistance, $c = 1$ pu, $b = 1$ pu, curve $A - R_2 = 1.76$ ohm, curve $B - R_2 = 2.11$ ohm, curve $C - R_2 = 2.64$ ohm.

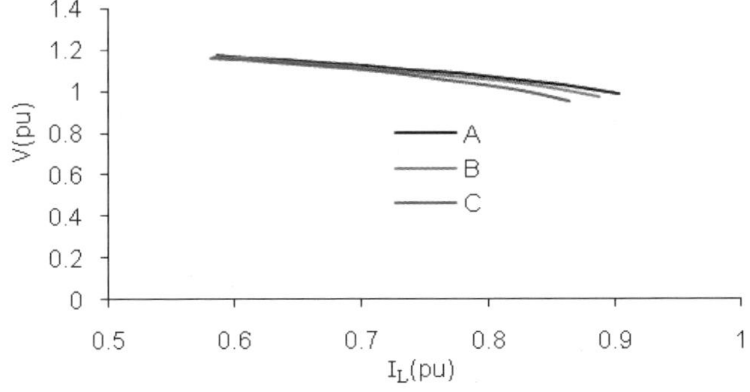

Fig. 26.35.   Effect of rotor reactance, $c = 1$ pu, $b = 1$ pu, curve $A - X_2 = 4.85$ ohm, curve $B - X_2 = 5.82$ ohm, curve $C - X_2 = 7.27$ ohm.

Table 26.1. Effect of rotor reactance on generated frequency on machine 1, $c = 1$ pu, $b = 1$ pu.

Load current in pu	Frequency (pu)		
	$X_2 = 4.85\,\Omega$	$X_2 = 5.82\,\Omega$	$X_2 = 7.27\,\Omega$
0.586	0.976	0.976	0.976
0.612	0.975	0.975	0.975
0.639	0.973	0.973	0.973
0.669	0.972	0.972	0.972
0.702	0.971	0.971	0.971
0.738	0.969	0.969	0.969
0.776	0.968	0.968	0.968
0.816	0.966	0.966	0.966
0.859	0.963	0.963	0.963
0.902	0.961	0.961	0.961

26.5.9 *Voltage and frequency control*[2,17–19]

A self-excited induction generator may be operated for the conversion of wind energy to electrical energy. Wind turbines are used to run the induction generators through the controllers to convert the wind energy to electrical energy. With the help of these controllers the induction generators may be operated either at a constant or variable speed under certain limitations. A standard wind energy converter of today has a constant turbine speed of 30 to 50 rpm and uses a gearbox to run a four or six pole induction generator.

In this section an attempt has been made to look into the performance of the generator, when the operating speed of the prime-mover running the machine is controllable by any means. Depending upon the mode of control adopted, the machine may be operated at a constant speed or at variable but desired speed. Here it is proposed to analyze the behavior of the self-excited induction generator for the following operations:

- Constant Speed Operation.
- Variable Speed Operation.

A. Constant speed operation

In case it is possible to run the generator at a constant speed, which results in the required terminal voltage and frequency at the rated load of the machine, the

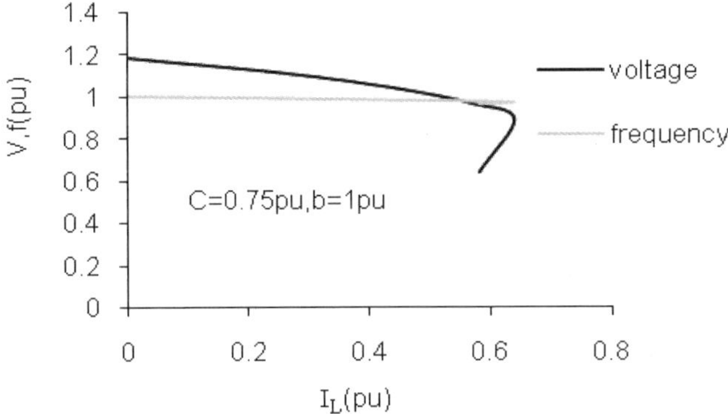

Fig. 26.36. Variation of voltage and frequency.

operation is called a constant speed controlled operation. Practically it is not possible to run the generator at the rated load throughout, especially if the machine is supplying a load distributed among number of consumers such as lighting load, etc. In such cases the load on the generator will vary depending upon the load cycle. Where, the load cycle is totally dependent upon the nature of load and load requirement by the consumers. Figure 26.36 gives the variation in the terminal voltage and frequency for machine 1, with such operating conditions.

It is observed that for a given value of the excitation capacitance and rotor speed the terminal voltage and frequency falls when the generator is loaded. The effect is opposite in case load decreases. Thus the generator operation at different load conditions certainly affects the quality of the supply and performance of the equipment at the consumer end, if these are voltage sensitive devices. The constant speed operation is satisfactory only, when the load on the generator remains undisturbed throughout. Otherwise it is required to minimize the effects of variation of the load using some control scheme. The different ways to compensate the voltage changes under such conditions are:

A.1. Load Adjustment.
A.2. Excitation Control.
A.3. Compensation.

A.1. Load adjustment[2]

During the constant speed operation the load on the machine must be such that it gives the required voltage and frequency at the load terminals for a given value of excitation capacitance and rotor speed. Or the machine may be run at a particular speed which along with a selected value of terminal capacitance results in the rated voltage and

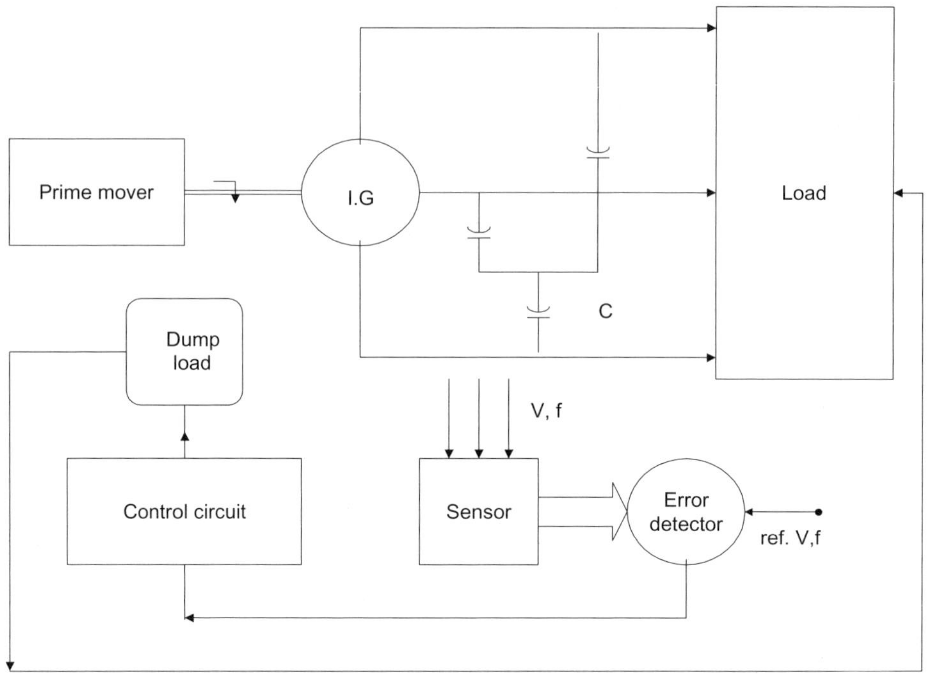

Fig. 26.37.   Automatic control system.

frequency for the full load condition. During such operations the machine becomes under-loaded in case the consumers disconnect some portion of the load on the machine. Under such conditions, in order to maintain the total load on the generator as rated load, additional load may be switched on to the machine as soon as some portion of the total load gets disconnected.

An automatic control system as shown in Fig. 26.37 may be used to maintain the load on the machine as;

I. The sensor may be used to detect the load on the machine and in case of any error the signal may be fed to the controller through the amplifier.
II. The controller will adjust the load on the machine in such a manner to reduce the error given by the error detector.

Under such circumstances additional load may be used for heating, etc.

This is the simplest way to control the output of the machine.

A.2. Excitation control

In the previous sections, it was found that any change in the excitation capacitance of the self-excited induction generator affects the terminal voltage. However the

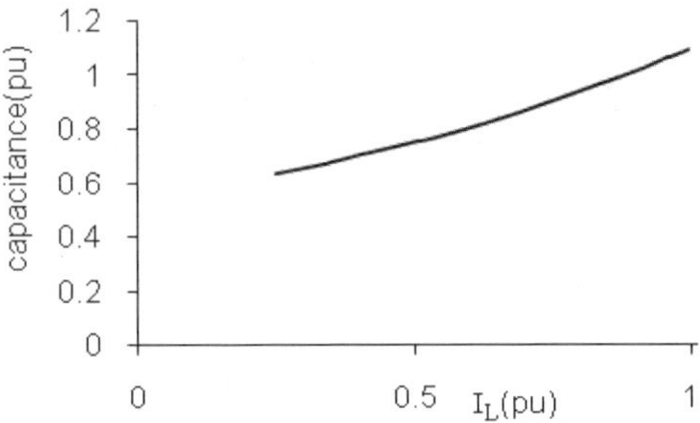

Fig. 26.38. Variation of capacitance with load current, $b = 1$ pu, $V = (1.001–1.008)$ pu.

effect on the generated frequency is found to be insignificant. It is observed that any increase in the excitation capacitance will reduce the capacitive reactance X_c, and this in turn shifts the operating point on the magnetization curve which results in an increase in the terminal voltage. Such an effect is very useful to control the voltage developed by the generator at different load conditions. By proper variation of the excitation capacitance called "capacitor switching" the voltage across the machine terminals may be maintained within the limits as shown in Fig. 26.38. The capacitance across the stator terminals of machine 1 may be switched accordingly to achieve the terminal voltage between the limits 1.00 pu to 1.008 pu. Although the "switched capacitor" scheme seems to be very simple but practically it is observed that such a process due to the additional control circuit becomes expensive and complex.

A.3. Compensation

It is well known that the terminal voltage of the self-excited induction generator falls considerably as the machine is loaded to the rated value. Thus the voltage regulation of the induction generator when operating in isolation is very poor. This is the major drawback, which puts a question mark on the use of this machine as a self-excited induction generator.

As the load on the induction generator increases, it will result in an increase in the operating slip to meet the requirement. Thus for a constant speed operation any increase in the operating slip results in a reduction of generated frequency. Since the capacitive reactance is inversely proportional to the operating frequency, it increases due to any reduction in the operating frequency of the system. As the value of "X_c" increases, the terminal voltage decreases, resulting in a poor voltage regulation. So if by any means, it is possible to maintain the excitation reactance, i.e., by additional generation of VAr, the voltage regulation may be improved to a great extent. Many researchers recommended the use of voltage regulators to improve the voltage regulation of the machine. But the complex system configuration, control circuit, switching transients, cost, etc., puts some restrictions on the use of self-excited induction generator along with such regulators.

The other simple and cheap method to improve the voltage regulation of self-excited induction generators is by using series capacitors. Such a process involving a series capacitor is called compensation[17–19] for the induction generators working in isolations.

Many researchers tried to look for the voltage regulation of the machine under two types of compensations[19] depending upon the location of series capacitor. These are:

- Short Shunt Compensation.
- Long Shunt Compensation.

A.3.1. *Short shunt compensation*

Figure 26.39 gives the per phase equivalent circuit representation for a self-excited induction generator with short shunt connections.

The variation of the load and terminal voltage may be plotted as a function of load current for different values of compensation factor "Cf", where Cf is defined as the ratio of series capacitive reactance "X_{se}" to shunt capacitive reactance "X_{sh}" as

$$Cf = \frac{X_{se}}{X_{sh}} = \frac{C_{sh}}{C_{se}}. \tag{26.30}$$

As the series capacitor does not come into the picture when the machine is operating at no load, it is only the shunt capacitor which will decide the no-load voltage across the machine terminals. However the series capacitor which carries the load

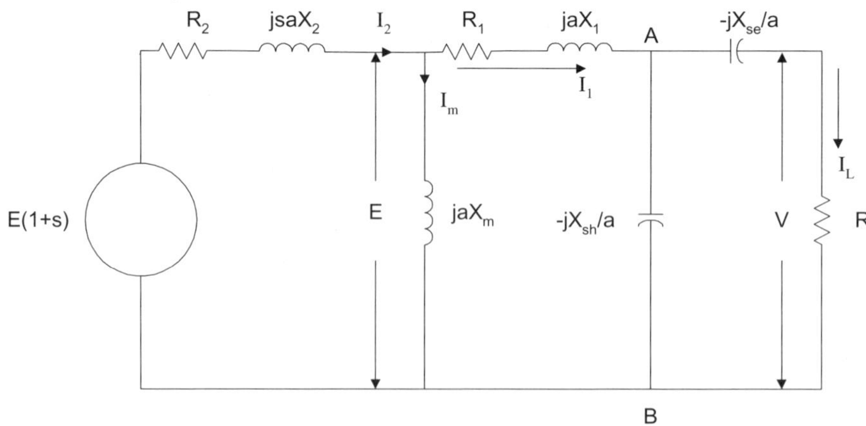

Fig. 26.39.   Short shunt compensation.

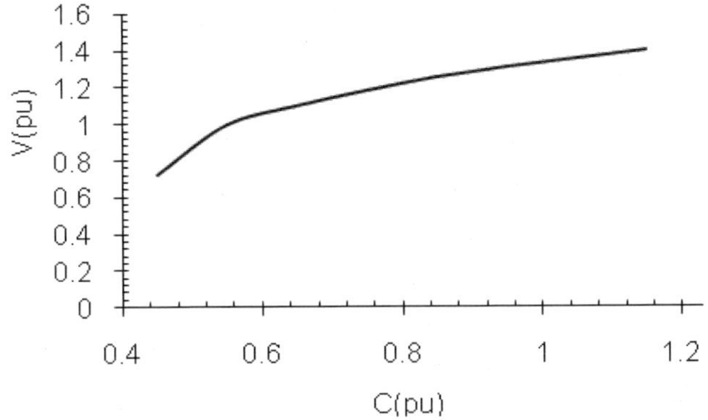

Fig. 26.40.　Variation of no-load terminal voltage with capacitance, $b = 1$ pu.

current is effective during the load conditions. A suitable methodology is required for the selection of shunt and series capacitors, which results in a minimum voltage regulation from no load to full load conditions.

The value of the shunt capacitor may be obtained from the variation of no-load voltage of the machine as a function of excitation capacitance "C" as shown in Fig. 26.40.

C_{sh} may be selected as capacitance C sufficient to generate the required voltage across the machine terminals at no load. For the selection of series capacitor "C_{se}" Eq. (26.30) may be used for different values of "compensation factor". The value of "Cf" for which the variations in the load voltage are minimum may be used to determine the value of series capacitor for a short shunt connection.

Figure 26.41 gives the variation of the load voltage of machine 1 as a function of load, for different values of compensation factor "Cf". From this the value of Cf may be obtained which almost results in the flat load voltage curve.

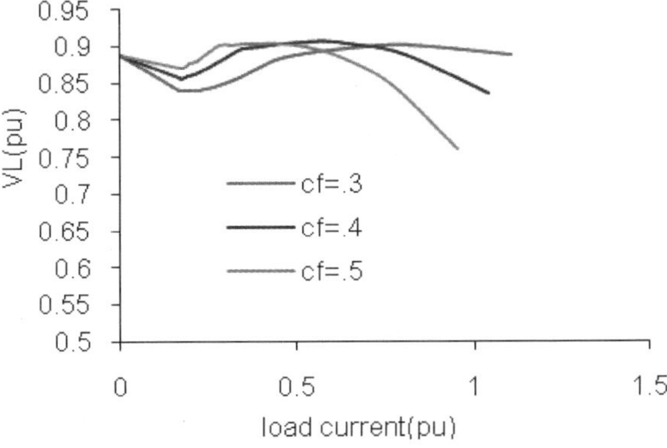

Fig. 26.41.　Short shunt compensation with $C_{sh} = .5$ pu, $b = 1$ pu.

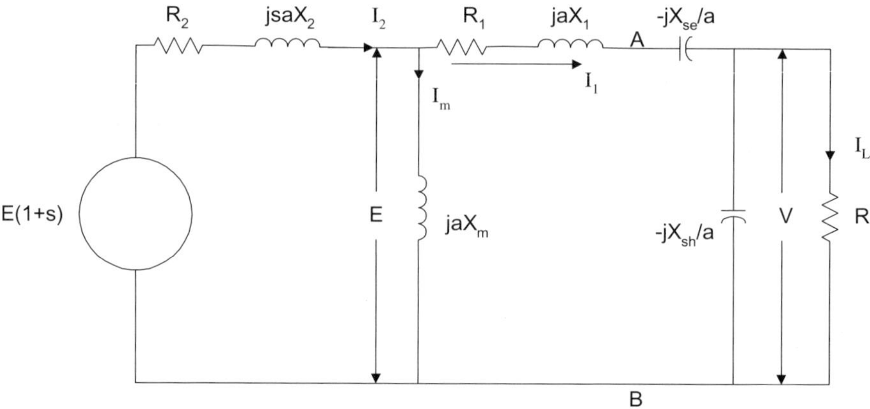

Fig. 26.42. Long shunt compensation.

A.3.2. *Long shunt compensation*

Figure 26.42 shows the per phase equivalent circuit representation for a self-excited induction generator with long shunt compensation. From the circuit it is clear that the series capacitor "C_{se}" carries stator current under all operating conditions and hence affects the generator's performance for any value of load. However at no load when R is ∞ this capacitor "C_{se}" comes in series with the shunt capacitor "C_{sh}" and thus causes a reduction in the net capacitance across stator terminals. This results in a low voltage across the machine terminals. Further it is noticed that the main difference between the long and short shunt compensation is the position of the series capacitor. For a long shunt compensation it is in series with the stator of the machine, hence carries a stator current for all operating conditions. On the other hand in short shunt compensation the series capacitor is placed in series with the load resistance, hence it carries load current only. Due to this difference of currents carried by the series capacitors in the two cases, the procedure to select the proper values of shunt and series capacitors in two cases becomes totally different.

The following procedures may be adopted to select the values of shunt and series capacitors in case of long shunt compensation.

Procedure 1

- As the no-load voltage developed by the machine is affected by the series capacitor, so it is necessary to include its effect on the no-load voltage of the generator. A curve between the no-load voltage generated by the machine against the capacitance, as shown in Fig. 26.40 gives the net capacitance C, where "C" is the combined effect of C_{se} and C_{sh} at no load and it may be written that

$$\frac{1}{C} = \frac{1}{C_{\text{sh}}} + \frac{1}{C_{\text{se}}}$$

or

$$C = \frac{C_{\text{sh}} C_{\text{se}}}{C_{\text{sh}} + C_{\text{se}}}. \tag{26.31}$$

- Cf is the compensating factor already defined in the previous section as

$$Cf = \frac{X_{\text{se}}}{X_{\text{sh}}} = \frac{C_{\text{sh}}}{C_{\text{se}}}. \tag{26.32}$$

- Using Eqs. (26.31) and (26.32)

$$C = \frac{Cf C_{\text{se}}^2}{Cf C_{\text{se}} + C_{\text{se}}} = \left(\frac{Cf}{1 + Cf} \right) C_{\text{se}}.$$

Using the above expression with Eq. (26.32) gives

$$\left. \begin{aligned} C_{\text{se}} &= C \left(\frac{1 + Cf}{Cf} \right) \\ C_{\text{sh}} &= C(1 + Cf) \end{aligned} \right] . \tag{26.33}$$

Equation (26.33) may be used to determine the value of the series and shunt capacitor for a particular value of compensation factor, provided the net capacitance C required to produce the required maximum tolerable no-load voltage across the machine terminals is known. Thus the curve of Fig. 26.40 may be used to select the value of C and then for different values of "Cf" the load voltage variation may be obtained as a function of load current. A particular value of Cf, which results in the minimum voltage variations from no load to full load conditions, may be selected for compensation. This value of Cf will result in the proper selection of series and shunt capacitors as per Eq. (26.33).

Procedure 2

- The value of shunt capacitor C_{sh} is kept constant as in case of short shunt compensation as "C".
- The value of "C_{se}" for different values of compensation factor may be obtained using the Eq. (26.32) as $C_{\text{se}} = C_{\text{sh}}/Cf$.

Thus for different values of compensation factor "Cf", the load and terminal voltage variations are plotted. The value of Cf which results into minimum voltage variations may be selected to obtain the values of series and shunt capacitors.

Application of procedure 1 to machine 1 gives the variation of load voltage for different values of compensation factor as shown in Fig. 26.43. From these curves the value of Cf of 0.5 may be selected which results in a better voltage regulation.

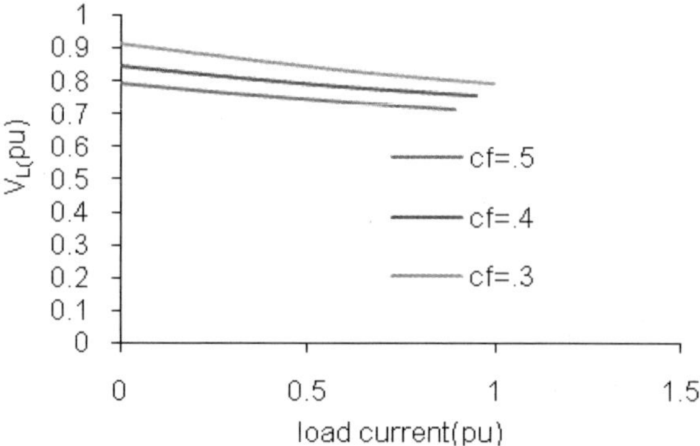

Fig. 26.43. Variation of load voltage with long shunt compensation (Procedure 1), $C_{sh} = .75$ pu, speed = 1 pu.

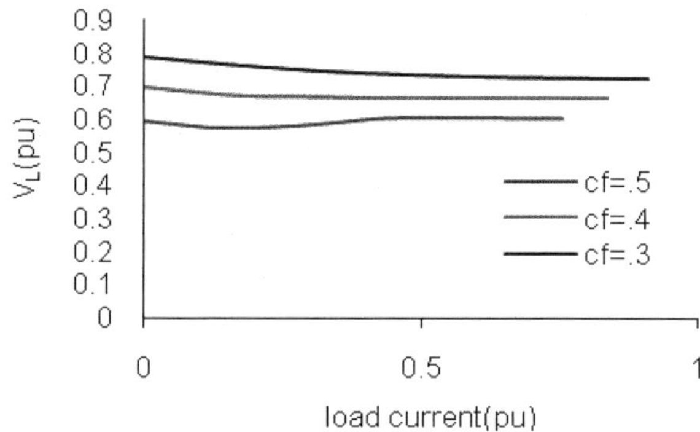

Fig. 26.44. Variation of load voltage with long shunt comprensation (Procedure 2), $c = 1$ pu, speed = 1 pu.

Similarly the selection of C_{sh} and C_{se} as per procedure 2 results in the machine's load voltage variations with different values of compensation factor, is shown in Fig. 26.44. Again a selection of Cf of 0.5, results in minimum voltage regulation.

B. Variable speed operation

Just like excitation capacitance, the other way to control the output of the self-excited induction generator is its operating speed. An increase in speed results in a higher terminal voltage for the same load resistance. This phenomenon may be utilized to control the terminal voltage of the machine. But it is noticed that any change in the operating speed affects the generated frequency more than in comparison to the excitation control. The generated frequency almost increases linearly with an increase in the operating speed.

26.6 Conclusions

On the basis of discussions as per previous sections, the following conclusions may be drawn.

- The generated voltage and frequency of a grid connected induction generator remains grid values irrespective of its operating conditions. The power flow can be maintained by proper control of its operating slip. However the main drawback of this machine is its limited capability to operate as a generator, in terms of operating speed.
- The main advantage of a self-excited induction generator is its capability to operate as a generator for large variations of operating speeds. However the generated voltage and frequency of this machine are dependent upon its operating conditions, i.e., load, speed and excitation capacitance. Hence there is a need to control the terminal conditions.
- As discussed earlier the main drawback of a self-excited induction generator is its poor voltage regulation. For the improvement of voltage regulation many schemes such as load adjustment, capacitor switching, variable speed operation and compensation has been discussed. Although load adjustment and capacitor switching seem to be very simple, it is observed that their application becomes expensive and complicated due to the control circuit involved. On the other hand, the variable speed operation gives a limited control due to the speed limitations of the wind turbine. Thus the series compensation is the only alternative which, when implemented, makes the self-excited induction generator more attractive for its autonomous operation.

Appendix

Specifications of Machine 1.

3-phase, 4-pole, 50 Hz, delta-connected, squirrel cage induction machine.

2.2 kW/3.0 hp, 230 V, 8.6 A, $R_1 = 3.35$ ohm, $R_2 = 1.76$ ohm, $X_1 = X_2 = 4.85$ ohm.

Base voltage $= 230$ V.

Base current $= 4.96$ A.

Base impedance $= 46.32$ ohm.

Base capacitance $= 68.71 \, \mu$F.

Base frequency $= 50$ Hz.

Base speed $= 1500$ rpm.

Variation of air gap voltage with magnetizing reactance is;

$$X_m < 82.292 \quad E = 344.411 - 1.61 X_m$$
$$95.569 > X_m \geq 82.292 \quad E = 465.12 - 3.077 X_m$$
$$108.00 > X_m \geq 95.569 \quad E = 579.897 - 4.278 X_m$$
$$X_m \geq 108.00 \quad E = 0.$$

References

1. www.wwindea.org.
2. K.S. Sandhu, "Behaviour of self-excited induction generators under different operating conditions," Ph.D. Thesis, Kurukshetra University, Kurukshetra (2001).
3. A.S. Langsdorf, *Theory of Alternating-Current Machinery*, 2nd edn. (McGraw-Hill, New Delhi).
4. M.G. Say, *Performance and Design of Alternating Current Machines*, 3rd edn. (PITMAN Publishing Corporation, 1961).
5. K.S. Sandhu and S. Vadhera, "Reactive power requirements of grid connected induction generator in a weak grid," *WSEAS Trans. Circuits and Systems* **7** (2008) 150.
6. L. Ouazene and G. McPherson, "Analysis of the isolated induction generator," *IEEE Trans. on Power Apparatus and Systems* **PAS-102** (1983) 2793.
7. K.S. Sandhu and S.K. Jain, "Operational aspects of self-excited induction generator using a new model," *Electric Machines and Power Systems* **27** (1999) 169.
8. K.S. Sandhu, "Iterative model for the analysis of self-excited induction generators," *Electric Power Components and Systems* **31** (2003) 925.
9. D. Joshi, K.S. Sandhu and M.K. Soni, "Performance analysis of three-phase self-excited induction generator using GA," *Electric Power Components and Systems* **34**, (2006) 461.
10. T.F. Chan, "Capacitance requirements of self-excited induction generators," *IEEE Trans. Energy Conversion* **8** (1993) 304.
11. T.F. Chan, "Self-excited induction generators driven by regulated and unregulated turbines," *IEEE Trans. Energy Conversion* **11** (1996) 338.
12. S.S. Murthy, O.P. Malik, and A.K. Tandon, 'Analysis of self-excited induction generators," *Proc. IEE, Pt. C* **129** (1982) 260.
13. G. Raina and O.P. Malik, "Wind energy conversion using a self-excited induction generator," *IEEE Trans. Power Apparatus and Systems* **PAS-102** (1983) 3933.
14. A.K. Tandon, S.S. Murthy and G.J. Berg, "Steady state analysis of capacitor self-excited induction generators," *IEEE Trans. Power Apparatus and Systems* **PAS-103** (1984) 612.
15. A.K. Tandon, S.S. Murthy and C.S. Jha, "New method of computing steady state response of capacitor self-excited induction generator," *IE (1) J.-EL* **65** (1985) 196.
16. N.H. Malik and S.E. Haque, "Steady state analysis and performance of an isolated self-excited induction generator," *IEEE Trans. Energy Conversion* **EC-1** (1986) 134.
17. E. Bim, J. Szajner and Y. Burian, "Voltage compensation of an induction generator with long-shunt connection," *IEEE Trans. Energy Conversion* **4** (1989) 526.
18. L. Shridhar, B. Singh and C.S. Jha, "A step towards improvements in the characteristics of self-excited induction generator," *IEEE Trans. Energy Conversion* **8** (1993) 40.
19. T.F. Chan, "Analysis of self-excited induction generators using an iterative method," *IEEE Trans. Energy Conversion* **10** (1995) 502.

Chapter 27

Control of Doubly Fed Induction Generators under Balanced and Unbalanced Voltage Conditions

Oriol Gomis-Bellmunt[*] and Adrià Junyent-Ferré[‡]

[*]*IREC Catalonia Institute for Energy Research*
Josep Pla, B2, Pl. Baixa. E-08019 Barcelona, Spain
oriol.gomis@upc.edu

[‡]*Centre d'Innovació Tecnològica en Convertidors*
Estàtics i Accionaments (CITCEA-UPC)
Departament d'Enginyeria Elèctrica
Universitat Politècnica de Catalunya.
ETS d'Enginyeria Industrial de Barcelona,
Av. Diagonal, 647, Pl. 2. 08028 Barcelona, Spain

The present chapter presents a control technique to deal with the control of doubly fed induction generators under different voltage disturbances. Certain current reference values are chosen in the positive and negative sequences so that the torque and the DC voltage are kept stable during balanced and unbalanced conditions. Both rotor-side and grid-side converters are considered, detailing the control scheme of each converter while considering the effect of the crow-bar protection. The control strategy is validated by means of simulations.

27.1 Introduction

Power electronics have motivated an important change in the conception of wind farms and have forced experts to start thinking about wind power plants. Modern wind power plants are based on doubly fed induction Generators (DFIG) or synchronous generators with full power converters (FPC) while they are required to provide support to the grid voltage and frequency by the different power system operators worldwide.

The current grid codes of most countries where wind power is being massively integrated do not not allow wind farms to disconnect when faults in the main grid occur. Moreover, in some countries support to the main grid in terms of reactive power is demanded during faults.

The task of controlling the Doubly Fed Induction Generator (DFIG) during a voltage sag is specially challenging, since the stator is directly connected to the grid and therefore the stator voltage cannot be prevented to drop suddenly. In the case of unbalanced voltage sags, there also appears a negative sequence which provokes severe power oscillations. The present chapter describes a control technique[1] to deal with such balanced and unbalanced voltage perturbations.

27.2 Nomenclature

The chapter employs the following nomenclature. Vectors are expressed as

- $\mathbf{i}_x$: Current vector $i_{xd} + ji_{xq}$
- $\mathbf{v}_x$: Voltage vector $v_{xd} + jv_{xq}$
- $\mathbf{S}_x$: Power vector $P_x + jQ_x$

Scalar quantities:

- λ: Flux linkage
- Γ: Torque
- E: DC bus voltage
- t: Time
- ω_e: Electrical angular velocity
- ω_r: Rotor electrical angular velocity
- ω_m: Mechanical angular velocity
- θ: Angle
- s: Slip
- P: Generator number of poles
- f: Frequency
- R_r: Rotor resistance
- R_s: Stator resistance
- L_r: Rotor inductance
- L_s: Stator inductance
- M: Mutual inductance

The first subscript:

- s: Stator
- r: Rotor
- c: Rotor-side converter
- l: Grid-side converter

- z: Grid
- f: Filter

The second subscript:

- d: d-axis
- q: q-axis
- 0: Non-oscillating component
- *sin*: sin oscillating component
- *cos*: cos oscillating component

Superscripts:

- $*$: Set-point
- p: Positive sequence
- n: Negative sequence

27.3 General Considerations

27.3.1 *System under study*

The analyzed system is illustrated in Fig. 27.1. The Doubly Fed Induction Generator (DFIG) is attached to the wind turbine by means of a gearbox separating the fast axis connected to the generator to the slow axis attached to the turbine. The DFIG stator windings are connected directly to the wind farm transformer while the rotor windings are connected to a back-to-back converter shown in Fig. 27.2. The converter is composed by the grid-side converter connected to the grid and the rotor-side converter connected to the wound rotor windings.

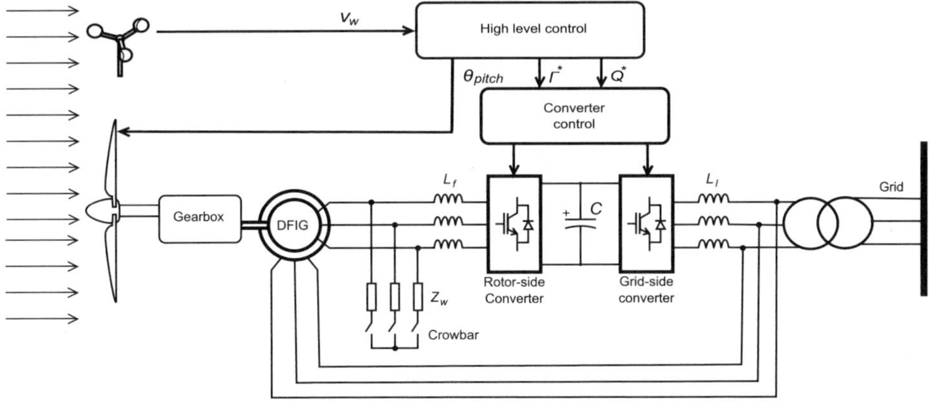

Fig. 27.1.   General system scheme.

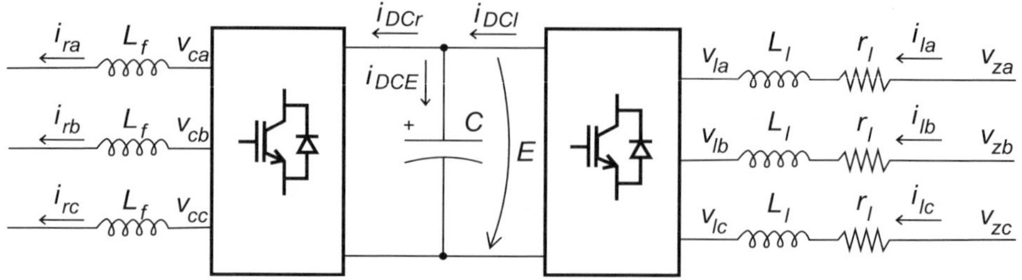

Fig. 27.2.   Back-to-back converter.

The converter set-points are established by the so-called high level controller. It uses the knowledge of the wind speed and the grid active and reactive power requirements, to determine the optimum turbine pitch angle and the torque and reactive power set-points referenced to the converter. The rotor-side converter controls torque and reactive power, while the grid-side converter controls the DC voltage and grid-side reactive power.

Although the back-to-back converter can control both the reactive power injected by the stator by controlling the rotor currents and the reactive power injected directly to the grid with the grid-side converter, it is a common practice to deliver most of the referenced reactive power through the stator while keeping a low or null reactive power set-point in the grid-side converter.

27.4   Control of the Doubly Fed Induction Generator under Balanced Conditions

In this section, the classical DFIG control scheme[2] is presented. It was designed for balanced voltage supply.

27.4.1   *Analysis*

27.4.1.1   *Grid-side converter*

In the grid-side converter, the DC bus voltage and reactive power references determine the current references, which determine the voltages to be applied in the grid-side.

In a synchronous reference frame, the grid-side voltage equations can be written as:

$$\left\{\begin{array}{c} v_{zq} \\ v_{zd} \end{array}\right\} - \left\{\begin{array}{c} v_{lq} \\ v_{ld} \end{array}\right\} = \left[\begin{array}{cc} R_l & -L_l\omega_e \\ L_l\omega_e & R_l \end{array}\right] \left\{\begin{array}{c} i_{lq} \\ i_{ld} \end{array}\right\} + \left[\begin{array}{cc} L_l & 0 \\ 0 & L_l \end{array}\right] \frac{d}{dt} \left\{\begin{array}{c} i_{lq} \\ i_{ld} \end{array}\right\}. \qquad (27.1)$$

Active and reactive power provided by the grid-side converter can be written as $P_z = \frac{3}{2}(v_{zq}i_{lq} + v_{zd}i_{ld})$ and $Q_z = \frac{3}{2}(v_{zq}i_{ld} - v_{zd}i_{lq})$.

The DC bus voltage can be expressed as:

$$E = E_0 + \frac{1}{C}\int_0^t (i_{DCl} - i_{DCr})dt. \tag{27.2}$$

27.4.1.2 *Machine-side converter*

In the rotor side converter, the referenced torque and reactive power determine the current references, which determine the voltages to be applied in the rotor-side.

It is usually assumed that stator and rotor windings are placed sinusoidally and symmetrically, the magnetic saturation effects and the capacitance of all the windings are negligible. The relations between voltages and currents on a synchronous reference qd can be written as:

$$\begin{Bmatrix} v_{sq} \\ v_{sd} \\ v_{rq} \\ v_{rd} \end{Bmatrix} = \begin{bmatrix} L_s & 0 & M & 0 \\ 0 & L_s & 0 & M \\ M & 0 & L_r & 0 \\ 0 & M & 0 & L_r \end{bmatrix} \frac{d}{dt} \begin{Bmatrix} i_{sq} \\ i_{sd} \\ i_{rq} \\ i_{rd} \end{Bmatrix}$$

$$+ \begin{bmatrix} R_s & L_s\omega_e & 0 & M\omega_e \\ -L_s\omega_e & R_s & -M\omega_e & 0 \\ 0 & sM\omega_e & R_r & sL_r\omega_e \\ -sM\omega_e & 0 & -sL_r\omega_e & R_r \end{bmatrix} \begin{Bmatrix} i_{sq} \\ i_{sd} \\ i_{rq} \\ i_{rd} \end{Bmatrix}. \tag{27.3}$$

Linkage fluxes can be written as:

$$\begin{Bmatrix} \lambda_{sq} \\ \lambda_{sd} \\ \lambda_{rq} \\ \lambda_{rd} \end{Bmatrix} = \begin{bmatrix} L_s & 0 & M & 0 \\ 0 & L_s & 0 & M \\ M & 0 & L_r & 0 \\ 0 & M & 0 & L_r \end{bmatrix} \begin{Bmatrix} i_{sq} \\ i_{sd} \\ i_{rq} \\ i_{rd} \end{Bmatrix}. \tag{27.4}$$

The torque can expressed as:

$$\Gamma_m = \frac{3}{2}PM(i_{sq}i_{rd} - i_{sd}i_{rq}). \tag{27.5}$$

The reactive power yields:

$$Q_s = \frac{3}{2}(v_{sq}i_{sd} - v_{sd}i_{sq}). \tag{27.6}$$

27.4.2   Control scheme

27.4.2.1   Grid-side converter

The grid-side converter control reactive power and DC bus voltage. The q axis may be aligned to the grid voltage allowing active and reactive decoupled control. To control the reactive power, a i_{ld} reference is computed as:

$$i_{ld}^* = \frac{2Q_z^*}{3v_{zq}}.$$
(27.7)

The active power, which is responsible for the evolution of the DC bus voltage is controlled by the i_{lq} component. A linear controller is usually designed to control the DC bus voltage.

The current control is done by the following state linearization feedback:[3]

$$\left\{ \begin{array}{c} v_{lq} \\ v_{ld} \end{array} \right\} = \left\{ \begin{array}{c} -\hat{v}_{lq} + v_{zq} - L_l\omega_e i_{ld} \\ -\hat{v}_{ld} + L_l\omega_e i_{lq} \end{array} \right\},$$
(27.8)

where the $\hat{v}_{lq}$ and $\hat{v}_{ld}$ are the output voltages of the current controller. The decoupling leads to:

$$\frac{d}{dt}\left\{ \begin{array}{c} i_{lq} \\ i_{ld} \end{array} \right\} = \left[\begin{array}{cc} -\dfrac{R_l}{L_l} & 0 \\ 0 & -\dfrac{R_l}{L_l} \end{array} \right] \left\{ \begin{array}{c} i_{lq} \\ i_{ld} \end{array} \right\} + \left[\begin{array}{cc} \dfrac{1}{L_l} & 0 \\ 0 & \dfrac{1}{L_l} \end{array} \right] \left\{ \begin{array}{c} \hat{v}_{lq} \\ \hat{v}_{qd} \end{array} \right\}.$$
(27.9)

27.4.2.2   Machine-side converter

$$i_{sq} = \frac{\lambda_{sq} - Mi_{rq}}{L_s},$$
(27.10)

$$i_{sd} = \frac{\lambda_{sd} - Mi_{rd}}{L_s} = -\frac{Mi_{rd}}{L_s}.$$
(27.11)

Thus:

$$\Gamma_m = \frac{3}{2}\frac{PM}{L_s}\lambda_{sq}i_{rd},$$
(27.12)

$$Q_s = \frac{3}{2L_s}(-v_{sq}Mi_{rd} - v_{sd}\lambda_{sq} + Mv_{sd}i_{rq}).$$
(27.13)

Orientating the synchronous reference qd with the stator flux vector so that $\lambda_{sd} = 0$, the rotor current references can be computed as:

$$\left\{ \begin{array}{c} i_{rq}^* \\ i_{rd}^* \end{array} \right\} = \left\{ \begin{array}{c} \dfrac{\frac{2}{3}L_s Q_s^* + Mv_{sq}i_{rd} + v_{sd}\lambda_{sq}}{Mv_{sd}} \\ \dfrac{2L_s\Gamma_m^*}{3PM\lambda_{sq}} \end{array} \right\}.$$
(27.14)

The control of the current is done by linearizing the current dynamics using the following state feedback:

$$\begin{Bmatrix} v_{rq} \\ v_{rd} \end{Bmatrix} = \begin{Bmatrix} \hat{v}_{rq} + M(\omega_e - \omega_r)i_{sd} + L_r(\omega_e - \omega_r)i_{rd} \\ \hat{v}_{rd} - M(\omega_e - \omega_r)i_{sq} - L_r(\omega_e - \omega_r)i_{rq} \end{Bmatrix}. \tag{27.15}$$

By neglecting stator current transients, the decoupling leads to:

$$\frac{d}{dt}\begin{Bmatrix} i_{rq} \\ i_{rd} \end{Bmatrix} = -\begin{bmatrix} \dfrac{R_r}{L_r} & 0 \\ 0 & \dfrac{R_r}{L_r} \end{bmatrix}\begin{Bmatrix} i_{rq} \\ i_{rd} \end{Bmatrix} + \begin{bmatrix} \dfrac{1}{L_r} & 0 \\ 0 & \dfrac{1}{L_r} \end{bmatrix}\begin{Bmatrix} \hat{v}_{rq} \\ \hat{v}_{rd} \end{Bmatrix}. \tag{27.16}$$

27.4.3 *Current controllers design*

Current controllers can be been designed using the so-called Internal Model Control (IMC) methodology.[4] The parameters of a Proportional Integral (PI) controller to obtain a desired time constant τ are obtained as:

$$K_p = \frac{L}{\tau}, \quad K_i = \frac{R}{\tau}. \tag{27.17}$$

The currents and voltages have been limited according to the converter operating limits. PI controllers have been designed with anti-windup in order to prevent control instabilities when the controller exceed the limit values.

Example current loop responses to voltage disturbances are shown in Fig. 27.3.

27.4.4 *Crowbar protection*

The so-called crow-bar is connected to avoid overvoltages in the DC bus due to excessive power flowing from the rotor inverter to the grid-side converter, guaranteeing ride-through operation of the generator when voltage sags or other disturbances occur.

The crow-bar is triggered when the DC voltage reaches a threshold $v_{\text{crow}-c}$ and disconnects when it goes below another threshold $v_{\text{crow}-d}$.

During its operation, the rotor-side converter may be disconnected[5] or kept connected[6] to avoid losing control over the machine.

A DC bus voltage evolution example is shown in Fig. 27.4. It can be seen that the crow-bar protection is connected when the overvoltage occur and that after a transient the DC bus voltage can return to the reference value. The threshold $v_{\text{crow}-c}$ is located at 1180 V and $v_{\text{crow}-d}$ at 1140 V.

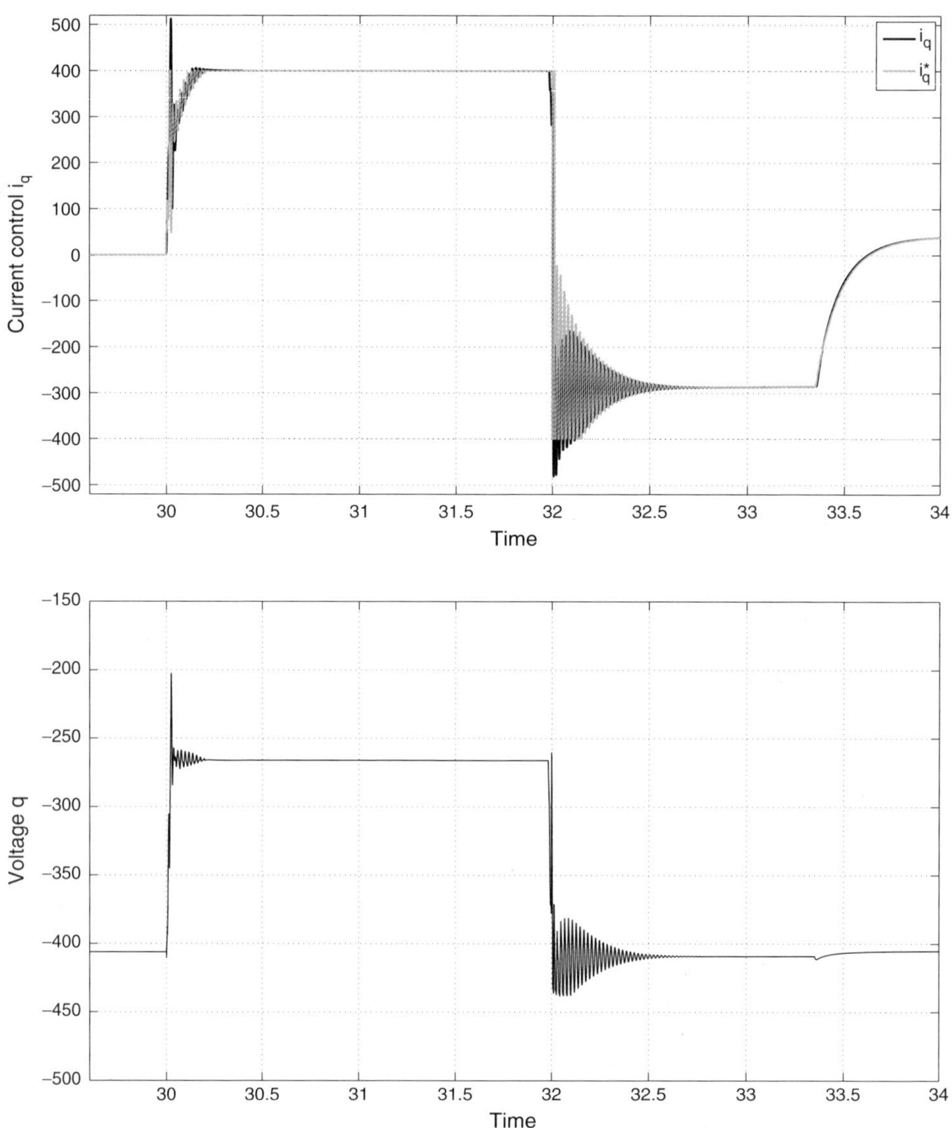

Fig. 27.3. Current loop example.

27.5 Control of the Doubly Fed Induction Generator under Unbalanced Conditions

27.5.1 *Analysis*

In this section non symmetrical voltage sags are considered. Such unbalanced sags imply negative sequence components in all the relevant quantities. Therefore,

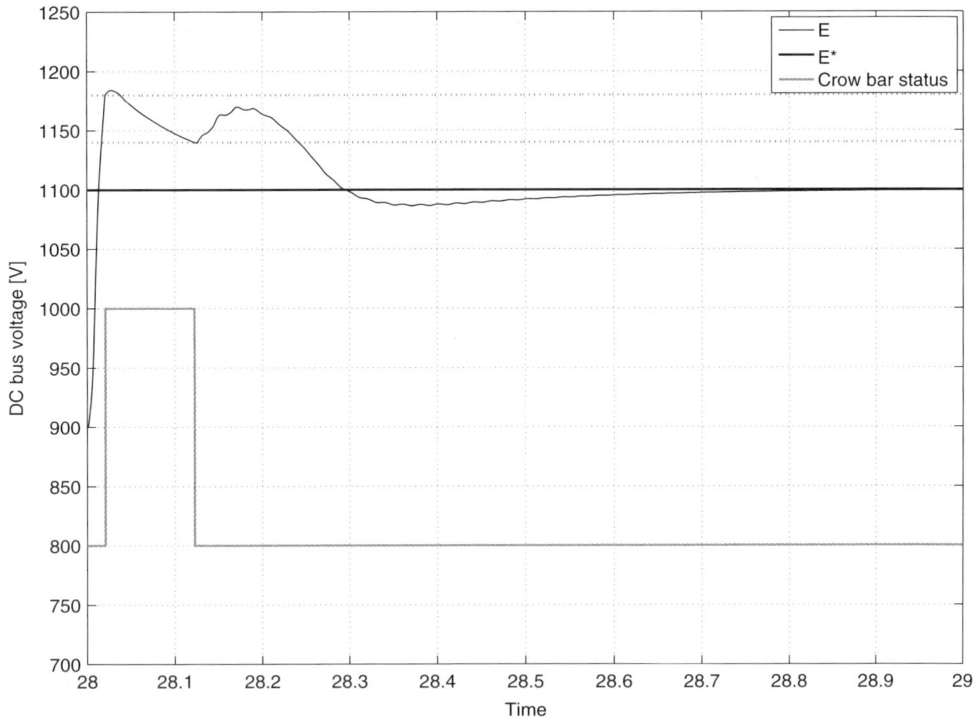

Fig. 27.4.   DC bus voltage evolution example.

important oscillations appear in torque, active and reactive power. Such oscillations have a pulsation of $2\omega_e$. Examples of power and torque oscillations employing the control scheme of the previous section are shown in Fig. 27.5.

In order to mitigate such oscillations, an approach taking into account the negative sequence quantities is required. The present section analyzes a whole back-to-back converter taking into account both the positive and negative sequence components, and proposes a technique to control optimally both the DC bus voltage and the torque when unbalanced voltage sags occur.

As far as unbalanced systems are concerned, it is useful to express three-phase quantities $x_{abc} = \{x_a, x_b, x_c\}^T$ in direct and inverse components as:

$$\mathbf{x} = e^{j\omega_e t + j\theta_0}\mathbf{x}^p + e^{-j\omega_e t - j\theta_0}\mathbf{x}^n, \tag{27.18}$$

where $\mathbf{x} = \frac{2}{3}(x_a + ax_b + a^2 x_c)$, $a = e^{j2\pi/3}$, $\mathbf{x}^p = x_d^p + jx_q^p$ and $\mathbf{x}^n = x_d^n + jx_q^n$.

In the present section, voltages, currents and fluxes are regarded as a composition of such positive and negative sequences.

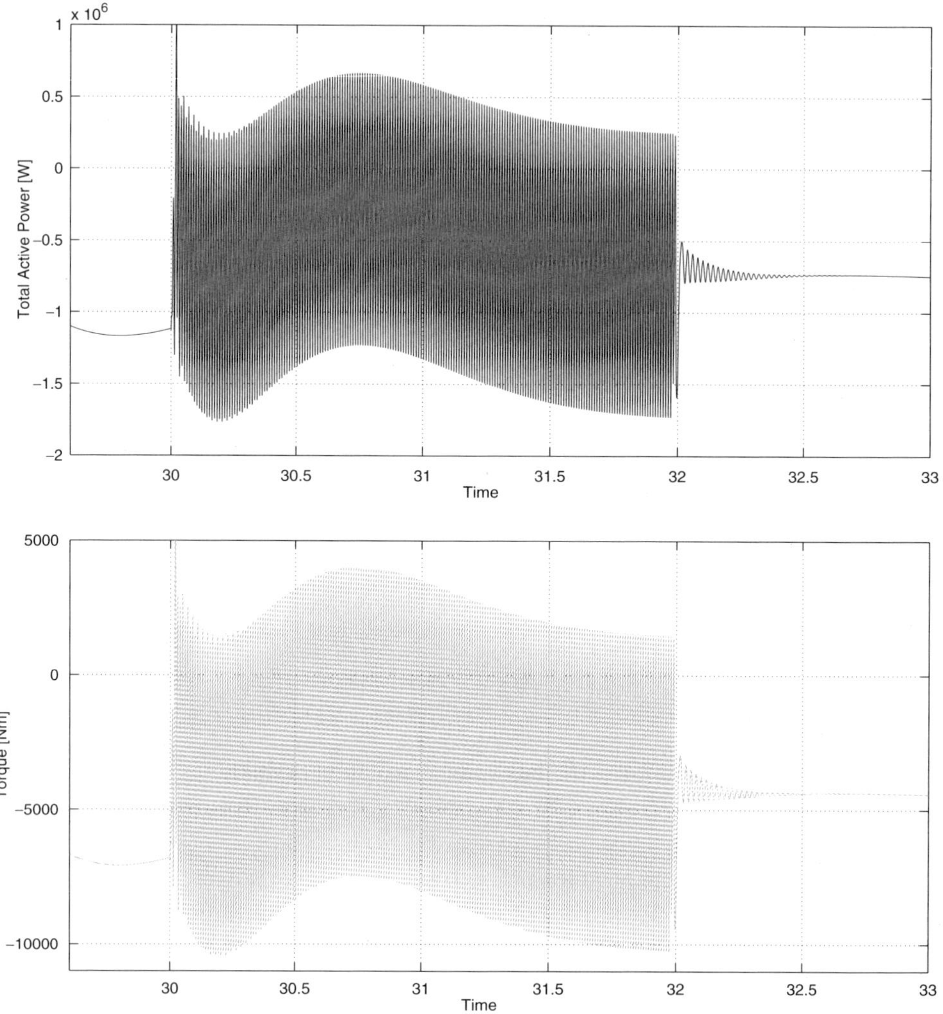

Fig. 27.5.   Generator power and torque during an unbalanced voltage sag using conventional control.

27.5.1.1   *Grid-side converter*

Considering two rotating reference frames at $+\omega_e$ and $-\omega_e$, the voltage equations for the positive and negative sequence yield:

$$\mathbf{v}_{zqd}^{p} - \mathbf{v}_{lqd}^{p} = (R_l + j\omega_e L_l)\mathbf{i}_{lqd}^{p} + L_l \frac{d\mathbf{i}_{lqd}^{p}}{dt}, \tag{27.19}$$

$$\mathbf{v}_{zqd}^{n} - \mathbf{v}_{lqd}^{n} = (R_l - j\omega_e L_l)\mathbf{i}_{lqd}^{n} + L_l \frac{d\mathbf{i}_{lqd}^{n}}{dt}. \tag{27.20}$$

Active and reactive power can be written:[7]

$$P_l = \frac{3}{2}[P_{l0} + P_{l\cos}\cos(2\omega_e t) + P_{l\sin}\sin(2\omega_e t)], \tag{27.21}$$

$$Q_l = \frac{3}{2}[Q_{l0} + Q_{l\cos}\cos(2\omega_e t) + Q_{l\sin}\sin(2\omega_e t)], \tag{27.22}$$

where

$$
\begin{Bmatrix} P_{l0} \\ P_{l\cos} \\ P_{l\sin} \\ Q_{l0} \\ Q_{l\cos} \\ Q_{l\sin} \end{Bmatrix} =
\begin{bmatrix}
v_{zd}^p & v_{zq}^p & v_{zd}^n & v_{zq}^n \\
v_{zd}^n & v_{zq}^n & v_{zd}^p & v_{zq}^p \\
v_{zq}^n & -v_{zd}^n & -v_{zq}^p & v_{zd}^p \\
v_{zq}^p & -v_{zd}^p & v_{zq}^n & -v_{zd}^n \\
v_{zq}^n & -v_{zd}^n & v_{zq}^p & -v_{zd}^p \\
-v_{zd}^n & -v_{zq}^n & v_{zd}^p & v_{zq}^p
\end{bmatrix}
\begin{Bmatrix} i_{ld}^p \\ i_{lq}^p \\ i_{ld}^n \\ i_{lq}^n \end{Bmatrix}. \tag{27.23}
$$

It can be noted that both active and reactive power quantity have three different components each, and hence with the four regulable currents i_{ld}^p, i_{lq}^p, i_{ld}^n and i_{lq}^n only four of such six powers can be controlled.

27.5.1.2 *Machine-side converter*

Voltage equations. Considering two rotating reference frames at $+\omega_e$ and $-\omega_e$, the voltage equations for the positive and negative sequence can be obtained as:

$$
\begin{Bmatrix} \mathbf{v}_s^p \\ \mathbf{v}_r^p \end{Bmatrix} =
\begin{bmatrix} L_s & M \\ M & L_r \end{bmatrix} \frac{d}{dt} \begin{Bmatrix} \mathbf{i}_s^p \\ \mathbf{i}_r^p \end{Bmatrix}
$$
$$
+ \begin{bmatrix} R_s + jL_s\omega_e & jM\omega_e \\ jM(\omega_e - \omega_r) & R_r + jL_r(\omega_e - \omega_r) \end{bmatrix} \begin{Bmatrix} \mathbf{i}_s^p \\ \mathbf{i}_r^p \end{Bmatrix}, \tag{27.24}
$$

$$
\begin{Bmatrix} \mathbf{v}_s^n \\ \mathbf{v}_r^n \end{Bmatrix} =
\begin{bmatrix} L_s & M \\ M & L_r \end{bmatrix} \frac{d}{dt} \begin{Bmatrix} \mathbf{i}_s^n \\ \mathbf{i}_r^n \end{Bmatrix}
$$
$$
+ \begin{bmatrix} R_s - j\omega_e L_s & -j\omega_e M \\ +jM(-\omega_e - \omega_r) & R_r + jL_r(-\omega_e - \omega_r) \end{bmatrix} \begin{Bmatrix} \mathbf{i}_s^n \\ \mathbf{i}_r^n \end{Bmatrix}. \tag{27.25}
$$

Stator power expression. The apparent stator power can be expressed as:

$$\mathbf{S}_s = P_s + jQ_s = \frac{3}{2}\mathbf{v}_s\mathbf{i}_s^*. \tag{27.26}$$

Using Eq. (27.18):

$$\mathbf{S}_s = (e^{j\omega_e t + j\theta_0}\mathbf{v}_s^p + e^{-j\omega_e t - j\theta_0}\mathbf{v}_s^n)((e^{j\omega_e t + j\theta_0})^*\mathbf{i}_s^{p*} + (e^{-j\omega_e t - j\theta_0})^*\mathbf{i}_s^{n*}),$$

$$\mathbf{S}_s = \mathbf{v}_s^p\mathbf{i}_s^{p*} + \mathbf{v}_s^n\mathbf{i}_s^{n*} + e^{j2\omega_e t + j2\theta_0}\mathbf{v}_s^p\mathbf{i}_s^{n*} + e^{-j2\omega_e t - j2\theta_0}\mathbf{v}_s^n\mathbf{i}_s^{p*}. \tag{27.27}$$

Taking into account $\mathbf{x}_s^i = x_{sd}^i + jx_{sq}^i$, and rearranging, it can be written as $\mathbf{S}_s = P_s + jQ_s$, with

$$P_s = \frac{3}{2}[P_{s0} + P_{s\cos}\cos(2\omega_e t + 2\theta_0) + P_{s\sin}\sin(2\omega_e t + 2\theta_0)], \tag{27.28}$$

$$Q_s = \frac{3}{2}[Q_{s0} + Q_{s\cos}\cos(2\omega_e t + 2\theta_0) + Q_{s\sin}\sin(2\omega_e t + 2\theta_0)], \tag{27.29}$$

where

$$\begin{Bmatrix} P_{s0} \\ P_{s\cos} \\ P_{s\sin} \\ Q_{s0} \\ Q_{s\cos} \\ Q_{s\sin} \end{Bmatrix} = \begin{bmatrix} v_{sd}^p & v_{sq}^p & v_{sd}^n & v_{sq}^n \\ v_{sd}^n & v_{sq}^n & v_{sd}^p & v_{sq}^p \\ v_{sq}^n & -v_{sd}^n & -v_{sq}^p & v_{sd}^p \\ v_{sq}^p & -v_{sd}^p & v_{sq}^n & -v_{sd}^n \\ v_{sq}^n & -v_{sd}^n & v_{sq}^p & -v_{sd}^p \\ -v_{sd}^n & -v_{sq}^n & v_{sd}^p & v_{sq}^p \end{bmatrix} \begin{Bmatrix} i_{sd}^p \\ i_{sq}^p \\ i_{sd}^n \\ i_{sq}^n \end{Bmatrix}. \tag{27.30}$$

Substituting stator currents in Eq. (27.30):

$$\begin{Bmatrix} P_{s0} \\ P_{s\cos} \\ P_{s\sin} \\ Q_{s0} \\ Q_{s\cos} \\ Q_{s\sin} \end{Bmatrix} = \frac{1}{L_s} \begin{bmatrix} v_{sd}^p & v_{sq}^p & v_{sd}^n & v_{sq}^n \\ v_{sd}^n & v_{sq}^n & v_{sd}^p & v_{sq}^p \\ v_{sq}^n & -v_{sd}^n & -v_{sq}^p & v_{sd}^p \\ v_{sq}^p & -v_{sd}^p & v_{sq}^n & -v_{sd}^n \\ v_{sq}^n & -v_{sd}^n & v_{sq}^p & -v_{sd}^p \\ -v_{sd}^n & -v_{sq}^n & v_{sd}^p & v_{sq}^p \end{bmatrix} \begin{Bmatrix} \lambda_{sd}^p - Mi_{rd}^p \\ \lambda_{sq}^p - Mi_{rq}^p \\ \lambda_{sd}^n - Mi_{rd}^n \\ \lambda_{sq}^n - Mi_{rq}^n \end{Bmatrix}. \tag{27.31}$$

It can be noted that both active and reactive power quantity have three different components each, and therefore with the four regulable currents i_{rd}^p, i_{rq}^p, i_{rd}^n and i_{rq}^n only four of the six power quantities can be controlled.

Rotor power expression. The apparent rotor power can be expressed as:

$$\mathbf{S}_r = P_r + jQ_r = \frac{3}{2}\mathbf{v}_r\mathbf{i}_r^*, \tag{27.32}$$

$$\mathbf{S}_r = \frac{3}{2}(e^{j(\omega_e - \omega_r)t + j\theta_{r0}}\mathbf{v}_r^p + e^{-j(\omega_e + \omega_r)t - j\theta_{r0}}\mathbf{v}_r^n)$$

$$\times (e^{j(\omega_e - \omega_r)t + j\theta_{r0}}\mathbf{i}_r^p + e^{-j(\omega_e + \omega_r)t - j\theta_{r0}}\mathbf{i}_r^n)^*. \tag{27.33}$$

Using Eq. (27.18):

$$\mathbf{S}_r = \frac{3}{2}[\mathbf{v}_r^p \mathbf{i}_r^{p*} + \mathbf{v}_r^n \mathbf{i}_r^{n*} + e^{j2\omega_e t + 2j\theta_{r0}}\mathbf{v}_r^p \mathbf{i}_r^{n*} + e^{j-2\omega_e t - j2\theta_{r0}}\mathbf{v}_r^n \mathbf{i}_r^{p*}]. \quad (27.34)$$

Taking into account $\mathbf{x}_s^i = x_{sd}^i + jx_{sq}^i$, and rearranging, and analyzing the active rotor power:

$$P_r = \frac{3}{2}[P_{r0} + P_{r\cos}\cos(2\omega_e t + 2\theta_{r0}) + P_{r\sin}\sin(2\omega_e t + 2\theta_{r0})], \quad (27.35)$$

where

$$\left\{\begin{array}{c} P_{r0} \\ P_{r\cos} \\ P_{r\sin} \end{array}\right\} = \left[\begin{array}{cccc} v_{cd}^p & v_{cq}^p & v_{cd}^n & v_{cq}^n \\ v_{cd}^n & v_{cq}^n & v_{cd}^p & v_{cq}^p \\ v_{cq}^n & -v_{cd}^n & -v_{cq}^p & v_{cd}^p \end{array}\right]\left\{\begin{array}{c} i_{rd}^p \\ i_{rq}^p \\ i_{rd}^n \\ i_{rq}^n \end{array}\right\}. \quad (27.36)$$

Torque expression. Analogously, electrical torque can be expressed as:

$$\Gamma = \frac{P}{2}\frac{3}{2}[\Gamma_0 + \Gamma_{\sin}\sin(2\omega_e t) + \Gamma_{\cos}\cos(2\omega_e t)], \quad (27.37)$$

where

$$\left\{\begin{array}{c} \Gamma_0 \\ \Gamma_{\cos} \\ \Gamma_{\sin} \end{array}\right\} = \frac{M}{L_s}\left[\begin{array}{cccc} -\lambda_{sq}^p & \lambda_{sd}^p & -\lambda_{sq}^n & \lambda_{sd}^n \\ \lambda_{sd}^n & \lambda_{sq}^n & -\lambda_{sd}^p & -\lambda_{sq}^p \\ -\lambda_{sq}^n & \lambda_{sd}^n & -\lambda_{sq}^p & \lambda_{sd}^p \end{array}\right]\left\{\begin{array}{c} i_{rd}^p \\ i_{rq}^p \\ i_{rd}^n \\ i_{rq}^n \end{array}\right\}. \quad (27.38)$$

27.5.2 *Control scheme*

27.5.2.1 *General control structure*

Since there are eight degrees of freedom (the rotor-side currents i_{rd}^p, i_{rq}^p, i_{rd}^n, i_{rq}^n and the grid-side currents i_{ld}^p, i_{lq}^p, i_{ld}^n, i_{lq}^n), eight control objectives may be chosen. This implies that it is not possible to eliminate all the oscillations provoked by the unbalance. In this work, the main objective is to ride through voltage dips. Hence, it is important to keep the torque and DC bus voltage as constant as possible and to keep reasonable values of reactive power. To this end it has been chosen to determine the currents to keep certain values of Γ_0^*, $\Gamma_{\cos}^*$, $\Gamma_{\sin}^*$, Q_{s0}^* for the rotor-side converter and P_{l0}^*, $P_{l\cos}^*$, $P_{l\sin}^*$ and Q_{l0}^* for the grid-side converter. It can be noted that P_{l0}^*, $P_{l\cos}^*$ and $P_{l\sin}^*$ are directly linked to the DC bus voltage.

The DC voltage E is regulated by means of a linear controller whose output is the power demanded to the grid-side converter. Considering the power terms P_{r0}, $P_{r\cos}$ and $P_{r\sin}$ in the rotor side converter, P_{r0} can be regarded as the average power delivered, while $P_{r\cos}$ and $P_{r\sin}$ are the rotor power oscillating terms. Such terms will cause DC voltage oscillations, and hence they can be canceled by choosing:

$$
\begin{aligned}
P_{l\cos}^* &= P_{r\cos}, \\
P_{l\sin}^* &= P_{r\sin}.
\end{aligned}
\tag{27.39}
$$

P_{l0} can be computed as:

$$
P_{l0}^* = P_{r0} + P_E^*,
\tag{27.40}
$$

where P_E^* is the output of the DC voltage linear controller.

The grid reference currents can be computed from Eqs. (27.23), (27.36), (27.39) and (27.40) as:

$$
\begin{Bmatrix} i_{ld}^{p*} \\ i_{lq}^{p*} \\ i_{ld}^{n*} \\ i_{lq}^{n*} \end{Bmatrix} = \begin{bmatrix} v_{zd}^{p} & v_{zq}^{p} & v_{zd}^{n} & v_{zq}^{n} \\ v_{zd}^{n} & v_{zq}^{n} & v_{zd}^{p} & v_{zq}^{p} \\ v_{zq}^{n} & -v_{zd}^{n} & -v_{zq}^{p} & v_{zd}^{p} \\ v_{zq}^{p} & -v_{zd}^{p} & v_{zq}^{n} & -v_{zd}^{n} \end{bmatrix}^{-1} \left(\begin{Bmatrix} P_E \\ 0 \\ 0 \\ Q_{l0}^* \end{Bmatrix} \right.
$$
$$
\left. + \begin{bmatrix} v_{cd}^{p} & v_{cq}^{p} & v_{cd}^{n} & v_{cq}^{n} \\ v_{cd}^{n} & v_{cq}^{n} & v_{cd}^{p} & v_{cq}^{p} \\ v_{cq}^{n} & -v_{cd}^{n} & -v_{cq}^{p} & v_{cd}^{p} \\ 0 & 0 & 0 & 0 \end{bmatrix} \begin{Bmatrix} i_{rd}^{p} \\ i_{rq}^{p} \\ i_{rd}^{n} \\ i_{rq}^{n} \end{Bmatrix} \right).
\tag{27.41}
$$

The rotor reference currents can be computed from Eqs. (27.38) and (27.31) as:

$$
\begin{Bmatrix} i_{rd}^{p*} \\ i_{rq}^{p*} \\ i_{rd}^{n*} \\ i_{rq}^{n*} \end{Bmatrix} = \begin{bmatrix} -\lambda_{sq}^{p} & \lambda_{sd}^{p} & -\lambda_{sq}^{n} & \lambda_{sd}^{n} \\ \lambda_{sd}^{n} & \lambda_{sq}^{n} & -\lambda_{sd}^{p} & -\lambda_{sq}^{p} \\ -\lambda_{sq}^{n} & \lambda_{sd}^{n} & -\lambda_{sq}^{p} & \lambda_{sd}^{p} \\ -v_{sq}^{p} & v_{sd}^{p} & -v_{sq}^{n} & v_{sd}^{n} \end{bmatrix}^{-1}
$$
$$
\times \begin{Bmatrix} \dfrac{2}{P}\dfrac{2}{3}\dfrac{L_s}{M}\Gamma_0^* \\[2mm] \dfrac{2}{P}\dfrac{2}{3}\dfrac{L_s}{M}\Gamma_{\cos}^* \\[2mm] \dfrac{2}{P}\dfrac{2}{3}\dfrac{L_s}{M}\Gamma_{\sin}^* \\[2mm] \dfrac{1}{M}[L_s Q_{s0}^* - \lambda_{sd}^{p} v_{sq}^{p} + \lambda_{sq}^{p} v_{sd}^{p} - \lambda_{sd}^{n} v_{sq}^{n} + \lambda_{sq}^{n} v_{sd}^{n}] \end{Bmatrix}.
\tag{27.42}
$$

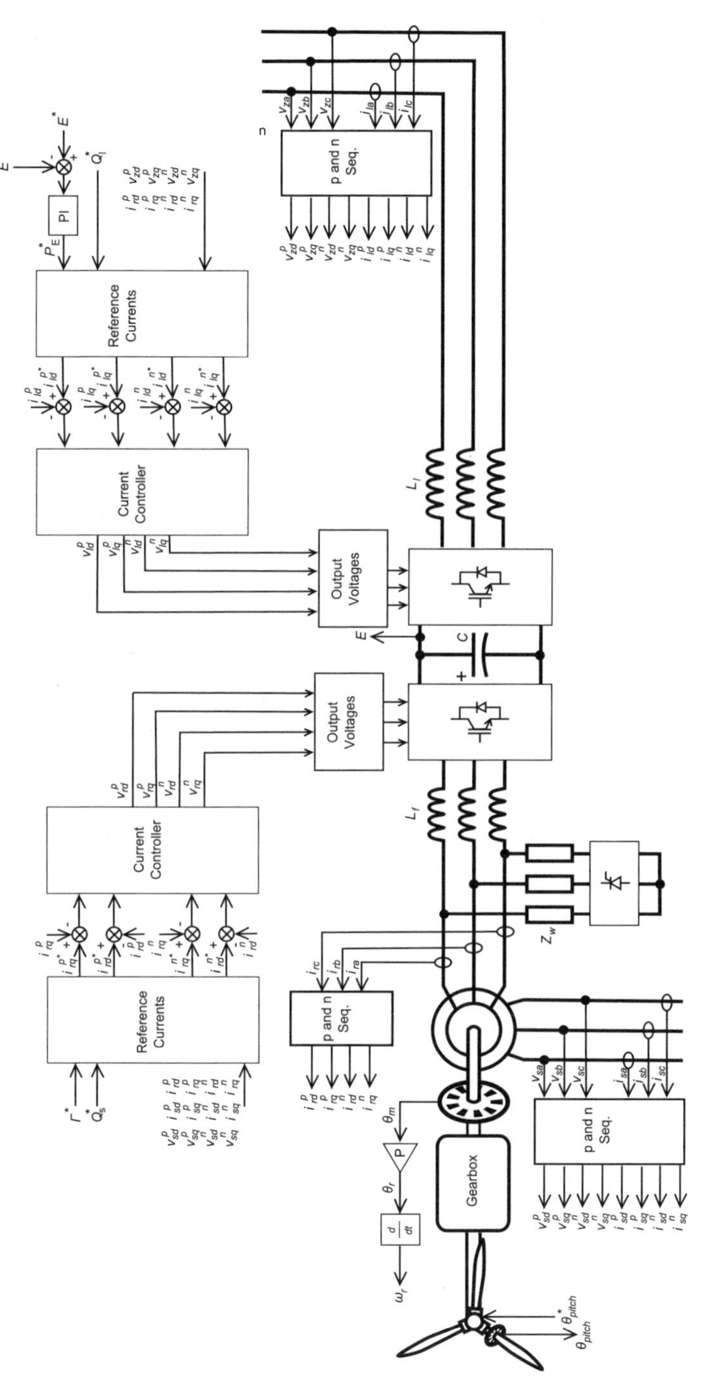

Fig. 27.6. General control scheme.

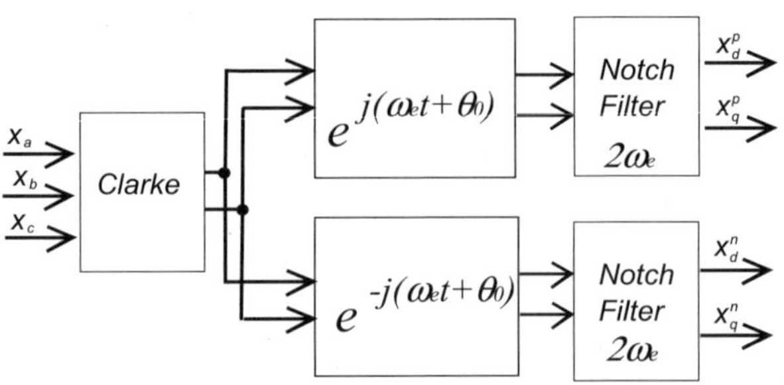

Fig. 27.7.   Positive and negative components calculation.

27.5.2.2 *Positive and negative components calculation*

The positive and negative sequence components calculation is done by doing the Clarke transformation, rotating either $e^{j\omega_e t}$ or $e^{-j\omega_e t}$ and finally applying a notch-filter at $2\omega_e$ to eliminate the opposite sequence. The technique is exemplified in Fig. 27.7. For the rotor voltages and currents, the rotation applied is either $e^{j(\omega_e - \omega_r)t}$ or $e^{j(-\omega_e - \omega_r)t}$.

27.5.2.3 *Reference orientation*

The rotating references have been aligned with the stator voltage so that $v_{sq}^p = 0$. Nevertheless, v_{sq}^p has not been substituted in previous expressions for the sake of describing general results. Orientation may be done computing the required θ_0 assuming a constant ω_e or using a Phase Locked Loop (PLL) to determine both ω_e and θ_0.

27.5.2.4 *Controllers linearization and tuning*

Grid-side. Similarly to the balanced case the control of the current is done by linearizing the current dynamics using:

$$\hat{\mathbf{v}}_{zqd}^p = \mathbf{v}_{zqd}^p - \mathbf{v}_{lqd}^p - j\omega_e L_l \mathbf{i}_{lqd}^p, \tag{27.43}$$

$$\hat{\mathbf{v}}_{zqd}^n = \mathbf{v}_{zqd}^n - \mathbf{v}_{lqd}^n + j\omega_e L_l \mathbf{i}_{lqd}^n. \tag{27.44}$$

The decoupled system yields:

$$\frac{d\mathbf{i}_{lqd}^p}{dt} = \frac{\hat{\mathbf{v}}_{zqd}^p - R_l \mathbf{i}_{lqd}^p}{L_l}, \tag{27.45}$$

$$\frac{d\mathbf{i}_{lqd}^n}{dt} = \frac{\hat{\mathbf{v}}_{zqd}^n - R_l \mathbf{i}_{lqd}^n}{L_l}. \tag{27.46}$$

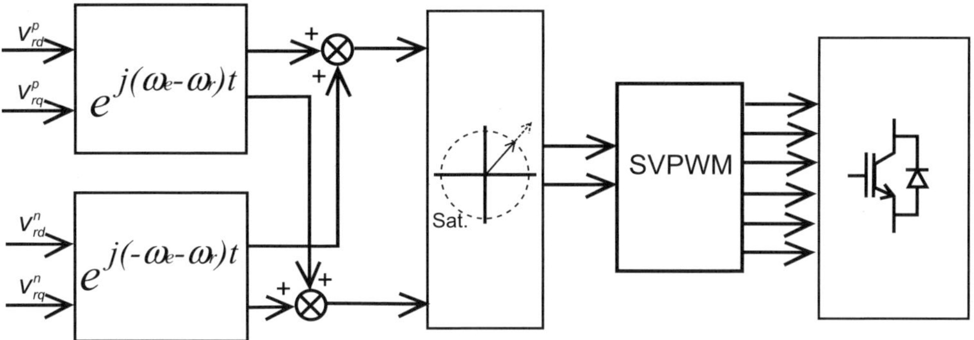

Fig. 27.8.   Output voltage calculation: Rotor-side converter example.

Rotor-side. Analogously to the balanced case

$$\hat{\mathbf{v}}_r^p = \mathbf{v}_r^p - jM(\omega_e - \omega_r)\mathbf{i}_s^p - jL_r(\omega_e - \omega_r)\mathbf{i}_r^p, \tag{27.47}$$

$$\hat{\mathbf{v}}_r^n = \mathbf{v}_r^n - jM(-\omega_e - \omega_r)\mathbf{i}_s^n - jL_r(-\omega_e - \omega_r)\mathbf{i}_r^n. \tag{27.48}$$

Neglecting the derivative of stator currents, the decoupled system yields:

$$\frac{d\mathbf{i}_r^p}{dt} = \frac{\hat{\mathbf{v}}_r^p - R_r\mathbf{i}_r^p}{L_r}, \tag{27.49}$$

$$\frac{d\mathbf{i}_r^n}{dt} = \frac{\hat{\mathbf{v}}_r^n - R_r\mathbf{i}_r^n}{L_r}. \tag{27.50}$$

27.5.2.5   *Output voltage calculation*

The output voltages calculation is done by summing the resulting positive sequence and negative sequence voltages in the stationary reference frame. For the line-side:

$$\mathbf{v}_l = e^{j\omega_e t}\mathbf{v}_l^p + e^{-j\omega_e t}\mathbf{v}_l^n. \tag{27.51}$$

For the rotor-side:

$$\mathbf{v}_r = e^{j(\omega_e - \omega_r)t}\mathbf{v}_r^p + e^{j(-\omega_e - \omega_r)t}\mathbf{v}_r^n. \tag{27.52}$$

The resulting voltages are limited according to the converter rating. The final voltages can be applied using standard SVPWM techniques. The technique is exemplified for rotor-side converter case in Fig. 27.8.

27.6   Simulation Results

The proposed control scheme have been evaluated by means of simulations with one balanced and one unbalanced voltage sag.

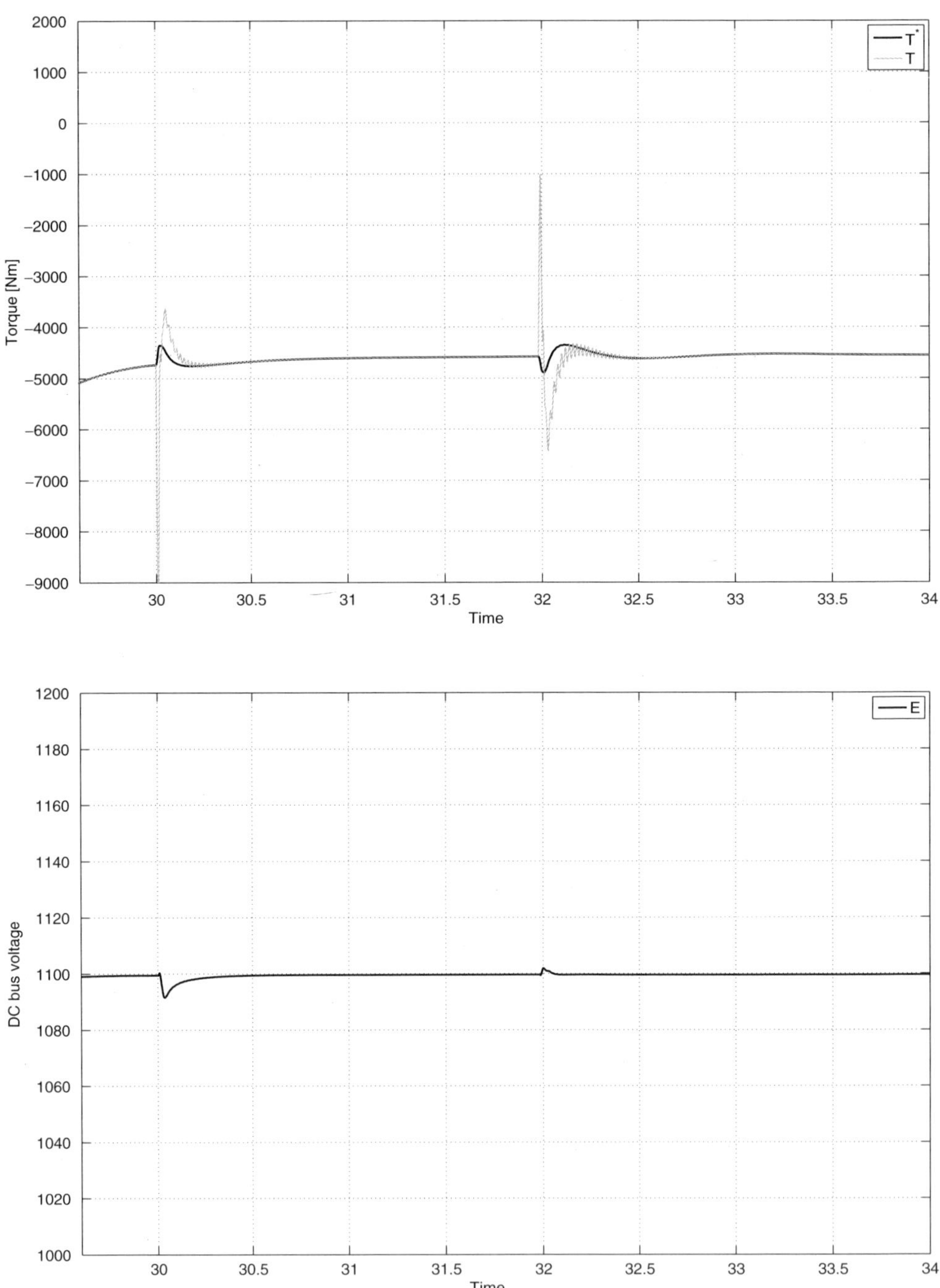

Fig. 27.9. Torque and DC bus response to a balanced voltage sag.

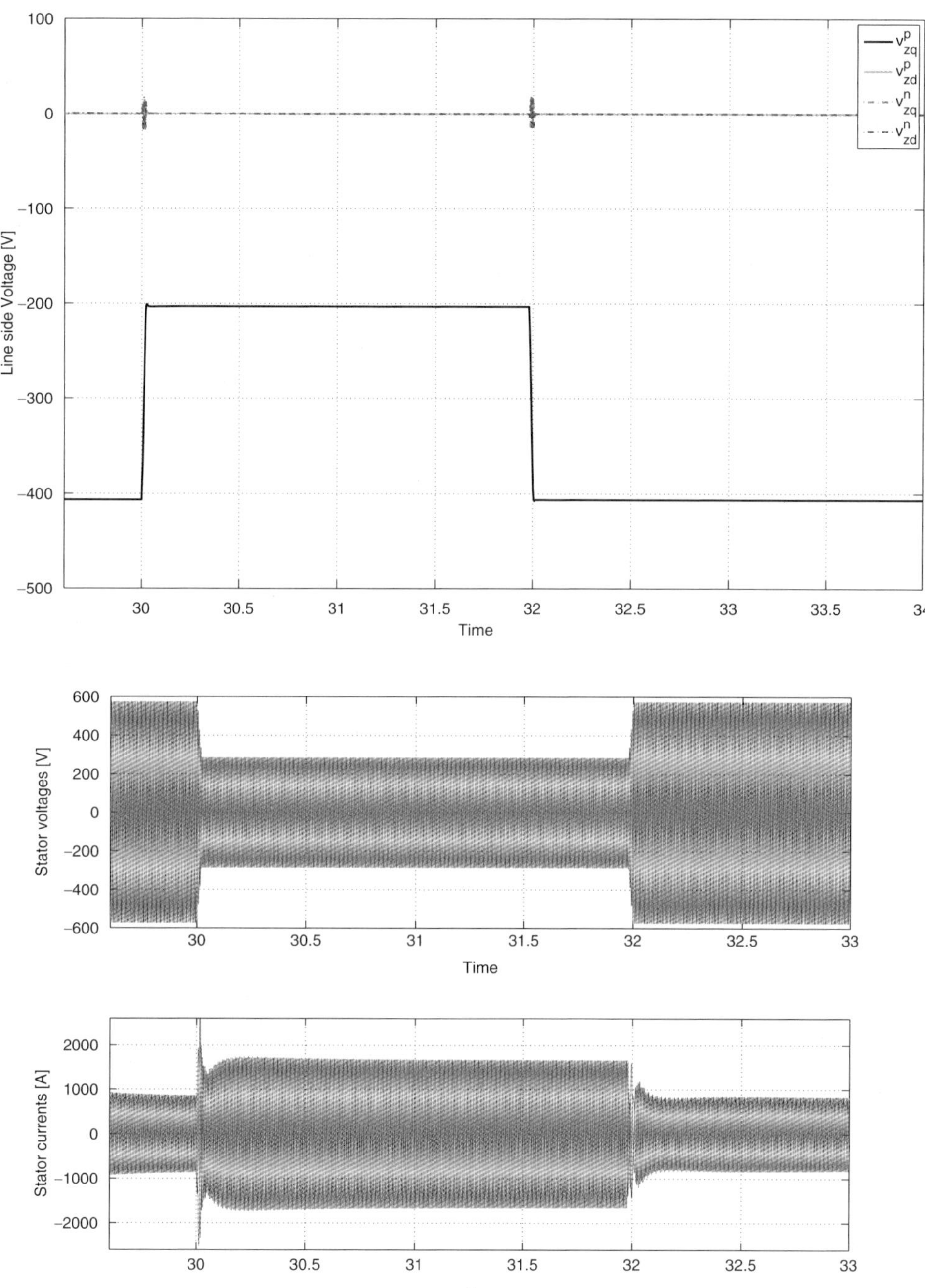

Fig. 27.10. Stator voltage in positive and negative sequence (top) and stator abc voltages and currents (bottom).

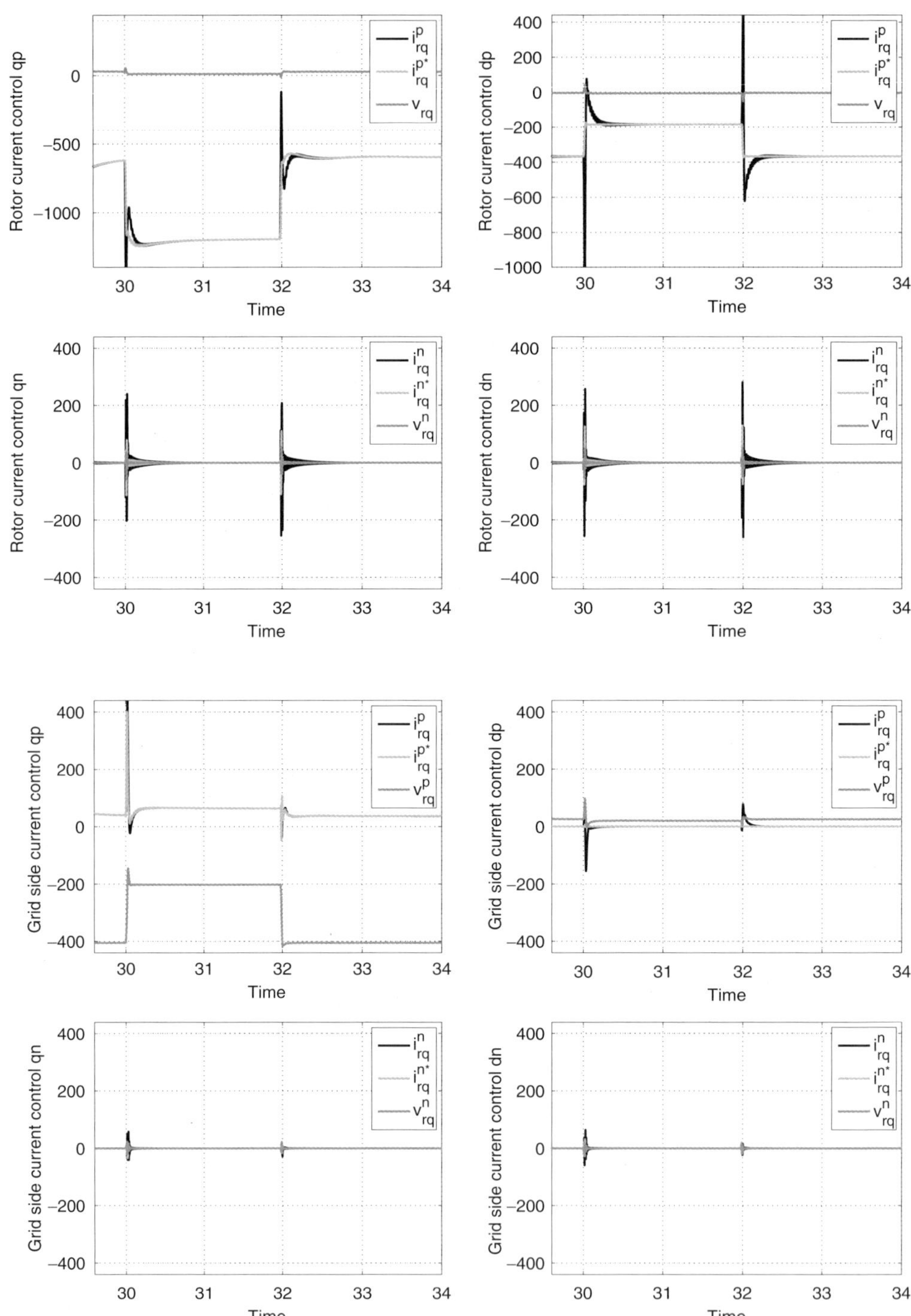

Fig. 27.11.   Rotor-side and grid-side converter current loops.

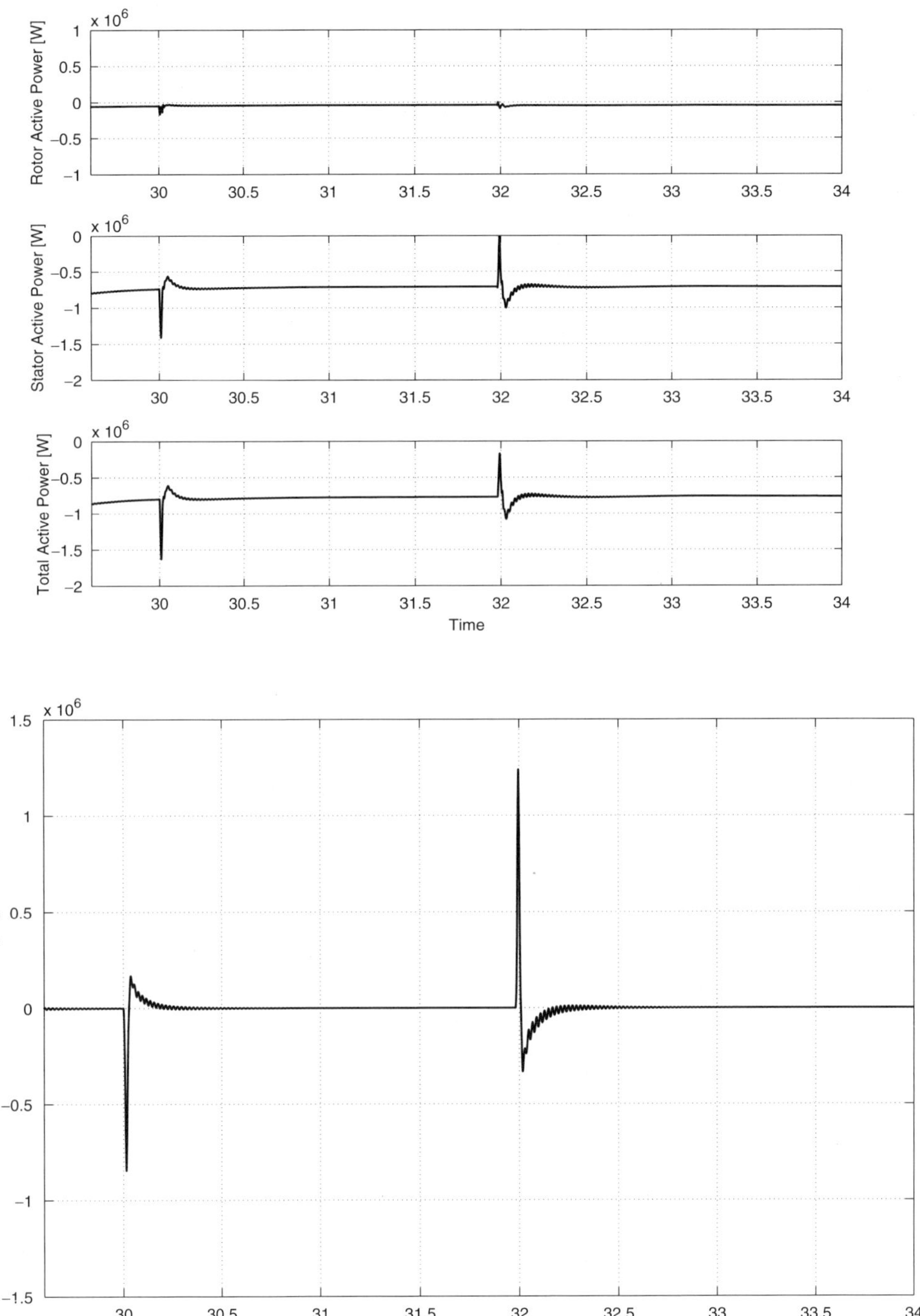

Fig. 27.12. Rotor, stator and total active power and stator reactive power.

 O. Gomis-Bellmunt and A. Junyent-Ferré

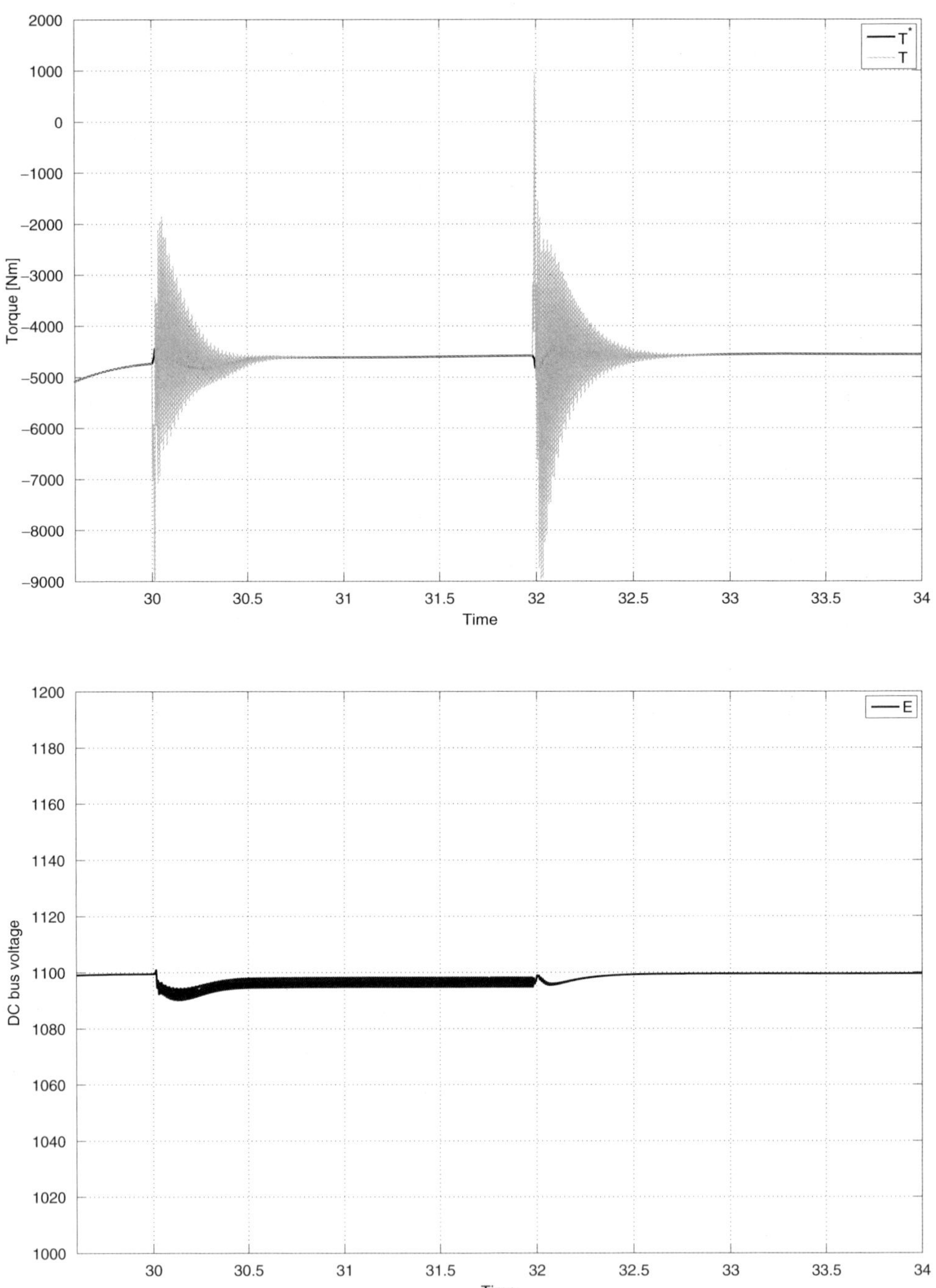

Fig. 27.13.   Torque and DC bus response to an unbalanced voltage sag.

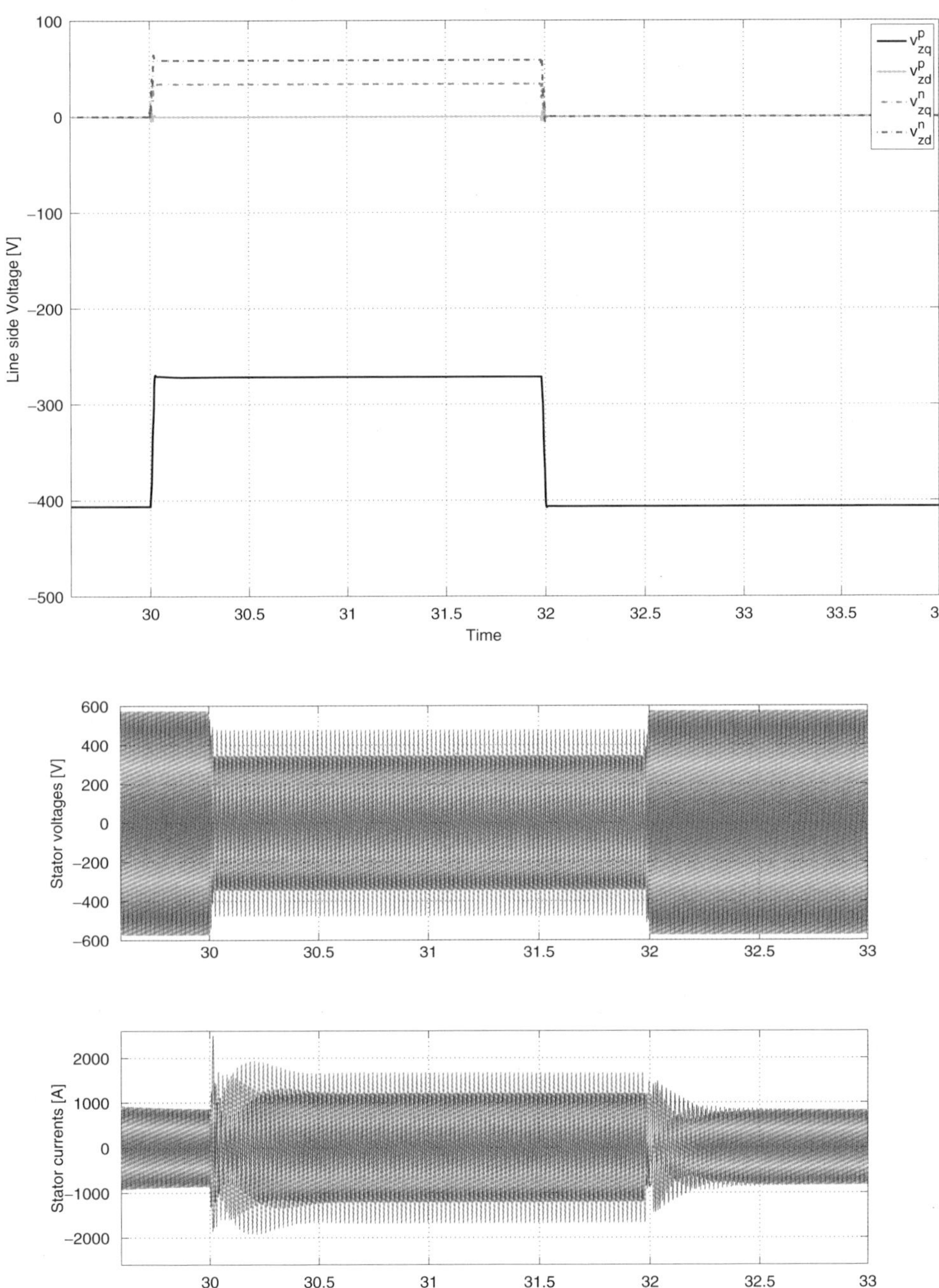

Fig. 27.14.   Stator voltage in positive and negative sequence (top) and stator abc voltages and currents (bottom).

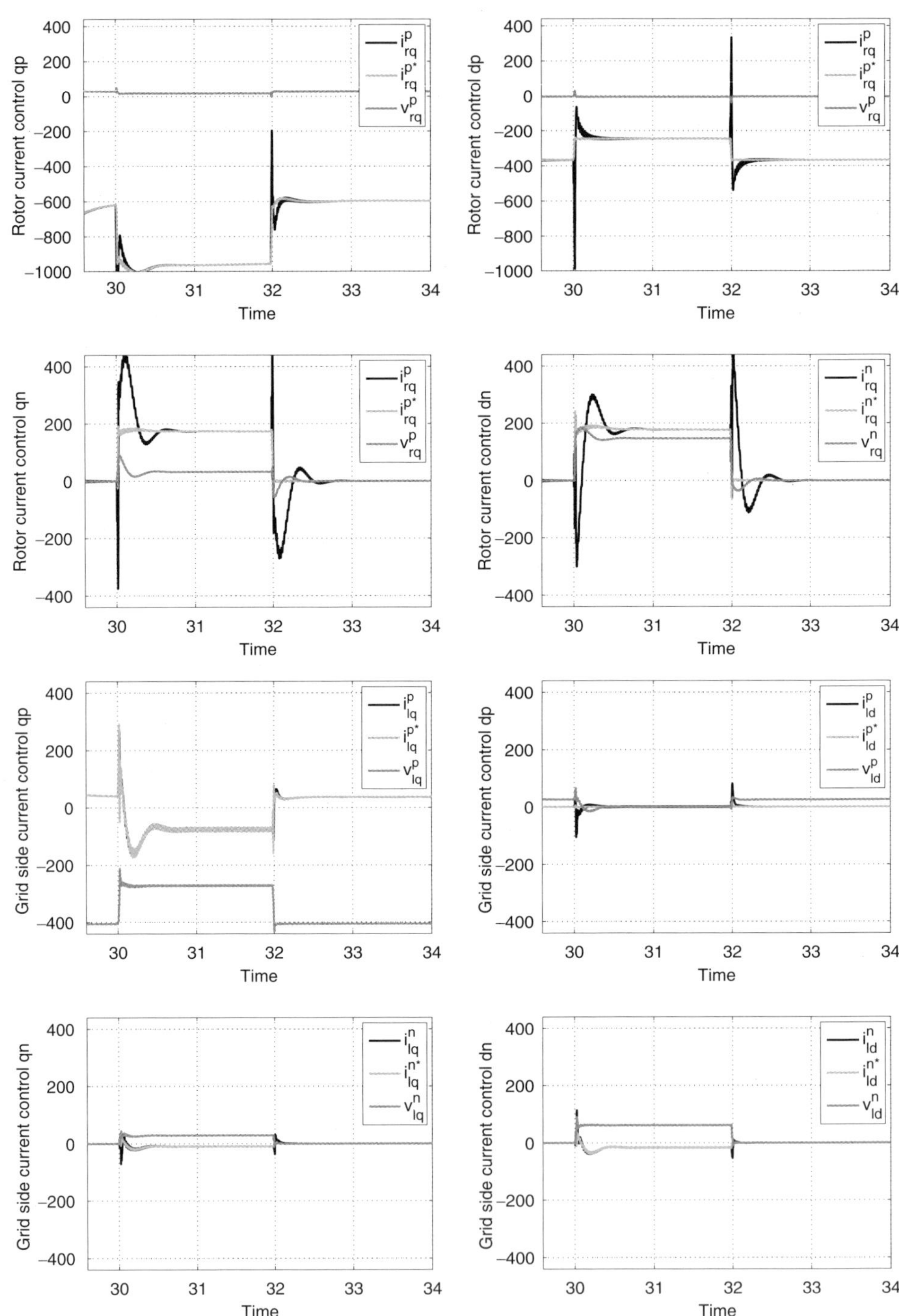

Fig. 27.15.   Rotor-side and grid-side converter current loops.

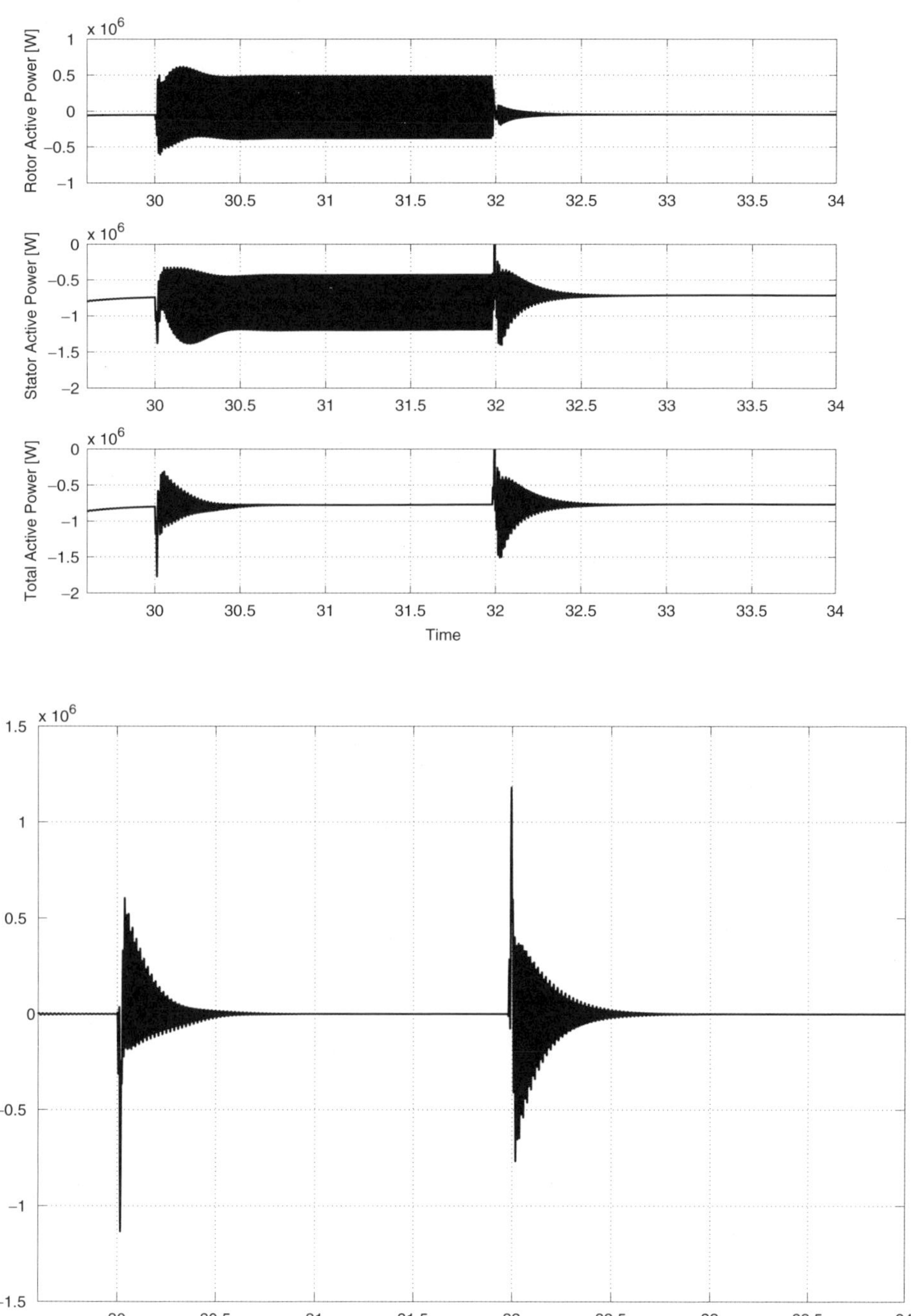

Fig. 27.16.   Rotor, stator and total active power and stator reactive power.

27.6.1 *Balanced voltage sag*

A 50% voltage sag of 2 seconds have been applied when a 2 MW wind turbine generating a wind of 8.2 m/s.

The generator torque and DC bus voltage response are illustrated in Fig. 27.9. Stator voltages in positive and negative sequence along with abc stator voltages and currents are shown in Fig. 27.10. Rotor-side and grid-side converter current loops are shown in Fig. 27.11. Active and reactive powers are illustrated in Fig. 27.12.

27.6.2 *Unbalanced voltage sag*

A 50% voltage sag have been applied to two phases leaving the third phase undisturbed. The disturbance has been analyzed in a 2 MW wind turbine generating a wind of 8.2 m/s.

The generator torque and DC bus voltage response are illustrated in Fig. 27.13. It can be seen that although the inverse sequence provokes an oscillating flux, an almost constant torque can be achieved after a transient. The DC voltage response is shown in Fig. 27.13. As it has been stated, the constant torque implies oscillating rotor power which can be compensated with oscillating grid-side converter power. Using the proposed technique, the resulting DC bus voltage has minimized the oscillations.

The stator voltages in positive and negative sequence and the abc stator voltages and currents are illustrated in Fig. 27.14. Rotor-side and grid-side converter current loops are shown in Fig. 27.15.

Active and reactive power are illustrated in Fig. 27.16. It can be seen that while the total power (depending on the torque) is almost constant, stator and rotor active power are of oscillating nature.

27.7 Conclusions

This present chapter presents a control technique for doubly fed induction generators under different voltage disturbances. The current reference are chosen in the positive and negative sequences so that the torque and the DC voltage are kept stable during balanced and unbalanced conditions. Both rotor-side and grid-side converters have been considered, detailing the control scheme of each converter while considering the effect of the crow-bar protection. The control strategy has been validated by means of simulations for balanced and unbalanced voltage sags.

References

1. O. Gomis-Bellmunt, A. Junyent-Ferre, A. Sumper and J.Bergas-Jane, "Ride-through control of a doubly fed induction generator under unbalanced voltage sags," *IEEE Trans. Energy Conversion* **23** (2008) 1036–1045.
2. R. Pena, J.J.C. Clare and G. Asher, "Doubly fed induction generator using back-to-back PWM convertersand its application to variable-speed wind-energy generation," *IEE Proc. Electric Power Applications* **143** (1996) 231–241.
3. A. Junyent-Ferré, "Modelització i control d'un sistema de generació elèctrica de turbina de vent," Master's thesis, ETSEIB-UPC (2007).
4. L. Harnefors and H.-P. Nee, "Model-based current control of ac machines using the internal model control method," *IEEE Trans. Industry Applications* **34** (1998) 133–141, doi: 10.1109/28.658735.
5. A.D. Hansen and G.Michalke, "Fault ride-through capability of DFIG wind turbines," *Renewable Energy* **32** (2007) 1594–1610, doi: 10.1016/j.renene.2006.10.008.
6. J. Morren and S. de Haan, "Ridethrough of wind turbines with doubly-fed induction generator during a voltage dip," *IEEE Trans. Energy Conversion* **20** (2005) 435–441, doi: 10.1109/TEC.2005.845526.
7. H.-S. Song and K. Nam, "Dual current control scheme for PWM converter under unbalanced input voltage conditions," *IEEE Trans. Industrial Electronics* **46** (1999) 953–959, doi: 10.1109/41.793344.

Chapter 28

Power Quality Instrumentation and Measurement in a Distributed and Renewable Environment

Mario Manana[*], Alfredo Ortiz, Carlos J. Renedo, Severiano Perez and Alberto Arroyo

*Electrical and Energy Engineering Department,
University of Cantabria, Avda. Los Castros,
s/n, 39005 Santander, Cantabria, Spain*
[*]*mananam@unican.es*

This chapter provides a basic review of the architecture and features of a modern power quality meter, considering its application to renewable energy generation. Power quality monitoring of renewable energy facilities has to consider not only voltage and current but also other parameters like grid impedance and wind speed. In addition, the power quality survey has to be extended to include the grid topology and other operational information like resource distribution. The chapter also details the basic structure of the IEC standards related with power quality monitoring.

28.1 Introduction

Power quality (PQ) instrumentation has evolved significantly during the last few years.[1] From general purpose oscilloscopes and voltmeters to specialized transient recorders, these types of devices have introduced numerous improvements related with selective disturbance detection and automatic report generation. The state of the art on power quality instrumentation includes the latest standards related with power quality according to the European Union[2,3] and the USA.[4] The increase in electricity generation based on renewable energies has produced new power quality problems that have to be measured and analyzed.[5] This is due to:

- The energy vector is not constant.
- Renewable energy power plants usually include an electronic power converter.
- The increasing ratio between the nominal power of the renewable energy and classical generation.

The integration of renewable energy systems into the power grid introduce changes in the behavior and characteristics of the system. These include, among others:

- The short-circuit power is modified.
- The voltage profile suffers variations due to the variable energy vector.
- The voltage variations produce fluctuations, flicker, imbalance, harmonics and subharmonics.

In addition, power quality monitoring of renewable energy facility has to consider not only voltage and current but also other parameters like grid impedance and wind speed. Some research groups[6,7] have developed a power quality meter specifically designed to fulfil both the IEC 61000-4-15[8] and the IEC 61400-21.[9]

From a general point of view, PQ measurements should answer some basic question[10–13]:

- When to do the PQ survey? Most power quality surveys are programmed after the problem is detected.
- Where to put the PQ meters? The choice of the best point or set of points to install the power quality meters is not an easy question. The answer should address topics like system topology, sensitive loads, disturbance generators, grounding, etc.
- What PQ meter to use? The use of hand-held, portable or fixed power quality equipment has to be determined based on various parameters like: physical place where the power quality meter has to be installed, recording period, number of channels, kinds of disturbances, remote control, etc.
- What magnitudes should be monitored? General purpose power quality surveys should include all the usual parameters, considering both voltage and current. If the power quality survey is devoted to a wide area, current is not considered. If the survey is related with a final user, current should also be considered.
- How to process the registered data? Once the power quality survey has concluded, it is necessary to generate a standardized report considering conclusions and recommendations.

28.2 Regulatory Framework

Wikipedia[14] defines a technical standard as "*an established norm or requirement. It is usually a formal document that establishes uniform engineering or technical criteria, methods, processes and practices.*"

Standards provide a common reference framework that allows us to compare the qualities of products that we, as consumers, use in our daily lives. The set of standards that regulate power-quality measurements belongs to various groups of documents. The first group deals with the overall regulation that defines the technical

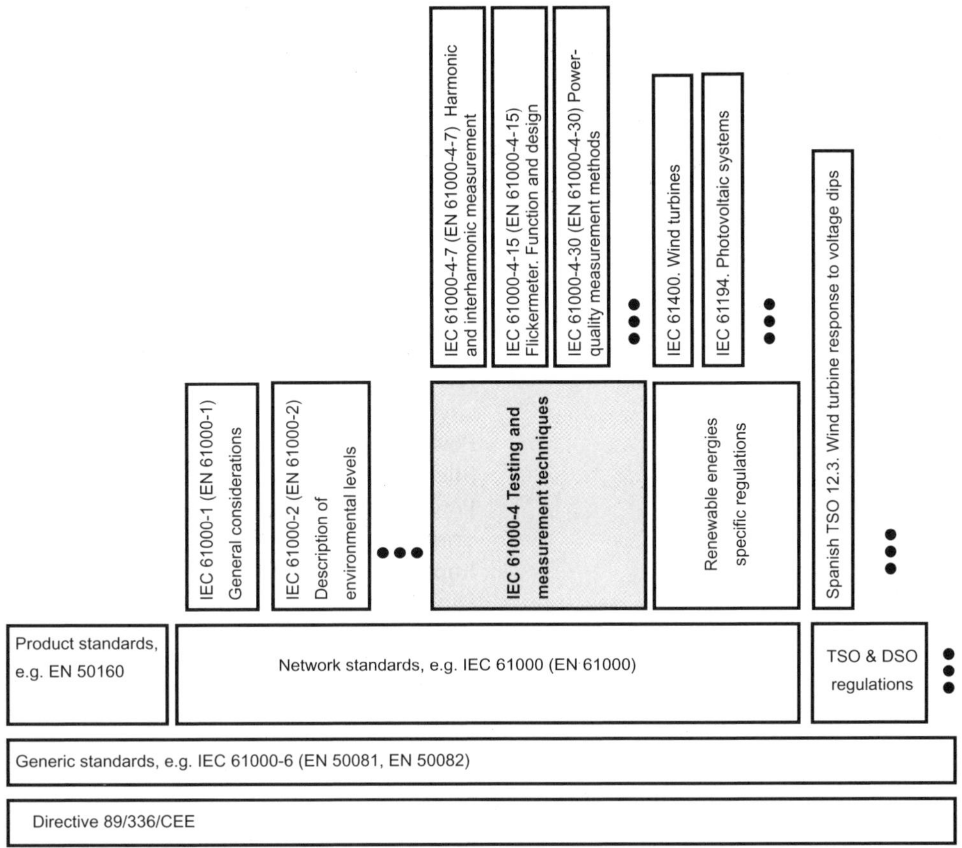

Fig. 28.1.   Basic structure of the IEC standards related to PQ.

characteristics of the instrumentation. The second group is devoted to measuring procedures. The last one defines the limits of the power-quality indices.

One of the main difficulties of dealing with standards is the existence of multiple regional, national and international standards organizations. This coexistence of multiple reference documents in a globalized world was solved with a coordinated action that ended with the adoption of the existing standard or with the definition of a new one.

Figure 28.1 summarizes the basic structure of the IEC standards related with PQ measurements. The IEC 61400 series focuses on wind turbines and the IEC 61194 on photovoltaic systems.

28.3   State-of-the-art

From a general point of view, power quality instrumentation can be classified according to several criteria such as kind of application, graphical user interface,

Table 28.1.   Typical parameters of a power-quality meter.

Type of application	Hand-held
	Portable
	Fixed installation
User Interface	Alphanumeric
	Graphic
	Oscilloscope
	Text
	Blackbox
Measured parameters	DC voltage and current
	Harmonics and interharmonics
	Ground resistivity
	Power factor
	Flicker
	Power and energy
	Transients ($>$ 200 us)
	Impulses ($<$ 200 us)
	Dips
	Overvoltages
	Imbalance
	Frequency
	Other disturbances
Type of meter	Trends
	Energy
	Spectrum analyser
	Transients recorder
Type of communication	RS-232 and/or RS-485
	Ethernet
	TCP/IP
	Modem
	Power-Line Communication
	Other

etc.[1] Table 28.1 summarizes the most important parameters that have to be considered in order to compare power quality meters.[15–18]

In general, this type of instrument makes it possible to carry out the evaluation of any type of conducted disturbance: variations of the nominal frequency of supply, variations in the magnitude of the voltage supply, transients, flicker, imbalance, harmonics, interharmonics, dips and interruptions as defined by the standard EN 50160.[19] In many cases the PQ meters have, in addition, specialized software able to analyze the stored measurements.

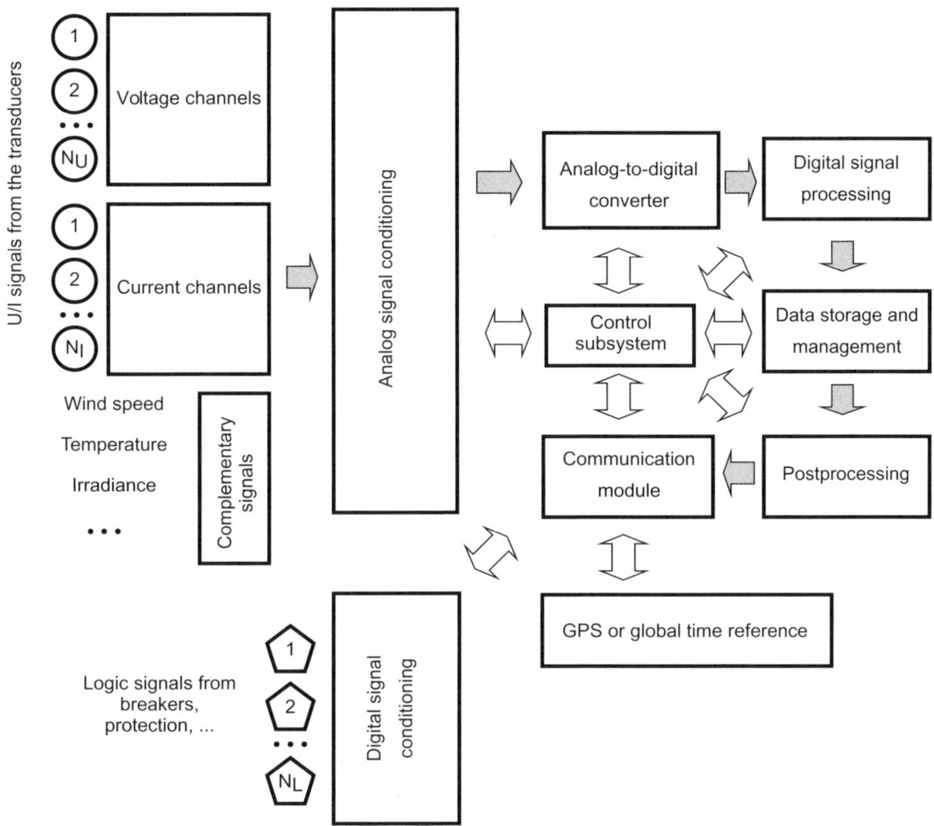

Fig. 28.2.   Basic architecture of a PQ meter.

28.4   Instrumentation Architecture

A PQ meter has a basic architecture that includes the blocks shown in Fig. 28.2. This figure does not consider voltage and current transformers, if needed. More detailed information about this topic can be found in the literature.[1,20,21] The signal conditioning module provides signal isolation, antializing filtering and the maximum available signal-to-noise ratio. The analog-to-digital conversion module is resposible for the analog to digital translation. Then, the digital representation of the signal can be processed by the digital signal processing module. The processed data is stored or sending to a remote control system. The time reference is provided by the GPS[22,23] or global time reference.

28.4.1   *Number of channels*

The number of channels is a key question because the instrumentation has to measure not only voltage and current but also meteorological parameters like wind speed, irradiance and temperature. From the point of view of the electrical magnitudes, and due to economic criteria, the choice of the number of channels is based on a specification of minimal needs.[1]

28.4.2 *Transducers*

Voltage and current transducers have to provide a small voltage signal (with typical amplitudes between -10 and 10 V), which is proportional to the levels of voltage and current that have to be measured. In general, correct utilization of the transducers means that their range and frequency responses have to be completely characterized, both in terms of magnitude and phase. Most meters can handle voltages up to 700 volts without any additional element. Voltages over this value should be measured using a voltage transformer. Table 28.2 summarizes the basic characteristics of various types of current transducers covering the range from a few mA to thousands of amps.

28.4.3 *Data acquisition module*

The Data Acquisition Module includes two main elements:

- Signal conditioning module (SCM). This subsystem acts as an interface between the transducers and the analog-to-digital converter. It can be seen as a glue element between the analog and digital worlds. The main objectives of the SCM can be summarized as follows:

  — To provide galvanic isolation between the power system and the user of the instrumentation.
  — To adequately amplify and/or to attenuate the voltage or current signals in order to obtain the maximum dynamic signal range, guaranteeing in this way a high signal-to-noise ratio.
  — To avoid aliasing problems by means of low-frequency filters.

- Analog-to-digital converter (ADC). This module has the responsibility of translating the signal from the analog to the digital domain. It carries out two basic tasks:

  — Signal discretization. This is the process of periodic sampling of the signal using the Nyquist–Shannon theorem. It allows us to obtain a sequence of samples $x[n]$ obtained from a continuous-time signal $x(t)$.
  — Signal quantization. This is a mathematical transformation that assigns a fixed-point binary number to a sampled signal value that belongs to the real numbers.

The resolution of analog-to-digital converters in power-quality instrumentation has values between 8 and 24 bits. Other values can also be found, but are not

Table 28.2.   Current transducer comparison.

	Input range	Overcurrent	Accuracy	Sensitiviy	Bandwith	Security
Resistive *shunt*	✓	✓	✓✓✓✓ Better than 0.01%	✓✓✓✓	✓✓✓✓ dc–100 MHz	— No isolation
Inductive	✓✓✓✓	✓✓✓	✓✓ ±0.5%, 0.5 degrees	✓✓	✓✓ 20 Hz–1 kHz	✓✓ Single isolation
Hall effect	✓✓✓✓	✓✓✓	✓ ±1%, 1.5 degrees	✓✓ 1.5–3.0 V/FSA	✓✓✓ dc–50 kHz	✓✓✓✓ Double isolation
Rogowski coil	✓✓✓✓	✓✓✓	✓ ±1%, 1.5 degrees	✓ 0.2 V/FSA	✓✓✓ 1 Hz–10 MHz	✓✓✓✓ Double isolation
Optical	✓✓✓✓✓	✓✓✓✓✓	✓✓	✓	✓✓✓✓	✓✓✓✓✓ Double isolation

typical. From a numerical point of view, quantification is a nonlinear transformation that produces an error. The error range depends on the resolution of the ADC and the input range.

28.4.4 *Signal processing module*

The signal-processing module is the core of the instrumentation. It includes the task manager, which controls both the device and the set of algorithms that computes the power-quality indexes, the file manager and the communication facilities. The main features of this module are:

- Reprogrammed firmware. This characteristic allows us to programme and to reschedule the functionality of the system without hardware modification, which reduces the development costs enormously. For instance, a system can be updated to measure harmonics with several algorithms or standards without changing the hardware.
- System stability. This involves repeatability of the implementation and of the response. It is easy to understand that a digital system provides the same answer to the same question. This is not so with analog systems, where the response is a function of the temperature, age, humidity, etc.
- Suitable for implementing adaptive algorithms and special functions such as linear phase filters. The utilization of numerical algorithms allows external errors, due to changes in the operating conditions, to be corrected dynamically.
- Able to compress and store measurement data. It is not necessary to highlight here the importance of computers in the storage, treatment and recovery of information. The IEEE 1159.3[24] defines a file format suitable for exchanging power quality-related measurement and simulation data in a vendor-independent manner.
- User-friendly interfaces. The utilization of graphical user interfaces (GUIs) facilitates user interaction with the instrument.
- Low power consumption. These kinds of devices have a power consumption of less than 3 VA.

28.5 PQ Monitoring Surveys in Distributed and Renewable Environments

Power Quality surveys should be planned to include, at least, the following steps[25].

1. The complaint. Power quality complaints can generally be classified according to three main sets:

   - Physical damage to information technology (IT) hardware and/or other equipment.

Customer PQ recording form

Date of disturbance: Time of disturbance:
Company:
Address:
Contact name:
Phone number: Fax number:
email:
Description of disturbance:

Basic electrical schematic diagram:

Equipment type	Manufacturer	Equipment Rating	Cost
PQ meter	Configuration	Placement	Period

Fig. 28.3.   PQ form modified from Ref. 4.

- Damage and/or corruption of stored data and/or data that was being processed. Corrupted data can be probed using printouts.
- Loss of production.

2. Visiting the site. Once the complaint has been reviewed it is important to visit the site with a "walk-through". This visit should include interviews with equipment operators and other personnel related with the electrical installation. Figure 28.3 shows a basic form that can be completed during the visit. It is important to collect data related with the following questions, among others:

- Magnitude, frequency and duration of the events.
- To establish if there is a constant frequency pattern.
- To identify the loads that are electrically close to the disturbed one and can be considered potential disturbance generators.

3. Collecting data. Relevant data includes not only electrical parameters but also other important additional information like meteorological data and the status of the breakers. If there is any damaged equipment, it is important to retain it in order to be examined by an expert.

4. Planning the survey. This is not an easy task because planning the survey should define what places are the optimal ones in which to install the PQ meter or meters. In addition, it is necessary to define the number of channels, the monitoring period, the PQ parameters to measure, the communication protocol if the meters are installed in a remote place, etc.

5. Defining PQ instrumentation. Voltage and current transformers and transducers have to be chosen to match the installation requirements.
6. Installing the PQ monitors. The installation of the PQ meters has to be done by specialized technicians and by following all the security guidelines. This is especially important if the installation has to be done without voltage interruption.
7. Handling[26] and postprocessing the recorded data.
8. Evaluating results and defining cost-effective solutions.

It is important to reduce response time as much as possible. A fast response can avoid problems relating to sensitive equipment, loss of data and loss of production. In addition, a rapid response can prevent a complaint from turning into a lawsuit.

Communication with all the parties involved in the complaint or PQ problem (utility, customer, administration, etc.) is essential to find a fast and appropriate solution. It is important to avoid providing partial conclusions before finishing the study. A fault in communication can transform one or more of the parties into adversaries.

The availability of power quality and weather information allows to correlate electrical and physical parameters. In the literature[27] it is possible to find some graphs that summarize the correlation between voltage, or flicker, and wind speed in wind parks.

A sample PQ survey report is defined in IEEE 1159.[4] Figure 28.3 includes the basic data defined in the PQ survery report and other additional information.

28.5.1 *PQ monitoring in a wind park*

As an application example of Sec. 28.5, this subsection summarizes a PQ survey in a 35 MW wind park. Figure 28.4 shows the electrical diagram of the wind park. This PQ survey includes four monitoring places. Places 1 and 2 register PQ data at the substation level. It is important to control PQ at the grid connection point. The grid can be considered to have unlimited capacity with constant voltage and frequency. The main problem, from a PQ point of view, is due to the fact that this wind park is in a remote place and the connection has been made using a dead-end feeder. The short circuit power in the connection point can be considered small considering the wind park nominal power. Places 3 and 4 register PQ data at the wind generator level.

Figures 28.5–28.9 sumarize the results of the PQ survey. Figures 28.5 and 28.6 show the timeplot and the probability distribution function of the voltage at place 1. It can be seen that voltage rms values remain almost constant during the survey, with the exception of some voltage sags. Figure 28.7 summarizes the probability distribution function of the flicker P_{st} at place 1. The flicker values range from 0.05 to 6.8, with an average value of 0.52. Figure 28.8 summarizes voltage THD

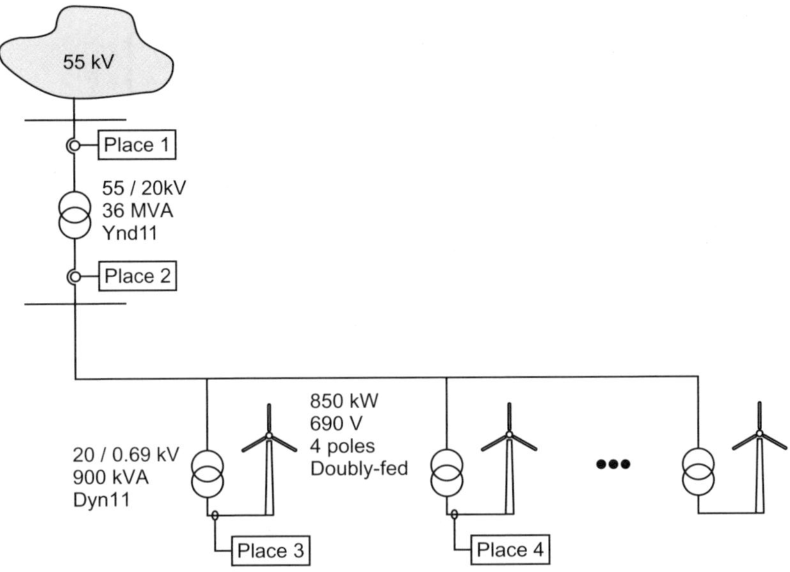

Fig. 28.4.   PQ survey in a 35 MW wind park.

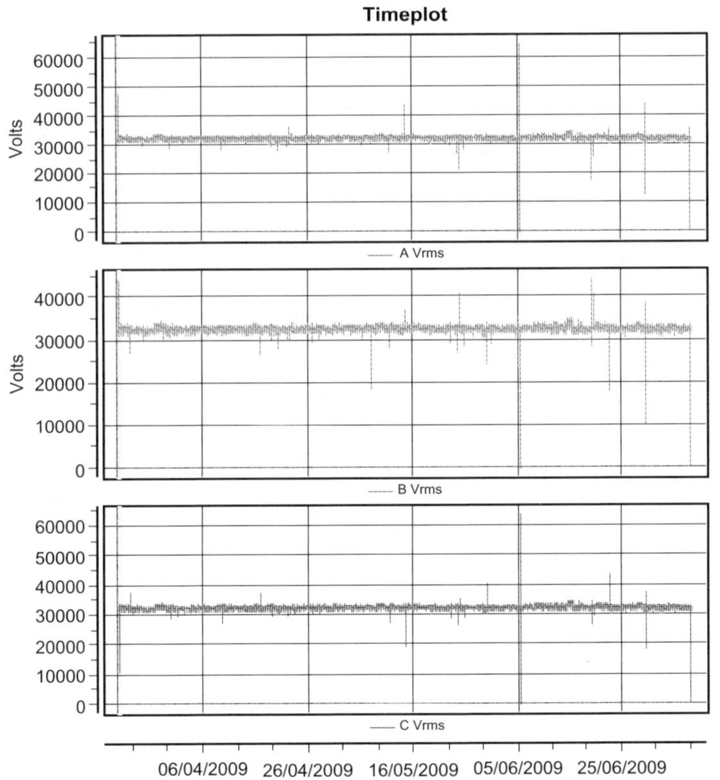

Fig. 28.5.   Rms voltage evolution at place 1 during survey.

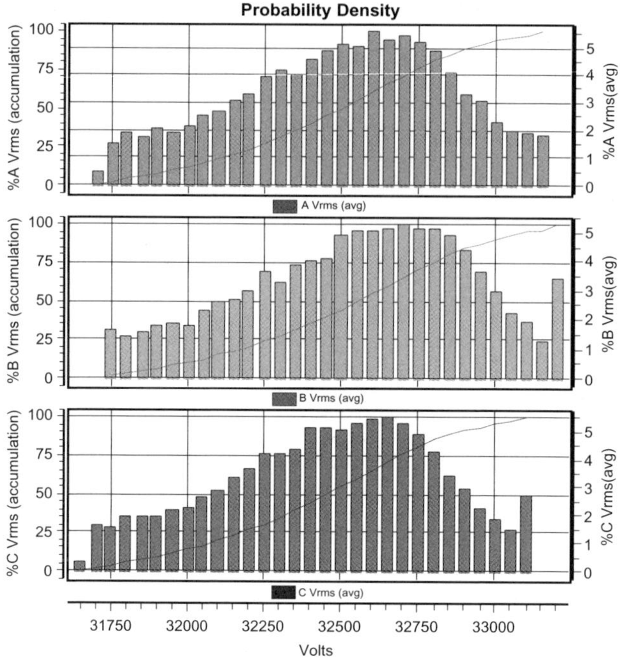

Fig. 28.6.   Probability distribution function of voltage at place 1.

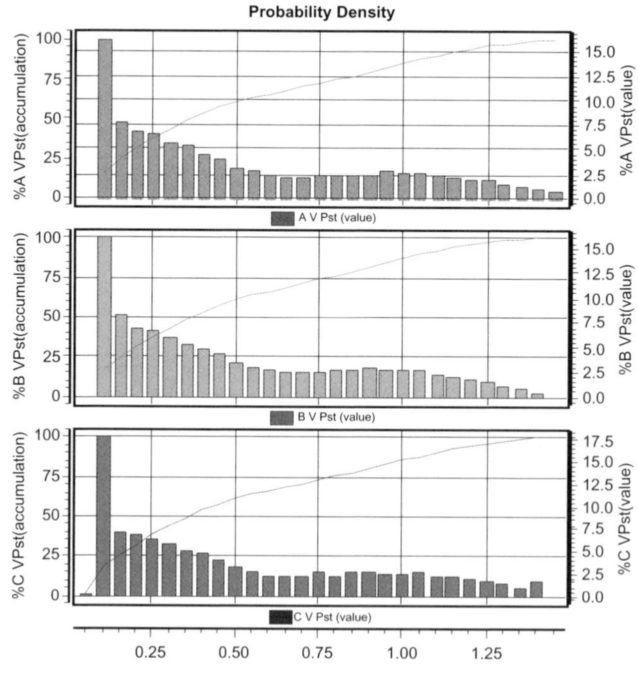

Fig. 28.7.   Probability distribution function of flicker at place 1.

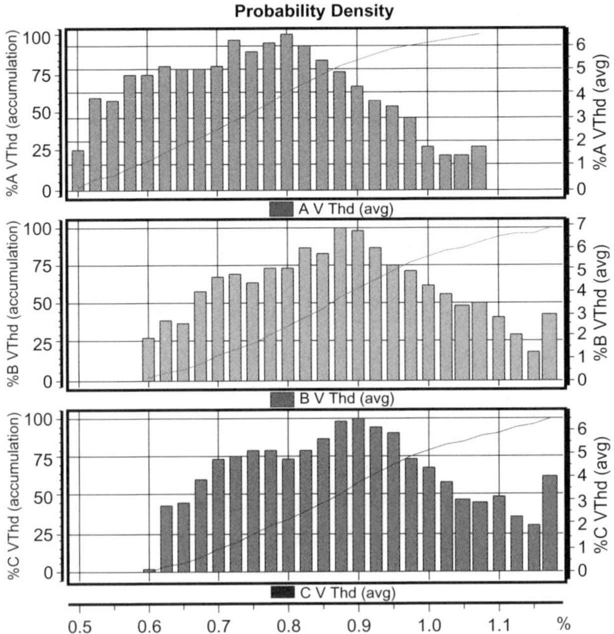

Fig. 28.8.   Probability distribution function of voltage THD at place 1.

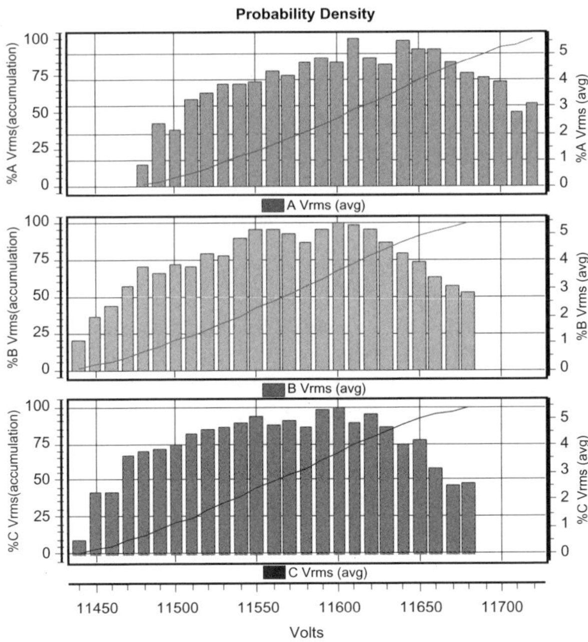

Fig. 28.9.   Probability distribution function of voltage at place 2.

at place 1. Figure 28.9 shows the probability distribution function of voltage at place 2.

At places 3 and 4, the general data exhibits similar behavior with greater voltage variations at places 1 and 2. The aggregation of individual generators contributes to smooth the variations.

28.6 Summary

This chapter has provided a basic review of the architecture of a modern power quality meter considering its application to renewable energy generation. The text also focuses on the importance of considering not only the generation facility but also the integration of the facility into the grid. Governments have established new regulatory frameworks that have to be carefully considered in order to fulfil the standards. In conclusion, it is necessary to underline the fact that the study of renewable energy resources should cover not only technical problems related with the generation, transformation and final use of the energy but also other technical aspects like grid effects in terms of changes related with the short-circuit power, voltage fluctuations and other disturbances, grid reliability, etc. All the above considerations can be summarized as power quality issues related to renewable energy resources.

References

1. A. Moreno-Munoz (ed.) *Power Quality: Mitigation Technologies in a Distributed Environment* (Springer-Verlag London Limited, 2007).
2. European Union. "Directive emc 89/336/eec of 23 May 1989 on the approximation of the laws of the member states relating to electromagnetic compatibility," no. 1 139/19 (1989).
3. European Union. "Directive 92/31/eec of 28 April 1992 amending directive 89/336/eec on the approximation of the laws of the member states relating to electromagnetic compatibility," no. 1 126/11 (1992).
4. IEEE. "Recommended practice for monitoring electric power quality," *IEEE Std 1159-2009 (Revision of IEEE Std 1159-1995)* (2009), pp. c1–81, doi:10.1109/IEEESTD.2009.5154067.
5. S. Heier, *Grid Integration of Wind Energy Conversion Systems* (John Wiley & Sons Ltd, 2006).
6. C. Gherasim, T. Croes, J. Van den Keybus, J. Driesen, and R. Belmans, "Development of a flickermeter for grid connected wind turbines using a DSP based prototyping system," *Instrumentation and Measurement Technology Conf., 2004. IMTC 04. Proc. 21st IEEE*, Vol. 3 (2004), pp. 2015–2019, doi:10.1109/IMTC.2004.1351484.
7. J. Gutierrez, J. Ruiz, L. Leturiondo, and A. Lazkano, "Flicker measurement system for wind turbine certification," *IEEE Trans. Instrumentation and Measurement* **57** (2008) 375–382, doi: 10.1109/TIM.2008.928871.
8. IEC. "IEC 61000-4-15. Electromagnetic compatibility (emc) — part 4: Testing and measurement techniques — section 15: Flickermeter — functional and design specifications," (2003).
9. IEC. "IEC 61400-21. Wind turbines — part 21: Measurement and assessment of power quality characteristics of grid connected wind turbines," (2008).

10. E. Gunther, D. Sabin, and H. Mehta, "Update on the EPRI distribution power quality monitoring project," *PQA'93* (1993).

11. E. Gunther, J. Thompson, R. Dwyer, and H. Mehta, "Monitoring power quality levels on distribution systems," *PQA'92* (1992).

12. P. Ribeiro and R. Celio, "Advanced techniques for voltage quality analysis: Unnecessary sophistication or indispensable tools," *PQA'94*, Vol. Ref. A-2.06 (1994).

13. D. Sabin, T. Grebe and A. Sundaram, "Preliminary results for eighteen months of monitoring from the EPRI distribution power quality project," *PQA'95* (1995).

14. Wikipedia, http:\\wikipedia.org.

15. J. Balcells, V. Parisi and D. Gonzalez, "New trends in power quality measuring instruments," *IECON 02, IEEE 2002 28th Annual Conf. of the Industrial Electronics Society*, Vol. 4 (2002), pp. 3344–3349, doi:10.1109/IECON.2002.1182934.

16. R. Bingham, "Measurement instruments for power quality monitoring," *Transmission and Distribution Conference and Exposition, 2008. T&D. IEEE/PES* (2008), pp. 1–3, doi: 10.1109/TDC.2008.4517281.

17. R. Simpson, "Instrumentation, measurement techniques, and analytical tools in power quality studies," *IEEE Trans. Industry Applications* **34** (1998) 534–548, doi:10.1109/28.673724.

18. D. Brasil, J. Medeiros, P. Ribeiro, J. Oliveira, and A. Delaiba, "Considerations on power quality measurements and measurement instrumentation," *Transmission and Distribution Conf. and Exposition: Latin America, 2004 IEEE/PES* (2004), pp. 113–116.

19. CENELEC. "EN 50160. voltage characteristics of electricity supplied by public distribution systems," (2007).

20. L. Eguiluz, M. Manana, P. Lara, J. Lavandero, and P. Benito, "Mepert i: Electric disturbance and energy meter," *Int. Conf. on Industrial Metrology (ICIM'95)*, (1995).

21. M. Manana, "Contributions to the representation, detection and classifications of power disturbances," Ph.D. thesis, University of Cantabria (in Spanish).

22. M. Caciotta, A. Di Liberto, F. Leccese, S. Pisa, and E. Piuzzi, "A system for obtaining a frequency reference constantly locked to 11 gps carrier for distributed power quality assessment," *Electronic Measurement & Instruments, 2009. ICEMI '09. 9th Int. Conf. on* (2009), pp. 2–7–2–10, doi: 10.1109/ICEMI.2009.5274488.

23. A. Miller and M. Dewe, "Multichannel continuous harmonic analysis in real-time," *IEEE Trans. Power Delivery* **7** (1992) 1813–1819, doi:10.1109/61.156983.

24. IEEE. "IEEE recommended practice for the transfer of power quality data," *IEEE Std 1159.3-2003* (2004), pp. 1–119.

25. A. Baggini (ed.) *Handbook of Power Quality* (Wiley, 2008).

26. E. Gunther, "On creating a new format for power quality and quantity data interchange," *Transmission and Distribution Conf. and Exhibition, 2005/2006 IEEE PES* (2006), pp. 354–358, doi:10.1109/TDC.2006.1668517.

27. G. Chicco, P. Di Leo, I.-S. Ilie, and F. Spertino, "Operational characteristics of a 27-MW wind farm from experimental data," *Electrotechnical Conf., 2008. MELECON 2008. The 14th IEEE Mediterranean* (2008), pp. 520–526, doi:10.1109/MELCON.2008.4618488.

Chapter 29

Energy Resource Allocation in Energy Planning

Sandip Deshmukh

Centre for Environment Strategy,
University of Surrey, Guildford
s.deshmukh@surrey.ac.uk

The gap in demand and supply of energy can only be met by optimal allocation of energy resources and is need of the day for developing countries like India. For the socio-economic development of India, energy allocation at the rural level is gaining importance these days.

Integrated Renewable Energy System (IRES) in rural context aims at optimal resource allocation, thereby reducing dependence on commercial energy and reducing associated environmental hazards, and opening new avenues for employment generation. The chapter aims at finding optimal energy resource allocation in the energy planning process. To demonstrate the development of IRES, a survey was conducted in villages of Jhunjhunu district of Rajasthan for estimating the end-use energy requirements and energy resource availability. The Goal Programming model developed is used for determining energy resource allocation for meeting present energy needs. Six different scenarios are developed by considering alternative priorities to the objective functions. The developed scenarios are evaluated on the basis of associated cost and emissions, and optimal scenario, wherein cost and employment generation is priority, is suggested for implementation.

29.1 Introduction to Energy Planning Process

The local energy resource utilization is a response to the need, which has now become very pressing, to limit the dependence of the regions on the conventional power grid with a view to improving their security of energy supply. The existing approach for planning and implementing energy programmes in India is top-down and sectoral. Ministry and State Nodal Agencies (SNAs) are responsible for the supply of different energy resources and programmes in rural areas. These SNAs prepare and implement their own plans and programmes. The targets for these programmes are imposed from the top-down (e.g., pump-sets to be energized; villages to be electrified; hectares for afforestation; biogas plants to be installed; improved stoves to be promoted, etc.) through uniform directives which go down to inadequately trained employees in these departments, or to the district or other administrative agencies at the grassroots level.[35] Moreover, the targets for the different rural energy programmes and options do not have any relationship with one another. Therefore, there is a need to develop a region dependent Integrated Renewable Energy System (IRES), which incorporates proper utilization of renewable energy to supplement conventional energy sources.

In developing countries, constant endeavor is to evaluate various investment alternatives for meeting increasing energy demand from a techno-economic point of view. A large number of models have been developed for energy planning. However, these models are suitable for a centralized energy supply system using mainly conventional sources. It has resulted in inequities, external debt and environmental degradation.[20] In this system, the power supply to the region remains inadequate, erratic and unreliable due to increasing pressure from urban centres. As a result, development of economically productive activities in rural areas has been far slower than in the urban areas.

The role of centralized (macro) energy planning is questionable when it comes to addressing the variations in socio-economic and ecological factors of a region.[3] Decentralized (micro) energy planning is in the interest of efficient utilization of resources. The regional planning mechanism takes into account various available resources and demands in a region.

The main objective of energy planning activity is to develop an optimal plan for energy resource allocation to various applications over 5 to 20 years with the consideration of future energy requirement at minimum costs and environmental emissions, maximum employment generation, social acceptance, reliability and system efficiency, maximum use of local energy resources and minimum use of petroleum products.

29.1.1 *Need of energy resource allocation*

Energy planning has always been important, though it caught the attention of planners only after the oil crisis in the years of early 1970s. In the post gulf-crisis era, sufficient attention has been given to critical assessment of fuel reserves, rational use and conservation of energy resources, and long term energy planning. Energy planning process usually includes a study of sectoral demand and supply, forecasts of the trends based on economics and technological models, and a list of actions to achieve the objectives of the energy plan. The action plan is addressed to specific strategies and interventions, which are able to match demand and supply in the best possible way, considering associated constraints and factors. The constraints chosen in energy planning process are mainly cost and efficiency of the system. The energy planning also takes into account factors like political, social and environmental considerations, and is carried out taking into account the historical data collected in the previous energy plans of the country or region under examination.[4,57]

Energy planning methods are broadly classified into three categories: (i) planning by models, (ii) planning by analogy and (iii) planning by inquiry. The accuracy of these methods depends on the intended time interval for implementation of energy plan, viz., short-term and medium-term, up to 10 and 20 years, long-term beyond 20 years.[27]

The energy planning by the models methodology includes the optimization and econometric models. The optimization model is the most widely used tool for energy planning. Optimization models yield the optimal solution depending on a goal or objectives set in the energy planning process. The optimization model achieves optimization through minimization or maximization of the goal parameters. The optimization is carried out by using a single or multi objective linear programming technique.[10] The econometric model uses mathematical and statistical methods such as regression analysis, to model economic systems. In particular, econometric models aim at empirical validation of theoretical models and at computation of values of operational parameters for economic operation of the system.[36,37,54] All the econometric models use empirical statistical data. The econometric model can also be used for energy system modeling. Energy system modeling involves one or several energy forms, different energy sectors and energy uses.

The energy planning by analogy allows the prediction of supply and demand of energy over a period of times.[19] In developing countries, energy planning is carried out on the basis of a future scenario, also referred to as reference scenario. In developed countries, energy planning is carried out on the basis of the knowledge of recorded trends of energy supply and demand, and energy source potential. The analogy approach is often used to check and compare results obtained by using energy planning models.[24]

The energy planning by inquiry includes statistical evaluation of the responses of a selected panel of experts, to questionnaires, in order to formulate an accurate action plan for the future. The questionnaire is designed to seek the opinion on developed scenarios in a multi criteria context.[39]

29.1.2 *Classification of energy planning models*

Energy planning models are classified on the basis of methodology adopted, spatial coverage, sectoral coverage and temporal coverage. The energy planning model under each class is discussed in this section with an aim to identify the approach for micro-level energy planning in developing countries like India.

29.1.2.1 *Methodological paradigm*

Depending on the impact of energy supply and demand on economic issues, the planning methodology approach can be classified as bottom-up or top-down. The bottom-up approach entails detailed consideration of the energy resources, technologies, and energy demand. The bottom-up approach with detailed consideration allows assessment of implications of policy options such as technology mix, fuel mix, logistics and emissions in the energy sector at local, regional and national levels. The bottom-up approach to energy planning is useful for energy sector in isolation without consideration of its linkage with other sectors of economy.[29]

The top-down approach to energy planning, allows consideration of all the sectors of the national economy along with their cross linkages. In such cases, the effect of the energy plan on macroeconomic indicators such as GDP or GNP and national level emissions can be investigated. Top-down modeling approach is useful in the cases of developed countries wherein technological efficiencies and rate of capital investments have already reached close to saturation levels.

29.1.2.2 *Spatial coverage*

The energy planning model can also be classified in terms of its spatial coverage. The coverage can be for local, national, and global regions. For environmental planning, the spatial coverage of the model is usually global or national. For energy planning the spatial coverage can be local, regional and national. Local, regional and national models adopt the bottom-up modeling approach while national and global models adopt the top-down modeling approach.[38]

29.1.2.3 *Sectoral coverage*

Based on the sectoral coverage, models can be classified as economy-wide, sectoral, and sub-sectoral. Sectoral models address concerns within an economic sector such as energy, industry, transport, and agriculture. Sub-sector models are models which are used in coal sector, power sector, petroleum sector, steel industry, or railways. However, the scope for sectoral and sub-sectoral models can be regional or national, i.e., economy-wide. Most of these models follow the bottom-up approach, since other sectors of economy are not considered. Most economy-wide models follow the top–down approach, and address policy concerns at national or global level.[15,16]

29.1.2.4 *Temporal coverage*

Models can also be classified on the basis of time scale considered for plan implementation. Accordingly, on the basis of temporal coverage of planning, the models can be classified as short-term (up to 10 years), medium-term (up to 20 years), and long-term (beyond 20 years). The model addressing energy plan at local, regional and national level can be short-term, medium-term or long-term. Short and medium-term models follow the bottom-up modeling approach. Long-term models consider either the bottom-up or top-down modeling approach. Very-long-term models (100 to 300 years) always adopt the top-down modeling approach, as for global models assessing impacts of global green house gas emission and atmospheric chemistry.

29.1.3 *Review of energy planning methods*

A review of renewable energy planning models is presented in this section with an aim to identify the energy planning methodology, objectives and constraints considered for macro and micro-level planning. Several researchers have reported use of computer-based optimization and simulation models in renewable energy planning. Also, a number of optimization and simulation models have been developed for renewable energy allocation at both the macro (national level) and micro (local level) level of energy planning.

29.1.3.1 *Macro-level energy planning*

Mezher *et al.*[33] has developed a macro level energy planning model for energy resource allocation. The energy resource allocation has been carried out by a multi objective goal programming technique for Lebanon from two points of view: economy and environment. The economic objectives considered were costs, efficiency, energy conservation, and employment generation. The environmental objectives considered were environmental friendliness factors. The objective functions were expressed as mathematical expressions and multi objective allocation was carried out using a pre-emptive goal programming technique. Later the authors demonstrated the use of the fuzzy programming approach for energy resources allocation.[8] The similar fuzzy multi objective approach for energy allocation for cooking in UP households (India) was also demonstrated by Agrawal and Singh.[1] Suganthi and Samuel[55] presented a macro level energy forecasting model for energy, economy, and environment considerations. Their model was based on a two stage least square principle to calculate future energy requirement. The requirement calculated was then used in the MPEEE (Mathematical Programming Energy-Economy-Environment) model developed by the authors. The developed model seeks to maximize the GNP-energy ratio subjected to the constraints such as limits imposed on the emissions of CO_2, SO_2, NO_2, total suspended particles, CO and volatile organic compounds. The authors concluded that the GNP-energy ratio is closely related to energy efficiency. Iniyan and Sumathy (2000) presented the top-down approach based on the optimal renewable energy model (OREM) to minimize the cost/efficiency ratio. The potential of renewable energy sources, energy demand, reliability of energy system, and social acceptance of energy sources were considered as constraints for optimization. The performance and reliability of the wind energy system and its effects on the OREM model has been presented by Iniyan *et al.*.[22] Ghosh *et al.*[18] developed a methodology by using the bottom-up modeling approach. The approach takes into account the linkage between renewable energy and carbon emissions. The authors have assessed the potential for mitigation of CO_2 emissions in the power sector. Mihalakakou *et al.*[34] presented the scope for the utilization of renewable energy sources in the Greek islands of the South Aegean Sea area. The authors have developed alternative scenarios such as (i) energy conservation scenarios and (ii) exploitation of renewable energy sources, for meeting the future energy demand. The energy conservation scenarios emphasize rational use of energy in all sectors of economy and promotion of combined heat and power systems. In renewable energy sources scenarios, maximizing the use of renewable energy sources is considered. Zouros *et al.*[63] presented an integrated tool for the comparative assessment of alternative regulatory policies, along with a methodology for decision-making, on the basis of evaluation of alternative scenarios for social welfare functions.

29.1.3.2 *Micro-level energy planning*

Ramakumar *et al.*[41] presented a micro level linear programming approach for the design of integrated renewable energy systems for developing countries. The objectives of energy resource allocation were-cost of energy and energy conversion efficiencies. Later the same authors modified their approach by considering the prediction of energy resources for utilization.[2] Optimization models have been modified by researchers to Indian conditions, for modeling renewable energy systems. Sinha and Kandpal[51] had developed a linear programming model for determining an optimal mix of technologies for domestic cooking in the rural areas of India. Similar exercises have been done for irrigation[52] and lighting.[53] Minimizing cost was chosen as the objective in all cases. Joshi *et al.*[25] developed a linear programming model for decentralized energy planning for three villages in Nepal. The authors have presented results on optimizing the use of energy sources subject to constraints of energy conversion efficiency of different end-uses, resource availability, and the cost. Ramanathan and Ganesh (1993) developed a multi objective goal programming model for energy resource allocation at the micro level for Madras city in India. The objective functions chosen for the energy resource allocation were: minimization of cost, use of petroleum products, CO_x, NO_x and SO_x emissions, and maximization of system efficiency, use of locally available resources, and employment generation. Using a pre-emptive goal programming technique, the optimal energy resource allocation was carried out. Later the same authors used the Analytical Hierarchical Process (AHP) technique to allocate priorities to the objective functions.[43] The authors considered the aggregate option of economists, environment analysts and local people to allocate priority to the objective functions. Srinivasan and Balachandra presented a micro-level, bottom approach based linear programming model for Bangalore North Taluk. Devdas[12–14] presented an approach for renewable energy planning at the micro level in a Kanyakumari District of Tamilnadu, India, and also identified the important parameters which control the economy of rural system, particularly in relation with energy inputs and outputs. Cormio *et al.*[10] presented a bottom-up approach for formulating energy planning policies to promote use of renewable energy sources. Weisser[60] evaluated the costs of renewable electricity under various scenarios. The scenarios considered for the study were (i) business as usual, based on the power generation capacity expansion plan, (ii) hybrid, based on the assumption that the installed capacity of conventional power generation will remain at the same level, and to meet estimated future electricity demand co-fired biomass/waste combustion burners will be used, and (iii) renewable energy technology, based on the incorporation of renewable energy technologies in the existing electric power systems in Rodrigues, Mauritius.

29.1.4 *Review of energy planning models*

There are various models available to assess economic and environmental benefits of different supply and demand options in energy planning, both at the macro and micro level. The most widely used energy planning models are classified in the Table 29.1.

Cosmi *et al.*[11] evaluated the feasibility of use of renewable energy sources on a local scale, as per the European Union energy policies, which foster their utilization in member states. The authors presented an application of the R-MARKAL model to investigate the feasibility of renewable energy use for electricity and thermal energy generation. Energy planning models are also used in the Indian context. Market Allocation, (MARKAL) model has been applied by Shukla[49] to Indian renewable energy systems. The author has used MARKAL to analyze the technologies, peak electricity demand, carbon taxes and different energy scenarios. Carbon taxes and emission permits were analyzed using SGM. Mathur[32] presented the use of MARKAL for the energy-environment analysis of the Indian power sector. Several scenarios were developed such as (i) base case, by assuming unconstrained

Table 29.1.   Recommended paradigms for addressing issues in energy planning.

Paradigm	Issues addressed in energy planning	Spatial coverage	Sectoral coverage	Temporal coverage
Top-down simulation	Impact of market measures and trade policies on cost to economies and global/ national emissions	Global, national	Macro-economy/ Energy	Long-term
	Impact of market structure, competition and uncertainties on capacity investment, technology-mix, cost to consumers and emissions			
Bottom-up optimization/ Accounting	Impact of market measures and other policies such as regulations on technology-mix, fuel-mix, emissions and cost to energy systems; capacity investment planning	National, regional	Energy	Long-term
Bottom-up optimization/ Accounting	Impact of sectoral policies on sectoral technology-mix, fuel-mix, cost and emissions; planning for generation mix; unit scheduling; logistics	National, regional, local	Energy	Medium-term, short-term

development in power generation capacity except for presence of upper limits for renewable energy technologies as per their potential, (ii) bound growth, by the form of technological innovations, (iii) learning technologies, in the form of past experience, and (iv) bound growth with learning technologies. The developed scenarios were analyzed by cost minimization and by CO_2 taxations.

The review presented above show that a micro-level planning model is suitable for energy planning within a region. It is found that renewable energy optimization models generally deal with maximizing output, income, utilization of energy resources, profit, demand, performance of energy system or energy production. In the case of minimization, overall cost, energy system cost or capital investment are to be considered. The constraints considered are limitation of technology, supply, demand, efficiency, resource availability or installed system capacity. In addition, to above, in recent years, there are certain other factors gaining importance in favor of large-scale utilization of renewable energy sources. For instance, there is a certain amount of emission from renewable energy utilization. Secondly, when it comes to large-scale utilization, the installation of renewable energy sources needs a workforce for construction and maintenance, etc. Hence, the bottom-up approach for developing the multi-objective linear programming model is demonstrated considering the above parameters, critical for utilization of renewable energy sources in existing energy system.

29.1.5 *Review of energy decision-making methods*

A review of renewable energy decision-making methods is discussed in this section, aiming at identification of appropriate decision-making methodology suitable for micro-level planning energy planning.

Multi Criteria Decision-Making (MCDM) is a branch of operations research models which deal with decision-making subject to decision criteria. The MCDM methods are classified into two categories: multi-objective decision-making (MODM) approach and multi-attribute decision-making (MADM) approach.[9,58] The MCDM techniques have been widely used in renewable energy planning for ranking of developed scenarios.

The decision problem in MODM is solved using multi-objective linear or non-linear mathematical programming models in which several objective functions are considered and optimized, subject to a set of constraints. In MADM, each planning or design strategy is associated with a set of attributes whereby various planning or design strategies can be compared. MADM is preferred when the criteria are qualitative in nature and MODM is preferred when criteria can be quantified. MODM problems are defined and solved by several alternative optimization models, such as compromising programming,[28,30,59] constraint method, goal programming, and fuzzy multi-objective programming.[62] For MADM problems, the utility function method,[26] trade-off analysis method[17,61] and analytical hierarchy process method.[41,40,46,47]

29.2 Energy Requirement and Energy Resource Estimations

Energy use patterns are closely linked to agro-climatic and socio-economic conditions. Energy problems in rural areas are also closely linked to soil fertility, landholding, livestock holding, etc. Energy planning of any region should be based on the existing levels of energy consumption. However, the information available in published form is either at the state level or at the national level. Devdas[12] highlighted that the regional developmental activities have to be based on detailed information from each sector. Hence, a detailed energy survey was conducted by visiting and consulting local people, to understand the household and agricultural energy use patterns in various socio-economic zones. For this purpose, a survey was conducted to investigate household and agricultural energy consumption due to cooking, lighting, pumping, cooling, heating and appliances in the identified villages.

29.2.1 *Need and methodology of survey*

The Jhunjhunu district located in Northern part of Rajasthan, India, is one of the prosperous districts of Rajasthan with an area of 5929 sq. kms (latitude 28.060 N, longitude 75.200 E). Most parts of the district fall under the semi-arid zone of desert.[6] The main energy resources in the Jhunjhunu district are traditional fuels, mainly fuelwood, agricultural residue and dung. The households in rural areas of the Jhunjhunu district possess their own land, cultivate mustard and wheat as the major crops and are dependent on agricultural revenue. Preliminary investigations were conducted in six villages from the Jhunjhunu district for household and agricultural energy needs, such as energy resource availability, accessibility, technological support and local cooperation/support by visiting the villages to choose the region for investigation. After preliminary investigations, *Panthadiya* village, a village in *Morva* Panchayat, is identified as a study village for estimating household and agrivculture end-use behavior. The *Panthadiya* village is also considered to study intra-village energy use mix with neighboring villages namely *Bisanpura 1st*, *Bisanpura 2nd*, and *Morva*. The detailed energy survey was then conducted in *Panthadiya*, consisting mainly of secondary and primary data. The secondary data is collected from the respective government offices and is used to prepare the framework for the primary survey. The energy needs were estimated for various household and agricultural end-uses such as cooking, pumping, heating, cooling, lighting and appliances.

29.2.1.1 *Collection of secondary data*

Most of the secondary data such as landholding, demography, and livestock population, was collected from government offices. Landholding particulars for each household were collected from the Village Accountants' offices (Patwari). Data on village wise demography and occupational and infrastructural facilities was collected from the Tahsildar's office at Chirawa. Data on livestock population was collected from the veterinary departments of the sub-tahsil at Surajgadh and tahsil at Chirawa. Table 29.2 shows information-related to demography and livestock of the surveyed villages.

The secondary data available with government offices is compiled over a period of several years by the concerned officials. Furthermore, data regarding several aspects having an important bearing on rural energy planning are not readily available in the published statistics. Hence, the survey was conducted for the household energy needs using multi-stage schedules for the present investigation.

Table 29.2. Demographic information and livestock population of surveyed villages (Source: Census, 2001).

Name of the village	Population	Total land (in hectors)	Irrigated land (in hectors)	Number of cows	Number of buffalos	Number of camels
Panthadiya	1483	522	481	366	340	36
Bisanpura 1st	647	190	169	239	111	11
Bisanpura 2nd	717	185	167	112	242	20
Morva	2477	857	733	341	672	60

The secondary data was analyzed to select households by stratified sampling, based on landholdings and community, for the energy survey. This survey was conducted during December 2004 to October 2005, which is considered to be the base year for this case (2004–2005). Table 29.3 shows the variation in secondary and primary data for a number of households in the surveyed villages. The variation in the data is attributed to the methodology adopted for estimating the number of households. The secondary data for the number of households as reported in the Census 2001, is calculated on the basis of landholdings as per the government records. It is observed that, in most of the families, operational landholdings records available in government offices are not updated. Therefore, primary data on the actual number of households is estimated by consulting Sarpanch and senior citizens of respective villages. The estimated primary data on the number of households is then used to conduct the survey.

It is observed that the population increases exponentially (Population reference bureau, 2005). In order to estimate the population of villages in the basis year, i.e., 2004–2005 for energy resource allocation, exponential regression analysis for the population in the year 1991 and 2001 is performed. Exponential regression analyses

Table 29.3. Variation in secondary and primary data for number of households.

Name of the village	Number of households as reported in Census 2001	Number of households estimated after survey	Population as reported in Census 1991	Population as reported in Census 2001
Panthadiya	182	240	1154	1483
Bisanpura 1st	95	101	538	647
Bisanpura 2nd	86	135	670	717
Morva	327	355	2088	2477

resulted in correlations. The exponential growth observed in the last decade, i.e., 1991–2001 is used as a basis for estimating present (i.e., 2005) and future (i.e., 2010 and 2015) populations of surveyed villages.

29.2.1.2 *Collection of primary data*

The secondary information was analyzed to select households for stratified sampling (based on landholdings and community) for the energy survey. Households in the village were categorized into landless, small, medium and large farmers based on the landholdings. Under each category, households were grouped community wise, and samples were selected from each category.

29.2.2 *Investigation of energy consumption patterns*

The classification adopted for the survey based on landholding is: (i) landless, (ii) small farmers (0 ± 1 ha), (iii) medium farmers (1 ± 2.5 ha), (v) large farmers (2.5 ± 5 ha) and (vi) very large farmers (> 5 ha), keeping in view the fragmented landholding scenario of the village. Table 29.4 shows the demographic information of the surveyed village and distribution according to socio-economic distribution of the village. The number of households is estimated by consulting Sarpanch and senior citizens of respective villages. Population to cattle ratio, as observed in Census 2001, is used as a basis to estimate the number of cattle available in the base year, i.e., 2005. The estimated number of cattle in the surveyed villages is shown in the table. It can be seen that the percentage of small and medium farmers is larger, followed by large farmers, very large farmers, and landless farmers in the surveyed villages. The percentage of small and medium farmers are more in the surveyed villages due to their separation from main family.

29.2.2.1 *Factors influencing economy in the surveyed villages*

29.2.2.1.1 Distribution of households

Figure 29.1 shows the distribution of number of households against the size of the family for *Panthadiya* village. It is seen that the average size of the family is about five in all categories. The relatively flat family size distribution for very large and landless farmers is possibly due to their population in the village.

29.2.2.1.2 Population distribution and landholding

Population is an important parameter having direct impact on energy consumption, demand and supply of energy in rural regions. Agriculture being the major source of income in the study area, the size of the operational landholdings is an important

Table 29.4.   Demography of surveyed villages.

Name of the village	Landless	Small farmers $(0 \pm 1\,ha)$	Medium farmers $(1 \pm 2.5\,ha)$	Large farmers $(2.5 \pm 5\,ha)$	Very large farmers $(> 5\,ha)$	Number of cattle
Panthadiya	9 (3.75%)	81 (33.75%)	103 (42.92%)	37 (15.41%)	10 (4.17%)	820
Bisanpura 1st	0 (0.00%)	58 (57.43%)	33 (32.67%)	8 (7.92%)	2 (1.98%)	389
Bisanpura 2nd	4 (2.96%)	48 (35.56%)	54 (40.00%)	19 (14.07%)	10 (7.41%)	384
Morva	5 (1.41%)	128 (36.06%)	121 (34.08%)	77 (21.69%)	24 (6.76%)	1149

Fig. 29.1. Households distribution by family size for *Panthadiya* Village.

parameter, which determines the demand and supply of energy, and the distribution of energy consumption. Hence, it is essential to consider the population distribution in the rural region in terms of the size of farms. For the present case, the sample households were divided into five categories, (i) landless, (ii) small farmers (0±1 ha), (iii) medium farmers (1 ± 2.5 ha), (iv) large farmers (2.5 ± 5 ha) and (v) very large farmers (>5 ha). Figure 29.2 shows the population distribution according to landholding. It has been observed during the survey that, households having larger operational landholding are found to consume a larger quantity of energy while the reverse is the case with households having marginal operational landholding.

29.2.2.1.3   Cropping pattern and irrigation intensity

Irrigation intensity for cropping is an important factor which determines the energy requirement in agricultural operations. In the study area, the field crops are cultivated

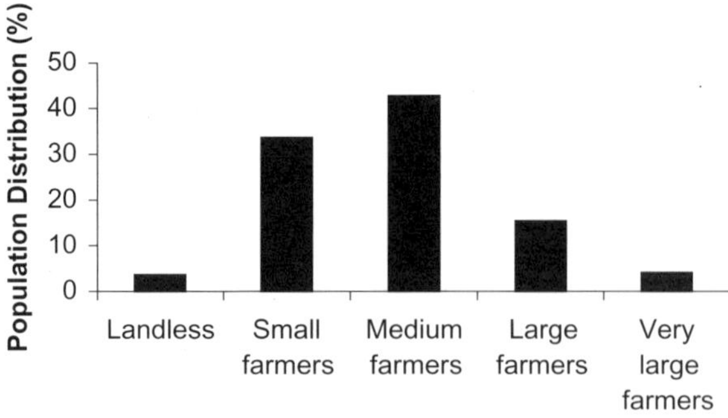

Fig. 29.2. Population distributions according to landholdings for *Panthadiya* Village.

in one, two or three seasons depending upon the infrastructure facilities. For a farm cultivating field crops in more than one season, the energy demand would increase accordingly. Energy consumption for irrigation purely depends on the nature of the water source and the irrigation methods employed. The major source of irrigation in the study area is ground water. Since, the ground water is available below 250 feet, installation of electric pump sets lead to an increased consumption of commercial energy. The electric pumps which are used in the region are of 12.5 hp capacity.

29.2.2.1.4   Crop residues

The ratio of the main product to the by-product varies due to differences in verities of crop, cultivation practices, and application of different types of technologies at the farm level (Ramchandra, 2001). The ratio for crop residues is calculated on the basis of the production of the main products and their by-products for analysis. The main crops in the study area are mustard, and wheat. It is observed that residue of mustard is used for cooking and heating end-uses.

29.2.2.2   *Equations employed for calculating the end-use energy requirements*

The detailed survey questionnaire is developed to collect relevant data for various end-use energy requirements per household. To calculate energy requirement per household, a primary survey is conducted by visiting these villages. End-use energy requirements are categorized as cooking, lighting, pumping, heating, cooling, and appliances.

The primary survey has considered only six important end-uses and for each end-use commonly used devices have been considered. The equations used to compute the energy requirements for device-end use combination are as follows:

Energy consumption = (Number of devices used) × (energy consumed for 1 hour of usage) × (Average number of hours of usage of the device) × (Number of days of usage in a year).

Computation of Per Capita Energy Consumption (PCEC)
Per capita energy consumption is calculated by using following formula.

$$PCEC = EC/p$$

where

EC = energy consumed per day and p = number of adult equivalents, for whom the energy is used.

The data is grouped based on landholding category. Then the average values for each end-use are calculated.

29.2.2.3 *End-use energy requirements of surveyed village*

The average estimated energy requirement per person for each end-use in the surveyed villages is calculated using equations given in the previous section and the results of analysis is shown in Table 29.5.

29.2.3 *Estimation of energy resource availability*

The survey questionnaire was also designed to estimate energy resource availability of the villages. Biogas availability in the village is calculated on the basis of number of cattle and the dung available. It is assumed that on an average, two cattle can provide 25 kg of dung or 1 m^3 of biogas or 20.14 MJ of energy per day.[43] During the field visits, it is also observed that in most of the families 10–25% dung available is used for making dung cakes and 75–90% is then used in agriculture. But the cooking pattern of the region indicates that the dung cakes are not fully consumed for the cooking and heating applications. Therefore, it is assumed that 15% of the dung cakes produced are used for cooking and heating applications. The remaining 85% of dung available can be utilized for biogas production. Table 29.6 shows the estimated dung available, dung cake consumption and biogas availability per day for the surveyed villages.

Biomass energy resource is calculated on the basis of amount of firewood and agricultural residue consumed in a socio-economic group and its corresponding calorific value. The availability of firewood is calculated on the basis of the number of *Jati* trees in the area and their annual yield. It is noted that one *Jati* tree yields is around 50 kg/year and there are on an average 25 trees/hector. Therefore, the number of *Jati* trees in the village can be estimated and their yield is shown in the table. It is observed that along with the firewood, agricultural residue of Mustard is also used for cooking and heating endues. The Mustard is cultivated in 35% of the land available in the region and residue produced per hector of Mustard cultivated is 200 kg. It is also observed that most of the times the biomass requirement is met locally. On an average, 1 kg (dry) of biomass produces 1.66 to 2.101 m^3 of producer gas in a circulating fluidized bed gasifier and the calorific value of the gas generated varies from 6.94 to 7.26 MJ/m^3. Table 29.7 shows the estimated biomass energy availability per year for the surveyed villages.

The estimated actual energy requirement for household and agriculture end-uses and energy resource availability of the surveyed villages is used further for energy resource allocation.

Table 29.5.   Estimated actual end-use energy requirements for surveyed villages.

	Panthadiya		Bisanpura 1st		Bisanpura 2nd		Morva	
Energy end-use	Energy reqd per person, per day, kWh	Annual energy reqd MWh/yr	Energy reqd person, per day, kWh	Annual energy reqd MWh/yr	Energy reqd per person, per day, kWh	Annual energy reqd MWh/yr	Energy reqd perperson, per day, kWh	Annual energy reqd MWh/yr
Cooking	1.495	0.895×10^3	1.499	0.381×10^3	1.463	0.394×10^3	1.455	1.409×10^3
Lighting	0.10	0.060×10^3	0.14	0.036×10^3	0.12	0.032×10^3	0.11	0.107×10^3
Pumping	—	0.790×10^3	—	0.286×10^3	—	0.327×10^3	—	0.980×10^3
Heating	0.0002	0.120	0.0002	0.051	0.0002	0.054	0.0002	0.194
Cooling	0.212	0.127×10^3	0.344	0.088×10^3	0.274	0.074×10^3	0.207	0.200×10^3
Appliances	0.055	0.033×10^3	0.082	0.021×10^3	0.061	0.016×10^3	0.052	0.050×10^3

Table 29.6. Estimated dung-cake consumption and biogas availability in surveyed villages.

Name of the village	Number of cattle	Dung available kg/year	Dung cake consumption MJ/year	Biogas availability MJ/year
Panthadiya	820	3.74×10^6	5.61×10^6	2.56×10^6
Bisanpura 1st	389	1.78×10^6	2.67×10^6	1.22×10^6
Bisanpura 2nd	384	1.75×10^6	2.63×10^6	1.20×10^6
Morva	1149	5.24×10^6	7.86×10^6	3.59×10^6

Table 29.7. Estimated biomass energy available in surveyed villages.

Name of the village	Total land (in hectors)	Irrigated land (in hectors)	Firewood availability, tons/year	Agricultural residue, tons/year	Biomass energy available, MJ/year
Panthadiya	522	481	602	58.92	9.26×10^6
Bisanpura 1st	190	169	212	20.70	3.26×10^6
Bisanpura 2nd	185	167	209	20.46	3.21×10^6
Morva	857	733	917	89.79	14.10×10^6

29.3 Energy Resource Allocation

Renewable energy resources play a significant role in supplying the energy needed in the rural region of the developing countries for improving the living environment and for economic development. To design Integrated Renewable Energy System (IRES) at micro-level, the region in which energy needs are both for thermal and electrical applications are particularly suitable for design of IRES at micro-level. Moreover region should be rich in resources both renewable and conventional.

29.3.1 *Integrated energy system model development*

The methodology adopted involves the development of a model for optimal energy resource allocation for different end uses. The resource, end use, and their combination are chosen on the basis of the availability of data and the feasibility of resource utilization in the surveyed region in Northern parts of Rajasthan. In all, eleven energy resources and six end-uses have been considered, and forty-one resource-end-use combinations have been chosen as shown in Table 29.8.

Table 29.8.   Energy Resource — end-use combinations.

Energy Resources	Cooking	Lighting	Pumping	Heating	Cooling	Appliances
Dung cake	1	—	—	23	—	—
Biomass	2	—	—	24	—	—
LPG	3	—	—	—	—	—
Kerosene	4	12	—	—	—	—
Biogas	5	13	—	25	—	—
Solar Thermal	6	—	—	26	—	—
Biogas electricity	7	14	19	27	32	37
Biomass electricity*	8	15	20	28	33	38
PV electricity	9	16	–	29	34	39
Diesel electricity	10	17	21	30	35	40
Grid electricity	11	18	22	31	36	41

The optimization model aims at minimization of cost, usage of petroleum products, CO_x SO_x, NO_x emissions and maximizes system efficiency, use of local resources, employment generation, social acceptance of resources, and reliability of the system. The constraints are available potential of energy resources and end-use energy requirements in the form of cooking, lighting pumping, heating, cooling, and appliances. In addition to these constraints operational constraints are also considered for the use of solar thermal and dung cakes for cooking end-use. In general the mathematical representation of model includes defining objectives as minimization or maximization and represented as:

$$\sum (R_i X_i). \qquad (29.1)$$

29.3.2 *Data requirement in the model*

The unit cost of energy used in the optimization model is taken from current published data. The cost of solar photovoltaic electric conversion is estimated to be Rs. 15/kWh and diesel electricity is estimated to be Rs. 15/kWh (based on present cost of diesel) and the cost of grid electricity for household and agriculture applications is estimated to be Rs. 3/KWh and Rs. 0.75/kWh, respectively; the cost of the biomass gasifier electric conversion and biogas electric conversion systems is estimated to be Rs. 2.50/kWh and Rs. 1.25/kWh, respectively; the cost of dung, biomass, LPG, and kerosene is estimated as Rs. 1/kg, Rs. 3/kg, Rs. 21.8/kg, and Rs. 10/lit, respectively. The energy system efficiency is calculated by multiplying the external efficiency of energy source and the end-use device efficiency. The external efficiency is the energy source efficiency just before the end-use point. The energy system efficiencies used in the model are: 12% for solar photovoltaic system, 23% diesel electric system, 18.40% grid power system, 21.89% biomass gasifier conversion system, 28.16% for the biogas electric conversion system, 16.15% biomass direct combustion, 40% solar direct thermal, 44% biogas system, 32.40% kerosene system and 36% for LPG system. The reliability factor of 0.1 at 10,000 hours for solar photovoltaic system, 0.9 at 10,000 hours for biomass energy and 0.9 at 10,000 hours for biogas energy system is used in the model as reported by Iniyan *et al.*[22] The number of people employed in developing various energy resources, along with the total consumption of energy resources is used for estimation. The social acceptance factor for solar, biomass/biogas, and commercial energy sources are 7.12, 10.49, and 74.49, respectively, is used in the model. Stoichiometry quantity of pollutants per weight of fuel is used to estimate the emission rates in kg/kWh.

Based on the objectives and constraints the multi objective goal programming model has been built and is discussed below:

$$\text{Minimize} \sum d_j^- + d_j^+, \quad (j = 1, 2, \ldots, 10). \tag{29.2}$$

Subject to,

$$Objective\,Function_j + w_j d_j^- - w_j d_j^+ = b_j, \tag{29.3}$$

where

d_j^- and d_j^+ are the underachievement and over-achievement of the goal, respectively.

Each of the objective function is referred as the goal for the optimization. First, all the objectives are individually optimized, and the optimum value for each of the objectives are fixed as the corresponding goal b_j. The worst possible value, i.e., minimum value for the maximization objectives and maximum value for the minimization objective for the objective function is calculated and referred as L_j. Then the weighing factor w_j for each of the goal is calculated as the difference in the value of goals and the worst value of the goals.

29.3.3   *Development and selection of scenario for implementation*

The optimal scenario is described in terms of goal values for individual objective functions by maximization or minimization. The goal value for an objective function is obtained by optimizing each objective function individually by linear programming technique. Next, the multi-objective optimal scenario is obtained by optimizing all objective functions simultaneously by the pre-emptive goal programming method. In this method, weighting factors for individual objective function are determined. The weighting factor for an objective function is the difference between

Table 29.9.　Present energy resource consumption pattern for various end-uses.

	End-uses					
Scenario	Cooking	Lighting	Pumping	Heating	Cooling	Appliances
Present energy consumption scenario	1. Dung cake (15%) 2. Biomass (70%) 3. LPG (15%)	Grid electricity (100%)	Grid electricity (100%)	1. Dung cake (20%) 2. Biomass (80%)	Grid electricity (100%)	Grid electricity (100%)

the goal value and the goal obtained by reversal of optimization, i.e., for maximization to minimization or minimization to maximization. The goal value obtained by reversal of optimization from maximization or minimization is called the worst value. Table 29.9 shows the present energy consumption pattern in the study village and is used for the comparison.

29.3.3.1　*Alternate energy scenarios*

Scenario 1 — *Equal Priority Scenario*: In this scenario, all the objective functions are taken into account while arriving at the energy resource allocation. This scenario is developed without assigning priority to objective functions. The optimal energy resource allocation pattern is shown in Table 29.10. The results of optimization without assigning priority to objective function show that use of biomass, LPG, solar thermal and PV electricity should be promoted for cooking end-use, PV electricity for lighting, cooling and appliance end-uses, biomass electricity for pumping end-use, and solar thermal for heating end-use. Energy resource allocations in scenario 1 also show that biomass can meet 30%, LPG can meet 37.65% and PV electricity can met 12.35% of total cooking energy requirement. Similarly, PV electricity can meet 100% of lighting, cooling and appliance end-use requirement.

The associated cost and emission with scenario 1 are tabulated for comparing it with the present energy scenario as shown in Table 29.11. The comparison of the present and scenario 1 show that the cost associated with scenario 1 is almost two and half times the present cost of energy consumption, and the associated emissions are reduced. The selection of energy scenario is primarily guided by the cost incurred, and also by avenues for higher employment generation, use of local resources, and associated emissions. Since, the cost associated is higher with this scenario, therefore scenario 1 should not be promoted for implementation. In order to implement this scenario, the cost of PV electricity should be decreased. Hence, different scenarios

Table 29.10.   Energy resource allocation for *Panthadiya* village in different scenarios.

End-uses	Present energy consumption scenario	Energy Consumption scenarios					
		Scenario 1	Scenario 2	Scenario 3	Scenario 4	Scenario 5	Scenario 6
Cooking	1. Dung cake (15%) 2. Biomass (70%) 3. LPG (15%)	1. Biomass (30.00%) 2. LPG (37.65%) 3. Solar Thermal (20%) 4. PV electricity (12.35%)	1. Dung cake (13.30%) 2. Biomass (48.83%) 3. PV electricity (37.87%) 1. Biomass (27.60%) 2. Solar thermal (20.00%) 3. PV electricity (52.40%) 1. LPG (33.41%) 2. Solar Thermal (20%) 3. Biomass electricity (39.44%) 4. PV electricity (7.15%)	1. Biomass (6.59%) 2. Solar Thermal (20%) 3. Biomass electricity (73.41%) 1. Solar thermal (20%) 2. Biomass electricity (80%) 1. Solar Thermal (20%) 2. Biomass electricity (80.00%)	1. Biomass (22.24%) 2. LPG (57.76%) 3. Solar thermal (20%) 1. Biomass (22.24%) 2. LPG (57.76%) 3. Solar thermal (20%) 1. Biomass (22.24%) 2. LPG (18.10%) 3. Biogas (39.66%) 4. Solar thermal (20%)	1. Biomass (17.32%) 2. LPG (22.96%) 3. Biogas (39.72%) 4. Solar thermal (20%)	1. Biomass (17.32%) 2. LPG (22.96%) 3. Biogas (39.72%) 4. Solar thermal (20%) 1. Biomass (17.32%) 2. LPG (22.96%) 3. Biogas (39.72%) 4. Solar thermal (20%)

Table 29.10. (*Continued*)

		Energy Consumption scenarios					
End-uses	Present energy consumption scenario	Scenario 1	Scenario 2	Scenario 3	Scenario 4	Scenario 5	Scenario 6
Lighting	Grid elect. (100%)	PV elect. (100%)	PV elect. (100%) PV elect. (100%) PV elect. (100%)	Biomass electricity (100%) Biomass elect. (100%) Biomass electricity (100%)	Biomass elect. (100%) Biomass elect. (100%) Biomass elect. (100%)	Biomass elect. (100%)	Biomass elect. (100%) Biomass elect. (100%)
Pumping	Grid elect. (100%)	Biomass electricity (100%)	Grid electricity (100%) Grid electricity (100%) Biomass electricity (100%)	Biomass electricity (100%) Biomass electricity (100%) Biomass electricity (100%)	Biomass electricity (100%) Biomass electricity (100%) Biomass electricity (100%)	Biomass elect. (100%)	Biomass elect. (100%) Biomass elect. (100%)
Heating	1. Dung cake (20%) 2. Biomass (80%)	Solar Thermal (100%)	Solar thermal (100%)	Solar thermal (100%)	Solar thermal (100%)	Solar thermal (100%)	Solar thermal (100%)

Table 29.10. (*Continued*)

Energy Consumption scenarios

End-uses	Present energy consumption scenario	Scenario 1	Scenario 2	Scenario 3	Scenario 4	Scenario 5	Scenario 6
			Solar thermal (100%)	Solar thermal (100%)	Solar thermal (100%)		Solar thermal (100%)
			Solar thermal (100%)	Solar thermal (100%)	Solar thermal (100%)		
Cooling	Grid elect. (100%)	PV electricity (100%)	PV electricity (100%)	Biomass elect. (100%)	Biomass elect. (100%)	Biomass elect. (100%)	Biomass elect. (100%)
			PV electricity (100%)	Biomass elect. (100%)	Biomass elect. (100%)		Biomass elect. (100%)
			PV electricity (100%)	Biomass elect. (100%)	Biomass elect. (100%)		
Appliances	Grid elect. (100%)	PV elect. (100%)	PV elect. (100%)	Biomass elect. (100%)	Biomass electricity (100%)	Biomass elect. (100%)	Biomass elect. (100%)
			PV elect. (100%)	Biomass elect. (100%)	Biomass electricity (100%)		Biomass elect. (100%)
			PV elect. (100%)	Biomass elect. (100%)	1. Biomass elect. (99%) 2. PV elect. (1%)		

Table 29.11. Comparison of cost and emission for present energy consumption scenario and different scenarios.

		Cost associated, million Rs/year	Emissions associated		
			CO_x, Tons/year	SO_x, Tons/year	NO_x, Tons/year
Present energy consumption scenario		5.38	3829.05	3.66	27.05
Scenario 1	Case 1	13.16	1630.82	0.02	3.02
		15.43	2679.47	2.49	20.67
Scenario 2	Case 2	16.91	1405.79	1.58	7.04
	Case 3	11.43	618.83	0.02	1.04
	Case 1	5.89	1096.68	—	2.21
Scenario 3	Case 2	5.83	864.00	—	1.75
	Case 3	5.83	864.00	—	1.75
	Case 1	6.20	1615.20	0.04	2.86
Scenario 4	Case 2	6.20	1615.20	0.04	2.86
	Case 3	5.17	1621.72	29.68	2.86
Scenario 5		5.06	1445.11	29.74	2.48
Scenario 6	Case 1	5.06	1445.11	29.74	2.48
		5.06	1445.11	29.74	2.48

are developed by varying the priority of objective functions to reduce the associated cost.

Scenario 2 — *Priority scenario*: In this scenario, the objective functions are divided into three categories: economic, security-acceptance and environmental. Under economic objectives cost of energy, system efficiency, reliability of energy system, and employment generation are considered; while under security-acceptance, minimization of imported petroleum products, maximization of local resources and social acceptance are considered. The environment related objectives include the minimization of CO_x, SO_x and NO_x. In this scenario, the priority of environment emissions is varied from one to three and the economic objectives have always been given higher priority as compared to security-acceptance objectives. The results of energy resource allocation are shown in Table 29.10.

Case 1: The results of optimization show that when environment objectives are given higher priority, PV electricity should be promoted for lighting, cooling and appliance end-uses since the energy source is emission free. There are no constraints on the availability of the solar energy in the village, since it is available most of the time during a year and is observed to be available for more than 270 days in a year. The analysis results show that grid electricity is only to be preferred for pumping end-use from the point of view of present subsidized prices.

Case 2: The results are almost similar as observed in case 1 except for the allocation of dung cake for cooking end-use. In this case biomass energy share in cooking energy requirement is reduced from 48.83% to 27.60%. Solar thermal and PV electricity is also allocated for cooking end-use due to decrease in the environmental priority from one to two.

Case 3: The results of optimization when economic objectives are given higher priority than security-acceptance and environment objectives show that large portions of LPG (33.41%) and biomass electricity (39.44%) is to be promoted for cooking. The results of optimization show that PV electricity (7.15%) should also be allocated for cooking end-use. Solar thermal with its low cost, will meet 20% of the cooking energy requirement, and total heating energy requirement. Biomass electricity should be promoted for pumping, cooling and appliance end-uses and PV electricity for cooling end-use due to increase in priority to social-acceptance objectives.

The associated cost and emission with scenario 2 are tabulated for comparing it with the present energy scenario as shown in Table 29.11. The comparison of cost associated in present energy consumption scenario and scenario 2 show that the cost increases by many folds, when environment emissions are given priority and can be observed in case 1 and 2 as shown in the table. If the security-acceptance objectives are given more priority it also results in higher cost than the present energy consumption scenario. Therefore, these scenarios should only be preferred only when the reduction in environment emissions is the priority. In order to implement these scenarios, the cost of PV electricity should be decreased. Different scenarios are again developed to reduce the associated cost by varying the priority of economic objectives as discussed in scenario 3.

Scenario 3 — *Economic objective scenario*: In this scenario, changes are made within the priorities of economic objectives. Priority to cost objective function is varied from one to three, and the employment generation is always given higher priority as compared to efficiency and reliability. In this scenario, the other objective-functions have been given lowest priority. The results of the energy resource allocation are shown in Table 29.10.

Case 1: The results of optimization when energy cost is assigned the highest priority show that biomass and biomass electricity for cooking; and solar thermal for cooking and heating should be preferred, due to their low cost and higher potential for employment. Biomass electricity is to be promoted for lighting, cooling and appliance end-uses due to its low cost (Rs. 2.50/kWh) compared with other energy resources. Biomass electricity should be promoted for pumping end-use due the lower costs as Rs. 2.50/kWh and is local energy resource.

Case 2: The results of optimization when employment generation is assigned the higher priority than cost, results in the almost same energy resources allocation for the end-uses, except the use of biomass electricity for cooking in place of biomass. Therefore, a decrease in the priority of the cost function from one to two does not change the energy resource allocation.

Case 3: The results of optimization when employment generation is assigned the highest priority and cost is given the lower priority, as in case 3, show the similar energy resources allocation as observed in case 2. The biomass electricity is to be allocated for different end-uses, due to high employment potential in bio-energy resources at lesser cost. Therefore, a decrease in the priority of the cost function from one to three does not change the energy resource allocation.

The associated cost and emission with scenario 3 are tabulated for comparing it with the present energy scenario as shown in Table 29.11. The comparison of costs associated with the present energy consumption scenario and scenario 3, show that the cost and environmental emissions are reduced for all the cases. In all the cases, biomass electricity is to be promoted for lighting, pumping, cooling and appliance end-uses, which is due to the availability of biomass in the village. Therefore, the case 2 scenario should be preferred for implementation which will have higher employment generation potential due to the use of local available resources at the optimal cost. When the employment generation is assigned higher priority than reliability and efficiency of energy system, the cost associated in achieving scenario 3 increases as compared to the present energy consumption scenario. Therefore, this scenario should only be preferred when the employment generation is the priority. Different scenarios are again developed by varying the priority of security-acceptance objectives to find an acceptable scenario at the lower cost and higher employment generation options as discussed in scenario 4.

Scenario 4 — *Security-acceptance scenario*: In this scenario, the security-acceptance objectives functions are given the higher priorities and other objective-functions are given the lowest priority. The results of energy resource allocation are shown in Table 29.10. The results of optimization in case 1 and case 2 show that LPG and solar thermal is to be promoted for cooking energy requirements, since minimum use of petroleum products leads to maximum use of local resources. All the cases result in an almost similar energy resources allocation pattern for the end-uses, except for the use of biomass and biogas for cooking end-use. Therefore, an increase in the priority of the social acceptance factor from three to one does not greatly change the energy resource allocation.

The associated cost and emission with scenario 4 are tabulated for comparing it with the present energy scenario as shown in Table 29.11. The comparison of the present energy consumption scenario and scenario 4 shows that the cost associated in the cases 1 and 2 are higher than in the reference scenario, i.e., present energy consumption scenario, and the associated emissions are reduced. Therefore, these scenarios should only be preferred when the maximum use of local resources is the objective. The results of optimization when social acceptance and use of local resources objective is given higher priority than the use of petroleum products objective show the reduction in associated cost and environment emissions. It can be seen that the Sox emissions increases from 3.66 to 29.68 tons/year due to the allocation of biogas for cooking. Therefore, case 3 of the security-acceptance scenario should only be preferred when the social acceptance and use of local resources is the priority. Different scenarios are developed by assigning higher priority to cost and employment generation to find an acceptable scenario at the lower cost, and higher employment generation options as discussed in scenario 5.

Scenario 5 — *Cost-employment generation scenario*: In this scenario, cost and employment generation objective functions are given higher priority as compared to other objective functions. This scenario is important, where the objective of energy resource allocation is socio-economic development. The results of energy resource allocation are shown in Table 29.10. The results of optimization show that biomass, biogas and solar thermal should be promoted for cooking, and solar thermal for heating end-use. LPG (22.96%) is to be allocated for cooking due to the constraint of biogas and biomass energy resource potential. Biomass electricity is to be promoted for lighting, pumping, cooling, and appliance end-uses due to their high employment generation potential at the lower costs.

The associated cost and emission with scenario 5 are tabulated for comparing it with the present energy scenario as shown in Table 29.11. The comparison of the present energy consumption scenario and scenario 5 shows that the cost associated is lower than the present cost of utilization, and the associated emissions are reduced. Therefore, this scenario should be preferred only when the maximum use of local resources and employment generation are the objectives.

Scenario 6 — *Efficiency scenario*:

Case 1: In this scenario, the maximization of system efficiency is given the first priority and the other objective functions are given a priority of two. The results of energy resource allocation are shown in Table 29.10. The results of optimization resulted in similar results as observed in scenario 5 and show that, biomass electricity is to be promoted for lighting, pumping, cooling and appliance end-uses. The energy allocation is due to the system efficiency of 21.89%. Biomass and biogas for cooking, and solar thermal for cooking and heating is to be allocated due to their resource availability and associated system efficiency of 16.15%, 44% and 40%, respectively.

Case 2: In this case, due to technological advancement, an increase of 25% is assumed for all renewable energy sources. The optimization problem is carried for new values of system efficiency. The optimization results for energy resource allocation are shown in Table 29.10. The results of optimization for case 2 show that even though there is a 25% increase in system efficiency of renewable energy sources, it does not change the energy resource allocations as observed in case 1. Thus, the solution is found to be not sensitive to a 25% increase in the system efficiency.

The associated cost and emission with scenario 6 are tabulated for comparing it with the present energy scenario as shown in Table 29.11. The comparison of the present energy consumption scenario and scenario 6 shows that the cost and emissions associated, in all the cases, is still higher than the present cost of utilization. Therefore, these scenarios should not be preferred for implementation.

29.4   Region Dependent Development in Energy Planning

Energy resource allocation is important in IRES design and development. The energy resource allocation for the *Panthadiya* village (study village) is discussed in the previous section. There it is shown that the present cost of energy consumption and emissions can be reduced by implementing scenario 5. It is observed that biomass energy of the 10.04% of biomass energy is unutilized in scenario 5 and dung cake energy is not allocated in present energy resource allocation. Thus, unutilized biomass and dung cake energy is available for allocating to neighboring villages to identify regions for fast track development of IRES.

29.4.1   *Intra village-mix for IRES development*

The methodology adopted for energy resource allocation is the same as discussed in the previous section. The optimal energy scenarios are generated for *Bisanpura 1st, Bisanpura 2nd* and *Morva* villages. The estimated end-use energy requirements of the villages are shown in Table 29.12.

Also, the estimated biomass, dung cake and biogas energy resource availability in neighboring villages are shown in Table 29.13. It can be seen that the number of

Table 29.12.   Estimated end-use energy requirement for neighboring villages.

End-use	Energy requirement per person, per day, kWh for Bisanpura 1st	Annual energy requirement MWh/yr for Bisanpura 1st	Energy requirement per person, per day, kWh for Bisanpura 2nd	Annual energy requirement MWh/yr for Bisanpura 2nd	Energy requirement per person, per day, kWh for Morva	Annual energy requirement MWh/yr for Morva
Cooking	1.499	0.381×10^3	1.463	0.394×10^3	1.455	1.409×10^3
Lighting	0.14	0.036×10^3	0.12	0.032×10^3	0.11	0.107×10^3
Pumping	—	0.286×10^3	—	0.327×10^3	—	0.980×10^3
Heating	0.0002	0.051	0.0002	0.054	0.0002	0.194
Cooling	0.344	0.088×10^3	0.274	0.074×10^3	0.207	0.200×10^3
Appliances	0.082	0.021×10^3	0.061	0.016×10^3	0.052	0.050×10^3

Table 29.13. Estimated biomass, dung cake, biogas energy potential in neighboring villages.

Name of the village	Population	Irrigated Land (in hectors)	Biomass energy available, MJ/year	Number of Cattle	Dung cake consumption, MJ/year	Biogas availability, MJ/year
Bisanpura 1st	697	169	3.26×10^6	389	2.67×10^6	1.22×10^6
Bisanpura 2nd	737	167	3.21×10^6	384	2.63×10^6	1.20×10^6
Morva	2652	733	14.10×10^6	1149	7.86×10^6	3.59×10^6

cattle to population ratio is 0.59 and 0.52, which is higher than the ratio observed in *Panthadiya* village (0.50), due to which the biogas energy potential is higher in *Bisanpura 1st* and *Bisanpura 2nd* villages.

The results of optimization for neighboring villages show that biomass, biogas and solar thermal should be promoted for cooking, and solar thermal for heating end-use. LPG (29.425% in *Bisanpura 1st*, 33.86 in *Bisanpura 2nd* and 25.71% in *Morva* village) is to be allocated for cooking due to the constraint of biogas and biomass energy resource potential and is shown in Table 29.14. Biomass electricity is to be promoted for lighting, pumping, cooling, and appliance end-uses due to their high employment generation potential at the lower costs in all the neighboring villages.

Table 29.15 shows the unutilized energy resource in implementing an optimal scenario for neighboring villages in the base year (2005–2006).

It can be observed that in the *Bisanpura 1st* village the dung cake (100%) and biomass energy resource (15.34%) is not utilized in the optimal scenario. In *Bisanpura 2nd* village, the dung cake (100%) and biomass energy resource (18.69%) is not utilized in the optimal scenario. In *Morva* village dung cake (100%) and biomass energy resource (11.21%) is not utilized in the optimal scenario. Therefore, *Bisanpura 1st*, *Bisanpura 2nd* and *Morva* village is considered to study intra-village energy mix with *Panthadiya* village.

The results of optimization for the intra village mix show similar results and are shown in Table 29.16. The scenario analysis shows that biomass, biogas and solar thermal should be promoted for cooking, and solar thermal for heating end-use. LPG (21.40% to 26.42%) is to be allocated for cooking due to the constraint of biogas and biomass energy resource potential. The contribution of LPG for cooking end-use is observed to be proportional to the use of biogas. Biomass electricity is to be promoted for lighting, pumping, cooling, and appliance end-uses due to their high employment generation potential at the lower costs.

Table 29.14. Energy resource allocation pattern in optimal scenario for implementation in surveyed villages.

Optimal scenario	End-uses					
	Cooking	Lighting	Pumping	Heating	Cooling	Appliances
Panthadiya Village	1. Biomass (17.32%) 2. LPG (22.96%) 3. Biogas (39.72%) 4. Solar thermal (20%)	Biomass electricity (100%)	Biomass electricity (100%)	Solar thermal (100%)	Biomass electricity (100%)	Biomass electricity (100%)
Bisanpura 1st Village	1. Biomass (6.09%) 2. LPG (29.42%) 3. Biogas (44.49%) 4. Solar thermal (20%)	Biomass electricity (100%)	Biomass electricity (100%)	Solar thermal (100%)	Biomass electricity (100%)	Biomass electricity (100%)
Bisanpura 2nd Village	1. Biomass (3.75%) 2. LPG (33.86%) 3. Biogas (42.39%) 4. Solar thermal (20%)	Biomass electricity (100%)	Biomass electricity (100%)	Solar thermal (100%)	Biomass electricity (100%)	Biomass electricity (100%)
Morva Village	1. Biomass (18.91%) 2. LPG (25.71%) 3. Biogas (35.38%) 4. Solar thermal (20%)	Biomass electricity (100%)	Biomass electricity (100%)	Solar thermal (100%)	Biomass electricity (100%)	Biomass electricity (100%)

Table 29.15.   Estimated unutilized energy resources in optimal scenario for surveyed villages.

Name of the village	Estimated dung cake availability, MJ/year	Unutilized dung cake, MJ/year	Estimated biomass energy availability, MJ/year	Unutilized biomass energy availability, MJ/year
Panthadiya	5.61×10^6	5.61×10^6	9.26×10^6	0.93×10^6
Bisanpura 1st	2.67×10^6	2.67×10^6	3.26×10^6	0.50×10^6
Bisanpura 2nd	2.63×10^6	2.63×10^6	3.21×10^6	0.60×10^6
Morva	7.86×10^6	7.86×10^6	14.10×10^6	1.58×10^6

Table 29.17 shows the unutilized energy resource in implementing optimal energy scenario for intra village mix in the base year (2005–2006). It can be observed that the dung cake (100%) is not allocated in any intra village mix. Biomass energy resource 11.42%, 11.95%, 10.53%, 11.08%, 11.93% and 11.86% is not allocated in the optimal scenario for *Panthadiya–Bisanpura 1st*, *Panthadiya–Bisanpura 2nd*, *Panthadiya–Morva*, *Panthadiya–Bisanpura 1st-Bisanpura 2nd*, *Panthadiya–Bisanpura 1st Morva*, *Panthadiya–Bisanpura 2nd-Morva* and *Panthadiya–Bisanpura 1st-Bisanpura 2nd-Morva* village-mix, respectively. The dung cake energy resource is not allocated for cooking end-use due to its higher associated emissions (0.633 kg/kWh, 0.0013 kg/kWh and 1.709×10^{-2} kg/kWh for Carbon, Sulphur and Nitrogen, respectively). The biomass energy is unutilized due to more potential of energy resource.

In micro-level energy planning, the energy scenario when implemented is required to fulfill the objective of meeting energy requirements subject to constraints. These constraints correspond to resource availability, technology options, cost of utilization, environmental impact, socio-economic impact, employment generation, subject to present as well as future considerations. The success of the energy scenario depends on an accurate estimation of energy resource, energy demand and unutilized local energy resource. The quantum of unutilized local energy resource will make the plan successful.

The region for fast track IRES development is defined with respect to available energy sources and energy demand. The results of the intra-village mix for present energy requirements for different end-uses show that the dung cake energy resource is not preferred in optimal energy resource allocation. Moreover, the unutilized energy resource potential of biomass energy can be observed from Table 3.17 for the intra village-mix with *Panthadiya* village. The results of the intra-village mix show that *Panthadiya–Bisanpura 2nd* village-mix has maximum unutilized local

Table 29.16. Energy resource allocation pattern in optimal scenario for implementation in *Panthadiya–Bisanpura 1st village-mix*.

Optimal Scenario for	End-uses					
	Cooking	Lighting	Pumping	Heating	Cooling	Appliances
Panthadiya–Bisanpura 1st Village-mix	1. Biomass (13.99%) 2. LPG (24.87%) 3. Biogas (41.14%) 4. Solar thermal (20%)	Biomass electricity (100%)	Biomass electricity (100%)	Solar thermal (100%)	Biomass electricity (100%)	Biomass electricity (100%)
Panthadiya–Bisanpura 2nd Village-mix	1. Biomass (13.24%) 2. LPG (26.42%) 3. Biogas (40.34%) 4. Solar thermal (20%)	Biomass electricity (100%)	Biomass electricity (100%)	Solar thermal (100%)	Biomass electricity (100%)	Biomass electricity (100%)
Panthadiya–Morva Village-mix	1. Biomass (18.39%) 2. LPG (24.50%) 3. Biogas (37.11%) 4. Solar thermal (20%)	Biomass electricity (100%)	Biomass electricity (100%)	Solar thermal (100%)	Biomass electricity (100%)	Biomass electricity (100%)
Panthadiya–Bisanpura 1st-Bisanpura 2 Village-mix	1. Biomass (17.28%) 2. LPG (21.40%) 3. Biogas (41.32%) 4. Solar thermal (20%)	Biomass electricity (100%)	Biomass electricity (100%)	Solar thermal (100%)	Biomass electricity (100%)	Biomass electricity (100%)

Table 29.16. (*Continued*)

Optimal Scenario for	End-uses					
	Cooking	Lighting	Pumping	Heating	Cooling	Appliances
Panthadiya–Bisanpura 1st-Morva Village-mix	1. Biomass (16.63%) 2. LPG (25.19%) 3. Biogas (38.18%) 4. Solar thermal (20%)	Biomass electricity (100%)	Biomass electricity (100%)	Solar thermal (100%)	Biomass electricity (100%)	Biomass electricity (100%)
Panthadiya–Bisanpura 2nd-Morva Village-mix	1. Biomass (16.04%) 2. LPG (26.15%) 3. Biogas (37.81%) 4. Solar thermal (20%)	Biomass electricity (100%)	Biomass electricity (100%)	Solar thermal (100%)	Biomass electricity (100%)	Biomass electricity (100%)
Panthadiya–Bisanpura 1st-Bisanpura 2nd-Morva Village-mix	1. Biomass (15.05%) 2. LPG (26.30%) 3. Biogas (38.65%) 4. Solar thermal (20%)	Biomass electricity (100%)	Biomass electricity (100%)	Solar thermal (100%)	Biomass electricity (100%)	Biomass electricity (100%)

Table 29.17. Estimated unutilized energy resources in optimal scenario for village-mix.

Name of the village	Estimated dung cake availability, MJ/year	Unutilized dung cake, MJ/year	Estimated biomass energy availability, MJ/year	Unutilized biomass energy availability, MJ/year
Panthadiya–Bisanpura 1st	8.28×10^6	8.28×10^6	12.52×10^6	1.43×10^6
Panthadiya–Bisanpura 2nd	8.24×10^6	8.24×10^6	12.47×10^6	1.49×10^6
Panthadiya–Morva	13.47×10^6	13.47×10^6	23.36×10^6	2.46×10^6
Panthadiya–Bisanpura 1st-Bisanpura 2nd	10.91×10^6	10.91×10^6	15.73×10^6	—
Panthadiya–Bisanpura 1st-Morva	16.14×10^6	16.14×10^6	26.62×10^6	2.95×10^6
Panthadiya–Bisanpura 2nd-Morva	16.10×10^6	16.10×10^6	26.57×10^6	3.17×10^6
Panthadiya–Bisanpura 1st-Bisanpura 2nd-Morva	18.77×10^6	18.77×10^6	29.83×10^6	3.54×10^6

energy potential. Therefore, *Panthadiya–Bisanpura 2nd* village-mix is identified as a region for energy planning.

29.4.2 *Energy planning for the region*

As discussed in a previous article the *Panthadiya–Bisanpura 2nd* village-mix is identified as a region for IRES design for future energy requirements. In a IRES design for the region, energy plans are developed for short and medium term objectives. In short term energy planning, objectives to be achieved are minimum cost of energy utilization, maximum employment generation and maximum use of local resource. In medium term planning, objectives to be achieved are minimum cost of energy utilization, maximum employment generation, maximum use of local resource and minimum environment emissions. The short and medium term plans are generated for the projected end-use energy requirements and estimated energy resource availability in the region.

29.4.2.1 *Short-term planning*

Short term energy plans are developed for the region by considering the expected end-use energy requirements in the year 2010–2011. The average per capita energy consumption in *Panthadiya* and *Bisanpura 2nd* is used to project future end-use energy requirements. For pumping end-use, it is assumed that 2 tube wells will be added in the agricultural application on a yearly basis, due to higher initial cost for constructing a tube well. The cost of energy utilization is computed by considering an inflation rate of 6% and escalation rate of 6% for petroleum products.

Scenario 1 — *Base-case energy scenario*

In Scenario 1, it is assumed that the biomass energy resource availability in 2010–2011 will remain the same as that of the base year (2005–2006), and for biogas energy resource, the number of cattle to population ratio (0.51) is assumed to remain constant. In scenario 1, it assumed that system efficiency, reliability of the energy source and system, social acceptance factors, employment generation rate and environment emission rates will remain the same as observed in the year base year (2005–2006). The multi-objective optimization problem is solved for the projected end-use energy requirements and resource availability in the region.

In scenario 1, energy resource allocation is carried out with respect for short term objectives of energy planning. In scenario 1, cost, employment generation and use of local resources objective functions are assigned higher priority as compared to other objective functions. The results of energy resource allocation are shown in Table 29.18.

The results of optimization show that LPG, biogas and solar thermal should be promoted for cooking, and solar thermal for heating end-use. LPG (39.18%) is to be allocated for cooking due to the constraint of biogas and biomass energy resource potential, and is preferred over dung cake due to the lower associated emissions and higher system efficiency. Biomass electricity is to be promoted for lighting, pumping, cooling, and appliance end-uses due to their high employment generation potential at the lower costs. The cost associated with scenario 1 is Rs. 11.60 millions and the associated emissions are 1331.85, 1.84, 48.54 Tons/year for CO_x, NO_x and SO_x, respectively.

Scenario 2 — *Biogas energy scenario*

In scenario 2, it is assumed that the population to cattle ratio increases from the present, i.e., 0.51 by 20% in the future to 0.612 and biomass energy resource remains unchanged due to more dependence on firewood collection from the region as compared to agriculture residue. In scenario 2, it assumed that system efficiency, reliability of the energy source and system, social acceptance factors and employment generation rate and environment emission rates will remain the same as observed in the year base year (2005–2006). The multi-objective optimization problem is solved for the projected end-use energy requirements and resource availability in the region. In scenario 2, energy resource allocation is carried out with respect to the short term objectives of energy planning. In scenario 2, cost, employment generation and use

Table 29.18. Short-term energy resource allocation for 2010–2011.

Objective function	End-uses					
	Cooking	Lighting	Pumping	Heating	Cooling	Appliances
Scenario 1	1. LPG (39.18%) 2. Biogas (40.82%) 3. Solar thermal (20%)	Biomass electricity (100%)	Biomass electricity (100%)	Solar thermal (100%)	Biomass electricity (100%)	Biomass electricity (100%)
Scenario 2	1. LPG (31.04%) 2. Biogas (56.18%) 3. Solar thermal (20%)	Biomass electricity (100%)	Biomass electricity (100%)	Solar thermal (100%)	Biomass electricity (100%)	Biomass electricity (100%)
Scenario 3	1. LPG (39.25%) 2. Biogas (40.75%) 3. Solar thermal (20%)	Biomass electricity (100%)	Biomass electricity (100%)	Solar thermal (100%)	Biomass electricity (100%)	Biomass electricity (100%)

of local resources objective functions are assigned a higher priority as compared to other objective functions. The results of the energy resource allocation are shown in Table 29.18.

The scenario results indicate that biomass, biogas and solar thermal should be promoted for cooking, and solar thermal for heating end-use. LPG (31.04%) is to be allocated for cooking due to the constraint of biogas and biomass energy resource potential. Biomass electricity is to be promoted for lighting, pumping, cooling, and appliance end-uses due to their high employment generation potential at the lower costs. The cost associated with scenario 2 is Rs. 10.97 million and the associated emissions are 1333.97, 1.84, 58.20 Tons/year for CO_x, NO_x and SO_x, respectively.

Scenario 3 — *Biomass energy scenario*
In scenario 3, it is assumed that the biomass energy potential increases by 20% in the future and for biogas energy resource, the number of cattle to population ratio (0.51) is assumed to remain the same as that estimated in the base year (2005–2006). In scenario 3, it assumed that reliability of the energy source and system, social acceptance factors and employment generation rate and environment emission rates will remain the same as observed in the year base year (2005–2006). The multi-objective optimization problem is solved for the projected end-use energy requirements and resource availability in the region. In scenario 3, energy resource allocation is carried out with respect to the short term objectives of energy planning. In scenario 3, cost, employment generation and the use of local resources objective functions are assigned a higher priority as compared to other objective functions. The results of energy resource allocation are shown in Table 29.18.

The scenario indicates that biomass, biogas and solar thermal should be promoted for cooking, and solar thermal for heating end-use. LPG (39.25%) is to be allocated for cooking due to the constraint of biogas and biomass energy resource potential. Biomass electricity is to be promoted for lighting, pumping, cooling, and appliance end-uses due to their high employment generation potential at the lower costs. The cost associated with scenario 3 is Rs. 11.60 million and the associated emissions are 1331.82, 1.84, 48.46 Tons/year for CO_x, NO_x and SO_x, respectively.

29.4.2.2 *Medium-term planning*

Medium term energy plans are developed for the region considering expected energy demand for the year 2015. The average per capita energy consumption in *Panthadiya* and *Bisanpura 1st* villages is used to project future energy requirements. For pumping end-use it is assumed 2 tube wells will be added in the agricultural application on a yearly basis. The cost associated with constructing a tube well depends on the water table available. The lower rate of increase is assumed due to the lower water table observed in the region, which is below 300 feet. The cost of energy utilization is computed by considering an inflation rate of 6% and escalation rate of 6% for petroleum products. The projected unit cost of energy (Rs/kWh).

Scenario 1 — *Base-case energy scenario*
In scenario 1, it is assumed that the biomass energy resource availability in 2015–2016 will remain the same as that of the base year (2005–2006) and for biogas energy resource, the number of cattle to population ratio (0.51) is assumed to remain constant. In scenario 1, it assumed that system efficiency, reliability of the energy source and system, social acceptance factors and employment generation rate and environment emission rates will remain the same as observed in the year base year (2005–2006). The multi-objective optimization problem is solved for the projected end-use energy requirements and resource availability in the region. In scenario 1, energy resource allocation is carried out with respect to medium term objectives of energy planning. In scenario 1, cost, employment generation, use of local resources and environment emissions objective functions are assigned a higher priority as compared to other objective functions. The results of the energy resource allocation are shown in Table 29.19.

The results of optimization show that biomass, biogas and solar thermal should be promoted for cooking, and solar thermal for heating end-use. LPG (28.99%) is to be allocated for cooking due to the constraint of biogas and biomass energy resource potential. Biomass electricity is to be promoted for lighting, pumping, cooling, and appliance end-uses due to their high employment generation potential at the lower costs. The cost associated with scenario 1 is Rs. 18.33 million and the associated emissions are 2118.062, 6.00, 53.57 Tons/year for CO_x, NO_x and SO_x, respectively.

Scenario 2 — *Biogas energy scenario*
In scenario 2, it is assumed that the population to cattle ratio increases from the present, i.e., 0.51, by 20% in the future to 0.612 and the biomass energy resource remains unchanged due to more dependence on firewood collection from the region as compared to agriculture residue. In scenario 2, it assumed that system efficiency, reliability of the energy source and system, social acceptance factors and employment generation rate and environment emission rates will remain the same as observed in the year base year (2005–2006). The multi-objective optimization problem is solved for the projected end-use energy requirements and resource availability in the region. In scenario 2, energy resource allocation is carried out with respect to medium term objectives of energy planning. In scenario 2, cost, employment generation, use of local resources and environment emissions objective functions are assigned a higher priority as compared to other objective functions. The results of the energy resource allocation are shown in Table 29.19.

Table 29.19. Medium-term energy resource allocation for 2015–2016.

Objective function	End-uses					
	Cooking	Lighting	Pumping	Heating	Cooling	Appliances
Scenario 1	1. Biomass (10.25%) 2. LPG (28.99%) 3. Biogas (40.76%) 4. Solar thermal (20%)	Biomass electricity (100%)	Biomass electricity (100%)	Solar thermal (100%)	Biomass electricity (100%)	Biomass electricity (100%)
Scenario 2	1. Biomass (10.25%) 2. LPG (20.75%) 3. Biogas (49.00%) 4. Solar thermal (20%)	Biomass electricity (100%)	Biomass electricity (100%)	Solar thermal (100%)	Biomass electricity (100%)	Biomass electricity (100%)
Scenario 3	1. Biomass (17.00%) 2. LPG (13.97%) 3. Biogas (49.03%) 4. Solar thermal (20%)	Biomass electricity (100%)	Biomass electricity (100%)	Solar thermal (100%)	Biomass electricity (100%)	Biomass electricity (100%)

The results of optimization show that biomass, biogas and solar thermal should be promoted for cooking, and solar thermal for heating end-use. LPG (20.75%) is to be allocated for cooking due to the constraint of biogas and biomass energy resource potential. Biomass electricity is to be promoted for lighting, pumping, cooling, and appliance end-uses due to their high employment generation potential at the lower costs. The cost associated with scenario 2 is Rs. 17.03 million and the associated emissions are 2039.44, 6.00, 64.38 Tons/year for CO_x, NO_x and SO_x, respectively.

Scenario 3 — *Biomass energy scenario*
In Scenario 3, it is assumed that the biomass energy potential increases by 20% in the future and for biogas energy resource, the number of cattle to population ratio (0.51) is assumed to remain the same as that estimated in the base year (2005–2006). In scenario 3, it assumed that the reliability of the energy source and system, social acceptance factors and employment generation rate and environment emission rates will remain the same as observed in the year base year (2005–2006). The multi-objective optimization problem is solved for the projected end-use energy requirements and resource availability in the region. In scenario 3, energy resource allocation is carried out with respect to medium term objectives of energy planning. In scenario 3, cost, employment generation, use of local resources and environment emissions objective functions are assigned higher priority as compared to other objective functions. The results of the energy resource allocation are shown in Table 29.19.

The results of the optimization show that biomass, biogas and solar thermal should be promoted for cooking, and solar thermal for heating end-use. LPG (13.97%) is to be allocated for cooking due to the constraint of biogas and biomass energy resource potential. Biomass electricity is to be promoted for lighting, pumping, cooling, and appliance end-uses due to their high employment generation potential at the lower costs. The cost associated with scenario 3 is Rs. 16.67 million and the associated emissions are 2464.61, 4.37, and 64.37 Tons/year for CO_x, NO_x and SO_x, respectively.

29.5   Conclusions

This chapter presents detailed methodology of a multi objective goal programming model and its application for the energy resource allocation. This model gives decision-makers a tool to use in making strategic decisions on matters related to energy policy. The objective of this work was to determine the optimum allocation of energy resources to six end-uses in the household and agriculture sector in *Panthadiya* village. Eleven different energy resources were selected, based on either their present or potential availability in village. The results of analysis show the followings:

— To meet the cooking energy demands: biomass, LPG, biogas and solar thermal should be promoted.
— To meet the lighting energy demands: biomass electricity should be promoted.
— To meet the pumping energy demands: biomass electricity should be promoted.
— To meet the heating energy demands: solar thermal should be promoted.
— To meet the cooling energy demands: biomass electricity should be promoted.
— To meet the appliances energy demands: biomass electricity should be promoted.

Biomass electricity generation should be encouraged for all end uses. Grid electricity for all end-uses should be discouraged. Solar photovoltaic can be used for small-scale applications, where the connections from the grid are expensive and there are no other economically competing technologies. This resource will become more prominent in the near future, especially when environmental quality receives a higher importance.

The generalized goal programming model is also used to identify regions for fast track development. The region is identified on the basis of scope of utilization of energy resources. An attempt has been made to incorporate the needs of group of villages in micro level energy planning. Optimal scenarios are developed for the study as well as neighboring villages by assigning higher priority to cost and employment generation objective as compared to other objectives. It was found that mix of two villages for micro level energy planning results in better utilization of available energy sources compared to each individual village. However, the success of intra village mix for micro level energy planning depends on socio-economic and local political will, to implement such intra village mix micro plans.

References

1. R.K. Agrawal and S.P. Singh, "Energy allocation for cooking in UP households (India): A fuzzy multi objective analysis," *Energy Conservation and Management* **42** (2001) 2139–2154.
2. K. Ashenayi and R. Ramakumar, "IRES — A program to design integrated renewable energy systems," *Energy* **15** (1990) 1143–1152.
3. M. Beccali, M. Cellura and M. Mistretta, "Decision-making in energy planning. Application of the Electre method at regional level for the diffusion of renewable energy technology," *Renewable Energy* **28** (2003) 2063–2087.
4. J. Byrne, B. Shen and Wallace W, "The economics of sustainable energy for rural development: A study of renewable energy in rural China," *Energy Policy* **26** (1998) 45–54.
5. Census of Rajasthan, Ministry of Home Affairs, Government of India (2001).
6. Central Arid Zone Research Institute (CAZRI), Jodhpur, Rajasthan. Information Brochure (2005).

7. R. Chedid, H. Akiki and S. Rahman, "A decision support technique for the design of hybrid solar — wind power systems," *IEEE Trans. Energy Conversion* **13** (1998) 76–83.

8. R. Chedid, T. Mezher and C. Jarrouche, "A fuzzy programming approach to energy resource allocation," *International Journal of Energy Research* **23** (1999) 303–317.

9. J. Climaco, *Multicriteria Analysis* (Springer-Verlag, New York, 1997).

10. C. Cormio, M. Dicorato, A. Minoia and M. Trovato, "A regional energy planning methodology including renewable energy sources and environmental constraints," *Renewable and Sustainable Energy Reviews* **7** (2003) 99–130.

11. C. Cosmi, M. Macchiato, L. Mangiamele, G. Marmo, F. Pietrapertosa and M. Salvia, "Environmental and economic effects of renewable energy sources use on a local case study," *Energy Policy* **31** (2003) 443–457.

12. V. Devdas, "Planning for rural energy systems: part I," *Renewable and Sustainable Energy Reviews* **5** (2001a) 203–226.

13. V. Devdas, "Planning for rural energy systems: part II," *Renewable and Sustainable Energy Reviews* **5** (2001b) 227–270.

14. V. Devdas, "Planning for rural energy systems: part III," *Renewable and Sustainable Energy Reviews* **5** (2001c) 271–297.

15. R. Dossani, "Reorganization of the power distribution sector in India," *Energy Policy* **32** (2004) 1277–1289.

16. I. Dyner and E.R. Larsen, "From planning to strategy in the electricity industry," *Energy Policy* **29** (2001) 1145–1154.

17. E.S. Gavanidou and A.G. Bakirtzis, "Design of a stand alone system with renewable energy sources using trade off methods," *IEEE Trans. Energy Conservation* **7** (1992) 42–48.

18. D. Ghosh, P.R. Shukla, A. Garg and P.V. Ramana, "Renewable energy technologies for the Indian power sector: Mitigation potential and operation strategies," *Renewable and Sustainable Energy Reviews* **6** (2002) 481–512.

19. S.W. Hadley and W. Short, "Electricity sector analysis in the clean energy futures study," *Energy Policy* **29** (2001) 1285–1298.

20. R.B. Hiremath, S. Shikha and N.H. Ravindranath, "Decentralized energy planning; modeling and application — a review," *Renewable and Sustainable Energy Reviews*, Article in Press.

21. J.P. Huang, K.L. Poh and B.W. Ang, "Decision analysis in energy and environmental modeling," *Energy* **20** (1995) 843–855.

22. S. Iniyan, L. Suganthi and T.R. Jagadeesan, "Critical analysis of wind farms for sustainable generation," Solar World Congress, Korea (1997).

23. S. Iniyan and K. Sumanthy, "An optimal renewable energy model for various end-uses," *Energy* **25** (2000) 563–575.

24. S. Jebaraja and S. Iniyan, "A review of energy models," *Renewable and Sustainable Energy Reviews* **10** (2006) 281–311.

25. B. Joshi, T.S. Bhatti, N.K. Bansal, K. Rijal and P.D. Grover, "Decentralized energy planning model for optimum resource allocation with a case study of the domestic sector of rurals in Nepal," *Int. J. Energy Research* **15** (1991) 71–78.

26. R.L. Keeney and H. Raiffa, *Decisions with Multiple Objectives: Preferences and Value Trade-offs* (John Wiley & Sons, New York, 1995).

27. M. Kleinpeter, *Energy Planning and Policy* (John Wiley & Sons, New York, 1995).

28. S. Koundinya, "Incorporating qualitative objectives in integrated resource planning: Application of analytical hierarchy process and compromising programming," *Energy Sources* **17** (1995) 565–581.

29. A.S. Kydes, S.H. Shaw and D.F. McDonald, "Beyond the horizon: Recent directions in long-term energy modelling," *Energy* **20** (1995) 131–149.

30. D.M. Logan, "Decision analysis in engineering-economic modeling," *Energy* **15** (1990) 677–696.

31. N.M. Maricar, "Efficient resource development in electric utilities planning under uncertainty," Doctoral Thesis Virginia Polytechnic Institute and State University, Virginia (2004).

32. J. Mathur, N.K. Bansal and H.J. Wagner, "Dynamic energy analysis to assess maximum growth rates in developing power generation capacity: Case study of India," *Energy Policy* **32** (2004) 281–287.

33. T. Mezher, R. Chedid and W. Zahabi, "Energy resource allocation using multi-objective goal programming: The case of Lebanon," *Applied Energy* **61** (1998) 175–192.

34. G. Mihalakakou, B. Psiloglou, M. Santamouris and D. Nomidis, "Application of renewable energy sources in the Greek islands of the South Aegean Sea," *Renewable Energy* **26** (2002) 1–19.

35. R.C. Neudoerffer, P. Malhotra and P.V. Ramana, "Participatory rural energy planning in India — a policy context," *Energy Policy* **29** (2001) 371–381.

36. G. Notton, M. Muselli and A. Louche, "Autonomous hybrid photovoltaic power plant using a back-up generator: A case study in a Mediterranean Island," *Renewable Energy* **7** (1996) 371–391.

37. G. Notton, M. Muselli, P. Poggi and A. Louche, "Decentralized wind energy systems providing small electrical loads in remote areas," *Int. J. Energy Research* **25** (2001) 141–164.

38. R. Pandey, "Energy policy modelling: Agenda for developing countries," *Energy Policy* **30** (2002) 97–106.

39. S.D. Pohekar and M. Ramachandran, "Assessment of solar cooking technology and its dissemination in India," *Energy and Fuel Users J.* (2004).

40. S. Rahman and L.C. Frairm, "A hierarchical approach to electric utility planning," *Energy Research* **8** (1984) 185–196.

41. R. Ramakumar, P.S. Shetty and K. Ashenayi, "A linear programming approach to the design of integrated renewable energy systems for developing countries," *IEEE Trans. Energy Conversion* **EC-I** (1986) 18–24.

42. R. Ramanathan and L.S. Ganesh, "A multiobjective programming approach to energy resource allocation problems," *Int. J. Energy Research* **17** (1993) 105–119.

43. R. Ramanathan and L.S. Ganesh, "Energy alternatives for lighting in households: An Evaluation using an integrated goal programming-AHP model," *Energy* **20** (1995) 63–72.

44. R. Ramanathan and L.S. Ganesh, "Energy resource allocation incorporating qualitative and quantitative criteria: An integrated model using goal programming and AHP," *Int. J. Socio-economic Planning Science* **29** (1995) 197–218.

45. R. Ramanathan and L.S. Ganesh, "Using AHP for resource allocation problems," *European J. Operational Research* **80** (1995) 410–417.

46. T.L. Saaty and K.P. Kearns, *Analytic Planning: The Organization of Systems* (Pergamon Press, New York, 1985).

47. T.L. Saaty, *The Analytic Hierarchy Process: Planning, Priority Setting, Resource Allocation* (McGraw-Hill, New York, 1980).

48. Y. Sarafidis, D. Diakoulaki, L. Papayannkis and A. Zervos, "A regional planning approach for the promotion of renewable energies," *Renewable Energy* **18** (1999) 317–330.

49. P.R. Shukla, "The modelling of policy options for greenhouse gas mitigation in India," *AMBIO* **25** (1996) 240–248.

50. C.S. Sinha and T.C. Kandpal, "Decentralized vs grid electricity for rural India," *Energy Policy* (1991), pp. 441–448.

51. C.S. Sinha and T.C. Kandpal, "Optimal mix of technologies for rural India — the cooking sector," *Int. J. Energy Research* **15** (1991a) 85–100.

52. C.S. Sinha and T.C. Kandpal, "Optimal mix of technologies for rural India — the irrigation sector," *Int. J. Energy Research* **15** (1991b) 331–346.

53. C.S. Sinha and T.C. Kandpal, "Optimal mix of technologies for rural India — the lighting sector," *Int. J. Energy Research* **15** (1991c) 653–665.

54. R. Sontag and A. Lange, "Cost effectiveness of decentralized energy supply systems taking solar and wind utilization plants into account," *Renewable Energy* **28** (2003) 1865–1880.

55. L. Suganthi and A.A. Samuel, "An optimal energy forecasting model for the economy environment match," *International Journal of Ambient Energy* **20** (1999) 137–148.

56. L. Suganthi and A. Williams, "Renewable energy in India — a modeling study for 2020–2021," *Energy Policy* **28** (2000) 1095–1109.

57. J.W. Sun, "Is GNP — energy model logical?" *Energy Policy* **29** (2001) 949–950.

58. E.B. Triantaphyllou, S. Shu, T. Sanchez and T. Ray, "Multicriteria decision making: An operations research approach," in *Encyclopedia of Electrical and Electronics Engineering* (John Wiley & Sons, New York, 1998).

59. G.H. Tzeng, T. Shiau and C.Y. Lin, "Application of multicriteria decision making to the evaluation of new energy system development in Taiwan," *Energy* **17** (1992) 983–992.

60. D. Weisser, "Costing electricity supply scenarios: A case study of promoting renewable energy technologies on Rodriguez, Mauritius," *Renewable Energy* **29** (2004) 1319–1347.

61. M. Yehia, S. Karaki, R. Chedid, A.E. Tabors and P. Shawcross, "Tradeoff approach in power system planning case study: Lebanon," *Electrical Power & Energy Systems* **17** (1995) 137–141.

62. J. Zhu and M. Chow, "A review of emerging techniques on generation expansion planning," *IEEE Trans. Power Systems* **12** (1997) 1722–1728.

63. N. Zouros, G.C. Contaxis and J. Kabouris, "Decision support tool to evaluate alternative policies regulating wind integration into autonomous energy systems," *Energy Policy* **33** (2005) 1541–1555.

Index